Chemistry in Context
Third Edition

Books are to be returned on or before the last date below.

Chemistry in Context
Third Edition

Graham C. Hill MA (Cantab), FRSC
Deputy Headmaster,
Dr Challoner's Grammar School, Amersham,
formerly Head of Science,
Bristol Grammar School

John S. Holman MA (Cantab), FRSC
Head of Science,
Watford Grammar School

Nelson

Thomas Nelson and Sons Ltd
Nelson House   Mayfield Road
Walton-on-Thames   Surrey
KT12 5PL   UK

51 York Place
Edinburgh
EH1 3JD   UK

Thomas Nelson (Hong Kong) Ltd
Toppan Building 10/F
22A Westlands Road
Quarry Bay   Hong Kong

Thomas Nelson Australia
102 Dodds Street
South Melbourne
Victoria 3205 Australia

Nelson Canada
1120 Birchmount Road
Scarborough   Ontario
M1K 5G4   Canada

© G C Hill, Js Holman 1978, 1983, 1989
First published by Thomas Nelson and Sons Ltd 1978
2nd edition published 1983
3rd edition published 1989
ISBN 0-17-438401-7
NPN 9 8 7 6 5 4

Acknowledgement is due to the following for
permission to use photographs:

Adrian Meredith Photography (p. 257 *top*)
Aerofilms Ltd (pp. 59, 246 *top*)
Agricultural Research Council Weed Research
   Organisation (p. 429)
All-Sport UK Ltd (p. 278)
All-Sport UK Ltd/Gerard Vandystadt (p. 284 *top*)
Ann Ronan Picture Library (pp. 186 *top*, 574)
Associated Octel Company Ltd (p. 212)
Associated Press Ltd (pp. 225, 458)
Atomic Energy Research Establishment, Harwell
   (p. 58)
BASF (pp.362, 363 *both*, 365 *top*, 581)
Banana Rite Ltd (p. 472)
BBC Hulton Picture Library (pp. 321 *both*, 494 *top*,
   506, 539 *bottom*)
Blue Circle Cement (p.338)
Bridgeman Art Library (p. 184)
British Alcan Aluminium plc (pp. 260, 261 *bottom*)
British Gypsum Ltd (p. 252)
British Leyland (pp.257 *bottom*, 465 *bottom*)
British Museum (Natural History) (pp. 96, 256)
British Petroleum Chemicals Ltd (pp. 483, 485, 518
   bottom, 591)
British Steel Corporation (pp. 312 *top and bottom*, 314,
   318 *bottom*)
British Tourist Authority (p. 328 *bottom*)
Bruce Coleman Ltd (pp. 279, 336)
Bryn Campbell (p. 29)
Camera Press (pp. 79 *right*, 89 *bottom*, 152, 507, 573
   top)
Cavendish Laboratory, University of Cambridge
   (pp. 54, 61)
CERN (p. 88)
Ciba-Geigy (p. 284 *bottom*)
Claridge Communications (p. 549)
Courtaulds Ltd (p. 592)
Daily Telegraph Colour Library (pp. 90, 103 *left*, 173)
Daily Telegraph Colour Library/Clive Coote (p. 253)
Daily Telegraph Colour Library/Wilhelm Schmidt
   (p. 451)
De Beers Industrial Diamond Division (Pty) Ltd
   (pp. 138 *bottom*, 139 *right*, 282 *all photos*, 283 *centre*)
Distillers Company (p. 290)
Dr D.W. Thompson (p. 2)
Dr J.D. Dodge (p. 232)
Dr J.H. Holloway (pp. 40, 78)
Ed Barber (pp. 5, 272, 293, 394, 403, 438 *left*, 441, 454,
   463, 467, 477 *left*, 514, 539 *top*, 555 *left*, 567 *bottom*,
   573 *bottom*)
Enraf-Nonius Ltd (p. 133 *top*)
Fotofirst (pp. 319, 320 *top*)
Fox Photos Ltd (p. 526)
Gallenkamp (p. 169)
Gareth J. Jones (pp. 488, 547)
General Electric Co., USA (p. 283 *left*)
GeoScience Picture Library (pp. 250, 254 *top*)
Glaxo Operations UK Ltd (pp. 342, 518 *centre*)
Godfrey New Photographics (p. 246 *centre*)
G.S.F. Picture Library (p. 245 *top*)
Heather Angel (p. 376 *both*)
Homer Sykes (pp. 312 *centre*, 334)
Image Bank (p. 16 *right*)
Imperial Chemical Industries Agrochemicals (p. 597)
Imperial Chemical Industries Ltd (pp. 25, 366, 367
   right, 411, 484 *bottom photos*, 515)
International Gold Corporation (p. 248)
J. Allan Cash Ltd (pp. 10, 268 *top*, 305, 375)
James Austin (pp. 285)
John Freeman (by courtesy of the Royal Institution)
   (pp. 22 *bottom*, 33, 34)
John Hillelson Agency (pp. 245 *bottom left*, 266, 267,
   457 *bottom*)
John Topham Picture Library (131 *Top*)
John U. Clark (p.138 *centre*)

John Walmsley (pp. 283 *right*, 428, 438 *right*, 464)
Johnathan Ledger (p. 538 *top*)
Keystone Press Agency Ltd (pp. 6, 55, 185 *bottom*,
   187, 315 *bottom*, 408 *right*, 409, 499, 504, 571 *top*,
   582)
Leslie Bishop Company Ltd (p. 408 *right*)
London Fire Brigade (p. 480)
Mansell Collection (pp. 1 *centre*, 21 *left and right*, 156,
   249 *top*, 269, 513 *top*)
Mark Edwards (p. 335)
Mary Evans Picture Library (pp. 150, 185 *top*, 268
   bottom, 361, 368, 532)
Metropolitan Museum of Art (p. 273)
National Aeronautics and Space Administration
   (p. 172)
National Museum of Science and Industry (p. 246
   bottom)
National Portrait Gallery (p. 580)
Nicholas Horne (p. 367 *left*)
Nitrate Sales International N.V. (p. 264)
Novosti Press Agency (p. 37)
Oxford Scientific Films (p. 79 *left*, 122)
Panos Pictures (p. 245 *bottom right*)
Penni Bickle (p. 541 *centre*)
Peter Thorne Ltd Photographic (p. 465 *centre*)
Philip Harris (p. 99)
Professor F.G.A. Stone (p. 38)
Punch Publications (pp. 457 *top*, 541 *top*)
Redland Photographic Library (p. 453 *bottom*)
Rio Tinto Zinc Photographic Library (pp. 16 *left and
   centre*, 311 *top*, 320 *bottom*)
Robert Harding Picture Library (p. 465 *top*)
Ron Chapman (pp. 24, 69 *top*)
Royal College of Surgeons (p. 533)
Science Museum (pp. 1 *right*, 2, 22 *top*, 26 *bottom*, 56,
   87, 88 *top*, 103 *right*, 131 *centre*, 153 *bottom*, 261 *top*,
   310 *bottom*, 311 *bottom*, 313, 315 *top*, 328 *top*, 447
   both, 494 *bottom*, 495, 498 *top*, 527, 590)
Science Photo Library (pp. 62, 92 *left and right*, 93
   bottom, 121, 130, 186 *centre*, 286, 318 *top*, 442, 513
   bottom, 554)
Science Photo Library/Mike McNamee (p. 137)
Scott Polar Research Institute (p. 395)
Shell Photographic Library (pp. 26 *top*, 453 *top*, 466,
   498 *bottom*)
Sim-chem and UKF Fertilizers Ltd (p. 346)
Spectrum Colour Library (p. 249 *bottom*)
Stewart Keating (p. 310 *top*)
Stoe & Cie GmbH (p. 132)
Sunrise Medical Ltd (p. 224)
Time-Life Inc. (p. 484 *top*)
Tony Othen (pp. 530, 538 *bottom*, 555 *right*, 561, 571
   bottom)
Treak Cliff Cavern (p. 263)
Trustees of the British Museum (p. 291)
Ullstein (p. 82)
Unilever (p. 566)
United Kingdom Atomic Energy Authority (pp. 93
   top, 94, 165, 247)
University of California, San Diego (p. 89 *top*)
Van den Berghs & Jurgens (p. 477 *right*)
Will Green (p. 567)
Zefa Picture Library (UK) Ltd (pp. 186 *bottom*, 254
   bottom)

Cover: All-Sport UK Ltd/Gerard Vandystadt *top left*;
Daily Telegraph Colour Library/Wilhelm Schmidt
bottom; Image Bank *top right*; Science Photo Library/
Mike McNamee *background*

The publishers would also like to acknowledge the
assistance of Barclays Bank plc and Butterworth
Laboratories, Teddington.

Preface

This book is about chemistry and chemists and the contribution they make towards science, industry and society. It is designed primarily for students studying chemistry to Advanced level. In planning this edition, we have been particularly influenced by:

— the Nuffield Advanced level chemistry project and its revisions (with which both of us have been associated);
— the common core statements for Advanced level chemistry syllabuses published by the Standing Conference on University Entrance (SCUE) and the examination boards;
— the development of GCSE syllabuses and courses.

Between us, we have worked in various capacities for several A-level examining boards, including the University of Cambridge Local Examinations Syndicate, the Joint Matriculation Board, the University of London Examinations Board, the Associated Examining Board and the Southern Universities Joint Board. Nevertheless, the book is not written specifically for any one syllabus. Indeed, it is clear from the success of the first and second editions that it has been helpful to students studying for all Advanced level examinations.

Human interest in chemistry is essentially practical. We are interested in the way materials in the universe behave and react under different conditions and the ways in which we can use these materials for our own purposes. Complementary to this practical approach is a logical scientific method of working, the discovery of laws to summarise information and the development of ideas and theories to interpret and explain observations.

With these important points in mind, *Chemistry in Context* fulfils certain important criteria. In a single concise volume, it presents chemistry as a unified and integrated subject using physical principles as a basis for the inorganic and organic sections which follow. Our aim has been to show chemistry in its wider context as a relevant and developing science which makes an essential contribution towards society, industry and civilisation. We have tried to show, both in the text and by carefully chosen photographs, how chemical knowledge and chemical industry have influenced human life and society.

The text is designed for use with the present modern syllabuses. It draws on experimental evidence in establishing laws and theories. Throughout the text, we have used experimental results and data to develop key ideas and concepts. The book is intended to complement a course of practical work and consequently space is not devoted to instructions or suggestions for experiments.

However, the companion volume *Chemistry in Context — Laboratory Manual and Study Guide* (2nd edn) will be found a most helpful source of practical work.

Evidence from the earlier editions shows that *Chemistry in Context* is suitable for use by Advanced level students of all abilities, not just the most able. We have borne in mind that many students starting an Advanced level course have little experience of using advanced texts. Consequently, each chapter is divided into fairly short sections and concludes with a summary of the more important facts and ideas. In addition, there are frequent questions in the text which will encourage the student to read more carefully and critically. It is important that an effort is made to answer these questions, as they often form an integral part of the text.

The study questions at the end of each chapter vary in style and may be used in tests, in class discussion, in revision or for homework.

Considerable care has been taken over the layout of each page so as to present the subject clearly and attractively using figures, tables, photographs, charts, flow diagrams and graphs to reinforce the text.

The first four chapters begin by revising and extending key ideas concerning the mole, redox and periodicity. The next ten chapters (5–14) develop the major themes of structure, bonding, energetics, periodicity and competition which are then used throughout the following chapters of inorganic chemistry (15–20). Here, emphasis is placed on patterns and on the use of the periodic table and the activity series to summarise information. Five chapters (21–25) on equilibria, entropy and reaction rates complete the development of principles and illustrate the importance of these factors in industrial processes. The book concludes with ten chapters (26–35) of organic chemistry. This is developed systematically in terms of functional groups, and the patterns of behaviour are related to structure, kinetics and energetics. Mechanistic explanations are introduced where appropriate.

The order in which topics are presented is not intended as the ideal learning or teaching sequence. It is, however, a convenient, consistent and systematic order of the topics within a book. Different teachers will wish to cover the material in different ways according to their own views and the interests and ability of their students.

As far as nomenclature, units and abbreviations are concerned, we have been guided by *Chemical Nomenclature, Symbols and Terminology for Use in School Science* (ASE, 3rd edn, 1985) and we have used the International System of Units (SI) with very few exceptions. Consequently, we have expressed volumes in cubic decimetres (dm^3) rather than litres, and concentrations are given as moles per cubic decimetre rather than moles per litre.

In general, we have used the SI unit of pressure (Nm^{-2}, sometimes called a pascal (Pa)), although atmospheres (atm) have been used in making comparisons with atmospheric pressure.

In a similar fashion, we have used degrees celsius (°C) to express temperature when this has seemed more suitable or more appropriate than kelvins (K).

The SI unit for the quantity referred to by international agreement as 'amount of substance' is 'mol'. Unfortunately, the word 'molecule' is often abbreviated to 'mol.' and to avoid ambiguity we have used the unit mole rather than mol for amount of substance. Thus, the units of concentrations are given as $mole\ dm^{-3}$ not $mol\ dm^{-3}$, and those for molar enthalpies are written as $kJ\ mole^{-1}$ not $kJ\ mol^{-1}$.

Modern nomenclature has been used throughout the book. In general, organic chemicals are named systematically following IUPAC Rules, though trivial names of common substances, such as ethanoic acid and phenylamine, are sometimes bracketed after the systematic name.

In the six years since the second edition of *Chemistry in Context* was published, some significant changes have influenced the study and teaching of chemistry at Advanced level. The most important of these changes is the introduction of GCSE. This has led to the following changes in science courses for 14 to 16 year olds:

— a reduction in content and an elimination of the most difficult concepts;
— a better balance between the knowledge and understanding that can be gained from a course and the skills which it can develop;
— a greater emphasis on social, economic, technological and industrial aspects;
— the introduction of internal (school-based) teacher assessment.

The advent of GCSE has resulted in a revision of A-level chemistry syllabuses. In general, exam groups have taken steps to reduce the content of A-level syllabuses and to spell out more carefully certain concepts to ensure progression from GCSE courses to A-level.

Another significant change relating to A-levels over the last 10 years or so is the wider ability range of the students. Nowadays, there are many more students pursuing A-level studies who will not continue their studies into higher education.

With these changes in mind we have made some important improvements to this third edition of *Chemistry in Context*.

First, we have revised the text so that it is appropriate for a wider ability range, while retaining the capacity to stretch the most able students.

Secondly, we have removed certain topics (such as quantitative aspects of mass spectrometry and the Nernst Equation) which are no longer in A-level syllabuses.

Thirdly, we have added some important new material, such as that on the mole in chapter 2, to ensure a better articulation with GCSE.

Fourthly, we have further emphasised the social, economic, technological and industrial aspects of chemistry by extending the discussion at some points and by the introduction of full-colour sections with carefully chosen photographs and illustrations.

Finally, we have incorporated important material formerly in appendices into the main text, and we have enhanced the treatment of organic reaction mechanisms.

Many friends and colleagues have influenced this and earlier editions of *Chemistry in Context*. Countless students at different schools have influenced our styles of teaching and the way in which topics are introduced and developed in the book. This edition includes a number of improvements suggested by users of the first two editions.

To all these people, to our publishers and especially to our wives Elizabeth and Wendy, we owe our sincere thanks.

Graham Hill
John Holman
January 1989

Contents

x

To Elizabeth and Wendy

xii

1 Atoms, Atomic Masses and Moles

1.1 Atoms

The first chemist to use the name '**atom**' was John Dalton (1766–1844). Dalton used the word 'atom' to mean the smallest particle of an element. He then went on to explain how atoms could react together to form **molecules** which he called 'compound atoms'.

Right. John Dalton (1766–1844), the quiet genius. Dalton was born in the remote village of Eaglesfield in Cumbria. He was the son of a handloom weaver. For most of his life, Dalton lived and worked in Manchester at what was then the Presbyterian College.

Far right. Dalton's symbols for the elements. In 1803, Dalton published his Atomic Theory. He suggested that all matter was composed of small particles which he called 'atoms'. Later, Dalton went on to suggest symbols for the atoms of different elements and these are shown in the photograph. What do we now call 'Azote'? Which substances in Dalton's list are not elements, but compounds?

Do not confuse the terms atom and molecule.

An atom is the smallest part of an element which can ever exist.
A molecule is the smallest part of an element or a compound which can exist alone under ordinary conditions.

For example, chlorine consists of particles of Cl_2 under ordinary conditions, but at very high temperatures these split up to form particles of Cl. So, molecules of chlorine are written as Cl_2 and atoms of chlorine are written as Cl.

1

An electronmicrograph of copper phthalocyanine, a complex copper compound. The parallel dark lines indicate rows of copper atoms which have been magnified 583 330 times.

■ Name an element (other than chlorine) whose molecules consist of two atoms under normal conditions.
■ Name an element whose molecules consist of one atom under normal conditions.
■ Is it possible to have an atom of a compound?

Most atoms have a radius of about 10^{-10} m.

Nowadays, the unit used in measuring atomic distances is usually the nanometre (nm)

$$
\begin{aligned}
1\,\text{m} &= 10^9 \text{ nanometres} \\
10^{-9}\,\text{m} &= 1\,\text{nm} \\
\Rightarrow 10^{-10}\,\text{m} &= 0.1\,\text{nm}
\end{aligned}
$$

so, the radii of atoms are about 0.1 nm.

At one time, the unit used for atomic distances was the Ångstrom

$$1\,\text{Å} = 10^{-10}\,\text{m} = 0.1\,\text{nm}$$

Atoms, of course, are far too small to be seen even with the most powerful light microscope. However, chemists have discovered that it is possible to identify single atoms of certain elements using electron microscopes. In 1958, scientists in the USSR reported that they could pick out atoms of barium (which have a diameter of about 0.4 nm). They used an electron microscope with a magnification of 2 000 000. In 1967, Japanese scientists provided evidence for their identification of particles as small as 0.1 nm.

1.2 Comparing the masses of atoms

Atoms are too small to be weighed. Nevertheless, early in the nineteenth century Dalton suggested that it should be possible *to compare* the masses of atoms even though the atoms themselves could not be weighed. During the nineteenth century various methods were devised for obtaining approximate values for the relative masses of different atoms.

In 1919, Aston invented the mass spectrometer. This gave chemists a reliable and accurate method of comparing the relative masses of atoms. At one time, the relative masses of atoms were known as atomic weights, but nowadays we refer to them as **relative atomic masses.**

The basic idea of a mass spectrometer can be demonstrated using the apparatus in figure 1.1. Wooden balls of a *different* size, but with *identical* iron cores, roll down a sloping plane. At the bottom of the slope, a powerful magnet attracts the iron cores and the moving balls are deflected.

collecting compartments — sloping plane — large powerful magnet — increasing deflection

Figure 1.1
A simple model to illustrate the working of a mass spectrometer.

■ Why does the magnet have the same attraction for all the balls?
■ Which size of ball will be deflected the most? Why?

As the balls have identical iron cores, they are all attracted equally by the magnet. But, the smaller balls are lighter and therefore they are deflected the most. The balls collect in different compartments depending on their mass. All balls of the same mass collect in the same compartment. Using this simple

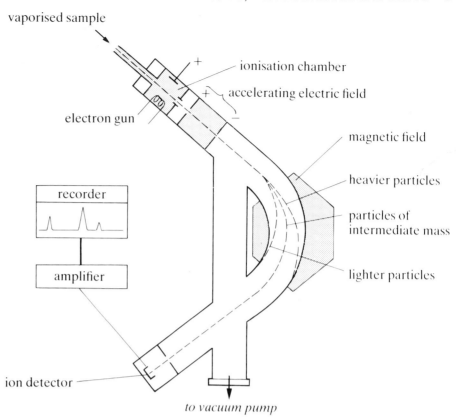

vaporised sample

ionisation chamber

accelerating electric field

electron gun

magnetic field

heavier particles

particles of intermediate mass

lighter particles

recorder

amplifier

ion detector

to vacuum pump

Figure 1.2
A diagram of a mass spectrometer.

Aston's mass spectrometer. A vaporised sample of the element was bombarded by electrons in the large glass bulb on the left. The ions produced were then accelerated by an electric field towards the magnetic field on the right (produced by the many coils of the electromagnet).

apparatus, it is possible to separate the different-sized balls according to their mass and to find the relative numbers of each present.

A real mass spectrometer works in a similar fashion to this simple model. It separates atoms according to their mass and shows the relative numbers of the different atoms present. Before the atoms can be deflected and separated, they must be converted to positively charged ions.

Figure 1.2 shows a simple mass spectrometer. There are five main stages.

1 Vaporisation—the sample of element is vaporised.

2 Ionisation—positive ions are obtained from the vapour.

3 Acceleration—the positive ions are accelerated by an electric field.

4 Deflection—the positive ions are deflected by a magnetic field.

5 Detection—the ions are detected and a record is made.

Vaporisation

Gases, liquids and volatile solids are injected into the instrument just before the ionisation chamber. Less volatile solids must be pre-heated to vaporise them.

Ionisation

After vaporisation, the element passes into the ionisation chamber. Here, atoms of the element are bombarded with a stream of high-energy electrons. This causes ionisation. One or occasionally two electrons are knocked out of the atoms leaving positive ions. Thus, for an atom X we have:

$$e^- + X \rightarrow X^+ + e^- + e^-$$

high-energy electron   atom   positive ion   electron knocked out of X   high-energy electron retreating

or occasionally $e^- + X \rightarrow X^{2+} + e^- + e^- + e^-$

2 electrons knocked out of X

Figure 1.3
A mass spectrometer trace for naturally occurring magnesium.

Acceleration

These positive ions, such as X^+ and X^{2+}, now pass through holes in parallel plates to which an electric field is applied. The electric field accelerates the ions into the instrument towards the magnetic field.

Deflection

As the ions pass through the magnetic field, they are deflected according to their mass and their charge Notice that particles can only pass through the instrument if they are positively charged.

Detection

If the accelerating electric field and the deflecting magnetic field stay constant, ions of only one particular mass/charge ratio will hit the ion detector at the end of the apparatus. Ions of smaller mass/charge ratio will be deflected too much. Ions of greater mass/charge ratio will be deflected too little.

The ion detector is usually linked through an amplifier to a recorder. As the strength of the magnetic field is slowly increased, ions of increasing mass will be detected and a mass spectrum similar to that shown in figure 1.3 is traced out by the recorder. The relative heights of the peaks in the mass spectrum give a measure of the relative amounts of the different ions present. (Strictly speaking, it is the areas under the peaks and not the peak heights which give the relative amounts or abundancies.)

In practice, a reference peak using a known substance is first obtained on the mass spectrum. The relative masses of other particles can then be obtained by comparison with this. Look closely at figure 1.3.

■ How many different ions are detected in the mass spectrum of naturally occurring magnesium?
■ What are the relative masses of these different ions?
■ What are the relative proportions of these different ions?

The interpretation of mass spectrometer traces and the calculation of relative atomic masses is taken up again in sections 5.5 and 26.4.

1.3 Relative atomic masses—the relative atomic mass scale

Chemists use a **relative atomic mass scale** to compare the masses of different atoms. This scale is sometimes called the **relative atomic weight scale**. At first the element hydrogen was chosen as the standard against which the masses of other atoms were compared. Hydrogen was chosen initially because chemists realised that it had the smallest atoms which could conveniently be assigned a relative atomic mass of one (H = 1). At a later date, when relative atomic masses could be obtained more accurately and when scientists realised that one element could contain atoms of different mass (**isotopes**—see section 5.4), it became necessary to chose a single isotope as the reference standard for relative atomic masses.

In 1961, carbon-12 (^{12}C) was chosen as the new standard. This is the isotope of carbon with a relative atomic mass of 12. An isotope of carbon was chosen rather than an isotope of hydrogen because carbon is a very common element. Being a solid, it is also easier to store and transport than hydrogen, which is a gas.

On the carbon-12 scale, atoms of the isotope ^{12}C are assigned a relative atomic mass of 12. The relative masses of all other atoms are obtained by comparison with the mass of a carbon-12 atom. A few relative atomic masses are listed in table 1.1. For example, the relative atomic mass of magnesium, on the carbon-12 scale, is 24.312. This means that the average mass of a magnesium atom and the mass of a carbon-12 atom are in the ratio 24.312:12.000. Notice in table 1.1 that the relative atomic mass of carbon is 12.011, i.e. the average mass of a carbon atom is 12.011, not 12.000. This is because naturally occurring carbon contains a

Table 1.1
The relative atomic masses of some elements

Element	Symbol	Relative atomic mass
Carbon-12	^{12}C	12.000
Carbon	C	12.011
Chlorine	Cl	35.453
Copper	Cu	63.540
Hydrogen	H	1.008
Iron	Fe	55.847
Magnesium	Mg	24.312
Sulphur	S	32.064

Chemists are not the only people who 'count by weighing'.

few atoms of carbon-13 and carbon-14 mixed in with those of carbon-12.

Use table 1.1 to answer the following questions.

■ Roughly how many times are magnesium atoms heavier than ^{12}C atoms?
■ Approximately how many times are carbon atoms heavier than hydrogen atoms?
■ Which element has atoms approximately twice as heavy as sulphur atoms?

A complete list of relative atomic masses can be found on page 602.

1.4 Using relative atomic masses to count atoms—the mole

Since one atom of carbon is 12 times as heavy as one atom of hydrogen, it follows that 12 grams of carbon and one gram of hydrogen contain the same number of atoms. In the same way, since an atom of sulphur is 32 times as heavy as an atom of hydrogen, it follows that 32 grams of sulphur will also contain the same number of atoms as one gram of hydrogen. In fact, *the relative atomic mass in grams of all elements will contain the same number of atoms*. Experiments show that this number is 6.0×10^{23}. Written out in full, this is 600 000 000 000 000 000 000 000. It is called the **Avogadro constant**. At one time, this number was referred to as the Avogadro number, but the term Avogadro constant (rather than number) is now used since its value is constant for all elements.

The Avogadro constant, represented by the symbol L, is now defined as *the number of atoms in exactly 12 grams of ^{12}C*.

Since 12 g of carbon contains 6.0×10^{23} atoms,

$$1 \text{ g of carbon contains } \frac{6.0 \times 10^{23}}{12} \text{ atoms, and}$$

$$5 \text{ g of carbon contains } 5 \times \frac{6.0 \times 10^{23}}{12} \text{ atoms.}$$

In this way, it is easy for chemists to count the number of atoms in a sample of an element by weighing. Chemists are not the only people who 'count by weighing'. Bank clerks use the same idea when they count money (silver or bronze) by weighing it. Since 100 pennies weigh 356 grams, it is quicker to take one hundred 1p coins by weighing out 356 grams of them than by counting.

■ What is the mass of 6×10^{23} atoms of copper?
■ What is the mass of 6×10^{24} atoms of magnesium?
■ How many atoms are there in 6 g of carbon?
■ How many atoms are there in 20 g of hydrogen?
■ How heavy is one atom of carbon?

Since the atomic mass in grams of all elements contains 6.0×10^{23} atoms, chemists refer to 6.0×10^{23} atoms of an element as one **mole** of atoms. They also use the term mole to mean *the relative atomic mass in grams*. Strictly speaking, the mole is defined as the amount of substance which contains the same number of particles (atoms, ions or molecules) as there are atoms in exactly 12 grams of ^{12}C. Thus, one mole of any substance contains 6×10^{23} particles.

In equations, the symbol for an element can be used to represent one mole of the element as well as one particle of the element. Thus, C represents one mole of carbon atoms (12 g of carbon) and $\frac{1}{10}$ C represents one-tenth of a mole of carbon atoms (1.2 g of carbon). O represents one mole of oxygen atoms (16 g of oxygen) but O_2 represents one mole of oxygen molecules (32 g of oxygen).

Notice how important it is to specify precisely which particles you mean in discussing the number of moles of different substances. For example, the statement 'one mole of oxygen' is ambiguous. It could mean one mole of oxygen

Sir John Cockroft beside the apparatus with which he and Dr E. T. Walton first 'split the atom' in 1932. Protons were accelerated down the large central tube towards a lithium target. The bombarding protons split some of the lithium atoms into helium.

atoms (O), i.e. 16g; or it could mean one mole of oxygen molecules (O_2), i.e. 32g. Because of this ambiguity, it is important to state the formula of the substance involved.

In considering giant structures such as sodium chloride, Na^+Cl^-, it is again important to quote the formula in discussing amounts of substance. Thus, one mole of sodium chloride, $Na^+ Cl^-$, will have a mass equal to the formula weight in grams, i.e. 58.5g. It contains one mole of sodium ions, Na^+ (23.0g), and one mole of chloride ions, Cl^-(35.5g). On the other hand, one mole of magnesium chloride, $Mg^{2+} (Cl^-)_2$ (95.3g), contains one mole of Mg^{2+} ions (24.3g) and two moles of Cl^- ions (2 × 35.5g).

- What is the mass of one mole of hydrogen molecules, H_2?
- What is the mass of one mole of sulphur dioxide, SO_2?
- What is the mass of 0.5 moles of SO_2?
- How many moles of SO_2 are there in 16g of SO_2?
- How many molecules of SO_2 are there in 16g of SO_2?

From the discussion you will see that it is more sensible for chemists to consider amounts which contain the same number of particles (atoms, ions, molecules or formula units) rather than amounts which contain the same number of grams. Thus, in making one mole of iron (II) sulphide (88g of FeS) from iron and sulphur, it is necessary to weigh out 56 grams of Fe and 32 grams of S (and not 44g of Fe and 44g of S).

1.5 Using relative atomic masses to find the formulae of compounds

The formula of a compound shows the relative number of atoms of each element in it. The formula can be calculated once we know the actual masses and relative atomic masses of the different elements in a sample of the compound.

The following example shows how the formula of a compound can be obtained after its composition has been determined by experiment.

When ethene is analysed it is found to contain 85.72% carbon and 14.28% hydrogen.

	C	H
masses reacting	85.72 g	14.28 g
mass of 1 mole	12 g	1 g
∴ moles present	$\dfrac{85.72}{12}$ $= 7.14$	$\dfrac{14.28}{1}$ $= 14.28$
ratio of moles	1	2
∴ ratio of atoms	1	2
⇒ formula	C	H_2

This formula for ethene shows only the simplest ratio of carbon atoms to hydrogen atoms. The actual formula showing the correct number of carbon atoms and hydrogen atoms in one molecule of ethene could be CH_2, C_2H_4, C_3H_6, C_4H_8, etc., since all these formulae give CH_2 as the simplest ratio of atoms.

The simplest formula for a substance, such as CH_2 for ethene, is called the **empirical formula**. This shows the simplest whole-number ratio for the atoms of different elements in the compound. In the case of molecular compounds such as ethene, it is generally better to use the **molecular formula**. This shows the actual number of atoms of the different elements in one molecule of the compound. The molecular formula for ethene is not CH_2 but C_2H_4.

In substances composed of giant lattices, whether these are giant covalent structures, such as silicon (IV) oxide, SiO_2, or giant ionic structures, such as sodium chloride, Na^+Cl^-, it is of course meaningless to talk of molecules and molecular formulae. In this case, the formulae we use are empirical formulae. Experimental methods of determining the structures and formulae of compounds are described in sections 8.5, 10.2, 10.3, 11.10, 11,10 and 26.4.

1.6 Solutions and moles

When two solutions react together there is usually a simple ratio between the number of moles of solute in one solution, and the number of moles of solute with which it reacts in the other solution. For this reason, it is convenient to represent the concentration of a solute in terms of the number of moles of it in a particular volume of solution. Chemists usually measure the concentration of solutions as **moles per cubic decimetre**. The cubic decimetre (dm^3) is now the agreed international name for the unit of volume also known as the litre. In this book, we shall use cubic decimetres (not litres) and cubic centimetres, cm^3 (not millilitres, ml).

$$1 \text{ decimetre (dm)} = \tfrac{1}{10} \text{ metre} = 10 \text{ cm}$$

$$\therefore 1 \text{ dm}^3 = (10 \text{ cm})^3 = 1000 \text{ cm}^3 = 1 \text{ litre}$$

A solution of HCl(aq) containing 1.0 mole per cubic decimetre will have 36.5 g of HCl in 1 dm^3 of solution. A solution with 2.0 moles of HCl per cubic decimetre contains 73 g (2×36.5 g) in 1 dm^3 of solution.

Notice that these concentrations are expressed as the number of moles in 1 dm^3 of *solution*, not 1 dm^3 of *solvent*.

At one time, a solution containing 1.0 mole of solute per dm^3 was described as 'one molar', but the words *molar* and *molarity* are now obsolete and it is wrong to use them in the context of solution concentrations. The units associated with the word 'molar' are 'per mole', so it is quite incorrect to refer to a solution containing 0.1 moles of solute per dm^3 as '0.1 molar'. Nevertheless, the abbreviation '0.1 M' for '0.1 moles per cubic decimetre' is still used even though the *word* 'molar' is no longer appropriate. Thus, the symbol, M, can be used to mean 'moles per cubic decimetre', but the word 'molar', should *not* be used to describe the concentrations of solutions.

■ How many moles of H_2SO_4 are there in 9.8 g?
■ What is the concentration, in moles per dm^3, of a solution containing 9.8 g of H_2SO_4 in 100 cm^3?
■ How many moles of NaCl are there in 5 dm^3 of a 1.0 M solution?
■ How many moles of NaCl are there in 200 cm^3 of 2.0 M solution?

Summary

1   An atom is the smallest part of an element which can ever exist.

2   A molecule is the smallest part of an element or a compound which can exist alone under ordinary conditions.

3   Most atoms have a radius between 0.07 and 0.20 nanometres (nm). 1 nanometre $\equiv 10^{-9}$ m.

4   Chemists use the relative atomic mass scale to compare the masses of different atoms. Atoms of the isotope ^{12}C are assigned a relative atomic mass of 12 and the relative masses of all other atoms are obtained by comparison with the mass of a carbon-12 atom.

5 The relative atomic mass in grams of any element contains 6×10^{23} atoms.

6 The relative molecular mass in grams of any compound contains 6×10^{23} molecules.

7 The Avogadro constant (6.0×10^{23}) is defined as the number of atoms in exactly 12 g of ^{12}C.

8 The mole is the amount of substance which contains the same number of particles (atoms, ions, molecules or formula units) as there are atoms in exactly 12 g of ^{12}C.

9 An empirical formula shows the simplest whole-number ratio for the atoms of different elements in a compound.

10 A molecular formula shows the actual number of atoms of the different elements in one molecule of a compound.

11 The concentrations of solutes in solutions are usually expressed in moles per cubic decimetre (moles dm^{-3}).

12 The symbol M can be used to mean 'moles per cubic decimetre', but the word 'molar' should not be used to describe the concentrations of solutions.

Study questions

1 What is the mass of
   (a) 6×10^{23} atoms of O,
   (b) 6×10^{23} atoms of P,
   (c) 6×10^{23} molecules of O_2,
   (d) 6×10^{23} molecules of P_4,
   (e) one mole of carbon dioxide, CO_2,
   (f) two moles of silver, Ag,
   (g) 0.2 moles of sulphur dioxide, SO_2,
   (h) NaOH in 2 dm^3 of 1.5 M solution?

2 How many moles of
   (a) Cl_2 are there in 7.1 g of chlorine,
   (b) $CaCO_3$ are there in 10.0 g of calcium carbonate,
   (c) Ag are there in 10.8 g of silver,
   (d) NH_3 are there in 3.4 g of ammonia,
   (e) S are there in 32 g of sulphur,
   (f) S_8 are there in 32 g of sulphur,
   (g) NaOH are there in 1 dm^3 of 3.0 M solution,
   (h) NaOH are there in 20 cm^3 of 0.1 M solution?

3 How many atoms are there in
   (a) two moles of iron, Fe,
   (b) 0.1 moles of sulphur, S,
   (c) 18 g of water, H_2O,
   (d) 0.44 g of carbon dioxide, CO_2?

4 Read section 1.2 again.
   (a) Why must a mass spectrometer be evacuated to a very low pressure before being used?
   (b) How would the accelerating field differ in its effect on X^+ and X^{2+}?
   (c) How would the deflecting magnetic field differ in its effect on X^+ and X^{2+}?
   (d) Relative atomic masses are not really masses. What are they?
   (e) What units do relative atomic masses have?

5 15.3g of element X (X = 27) will combine with 13.6g of oxygen to form an oxide.
(a) Express this in moles.
(b) How many moles of oxygen, O, will combine with one mole of X?
(c) What is the simplest formula for the oxide?

6 5.34g of a salt of formula M_2SO_4 (where M is a metal) were dissolved in water. The sulphate ion was precipitated by adding excess barium chloride solution when 4.66g of barium sulphate ($BaSO_4$) were obtained.
(a) How many moles of sulphate ion were precipitated as barium sulphate?
(b) How many moles of M_2SO_4 were in the solution?
(c) What is the formula mass of M_2SO_4?
(d) What is the relative atomic mass of M?
(e) Use a table of relative atomic masses to identify M.

7 2.4g of a compound of carbon, hydrogen and oxygen gave, on combustion, 3.52g of CO_2 and 1.44g of H_2O. The relative molecular mass of the compound was found to be 60.
(a) What are the masses of carbon, hydrogen and oxygen in 2.4g of the compound?
(b) What are the empirical and molecular formulae of the compound?

8 Figure 1.4(a) shows the mass spectrum of HCl. The peak at mass 36 corresponds to the molecular ion, $(H^{35}Cl)^+$.
(a) What particle is responsible for the prominent peak at mass 38?
(b) What particles are responsible for the two lower peaks?
(c) How do you explain the relative heights of the peaks at mass 36 and 38?
Figure 1.4(b) shows the mass spectrum of methane. The peak at mass 16 corresponds to the molecular ion $(CH_4)^+$.
(d) How do you explain the peaks of relative mass 1, 2, 12, 13, 14, 15 and 17?

Figure 1.4
Mass spectra for (a) hydrogen chloride and (b) methane.

2 Reacting Quantities and Equations

Formulae and equations help us to calculate how much product we can get from a given amount of reactant. This is important in processes such as the manufacture of lime from limestone. This was once carried out in kilns like the one shown above.

2.1 What is an equation?

In your early studies of chemistry you have used word equations to summarise the reactions between substances. For example, when shiny aluminium tarnishes

$$\text{aluminium} + \text{oxygen} \rightarrow \text{aluminium oxide}$$

However, it is more convenient to use the formula of a substance in equations as a summary of its name. For example, in the word equation above we could write aluminium as Al, oxygen as O_2 and aluminium oxide as Al_2O_3. The symbols Al_2O_3 can be used to represent 1 mole of aluminium oxide or 1 particle of Al_2O_3. In the same way, H_2O can be used to represent 1 mole of water or 1 molecule of water.

When iron (Fe) reacts with sulphur (S), experiments show that the product is iron (II) sulphide (FeS). This can be summarised in a word equation as

$$\text{iron} + \text{sulphur} \rightarrow \text{iron (II) sulphide}$$

or more simply using symbols and formulae as

$$\text{Fe} + \text{S} \rightarrow \text{FeS}$$

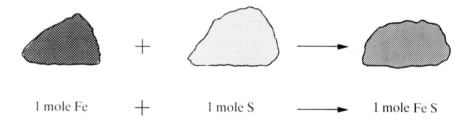

1 mole Fe $+$ 1 mole S $\longrightarrow$ 1 mole Fe S

2.2 Relative numbers of moles of reactants and products

It would be very helpful if we could summarise all reactions using balanced equations. However, these equations cannot be written simply by looking at the formulae of the reactants.

A drawing of the original balance with which John Dalton studied the reacting quantities of different substances.

In order to write an equation, we must know:

—the formulae of the reactants and products;
—how many moles of each reactant are involved;
—how many moles of each product are formed.

Various methods can be used to find the number of moles of substances which react. In acid/alkali reactions, we can use indicators to decide how much acid reacts with a given volume of alkali.

In precipitation reactions, we can investigate the volumes of solutions which react by adding increasing amounts of one liquid to a fixed volume of the other liquid until no more precipitate forms.

For reactions in which heat is evolved (or absorbed) we can find how much of one substance reacts with another by measuring temperature changes.

For reactions involving gases we can measure pressure or volume changes.

2.3 Equations for reactions involving precipitates

When colourless solutions of lead nitrate and potassium iodide are mixed together, a bright yellow precipitate forms immediately.

■ What is the formula of lead nitrate?
■ What is the formula of potassium iodide?
■ What is the yellow precipitate?

Careful analysis shows that the formula of the yellow precipitate is PbI_2. If the solution left behind after filtering off the precipitate is evaporated to dryness, a white solid remains. This is potassium nitrate (KNO_3).

The numbers of moles of lead nitrate (($PbNO_3$)$_2$) and potassium iodide (KI) which react together can be studied as follows. Put $5 cm^3$ of 1.0 M KI into each of eight test tubes. Then add $0.5 cm^3$ of 1.0 M $Pb(NO_3)_2$ to the first test tube, $1.0 cm^3$ to the second, $1.5 cm^3$ to the third and so on. A yellow precipitate of lead iodide forms in each tube. Centrifuge each tube for the same length of time and measure the depth of precipitate in each tube. The depth of precipitate is a measure of the mass of the solid produced. Figure 2.1 shows a typical set of results.

Figure 2.1

The depth of the precipitate produced when different volumes of 1.0 M $Pb(NO_3)_2(aq)$ are added to $5 cm^3$ of 1.0 M KI(aq).

volume of 1.0M KI/cm³	5	5	5	5	5	5	5	5
number of moles of KI	5×10^{-3}	5×10^{-3}	5×10^{-3}	5×10^{-3}	5×10^{-3}	5×10^{-3}	5×10^{-3}	5×10^{-3}
volume of 1.0M Pb(NO₃)₂/cm³	0.5	1.0	1.5	2.0	2.5	3.0	3.5	4.0
number of moles of Pb(NO₃)₂	0.5×10^{-3}	1.0×10^{-3}	1.5×10^{-3}	2.0×10^{-3}	2.5×10^{-3}	3.0×10^{-3}	3.5×10^{-3}	4.0×10^{-3}
depth of precipitate/mm	5	6	8	10	12	12	12	12

depth of precipitate/mm

volume of 1.0M Pb(NO₃)₂/cm³

Notice that $1.0\,\text{cm}^3$ of $1.0\,\text{M}$ $Pb(NO_3)_2$ produces more precipitate than $0.5\,\text{cm}^3$. $1.5\,\text{cm}^3$, $2.0\,\text{cm}^3$ and $2.5\,\text{cm}^3$ of $1.0\,\text{M}$ $Pb(NO_3)_2$ cause further increases in the amount of lead iodide produced. But, after $2.5\,\text{cm}^3$ of $1.0\,\text{M}$ $Pb(NO_3)_2$ has been added, there is no further increase in the amount of precipitate. This means that:

$5.0\,\text{cm}^3$ of $1.0\,\text{M}$ KI just react with $2.5\,\text{cm}^3$ of $1.0\,\text{M}$ $Pb(NO_3)_2$

These amounts are in the proportion of 2 moles of potassium iodide to 1 mole of lead nitrate, so the left-hand side of the equation is

$$2KI + Pb(NO_3)_2 \rightarrow$$

The products of the reaction are lead iodide, PbI_2, and potassium nitrate, KNO_3, so the overall balanced equation is

$$2KI(aq) + Pb(NO_3)_2(aq) \rightarrow PbI_2(s) + 2KNO_3(aq)$$

All four substances in this equation are ionic compounds. The ions are free to move apart in the aqueous solutions, but in the precipitate of lead iodide the Pb^{2+} ions are strongly attached to I^- ions, forming insoluble PbI_2. We can write an ionic equation for the reaction as

$$2K^+(aq) + 2I^-(aq) + Pb^{2+}(aq) + 2NO_3^-(aq) \rightarrow Pb^{2+}(I^-)_2(s) + 2K^+(aq) + 2NO_3^-(aq)$$

If we cancel the spectator ions ($2K^+(aq)$ and $2NO_3^-(aq)$) on both the left-hand side and right-hand side of this ionic equation, we get

$$Pb^{2+}(aq) + 2I^-(aq) \rightarrow Pb^{2+}(I^-)_2(s)$$

- How many moles are there in $5\,\text{cm}^3$ of $1.0\,\text{M}$ KI?
- How many moles are there in $2.5\,\text{cm}^3$ of $1.0\,\text{M}$ $Pb(NO_3)_2$?
- $5.0\,\text{cm}^3$ of $1.0\,\text{M}$ KI just react with $2.5\,\text{cm}^3$ of $1.0\,\text{M}$ $Pb(NO_3)_2$ and neither is in excess. How could you test that neither substance is in excess?

This nineteenth-century engraving shows water being electrolysed to form hydrogen and oxygen. The apparatus can be used to find the volumes of hydrogen and oxygen that react to form water.

2.4 Equations for reactions involving gases

When a small piece of magnesium is added to dilute hydrochloric acid a vigorous effervescence occurs, showing that a gas is produced. This gas is hydrogen, H_2.

We can investigate how many moles of hydrochloric acid react with 1 mole of magnesium by pipetting $25\,\text{cm}^3$ of $1.0\,\text{M}$ HCl into a small beaker, and adding $0.4\,\text{g}$ of magnesium ribbon. A vigorous reaction occurs and hydrogen is evolved. When the effervescence stops, all the acid has been used up and excess unreacted magnesium is left. The unreacted magnesium is washed, dried and weighed. Here are some typical results:

$$\text{volume of } 1.0\,\text{M HCl taken} = 25.0\,\text{cm}^3$$
$$\text{mass of magnesium before reaction} = 0.41\,\text{g}$$
$$\text{mass of magnesium unreacted} = 0.11\,\text{g}$$
$$\Rightarrow \text{mass of magnesium reacting} = 0.30\,\text{g}$$

$\therefore 25\,\text{cm}^3$ of $1.0\,\text{M}$ HCl reacts with $0.30\,\text{g}$ of Mg

$25\,\text{cm}^3$ of $1.0\,\text{M}$ HCl contains $\dfrac{25}{1000} \times 1.0$ moles HCl

$$= 0.025 \text{ moles HCl}$$

$0.30\,\text{g}$ of Mg $= \dfrac{0.30}{24}$ moles Mg $\qquad = 0.0125$ moles Mg

$\therefore 0.0125$ moles of Mg react with 0.025 moles of HCl
$\Rightarrow 1$ mole of Mg reacts with 2 moles of HCl

So, the left-hand side of the equation for the reaction is

$$Mg(s) + 2HCl(aq) \rightarrow$$

The next question to ask is 'how many moles of hydrogen are produced from 1 mole of magnesium?' We can find the answer to this question by adding $0.03\,\text{g}$ of

Figure 2.2
Measuring the volume of hydrogen produced when magnesium reacts with hydrochloric acid.

magnesium to about $5\,cm^3$ (an excess) of $2\,M$ HCl. The volume of hydrogen produced can be measured by connecting the tube in which the reaction takes place to a syringe (figure 2.2). Here are some typical results:

mass of magnesium taken $=$ $0.030\,g$
volume of hydrogen produced $=$ $30\,cm^3$

■ How many moles of magnesium is $0.030\,g$?
■ What is the mass of $30\,cm^3$ of hydrogen? ($1\,dm^3$ ($1000\,cm^3$) weighs $0.084\,g$ at room temperature.)
■ How many moles of $H_2(g)$ is this mass of hydrogen?
■ How many moles of $H_2(g)$ are produced from 1 mole of magnesium?

The results of this experiment show that 1 mole of magnesium reacts with hydrochloric acid to produce 1 mole of hydrogen (H_2). If the remaining solution was evaporated to dryness, pure white solid magnesium chloride, $MgCl_2$, would remain. Thus, the overall equation for this reaction is

$$Mg(s) + 2HCl(aq) \rightarrow H_2(g) + MgCl_2(aq)$$

Both hydrochloric acid and magnesium chloride are ionic substances. They consist of ions which are free to move apart in the aqueous solutions. The hydrochloric acid, HCl(aq), contains H^+ and Cl^- ions, and the aqueous magnesium chloride, $MgCl_2$(aq), contains Mg^{2+}(aq) and Cl^-(aq) ions. An ionic equation for the reaction is

$$Mg(s) + 2H^+(aq) + 2Cl^-(aq) \rightarrow H_2(g) + Mg^{2+}(aq) + 2Cl^-(aq)$$

Notice that neither magnesium nor hydrogen are written as ions. *The only ionic compounds are salts, bases and aqueous solutions of acids. Metals, like magnesium, consist of atoms. Non-metals, like hydrogen, and non-metal/non-metal compounds, like water, consist of molecules.* Both atoms and molecules are uncharged particles. For example, magnesium consists of uncharged magnesium atoms (Mg) and hydrogen gas consists of uncharged hydrogen molecules (H_2).

2.5 Reacting volumes of gases

The volume occupied by 1 mole of any gas at s.t.p. is $22.4\,dm^3$. This volume is usually called the **Molar Volume**. Experiments which lead to this result for the molar volume are described in section 11.6. Remember that s.t.p. means standard temperature and pressure. This is 0°C (273 K) and 1 atmosphere pressure. At room temperature (say 25°C) and pressure (1 atmosphere), the volume of 1 mole of any gas is about $24.4\,dm^3$.

Using the idea of molar volumes, it is possible to calculate the volumes of gases which react. For example, consider the manufacture of ammonia from nitrogen and hydrogen:

$$\text{nitrogen} + \text{hydrogen} \rightarrow \text{ammonia}$$
$$N_2(g) + 3H_2(g) \rightarrow 2NH_3(g)$$

From the equation:

$$1 \text{ mole of } N_2 + 3 \text{ moles } H_2 \rightarrow 2 \text{ moles } NH_3$$

Using the idea that 1 mole of each gas will occupy $22.4 \, dm^3$ at s.t.p., we can write:

$$22.4 \, dm^3 \, N_2 + (3 \times 22.4) dm^3 \, H_2 \rightarrow 2 \times 22.4 \, dm^3 \, NH_3$$
$$\Rightarrow 1 \, dm^3 \, N_2 + 3 \, dm^3 \, H_2 \rightarrow 2 \, dm^3 \, NH_3$$
$$\Rightarrow 1 \text{ volume } N_2 + 3 \text{ volumes } H_2 \rightarrow 2 \text{ volumes } NH_3$$

So, if we had $10 \, cm^3$ of nitrogen it would react with 3 volumes of hydrogen ($30 \, cm^3$) to give 2 volumes of ammonia ($20 \, cm^3$).

Methane burns in oxygen to form carbon dioxide and water:

$$CH_4(g) + 2O_2(g) \rightarrow CO_2(g) + 2H_2O(g)$$

■ How many moles of CH_4, O_2, CO_2 and H_2O does the equation show?
■ What are the relative volumes of the reactants and products in the equation?
■ How much oxygen is needed to react with $10 \, dm^3$ of methane?
■ A biogas digester produces $1 \, dm^3$ of methane every hour. How much methane collects in one day? How much carbon dioxide is produced when this methane burns?

2.6 Predicting and writing equations

You will realise by now that it is impossible to write an equation unless you know:

—the formulae of the reacting substances and products;
—the relative numbers of moles of reactants.

However, if you know the names of all the reactants and products, it is possible to predict their formulae and a possible equation. It is very convenient to predict equations in this way, instead of carrying out an experiment to establish the equation for every reaction.

There are three stages in writing equations.

1 Write out the equation in words. For example

$$\text{copper} + \text{oxygen} \rightarrow \text{copper (II) oxide}$$

2 Write symbols for elements and formulae for compounds in the equation. This gives

$$Cu + O_2 \rightarrow CuO$$

Remember that hydrogen, oxygen, nitrogen and the halogens exist as molecules containing two atoms when they are not combined with other elements. They are called **diatomic molecules**. Thus they are represented in equations as H_2, O_2, N_2, Cl_2, Br_2 and I_2.

All other elements are regarded as single atoms and represented by a monatomic symbol in equations.

3 Balance the equation by making the number of atoms of each element the same on both sides. In the last equation, there are two oxygen atoms on the left and only one on the right. Therefore CuO on the right must be doubled

(2CuO). There are now two copper atoms on the right, so two copper atoms must be shown on the left (2Cu). The balanced equation is

$$2Cu + O_2 \rightarrow 2CuO$$

Remember that formulae must *never* be altered in balancing an equation. Atoms can only be balanced by putting a number in front of a formula, thus doubling or trebling, etc., the whole formula.

Here are two more examples to show how equations are written:

1 The reaction between hydrogen and chlorine:
   (i) Write the equation in words:

$$hydrogen + chlorine \rightarrow hydrogen\ chloride$$

   (ii) Write symbols for elements and formulae for compounds:

$$\Rightarrow H_2 + Cl_2 \rightarrow HCl$$

   Notice that hydrogen and chlorine are written as H_2 and Cl_2 since they exist as diatomic molecules.
   (iii) Balance the equation. There are two hydrogen atoms on the left and only one on the right. There are also two chlorine atoms on the left and only one on the right. By doubling the formula of HCl (i.e. 2HCl), both hydrogen and chlorine are balanced:

$$\Rightarrow H_2 + Cl_2 \rightarrow 2HCl$$

2 The reaction between magnesium and nitric acid:
   (i) Write the equation in words:

$$magnesium + nitric\ acid \rightarrow magnesium\ nitrate + hydrogen$$

   (ii) Write symbols for elements and formulae for compounds:

$$Mg + HNO_3 \rightarrow Mg(NO_3)_2 + H_2$$

   (iii) Balance the equation. Groups of atoms that form a single ion, such as nitrate (NO_3^-), are regarded as a single unit in balancing equations. There are two nitrate ions on the right and only one on the left. There are two hydrogen atoms on the right and only one on the left. By writing $2HNO_3$ on the left, the whole equation is balanced:

$$Mg + 2HNO_3 \rightarrow Mg(NO_3)_2 + H_2$$

   These equations in which all substances are represented by their full chemical formulae are called **formal** or **conventional equations**.

Write equations for the reaction of
■ zinc with oxygen.
■ copper oxide with hydrochloric acid.
■ sodium with water.

2.7 What do equations tell us?

A balanced chemical equation tells us

—the reactants and products;
—the formulae of reactants and products;
—the relative numbers of moles of the reactants and products.
For example, the equation

$$2Cu + O_2 \rightarrow 2CuO$$

tells us that

—the formula of copper (II) oxide is CuO and that of oxygen is O_2;
—copper reacts with oxygen to form copper oxide;
—two moles of copper atoms react with 1 mole of oxygen molecules (O_2) to produce 2 moles of copper oxide (CuO).

Some of the most useful data we can obtain from balanced equations are the relative masses of reactants and products.

The mass of 1 mole of any substance can be calculated from the relative atomic masses of its constituent elements. So, from the last equation, we can calculate that

$$(2 \times 63.5)\text{g Cu} + 32\text{g O}_2 \rightarrow (2 \times 79.5)\text{g CuO}$$

i.e. 127g copper react with 32g oxygen to form 159g of copper (II) oxide (see figure 2.3).

(2 × 63.5) g of Cu

32g of oxygen (about 24 litres at room temperature and pressure; about 1 large biscuit tin)

(2 × 79.5)g of CuO

Figure 2.3
The mass of reactants and products when copper reacts with oxygen to form copper oxide.

Information concerning the masses of reactants and products is especially useful in industry. Industrial chemists need to know how much product they can obtain from a given amount of starting material. For example, in the manufacture of lime they can calculate the mass of lime (calcium oxide) that can be obtained from 1 tonne (1000 kg) of pure limestone (calcium carbonate), $CaCO_3$.

The equation for the reaction is

$$CaCO_3 \rightarrow CaO + CO_2$$
$$\Rightarrow 1 \text{ mole of } CaCO_3 \rightarrow 1 \text{ mole of } CaO + 1 \text{ mole of } CO_2$$
$$(40 + 12 + (3 \times 16))\text{g} \rightarrow (40 + 16)\text{g} + (12 + (2 \times 16))\text{g}$$
$$\Rightarrow 100 \text{g } CaCO_3 \rightarrow 56 \text{g } CaO + 44 \text{g } CO_2$$

∴ 1000 kg of limestone produces 560 kg of lime.

Copper is obtained from copper pyrites, $CuFeS_2$ (Cu = 64, Fe = 56, S = 32).

■ What is the relative formula mass of $CuFeS_2$?
■ How much copper can be obtained from 1 mole of $CuFeS_2$?
■ How much copper can be obtained from 1 tonne (1000 kg) of $CuFeS_2$?

The photographs below show three stages in the manufacture of copper. First the copper ore is mined. The ore is then crushed and separated by froth flotation (centre). The copper ore is then reduced and purified to obtain bars of copper similar to those in the third photograph. How much copper can be obtained from 100 tonnes of copper ore, $CuFeS_2$? (Cu = 64, Fe = 56, S = 32)

Equations are very useful, but we must remember their limitations. An equation will *not* tell us:

—the rate of the reaction—whether it is fast or slow, whether it will explode or whether it will take years to react;

—how or why a reaction takes place. The equation for the reaction of magnesium with oxygen is

$$2Mg(s) + O_2(g) \rightarrow 2MgO(s)$$

The equation does not show that the magnesium must be heated before a reaction will take place. Nor does it show that once the reaction has started it will continue vigorously of its own accord. The equation does not tell us why the magnesium reacts with oxygen or how the atoms and molecules behave in the reaction. Do oxygen molecules (O_2) each combine with two magnesium atoms, or do oxygen molecules split to form separate oxygen atoms which then combine with separate magnesium atoms? The equation cannot answer these questions.

Equations can only tell us about the amounts and states of the reactants and products. They cannot tell us anything about what happens on the route between the reactants and the products. Equations can only tell us the overall change which occurs.

2.8 Ionic equations

Many reactions involve ionic compounds. The part played by the separate ions of these compounds in a reaction can be shown in an ionic equation. Here are some examples of ionic equations.

1 When hydrochloric acid reacts with sodium hydroxide solution, the equation for the reaction is

$$HCl(aq) + NaOH(aq) \rightarrow NaCl(aq) + H_2O(l)$$

HCl(aq), NaOH(aq) and NaCl(aq) consist of ions, so we can write an ionic equation for the reaction as:

$$\underbrace{H^+(aq) + Cl^-(aq)}_{\substack{\text{hydrochloric} \\ \text{acid}}} + \underbrace{Na^+(aq) + OH^-(aq)}_{\substack{\text{sodium hydroxide} \\ \text{solution}}} \rightarrow \underbrace{Na^+(aq) + Cl^-(aq)}_{\substack{\text{sodium chloride} \\ \text{solution}}} + \underbrace{H_2O}_{\text{water}}$$

2 The combination of iron and sulphur can also be represented by an ionic equation:

$$Fe(s) + S(s) \rightarrow Fe^{2+}S^{2-}(s)$$

3 When molten sodium chloride is electrolysed, the reactions which take place at the electrodes can be summarised as follows:

at the anode (+)

$$2Cl^-(l) \quad \rightarrow \quad 2e^- + 2Cl \quad \rightarrow \quad Cl_2(g)$$
(chloride ions in liquid sodium chloride)   (electrons taken by positive anode)   (two chlorine atoms join together to form a chlorine molecule)

at the cathode (−)

$$2Na^+(l) \quad + \quad 2e^- \quad \rightarrow \quad 2Na(l)$$
(sodium ions in liquid sodium chloride)   (electrons taken from cathode)   (sodium atoms remain liquid at the high temperature of the process)

4 Earlier in this chapter we used an ionic equation to represent the precipitation of lead iodide during the reaction between lead nitrate and potassium iodide:

$$Pb^{2+}(aq) + 2I^-(aq) \rightarrow Pb^{2+}(I^-)_2(s)$$

5 Earlier in this chapter we also used an ionic equation to summarise the reaction between magnesium and hydrochloric acid:

$$Mg(s) + 2H^+(aq) + 2Cl^-(aq) \rightarrow Mg^{2+}(aq) + 2Cl^-(aq) + H_2(g)$$

These five examples illustrate the commonest types of reaction for which ionic equations can be used:

—The reactions between acids and bases (neutralisation).
—The reactions of metals with non-metals.
—Electrolyses.
—The precipitations of insoluble ionic solids.
—The reactions of metals with acids.

True or false? In chemical reactions:
- atoms may be converted to ions of the same element.
- atoms may be converted to atoms of a different element.
- molecules may be converted to ions.
- ions may be converted to molecules.
- there is sometimes a smaller number of molecules after the reaction than before.

2.9   Writing ionic equations

Ionic equations are often simpler and easier to write than formal equations. In order to obtain an ionic equation for a reaction:

1 Write ionic formulae for the ionic substances that take part in the reaction. The three main types of ionic substance are salts, bases and aqueous solutions of acids. For example

$$\underbrace{2H^+(aq) + SO_4^{2-}(aq)}_{\text{sulphuric acid}} + \underbrace{2Na^+(aq) + 2OH^-(aq)}_{\text{sodium hydroxide}} \rightarrow \underbrace{2Na^+(aq) + SO_4^{2-}(aq)}_{\text{sodium sulphate}} + \underbrace{2H_2O(l)}_{\text{water}}$$

2 Cancel spectator ions which appear unchanged on both sides of the equation and take no part in the reaction:

$$2H^+(aq) + \cancel{SO_4^{2-}(aq)} + \cancel{2Na^+(aq)} + 2OH^-(aq) \rightarrow \cancel{2Na^+(aq)} + \cancel{SO_4^{2-}(aq)} + 2H_2O(l)$$
$$\Rightarrow 2H^+(aq) + 2OH^-(aq) \rightarrow 2H_2O(l)$$

3 Check that ionic substances are written in the correct state.
    Aqueous solutions of ionic substances are written with a + sign between the different ions because the ions are free to move apart in the aqueous solution (e.g. $2Na^+(aq) + SO_4^{2-}(aq)$).
    The ions in molten (liquid) ionic substances are also free to move apart. Thus, liquid sodium sulphate is $2Na^+(l) + SO_4^{2-}(l)$.
    In solid ionic substances, the ions are held together by the attraction of their oppositely charged neighbours. The ions are *not* free to move apart and so they are written without a + sign between them (e.g. $(Na^+)_2SO_4^{2-}(s)$ for sodium sulphate).

4 Check that the equation is balanced with respect to atoms and with respect to positive and negative charges. Some ionic equations (in particular those for electrolyses) involve electrons which carry one unit of negative charge.

The main advantage of ionic equations is that they provide a clearer picture of what is happening in chemical reactions. It is possible to see which ions are taking part in the reaction and which are spectators.

Summary

1   An equation is a summary of the reactants and the products in a chemical reaction.

2   In order to write an equation we must know:
   (i)   the formulae of reactants and products, and
   (ii) the relative numbers of moles of reactants and products.

3   The rules for writing equations are
   (i)   write a word equation,
   (ii) write symbols for elements and formulae for compounds,
   (iii) balance the equation.

4   Equations can only tell us about the amounts and states of the reactants and products. They cannot tell us how or why the reaction occurs.

5   Equations in which all substances are represented by their full chemical formulae are called formal or conventional equations.

6   The part played by separate ions in a reaction can be shown in an ionic equation.

7   Ionic equations are useful in summarising:
   (i)   the reactions between acids and bases,
   (ii) the reactions between acids and metals,
   (iii) the reactions between metals and non-metals,
   (iv) electrolyses,
   (v)   the precipitation of insoluble ionic solids.

Study questions

1   0.27 g of aluminium reacts with 2.4 g of bromine.
   (a) How many moles are 0.27 g of aluminium?
   (b) How many moles are 2.4 g of bromine?
   (c) Write a possible formula for the compound of aluminium and bromine which is produced.
   (d) What is the relative molecular mass of a compound with this formula?
   (e) The *actual* relative molecular mass of the compound of aluminium with bromine is 534. What is the *actual* molecular formula of the compound?
   (f) Write an equation for the reaction of aluminium with bromine.

2   A compound of nitrogen and oxygen was found to contain 28 g of nitrogen to every 64 g of oxygen ($N = 14$, $O = 16$). Which of the following formulae could represent the compound?
   A N_2O B N_2O_2 C NO_2 D N_2O_4 E N_4O_2

3   (a) Explain what is meant by 'hydrogen peroxide has the molecular formula H_2O_2'.
   (b) $2H_2O_2 \rightarrow 2H_2O + O_2$
        What mass of oxygen can be formed from 17 g of H_2O_2?
   (c) What mass of oxygen can be formed from 17 g of H_2O_2 if 1 g of manganese dioxide catalyst is used?

4   When lithium perchlorate, $LiClO_4$, is heated, a gas is given off and a white solid is left. The gas is colourless and re-lights a glowing splint. The solid residue dissolves completely in water. When silver nitrate is added to this solution, a white precipitate is formed.
   (a) What is the name and formula of the gas produced?
   (b) What is the name and formula of the solid left on heating?

(c) Write an equation for the reaction which takes place when lithium perchlorate is heated.

(d) Write an equation for the reaction of the aqueous solution of the solid residue with silver nitrate.

(e) 2.13 g of lithium perchlorate were heated to constant weight. What weight of residue was produced?

5   By calculating relative molecular masses, find whether it is cheaper to buy washing soda (sodium carbonate) as the decahydrate ($Na_2CO_3.10H_2O$) at 4p per kg, or as the anhydrous salt (Na_2CO_3) at 8p per kg.

6   Write formal equations to summarise the following:

(a) When aqueous solutions of sodium chloride and silver nitrate are mixed, a white precipitate of silver chloride forms.

(b) The electrolysis of molten aluminium oxide produces aluminium and oxygen.

(c) Iron reacts with chlorine to form iron (III) chloride.

(d) Methane (CH_4) burns in oxygen to produce carbon dioxide and water.

(e) Copper (II) oxide reacts with hot dilute sulphuric acid to produce copper (II) sulphate and water.

Rewrite (a), (b) and (c) as ionic equations.

7   True or false? A balanced equation shows:

A   the formulae of the products.

B   the molar proportions in which the products are formed.

C   that a reaction can occur.

D   the relative numbers of atoms, ions and molecules which react.

E   that a reaction is exothermic.

8   Sulphuric acid reacts with 'thio' (sodium thiosulphate) producing a precipitate of sulphur. Samples of $10 cm^3$ of 0.1 M 'thio' were taken in similar tubes and varying quantities of 0.2 M acid were added. The sulphur precipitate was allowed to settle and its depth measured. The results are given in figure 2.4.

Volume of 0.2 M H_2SO_4 added/cm^3	1	3	5	7	9	11
Depth of precipitate/mm	1	3	5	4.5	5	5

Figure 2.4

(a) Draw a careful sketch graph (it is unnecessary to use graph paper) of the depth of precipitate against the volume of acid added.

(b) In what molar proportions do 'thio' and acid react?

9   A hydrate of potassium carbonate has the formula $K_2CO_3.xH_2O$. Ten grams of the hydrate leave 7.93 g of anhydrous salt on heating.

(a) What is the mass of anhydrous salt in 10 g of hydrate?

(b) What is the mass of water in 10 g of hydrate?

(c) How many moles of anhydrous salt are there in 10 g of hydrate?

(d) How many moles of water are there in 10 g of hydrate?

(e) How many moles of water are present in the hydrate for every 1 mole of K_2CO_3?

(f) What is the formula of the hydrate?

10   $25 cm^3$ of a solution of NaOH required $28 cm^3$ of 1.0 M H_2SO_4 to neutralise it.

(a) Write the equation for the reaction.

(b) How many moles of H_2SO_4 were needed?

(c) How many moles of NaOH were thus neutralised?

(d) How many moles of NaOH are there in $25 cm^3$ of solution?

(e) What is the molarity of the NaOH?

3 Redox

Above. Antoine Lavoisier (1743–1794), the father of modern chemistry. Lavoisier was the son of a rich Parisian lawyer. He was the first chemist to explain the redox reactions which occur during burning.

Right. In 1775, Lavoisier was appointed to a post at the French government munitions factory. Here he carried out experiments in combustion and respiration. This print shows Lavoisier collecting the exhaled air from a seated volunteer. His wife, Marie, assisted in the experiments and illustrated their many publications. She is shown here updating the laboratory records.

3.1 Introduction

The term '**redox**' is used by chemists as an abbreviation for the processes of **red**uction and **ox**idation. These two processes usually occur simultaneously. Redox reactions include processes such as burning, rusting and respiration. Originally, chemists had a very limited view of redox, using it to account for only the reactions of oxygen and hydrogen. Nowadays, our ideas of redox have been extended to include all electron-transfer processes. An important feature of several electron-transfer redox reactions is that the energy of the chemical reaction may be released in the form of electrical energy and harnessed to provide electricity. This is what happens in the dry cell of a small torch, in the button cell of a hearing aid and in the battery of a motor car. Commercial cells and batteries are discussed in some detail in section 14.7.

3.2 Redox processes in terms of electron transfer

When metals react with oxygen they form oxides. For example:

$$2Mg + O_2 \rightarrow 2Mg^{2+}O^{2-}$$
$$4Na + O_2 \rightarrow 2(Na^+)_2O^{2-}$$

The metal is oxidised and the oxygen is reduced. During this process the metal atoms lose electrons to form positive ions and oxygen gains electrons to form negative oxide ions, O^{2-}. The oxygen takes the electrons given up by the metal.

$$2Mg \rightarrow 2Mg^{2+} + \boxed{4e^-}$$
$$O_2 + \boxed{4e^-} \rightarrow 2O^{2-}$$

The Lavoisiers' apparatus for preparing oxygen. Red mercury(II) oxide was heated in the retort on the left. The oxygen which formed was collected in the bell jar above mercury.

This cartoon (published in the eighteenth century) pokes fun at Joseph Priestley (1733–1804). Priestley was a nonconformist English clergyman who carried out experiments on combustion. Priestley's discovery of oxygen provided the Lavoisiers with the vital information they needed for their redox theory of combustion.

Electron-transfer reactions such as this are called **redox reactions**. The separate equations showing which substance gains electrons and which substance loses electrons are known as **half-equations**.

In the half-equations above Mg loses electrons and is oxidised to Mg^{2+}; O_2 gains electrons and is reduced to O^{2-}.

Thus, **oxidation** is defined as the *loss of electrons* and **reduction** is defined as the *gain of electrons*. **Oxidising agents**, such as oxygen, are defined as substances which *accept electrons*; **reducing agents** are substances which *donate electrons*.

When metals react they lose electrons and form their ions. Thus, metals are oxidised and act as reducing agents in their reactions. Notice that the oxidised substance (magnesium in the above example) acts as the reducing agent and the reduced substance (oxygen in the above example) acts as the oxidising agent.

When redox is viewed in terms of electron transfer it is easy to see why oxidation and reduction always take place together. One substance cannot lose electrons and be oxidised unless another substance gains electrons and is reduced.

Do the following processes involve oxidation, reduction, both oxidation and reduction or none of these?

- $2H^+ + 2e^- \rightarrow H_2$
- $Cu^+ \rightarrow Cu^{2+} + e^-$
- $Mg \rightarrow Mg^{2+} + 2e^-$
- $Ag^+ + Cl^- \rightarrow AgCl$
- $NH_3 + H^+ \rightarrow NH_4^+$
- $2Cu^+ \rightarrow Cu^{2+} + Cu$

3.3 Electron transfer in redox reactions

When powdered zinc is added to copper(II) sulphate solution an exothermic reaction occurs. Copper ions are reduced to a deposit of red-brown copper and zinc goes into solution as zinc ions.

$$Zn(s) + Cu^{2+}(aq) \rightarrow Zn^{2+}(aq) + Cu(s)$$

The overall reaction can be separated into two simpler processes involving electron transfer.

and

The apparatus in figure 3.1 shows how these two half-reactions can be separated to demonstrate that electron transfer is occurring. As soon as the circuit is complete, the bulb lights, showing that electrons are flowing through the wire. An ammeter used in place of the bulb will measure the electric current.

Zinc dissolves from the zinc rod which loses weight, whilst copper is deposited on the copper rod. The overall reaction is

$$Zn(s) + Cu^{2+}(aq) \rightarrow Zn^{2+}(aq) + Cu(s)$$

This is identical to the reaction which occurs when zinc is added to copper (II) sulphate solution. In this experiment, however, the energy of the reaction is liberated as electrical energy (electron transfer), whereas on direct mixing the energy is liberated as heat.

Using the apparatus in figure 3.1 the overall process has been separated into two distinct half-reactions:

$$Zn(s) \rightarrow Zn^{2+}(aq) + 2e^- \quad \text{at the zinc rod}$$

and

$$Cu^{2+}(aq) + 2e^- \rightarrow Cu(s) \quad \text{at the copper rod}$$

At the zinc rod, zinc atoms give up electrons and form zinc ions which go into solution as $Zn^{2+}(aq)$. The electrons flow from the zinc rod through the external

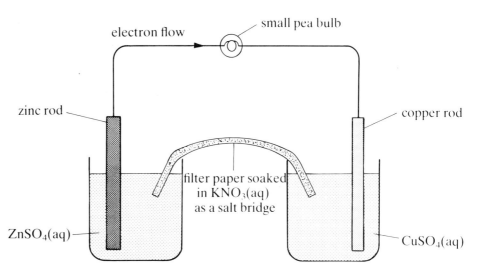

Figure 3.1
Electron transfer between zinc and copper(II) sulphate solution.

circuit including the bulb to the copper rod where they combine with Cu^{2+} ions to form copper atoms.

- Will electrons flow from positive to negative or from negative to positive?
- Will the zinc rod be positive or negative?
- What is the function of the salt bridge?
- What happens when the salt bridge is removed?

When the circuit in figure 3.1 is complete, the zinc rod dissolves away and the concentration of $Zn^{2+}(aq)$ in the left-hand beaker rises. If this net increase in positive charge is not 'neutralised', the reaction will soon stop because the excess positive charge in the solution will prevent any more Zn^{2+} ions from entering it. The excess positive charge from Zn^{2+} ions in the solution is reduced by nitrate ions, NO_3^-, moving out of the salt bridge into the solution or by zinc ions moving into the salt bridge.

In the right-hand beaker, copper ions are converted to copper atoms leaving an excess of sulphate ions, SO_4^{2-}, in solution. The excess negative charge in the right-hand half-cell is reduced by SO_4^{2-} ions moving into the salt bridge or potassium ions, K^+, diffusing out of the salt bridge into the solution.

Notice how the salt bridge serves two main functions:

(a) It completes the circuit by allowing ions carrying charge to move towards one half-cell from the other. When the salt bridge is removed, the current ceases since charge (whether it is ions or electrons) can no longer flow around the circuit.

(b) It provides cations and anions to replace those consumed at the electrodes and balance the charges on any ions formed from the electrodes. The movement of charge around the circuit is shown in figure 3.2.

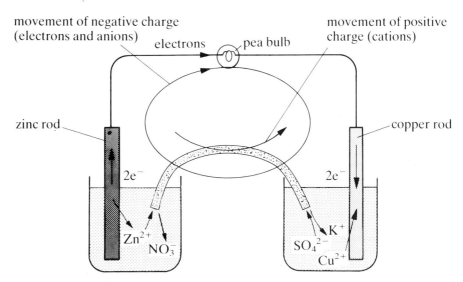

Figure 3.2
Movement of charge around a circuit.

3.4 Balancing redox reactions

Many substances can act as oxidisating agents (e.g. MnO_2, Cl_2, H_2O_2, Fe^{3+}); and there is a similar large number of reducing agents (e.g. I^-, Fe^{2+}, Zn). This means that many possible redox reactions can result from various combinations of oxidising agents and reducing agents. The following simple rules can be used to obtain a balanced equation for these redox processes.

1 Write down the oxidising and reducing agents and their products.

2 Write separate half-equations for the oxidation and reduction processes and balance these with respect to atoms and charge.

3 Combine the half-equations to obtain the overall equation.

We can use these simple rules to write an equation for the reaction between bromine and iron (II) ions.

1 The oxidising agent is Br_2 which forms Br^- on reaction. The reducing agent is Fe^{2+} which forms Fe^{3+} on reaction.

2 The balanced Br_2/Br^- half-equation is

$$Br_2 + 2e^- \rightarrow 2Br^-$$

Of course, one Br_2 can form two Br^- ions and consequently the Br^2 requires $2e^-$ to undergo reduction to $2Br^-$. The balanced Fe^{2+}/Fe^{3+} half-equation is

$$Fe^{2+} \rightarrow Fe^{3+} + e^-$$

3 During this reaction, Br_2 molecules take electrons from Fe^{2+} ions. In order to obtain a balanced overall equation, it is necessary to double the Fe^{2+}/Fe^{3+} equation so that $2Fe^{2+}$ gives up two electrons which are taken by one Br_2:

$$2Fe^{2+} \rightarrow 2Fe^{3+} + 2e^-$$
$$Br_2 + 2e^- \rightarrow 2Br^-$$

Adding these together gives the overall equation:

$$2Fe^{2+} + Br_2 \rightarrow 2Fe^{3+} + 2Br^-$$

As a second example consider the reaction between manganate (VII), MnO_4^-, and iodide (I^-) in acid solution.

1 The oxidising agent is MnO_4^-. In acid solution this is reduced to Mn^{2+} and H_2O. Many other oxyanions including $Cr_2O_7^{2-}$, BrO_3^-, IO_3^-, ClO^- and ClO_3^- act as oxidising agents in acid solution. Oxygen in the oxyanions combines with H^+ ions during the reaction to form water, and atoms such as Cr or Br in the oxyanion are reduced to stable ions.

2 The unbalanced MnO_4^-/Mn^{2+} half-equation is therefore

$$MnO_4^- + H^+ \rightarrow Mn^{2+} + H_2O$$

Since the MnO_4^- ion contains four oxygen atoms it can produce four H_2O molecules. This means that eight H^+ ions are required to balance the hydrogen atoms in four H_2O molecules. The equation is now balanced with respect to atoms:

$$MnO_4^- + 8H^+ \rightarrow Mn^{2+} + 4H_2O$$

We must now balance it with respect to charge.
On the left-hand side, charge $= -1 + 8 = +7$.
On the right-hand side, charge $= +2$.
This means that $5e^-$ must be added to the left-hand side to obtain a balanced half-equation:

$$MnO_4^- + 8H^+ + 5e^- \rightarrow Mn^{2+} + 4H_2O$$

The balance I^-/I_2 half-equation is

$$2I^- \rightarrow I_2 + 2e^-$$

An unusual redox reaction!

3 In order to obtain the overall equation, the MnO_4^-/Mn^{2+} half-equation must be multiplied by 2 and the I^-/I_2 half-equation must be multiplied by 5 so that 10 electrons are transferred.

$$2MnO_4^- + 16H^+ + 10e^- \rightarrow 2Mn^{2+} + 8H_2O$$
$$10I^- \rightarrow 5I_2 + 10e^-$$
$$\overline{2MnO_4^- + 16H^+ + 10I^- \rightarrow 2Mn^{2+} + 8H_2O + 5I_2}$$

Try to write balanced half-equations and an overall equation for the following:
■ The reaction between I_2 and $S_2O_3^{2-}$. $S_2O_3^{2-}$ is oxidised to $S_4O_6^{2-}$ during the reaction.
■ The reaction between H_2O_2 and I^- in acid solution.

3.5 Important types of redox reaction

1 The reactions of metals with non-metals

In this case, metals give up electrons to form positive ions and non-metals (O_2, Cl_2, S) take these electrons to form negative ions (O^{2-}, Cl^-, S^{2-}). For example:

$$Fe + S \rightarrow Fe^{2+}S^{2-} \quad \begin{cases} Fe \rightarrow Fe^{2+} + 2e^- \\ S + 2e^- \rightarrow S^{2-} \end{cases}$$

$$2Fe + 3Cl_2 \rightarrow 2FeCl_3 \quad \begin{cases} 2Fe \rightarrow 2Fe^{3+} + 6e^- \\ 3Cl_2 + 6e^- \rightarrow 6Cl^- \end{cases}$$

The metal is oxidised and the non-metal is reduced. Metals higher in the activity (electrochemical) series lose electrons more easily than the less reactive metals lower down. Thus, moving down the activity series metals become weaker reducing agents. The reactions of metals and the activity series are discussed more fully in section 14.2 and in chapter 20.

2 The reactions of metals with water

Metals at the top of the activity series (K, Na, Ca and Mg) are sufficiently reactive to form hydrogen with water.

$$Ca + 2H_2O \rightarrow Ca^{2+}(OH^-)_2 + H_2$$

The next few metals in the activity series (e.g. Al, Zn and Fe) do not react noticeably with water, but they will react with steam to form hydrogen.

In these reactions the metal atoms are oxidised to form positive ions. The electrons which they release are accepted by water molecules which are reduced to hydroxide ions (OH^-) and hydrogen (H_2).

$$Ca \rightarrow Ca^{2+} + 2e^-$$
$$2H_2O + 2e^- \rightarrow 2OH^- + H_2$$

3 The reactions of metals with acids

In this case, the metal atoms lose electrons which are taken by aqueous H^+ ions in the acid to form H_2. For example:

$$Zn(s) + 2H^+(aq) \rightarrow Zn^{2+}(aq) + H_2(g)$$

The half-equations are:

$$Zn \rightarrow Zn^{2+} + 2e^-$$
$$2H^+ + 2e^- \rightarrow H_2$$

The reactivity of metals with acids also depends on the ease with which the metal loses electrons to form its aqueous ions. Metals which form ions more easily than hydrogen will therefore react with acids releasing electrons, which are taken by H^+ ions in the acid.

Metals below hydrogen in the activity series, such as copper and silver, form ions less readily than hydrogen and so do not react with acids to form hydrogen.

Tutankhamun's gold funerary mask: gold is such an unreactive metal that it remains unoxidised and untarnished after centuries.

4   Reactions at the electrodes during electrolysis

During electrolysis cations are attracted to the negatively charged cathode where they gain electrons and are reduced. For example, during the electrolysis of molten lead(II) bromide:

at the cathode(−)

$$Pb^{2+} + 2e^- \rightarrow Pb$$

At the same time, Br^- anions are attracted to the positively charged anode and oxidised by the loss of electrons.

at the anode (+)

$$2Br^- \rightarrow Br_2 + 2e^-$$

Some important industrial processes involving electrolysis are discussed in sections 15.8, 16.6, 17.3 and 20.4.

3.6   Oxidation number

Many redox reactions involve a *complete* transfer of electrons from one substance to another. These redox processes usually have ions as either the reactants or the products. In some cases, the reactants and the products are both ions.

There are, however, some reactions which are regarded as redox processes in spite of the fact that they involve only molecules and do not have a *complete* transfer of electrons from one substance to another. For example, the reactions

$$2H_2 + O_2 \rightarrow 2H_2O$$
and
$$C + O_2 \rightarrow CO_2$$

clearly involve redox, yet there is no complete transfer of electrons from one substance to another.

In order to overcome this problem, the concept of **oxidation number** (or oxidation state) was introduced. This provided a similar, but alternative, definition of redox to that involving electron transfer. An oxidation number is a number assigned to an atom or an ion to describe its relative state of oxidation or reduction.

Using oxidation numbers it is possible to decide whether redox has occurred in processes involving either transfer or re-sharing of electrons. For example:

The oxidation numbers of the atoms in uncombined elements are given an oxidation number of zero. Thus, the oxidation number of Mg is 0, as is the oxidation number of chlorine atoms in Cl_2. For simple ions, the oxidation number is simply the charge on the ion. Thus, the oxidation numbers of Cl^-, Fe^{2+} and Fe^{3+} are −1, +2 and +3, respectively.

For compounds and complex ions, the oxidation numbers of the atoms within them are obtained by assuming the compounds and complex ions are *wholly ionic* and then working out the charge associated with each atom. For example:

$H_2O((H^+)_2O^{2-})$	Ox. No. of H in H_2O = +1
	Ox. No. of O in H_2O = −2
$HCl(H^+Cl^-)$	Ox. No. of H in HCl = +1
	Ox. No. of Cl in HCl = −1

In assigning oxidation numbers in this way it is necessary to assume that the electrons in each bond of the molecule or ion belong to the *more electronegative atom* (i.e. the atom with the greater attraction for electrons).

Electrical gear and batteries under the bonnet of an experimental Daf car powered by fuel cells.

Early Japanese workmen smelting copper ore. What redox processes are involved in this?

Since hydrogen is the least electronegative non-metal, the oxidation number of H in its compounds is usually $+1$. For example:

$$PH_3(P^{3-}(H^+)_3)$$

Ox. No. of H in PH_3	$= +1$
Ox. No. of P in PH_3	$= -3$

$$H_2S((H^+)_2S^{2-})$$

Ox. No. of H in H_2S	$= +1$
Ox. No. of S in H_2S	$= -2$

In metal hydrides, however, the metal is more electropositive than hydrogen and in this case the oxidation number of H is -1. For example:

$$Na^+H^-$$

Ox. No. of H in NaH	$= -1$
Ox. No. of Na in NaH	$= +1$

Apart from fluorine, oxygen is the most electronegative element. This means that the oxidation number of oxygen in its compounds is usually -2.

$$Na_2O((Na^+)_2O^{2-})$$

Ox. No. of O in Na_2O	$= -2$
Ox. No. of Na in Na_2O	$= +1$

$$CO_2(C^{4+}(O^{2-})_2)$$

Ox. No. of O in CO_2	$= -2$
Ox. No. of C in CO_2	$= +4$

However, in OF_2 the oxidation number of oxygen is $+2$ and in peroxides the oxidation number of oxygen is -1.

$$OF_2(O^{2+}(F^-)_2)$$

Ox. No. of O in OF_2	$= +2$
Ox. No. of F in OF_2	$= -1$

$$Na_2O_2((Na^+)_2(O_2)^{2-})$$

Ox. No. of O in Na_2O_2	$= -1$
Ox. No. of Na in Na_2O_2	$= +1$

Rules for assigning oxidation numbers

These points concerning oxidation numbers can be summarised in six simple rules.

1 The oxidation number of atoms in uncombined elements is 0.

2 In neutral molecules, the algebraic sum of the oxidation numbers is 0.

3 In ions, the algebraic sum of the oxidation numbers equals the charge on the ion.

4 In any substance, the more electronegative atom has the negative oxidation number, and the less electronegative atom has the positive oxidation number.

5 The oxidation number of hydrogen in all its compounds, except metal hydrides, is $+1$.

6 The oxidation number of oxygen in all its compounds, except in peroxides and in OF_2, is -2.

■ What are the oxidation numbers of each element in the following?

$$MgCl_2; \quad SO_2; \quad CO; \quad NaOH; \quad PCl_3; \quad SO_4^{2-}; \quad MnO_4^-$$

■ What are the oxidation numbers of sulphur in the following compounds?

$$NaHSO_4, \quad CS_2, \quad SO_2Cl_2, \quad Na_2S, \quad S_2Cl_2$$

Some elements have five or more possible oxidation states. The principal oxidation states of sulphur are shown in an oxidation number chart in figure 3.3.

Oxidation numbers and nomenclature

The oxidation number concept is often important in naming compounds.

Simple molecular compounds containing only two elements, such as CO_2, $SiCl_4$ and S_2Cl_2, are usually named by reference to the numbers of their different

Oxidation number	Examples
$+6$	H_2SO_4, SO_4^{2-}, SO_3
$+4$	H_2SO_3, SO_3^{2-}, SO_2
$+2$	SCl_2
0	S
-2	H_2S, S^{2-}

Figure 3.3
An oxidation number chart for sulphur.

atoms without stating oxidation numbers. Hence, CO_2—carbon *dioxide*, $SiCl_4$—silicon *tetra*chloride and S_2Cl_2—*di*sulphur *di*chloride.

The systematic names of more complex molecular compounds, such as HClO and H_2CrO_4, are obtained by reference to the oxidation numbers of constituent elements which can have variable oxidation numbers.

For example:

	Recommended name	*Common (trivial) name*
HClO	chloric(I) acid	hypochlorous acid
H_2CrO_4	chromic(VI) acid	chromic acid

However, the names sulphuric, sulphurous, sulphate, sulphite, nitric, nitrous, nitrate and nitrite are recommended by IUPAC (The International Union of Pure and Applied Chemistry) in preference to sulphuric(VI), sulphuric(IV), sulphate(VI), sulphate(IV), nitric(V), nitric(III), nitrate(V) and nitrate(III), respectively.

The systematic names of giant ionic compounds and giant molecular compounds usually include the oxidation numbers of those constituent elements which have a variable oxidation state.

For example:

$FeSO_4$	iron(II) sulphate
$KMnO_4$	potassium manganate(VII)
$FeCl_3$	iron(III) chloride
NaClO	sodium chlorate(I)
$Cu(NO_3)_2$	copper(II) nitrate

Remember that Roman numerals are used in writing the oxidation number of an element within a compound to prevent any confusion with the real charge on an ion. Thus, CuO is named copper(II) oxide *not* copper(2) oxide and MnO_4^- is named manganate(VII) not manganate(7).

3.7 Explaining redox in terms of oxidation number

We are now in a position to define oxidation and reduction in terms of oxidation number. This definition of redox is an alternative to that involving electron transfer, although the two definitions are quite closely related.

An atom is said to be oxidised when its oxidation number increases and reduced when its oxidation number decreases.

Consider the following reactions.

(a)
$$\overset{0}{2\,Na} + \overset{0}{Cl_2} \rightarrow 2\overset{+1\ -1}{NaCl}$$

The oxidation number of sodium has increased from 0 to +1: it has been oxidised. The oxidation number of chlorine has decreased from 0 to −1: it has been reduced.

(b)
$$\overset{+1}{H^+} + \overset{-2+1}{OH^-} \rightarrow \overset{+1\ -2}{H_2O}$$

This ionic equation summarises the neutralisation of an acid with an alkali. Notice that the oxidation number of each element remains unchanged and so the reaction does not involve redox.

(c) Another important ionic reaction which does not involve redox is precipitation. Here again the oxidation number of each element remains unaltered during the reaction.

$$\overset{+1}{Ag^+}(aq) + \overset{-1}{Cl^-}(aq) \rightarrow \overset{+1\ -1}{Ag\,Cl}(s)$$

$$\overset{+2}{Ba^{2+}}(aq) + \overset{+6\ -2}{SO_4^{2-}}(aq) \rightarrow \overset{+2+6-2}{BaSO_4}(s)$$

(d)
$$\overset{+2-0}{CO} + \overset{0}{\tfrac{1}{2}O_2} \rightarrow \overset{+4-2}{CO_2}$$

In this case, the oxidation number of carbon rises from +2 to +4: it has been oxidised. The oxidation number of the oxygen atom in CO remains unchanged, but the elemental oxygen is reduced: its oxidation number falls from 0 to −2.

Which of the following reactions involve redox?
■ $Cl_2 + 2OH^- \rightarrow Cl^- + ClO^- + H_2O$
■ $Cu^{2+} + 2OH^- \rightarrow Cu(OH)_2$
■ $H_2O + SO_3 \rightarrow H_2SO_4$
■ $2CrO_4^{2-} + 2H^+ \rightarrow Cr_2O_7^{2-} + H_2O$

3.8 The advantages and disadvantages of the oxidation number concept

Oxidation numbers can help us decide whether or not redox is involved in a particular process. Oxidation numbers show that neutralisation and precipitation are not redox reactions, even though they involve ions. This point highlights the importance of oxidation numbers as an electron book-keeping device that allows us to recognise redox processes.

The second advantage in using oxidation numbers is that they allow us to see exactly which part of a molecule or a complex ion is reduced or oxidised. For example, the half-equation

$$MnO_4^- + 8H^+ + 5e^- \rightarrow Mn^{2+} + 4H_2O$$

shows that MnO_4^- and H^+ ions are reduced to Mn^{2+} and $4H_2O$. But, which element or elements in MnO_4^- and H^+ is reduced? If oxidation numbers are assigned to the atoms in the half-equation,

$$\overset{+7\ -2}{MnO_4^-} + \overset{+1}{8H^+} + 5e^- \rightarrow \overset{+2}{Mn^{2+}} + \overset{+1\ -2}{4H_2O}$$

it is clear that manganese is the reduced element because its oxidation number changes from +7 to +2.

The main disadvantage of the oxidation number concept is that it can cause a misunderstanding about the structure of molecular substances. No physical or structural significance can be attached to oxidation numbers of atoms in molecular substances. The oxidation number of carbon in CO_2 is +4, but it must *not* be supposed that there is a charge of +4 on the carbon atom.

In a few cases ambiguities can arise with oxidation numbers. For example, the rules for assigning oxidation numbers suggest that each sulphur atom in the thiosulphate ion, $S_2O_3^{2-}$, has an oxidation number of +2. However, the structure of the $S_2O_3^{2-}$ ion shows that the two sulphur atoms in it are quite different. One sulphur atom is at the centre of a tetrahedron bonded to the other four atoms (one S and three O atoms) similar to the S atom in the SO_4^{2-} ion. With this in mind, we could assign an oxidation number of +6 to the central S atom in $S_2O_3^{2-}$ (similar to the central S atom in SO_4^{2-}) and an oxidation number of −2 to each of the surrounding atoms, including the second sulphur atom.

Two further problems with oxidation numbers concern their use with organic compounds.

■ What is the oxidation number of carbon in

(i) CH_4; (ii) C_2H_6; (iii) C_3H_8?

The carbon atoms in CH_4, C_2H_6 and C_2H_8 all have four covalent bonds. In spite of this similarity, they have different oxidation numbers. The other problem (which C_3H_8 highlights) is that in some compounds, atoms can have oxidation numbers which are not whole numbers.

In spite of these disadvantages the concept of oxidation number is still very useful.

An exhausted competitor is given oxygen. Oxygen is essential for respiration, an important redox reaction.

Summary

1 There are two related definitions of redox, one involving electron transfer, the other involving the concept of oxidation number.

2 In terms of electron transfer, oxidation is defined as a loss of electrons, and reduction is defined as a gain of electrons.

3 Electron-transfer reactions are called redox reactions.

4 The energy evolved by many electron-transfer reactions can be released in the form of electrical energy and harnessed to provide electricity.

5 An oxidation number is a number assigned to an atom or ion to describe its relative state of oxidation or reduction. Using oxidation numbers it is possible to decide whether redox has occurred.

6 In terms of oxidation numbers, oxidation is defined as an increase in oxidation number and reduction is defined as a decrease in oxidation number.

Study Questions

1 What are the oxidation numbers of
(a) chlorine in HCl, $HClO$, ClO_3^-, PCl_3, Na_3AlCl_6, $POCl_3$;

(b) nitrogen in N_2O, NO, NO_2, NO_3^-, N_2H_4, HCN?

2 (a) Write the formulae for substances containing sulphur in which it shows the following oxidation states: $-2, -1, 0, +1, +2, +4, +6$.

(b) The following equations summarise redox reactions involving sulphur compounds. Deduce the oxidation number of all the atoms and ions in these equations and hence determine precisely which species is oxidised and which is reduced.

(i) $2MnO_4^- + 6H^+ + 5SO_3^{2-} \rightarrow 2Mn^{2+} + 3H_2O + 5SO_4^{2-}$

(ii) $2NaI + 3H_2SO_4 \rightarrow 2NaHSO_4 + 2H_2O + I_2 + SO_2$

(iii) $S_2O_3^{2-} + 2H^+ \rightarrow S + SO_2 + H_2O$

(iv) $SO_3^{2-} + H_2O + 2Ce^{4+} \rightarrow SO_4^{2-} + 2H^+ + 2Ce^{3+}$

(v) $2S_2O_3^{2-} + I_2 \rightarrow S_4O_6^{2-} + 2I^-$

3 Which of the following may be regarded as redox reactions? Explain your answers.
(a) $Cu^{2+} + 4NH_3 \rightarrow Cu(NH_3)_4^{2+}$

(b) $Cl_2 + 2OH^- \rightarrow Cl^- + ClO^- + H_2O$

(c) $Ca^{2+} + 2F^- \rightarrow CaF_2$

(d) $Ca + F_2 \rightarrow CaF_2$

(e) $2CCl_4 + CrO_4^{2-} \rightarrow 2COCl_2 + CrO_2Cl_2 + 2Cl^-$

(f) $NH_3 + H^+ \rightarrow NH_4^+$

4 Write redox half-equations for the following reactions:
(a) When copper is added to concentrated nitric acid the solution becomes pale blue and brown fumes of nitrogen dioxide are produced.
(b) When potassium iodide is added to acidified hydrogen peroxide a brown colour appears.

(c) Sodium sulphite reduces an acidified solution of orange dichromate(VI) ions, $Cr_2O_7^{2-}$, to a green solution containing Cr^{3+} ions.

(d) Manganese(IV) oxide will oxidise concentrated hydrochloric acid to chlorine.

(e) When zinc is added to silver nitrate solution, crystals of silver form on the zinc surface.

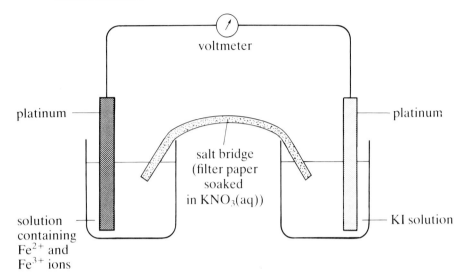

Figure 3.4
Electron transfer between KI(aq) and a solution containing Fe^{2+} and Fe^{3+} ions.

5   Look at the apparatus in figure 3.4. When the circuit is complete a brown colour appears around the platinum in the right-hand beaker.

(a) Write a half-equation to summarise the reactions at each of the platinum terminals.

(b) In which direction do electrons flow?

(c) Explain the function of the salt bridge, stating which ions are moving into it and out of it in each beaker.

(d) Do you think the voltage of the cell will increase, decrease or remain the same if the concentration of KI in the right-hand beaker is increased? Explain your answer.

6   (a) What is the oxidation number of chromium in each of the substances A–F in the reaction scheme in figure 3.5?

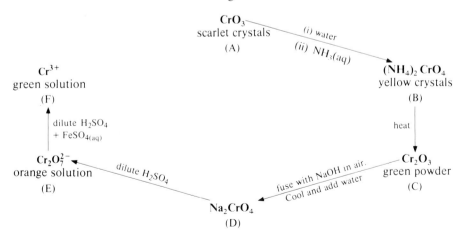

Figure 3.5
A reaction scheme involving compounds of chromium.

(b) Which steps in the scheme involve redox?

(c) Write equations or half-equations to describe each step in the scheme.

7   Certain features of the element vanadium, V, are presented in figure 3.6 on the next page. Consider the diagram carefully.

(a) What is the oxidation number of vanadium in the compounds A–I?

(b) What can you deduce about the oxidising power of the halogens towards vanadium?

(c) How does the action of chlorine on vanadium compare with the action of hydrogen chloride?

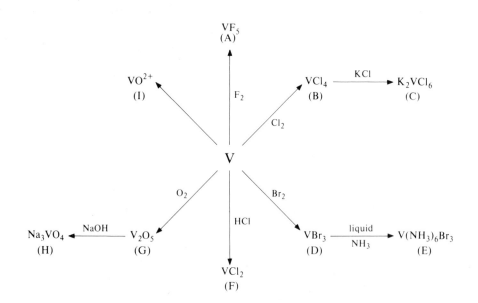

Figure 3.6
Products from vanadium.

8 What do you understand by the terms 'oxidation' and 'reduction'? In each of the following reactions say what (if anything) has been reduced and what has been oxidised.
   Write electron-transfer equations to explain your answers:

(a) $2FeCl_2 + Cl_2 \quad \rightarrow 2FeCl_3$

(b) $CuO + H_2 \quad \rightarrow Cu + H_2O$

(c) $3Cu + 8HNO_3 \quad \rightarrow 3Cu(NO_3)_2 + 4H_2O + 2NO$

(d) $2Na + H_2 \quad \rightarrow 2NaH$

(e) $AgNO_3 + NaCl \rightarrow AgCl + NaNO_3$

9 0.50g of hydrated iron(II) sulphate ($FeSO_4.7H_2O$) was dissolved in dilute sulphuric acid and titrated with 0.1M potassium manganate(VII) (potassium permanganate). What volume of the potassium manganate(VII) is required to complete the titration? What assumptions have you made in the calculations?

$$Fe = 55.8; \quad S = 32; \quad O = 16; \quad H = 1$$

10 10g of an impure iron(II) salt were dissolved in water and made up to 200cm³ of solution. 20cm³ of this solution, acidified with dilute sulphuric acid, required 25cm³ of 0.04M $KMnO_4$(aq) before a faint pink colour appeared.

(a) Write a balanced ionic equation (or half-equations) for the reaction of acidified manganate(VII) (permanganate) ions with iron(II) ions.

(b) How many moles of iron(II) ions react with one mole of MnO_4^- ions?

(c) How many moles of Fe^{2+} react with 25cm³ of 0.04M $KMnO_4$(aq)?

(d) How many grams of Fe^{2+} are there in the 200cm³ of original solution? (Fe = 56)

(e) Calculate the % by mass of iron in the impure iron(II) salt.

4 Patterns and Periodicity

Johann Wolfgang Döbereiner.

Born 1780 at Hof; died 1849 at Jena. Döbereiner studied as a pharmacist at Münchberg and then studied chemistry at Strasbourg. In 1810 he became professor of chemistry and pharmacy at Jena. Döbereiner made the first observations on platinum as a catalyst and on the periodic system of the elements. He also contributed to technical questions, and was, for example, a keen supporter for the building of gas-works. Döbereiner and the poet Goethe were close friends and colleagues.

Li
6.9
Na
23.0
K
39.1

Figure 4.1
Döbereiner's triad of alkali metals.

4.1 Introduction

Chemists have always searched for patterns and similarities in the properties and reactions of substances. From your own studies of chemistry you will already appreciate the classification of elements as metals and non-metals and you will have looked at the patterns in reactivity of different metals with oxygen, water and acids. Unfortunately, there are some elements that cannot be classified easily and unambiguously as either a metal or a non-metal. Non-metals are usually volatile and non-conductors of electricity, but graphite (carbon) and silicon (see p.47), usually classed as non-metals, have very high melting points and very high boiling points, and both conduct electricity. Mercury is untypical of metals in being liquid at room temperature. Can you think of other elements which do not fit neatly into the classification of elements as metals or non-metals?

4.2 Families of similar elements

Because of limitations to the overall classification of elements as metals or non-metals, chemists began to search for trends and similarities in the properties of much smaller groups of elements. Pairs of similar elements, such as sodium and potassium, calcium and magnesium, and chlorine and bromine, will no doubt spring to your mind.

To what extent do sodium and potassium resemble each other?

- Do they have similar properties (e.g. melting point, boiling point, hardness, density, lustre)?
- Do they react in the same way with oxygen in the air?
- What colour are their oxides, chlorides, sulphates, etc?
- How do sodium and potassium react with water? Write equations for their reactions with water. Are their reactions and products similar?
- Are the formulae of their compounds similar? Do sodium and potassium have the same oxidation number in their compounds?

In this chapter, we shall follow the search for patterns amongst elements leading to the development of the periodic table. Then, we shall study the physical and chemical periodicity of the elements in periods 2 and 3. In chapter 13, these periodic trends are explained in terms of atomic properties such as electronic structure, bonding and electronegativity.

Early in the nineteenth century, the German chemist Döbereiner pointed out that many of the known elements could be arranged in groups of three similar elements. He called these families of three elements 'triads'. Two of Döbereiner's triads were lithium, sodium and potassium (alkali metals—figure 4.1) and chlorine, bromine and iodine (halogens—figure 4.2). Döbereiner showed that

Cl
35.5
Br
79.9
I
126.9

Figure 4.2
Döbereiner's triad of halogens.

when the three elements in each triad were written in order of relative atomic mass, the middle element had properties intermediate between those of the other two elements. What is more, the relative atomic mass of the middle element was very close to the average of the relative atomic masses of the other two elements. In the triad of chlorine, bromine and iodine, the relative atomic mass of bromine (79.9) is close to the average of the relative atomic masses of chlorine and iodine ((35.5 + 126.9)/2 = (162.4)/2 = 81.2).

■ How close to the relative atomic mass of sodium is the average of the relative atomic masses of lithium and potassium?

4.3 Newlands' Octaves

The relationship that Döbereiner had discovered encouraged other chemists to look for trends between properties of elements and their relative atomic masses.

In 1866, John Newlands, an English chemist, read an important paper to the Chemical Society. Newlands suggested that when the elements were arranged in order of their relative atomic masses (atomic weights), any one element had properties similar to those of the elements eight places in front of it and eight places behind it in the list.

Newlands called this a '**Law of Octaves**'. He claimed that the eighth element was a kind of repetition of the first, like the eighth note of an octave in music.

Table 4.1 shows the first five of these 'octaves'. Although Newlands' 'octaves' showed similar elements in the same column, his ideas met with considerable scepticism.

Table 4.1
Newlands' octaves

H		Li	Be	B		C	N		O
F		Na	Mg	Al		Si	P		S
Cl		K	Ca	Cr		Ti	Mn		Fe
Co and Ni		Cu	Zn	Y		In	As		Se
Br		Rb	Sr	Ce and La		Zr	Di and Mo		Ro and Ru

■ Can you find any groups of similar elements in the columns of table 4.1?
■ Are Döbereiner's triads included in the columns of Newlands' octaves? Find one triad in one of the columns.
■ Why are the noble gases (He, Ne, Ar, etc.) missing from Newlands' octaves?
■ Why do you think chemists were sceptical about Newlands' octaves?

The periodic repetition of similar elements at regular intervals in Newlands' octaves led to the name **periodic table**. Newlands' classification was criticised for three important reasons.

1 It assumed that all elements had been discovered. The discovery of one more element could throw out the whole idea of 'octaves'. This was a strong criticism as four elements (thallium, indium, caesium and rubidium) had been discovered in the few years previous to Newlands' suggestions.

2 In order to ensure repeating octaves, Newlands had found it necessary to place two elements (e.g. Co and Ni) into only one space in certain cases.

3 Newlands' classification grouped together some elements which were very dissimilar. For example, cobalt and nickel were placed in the same family as fluorine, chlorine and bromine; copper was placed in the same family as lithium, sodium, potassium and rubidium.

For these reasons, Newlands' ideas were rejected and even ridiculed by his fellow scientists. One of his more sceptical critics enquired whether 'Mr

PROCEEDINGS OF SOCIETIES.

CHEMICAL SOCIETY.
Thursday, March 1.
Professor A. W. WILLIAMSON, Ph.D., F.R.S., Vice-President, in the Chair.

Mr. JOHN A. R. NEWLANDS read a paper entitled "*The Law of Octaves, and the Causes of Numerical Relations among the Atomic Weights.*" The author claims the discovery of a law according to which the elements analogous in their properties exhibit peculiar relationships, similar to those subsisting in music between a note and its octave. Starting from the atomic weights on Cannizzaro's system, the author arranges the known elements in order of succession, beginning with the lowest atomic weight (hydrogen) and ending with thorium (= 231·5); placing, however, nickel and cobalt, platinum and iridium, cerium and lanthanum, &c., in positions of absolute equality or in the same line. The fifty-six elements so arranged are said to form the compass of eight octaves, and the author finds that chlorine, bromine, iodine, and fluorine are thus brought into the same line, or occupy corresponding places in his scale. Nitrogen and phosphorus, oxygen and sulphur, &c., are also considered as forming true octaves. The author's supposition will be exemplified in Table II., shown to the meeting, and here subjoined :—

Table II.—Elements arranged in Octaves.

No.		No.		No.		No.		No.		No.		No.		No.
H	1	F	8	Cl	15	Co & Ni	22	Br	29	Pd	36	I	42	Pt & Ir	50
Li	2	Na	9	K	16	Cu	23	Rb	30	Ag	37	Cs	44	Os	51
G	3	Mg	10	Ca	17	Zn	24	Sr	31	Cd	38	Ba & V	45	Hg	52
Bo	4	Al	11	Cr	19	Y	25	Ce & La	33	U	40	Ta	46	Tl	53
C	5	Si	12	Ti	18	In	26	Zr	32	Sn	39	W	47	Pb	54
N	6	P	13	Mn	20	As	27	Di & Mo	34	Sb	41	Nb	48	Bi	55
O	7	S	14	Fe	21	Se	28	Ro & Ru	35	Te	43	Au	49	Th	56

Dr. GLADSTONE made objection on the score of its having been assumed that no elements remain to be discovered. The last few years had brought forth thallium, indium, cæsium, and rubidium, and now the finding of one more would throw out the whole system. The speaker believed there was as close an analogy subsisting between the metals named in the last vertical column as in any of the elements standing on the same horizontal line.

Professor G. F. FOSTER humorously inquired of Mr. Newlands whether he had ever examined the elements according to the order of their initial letters? For he believed that any arrangement would present occasional coincidences, but he condemned one which placed so far apart manganese and chromium, or iron from nickel and cobalt.

Mr. NEWLANDS said that he had tried several other schemes before arriving at that now proposed. One founded upon the specific gravity of the elements had altogether failed, and no relation could be worked out of the atomic weights under any other system than that of Cannizzaro.

A note concerning Newlands' octaves from the magazine *Chemical News*, published by the Chemical Society in March 1866.

Newlands had ever examined the elements according to the order of their initial letters'.

4.4 Patterns and periodicity

Despite the criticism of Newlands' suggestions, many scientists continued to search for a pattern relating the properties of elements and their relative atomic masses.

Lother Meyer's Curves

In 1869, the German chemist, Julius Lothar Meyer and the Russian chemist, Dmitri Mendeléev, separately published results which supported the ideas of periodicity. Lothar Meyer plotted various physical properties (melting point, boiling point, density) of the known elements against their relative atomic masses (atomic weights). Two of Lothar Meyer's curves are shown in figures 4.3 and 4.4

The graph in figure 4.3 shows the periodic repetition obtained by plotting the **molar volume** (the volume occupied by one mole of the element in the solid state) against the relative atomic mass. Notice that elements in the same chemical family occur at similar points on the curves. For example, lithium,

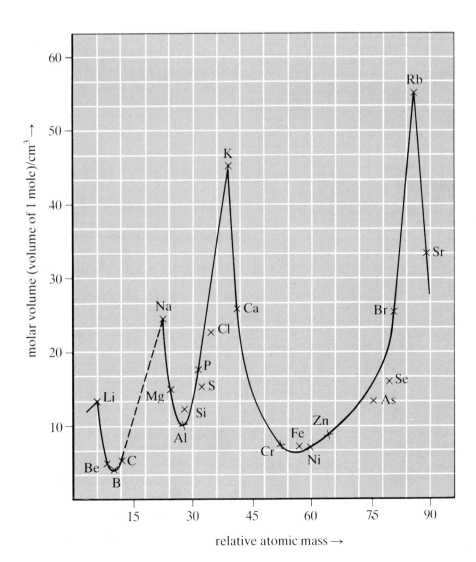

Figure 4.3
Lothar Meyer's curve of molar volume against relative atomic mass.

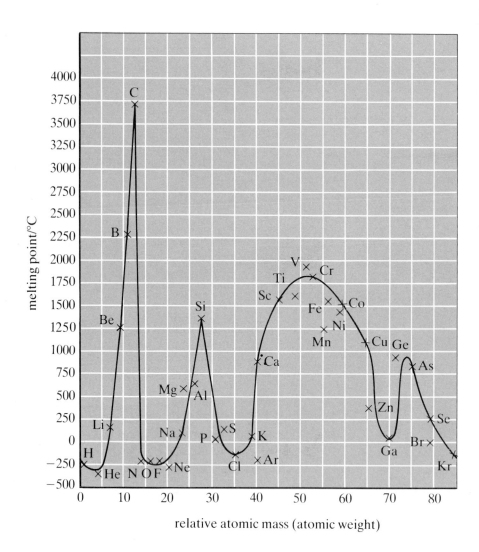

Figure 4.4
Lothar Meyer's curve of melting point against relative atomic mass.

sodium and potassium (alkali metals) appear on the peaks, whilst fluorine, chlorine and bromine (halogens) occur just before the peaks.

The graph in figure 4.4 shows the periodic variation in melting point with relative atomic mass. As in figure 4.3, similar elements fall at similar points on the curves.

■ Where do the alkali metals appear on the curves in figure 4.4? Do they occupy similar positions?
■ Where do the halogens appear on the curves in figure 4.4? Do they occupy similar positions?
■ Which elements appear on the peaks in figure 4.4? Are these elements alike in their properties?

Mendeléev's periodic table

Although Lothar Meyer's curves showed a periodic repetition of properties with relative atomic mass, most of the credit for arranging the elements in a periodic table is given to Mendeléev. Mendeléev arranged all the elements known to him in order of increasing relative atomic mass and showed that elements with similar properties recurred at regular intervals. Figure 4.5 shows part of Mendeléev's periodic table. Elements with similar properties fall in the same vertical column. These vertical columns of similar elements are called **groups** and the horizontal rows of elements are called **periods**.

	Group I	Group II	Group III	Group IV	Group V	Group VI	Group VII	Group VIII
Period 1	H							
Period 2	Li	Be	B	C	N	O	F	
Period 3	Na	Mg	Al	Si	P	S	Cl	
Period 4	K Cu	Ca Zn	* *	Ti *	V As	Cr Se	Mn Br	Fe Co Ni
Period 5	Rb Ag	Sr Cd	Y In	Zr Sn	Nb Sb	Mo Te	* I	Ru Rh Pd

Figure 4.5
Mendeléev's periodic table.

Why was Mendeléev's periodic table more successful than the one proposed by Newlands? In the first place, Mendeléev left gaps in his table in order that similar elements would fall in the same vertical column.

Secondly, he suggested that, in due course, elements would be discovered to fill these gaps. Thirdly, he predicted properties for some of the missing elements.

There are four gaps (indicated by *) in the portion of Mendeléev's periodic table reproduced in figure 4.5.

■ Which elements have since been discovered to fill these gaps?

Within Mendeléev's own lifetime some of these missing elements were discovered and their properties coincided very closely with his predictions.

Mendeléev also proposed that periods 4, 5, 6 and 7 should contain more than seven elements. In order to fit these longer periods into his pattern, he divided them into halves and placed the first half of the elements in the top left-hand corner of their space (e.g. K, Ca, etc., in period 4), and the second half in the bottom right-hand corners (e.g. Cu, Zn, etc.)

The accuracy of Mendeléev's predictions convinced scientists that his ideas were correct and his periodic table was accepted as a valuable overall summary of the properties of elements. Indeed, Mendeléev's concept of a periodic table has absorbed more and more new knowledge and the importance and usefulness of his original idea has been demonstrated many times.

Dmitri Ivanovich Mendeléev (1839–1907). In 1869, Mendeléev published his periodic table. This was the forerunner of modern periodic tables.

4.5 Modern forms of the periodic table

Mendeléev's periodic table arranged the elements in order of relative atomic mass. His periodic law stated that *the properties of the elements are a periodic function of their relative atomic masses*. This law fulfilled two important functions.

1 It summarised the properties of elements and classified them into groups with similar properties.

2 It enabled predictions to be made about the properties of known and unknown elements and led to considerable research activity.

Although Mendeléev used the order of relative atomic masses as a basis for his periodic table, he wrote the elements tellurium (Te = 127.6) and iodine (I = 126.9) in the reverse order. Mendeléev found that this was necessary if Te and I were to fall in their correct vertical groups. He argued that iodine must occupy the same group as chlorine and bromine.

Mendeléev's periodic table was proposed long before chemists could understand its fundamental relationship to electronic structure. Although the reverse order of Te and I worried scientists at the time of Mendeléev's proposals, it is now explained by modern forms of the periodic law. This states that *the properties of elements are a periodic function of their atomic numbers*. (The atomic number of Te is 52 and that of I is 53.)

If the elements are numbered along each period from left to right starting at period 1, then period 2, etc., the number given to each element is its **atomic number**. The real significance of atomic number is discussed in sections 5.2 and 5.4.

In modern periodic tables, all the elements are in strict order of their atomic numbers.

Although there are various forms of the periodic table suitable for one purpose or another, the 'wide form' shown in figure 4.6 is probably the most useful. In this figure the atomic numbers are shown below the symbols for each element.

Those groups with elements in periods 2 and 3 are numbered from I to VII followed by Group O. Some of these groups have names besides numbers. The most common names used for particular groups are:

Group number	Group name
I	alkali metals
II	alkaline-earth metals
VII	halogens
O	noble (inert) gases)

The most obvious difference between the periodic table in figure 4.6 and that proposed by Mendeléev is the removal of the **transition elements** from the simple groups. In period 4, for example, 10 transition elements (Sc, Ti, V, Cr, Mn, Fe, Co, Ni, Cu, Zn) have been taken out of the simple groupings suggested by Mendeléev and placed between Ca in Group II and Ga in Group III.

At the present time there are 106 known elements. The most recently discovered elements are kurchatovium (Ku, atomic number 104) which was first synthesised by Russian chemists in 1964, hahnium (atomic number 105) synthesised at Berkeley, USA, in 1970, and element 109 synthesised by West German scientists in 1983.

Besides dividing the periodic table into vertical groups of similar elements, it is also useful to split it into five blocks of elements with similar properties. These five blocks are shaded differently in figure 4.6.

An unusual form of the periodic table—emphasising the position of one element.

Figure 4.6
The modern periodic table (wide form).

* In 1983 West German scientists succeeded in synthesizing element 109 by bombarding $^{209}_{83}$Bi with $^{58}_{26}$Fe for one week. $^{209}_{83}$Bi + $^{58}_{26}$Fe → $^{?}_{109}$X

The reactive metals

The elements in Group I and Group II form a block of reactive metals. They are sometimes referred to as the **'s-block'** elements, since the outermost electrons in these metals are in *s* sub-shells (see section 6.6). These metals (including potassium, sodium, calcium and magnesium) are all high in the activity (electrochemical) series. They have lower densities, lower melting points and lower boiling points than most other metals and they form stable, involatile ionic compounds.

The transition metals

These elements occupy two rectangles between Group II and Group III. Some of the transition metals are called **'d-block'** elements, since electrons are being added to *d* sub-shells across this block of elements (section 6.6). These metals (including chromium, iron, copper, zinc and silver) are much less reactive than the metals in Groups I and II. In this block of elements there is also a marked horizontal similarity in properties as well as the usual vertical likeness.

How do the transition metals iron and copper compare with the reactive metals sodium and calcium in:

■ their melting points and boiling points,
■ their density,
■ their reaction with water,
■ the number of different oxidation states which they show in their compounds,
■ the colour of their compounds as solids and in aqueous solution?

The lanthanides and actinides form a second block of elements within the transition metals. Indeed, they are sometimes called the **inner transition elements**. Another name for these elements is the **'f-block'** elements, since electrons are being added to *f* sub-shells across this block of elements. The lanthanides consist of the 14 elements from cerium (Ce) to lutetium (Lu) which come immediately after lanthanum (La) in the periodic table. These elements resemble each other very much. In fact, the horizontal similarities across this block are so great that chemists experienced considerable difficulty in separating the lanthanides from one another. Lanthanum and the lanthanides are sometimes known as the rare earth elements. The actinides are the 14 elements from thorium (Th) to lawrencium (Lr) which follow actinium (Ac) in the periodic table. Only the first three elements in the actinide series (thorium (Th), protoactinium (Pa) and uranium (U)) are naturally occurring. All the elements beyond uranium have been synthesised by chemists since 1940.

The poor metals

The term 'poor metals' is not widely used, but it is a useful description for several metals including tin, lead and bismuth. These metals fall in a triangular block of the periodic table to the right of the transition metals. They are usually low in the activity (electrochemical) series and they have some resemblances to non-metals.

The non-metals

These elements also form a triangular block in the periodic table. The elements in this block and the last one are sometimes called the **'p-block'** elements since the outermost electrons in these elements are going into *p* sub-shells.

The noble gases

The atoms in these elements have completely filled *s* and *p* sub-shells of electrons. They are very unreactive and the first noble gas compound was not obtained until 1962. Because of their chemical unreactivity these elements were originally called 'inert gases'. Nowadays, several compounds of these elements (mainly oxides and fluorides of xenon and krypton) are known and the name 'inert' has been replaced by 'noble'.

The first crystals of xenon fluoride isolated at the Argonne National Laboratory, USA.

* The atomic conductance (atomic electrical conductivity) is the conductivity of a block of the substance 1 cm² in cross-section but long enough to contain one mole of atoms of the element. It is a measure of the conductivity of one mole of atoms of the element. Notice that the units of atomic conductance (atomic electrical conductivity) are ohm^{-1}cm^{-4}. The electrical conductivity of a substance is usually defined as the reciprocal of the electrical resistivity. This is the resistance of a section of the substance one square centimetre in cross-section and one centimetre long. The units of electrical resistivity are ohm cm so the units of electrical conductivity are therefore ohm^{-1}cm^{-1}. This compares equal volumes of substances. More appropriately, the atomic conductance (atomic electrical conductivity) compares the conductivity of equal numbers (i.e. 1 mole) of atoms and its value is obtained by dividing the electrical conductivity of an element by its molar volume. Hence, the units of atomic conductance are ohm^{-1}cm^{-4}.

4.6 Metals, non-metals and metalloids

Although the periodic table does not classify elements as metals and non-metals, there is a fairly obvious division between the two. *'Fairly obvious'* you will note, not *'clear cut'*. Separating the elements into either metals or non-metals is rather like trying to separate all the shades of grey into either black or white. It cannot be done without ambiguity. The fairly obvious division between metals and non-metals is shown by a thick stepped line in figure 4.6. The 20 or so non-metals are packed into the top right-hand corner above the thick stepped line. Some of the elements next to the thick steps, such as germanium, arsenic and antimony, have similarities to both metals and non-metals and it is difficult to place these in one class or the other. Chemists sometimes use the name **metalloid** (or **semi-metal**) for these elements which are difficult to classify one way or the other.

Figure 4.7 shows a classification of elements as metals, metalloids and non-metals on the basis of their electrical conductance. In this classification:

1 *Metals are good conductors of electricity* with atomic conductance* (atomic electrical conductivity) greater than 10^{-3}ohm^{-1}cm^{-4}. Their conductivity slowly falls as the temperature rises.

2 *Metalloids are poor conductors of electricity* with atomic conductance usually less than 10^{-3} but greater than 10^{-5}ohm^{-1}cm^{-4}. The conductivity of a metalloid increases as the temperature rises and is also considerably affected by the presence of impurity.

3 *Non-metals are virtually non-conductors* (insulators). Their atomic conductance is usually less than 10^{-10}ohm^{-1}cm^{-4}.

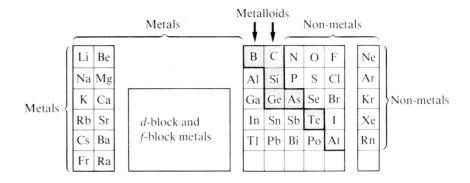

Figure 4.7
Classifying the elements as metals, metalloids and non-metals on the basis of their electrical conductance.

Notice in figure 4.7 that the cell for carbon has been shaded less heavily than those for the other metalloids. This is because carbon exists in two different solid forms called **allotropes**. One of these allotropes, graphite, is a poor conductor of electricity and would be classed as a metalloid; the other allotrope, diamond, is an insulator and therefore classed as a non-metal. In spite of problems such as this, the classification of elements into metals, metalloids and non-metals is useful and convenient.

One other criterion which has been used to classify elements as metals, metalloids and non-metals is the acidic/basic nature of their oxides.

A metal can be defined as an element whose oxide (with the group oxidation number) is basic. A metalloid is an element whose oxide is amphoteric and a non-metal is an element whose oxide is acidic.

■ Draw an outline of the periodic table as in figure 4.7 (omitting the *d*- and *f*-block elements) and shade in the cells for those elements which you consider to be metalloids on the criterion of amphoteric oxides.

■ How does this classification of metals, metalloids and non-metals compare with that in terms of electrical conductivity?

4.7　Periodic properties

In the following sections of this chapter we shall look at the variation in properties of the elements across the periodic table. Some of these features and other properties of the elements will be taken up again and extended in chapter 13. In chapters 15, 17 and 18, we shall look at the vertical trends in properties down the periodic table.

In this chapter we shall pay particular attention to elements in the second and third periods, Li, Be, B, C, N, O, F, Ne, and Na, Mg, Al, Si, P, S, Cl, Ar, respectively.

4.8　The periodicity of physical properties

Tables 4.2 and 4.3 show the values of various physical properties for the elements in the second and third rows, respectively, of the periodic table. Notice the following trends across the second and third periods.

1　The melting point and molar heat of fusion ($\Delta H_{\text{fus.}}$) rise to the element in Group IV and then fall to low values (figure 4.8).

Table 4.2

The values of various physical properties for the elements in the second row of the periodic table

	Li	Be	B	C (graphite)	C (diamond)	N	O	F	Ne
Melting point /°C	180	1280	2030	3700	3550	−210	−219	−220	−250
Heat of fusion/kJ mole^{-1}	3.0	11.7	22.2	—	—	0.36	0.22	0.26	0.33
Boiling point/°C	1330	2480	3930	sublimes	4830	−200	−180	−190	−245
Heat of vaporisation/kJ mole^{-1}	135	295	539	717 (sub)	—	2.8	3.4	3.3	1.8
Density*/g cm^3 (at 25°C)	0.53	1.85	2.55	2.25	3.53	0.81	1.14	1.11	1.21
Molar volume/cm^3 mole^{-1} (conditions as for density)	13.1	4.9	4.6	5.3	3.4	17.3	14.0	17.1	16.7
Atomic conductance × 1000 /ohm^{-1}cm^{-4}	8	51	—	0.14	—	—	—	—	—
Thermal conductivity /J cm^{-1} s^{-1} K^{-1} (at 25°C)	0.71	1.6	0.01	0.24	—	0.00025	0.00025	—	0.00042

Type of element	Metals		Metalloids		Non-metals				
	Li	Be	B	C graphite	C diamond	N	O	F	Ne
Type of structure	Giant metallic		Giant molecular			Simple molecular			
						N$_2$	O$_2$	F$_2$	Ne

* For those elements which are gaseous at 25°C the density quoted is that of the liquid at its boiling point.

2 The boiling point and molar heat of vaporisation (ΔH_{vap}) rise to the element in Group IV and then fall to low values (figure 4.9).

3 The density rises to the elements in Groups III and IV and then falls (figure 4.10).

4 The molar volume falls to the centre of the periods and then rises again (figure 4.10).

5 The electrical and thermal conductivity is relatively high for the metals on the left of each period, lower for metalloids in the centre of the periods and almost negligible for non-metals on the right of each period.

This periodicity of physical properties is, of course, related to a periodicity in the types of elements and their structures. Across a period, the elements change from metals through metalloids to non-metals. In period 3, sodium, the left-hand element, is a very reactive metal, whereas chlorine, next to the extreme right, is a very reactive non-metal. In between, the elements show a gradual transition from metals to non-metals. These periodic changes in the physical properties of the elements across the table are reflected in a periodic change in structure. The structure of the elements varies from metallic, through giant molecular in the metalloids to simple molecular structures in the non-metals.

Table 4.3
The values of various physical properties for the elements in the third row of the periodic table

	Na	Mg	Al	Si	P (white)	S (rhombic)	Cl	Ar
Melting point/°C	98	650	660	1410	44	119	−101	−189
Heat of fusion/kJ mole^{-1}	2.60	8.95	10.75	46.4	0.63	1.41	3.20	1.18
Boiling point/°C	890	1120	2450	2680	280	445	−34	−186
Heat of vaporisation/kJ mole^{-1}	89.0	128.7	293.7	376.7	12.4	9.6	10.2	6.5
Density*(at 25°C)/g cm^{-3}	0.97	1.74	2.70	2.33	1.82	2.07	1.57	1.40
Molar volume/cm^3 mole^{-1} (conditions as for density)	23.7	14.6	10.0	12.1	16.9	15.6	22.8	28.5
Atomic conductance × 1000 /ohm^{-1} cm^{-4}	10	16	38	4	10^{-16}	10^{-22}	—	—
Thermal conductivity /J cm^{-1} s^{-1} K^{-1} (at 25°C)	1.34	1.6	2.1	0.84	—	0.00029	0.00008	0.00017

Type of element	Metals			Metalloid	Non-metals			
	Na	Mg	Al	Si	P	S	Cl	Ar

Type of structure	Giant metallic			Giant molecular	Simple molecular			
	Na	Mg	Al	Si	P_4	S_8	Cl_2	Ar

* For those elements which are gaseous at 25°C the density quoted is that of the liquid at its boiling point.

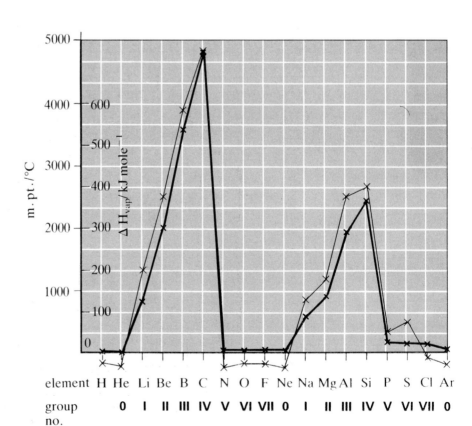

Figure 4.8
Variation of melting point (thick line) and $\Delta H_{fus.}$ (thin line) for the elements hydrogen to argon.

Figure 4.9
Variation of boiling point (thin line) and $\Delta H_{vap.}$ (thick line) for the elements hydrogen to argon.

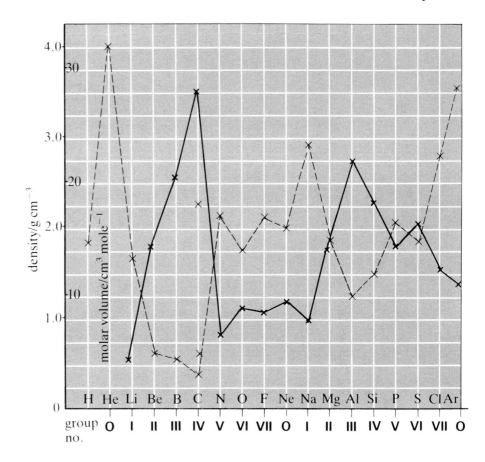

Figure 4.10
Variation of density (thick line) and molar volume (dashed line) for the elements hydrogen to argon.

- Which elements occur at or near the peaks on the graph in figure 4.9? What does this tell you about the strength of forces between their particles? In which groups of the periodic table are these elements found? What type of structure do these elements possess?
- Which elements occur at or near the troughs on the molar volume graph in figure 4.10? What does this tell you about the average distances between their atoms?
- Why is it that metals have high electrical and thermal conductivity?
- Why are the heat of fusion, heat of vaporisation and molar volume values in tables 4.2 and 4.3 given for one mole of the element and not one gram?

Notice that elements on the left-hand side of the periodic table exist as giant structures (metallic or giant molecular), whereas those on the right consist of small molecules (simple molecular). The graphs in figures 4.8 and 4.9 reflect this change in structure across periods 2 and 3 by the sharp drops from C to N and from Si to P. In period 2, carbon is the last element with a giant structure and nitrogen is the first element with simple molecules (N_2). In period 3, silicon is the last element with a giant structure and phosphorus is the first element with simple molecules (P_4).

4.9 Relating the properties of elements to their structure

Giant metallic structures

Metals usually have high melting points, high boiling points and high heats of fusion and vaporisation. These high values suggest that there are strong forces between the separate atoms in the metal. How does the structure of a metal

Figure 4.11
A model of metallic structure.

explain these and other typical metal properties? The physical properties of a metal can be explained using a model in which the outer shell electrons of the metal move randomly throughout a lattice of regularly spaced positive ions (figure 4.11). The moving electrons are sometimes described as a 'sea' or 'cloud' of fluctuating negative charge.

In the metal lattice, each positively charged ion is attracted to the 'cloud' of negative electrons and vice versa. These electrostatic attractions bind the entire crystal together as a single unit. In this model, one particular electron does not belong to one particular metal ion but is attracted to all the positive ions in the lattice.

Since the mobile outer shell electrons are responsible for the bonding in metals, it is not surprising that moving from sodium (one outer shell electron) through magnesium (two outer shell electrons) to aluminium (three outer shell electrons) the bonding gets gradually stronger. Thus, the melting point, the boiling point, and the heats of fusion and vaporisation rise from Na to Mg to Al. A similar trend is evident in the second period from Li → Be → B (see figures 4.8 and 4.9). The bonding in these elements is discussed more fully in sections 13.3 and 13.4.

The stronger bonding from Na → Mg → Al means that the atoms are pulled closer in Mg than Na and even closer in Al. This explains the increasing density (mass per unit volume) and the decreasing molar volume (volume of one mole of atoms) froms Na → Mg → Al. Similar trends are observed in the second period from Li → Be → B (see figure 4.10).

The mobile outer electrons in metals account for their high electrical and thermal conductivity. Under normal conditions the electrons will move fairly randomly throughout the lattice of positive ions in the metal crystal. When the metal is connected across a difference of potential there is an overall movement of electrons (superimposed on their random motion) away from the repelling negative electrode towards the attracting positive electrode. Thus, metals have good electrical conductivity.

The high thermal conductivity of metals can also be explained in terms of their mobile electrons. Electrons in the regions of high temperature (i.e. electrons with high kinetic energy) move rapidly and randomly towards the cooler regions of the metal, transferring their energy to other electrons throughout the metal.

Look at the atomic conductance (atomic electrical conductivity) and the thermal conductivity of elements in tables 4.2 and 4.3.

- Why do you think the electrical conductivity rises from Na → Mg → Al?
- Why do you think the thermal conductivity rises from Na → Mg → Al and from Li → Be?
- Why does B have a lower thermal conductivity than Be? What type of structure does B have?
- Why do metals have a shiny appearance? Use the model of metals with mobile outer electrons to explain their lustre.

The element silicon is used to make 'silicon chips' on which complex circuits can be made. This photograph shows the Ferranti F100, the first European-developed microprocessor. The surface area of the 'chip' is less than 0.5 cm² and it is only 0.05 cm thick. Even so, it contains about 7000 components.

Giant molecular structures

The metalloids (boron, graphite and silicon) and the non-metal, diamond, have giant molecular structures. Figure 4.12 shows the arrangement of atoms in the structure of silicon and diamond. Each atom can be imagined to be situated at the centre of a regular tetrahedron strongly bound by covalent bonds to four other atoms. The covalent linking in these elements extends from one atom to the next through the whole lattice forming a three-dimensional giant molecule (**macromolecule**). The strong covalent bonds hold each atom tightly in the crystal and it is extremely difficult to break one atom away from the lattice. Thus, these elements have very high melting points and boiling points. They also have very high heats of fusion and vaporisation and they are very hard. Indeed, they have even higher values than metals for their melting points, boiling points and heats of fusion and vaporisation (figures 4.8 and 4.9).

The strong covalent bonds pull the atoms together and result in high densities and small molar volumes (figure 4.10). Furthermore, the electrons within the covalent bonds are held much more tightly in these elements than in metals and consequently their thermal and electrical conductivities are lower. Compare the thermal and electrical conductivity of Si with the corresponding values for Na, Mg and Al in table 3.4

Figure 4.12
Models showing the arrangement of atoms in the structure of silicon and diamond. (a) The tetrahedral arrangement of atoms around a central atom in silicon and diamond. (b) An extended section of the structure of silicon and diamond.

Simple molecular structures

All the non-metals in periods 2 and 3 (except diamond) form simple molecular structures. Each of these elements consists of separate, small molecules, i.e. N_2, O_2, F_2, Ne, P_4, S_8, Cl_2, Ar. There are strong covalent bonds between the atoms within these molecules (i.e. between the two O atoms in an O_2 molecule), but only very weak molecular (Van der Waals') forces (sections 9.3 and 9.4) between the separate molecules (i.e. between one O_2 molecule and other O_2 molecules). Consequently, the small distinct molecules can be separated easily and these non-metals have low melting points, low boiling points and low heats of fusion and vaporisation (see figures 4.8 and 4.9.) The weak Van der Waals' forces between molecules such as N_2, O_2, P_4 and Cl_2 mean that their crystals are not packed tightly. So these non-metals have relatively low densities and high molar volumes (figure 4.10). There are no mobile electrons in the crystal structures of these elements and so they have very low conductivities.

4.10 The periodicity of chemical properties

Table 4.4 describes some reactions of the elements in period 3. Notice how the reactivities of the elements change across the period. Notice also that the trends in reactivity vary with different reagents. For example, the reactivity of the elements with chlorine gradually falls across the period from Na to Ar.

However, the reactivity with hydrogen falls at first to elements in the centre of the period and then rises for S and Cl_2. In spite of these differences, all the trends in table 4.4 can be related with considerable accuracy to the action of the elements as oxidising or reducing agents.

The metals (sodium, magnesium and aluminium) are strong reducing agents. They readily give up electrons to form their corresponding ions.

$$Mg \rightarrow Mg^{2+} + 2e^-$$

Table 4.4
Some reactions of the elements in the third period

Element	Heat element in dry chlorine	Heat element in dry oxygen	Heat element in dry hydrogen	Add cold dilute acid (e.g. H_2SO_4(aq)) to the element	Add conc. HNO_3 as oxidising agent
Na	very vigorous reaction forming Na^+ Cl^-	very vigorous reaction forming $(Na^+)_2O^{2-}$ + $(Na^+)_2(O_2^{2-})$	very vigorous reaction forming Na^+H^-	violent reaction $\rightarrow Na_2SO_4$(aq) + H_2(g)	explosive reaction $\rightarrow Na^+$ salt + H_2O + oxides of nitrogen
Mg	vigorous reaction forming $Mg^{2+}(Cl^-)_2$	very vigorous reaction forming $Mg^{2+}O^{2-}$	vigorous reaction forming $Mg^{2+}(H^-)_2$	very vigorous reaction $\rightarrow$ $MgSO_4$(aq) + H_2(g)	violent reaction $\rightarrow Mg^{2+}$ salt + H_2O + oxides of nitrogen
Al	vigorous reaction forming Al_2Cl_6	vigorous reaction at first forming $(Al^{3+})_2(O^{2-})_3$, then the oxide layer prevents further attack	no reaction	vigorous reaction (after oxide layer is removed)	oxide layer on Al reduces vigour of reaction
Si	slow reaction forming $SiCl_4$	slow reaction forming SiO_2	no reaction	no reaction with dilute acids	no reaction
P	slow reaction forming PCl_3 + PCl_5	vigorous reaction forming P_4O_6 and P_4O_{10}	no reaction	no reaction with dilute acids	vigorous reaction on heating forming P_4O_6 + P_4O_{10} + oxides of nitrogen
S	slow reaction forming SCl_2 + S_2Cl_2	slow reaction forming SO_2	very slow reaction forming H_2S	no reaction with dilute acids	slow reaction on heating forming a little SO_2
Cl	no reaction	no reaction	vigorous reaction in sunlight forming HCl	no reaction with dilute acids	no reaction
Ar	no reaction	no reaction	no reaction	no reaction with dilute acids	no reaction

Of course, these three metals never exist freely in nature, but only in compounds as Na^+, Mg^{2+} or Al^{3+} ions. Thus, these three metals react vigorously with non-metals and acids which are oxidising agents.

$$Cl_2 + 2e^- \rightarrow 2Cl^-$$

$$O_2 + 4e^- \rightarrow 2O^{2-}$$

$$H_2 + 2e^- \rightarrow 2H^-$$

$$2H^+ + 2e^- \rightarrow H_2$$

Silicon in Group IV is a very weak reducing agent. It will react slowly with Cl_2 and O_2 which are strong oxidising agents, but not with H_2 and acids which are weaker oxidising agents.

$$Si + 2Cl_2 \rightarrow SiCl_4$$

$$Si + O_2 \rightarrow SiO_2$$

The next two elements, phosphorus and sulphur, are weak reducing agents and weak oxidising agents. They react slowly with oxygen and chlorine (strong oxidising agents), moderately with concentrated HNO_3 on heating, but they have no reaction with dilute acids (weaker oxidising agents).

$$S + Cl_2 \rightarrow SCl_2$$

$$2S + Cl_2 \rightarrow S_2Cl_2$$

$$S + 2HNO_3 \rightarrow SO_2 + H_2O + NO + NO_2$$

Phosphorus does not react directly with hydrogen, but molten sulphur will react slowly with hydrogen to form hydrogen sulphide.

$$H_2 + S \rightarrow H_2S$$

Notice in this case that the sulphur is now acting as an oxidising agent and the hydrogen as a reducing agent.

Chlorine, at the other extreme to the metals in chemical reactivity, is a strong oxidising agent. Thus, it has no reaction with O_2, dilute acids or concentrated acids which are themselves oxidising agents. However, in sunlight, it will react violently with hydrogen (which can act as a reducing agent) to form hydrogen chloride.

$$H_2 + Cl_2 \rightarrow 2HCl$$

On the extreme right, argon (a noble gas) shows no reactivity with any of these reagents.

Other periods show a similar trend in reactivity to period 3. Excluding the noble gases, elements change from strong reducing agents on the extreme left to elements which are weak reducing agents and/or weak oxidising agents in the centre and finally to strong oxidising agents.

4.11 Periodicity in the structure and properties of chlorides

Look closely at table 4.5. It shows the formulas, states and boiling points of chlorides of the elements in the first three periods.

■ Work out the oxidation numbers of each element in the chlorides shown. How do the oxidation numbers of the elements vary across period 2 and period 3?
■ How does the volatility of the chlorides (as indicated by their boiling points) vary across period 2 and period 3?
■ The structures of the chlorides change from giant structures on the left of each period to simple molecular structures on the right. Which chlorides in periods 2 and 3 have giant structures and which have simple molecular structures?

We shall look at the trends in properties across a period more fully in chapter 13.

Table 4.5
The formulas, states and boiling points of chlorides of the elements in the first three periods

Period 1	H	He
Formula of chloride	HCl	No chloride
State of chloride (at 20°C)	g	—
B.pt. of chloride/°C	−85	—

Period 2	Li	Be	B	C	N	O	F	Ne
Formula of chloride	LiCl	BeCl$_2$	BCl$_3$	CCl$_4$	NCl$_3$	OCl$_2$ (O$_7$Cl$_2$)	FCl	No chloride
State of chloride (at 20°C)	s	s	g	l	l	g (g)	g	—
B.pt. of chloride/°C	1350	487	12	77	71	2 (decomposes)	−101	—

Period 3	Na	Mg	Al	Si	P	S	Cl$_2$	Ar
Formula of chloride	NaCl	MgCl$_2$	Al$_2$Cl$_6$	SiCl$_4$	PCl$_3$ (PCl$_5$)	SCl$_2$ (S$_2$Cl$_2$)	Cl$_2$	No chloride
State of chloride (at 20°C)	s	s	s	l	l (s)	l (l)	g	—
B.pt. of chloride/°C	1465	1418	423	57	74 (164)	59 (138)	−35	—

Summary

1   Most of the credit for arranging the elements in a periodic table is given to Mendeléev. Mendeléev arranged the elements in order of increasing relative atomic mass and showed that elements with similar properties recurred at regular intervals.

2   The vertical columns of similar elements are called groups and the horizontal rows of elements are called periods.

3   In modern periodic tables, all the elements are in strict order of atomic number. The modern periodic law states that the properties of elements are a periodic function of their atomic numbers.

4   Besides a division into vertical groups of similar elements, it is also useful to split the periodic table into five blocks of elements with similar properties:

   The reactive metals (elements in Groups I and II)
   The transition metals
   The poor metals (metals in Groups III, IV, V and VI)
   The non-metals in Groups III, IV, V, VI and VII
   The noble gases

5 In spite of its limitations, the classification of elements as metals, metalloids and non-metals is very useful. The commonest classification is based on electrical conductivity.

> Metals are good conductors of electricity.
> Metalloids are poor conductors of electricity.
> Non-metals are non-conductors.

6 Metals form giant metallic structures. They have high melting points, high boiling points, high density and good conductivity.

7 Metalloids and diamond form giant molecular structures. They have very high melting points, very high boiling points, high density but poor conductivity.

8 Non-metals (except diamond) form simple molecular structures. They have low melting points, low boiling points, low density and they are non-conductors.

9 From left to right across the periodic table, elements change from being reactive metals, through less reactive metals, metalloids, less reactive non-metals to reactive non-metals. (On the extreme right are the noble gases.)

10 These trends in metal/non-metal character are reflected in the changing redox behaviour across a period. Excluding the noble gases, the elements change from strong reducing agents (e.g. Na, Mg) through elements which are weak reducing agents (e.g. Si), then weak oxidising agents (e.g. S) to strong oxidising agents (e.g. Cl_2).

Study questions

1 Consider the elements (Li, Be, B, C, N, O, F, Ne) in the second period.
  (a) Which elements
    (i)   form cations,
    (ii) form a chloride of empirical formula, XCl_3,
    (iii) react together to form a compound of formula, XY,
    (iv) exist as diatomic molecules at room temperature?
  (b) How do each of the following properties vary for this sequence of elements?
    (i)   The boiling points of the elements.
    (ii) The oxidation numbers of the elements in their chlorides.
    (iii) The boiling points of the chlorides of the elements.
  (c) Give a brief explanation of the trends in oxidation number in terms of the electronic structures of the atoms of these elements.
  (d) How do you account for
    (i)   the trend in boiling points of the elements,
    (ii) the trend in boiling points of the chlorides of the elements?

2 Use a data book to plot (on the same graph) the melting points and boiling points of the elements from H to Ca against atomic number.
  (a) Which elements occur at or near the peaks on
    (i)   the melting point curves,
    (ii) the boiling point curves?
  (b) Which elements occur in the troughs on
    (i)   the melting point curves,
    (ii) the boiling point curves?
  (c) What type of structure is found in those elements which occur
    (i)   at or near the peaks,
    (ii) in the troughs?

(d) Make a note of elements which are liquid over
    (i)   unusually large temperature ranges,
    (ii) unusually small temperature ranges.
(e) What explanation can you give for the fact that some elements have
    (i)   unusually large liquid ranges,
    (ii) unusually small liquid ranges?

3   Table 4.6 shows the specific heat capacity for most of the elements from Li to Ca in the periodic table.
    (a) Plot the values of specific heat capacity against atomic number for the elements listed. Does the graph show a periodic variation?
    (b) Calculate the specific heat capacities per mole of atoms for those elements for which values are listed in table 4.6. Now plot these values of specific heat capacity per mole against atomic number. Does the graph show a periodic variation?

Table 4.6
The specific heat capacities for elements Li to Ca inclusive (all values are in $J\,g^{-1}K^{-1}$)

Li	Be	B	C	N	O	F	Ne
3.36	1.66	1.30	0.50				

Na	Mg	Al	Si	P	S	Cl	Ar
1.26	1.05	0.88	0.71	0.82	0.76	0.97	—

K	Ca
0.76	0.71

4   An element A reacts with another element B to form a compound of formula AB_2. The element B exists as molecules of formula B_2. Some properties of A, B_2 and AB_2 are tabulated below.

	A	B_2	AB_2
Melting point	High (in the range 700–1200°C)	Very low (less than −50°C)	Moderately high (in the range 400–700°C)
Electrical conductivity of the solid	High	Very low	Very low
Electrical conductivity of molten material	High	Very low	High
Electrical conductivity of aqueous solution of the material			High

(a) Which particles will move when a potential difference is applied across a sample of (i) solid A, (ii) molten AB_2?
(b) Explain why the electrical conductivity of molten AB_2 is high, whereas that of the solid is very low.
(c) Electrolysis of an aqueous solution of AB_2 with Pt electrodes gave A at the cathode and B_2 at the anode. Suggest possible names for the elements A and B consistent with all the above information.

5 The following table shows the melting point and conductivity of five substances.

	Melting point /K	Electrical conductivity in solid state	Electrical conductivity in molten state
Magnesium oxide, MgO	3173	poor	good
Sodium chloride, NaCl	1081	poor	good
Magnesium, Mg	923	good	good
Carbon dioxide, CO_2	217	poor	poor
Silicon(IV) oxide, SiO_2	1883	poor	poor

(a) Why is the electrical conductivity of MgO(l) good, but that of MgO(s) poor?

(b) Why is the melting point of MgO considerably higher than that of NaCl?

(c) Why is the electrical conductivity of magnesium good in *both* solid and liquid states?

(d) Why is the melting point of SiO_2 so much higher than that of CO_2 in spite of the fact that C and Si are both in Group IV of the periodic table?

6 Look closely at the boiling points of the elements Na to Ar in figure 4.9 or table 4.3.

(a) Why do the boiling points rise from Na → Si? Explain the trend in terms of the structures of the elements.

(b) Why are the boiling points of the second four elements (P to Ar) much lower than those of the first four?

(c) Why is there no clear trend in the boiling points of P, S, Cl and Ar? Plot a graph of the boiling points against the relative molecular mass of the simple molecules (P_4, S_8, Cl_2, Ar) of these elements to help you answer this question.

7 An element, X, reacts with hydrogen at 300°C to form a black compound, Y. When a sample of Y was heated to 1000°C, it decomposed forming 1.00 g X and 141 cm^3 of hydrogen at s.t.p.

If Y is heated with chlorine at 250°C, it forms a green chloride, Z, containing 37.3% by mass of chlorine and 62.7% by mass of X. Deduce

(a) the relative atomic mass of X,

(b) the element, X,

(c) the formulae of Y and Z.

(The molar volume of a gas at s.t.p. is 22.4 dm^3.)

5 Atomic Structure

5.1 Introduction

Less than a century ago, scientists believed that atoms were solid indestructible particles like minute snooker balls. Since then, they have built up a great deal of evidence concerning the detailed structure of atoms.

Experiments involving electrolysis suggested that certain compounds contained charged particles called ions. The formation of these ions from atoms could be explained by the loss or gain of negatively charged particles (electrons). This led to the idea that atoms consisted of a small positive nucleus surrounded by negative electrons. Later, scientists discovered that the nucleus is composed of two kinds of particle—protons and neutrons. But, what is the evidence for these particles? In this chapter we shall consider those experiments which have been important in shaping our ideas about atomic structure.

5.2 Evidence for atomic structure

1897—Thomson experiments with electrons

In 1874, G.J. Stoney suggested the name 'electron' for the tiny negative particles which made up an electric current. Stoney realised that the experiments involving electrolysis, which Faraday had carried out earlier in the nineteenth century, could be explained in terms of electrons.

J. J. Thomson (1856–1940) investigating the conductivity of electricity by gases. Using discharge tubes (similar to that on the right of this photograph), Thomson discovered the electron.

However, firm evidence for the existence of electrons was not found until 1897. In that year, J.J. Thomson was investigating the conductivity of electricity by gases at *very low pressure*. At ordinary pressures gases are electrical insulators, but when they are subjected to very high voltages at very low pressures (below 0.01 atm) they 'break down' and conduct electricity. When the gas pressure is lowered to 0.0001 atm and a potential difference of 5000 volts is applied across the tube containing the gas, the glass container begins to glow. Since then, the bright glows produced have found applications in neon advertising signs and sodium vapour street lamps.

When Thompson applied 15000 volts across the electrodes of a tube containing a trace of gas, a bright green glow appeared on the glass. The green glow results from the bombardment of the glass by rays travelling in straight lines from the cathode until they strike the anode or the glass walls of the tube. Thomson called these rays **cathode rays** (figure 5.1). Thomson also showed that when cathode rays were deflected onto an electrode of an electrometer, the instrument became negatively charged and the rays could be deflected by a magnetic or by an electric field (figure 5.2)

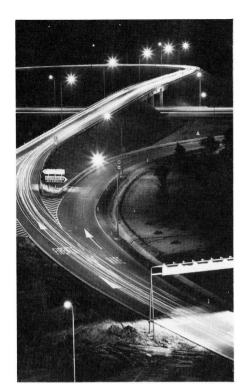

The bright glows produced in discharge tubes are now used for advertising signs and roadside lighting.

Figure 5.1
Cathode rays, deflecting plates uncharged.

Figure 5.2
Cathode rays, deflecting plates charged.

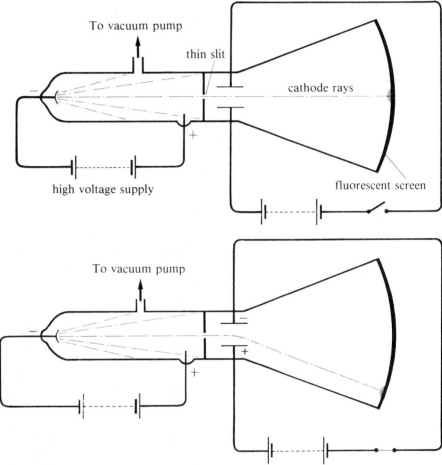

When the rays were deflected by an electric field across a pair of charged plates, the rays moved away from the negative plate towards the positive plate. This suggested that the cathode rays were negative. Thomson studied the bending of a thin beam of cathode rays by magnetic and electric fields and concluded that they consisted of electrons—tiny negatively charged particles. Later, electrons were shown to be 1840 times lighter than hydrogen atoms.

Now, 6×10^{23} H atoms weigh 1 g

$$\therefore \text{1 H atom weighs} \quad \frac{1}{6 \times 10^{23}} \text{ g}$$

$$\therefore \text{1 electron weighs} \quad \frac{1}{6 \times 10^{23}} \times \frac{1}{1840} \text{ g} = 9 \times 10^{-28}\text{g}$$

The same results were obtained with the cathode rays using different gases in the tube and with tubes and electrodes of different materials. This suggested that electrons were present in the atoms of all substances.

- Where do you think the electrons in the cathode rays have come from?
- Why do the cathode rays cause the glass to fluoresce?
- Which piece of Thomson's evidence showed that cathode rays could not be rays of light?

Thomson discovers positive rays

In his experiments with cathode rays, Thomson had noticed a red glow around the cathode. This red glow appeared on the opposite side to the anode. To investigate this red glow he designed a discharge tube with a central cathode which had a hole in it (figure 5.3)

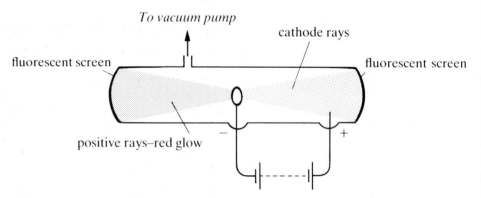

Figure 5.3
The apparatus used by Thomson to obtain positive rays.

This time he saw a red fluorescence as well as a green fluorescence in the tube. The green fluorescence was caused by cathode rays—electrons. The red glow was caused by rays which were deflected by magnetic and electric fields in the opposite direction to electrons. This showed that they contained positively charged particles. Thomson called these rays **positive rays**. They required much larger fields than electrons to cause their deflection. Also, the mass of the particles in them depended on the nature of the gas in the tube.

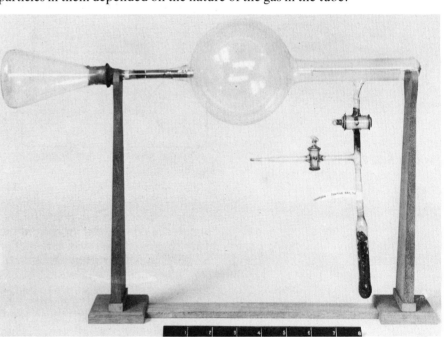

The apparatus with which Thomson investigated positive rays. Compare this with the diagram in figure 5.3.

For example, when the gas in the tube is hydrogen, the mass of the positive particles is almost identical to that of a hydrogen atom. If the gas in the tube is oxygen, the positive particles have a mass almost identical to that of oxygen atoms. Thomson concluded that the positive rays were produced when electrons in the cathode rays collided with molecules of gas in the tube. When a fast-moving electron, streaming away from the cathode, collides with a gas molecule it splits the molecule into atoms and strips off one or more electrons. The atoms are converted into positive ions by the removal of electrons.

$$e^- \quad + H_2 \longrightarrow e^- \quad + \underbrace{H^+ + H^+}_{} + \underbrace{e^- + e^-}_{}$$

| fast-moving electron | hydrogen molecule | retreating electron | hydrogen ions | electrons stripped from H atoms |

The positive ions are attracted towards the cathode and pass through the hole in it to the space behind.

The lightest positive ions are produced when the gas in the tube is hydrogen. Just as the unit of negative charge is the electron, so the unit of positive charge is that associated with a hydrogen ion, H^+. This unit of positive charge is called a **proton**.

1899—Thomson's model of atomic structure

As a result of his experiments, Thomson suggested that atoms consisted of rings of negative electrons embedded in a sphere of positive charge. The negative and positive charges balance and result in the atom being neutral. Thomson believed, wrongly of course, that the mass of the atom was due only to the electrons and, since one electron was only $\frac{1}{1840}$th of the mass of a hydrogen atom, Thomson concluded that each hydrogen atom must contain 1 840 electrons.

Thomson's model for the atom was compared to a Christmas pudding. The electrons in a sphere of positive charge were likened to currants in a spherical Christmas pudding.

1909—Geiger and Marsden explore the atom

How were the electrons arranged within the atom? Ernest Rutherford, who had been one of Thomson's research students at Cambridge University, had the idea of probing inside the atom using alpha particles. Rutherford used alpha-particles from radioactive substances as 'nuclear bullets'. Alpha-particles are helium ions, He^{2+}. They travel a few centimetres in air and produce a tiny pinpoint of light on striking a fluorescent screen.

Most of the experimental work was carried out by Geiger and Marsden, two of Rutherford's colleagues at Manchester University. Geiger and Marsden built the apparatus shown in figure 5.4. The source of alpha-particles was a piece of

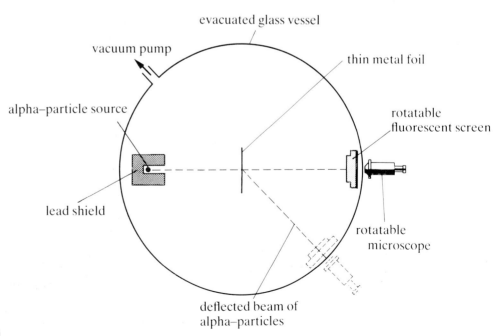

Figure 5.4

The apparatus used by Geiger and Marsden to investigate the deflection of α-particles by thin metal foil.

radium inside a protective lead shield. A narrow hole through the lead allowed the alpha-particles to travel in only one direction through the evacuated vessel towards a fluorescent zinc sulphide screen.

When an alpha-particle hits the fluorescent screen a tiny pinpoint of light can be observed through the microscope. Even when a thin metal foil was placed in the path of the alpha-particles, flashes were still seen on the screen.

Using Thomson's model of atomic structure, Rutherford expected that most of the very fast-moving alpha-particles would pass straight through the foil or be deviated a little. Most of the alpha-particles did pass straight through the foil, but when the detecting screen and microscope were rotated from the straight-on position flashes could still be seen. Clearly, some of the alpha-particles were deflected by the foil, but, to everyone's surprise, one particle in every 10000 appeared to rebound from the foil.

Here is Rutherford's account of the startling discovery.

'I remember Geiger coming to me in great excitement and saying, "We have been able to get some of the alpha-particles coming backwards . . .".

It was quite the most incredible event that has ever happened to me in my life. It was almost as incredible as if you fired a 15-inch shell at a piece of tissue paper and it came back and hit you. On consideration, I realised that this scattering backwards must be the result of a single collision . . . with a system in which the greater part of the mass of the atom was concentrated in a minute nucleus.'

1911—Rutherford explains the structure of the atom

Alpha-particles are positively charged. This led Rutherford to suggest that deflections and reflections could only be caused by the particles coming close to a concentrated region of positive charge. Only a very small fraction of the alpha-particles were deflected, so he concluded that the region of positive charge which caused the scattering would occupy only a small part of the atom (figure 5.5).

These two might look like Laurel and Hardy, but nothing could be further from the truth. Thomson (right) and Rutherford are unquestionably two of the most brilliant scientists of the twentieth century.

Figure 5.5
Most alpha-particles pass straight through the foil without coming near a nucleus. Others come close to a nucleus and are deflected. A few alpha-particles approach a nucleus head-on and are repelled back on the same side of the foil as they approached.

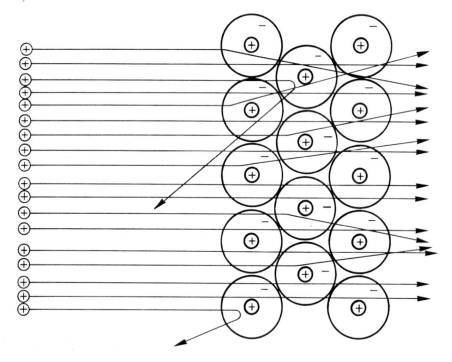

■ Why do most of the alpha-particles pass straight through the metal foil?
■ Why do some alpha-particles appear to rebound from the metal foil?
■ What factors will affect the extent of scattering of the alpha-particles?

Rutherford suggested that atoms in the metal foil consisted of a central positive nucleus composed of protons, where the mass of the atom was concentrated.

If we magnified an atom one million million times to the size of Wembley Stadium, the nucleus would be the size of a pea at the centre of the pitch.

This nucleus was surrounded by a much larger volume in which the electrons move.

From the angles through which alpha-particles are deflected, Rutherford calculated that the nucleus of an atom would have a radius of about 10^{-14} m. This is about one ten-thousandth of the size of the whole atom which has a radius of about 10^{-10} m. Thus, if we magnified an atom to the size of a football stadium (about 100 m across), the nucleus would be represented by a pea placed at the centre of the pitch.

Notice the difference between Rutherford's atomic model and that proposed by Thomson a few years earlier. Rutherford's atomic model has been compared to the solar system. Rutherford pictured each atom as a miniature solar system with electrons orbiting the nucleus, like planets orbiting the sun. He suggested that hydrogen, which had the smallest atoms, would have *one* proton in its nucleus, balanced by *one* orbiting electron. Atoms of helium which were next in size would have *two* protons in their nucleus, balanced by *two* orbiting electrons, etc., etc.

Although Rutherford's idea of planetary electrons has now been discarded, his idea of a small positive nucleus has been supported by many experiments.

1913—Moseley explores the nucleus

Calculations based on the results obtained by Geiger and Marsden showed that the number of positive charges (protons) in the nucleus of an atom was *about* half its relative atomic mass. At about the same time, it was noticed that the number of positive charges on the nucleus was equal to the atom's numbered position in the periodic table which had been given the symbol Z.

Initially, Z was thought to be simply a number corresponding to the order of the element in the periodic table. Later, experiments by Charles Moseley, a young research student at Oxford University, showed that Z could be obtained from properties of the element. This led to an appreciation of its fundamental significance as the number of protons in the nucleus.

Figure 5.6
An X-ray tube.

In 1913, Moseley bombarded various metallic elements with high-speed electrons and showed that X-rays were produced (figure 5.6). X-rays are electromagnetic waves like light, but with very short wavelength. Moseley found that the wavelength or frequency of the X-rays depended on the element he used. Whatever the conditions of his experiment, a particular target element always formed X-rays of the same frequency. What is more, the frequency of the X-rays, ν, was found to be given by a formula which involved Z:

$$\sqrt{\nu} = a(Z - b)$$

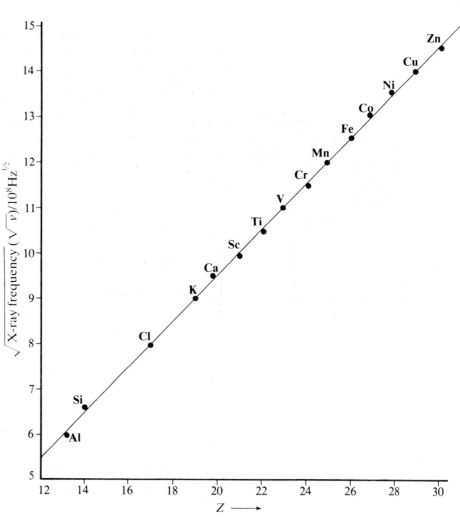

Figure 5.7
A graph of the square root of the frequency of X-rays against the atomic number of the target element.

Some of Moseley's results are shown in figure 5.7. The equation describing his results is of the form $y = mx + c$, in which $\sqrt{v}$ corresponds to y, and Z corresponds to x. a and b are constants where $a = m$, the gradient of the line and $-ab = c$, the intercept on the vertical axis.

Z is now known as the **atomic number**. It is the most important feature of an element's individuality because it represents

(i)   the number of protons in the nucleus,
(ii) the number of electrons in the neutral atom,
(iii) the order in which the element appears in the periodic table.

The atomic number of an element can therefore be predicted from its order in the periodic table; element number five in the periodic table has an atomic number of five, element 25 has an atomic number of 25, and so on.

Moseley's equation was also important in showing the values of Z for three undiscovered elements. Blank spaces occurred in Moseley's results between molybdenum (Mo) and ruthenium (Ru), between neodymium (Nd) and samarium (Sm), and between tungsten (W) and osmium (Os). Moseley's experiments sparked off a search for these missing elements which were found to occupy the blank spaces.

■ What are the names of the three missing elements?

Another valuable outcome of Moseley's work was that the frequency of the X-rays from a particular element provided an easy method of identifying the element.

1932—Chadwick discovers the neutron

In spite of the success of Rutherford and Moseley in explaining atomic structure, one major problem remained unsolved.

Part of the original apparatus with which Chadwick established the existence of neutrons.

Figure 5.8
The experiment in which Chadwick detected neutrons.

If the hydrogen atom contains one proton and the helium atom contains two protons, then the relative atomic mass of helium should be twice that of hydrogen. Unfortunately, the relative atomic mass of helium is four and not two.

Chadwick, one of Rutherford's collaborators, was able to show where the extra mass in helium atoms came from. Chadwick bombarded a thin sheet of beryllium with alpha-particles. The alpha-particles can be traced by an electric counter (figure 5.8(a)) which detects charged particles. When the beryllium is in place, the counter registers nothing (figure 5.8(b)) showing that the alpha-particles are being stopped by the beryllium. However, if a piece of paraffin wax is placed between the beryllium and the counter, charged particles are detected again (figure 5.8(c)).

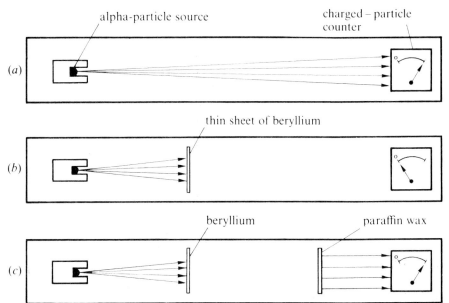

It seems that alpha-particles are stopped by the beryllium foil, yet charged particles are shooting out from the paraffin wax. How can this happen? What has caused charged particles to be ejected from the paraffin wax? Chadwick provided an explanation. He suggested that the alpha-particles striking the beryllium foil displaced *uncharged* particles called **neutrons** from the nuclei of beryllium atoms. These uncharged neutrons could not affect the charged-particle counter, but they could displace positively charged protons from the paraffin wax which would affect the counter. A summary of Chadwick's explanation is shown in figure 5.9. Further experiments showed that neutrons had almost the same mass as protons and Chadwick was able to explain the difficulty concerning the relative atomic masses of hydrogen and helium.

Figure 5.9
Chadwick's explanation of his experiment to detect neutrons.

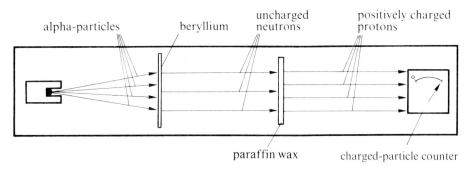

Hydrogen atoms have one proton, no neutrons and one electron. Since the mass of the electron is negligible compared to the masses of the proton and neutron, a hydrogen atom has a relative mass of one unit. Helium atoms have two protons, two neutrons and two electrons, so the relative mass of a helium atom is four units. This means that a helium atom is four times as heavy as a hydrogen atom, i.e. the relative atomic mass of helium = 4.

A bubble chamber photograph showing the tracks of sub-atomic particles. As the particles pass through the liquid, they lose energy and cause the liquid to vaporise, forming bubbles.

Lithium is the third element in the periodic table. Its relative atomic mass is very nearly seven.

- What is the atomic number of lithium?
- How many electrons does each lithium atom contain?
- How many neutrons does each lithium atom contain?
- How many protons, neutrons and electrons does one Li^+ ion contain?

5.3 Sub-atomic particles

As a result of the experiments described in section 5.2, scientists believe that all atoms are composed of three important particles—protons, neutrons and electrons.

The nucleus of an atom is composed of protons and neutrons. Both of these particles have a mass almost equal to that of a hydrogen atom. Virtually all the mass of the atom is concentrated in the nucleus which occupies only a small fraction of the total volume of the atom. The neutron has no charge, whereas the proton carries one positive charge. Electrons with one negative charge occupy the space outside the nucleus. The mass of an electron is 1 840 times less than that of a proton.

The nucleus of a fluorine atom has a diameter of about 10^{-12} cm and a mass of 3.1×10^{-23} g

$$\therefore \text{ The density of the fluorine nucleus} = \frac{\text{mass}}{\text{volume}}$$

$$= \frac{3.1 \times 10^{-23}}{\frac{4}{3}\pi(5 \times 10^{-13})^3} \text{ g cm}^{-3}$$

$$= 6 \times 10^{13} \text{ g cm}^{-3}.$$

The incredibly high density within the fluorine nucleus suggests that the particles in it are drawn very close together by extremely powerful forces. These forces are capable of overcoming the repulsion of the protons for each other. But, these powerful forces within the nucleus are effective over only a very short range because they do not pull the outer electrons into the nucleus.

Table 5.1 summarises the relative masses, the relative charges and the position within the atom of these sub-atomic particles. The atoms of all elements are built up from these three particles, different atoms having different numbers of the three particles. Other particles have been detected during atom-splitting experiments, but these other particles are less stable than protons, neutrons and electrons and only exist under extreme conditions.

Table 5.1
The relative masses, relative charges and positions within the atom of protons, neutrons and electrons

Particle	Mass (relative to that of a proton)	Charge (relative to that on a proton)	Position within the atom
Proton	1	+1	nucleus
Neutron	1	0	nucleus
Electron	$\frac{1}{1840}$	−1	shells

Figure 5.10
A mass spectrometer trace for chlorine.

5.4 Isotopes

In 1913 Thomson discovered that some elements could have atoms with different masses. Thomson's equipment was developed by Aston into a mass spectrometer which could compare the relative masses of atoms (section 1.2).

Thomson and Aston discovered that some naturally occurring elements contained atoms that were not exactly alike. When atoms of these elements were ionised and deflected in a mass spectrometer, the beam of ions separated into two or more paths. Since all ions in the beam have the same electrical charge, those in the separated paths must have different masses. Consequently, the atoms from which the ions were formed must also have different masses.

These atoms of the same element with different masses are called **isotopes**. All the isotopes of one particular element have the *same atomic number* because they have the same number of protons. But they have *different mass numbers* because they have different numbers of neutrons.

Figure 5.10 shows a mass spectrometer trace for chlorine. This shows that chlorine consists of two isotopes: each has an atomic number of 17, but they have different isotopic masses, 35 and 37, respectively.

Atoms of chlorine-35 have 17 protons and 18 neutrons, whereas atoms of chlorine-37 have 17 protons and 20 neutrons. The number of protons + the number of neutrons in an atom is called the **mass number**. Thus, chlorine-35 has a mass number of 35 (17 + 18) and chlorine-37 has a mass number of 37 (17 + 20).

We can write the symbol $^{35}_{17}Cl$ in order to specify the mass number and the atomic number of the chlorine-35 atom. Figure 5.11 shows both the chlorine isotopes represented in this way. The mass number is written at the top and to the left of the symbol. The atomic number is written at the bottom and to the left of the symbol.

Isotopes have the same number of electrons and hence the same chemical properties, because chemical properties depend upon the transfer and redistribution of electrons. But, isotopes have different numbers of neutrons, so they will have different masses and hence different physical properties. For example, pure $^{37}_{17}Cl_2$ has a higher density, higher melting point and higher boiling point than pure $^{35}_{17}Cl_2$.

Chlorine—35

Mass Number
= no. of protons +
    no. of neutrons ⟶ 35

Atomic Number ⟶ 17
= no. of protons

17 protons
18 neutrons
17 electrons

Chlorine—37

$^{37}_{17}Cl$

17 protons
20 neutrons
17 electrons

Figure 5.11
Specifying the mass number and atomic number with the symbol of an isotope.

5.5 Mass numbers, relative isotopic masses and relative atomic masses

The relative mass of a single isotope is called its **relative isotopic mass**. Relative isotopic masses relate to the same scale as relative atomic masses on which the isotopic $^{12}_6C$ is given a relative mass of 12.0000 units. (Relative atomic masses were introduced in section 1.3.) On this scale, the mass of the proton (1.0074 units) is almost the same as that of the neutron (1.0089 units), and the mass of the electron is very small in comparison (0.0005 units). Now since the *relative* masses of the proton and neutron are both very close to one and the electron has a negligible mass, it follows that all relative isotopic masses will be very close to whole numbers. In fact, the relative isotopic mass of an isotope will be very close to its mass number (number of protons + number of neutrons) and the two are assumed to be almost identical in all but the most accurate work.

However, naturally occurring elements often consist of a mixture of isotopes and this results in relative atomic masses which are not close to whole numbers. The relative atomic mass of an element represents the *average* mass of one atom, taking into account the different isotopes and their relative proportions. For example, chlorine consists of two isotopes—$^{35}_{17}Cl$ and $^{37}_{17}Cl$, with relative isotopic masses of 35 and 37, respectively. If chlorine consisted of $^{35}_{17}Cl$, its relative atomic mass would be 35. If it consisted of only $^{37}_{17}Cl$, its relative atomic mass would be 37. A 50:50 mixture of $^{35}_{17}Cl$ and $^{37}_{17}Cl$ would have a relative atomic mass of 36.

Table 5.2
Relating the relative atomic mass of chlorine to the composition of its isotopic mixture

$\% \, ^{35}_{17}Cl$	100	75	50	25	0
$\% \, ^{37}_{17}Cl$	0	25	50	75	100
Relative Atomic Mass	35.0	35.5	36.0	36.5	37.0

Look at figure 5.10. The relative heights of the peaks corresponding to $^{35}_{17}Cl$ and $^{37}_{17}Cl$ indicate that the isotopes occur in the ratio of about 3:1 (i.e. 75% chlorine-35 and 25% chlorine-37). This results in a relative atomic mass for chlorine of 35.5. Table 5.2 shows this more clearly. Can you see that *the relative atomic mass is a weighted mean of the relative isotopic masses* of the different isotopes; weighted, that is, in the proportions in which they occur?

∴ The relative atomic mass of chlorine

$$= \quad 35 \quad \times \quad \tfrac{75}{100} \quad + \quad 37 \quad \times \quad \tfrac{25}{100}$$

relative isotopic mass of chlorine-35   proportion of Cl-35   relative isotopic mass of chlorine-37   proportion of Cl-37

$$= \quad 26.25 + 9.25$$
$$= \quad 35.50$$

Figure 5.12 shows a mass spectrometer trace for the isotopes of neon. The atomic number of neon is 10.

■ What are the relative isotopic masses of the two neon isotopes?
■ What are the mass numbers of the two neon isotopes?
■ How many protons, neutrons and electrons has each of the neon isotopes?
■ What are the percentage abundancies of the two isotopes?
■ Calculate the relative atomic mass of neon.

Figure 5.12
A mass spectrometer trace for neon.

Summary

1 Cathode rays consist of tiny negatively charged particles called electrons. Electrons have a mass of $\tfrac{1}{1840}$th the mass of a hydrogen atom.

2 Rutherford pictured the atom as a miniature solar system in which electrons orbited the positively charged nucleus like planets orbiting the sun.

3 The nucleus is very small compared with the size of the atom as a whole. The volume of the nucleus is only about one million millionth (i.e. one 10^{-12}th) of the total volume of the atom.

4 The nucleus is composed of two particles:
protons with a mass of one relative to the hydrogen atom and a relative charge of $+1$, and
neutrons with a mass of one relative to the hydrogen atom but with no charge.

5 Atomic number = number of protons
              = order of element in the periodic table.

6 Mass number = number of protons + number of neutrons.

7 Isotopes are atoms with the same atomic number, but different mass numbers. Isotopes have the same number of electrons and therefore the same chemical properties.
Isotopes have different numbers of neutrons (i.e. different masses) and therefore different physical properties.

8 The relative isotopic mass of a single isotope is the relative mass of this isotope compared with an isotope of $^{12}_{6}C$ which is assigned a relative mass of 12.0000.

$$\text{Relative isotopic mass} = \frac{\text{mass of 1 isotope of the element}}{\frac{1}{12} \times \text{mass of 1 atom of } ^{12}_{6}C}$$

9 The relative atomic mass of an element is the average of the relative isotopic masses of the different isotopes balanced in the proportions in which they occur.

Study questions

1 (a) What are the atomic numbers of tellurium (Te) and iodine (I)?
  (b) What are the relative atomic masses (atomic weights) of Te and I?
  (c) What are the numbers of protons, neutrons and electrons in the commonest isotopes of tellurium (^{128}Te) and iodine (^{127}I)?
  (d) Why does Te come before I in the periodic table?
  (e) Te comes before I in the periodic table, but Te has a larger relative atomic mass than I. Explain.
  (f) Look closely at the periodic table and write down the names of two other *pairs* of elements which are placed in the periodic table in the reverse order to their relative atomic masses.

2 The five main proposals in Dalton's Atomic Theory of matter were:
  1 All matter is composed of tiny indestructible particles called atoms.
  2 Atoms cannot be created or destroyed.
  3 Atoms of the same element are alike in every way.
  4 Atoms of different elements are different.
  5 Atoms can combine together in small numbers to form molecules.
  (a) In the light of modern knowledge about atoms, isotopes, molecules and atomic structure comment on the truth of each of these proposals.
  (b) Why is Dalton's Atomic Theory still useful in spite of these limitations?

3 Assume that the fluorine atom ($^{19}_{9}F$) is a sphere of diameter 10^{-10}m and that its nucleus is a sphere of diameter 10^{-14}m.
  (a) What is (i) the atomic number.
              (ii) the mass number of $^{19}_{9}F$?
  (b) What is the *actual* mass of the nucleus in *one* $^{19}_{9}F$ atom?
      (6×10^{23} $^{1}_{1}H$ atoms have a mass of 1 g.)
  (c) What is the density of the nucleus in a $^{19}_{9}F$ atom?
  (d) What does the value in (c) suggest about the forces within the $^{19}_{9}F$ nucleus?
  (e) What is the ratio of the volume of the atom to the volume of the nucleus in a $^{19}_{9}F$ atom?

4 Natural silicon in silicon-containing ores contains 92% silicon-28, 5% silicon-29 and 3% silicon-30.
  (a) What is the atomic number of silicon?
  (b) What are the relative isotopic masses of the 3 silicon isotopes?
  (c) What is the relative atomic mass of silicon?
  (d) Samples of pure silicon obtained from natural silicon ores, mined in different parts of the world, have slightly different relative atomic masses. Explain.

5   Natural hydrogen contains two isotopes; ^1_1H and ^2_1H (sometimes called deuterium).
(a) Write down the possible formulae of a hydrogen molecule.
(b) What are the possible relative molecular masses for a hydrogen molecule?
(c) $1\,dm^3$ (1 litre) of a deuterium-enriched sample of hydrogen gas was found to weigh $0.10\,g$ at s.t.p. What is the isotopic composition of this gas?

6   The accurate relative isotopic masses of five isotopes are shown below.

$$^1_1H \quad ^2_1H\,(D) \quad ^{12}_6C \quad ^{14}_7N \quad ^{16}_8O$$

1.0078   2.0141   12.0000   14.0031   15.9949

(a) Calculate the accurate relative molecular masses (molecular weights) for

(i) N_2 (ii) DCN   (iii) CO   (iv) C_2H_4 (v) C_2D_2

(b) The relative molecular mass of a certain gas in a high-resolution mass spectrometer was 28.0171. What gas is probably under observation?

7   A compound containing only carbon, hydrogen and oxygen has peaks in its mass spectrum which correspond to relative mass of 60 (the molecular ion), 43, 31, 29 and 17. These masses are relative to the standard $^{12}_6C = 12$.
(a) What fragmented particles could be responsible for relative masses of 17, 29 and 31?
(b) Deduce the *two* possible structures for the compound.

8   The mass spectrum of dichlormethane shows peaks corresponding to 84, 86 and 88 atomic mass units (a.m.u.) with relative intensities of 9:6:1, respectively.
(a) What particles cause the three peaks?
(b) Why are their relative intensities 9:6:1? (H = 1.0; C = 12.0; Cl = 35.5.)

9   Suppose that you are a research chemist and that you have just discovered a new element.
(a) How would you identify its different isotopes and determine its relative atomic mass?
(b) How would you attempt to determine its atomic number?

6

Electronic Structure

6.1 Evidence for the electronic structure of atoms

When atoms react, electrons are redistributed. These electrons may be transferred from one atom to another or they may be shared between the reacting atoms in a different way. Remember that it is *only* electrons which are involved in chemical reactions: protons and neutrons take no part. As chemical reactions involve electrons, the similarities in chemical properties of certain elements (e.g. Na, K and Li; F, Cl and Br) suggest that these elements may have similar electronic structures.

It should be possible to obtain information about the arrangement of electrons in atoms by studying the ease with which atoms lose electrons. It would be useful to know the amount of energy which is needed to remove one electron from an atom. This energy is known as the **ionisation energy** and is given the symbol ΔH_i.

The first ionisation energy of an element is the energy required to remove one electron from each atom in a mole of gaseous atoms producing one mole of gaseous ions with one positive charge. Thus, the first ionisation energy of sodium is the energy required for the process

$$Na(g) \rightarrow Na^+(g) + e^- \qquad \Delta H = \Delta H_{i_1}$$

(The subscript number one in the symbol ΔH_{i_1} indicates the *first* ionisation energy.) Several methods can be used to determine the ionisation energy of an element. We shall consider two methods in this chapter. One of these methods involves bombarding the gaseous element with electrons. The other method involves a study of the spectra of the element.

6.2 Obtaining ionisation energies by electron bombardment

In this experiment the element is used as a gas at very low pressure inside a valve. Certain radio and television valves can be adapted and used in this way. Figure 6.1 shows a suitable circuit for measuring the ionisation energy of a noble gas.

When the cathode is heated from the 6.3 V a.c. supply, electrons are emitted from its surface. These electrons fall on the grid which is positive and a current flows in the cathode/grid circuit. As electrons stream from the cathode towards the grid they bombard atoms of the noble gas inside the valve.

Using the potentiometer device on the right-hand side in figure 6.1, it is possible to alter the voltage between the cathode and the grid. As the potential difference between the cathode and grid increases, electrons are accelerated through the valve at greater and greater speeds. (Notice in figure 6.1 that the *anode* is the most *negative* terminal inside the valve. We have retained the name

microammeter

valve anode

noble gas
at very low pressure

grid

3V dry
battery

cathode

6.3V a.c.
heater

30V

Figure 6.1
A circuit diagram of the apparatus
used to determine the ionisation
energy of a noble gas.

anode, because it is that part of the valve which is *normally* positive and used as the anode. Although this terminal is negative it cannot give off a stream of electrons because it is not heated.)

At first, the electrons travel slowly between the cathode and the grid. They collide gently with the noble gas atoms without causing any disruption. As the voltage between the cathode and grid is gradually increased, the kinetic energy of the bombarding electrons eventually becomes so great that, on collision with a gaseous atom, the most loosely bound electron is knocked off the atom forming a positively charged ion.

e^-	+	$X(g)$	$\rightarrow$	$X^+(g)$	+	e^-	+	e^-
fast-moving electron		gaseous atom		gaseous ion		electron knocked off the atom		retreating electron

As soon as positive ions ($X^+(g)$) appear they are attracted towards the anode (the most negative terminal inside the valve). This causes a flow of electrons from cathode to anode in order to combine with positive ions appearing on the anode surface. When ionisation first occurs, a current suddenly flows in the cathode/anode circuit and a rapid increase appears on the reading of the microammeter. When a sudden increase occurs in the microammeter reading the kinetic energy of the bombarding electrons is just sufficient to ionise the gaseous atoms inside the valve. The kinetic energy of the bombarding electrons can be determined from the voltage across the cathode and grid.

Figure 6.2 shows how the current in the microammeter varies with the voltage across the cathode and grid when the experiment is carried out with xenon and then argon.

- At what potential difference does the ionisation of xenon first occur?
- At what potential difference does the ionisation of argon first occur?

A similar experiment was carried out with helium, and the potential difference at which ionisation of the gas occurred was 22 volts.

When one electron (with a charge of 1.6×10^{-19}C) is accelerated through a potential difference of one volt, it acquires $1.6 \times 10^{-19} \times 1$ joules of energy.

When one electron is accelerated through 22 volts, it acquires $1.6 \times 10^{-19} \times 22$ joules of energy. This energy is sufficient to ionise one helium atom.

cathode/anode current/μA

Xenon Argon

p.d. between cathode and grid/volts

Figure 6.2
A graph of the current in the anode
circuit against the voltage across the
cathode and grid in experiments to
determine the first ionisation energy of
two noble gases.

∴ The first ionisation energy of helium

$$= 1.6 \times 10^{-19} \times 22 \text{ J per atom}$$

$$= 1.6 \times 10^{-19} \times 22 \times 6 \times 10^{23} \text{ J mole}^{-1}$$

$$= 211.2 \times 10^4 \text{ J mole}^{-1}$$

$$= 2112 \text{ kJ mole}^{-1}$$

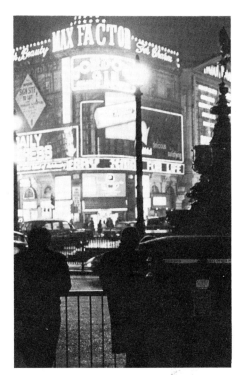

The bright lights! Certain gaseous materials such as neon emit light when they are subjected to high voltages in discharge tubes.

6.3 Obtaining ionisation energies from emission spectra

When sodium chloride is heated strongly in a bunsen, it gives off a brilliant yellow light. Some other substances also emit light when they are heated in this way. Have you noticed that compounds of one particular element always emit the same colour of light when they behave in this way?

What colour of light is emitted by:

- potassium compounds,
- calcium compounds,
- copper compounds?

Some gaseous materials emit light when they are subjected to large potential differences in electric discharge tubes. Neon advertising signs work in this way and the yellow sodium street lamps are discharge tubes containing sodium vapour.

If the light emitted by these substances is examined using a spectroscope, it does not consist of a continuous range of colours like the spectrum of white light or the colours in a rainbow. Instead, the light emitted by these substances is composed of separate lines of different colour. This kind of spectrum is called a **line emission spectrum**.

The photographs in figure 6.3 show the line emission spectra of sodium, calcium and barium in the visible region. Notice that each element has its own characteristic set of lines different from those of any other element. Consequently, elements can be identified by a study of their line emission spectra.

Figure 6.3
Photographs of the line emission spectra of sodium, calcium and barium in the visible region.

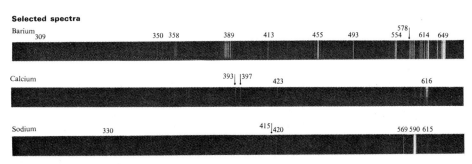

Each line in an emission spectrum corresponds to light of a particular frequency. The frequencies and colours of the prominent lines in the visible region of the spectrum of hydrogen are shown in figure 6.4. This particular series of lines is known as the **Balmer Series** for hydrogen.

What do you notice about the spacing of the lines? How do you explain the spacing of the lines? Why does an element only emit light of certain frequencies?

In order to answer these questions scientists assume that the electrons in an atom can exist only at certain energy levels. Under normal conditions, the electrons in an atom or ion fill the lowest energy levels first. When sufficient energy is supplied to the atom, it is sometimes possible to promote (excite) an electron from a lower energy level to a higher one. This process is called **excitation**. The electron is unstable in the higher energy level, so it will emit the excess energy as radiation and drop back into the lower energy level. Now, the energy difference between the higher and lower energy levels can have only certain fixed values because the energy levels themselves are fixed. This means

Figure 6.4
The colours and frequencies of the more prominent lines in the visible region of the atomic hydrogen spectrum.

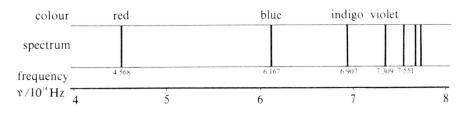

that the radiation emitted when an electron falls from a higher to a lower energy level can have only certain fixed frequencies (i.e. certain specific colours) because the frequency of any radiation is determined by its energy.

The small amount of radiation emitted by an electron when it falls from a higher to a lower energy level is referred to as a **quantum** of radiation.

The relationship between the energy (E) of a quantum of radiation and its frequency (v) is

$$E = h \times v$$

h is a constant, called Planck's constant. The value of h is 4×10^{-13} kJ s mole^{-1} or $4 \times 10^{-13}/6 \times 10^{23} = 6.66 \times 10^{-37}$ kJ s molecule^{-1}.

Figure 6.5 shows the energy levels in a hydrogen atom and the electron transitions which produce the lines in the visible region of the hydrogen spectrum.

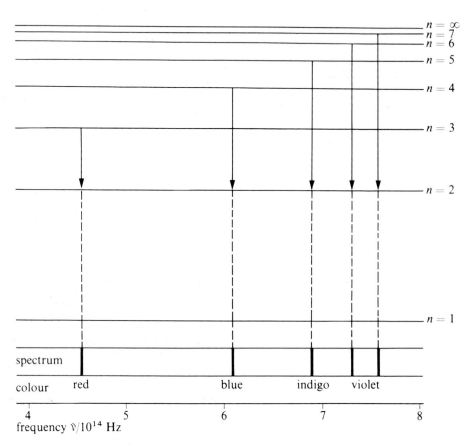

Figure 6.5
The energy levels in a hydrogen atom and the electron transitions which produce the lines in the visible region of the atomic hydrogen spectrum. Notice that the lines in the visible region are caused by transitions to the $n = 2$ level. Lines which result from transitions to the $n = 2$ level in an atom are known as the Balmer series of lines.

Notice the following points.

1 The electronic energy levels are numbered ($n = 1$, $n = 2$, $n = 3$, etc.). These numbers are sometimes referred to as the **principal quantum numbers** for the energy levels. The level of lowest energy is given the principal quantum number 1, the next lowest 2, and so on.

2 The colours in the visible region of the hydrogen spectrum are caused by electron transitions from higher levels to the level $n = 2$ and not to the lowest energy level ($n = 1$). For example, transitions from $n = 3$ to $n = 2$ result in a red line (at frequency 4.568×10^{14} Hz) in the hydrogen spectrum, while transitions from $n = 4$ to $n = 2$ produce a blue line.

3 As the energy levels get closer and eventually come together, it follows that the spectral lines also get closer and eventually come together.

When transitions occur from higher levels to the lowest energy level ($n = 1$), more energy is released than with transitions to the $n = 2$ level. Consequently, these lines appear at higher frequencies (i.e. higher energies) in the spectrum. In

Table 6.1

Frequencies of the Lyman Series for hydrogen

Frequency $v/120^{14}$ Hz	Transition to which frequency corresponds		
24.66	$n = 2$	to	$n = 1$
29.23	$n = 3$	to	$n = 1$
30.83	$n = 4$	to	$n = 1$
31.57	$n = 5$	to	$n = 1$
31.97	$n = 6$	to	$n = 1$
32.21	$n = 7$	to	$n = 1$
32.37	$n = 8$	to	$n = 1$

the case of hydrogen, transitions to the $n = 1$ level result in lines in the ultraviolet region of the spectrum.

If sufficient energy is given to an atom, it is possible to excite an electron just beyond the highest energy level. In this case the electron will escape and the atom becomes an ion. **Ionisation** has taken place.

In an atom, the highest possible energy level corresponds to the frequency at which the lines in the spectrum come together. So, by determining the frequency at which the converging spectral lines come together, we can find the ionisation energy of an element. This particular frequency is called the **'convergence limit'**.

The frequencies in table 6.1 relate to electron transitions to the $n = 1$ level in hydrogen. Lines corresponding to these frequencies occur in the ultraviolet region of the hydrogen spectrum. These lines which result from transitions to the lowest energy level in an atom are known as the **Lyman Series** of lines.

- Work out the difference in frequency (Δv) between successive lines in the Lyman Series for hydrogen.
- Plot a graph of Δv (vertically) against frequency, v. (Use the value of the lower frequency for plotting v.)
- Use your graph to estimate the frequency when Δv becomes 0.
- Δv becomes 0 when the difference in energy between the electronic energy levels becomes 0. Use the relationship $E = hv$ to find the energy which corresponds to the frequency when Δv becomes 0. This energy is the ionization energy for hydrogen.

An accurate value for the frequency at the convergence limit for hydrogen is 32.7×10^{14} Hz. Using $E = hv$, the energy of radiation with this frequency

$$= 4 \times 10^{-13} \times 32.7 \times 10^{14} \, \text{kJ mole}^{-1}$$

$$= 1308 \, \text{kJ mole}^{-1}$$

$\therefore$ The ionisation energy of hydrogen = 1308 kJ mole^{-1}.

6.4 Using ionisation energies to predict electronic structures—evidence for shells

If an atom containing several electrons is provided with sufficient energy it will lose one electron. Additional supplies of energy will result in the removal of a second electron, then a third, then a fourth, and so on. A succession of ionisations is therefore possible, each of which has its associated ionisation energy. For example, the first ionisation energy of sodium corresponds to

$$Na(g) \rightarrow Na^+(g) + e^- \qquad \Delta H_{i_1} = +494 \, \text{kJ mole}^{-1}$$

whereas the second ionisation energy of sodium corresponds to

$$Na^+(g) \rightarrow Na^{2+}e^- \qquad \Delta H_{i_2} = +4564 \, \text{kJ mole}^{-1}$$

The successive ionisation energies of an element are usually obtained from spectroscopic experiments.

Table 6.2 shows the successive ionisation energies of beryllium. Twice as much energy is needed to remove the second electron from beryllium than to remove the first, and eight times as much energy is required to remove the third electron than the second. The values for the different ionisation energies in table 6.2 suggest that beryllium has two electrons which are relatively easy to remove and two electrons which are very difficult to remove. Chemists have deduced that the beryllium atom has two electrons in a low energy level ($n = 1$) and therefore very difficult to remove; and two other electrons in a higher energy level ($n = 2$) and therefore easy to remove. This is represented on an energy level diagram in figure 6.6.

The electron arrangement in beryllium is written as 2,2. Lithium, with three electrons, has an electron arrangement of 2, 1 which shows two electrons in the $n = 1$ energy level and one electron in the $n = 2$ energy level.

Table 6.2
The successive ionisation energies of beryllium

| Element | Ionisation Energy/kJ mole^{-1} | | | |
	First	Second	Third	Fourth
Beryllium	900	1758	14905	21060
	2 electrons relatively easy to remove		2 electrons very difficult to remove	

In a beryllium atom, the two electrons in the lowest energy level ($n = 1$) spend most of their time closer to the nucleus than do the two electrons in the $n = 2$ level. These two electrons in the $n = 1$ level are said to occupy the first quantum **shell**. The two electrons in the second energy level ($n = 2$) spend most of their time further from the nucleus and are said to occupy the second quantum shell.

In figure 6.6 the electrons have been represented by arrows. When an energy level is filled the electrons have paired up with each other and in each of these pairs the electrons are spinning in opposite directions. Evidence that electrons pair up and spin in opposite directions is provided by spectroscopic and magnetic measurements. Chemists believe that paired electrons can only be stable when they spin in opposite directions so that the magnetic attraction which results from their opposite spins can counterbalance the electrical repulsion which results from their identical negative charges. In figure 6.6, the opposite spins of the paired electrons are shown by drawing arrows in opposite directions.

Figure 6.6
Energy levels and quantum shells for electrons in a beryllium atom.

6.5 How are the electrons arranged in larger atoms?

Figure 6.7 shows a graph of the logarithm to base ten ($\log_{10}$) for the successive ionisation energies of sodium.

Look closely at figure 6.7.

■ How many electrons does sodium have in the first shell close to the nucleus?
■ How many electrons does sodium have in the second shell?
■ How many electrons does sodium have in the third shell far away from the nucleus?
■ Write the electron structure for sodium showing the number of electrons in each shell. (The electron structure for beryllium is 2, 2.)
■ Draw an energy level diagram similar to figure 6.6 for the electrons in a sodium atom.

Figure 6.8(a) shows a sketch graph of the logarithm to base ten for the successive ionisation energies of potassium. This graph has been used to construct an energy level diagram (figure 6.8(b)) for the electrons in a potassium atom.

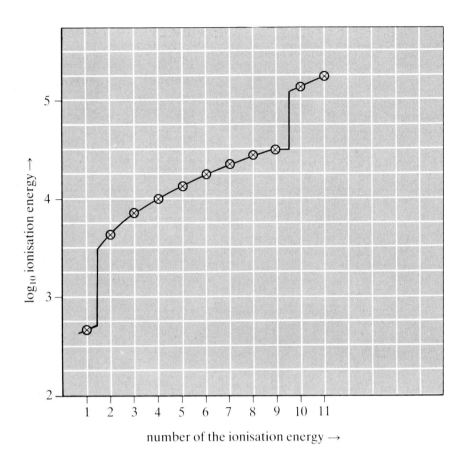

Figure 6.7
A graph of the logarithm to base ten of the successive ionisation energies for sodium.

The electron structure of potassium is written as 2, 8, 8, 1 which shows the number of electrons in each of the quantum shells as we move out from the nucleus. Using ionisation energy data, it is possible to predict the electron patterns in other elements. Table 6.3 shows the electron structures of the first 20 elements in the periodic table.

Notice that the first shell is filled when it contains two electrons and the second shell is filled when it contains eight electrons. Although the electron structures shown in table 6.3 suggest that the third shell can contain only eight electrons, it is possible for it to hold as many as 18 electrons.

Figure 6.8
(a) A sketch graph of the logarithm to base ten of the successive ionisation energies for potassium. (b) An energy level diagram for the electrons in a potassium atom.

Table 6.3
Electron structures of the first twenty elements in the periodic table

Atomic Number	Element	Symbol	Number of electrons in the first shell	Number of electrons in the second shell	Number of electrons in the third shell	Number of electrons in the fourth shell
1	Hydrogen	H	1	—	—	—
2	Helium	He	2	—	—	—
3	Lithium	Li	2	1	—	—
4	Beryllium	Be	2	2	—	—
5	Boron	B	2	3	—	—
6	Carbon	C	2	4	—	—
7	Nitrogen	N	2	5	—	—
8	Oxygen	O	2	6	—	—
9	Fluorine	F	2	7	—	—
10	Neon	Ne	2	8	—	—
11	Sodium	Na	2	8	1	—
12	Magnesium	Mg	2	8	2	—
13	Aluminium	Al	2	8	3	—
14	Silicon	Si	2	8	4	—
15	Phosphorus	P	2	8	5	—
16	Sulphur	S	2	8	6	—
17	Chlorine	Cl	2	8	7	—
18	Argon	Ar	2	8	8	—
19	Potassium	K	2	8	8	1
20	Calcium	Ca	2	8	8	2

6.6   Evidence for sub-shells of electrons

Figure 6.9 shows the first ionisation energies of the first 40 elements plotted against atomic number. The graph can be divided into sections ending with a noble (inert) gas whose ionisation energy is higher than those of all the elements between it and the previous noble gas. The large ionisation energies of noble gases are another indication of their stable electronic structures and unreactivity.

The graph between one noble gas and the next can also be divided into sub-sections. These sub-sections contain either two, six or 10 points on the graph. For example, after both He and Ne there are deep troughs in the graph followed by a small intermediate peak (Li to Be and Na to Mg), i.e. sub-sections

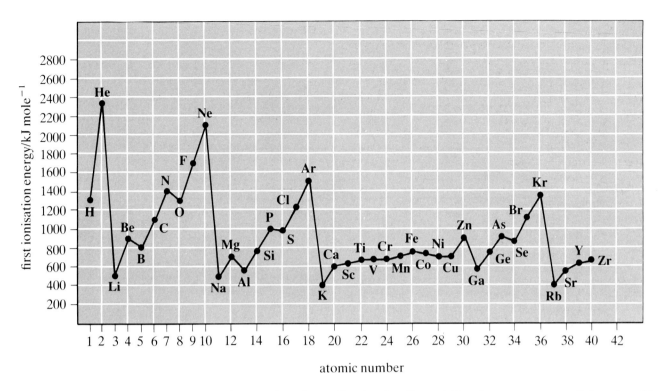

Figure 6.9
A graph of the first ionisation energies of the elements plotted against atomic number.

of two points. Immediately after Be and Mg, there are groups of six points (B to Ne and Al to Ar). Between Sc and Zn there is a sub-section containing 10 points.

In fact, the points on the graph in figure 6.9 between one noble gas and the next correspond to the filling of one shell with electrons. This means that the sub-sections of points correspond to **sub-shells** of electrons. By studying ionisation energies and atomic spectra in this way, scientists have concluded that:

the $n = 1$ shell can have 2 electrons in the same sub-level (sub-shell).

the $n = 2$ shell can have 2 electrons in one sub-level, and,
6 electrons in a slightly higher sub-level.

the $n = 3$ shell can have 2 electrons in one sub-level,
6 electrons in a slightly higher sub-level, and,
10 electrons in a still slightly higher sub-level.

the $n = 4$ shell can have 2 electrons in one sub-level,
6 electrons in a slightly higher sub-level,
10 electrons in a still slightly higher sub-level, and,
14 electrons in a still slightly higher sub-level.

The sub-shells (or sub-levels) containing 2 electrons are called s sub-shells (s electrons).
The sub-shells (or sub-levels) containing 6 electrons are called p sub-shells (p electrons).
The sub-shells (or sub-levels) containing 10 electrons are called d sub-shells (d electrons).
The sub-shells (or sub-levels) containing 14 electrons are called f sub-shells (f electrons).

The relative positions of these various sub-shells are shown on the left of figure 6.10 and the distribution of electrons amongst sub-levels in a potassium atom are shown on the right. Compare figure 6.10 with figure 6.8(*b*).

Notice that the *electrons always occupy the lowest available energy sub-level* and that *electrons 'pair-up' as soon as each sub-level is half-filled*.

The electron structure of an atom can therefore be described in terms of its sub-shells occupied by electrons. A number (1, 2, 3, etc.) is used to denote the quantum shell, a letter (*s, p, d* or *f*) to denote the sub-shell and a superscript to indicate the number of electrons in the sub-shell.

quantum
shell

sub-shell
notation

$n = 4$ ⠸⠂ ⎯⎯⎯⎯⎯⎯⎯⎯⎯⎯⎯⎯⎯⎯ $4d$
⎯⎯⎯⎯⎯⎯⎯⎯⎯⎯⎯⎯⎯⎯ $4p$
⎯⎯⎯⎯⎯⎯⎯⎯⎯ $3d$
⎯⎯⎯⎯⎯⎯⎯⎯⎯ $4s$

$n = 3$ ⠸⠂

⎯⎯⎯⎯⎯⎯⎯⎯⎯⎯⎯⎯ $3p$
⎯⎯⎯⎯⎯⎯⎯⎯⎯⎯⎯ $3s$

$n = 2$ ⠸⠂
⎯⎯⎯⎯⎯⎯⎯⎯⎯⎯⎯⎯ $2p$
⎯⎯⎯⎯⎯⎯⎯⎯⎯⎯⎯ $2s$

$n = 1$ ⎯⎯⎯⎯⎯⎯⎯⎯⎯⎯⎯⎯⎯ $1s$

(a)   (b)

Figure 6.10
(a) The positions of energy sub-levels in a potassium atom. (b) The distribution of electrons amongst the sub-levels in a potassium atom.

Thus, potassium has 2 electrons in the $1s$ sub-shell (i.e. $1s^2$)
2 electrons in the $2s$ sub-shell (i.e. $2s^2$)
6 electrons in the $2p$ sub-shell (i.e. $2p^6$)
2 electrons in the $3s$ sub-shell (i.e. $3s^2$)
6 electrons in the $3p$ subpshell (i.e. $3p^6$)
and 1 electron in the $4s$ sub-shell (i.e. $4s^1$).

This means that the electronic structure for potassium can be written in terms of energy levels as 2, 8, 8, 1; and more precisely in terms of energy sub-shells as

$$1s^{②}, 2s^{②} 2p^{⑥}, 3s^{②} 3p^{⑥}, 4s^{①}$$
$$\quad\ 2 \qquad\quad 8 \qquad\quad 8 \qquad\quad 1$$

Figure 6.10 shows that the $3d$ sub-shell is just above the $4s$ sub-shell, but just below the $4p$ sub-shell. This means that once the $4s$ level is filled (at element Ca in the periodic table) further electrons enter the $3d$ level, not the $4p$ level. Thus, scandium (the element after Ca in the periodic table) has the electronic structure $1s^2, 2s^2 2p^6, 3s^2 3p^6 3d^1, 4s^2$ and *not* $1s^2, 2s^2 2p^6, 3s^2 3p^6, 4s^2 4p^1$. The filling of the $3d$ sub-shell is very important in the chemistry of the elements from Sc to Zn. These are known as the **transition elements** and we shall deal with them in some detail in chapters 19 and 20.

■ Which element has the electronic structure $1s^2, 2s^2 2p^3$?
■ Write the electronic structure of this element in terms of energy levels.
■ Write the electronic structure of silicon in terms of energy levels.
■ Write the electronic structure of silicon in terms of energy sub-levels.
■ Draw an energy sub-level diagram for the electrons in a silicon atom.

6.7   Electrons and orbitals

We have described the electrons in atoms as occupying different quantum shells at increasing distances from the nucleus. But, suppose we could photograph the electron in a hydrogen atom at any given moment. The electron is moving in an unknown path at high speed, and if we took a second photograph an instant later, the electron would occupy a different position. If we superimposed millions of such photographs, the composite picture would resemble a cloud composed of an enormous number of dots, each dot representing one position of the electron (figure 6.11). Thus, in the hydrogen atom the electron can be

Figure 6.11
The charge cloud for the $1s$ electron in a hydrogen atom.

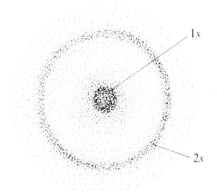

Figure 6.12
The charge clouds for electrons in a lithium atom

imagined to be a cloud of negative charge. The diffused cloud for the $1s$ electron in hydrogen is spherical in shape, but notice that the density of charge is not uniform throughout the cloud. Figure 6.12 shows the charge clouds for electrons in a lithium atom. The charge cloud for the two $1s$ electrons is again spherical as is the charge cloud for the $2s$ electron. Figure 6.12 also shows that the electron in the $2s$ level has a larger mean radius than electrons in the $1s$ level. The $2s$ electron can be imagined to occupy a 'spherical band'. In theory, there is no sharp boundary to the charge cloud, but in practice a boundary surface can be drawn to enclose almost all of the charge. Such regions which enclose almost all the charge cloud within them are called **orbitals**. Thus, orbitals are the regions in which there is the greatest probability of finding particular electrons, although the electrons are not confined to these regions.

Your own movements can be compared to those of an electron. At any one time you are most likely to be found either at home or in school, but fortunately you are not confined to these two places.

Each orbital can hold either one or a maximum of two electrons. If the orbital contains two 'paired-up' electrons they will be spinning in opposite directions.

Using complex calculations scientists have deduced that all s orbitals are spherical in shape; p orbitals are not spherical, but approximately 'dumb-bell' shaped with the nucleus located between the two halves of the 'dumb-bell'. In fact, each p sub-shell has three separate p orbitals, each of which can hold a maximum of two electrons, making a total of six electrons in a filled p sub-shell. Figure 6.13 shows the shapes of the three p orbitals. They are identical except for their axes of symmetry which, like the axes of a cartesian co-ordinate system, are mutually at right angles. Thus, it is convenient to distinguish between the p orbitals by labelling them p_x, p_y and p_z. Electrons will, of course, occupy the three p orbitals singly at first because of the repulsion between them. When the fourth electron is added to a p sub-shell, one orbital will contain a pair of electrons.

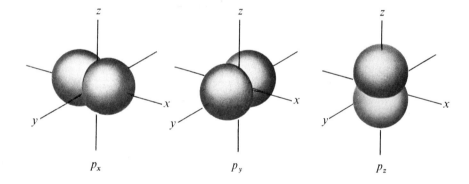

Figure 6.13
The shapes and relative positions of the three p orbitals in a p sub-shell.

The charge clouds for d electrons are even more complex than those of p electrons and we need not discuss them at this stage.

Figure 6.14 shows how the electronic structures of beryllium, carbon, nitrogen and oxygen can be represented to show the number of electrons in each orbital. Each box represents an orbital.

Figure 6.14
An orbital (electrons-in-boxes) representation of the electronic structures of beryllium, carbon, nitrogen and oxygen.

electron sub-shell	$1s$	$2s$	$2p$			s, p, d, f electron notation
element						
beryllium	↑↓	↑↓				$1s^2$ $2s^2$
carbon	↑↓	↑↓	↑	↑		$1s^2$ $2s^2$ $2p^2$
nitrogen	↑↓	↑↓	↑	↑	↑	$1s^2$ $2s^2$ $2p^3$
oxygen	↑↓	↑↓	↑↓	↑	↑	$1s^2$ $2s^2$ $2p^4$

6.8 Electronic structures and the periodic table

In the periodic table, elements are arranged in groups with similar chemical properties. The chemical similarities of elements in the same group arises because of their similar electron structures. Table 6.4 shows the electron structures of the first 20 elements in the periodic table. Note that elements in the same group have similar electron structures.

Table 6.4
The electron structures of the first 20 elements in the periodic table

Period 1				H 1				He 2
Period 2	Li 2, 1	Be 2, 2	B 2, 3	C 2, 4	N 2, 5	O 2, 6	F 2, 7	Ne 2, 8
Period 3	Na 2, 8, 1	Mg 2, 8, 2	Al 2, 8, 3	Si 2, 8, 4	P 2, 8, 5	S 2, 8, 6	Cl 2, 8, 7	Ar 2, 8, 8
Period 4	K 2, 8, 8, 1	Ca 2, 8, 8, 2						

Crystals of XeF_2 on the surface of a glass vessel. The crystals form when Xe and F_2 are mixed and exposed to sunlight (*Dr J. H. Holloway*).

Li 2,1
Na 2,8,1
K 2,8,8,1
Rb 2,8,18,8,1
Cs
Fr

Group 0: The noble (inert) gases

All of the noble gases have completely filled s and p orbitals in their outer shell: helium ($1s^2$), neon ($1s^2$, $2s^2 2p^6$), argon ($1s^2$, $2s^2 2p^6$, $3s^2 3p^6$), etc. Because of their complete and stable electronic structures, noble gases have large first ionisation energies. The atoms of these elements are very stable and exist as monatomic molecules. They do not combine with each other under any conditions to form diatomic molecules nor do they combine readily with the atoms of other elements.

The first noble gas compound was not prepared until 1962 and chemists have still not succeeded in preparing a single stable compound of helium or neon. The inert character of these elements is attributed to their stable electronic structures with filled s and p sub-shells.

As a result of their stable electronic structures, the atoms of noble gases have little interaction with each other and the forces of attraction between separate atoms are very small. Thus, noble gases have low melting points and low boiling points. Helium has the lowest melting point and the lowest boiling point of any known substance (3 K and 4 K, respectively). Radon, the heaviest inert gas, boils at −62°C (211 K).

Group I: The alkali metals

Each alkali metal follows a noble gas in the periodic table. Thus, alkali metals have one electron in their outermost shell which is fairly easily removed (low first ionisation energy). By losing their outermost electron, atoms of alkali metals form singly charged positive ions with stable electron structures like the previous noble gas (Li^+, Na^+, K^+, etc.). This means that all the alkali metals have an oxidation number of +1 in their compounds. They are very reactive because of the ease with which they lose the single electron in their outermost shell.

Above. A diver explores the sea bed. Strapped to his back are cylinders containing a mixture of oxygen/helium rather than oxygen/nitrogen. Helium is much less soluble in the blood than nitrogen. The diver is therefore less likely to suffer from 'bends' (pains caused by gases coming out of the blood) as he rises to the surface and the pressure falls.

Above right. Up she goes! An airship filled with helium gently floats upwards, though still held to the ground by ropes.

■ How many electrons have the atoms of Group VII elements in the outermost shell?

■ Write down the symbols with the correct charge for the ions of three elements in Group VII.

■ Why do elements in Group VII form ions with the same charge?

■ How many electrons have the atoms of Group II elements in the outermost shell?

■ Why do Group II elements have an oxidation number of +2 in all their compounds?

Summary

1 Each element has a characteristic emission spectrum which can be used to identify it.

2 The electrons in an atom can exist only at certain energy levels. If an atom is given sufficient energy, electrons can be promoted to higher energy levels. When such a promoted electron falls from a higher to a lower energy level, a quantum of radiation is emitted. This radiation can be detected as a line in the emission spectrum of the element.

3 The ionisation energy of an element can be obtained by electron bombardment in the gas phase or from the convergence limit of the Lyman Series of lines in its emission spectrum.

4 The principal quantum numbers (n = 1, 2, 3, etc.) are used to denote the shells of electrons in atoms.

5 In a stable atom, the electrons occupy the lowest available energy sub-levels. When each sub-level is half-filled, the addition of further electrons causes them to 'pair up'.

6 s sub-levels can hold a maximum of 2 electrons,
p sub-levels can hold a maximum of 6 electrons,
d sub-levels can hold a maximum of 10 electrons, and
f sub-levels can hold a maximum of 14 electrons.

7 The region in which an electron moves for most of the time is called an orbital. Each orbital can hold a maximum of two electrons. s sub-shells contain one orbital; p sub-shells contain three orbitals.

Study questions

1   (a) Draw a sketch graph of the logarithm to base 10 for the successive ionisation energies of phosphorus.
    (b) What conclusions could you draw from the graph concerning the electronic structure of phosphorus?
    (c) Write the electronic structure of phosphorus using s, p, d, f notation.

2   The following table shows the ionisation energies (in kJ mole^{-1}) of five elements lettered A, B, C, D and E.

Element	1st ionisation energy	2nd ionisation energy	3rd ionisation energy	4th ionisation energy
A	500	4600	6900	9500
B	740	1500	7700	10500
C	630	1600	3000	4800
D	900	1800	14800	21000
E	580	1800	2700	11600

   (a) Which of these elements is most likely to form an ion with a charge of 1+? Give reasons for your answer.
   (b) Which two of the elements are in the same group of the periodic table? Which group do they belong to?
   (c) In which group of the periodic table is element E likely to occur? Give reasons for your answer.
   (d) Which element would require the least energy to convert one mole of gaseous atoms into ions carrying two positive charges?

3   The electron energy levels of a certain element can be represented as $1s^2$, $2s^2$ $2p^6$, $3s^2$ $3p^4$.
    (a) Sketch a graph for the first seven ionisation energies of the element against the number of the ionisation.
    (b) What is the atomic number of the element?
    (c) What is the name of the element?
    (d) Draw an energy level diagram for the electrons in an atom of the element.

4   (a) List three factors which influence the size of the first ionisation energy of an element.
    (b) The first ionisation energies of the elements in Group I are shown below.

Element	First ionisation energy/kJ mole^{-1}
Li	520
Na	500
K	420
Rb	400
Cs	380

    Explain the variation in the first ionisation energy with atomic number.
    (c) Explain, with examples, how ionisation energies can provide evidence for the arrangement of electrons in shells.

5   Figure 6.15 shows a simplified diagram of the visible line spectrum of the hydrogen atom.

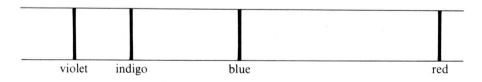

Figure 6.15
A simplified diagram of the visible line spectrum of the hydrogen atom.

(a) What quantity is represented along the horizontal axis in figure 6.15?
(b) Is this quantity increasing or decreasing from left to right?
(c) Explain why the atomic spectrum consists of a series of lines.
(d) To which energy level do the transitions corresponding to the visible lines in the spectrum of hydrogen relate?
(e) Why is the line spectrum of an element sometimes compared to the 'fingerprint of a criminal'?

6 The graph in figure 6.16 shows the first and second ionisation energies of the elements nitrogen to calcium.
(a) Why is there a sudden drop in the first ionisation energy after neon and after argon?
(b) Why is the first ionisation energy of magnesium greater than that for aluminium?
(c) Why is the second ionisation energy of each element greater than the corresponding first ionisation energy?
(d) Why do the maxima for the two graphs occur at different atomic numbers?

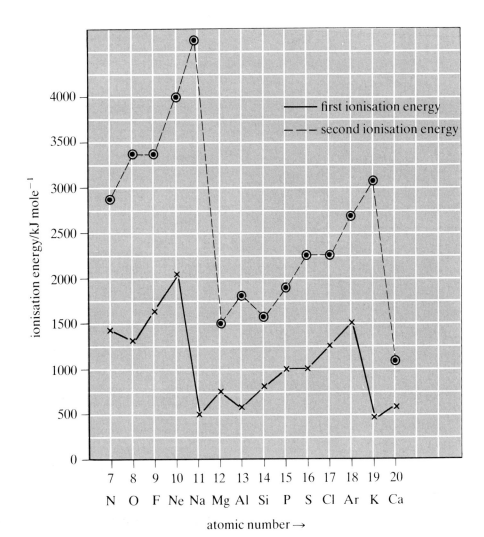

Figure 6.16
The first and second ionisation energies of the elements nitrogen to calcium.

7 Nuclear Structure and Radioactivity

Marie (1867–1934) and Pierre Curie (1859–1906) spent nearly 4 years isolating the radioactive elements polonium and radium from pitchblende. In 1903, they shared the Nobel Physics Prize with Becquerel. Then in 1911, Marie was awarded the Chemistry Prize—the first person ever to win two Nobel Prizes.

7.1 Introduction

In 1896, the Frenchman Henri Becquerel was investigating the possible production of X-rays from uranium salts exposed to bright sunlight. By chance, he left some crystals of a uranium salt on a photographic plate wrapped in black paper. Poor sunshine caused him to abandon his intended investigation, but, for interest's sake, he decided to develop the photographic plate. To his surprise, the plate was considerably darkened in the region where the crystals had been. Further experiments showed that the uranium salt was emitting a radiation which had quite different properties to X-rays, and Becquerel called this phenomenon **radioactivity**.

Intrigued by Becquerel's discovery, Marie and Pierre Curie examined the radioactivity of uranium salts in some detail. In particular they looked at a uranium ore called pitchblende. They found that pitchblende was much more radioactive than they had expected from its uranium content. In 1898, after a long and tedious extraction process, they isolated two other radioactive elements from pitchblende. They called these new radioactive elements polonium and radium. Radium was found to be about two million times more radioactive than uranium. By 1900, the elements thorium and actinium were also known to be radioactive.

The fogging of a photographic plate by radioactive materials still has practical applications. For example, the metabolism of CO_2 by a plant leaf can be studied by exposing the leaf to an atmosphere of radioactive $^{14}CO_2$ and then holding the leaf against a photographic plate. When the plate is developed the darkest parts indicate the presence of ^{14}C.

Before very long radioactive substances were shown to cause another effect which could detect their radiations. This was called scintillation. When a screen coated with zinc sulphide is placed near a radioactive radium salt and examined with a magnifying lens in the dark, tiny flashes of light appear on the surface of the zinc sulphide. Nowadays, crystals of zinc sulphide are combined with a photomultiplier. This can detect even the faintest of flashes and measure the radiations falling on the crystal. This arrangement of crystal and photomultiplier is known as a scintillation counter. It is used by many research scientists to detect alpha-particles (helium nuclei, $^4_2He^{2+}$).

The flashes of light on the zinc sulphide result from alpha-particles hitting the surface of the crystal. The alpha-particles emit the energy which they possess in the form of light. The alpha-particles are produced during the radioactive decay of radium atoms. From time to time, a radium atom 'explodes' and ejects an alpha-particle from its nucleus at great speed.

This spontaneous disintegration of atoms is responsible for radioactivity. Those isotopes which decay (break up) in this way are said to be **radioactive**.

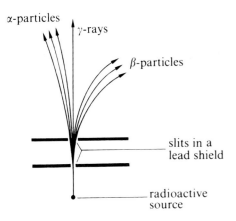

Figure 7.1
The effect of a magnetic field, perpendicular to the paper, on α-particles, β-particles and γ-rays.

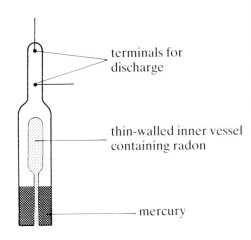

Figure 7.2
The experiment designed by Rutherford and Royds to investigate the nature of α-particles.

7.2 Alpha-particles, beta-particles and gamma-rays

When radium-226 decays it emits alpha-particles, but other radioactive substances behave differently when they disintegrate.

For example, potassium-43 emits much lighter beta-particles during decay and other materials emit gamma-rays.

Evidence for the nature of alpha-particles and beta-particles

Alpha-particles, beta-particles and gamma-rays can be separated easily using magnetic or electric fields (figure 7.1).

Alpha-particles are deflected in a direction indicating that they are positively charged. Beta-particles are deflected in the opposite direction and must be negatively charged. Gamma-rays are unaffected.

Rutherford determined the charge/mass ratio for alpha-particles by measuring their deflection in magnetic and electric fields. The value of charge/mass was half the value for $^1_1H^+$ ions (protons), showing that alpha-particles were either singly charged particles of atomic mass two, or doubly charged particles of atomic mass four, or triply charged particles of atomic mass six, etc. But, which of these suggestions is correct? The matter was settled by Rutherford and Royds using an ingenious yet simple experiment (figure 7.2).

Radioactive gaseous radon was allowed to decay for several days in a very thin-walled glass tube inside an evacuated thick-walled container. Alpha-particles from the decaying radon passed through the thin walls of the inner tube into the evacuated space outside. After a few days, the gas that had accumulated in the outer space was compressed into the top of the apparatus by raising the mercury level. It was then analysed by passing an electric discharge through it. When the radiation was analysed using a spectrometer it became clear that the gas collecting in the outer space was helium. This confirmed that alpha-particles had a mass of four units and thus a charge of +2. Alpha-particles are therefore doubly charged helium ions, $^4_2He^{2+}$.

Beta-particles were found to have similar properties to cathode rays. Measurement of charge/mass confirmed that beta-particles were electrons. The symbol $^0_{-1}e$ can be used to represent an electron since its mass number ($n_p + n_n$) is zero, and it carries a unit charge of -1.

Evidence for the nature of gamma-rays

■ What conclusion can you make from each of the following properties of gamma rays?

1 Gamma-rays are unaffected by an electric field.

2 Gamma-rays are unaffected by a magnetic field.

3 Gamma-rays can penetrate several centimetres of lead.

4 Gamma-rays can be diffracted by the lattice of a crystal.

5 When an atom emits gamma-rays there is no change in atomic number or mass.

This evidence suggests that gamma-rays are electromagnetic waves similar to light rays and X-rays. They have a wavelength of only 10^{-12} metres which is about one-tenth of the wavelength of X-rays and about one hundred-thousandth of the wavelength of visible light. The emission of gamma-rays enables a nucleus to lose surplus energy.

The nature and properties of alpha-particles, beta-particles and gamma-rays are summarised in table 7.1.

Table 7.1

The nature and properties of alpha-particles, beta-particles and gamma-rays

Emission	Nature	Relative effect of electric and magnetic fields	Distance travelled through air	Relative penetrating power
α-particles	helium nuclei ($_2^4\text{He}^{2+}$)	small deflection	a few cm	1
β-particles	electrons ($_{-1}^{\ 0}\text{e}$)	large deflection	a few m	100
γ-rays	electromagnetic waves	no deflection	a few km	10 000

7.3 Nuclear equations

When an atom disintegrates by losing an alpha-particle ($_2^4\text{He}^{2+}$), the remaining fragment will have a mass number four units less than the original atom and an atomic number two units lower. For example, when $_{92}^{238}\text{U}$ loses an alpha-particle (containing two protons and two neutrons), the breakdown product will have a mass number of 234 and an atomic number of 90. All isotopes (section 5.4) with an atomic number of 90 are atoms of thorium, Th. Thus, the products from the nuclear decay of uranium-238 are $_{90}^{234}\text{Th}$ and $_2^4\text{He}$ and the process can be summarised by the following nuclear equation.

$$_{92}^{238}\text{U} \rightarrow \, _{90}^{234}\text{Th} + \, _2^4\text{He}$$

Remember that both mass (indicated by the superscripts) and charge (indicated by the subscripts) must balance in nuclear equations.

Most of the radioactive isotopes of elements with atomic number greater than 83 (bismuth) undergo alpha-decay. Isotopes which decay by loss of alpha-particles include radium-226, plutonium-238, polonium-218 and radon-220.

■ Write nuclear equations for the alpha-decay of radium-226 and plutonium-238.
■ Use the periodic table in section 4.5 to find the symbols of the breakdown products.

Elements with an atomic number less than 83 do not usually exhibit alpha-decay. Instead, various isotopes of these elements emit beta-particles. For example, carbon-14 ($_6^{14}\text{C}$) decays by beta-particle emission forming $_7^{14}\text{N}$. The decay process can be represented by the equation

$$_6^{14}\text{C} \rightarrow \, _7^{14}\text{N} + \, _{-1}^{\ 0}\text{e}$$

Notice that both mass and charge are conserved in beta-decay as in alpha-decay.

■ What happens to the nucleus of the carbon-14 atoms during the beta-emission?
■ Where does the ejected electron come from?

During beta-decay, mass number remains constant, so the number of protons + neutrons stays constant. However, the atomic number increases by one which means that a neutron in the $_6^{14}\text{C}$ nucleus is converted to a proton + an electron:

$$_0^1n \rightarrow \, _1^1p + \, _{-1}^{\ 0}\text{e}$$

The proton remains in the nucleus (now $_7^{14}\text{N}$), but the electron is ejected as a beta-particle.

Thus, the overall effect of beta-decay causes an increase of one proton in an isotope and a corresponding decrease of one neutron.

■ Write nuclear equations for the beta-decay of $^{90}_{38}$Sr and $^{131}_{53}$I.

■ How do chemical reactions differ from nuclear reactions like alpha-decay and beta-decay?

During chemical reactions electrons are redistributed either by transfer from one atom to another or by sharing between atoms. Chemical reactions involve only the outer parts of atoms—the electrons.

During nuclear reactions one element may be converted to another either by radioactive decay or by atomic fission or fusion. Nuclear reactions involve the nucleus—the protons and neutrons.

Another important difference between chemical reactions and nuclear reactions is that the energy changes in nuclear reactions are usually much greater than those in chemical reactions.

7.4 Stable and unstable isotopes

At the beginning of this century detailed investigations were made of the disintegration of naturally occurring radioactive isotopes. Scientists were interested in:

1 the types of radioactive change,

2 the products of decay,

3 the rate of decay, and

4 the relative stability of different radioactive isotopes.

Experiments showed that the natural decay of a radioactive isotope could not be retarded nor accelerated by physical or chemical means. For example, extremes of temperature and pressure had virtually no effect on the rate of decay. Scientists found that the most convenient method of expressing the rate of disintegration of a radioactive element was its **half-life**. The half-life of a radioactive isotope is the time taken for the amount or concentration of the isotope to fall to half of its original value. Since the radioactivity of the isotope will also halve in this time, the half-life can also be defined as the time for the radioactivity of the isotope to be reduced to half its initial value. The half-life remains constant for each radioactive decay process. The half-life for $^{60}_{27}$Co is 5.2 years. This means that starting with one gram of cobalt-60, only a half gram would remain after 5.2 years. After another 5.2 years, the half gram would have decayed to one quarter gram and 5.2 years after that (i.e. 15.6 years from the start of the experiment) only one eighth of a gram would remain. Table 7.2 gives the half-lives of a few radioactive isotopes.

Table 7.2
The half-lives of some radioactive isotopes

Radioactive isotope	Half-life
Uranium-238, $^{238}_{92}$U	4.5×10^9 years
Carbon-14, $^{14}_{6}$C	5.7×10^3 years
Radium-226, $^{226}_{88}$Ra	1.6×10^3 years
Strontium-90, $^{90}_{38}$Sr	28 years
Iodine-131, $^{131}_{53}$I	8.1 days
Bismuth-214, $^{214}_{83}$Bi	19.7 minutes
Polonium-214, $^{214}_{84}$Po	1.5×10^{-4} seconds

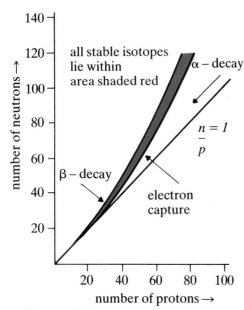

Figure 7.3
Variation in the number of neutrons
with the number of protons for stable,
non-radioactive isotopes.

The shorter the half-life, the faster the isotope decays and the more unstable it is. The longer the half-life, the slower the decay process and the more stable the isotope. Thus, the half-life of a radioactive isotope provides a quantitative measure of its relative stability.

Uranium-238 with a half-life of 4.5×10^9 years could be described as 'almost stable', but polonium-214 is quite the reverse.

Suppose you had one gram of $^{214}_{84}$Po.

■ How much of it would be left after 1.5×10^{-4} seconds?
■ How much of it would be left after 3×10^{-4} seconds?
■ How much of it would be left after 1.5×10^{-3} seconds?
■ How much of it would be left after 1.5 seconds?

7.5   The effect of n/p ratio on nuclear stability

For stable isotopes of small mass, the numbers of protons and neutrons are approximately equal. As the isotopic mass increases, the number of neutrons becomes increasingly larger than the number of protons.

The shaded area in figure 7.3 shows the variation in the number of neutrons with the number of protons for stable, non-radioactive isotopes. The straight line corresponds to equal numbers of neutrons and protons ($n/p = 1$). Figure 7.3 shows that the n/p ratio of a stable nucleus must lie within a restricted range for a given atomic number.

■ How does the n/p ratio for stable isotopes change as atomic number increases?

Beta-decay for isotopes with an n/p ratio above the stable value

During beta-decay a neutron splits up forming a proton and an electron.

$$^1_0n \rightarrow ^1_1p + ^{\ 0}_{-1}e$$

The proton remains in the nucleus, but the electron (beta-particle) is ejected. As a result of this, the number of neutrons in the isotope decreases by one and the number of protons increases by one. Thus, isotopes with an n/p ratio above the stable n/p ratio for the element try to stabilise themselves by beta-decay. Figure 7.3 shows how beta-decay enables an isotope with an n/p ratio above the stable value to move towards the stable n/p zone.

Isotopes which undergo beta-emission usually have more neutrons than stable isotopes of the same element. The heaviest isotopes of an element are, therefore, most likely to emit beta-particles in trying to increase their stability. Consequently, carbon-14 decays by beta-emission, but carbon-12 is stable. Hydrogen-3 is beta-active, but hydrogen-1 is stable.

Electron capture for isotopes with an n/p ratio below the stable value

Isotopes with an n/p ratio below the stable value try to stabilise themselves by electron capture. In this case, the unstable nucleus captures an electron from its innermost ($1s$) shell. This electron immediately combines with a proton in the nucleus to form a neutron.

$$^{\ 0}_{-1}e + ^1_1p \rightarrow ^1_0n$$

■ How does the number of neutrons change as a result of electron capture?
■ How does the number of protons change as a result of electron capture?
■ How does the n/p ratio change as a result of electron capture?

Figure 7.3 shows how an unstable isotope with a low n/p ratio can move towards the stable n/p zone as a result of electron capture. Isotopes which undergo electron capture usually have fewer neutrons than stable isotopes of the same element. For example, $^{38}_{19}$K undergoes electron capture, but $^{39}_{19}$K is stable.

Notice that electron capture is the opposite of beta-decay.

Alpha-decay for very heavy isotopes

Almost all the isotopes with an atomic number greater than 83 (Z>83) are unstable. Most of these isotopes are unstable because their nuclei are so heavy and their atomic mass is simply too large. Therefore, many of them try to become stable by ejecting an alpha-particle ($^4_2He^{2+}$) which contains two protons and two neutrons. For example:

$$^{226}_{88}Ra \rightarrow \, ^{222}_{86}Rn + \, ^4_2He$$

Figure 7.3 shows how alpha-decay enables a radioactive isotope to move towards or into the stable n/p zone.

At present, there are about 1600 known isotopes. Only one sixth of these are stable.

Results show that the decays of most radioactive isotopes can be summarised by the following simple rules:

1 Z≤83, n/p>stable n/p —beta-decay

2 Z≤82, n/p<stable n/p —electron capture

3 Z>83, element has no stable isotopes   —alpha-decay

7.6   An alchemist's dream—producing radioactive isotopes

Many of the alchemists who practised during the sixteenth and seventeenth centuries were little better than charlatans and magicians. They spent most of their time searching for a universal solvent which would dissolve all matter or trying to find a method of turning lead into gold. Their attempts at transmuting one element into another led to nothing but gloom and despair. Curiously, however, natural radioactive transmutations were happening all the time in nature.

The earliest evidence for the disintegration of an atomic nucleus was obtained in 1919. In that year, Rutherford bombarded nitrogen gas with alpha-particles

The alchemists of the sixteenth and seventeenth centuries were the first scientists to attempt the transmutations of elements.

One of the first cyclotrons designed by Lawrence. Positively charged particles enter the cyclotron through tube X and are then accelerated around the machine in an outward-spiralling path by a series of voltage kicks. Eventually, the particles emerge from the tube at the top as high-energy projectiles.

and discovered that protons were ejected. Later Blackett showed that the ejection of a proton was accompanied by the formation of an oxygen-17 isotope.

$$^{14}_{7}N + ^{4}_{2}He \rightarrow ^{1}_{1}p + ^{17}_{8}O$$

Within a few years, Rutherford, Chadwick and their collaborators had demonstrated several other transmutations. These included $F \rightarrow Ne$, $Na \rightarrow Mg$ and $Al \rightarrow Si$ using alpha-particles as projectiles.

A few years later, Cockcroft and Walton, working under Rutherford at Cambridge, developed equipment for bombarding atoms with protons rather than alpha-particles. Hydrogen gas was ionised to form electrons and protons. The protons were then accelerated to immense speeds by attracting them through a potential of more than 300 000 volts. In 1932, Cockcroft and Walton used their accelerator to bombard lithium atoms with protons and produce helium.

$$^{1}_{1}p + ^{7}_{3}Li \rightarrow ^{4}_{2}He + ^{4}_{2}He$$

During the next few years proton accelerators were followed by a variety of machines. These included cyclotrons, synchrotrons and cosmotrons, all designed to provide charged particles with greater and greater energy.

Both protons and alpha-particles have their limitations as 'nuclear bullets'. No matter what their energy may be, they will be repelled by the positive nucleus of any target atom they approach. Fortunately, the discovery of the neutron opened up new possibilities for nuclear transmutations.

■ What disadvantages do protons and alpha-particles have as nuclear projectiles?
■ What advantages do protons have over alpha-particles as nuclear projectiles?
■ Why are electrons ineffective as nuclear projectiles?
■ Why are neutrons superior to protons, alpha-particles and electrons as nuclear projectiles?

Radioactive $^{32}_{15}P$ is obtained by neutron bombardment of $^{31}_{15}P$.

$$^{1}_{0}n + ^{31}_{15}P \rightarrow ^{32}_{15}P$$

The $^{32}_{15}P$, being heavier than the stable isotope, undergoes beta-decay to $^{32}_{16}S$.

$$^{32}_{15}P \rightarrow ^{32}_{16}S + ^{0}_{-1}e$$

Radioactive $^{32}_{15}P$ has been used to study the uptake and metabolism of phosphorus by plants from phosphate fertilisers.

The reconstructed 600 MeV synchro-cyclotron of the European Organisation for Nuclear Research at Geneva. To the left is the rotary capacitor (No. 2) which controls the frequency of the accelerating electric field. The 2500-tonne electromagnet is at the rear. Protons are accelerated in a vacuum between the poles of this huge electromagnet. One of the two vacuum pumps is just behind the technicians.

Another radioactive isotope obtained by neutron bombardment is $^{14}_{6}C$.

$$^{1}_{0}n + ^{14}_{7}N \rightarrow ^{14}_{6}C + ^{1}_{1}H$$

$^{14}_{6}C$ has been extensively used in the elucidation of carbon pathways including photosynthesis and protein synthesis.

At first artificial transmutations involved elements with low relative atomic masses. Then in 1940 McMillan and Abelson succeeded in synthesising heavier elements from uranium, the heaviest naturally occurring element. Uranium-238 was bombarded with neutrons producing uranium-239.

$$^{1}_{0}n + ^{238}_{92}U \rightarrow ^{239}_{92}U$$

However, $^{239}_{92}U$ is radioactive, losing beta-particles in two stages to form first neptunium and then plutonium.

$$^{239}_{92}U \rightarrow ^{239}_{93}Np + ^{0}_{-1}e$$

$$^{239}_{93}Np \rightarrow ^{239}_{94}Pu + ^{0}_{-1}e$$

By 1955, another seven elements with atomic numbers from 95 to 101 had been synthesised by Seaborg and his co-workers. In the following 20 years evidence was obtained for the next four elements with atomic numbers 102–105.

All these relatively new elements beyond uranium in the periodic table are radioactive and most of them disintegrate rapidly into smaller atoms.

7.7 Atomic fission—atomic energy

Before 1938, scientists believed that all nuclear reactions involved the removal of small particles such as alpha-particles, electrons and protons from nuclei. In 1938, however, two German scientists, Hahn and Strassman, realised that isotopes of intermediate mass were formed when uranium was bombarded with neutrons. In this case, uranium nuclei had split into two similar halves.

This type of splitting during radioactive decay is very different from the previous examples we have studied, in which one large fragment and one very small fragment (either $^{4}_{2}He$, $^{0}_{-1}e$ or $^{1}_{1}p$) were produced. Thus, the term **atomic fission** or **nuclear fission** was used to describe the disintegration of an atom into two large fragments.

Hahn and Strassman noticed one other important feature of atomic fission—it could produce vast quantities of energy. Calculations suggested that one gram of natural uranium could provide as much energy as 10 000 grams of coal. Could the energy released during uranium fission be harnessed for the benefit of society? The answer, of course, was yes, but not without despair and destruction. In 1942, the first atomic reactor was built in America by the Italian scientist Enrico Fermi. Early in the 1940s, the possibility of nuclear warfare became obvious. The first 'atomic bombs' were dropped on the Japanese cities of Hiroshima and Nagasaki in 1945. The result was unbelievable. Vast areas of land were destroyed and thousands of inhabitants were killed. Many people died

On the left is Enrico Fermi, who built the first nuclear reactor in a Chicago squash court in 1942. Fermi is talking to Maria Goeppart-Mayer, who shared the Nobel Prize for physics in 1963.

Hiroshima after the bomb.

CND marchers demonstrating against the use of nuclear weapons

from the radiation emitted, not from the initial explosion. Nuclear warfare created worldwide unrest and indignation. During the 1950s and 1960s, large numbers of British scientists became supporters of a nuclear disarmament campaign which opposed the increased testing and development of nuclear weapons by the UK, the USA and the USSR. Fortunately, these countries began to exploit the peaceful uses of nuclear energy. In 1954, the first atomic power station supplying electrical energy was opened in the USSR, and in 1956, the world's first industrial-scale nuclear power station came into operation at Calder Hall, Cumbria.

During the 1960s, 1970s and early 1980s, the USSR and the USA continued to develop even more powerful and sophisticated nuclear weapons. This resulted in a re-emergence of the anti-nuclear movement. The movement is particularly strong in Europe, which would be the most likely theatre of war in any confrontation between the USA and the USSR. Fortunately the super-powers have now begun to reduce their stockpiles of nuclear weapons.

Uranium fission

Natural uranium contains two isotopes, $^{238}_{92}U$ and the much rarer $^{235}_{92}U$. Experiments showed that the rarer isotope of mass number 235 is mainly responsible for the vast release of energy during the fission. When natural uranium is bombarded by neutrons the $^{235}_{92}U$ isotopes absorb neutrons and then split up violently. The process releases two smaller nuclei, two or three neutrons and a large quantity of energy.
For example:

$$^{1}_{0}n + {}^{235}_{92}U \rightarrow [{}^{236}_{92}U] \rightarrow {}^{90}_{38}Sr + {}^{143}_{54}Xe + 3{}^{1}_{0}n$$

Although U-238 does not undergo fission during the neutron bombardment, it absorbs neutrons and disintegrates to neptunium and plutonium.

$$^{1}_{0}n + {}^{238}_{92}U \rightarrow {}^{239}_{92}U + \text{gamma rays}$$
$$^{239}_{92}U \rightarrow {}^{239}_{93}Np + {}^{0}_{-1}e$$
$$^{239}_{93}Np \rightarrow {}^{239}_{94}Pu + {}^{0}_{-1}e$$

The plutonium-239 which is formed undergoes atomic fission similar to U-235 and the newer atomic power stations, such as the one at Dounreay, use Pu-239 as fuel.

7.8 Chain reactions, atomic bombs and nuclear reactors

All of the neutrons released during the fission of U-235 are capable of splitting another uranium nucleus. Figure 7.4 shows how an exploding chain reaction can occur during uranium fission.

One reaction releases three neutrons, three neutrons release nine, then 27, 81, 243, etc. Provided the chain can build up rapidly and completely, there is an enormous release of energy—an atomic bomb.

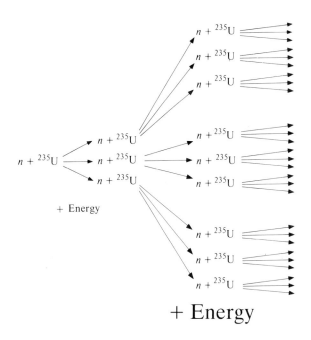

Figure 7.4
A simplified diagram to illustrate an exploding chain reaction during uranium fission.

This will only occur if the sample of U-235 is large enough. If it is too small the secondary neutrons will escape from the sample before they have caused further fissions and the chain reaction will be broken (figure 7.5(a)). Thus, a small piece of U-235 is quite safe, but a larger piece, within which the secondary neutrons cause further fissions before escaping, may explode (figure 7.5(b)). There is, therefore, a **critical size** below which U-235 is safe, but above which it may explode.

In an atomic bomb, two pieces of pure U-235 or Pu-239 each below the critical size are brought together. These form one piece larger than the critical size. Fission of U-235 or Pu-239 then proceeds in an uncontrolled manner as an exploding chain reaction.

Figure 7.5
A small piece of U-235 (as in (a)) is quite safe, but a piece larger than the critical size (as in (b)) will explode.

Figure 7.6
A simplified diagram of a nuclear reactor.

- Why must the U-235 used in an atomic bomb be pure?
- How can the fission of U-235 be controlled in an atomic reactor so that the energy is converted into electricity or heat?

Figure 7.6 shows a diagram of the commonest type of nuclear reactor. The atomic pile consists of rods of uranium inserted in channels surrounded by blocks of graphite. The neutrons resulting from fission of U-235 are slowed down, but not absorbed by collision with carbon atoms in the graphite. This ensures that as many neutrons as possible cause further fission of U-235 atoms before escaping from the pile or before capture by the more plentiful U-238 atoms. The temperature of the pile can be controlled by movable rods of boron or cadmium which absorb neutrons. The deeper these rods are inserted, the more neutrons are absorbed and the fewer fissions take place. By carefully adjusting the position of these rods a controlled chain reaction can be obtained. This will give a steady and continuous production of energy.

The heat produced during the fission process is removed by a stream of circulating carbon dioxide. The hot gas can be used to produce steam which drives a turbine.

- Why must the uranium rods be stored well apart before use in the reactor?
- Why do fewer fissions occur when the rods are removed from the pile even though they are the same distance apart?
- Why is it necessary to surround the pile with a thick concrete shield?

Nuclear energy: controlled and un-controlled! A nuclear explosion (left) and the Dounreay Nuclear Power Station (right).

Technicians loading cadmium absorbers into a small-scale test reactor at Harwell. The top shield plate of the reactor has been removed and is in the background.

Besides liberating energy, the atomic reactor also produces plutonium as a result of the neutron bombardment of U-238.

$$_0^1n + {}^{238}_{92}U \rightarrow {}^{239}_{92}U$$

$$_{92}^{239}U \rightarrow {}^{239}_{93}Np + {}^{\ 0}_{-1}e$$

$$_{93}^{239}Np \rightarrow {}^{239}_{94}Pu + {}^{\ 0}_{-1}e$$

The plutonium can be separated from the uranium and used as the atomic fuel in nuclear reactors such as the one at Dounreay.

Since plutonium requires faster neutrons for fission than U-235, no graphite moderator is necessary. The central core of plutonium is surrounded by natural uranium. The uranium captures any neutrons escaping from the central core and becomes converted to plutonium. By careful design and control, it is possible to produce more plutonium in the surrounding uranium than is used up in the central core. Consequently, this type of reactor is called a **fast-fission breeder reactor** since it uses fast neutrons and 'breeds' plutonium from uranium.

Although such reactors produce more plutonium than they use, we are not getting 'something for nothing'. There is no contradiction of the laws of Conservation of Matter and Energy, since uranium and plutonium are being converted into more plutonium. Check this yourself using the equations below:

$$_0^1n + {}^{239}_{94}Pu \rightarrow {}^{91}_{39}Y + {}^{146}_{55}Cs + 3{}^1_0n \qquad \text{Fission of } {}^{239}_{94}Pu$$

$$\left. \begin{array}{l} 2{}^1_0n + 2{}^{238}_{92}U \rightarrow 2{}^{239}_{92}U \\ 2{}^{239}_{92}U \rightarrow 2{}^{239}_{93}Np + 2{}^{\ 0}_{-1}e \\ 2{}^{239}_{93}Np \rightarrow 2{}^{239}_{94}Pu + 2{}^{\ 0}_{-1}e \end{array} \right\} \begin{array}{l} \text{Capture of neutron} \\ \text{by } {}^{238}_{92}U \text{ to} \\ \text{'breed' } {}^{239}_{94}Pu \end{array}$$

Overall: $2{}^{238}_{92}U + {}^{239}_{94}Pu \rightarrow 2{}^{239}_{94}Pu + {}^{91}_{39}Y + {}^{146}_{55}Cs + 4{}^{\ 0}_{-1}e$

The energy from disintegrating plutonium that destroyed Nagasaki is being tamed. For more than 20 years nuclear scientists have been trying to harness plutonium in 'fast-breeder' reactors as the fuel for the future. Many people, however, consider the dangers in using plutonium are too great. Atom bombs in the hands of terrorists or unscrupulous governments might prove catastrophic. On the other hand, the benefits of plutonium seem irresistible—cheap, abundant electricity for hundreds of years. Will our descendants curse us or bless us for the legacy of plutonium?

7.9 Using radioactive isotopes

Treatment of cancer

The structure of DNA within the genes of animals and plants can be changed by radiation. For reasons which are not yet clear, cancer cells are more susceptible than normal cells to destruction by radiation. Because of this, gamma-rays can be used in the treatment of cancer. Penetrating gamma-radiation from cobalt-60 (${}^{60}_{27}Co$) is used in treating inaccessible growths. Superficial cancers, such as skin cancer, can be treated by less penetrating radiation from ${}^{32}_{15}P$ or ${}^{90}_{38}Sr$ in plastic sheets strapped on the affected area.

Studying metabolic pathways

Radioactive isotopes can be used to trace the uptake and metabolism of various elements by animals and plants. For example, the uptake of phosphate and the metabolism of phosphorus by plants can be studied using a fertiliser containing ${}^{32}_{15}P$. Radioactive tracer studies using ${}^{14}_{6}C$ have helped in the elucidation of photosynthesis and protein synthesis. ${}^{131}_{54}I$ has been used in the diagnosis and treatment of thyroid diseases and in research into the working of the thyroid gland.

Radioactive cobalt-60 being used in the treatment of cancer. Cobalt-60 is the most common source of gamma-radiation.

A radiation thickness gauge using strontium-90 being used to control the thickness of tyre cord for the Avon Rubber Company at Melksham.

Thickness gauges and empty-packet detectors

The radiation passing through a material decreases as the material gets thicker. Thus, the amount of penetrating beta- or gamma-radiation can be used to estimate the thickness of various materials such as paper, metal or plastic. The advantage of using radiation for measuring the thickness of materials is that there need be no touching, marking or tearing of the article concerned. Thus, radiation thickness gauges can be used to control the thickness of sheet emerging from a high-speed rolling mill. Beta-rays can be used for steel thicknesses up to about 0.2 cm. More penetrating gamma-rays can be used with steel up to 10 cm thick.

A similar use of radioactive isotopes is made in level gauges and empty-packet detectors. The level of liquid in a closed vessel can be found by placing a source on one side of the container and a detector on the other. A sudden decrease in the detected radiation, when source and detector are moved down the vessel simultaneously, shows the level of liquid within the vessel. Level gauges of this type are used to measure the amount of liquid in fire extinguishers and gas cylinders. In a similar fashion, empty-packet detectors can be set to reject empty or insufficiently filled packets of biscuits or cigarettes.

Archaeological and geological uses

How old is the earth?

By knowing the rate at which an isotope decays, we can work out the time it takes for a certain amount of it to disappear. Thus, the decay of U-238 to lead-206 provides a method of dating rocks in the earth's crust. We must begin by assuming that the uranium-bearing rocks in the earth's crust originally contained uranium-238, but no lead-206. The present ratio of U-238 to Pb-206 in the rocks can then be used to calculate the time which has elapsed since the rocks formed. Using this method, the ages of rocks vary from forty million to four thousand million years. This second value is often taken by geologists as the age of the earth.

Dating ancient remains from carbon-14

The basis of carbon dating is the simultaneous production and disintegration of carbon-14($^{14}_{6}$C). $^{14}_{6}$C is continuously formed in the upper atmosphere by the collision of neutrons with nitrogen atoms:

$$^{14}_{7}N + {}^{1}_{0}n \rightarrow {}^{14}_{6}C + {}^{1}_{1}H$$

The colliding neutrons are produced when cosmic rays strike atoms and split them into fragments.

At the same time as $^{14}_{6}$C is being formed, the carbon-14 already present is decaying to $^{14}_{7}$N by the emission of electrons:

$$^{14}_{6}C \rightarrow {}^{14}_{7}N + {}^{0}_{-1}e$$

As a result of this simultaneous formation and decay of $^{14}_{6}$C, the atmosphere contains a constant concentration of $^{14}_{6}$C as $^{14}_{6}$CO$_2$. This $^{14}_{6}$CO$_2$ gets into plants via photosynthesis and from plants it is passed into animals. Thus, all living things have a constant proportion of their carbon in the form of carbon-14.

Now, when the animal or plant dies, replacement of $^{14}_{6}$C atoms ceases, but decay of $^{14}_{6}$C atoms continues. Just suppose that $^{14}_{6}$C makes up x% of all carbon in living things. About 5700 years after the plant or animal has died it will contain only half the percentage of $^{14}_{6}$C atoms (i.e. $\frac{x}{2}$%). After another 5700 years, the concentration of $^{14}_{6}$C will have halved again from $\frac{x}{2}$ to $\frac{x}{4}$%. By comparing the concentration of $^{14}_{6}$C in archaeological specimens with the concentration of $^{14}_{6}$C in similar materials living at the present time, it is possible to calculate the age of the specimen. In this way, carbon dating has been used to establish Egyptian chronology and to check the authenticity of ancient remains.

The best known forgery which was discovered by radioactive dating is probably Piltdown Man. In 1912, the archaeologist Charles Dawson claimed he had discovered the fossilised skull of a primitive apeman on Piltdown Common, near Lewes in England. In 1953, the remains were re-examined more closely. Results showed that the cranium and the jaw contained different percentages of fluorine, nitrogen, uranium and carbon. This made people very suspicious. In 1959, the age of the bones was determined from the radioactivity of carbon-14 in them. The remains were shown to be skilfully disguised fragments of a modern human cranium and an orang-utang jaw. Chemical tests revealed that Dawson had stained the fragments with chromium and iron(II) salts. Neither of these salts occurred in the region where the remains were found.

Summary

1 Radioactivity results from the spontaneous disintegration of nuclei.

2 Naturally occurring elements which undergo radioactive decay emit alpha-particles ($^4_2He^{2+}$ ions), beta-particles (electrons) and gamma-rays (electro-magnetic radiation).

3 Gamma-radiation is far more penetrating than alpha- and beta-radiation. It can penetrate thick metal sheets and is responsible for most of the damage caused by the disintegration of radioactive substances.

4 The half-life of a radioactive isotope is the time taken for the mass or concentration of the isotope (or for its rate of decay) to fall to half its initial value. The half-life is virtually unaffected by external conditions. It can be used as a measure of the relative stability of an isotope.

5 The n/p ratio for stable isotopes lies within a very restricted range for any given atomic number.

   Isotopes which are unstable due to an excess of protons over neutrons (n/p<stable value) can increase their stability by alpha-particle emission or electron capture.

   Isotopes which are unstable due to an excess of neutrons over protons (n/p>stable value) can increase their stability by beta-particle emission.

6 Atomic (nuclear) fission is the disintegration of an isotope into two large fragments.

7 Chemical reactions involve electrons in the outer parts of atoms, whereas nuclear reactions involve protons and neutrons in the nucleus.

Study questions

1   (a) What are the relative masses and charges of alpha- and beta-particles?
    (b) What is the nature of gamma-radiation?
    (c) What evidence can you provide to support your statements concerning the nature of alpha-, beta- and gamma-radiation?

2   A radioactive source with a very long half-life emits alpha-particles of energy 8×10^{-12} J at the rate of 4.2×10^{10} per second. The source is embedded in material sufficient to absorb the alpha-particles. The material loses heat at the rate of 8.4 J per minute for every 10°C its temperature is above that of its surroundings and eventually a steady temperature is attained.
    (a) What conditions apply when the steady temperature is attained?
    (b) How much has the temperature of the material risen above its surroundings when the steady state is achieved?
    (c) Mention two consequences or practical applications of this radioactive heating effect.

3 (a) How do the atomic number and the mass number of an isotope change when its nucleus loses
   (i) an alpha-particle,   (ii) a beta-particle,
   (iii) gamma-rays,   (iv) a neutron?

(b) $^{232}_{90}$Th emits a total of six alpha-particles and four beta-particles in its natural decay sequence. What is the atomic number, mass number and symbol of the final product?

(c) Write nuclear equations to summarise the following changes:
   (i) when a $^{7}_{3}$Li nucleus absorbs a colliding proton the product disintegrates into two exactly similar fragments.
   (ii) $^{232}_{90}$Th atoms undergo alpha-decay to form radium atoms.

(d) Scintillation counting shows that 1 mg of polonium emits 3×10^{18} alpha-particles in the course of complete decay. What is the relative atomic mass of polonium? (State the assumptions in your calculation.)

4 $^{214}_{83}$Bi has a half-life of 20 minutes.
(a) Plot a graph of the percentage of the $^{214}_{83}$Bi remaining (vertical) against time. What percentage of the $^{214}_{83}$Bi remains after 70 minutes?

(b) Rewrite the following nuclear equations substituting symbols and numbers for the question marks.
   (i) $^{14}_{?}? + ^{?}_{?}\text{He} \rightarrow ^{?}_{?}\text{O} + ^{1}_{1}?$
   (ii) $^{?}_{?}\text{H} + ^{56}_{?}\text{Fe} \rightarrow ^{57}_{?}\text{Co} + ^{1}_{0}?$

5 Outline the importance or uses of two of the following isotopes.

   $^{60}_{27}$Co, $^{32}_{15}$P, $^{131}_{54}$I, $^{239}_{94}$Pu

6 (a) How is carbon-14 produced in nature?
(b) Write an equation for the beta-decay of $^{14}_{6}$C.
(c) How is $^{14}_{6}$C used in 'dating' archaeological remains?
(d) How was radioactive 'dating' used to confirm the validity of the Dead Sea Scrolls and the forgery of Piltdown Man?

The reconstructured skull of Piltdown Man, a forgery uncovered by radioactive dating.

8 The Electronic Theory and Chemical Bonding

8.1 Introduction

Look at the information in table 8.1 showing the electron structures of the atoms and ions of the elements in period 3.

The first three elements in the period (Na, Mg and Al) *lose* electrons from their outermost shell to form positively charged ions (Na^+, Mg^{2+} and Al^{3+}). These ions have an electron structure like neon, the previous noble gas.

Elements at the end of the period in Groups VI and VII (S and Cl) *gain* electrons to form negatively charged ions (S^{2-} and Cl^-). These ions have the same electron structure as the next noble gas, argon.

But, what about elements in the middle of the period such as silicon and phosphorus? These elements do not usually form ions in their compounds, but do they obtain an electronic structure similar to that of a noble gas? The answer is 'yes', but they cannot obtain a noble gas structure by electron transfer, because this would result in their forming ions. Instead, silicon and phosphorus,

Table 8.1
Electronic structures of the atoms and ions of the elements in period 3

Element	Na	Mg	Al	Si	P	S	Cl	Ar
Electronic structure of the atom	2, 8, 1	2, 8, 2	2, 8, 3	2, 8, 4	2, 8, 5	2, 8, 6	2, 8, 7	2, 8, 8
No. of electrons in outermost shell of atom	1	2	3	4	5	6	7	8
Ion formed	Na^+	Mg^{2+}	Al^{3+}	—	—	S^{2-}	Cl^-	—
Electronic structure of the ion	2, 8	2, 8	2, 8	—	—	2, 8, 8	2, 8, 8	—

like other elements in Groups IV and V, achieve electron structures similar to noble gases by sharing electrons with other atoms.

In 1916, Kossel and Lewis realised that all the noble gases, except helium, had an outer shell containing eight electrons. They suggested that this arrangement was responsible for the stability and inertness of the noble gases. Thus, *when elements form compounds, they either lose or gain or share electrons so as to achieve stable (low-energy) electron configurations similar to the next higher or lower noble gas in the periodic table.* This simple idea forms the basis of the electronic theory of bonding.

Since Kossel and Lewis put forward their ideas in 1916, the noble gases have been shown to be more reactive than expected. We now know of many compounds in which the elements do not have a noble gas structure. For example, most of the ions of transition metals (e.g. Fe^{2+}, Fe^{3+}, Cu^{2+}) do not have an electron structure like a noble gas, nor does the sulphur atom in SO_2, SO_3 or SF_6. Nevertheless, the ideas suggested by Kossel and Lewis still form the basis of modern theories of bonding. These theories explain the formulae and structures of most compounds and the forces holding particles together.

8.2 Transfer of electrons—electrovalent (ionic) bonding

Typical ionic compounds are formed when metals in Group I or Group II react with non-metals in Group VI or Group VII. When the reaction occurs, electrons are transferred from the metal to the non-metal until the outer electron shells of the resulting ions are identical to those of a noble gas.

Figure 8.1 shows how the transfer of electrons from lithium to oxygen forms ions in lithium oxide. In figure 8.1, the nucleus of each atom is represented by its symbol and the electrons in each shell are represented by groups of dots or crosses around the symbol. Ions are shown in square brackets with the charge at the top right-hand corner.

■ What is the electron structure of: (i) the lithium ion, (ii) the oxide ion?
■ Which noble gas has an electron structure like Li^+?

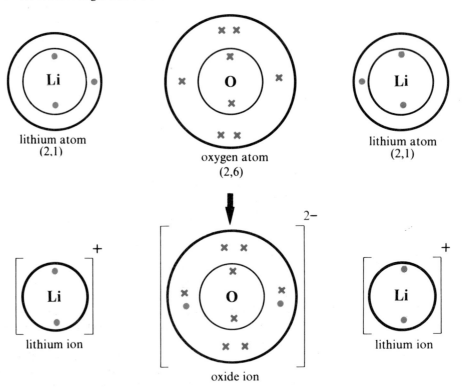

lithium atom (2,1)

oxygen atom (2,6)

lithium atom (2,1)

lithium ion

oxide ion

lithium ion

Figure 8.1
Transfer of electrons from lithium to oxygen in the formation of lithium oxide.

- Which noble gas has an electron structure like O^{2-}?
- Why is it that two lithium atoms react with only one oxygen atom?

Although the electrons of the different atoms in figure 8.1 are shown by dots and crosses, you must not think that electrons of lithium are any different from those of oxygen. All electrons are identical. They are shown differently in the diagram so that we can follow their transfer more easily.

Figure 8.2 shows the electron transfers which take place during the formation of sodium chloride and magnesium fluoride. Only those electrons in the outer shell of each atom are represented by dots or crosses round their symbol. The full electron structures are shown in brackets below the symbols.

- Draw similar 'dot/cross' diagrams to represent the electron transfers which take place in the formation of magnesium sulphide and potassium oxide.

Remember when writing 'dot/cross' diagrams that the dots and crosses are simply a means of counting electrons. They cannot show the precise location of electrons within the atom since electrons are distributed in space as diffuse negative charge clouds.

$$
\text{Na} \cdot \quad + \quad \overset{\text{x x}}{\underset{\text{x x}}{\text{x Cl x}}} \quad \longrightarrow \quad \left[\text{Na}\right]^{+} \quad \left[\overset{\text{x x}}{\underset{\text{x x}}{\text{x Cl x}}}\right]^{-}
$$

(2, 8, 1) (2, 8, 7)   (2, 8) (2, 8, 8)
Sodium atom Chlorine atom   Sodium ion Chloride ion

$$
\text{Mg} \vdots \quad + \quad \begin{array}{c} \overset{\text{x x}}{\underset{\text{x}}{\text{x F x}}} \\ (2,7) \\ \overset{\text{x x}}{\underset{\text{x}}{\text{x F x}}} \\ (2,7) \end{array} \quad \longrightarrow \quad \left[\text{Mg}\right]^{2+} \quad \begin{array}{c} \left[\overset{\text{x x}}{\underset{\text{x}}{\text{x F x}}}\right]^{-} \\ (2,8) \\ \left[\overset{\text{x x}}{\underset{\text{x}}{\text{x F x}}}\right]^{-} \\ (2,8) \end{array}
$$

(2, 8, 2)   (2, 8) (2, 8)
Magnesium 2 Fluorine   Magnesium 2 Fluoride
atom   atoms   ion   ions

Figure 8.2
Electron transfers during the formation of sodium chloride and magnesium fluoride.

The formation of ions in such compounds as lithium oxide, sodium chloride and magnesium fluoride involves a *complete transfer* of electrons. In the crystals of these substances ions are held together by **ionic (electrovalent) bonds**. These bonds involve an electrostatic attraction between oppositely charged ions. The structure, properties and bonding in ionic compounds are discussed in sections 10.7 and 10.8.

- Element X has two electrons in its outer shell and element Y has six electrons in its outer shell. What is the formula of the most likely compound formed from X and Y?

A photoelectric colorimeter which can be used to determine the concentration of a coloured solution. There is an explanation of the working of the instrument in section 25.4. This often involves the ionisation of caesium atoms.

8.3 Factors influencing the formation of ions

What factors determine whether an atom will form an ion?

(A) Cations—ionisation energies

In the case of metals, ionisation involves losing electrons. Thus, the ionisation energy of the metal provides a quantitative measure of the ease with which the metal atoms will form ions. In the case of doubly charged cations we must take the sum of the first and second ionisation energies into consideration. In sections 6.2 and 6.3, we discussed the measurement of ionisation energies. The influence of nuclear charge and atomic radius on the sizes of ionisation energies are discussed more fully in section 13.3.

The lower the ionisation energy of an atom or ion the more easily will it lose an electron. The further an electron is from the nucleus, the less firmly it is held and the easier it can be lost. Thus, in Group I, caesium, which has the largest atoms, forms ions most easily in spite of having the largest positive nuclear charge. In fact, caesium loses electrons so easily that it is used in photoelectric cells (section 19.6(b)). In general, it becomes easier to form positive ions on passing down a group from atoms of smaller to those of larger relative atomic mass.

Once an electron has been lost from an atom, the overall positive charge can hold the remaining electrons more firmly. So, although many metals form doubly charged cations, those with three units of charge are less common and ions with four positive charges are very rare. In fact, it requires four times as much energy to form $Mg^{2+}(g)$ from $Mg(g)$ than to form $Na^+(g)$ from $Na(g)$, and the process $Al(g) \rightarrow Al^{3+}(g)$ requires ten times as much energy as $Na(g) \rightarrow Na^+(g)$.

(B) Anions—electron affinity

Ionisation energies provide an indication of the ease with which atoms or ions can lose electrons. They are concerned with the formation of cations and are of no use in considering the formation of anions which involves the gain of an electron or electrons. A more appropriate measurement in the case of anions is the electron affinity, which gives the energy change for the process:

$$X(g) + e^- \rightarrow X^-(g)$$

Notice that the electron affinity for $X(g)$ is simply the reverse of the first ionisation energy for $X^-(g)$. Electron affinities can be obtained by valve measurements or by spectroscopic methods similar to those used in the determination of ionisation energies (sections 6.2 and 6.3). Table 8.2 shows some values of electron affinities. The more negative (i.e. more exothermic) the electron affinity, the more stable is the anion formed. Thus, Cl^- is more stable than Br^- with respect to their corresponding atoms. Likewise, Br^- is more stable than I^-.

Table 8.2
Electron affinities for some atoms and ions (values are in kJ mole^{-1})

$H \rightarrow H^-$			
-72			
$F \rightarrow F^-$	$Cl \rightarrow Cl^-$	$Br \rightarrow Br^-$	$I \rightarrow I^-$
-333	-364	-342	-295
$O \rightarrow O^-$	$O^- \rightarrow O^{2-}$	$S \rightarrow S^-$	$S^- \rightarrow S^{2-}$
-141	$+791$	-200	$+649$

■ Why are the electron affinities of halogen atoms more exothermic than those of the oxygen atom or the sulphur atom?
■ Why are the electron affinities O^- and S^- endothermic?

In general, electron affinities become more exothermic as a period is crossed from left to right because the incoming electron is attracted more strongly by the progressively smaller atoms with an increasing positive charge in their nucleus.

(C) Lattice energies

Although ionisation energies and electron affinities provide information about the ease with which atoms form ions, these processes are often endothermic. For example, the conversion of solid sodium to gaseous sodium ions ($Na(s) \rightarrow Na^+(g) + e^-$) is endothermic and this means that $Na^+(g)$ is less stable than $Na(s)$. However, the gaseous sodium ion (Na^+) is stabilised when it comes into contact with Cl^- ions forming solid NaCl. The process

$$Na^+(g) + Cl^-(g) \rightarrow Na^+Cl^-(s)$$

is highly exothermic ($\Delta H = -781\,\text{kJ mole}^{-1}$). The heat change for this reaction is known as the **lattice energy** for sodium chloride.

The more exothermic the lattice energy the more stable the ionic compound formed. Lattice energies are discussed in more detail in section 12.12.

8.4   Sharing electrons—covalent bonding

Look closely at figure 8.3. This shows an electron density map for the hydrogen molecule. (This can be compared with the electron density map of sodium chloride, a typical ionic compound, in figure 10.20). Lines on the map join points with the same electron density in the same way that contours on a map join points at the same height above sea-level. Notice that, although the highest concentration of electrons is near each nucleus, there is also a high concentration of electrons between the two nuclei. This suggests that in molecules such as H_2, electrons are shared by the two hydrogen atoms. The two atoms are held together by the strong attractions of their nuclei for the intervening electrons.

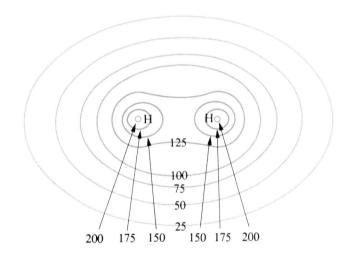

Figure 8.3
An electron density map for hydrogen (units for the contours are electrons per nm³). After C. A. Coulson, *Proc. Cam. Phil. Soc.*, **34**, 210.

Each hydrogen atom has only one electron and is very reactive. If, however, two hydrogen atoms come close together their electron orbitals can overlap. The pair of electrons is then attracted to each nucleus and shared by each atom. A schematic 'dot/cross' representation of this is shown in figure 8.4. Each hydrogen atom now has two electrons, which is the same electron structure as helium. Thus, the molecule H_2 is much more stable than the H atom.

The sharing of a pair of electrons between two atoms makes a covalent bond.

In a normal covalent bond, each atom contributes one electron to the shared pair. The shared pair help to fill the outermost shell of both atoms.

Covalent bonding can be used to explain the structures and formulae of non-metals (e.g. Cl_2, P_4, S_8) and also of non-metal/non-metal compounds (e.g. HCl, CH_4, CO_2). In these substances, *each atom usually gains a noble gas electron structure* as a result of electron sharing. Figure 8.5 shows how this happens in the case of chlorine, Cl_2. Each chlorine atom has the electronic structure 2, 8, 7 with seven electrons in its outermost shell. By sharing one pair of electrons they acquire an electron structure similar to argon (2, 8, 8). Figure 8.6 shows the electron 'dot/cross' diagrams for ammonia, NH_3; water, H_2O; carbon dioxide, CO_2; and nitrogen, N_2.

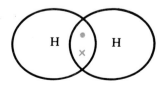

Figure 8.4
A 'dot/cross' representation of the electron structure of a hydrogen molecule.

■ What is the electron shell structure for a nitrogen atom?
■ Which noble gas has an electron structure similar to N in NH_3?

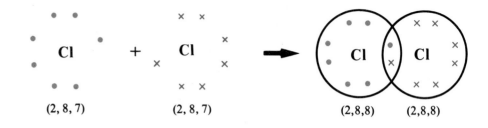

Figure 8.5
The formation of a covalent bond between two chlorine atoms.

■ How many double covalent bonds are there in one CO_2 molecule?
■ How many electrons are shared in each double covalent bond?
■ How would you describe the bond between nitrogen atoms in N_2?

In carbon dioxide, double bonds are formed by the sharing of four electrons, two contributed by each atom. Similarly, triple bonds can be formed by the sharing of six electrons. In this case, three electrons are contributed by each atom (see nitrogen in figure 8.6). Remember that 'dot/cross' diagrams are merely a convenient method for counting electrons. They do not represent the positions of electrons, nor do they give any indication of the shape of the molecules.

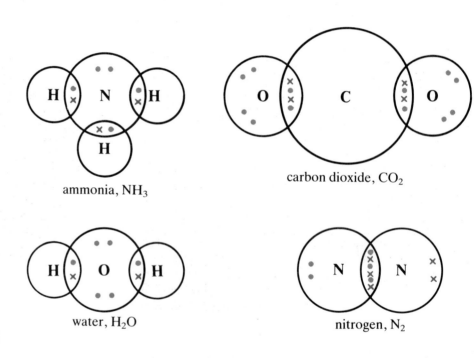

Figure 8.6
Electron 'dot/cross' diagrams for ammonia, water, carbon dioxide and nitrogen.

8.5 Determining the structures of simple molecules

When a beam of electrons is passed through a gas or vapour at low pressure, nuclei of the atoms in the gas molecules scatter (diffract) the electrons. From the pattern of the diffracted electrons, it is possible to determine the structure of the gas causing the diffraction. Figure 8.7 shows the experimental arrangement for electron diffraction studies. Electrons ejected by a heated filament are accelerated through about 50 000 volts and collimated into a thin beam. The gas under investigation is admitted through a jet towards a cold surface where it condenses. As the electron beam passes through the gas, diffraction takes place. The scattered electrons strike a photographic plate.

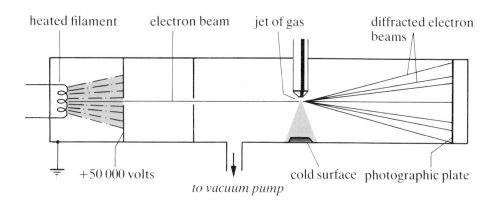

Figure 8.7
The experimental arrangement for electron diffraction studies.

Below right. An electron diffraction pattern produced by electrons passing through a very thin film of gold.

Below. An electron diffraction pattern produced by electrons passing through a very thin film of silver.

Photographs of the diffracted electron beams show a series of concentric circles. Each circle corresponds to the distance between atoms within one molecule. By comparing the observed diffraction pattern of the gas with the diffraction patterns calculated for various assumed structures, it is possible to deduce the correct structure of molecules in the gas. Furthermore, the diffraction pattern gives information from which bond lengths and bond angles can be determined.

Electron diffraction can be used with simple molecules in the gaseous or vapour state. It gives bond lengths to an accuracy of ± 0.001 nm or better. (The structure of solids can be investigated by X-ray diffraction (section 10.2).)

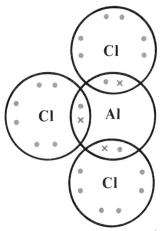

Figure 8.8
An electron 'dot/cross' diagram for $AlCl_3$.

8.6 Co-ordinate (dative covalent) bonding

In a normal covalent link, each atom provides one electron for the bond. However, in a few compounds a bond is formed by the sharing of a pair of electrons both of which are provided by one atom. Since both electrons in the bond are donated by one atom, the bonding is known as **dative covalent** or **co-ordinate**. Dative covalent bonding plays an important part in the formation of the addition compound, $NH_3.AlCl_3$.

In the vapour phase at high temperatures, aluminium chloride consists of simple molecules of $AlCl_3$ with covalent bonds between the aluminium and chlorine atoms (figure 8.8). This covalent bonding is very unexpected because metal/non-metal compounds like $AlCl_3$ are normally ionic. Notice that aluminium has only six electrons in its outer shell—two short of the noble gas structure for argon. Now look at the 'dot/cross' structure of ammonia in figure 8.6. The nitrogen atom has eight electrons in its outer shell, but two of these electrons are not shared with any other atom. When ammonia gas and aluminium chloride vapour are mixed, they react rapidly to form a solid with the formula $NH_3.AlCl_3$. In this compound, the NH_3 and $AlCl_3$ have formed a bond because the N atom in NH_3 can donate its unshared pair of electrons towards the Al atom in $AlCl_3$. This enables the Al atom to achieve a noble gas structure. Figure 8.9 shows the electron 'dot/cross' structure of the product and also the simpler notation using single lines to represent covalent bonds. The dative bond is represented by an arrow drawn from the atom donating the pair of electrons (in this case nitrogen) to the acceptor of the electron pair.

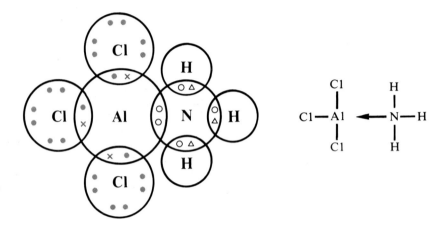

Figure 8.9
The electronic structure and bonding in $AlCl_3.NH_3$.

When gaseous aluminium chloride is cooled, $AlCl_3$ molecules dimerise to form molecules of Al_2Cl_6. Monomers of $AlCl_3$ are held together in the Al_2Cl_6 dimer by dative covalent bonding (figure 8.10).

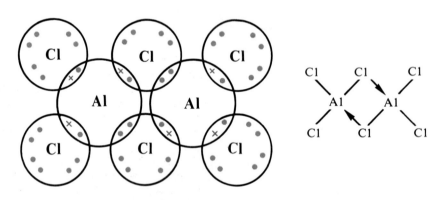

Figure 8.10
The electronic structure and bonding in Al_2Cl_6.

Compounds containing unshared pairs of electrons readily form dative covalent bonds. For example, when ammonia is mixed with gaseous hydrogen chloride, white clouds of ammonium chloride form. The reaction can be thought to result from the formation of a dative covalent bond between the unshared pair of electrons on the N atom in NH_3 and an H^+ ion from the HCl (figure 8.11). A similar reaction to this occurs when hydrogen chloride dissolves in water (figure 8.12).

Figure 8.11
An electron 'dot/cross' representation for the formation of NH_4Cl from NH_3 and HCl.

Figure 8.12
An electron 'dot/cross' representation for the reaction between hydrogen chloride and water.

Notice that both the ammonium ion and the oxonium ion (H_3O^+) have an overall charge of +1 which originates from the H^+ ions. In these ions, this charge is distributed over the whole structure. It is not localised on any one atom.

Once a dative bond has formed, it is indistinguishable from a covalent bond. The only difference between the two lies in the way we pictured the bond forming. Thus, although the NH_4^+ and H_3O^+ ions are sometimes represented as

$$\left[\begin{array}{c} H \\ | \\ H-N \rightarrow H \\ | \\ H \end{array} \right]^+ \quad \text{and} \quad \left[H-O \rightarrow H \atop \hspace{0.7em} | \atop \hspace{0.7em} H \right]^+$$

showing the co-ordinate bond as an arrow, it is important to remember that all four N—H bonds in NH_4^+ are identical as are the three O—H bonds in H_3O^+. There is no difference between any of the four hydrogen atoms in NH_4^+ and all four N—H bonds have the same length and the same strength.

The idea of dative covalent bonding has also been used to explain the existence of more than one oxide of the same non-metal. The structure of carbon dioxide can be explained purely in terms of covalent bonds. But, the structure of carbon monoxide must involve a dative covalent bond if both atoms are to attain a noble gas structure.

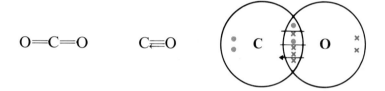

The strange properties of silicone bouncing putty ('potty putty') can be explained in terms of co-ordinate bonds. The structure of 'potty putty' resembles a silicone in which some silicon atoms have been replaced by boron (figure 8.13). These boron atoms have only six electrons in their outermost shell and readily form dative covalent bonds with the oxygen atoms in neighbouring silicone chains.

When a sample of 'potty putty' is pulled steadily, it extends like a piece of plasticine. The silicone chains slide over each other as the boron atoms form co-ordinate bonds to successive oxygen atoms along the same neighbouring chain. If, however, the 'potty putty' is pulled sharply, it breaks like a piece of crumbly cheese because the boron atoms are unable to progress smoothly by dative bonding from one oxygen to the next. Thus, 'potty putty' can be both plastic and brittle. Not surprisingly, it has been described as a schizophrenic material.

Figure 8.13
Co-ordinate bonding in 'potty putty'.

Co-ordinate (dative covalent) bond

8.7 The shapes of simple molecules

Look closely at the structures and 'dot/cross' diagrams of the simple molecules in figure 8.14. These structures have been determined by electron diffraction studies.

Name	Beryllium chloride	Boron trichloride	Methane
'Dot/cross' diagram	:Cl ˟ Be ˟ Cl:	(see structure)	H ˟ C ˟ H
Structure	Cl — Be — Cl	Cl Cl \ / B | Cl	H \ C / | \ H H H
Description of shape with respect to atoms	Linear	Trigonal planar	Tetrahedral

Figure 8.14
Electron 'dot/cross' diagrams and structures for $BeCl_2$, BCl_3 and CH_4.

Notice that the bonds in $BeCl_2$, BCl_3 and CH_4 spread out so as to be as far apart as possible. The three atoms in $BeCl_2$ are in a line. The shape is described as **linear**. The four atoms in BCl_3 are in the same plane with the chlorines at the corners of a triangle. The shape is described as **trigonal planar**. In CH_4, the four H atoms lie at the apices of a tetrahedron with the C atom at its centre. The shape is **tetrahedral**.

■ Why do the bonds get as far apart as possible?
■ What is unusual about the electron structures of Be and B in $BeCl_2$ and BCl_3, respectively?

Now look at the 'dot/cross' diagrams and structures of NH_3 and H_2O in figure 8.15. NH_3, in which the three hydrogen atoms form the base of a pyramid, is described as **pyramidal**. But why is ammonia pyramidal? Why is it not planar like BCl_3 which also has three atoms attached to the central atom? The answer lies in the non-bonded **lone pair of electrons** on the nitrogen atom. This lone pair on the nitrogen occupies the fourth tetrahedral position around the nitrogen atom in the NH_3 molecule. Each of the N—H bonds in ammonia is composed of a region of negative charge similar to the lone pair. The nitrogen atom is therefore surrounded by four regions of negative charge. These four negative

Name	Ammonia	Water
'Dot/cross' diagram	H ˟ N ˟ H (with H below)	H ˟ O: (with H below)
Structure	N (with xx) / | \ H H H	O (with dots) / \ H H
Description of shape with respect to atoms	Pyramidal	Bent (V-shaped)

Figure 8.15
Electron 'dot/cross' diagrams and structures for ammonia and water.

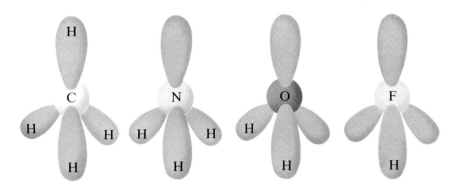

Figure 8.16
Electron charge cloud models of CH_4, NH_3, H_2O and HF.

charge clouds repel each other as far apart as possible. Consequently, the shape of ammonia, though pyramidal with respect to atoms, can be described as tetrahedral with respect to negative centres around the central nitrogen atom.

What about the water molecule? Can its **bent** or **V-shaped** structure be explained in a similar manner?

■ How many lone pairs does the O atom in water possess?
■ How many covalent bonds are there in the O atom in a water molecule?
■ How many centres of negative charge are there around the O atom in water?
■ What is the shape of water with respect to negative centres around the central O atom?

Figure 8.17
The electronic structure and bonding in HCN and SO_2.

Figure 8.16 shows the electron charge cloud models of CH_4, NH_3, H_2O and HF. You will realise from this discussion that it is the number of regions of negative charge (not the number of bonds) around the central atom which dictates the shape of a molecule. In predicting the shape of a molecule you must, therefore, count the number of negative centres around the central atom. Thus, CO_2 (figure 8.6) and HCN (figure 8.17) with two negative centres around their central C atoms are both linear with respect to negative centres and linear with respect to atoms. IN CO_2, each double covalent bond counts as a single negative centre as does the triple covalent bond in HCN. In contrast to CO_2, SO_2 has three centres of negative charge around the S atom, and is V-shaped with respect to atoms (figure 8.17). Our discussions concerning the shapes of molecules are summarised in table 8.3.

Table 8.3
The number of negative centres, the number of lone pairs and the shapes of some simple molecules

No. of negative centres around central atom	Angular separation of negative centres	Shape w.r.t. negative centres	Example	Shape w.r.t. atoms	No. of lone pairs
2	180°	linear	$BeCl_2$, CO_2	linear	0
3	120°	trigonal planar	BCl_3, BF_3	trigonal planar	0
3	120°	trigonal planar	SO_2	bent or v-shaped	1
4	109°	tetrahderal	CH_4, NH_4^+, BH_4^-, CCl_4	tetrahedral	0
4	109°	tetrahedral	NH_3, H_3O^+	pyramidal	1
4	109°	tetrahedral	H_2O, F_2O	bent or V-shaped	2
6	90°	octahedral	SF_6	octahedral	0

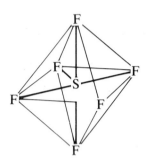

Figure 8.18
The electron 'dot/cross' diagram and the octahedral structure of SF₆.

The electron 'dot/cross' diagram and the **octahedral** structure of sulphur hexafluoride are shown in figure 8.18. There are six regions of negative charge around the central S atom in SF_6. These push each other as far apart as possible. This leaves four F atoms arranged in a square around the S atom and the other two F atoms above and below the square. If lines are drawn from each F atom to its nearest neighbours, the result is an octahedron (a solid with eight sides).

The fine structure of methane, ammonia and water

Look closely at table 8.4. Molecules of CH_4 are perfectly symmetrical. Consequently, the bond angle for H—C—H is that of a regular tetrahderon, 109°.

■ The bond angle for H—N—H in NH_3 is 107°, i.e. 2° less than the bond angle in CH_4. Does a bonded electron pair or a lone electron pair exert the greater repelling force?

■ The bond angle for H—O—H in water is 105°, i.e. 2° less than the bond angle in NH_3. Write the following repulsions in order of decreasing strength: bond pair/bond pair repulsion, bond pair/lone pair repulsion, lone pair/lone pair repulsion.

The region in space occupied by a lone pair of electrons is smaller and closer to the nucleus of an atom than a bonded pair. Bonded pairs of electrons are drawn out between the nuclei of the two atoms which they bind together. This means that a lone pair can exert a greater repelling effect than a bonded pair and this results in a decreasing bond angle from CH_4 to NH_3 to H_2O.

Table 8.4
Bond angles in methane, ammonia and water

	Methane CH_4	Ammonia NH_3	Water H_2O
HXH bond angle (X = C, N or O)	109°	107°	105°

Summary

1 In many cases, when elements form compounds they lose, gain or share electrons so as to achieve stable electron configurations similar to the next higher or lower noble gas in the periodic table.

2 An electrovalent bond involves the complete transfer of electrons from one atom to another forming oppositely charged ions. The ions are held together in the solid crystal by the electrostatic attraction between their opposite charges.

3 In general, cations form easiest when:
   (a) the resulting charge is small,
   (b) the radius of the atom is large.

4 In general, anions form easiest when:
   (a) the resulting charge is small,
   (b) the radius of the atom is small.

5 A covalent bond involves the sharing of a pair of electrons between two atoms, each atom contributing one electron to the shared pair.
   A double covalent bond involves the sharing of four electrons, two contributed by each atom.

6 A co-ordinate (dative covalent) bond involves the sharing of a pair of electrons between two atoms, both electrons in the bond being donated by one atom.

7 The shape of a simple molecule is dictated by the number of regions of negative charge around the central atom. The regions of negative charge could be a non-bonded (lone) pair of electrons, a shared pair of electrons, four shared electrons in a double covalent bond or even six electrons in a triple bond.

8 A lone pair of electrons exerts a greater repelling effect than a bonded pair.

Study questions

1 Draw 'dot/cross' electron structures showing electrons in the outermost shell of each atom in the following compounds. Show the overall charge on each ion in the ionic compounds. (None of the compounds involve co-ordinate bonding except those in part (d).)
  (a) Cl_2, H_2O_2, O_2, C_2H_4, C_2H_2, HCN
  (b) LiF, $CaCl_2$, Na_3P, Al_2S_3
  (c) PCl_3, Cl_2O, HNO_2, H_2CO_3, C_2H_5OH
  (d) SO_3, undissociated HNO_3, undissociated H_2SO_4.

2 (a) How is the type of bonding in the chlorides of the elements Na, Mg, Al, Si, P and S related to
    (i) their position in the periodic table
    (ii) the number of electrons in the outermost shells of these elements?
  (b) Explain the terms ionic, covalent and co-ordinate as applied to bonds in compounds. Illustrate your answer by clear 'dot/cross' diagrams to show the positions of all electrons in the outer shells of each atom or ion in four of the following compounds: hydrogen chloride, ammonia, ammonium chloride, aluminium chloride, sodium hydride.

3 Draw 'dot/cross' diagrams and predict the shapes with respect to atoms for molecules of the following compounds:

$$SF_6, \quad POCl_3, \quad SOCl_2, \quad H_3O^+, \quad BF_3, \quad C_2H_2$$

4 (a) State the electronic configurations of the following atoms (e.g. Be would be 2, 2):

$$C, \quad N, \quad O, \quad F$$

  (b) Draw a series of 'dot/cross' diagrams to show the structures of the simplest hydrides formed by carbon, nitrogen, oxygen and fluorine.
  (c) Sketch and describe the shapes of the molecules in part (b) and discuss the influence of any lone pairs of electrons on these shapes.
  (d) What shape would you predict for the following:

$$NH_4^+, \quad NH_3, \quad NH_2^-?$$

5 X, Y and Z represent elements of atomic numbers, 9, 19 and 34.
  (a) Write the electronic structures for X, Y and Z (e.g. Be would be 2, 2).
  (b) Predict the type of bonding which you would expect to occur between
    (i) X and Y, (ii) X and Z, (iii) Y and Z.
  (c) Draw 'dot/cross' diagrams for the compounds formed, showing only the electrons in the outermost shell for each atom.
  (d) Predict, giving reasons, the relative
    (i) volatility,
    (ii) electrical conductance and
    (iii) solubility in water
    of the compound formed between X and Y compared with that formed between X and Z.

6 Explain the following.
  (a) Tin and lead form the ions Sn^{4+} and Pb^{4+}, respectively, but carbon and silicon, elements in the same group, do not form C^{4+} and Si^{4+} ions.
  (b) The carbon–oxygen bond length in methoxymethane (dimethyl ether, $CH_3, O.CH_3$) is 0.14 nm, while that in carbon dioxide is 0.12 nm and that in carbon monoxide is only 0.11 nm.
  (c) The C—O bond lengths in $CO_3{}^{2-}$ are all identical.
  (d) Aluminium fluoride has a much higher melting point than aluminium chloride.

7 The first electron affinities of the elements in period 3 are given below.

Element	Na	Mg	Al	Si	P	S	Cl	Ar
First electron affinity/kJ mole^{-1}	-20	$+67$	-30	-135	-60	-200	-364	—

  (a) What is the general trend in the value of the electron affinity from Na to Cl?
  (b) Explain the general trend noticed in part (a).
  (c) Why is the first electron affinity of Mg more positive than one might expect from the general trend in the values above?
  (d) Why is the first electron affinity of silicon more exothermic than that of phosphorus?

9 Intermolecular Forces

9.1 Polar and non-polar molecules

Fill a burette with water. Open the tap and bring a charged ebonite rod close to the stream of water running from the jet. The water is deflected from its vertical path towards the charged rod (figure 9.1). Why is this?

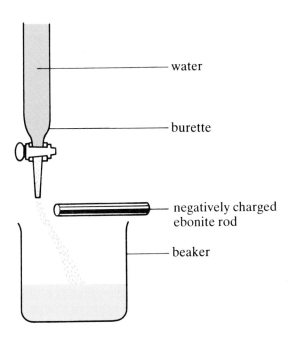

Figure 9.1
The effect of a charged rod on a thin stream of water.

water

burette

negatively charged
ebonite rod

beaker

Table 9.1
Testing the deflection of a jet of liquid using a charged rod

Liquids showing a marked deflection	Liquids showing no deflection
Water	
Trichloromethane	Tetrachloromethane
Propanone (acetone)	
Ethoxyethane (ether)	
Nitrobenzene	Benzene
Cyclohexene	Cyclohexane
Ethanol	

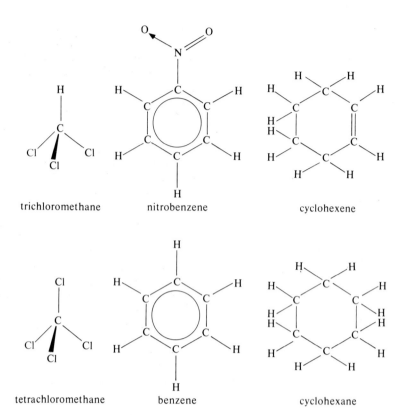

Figure 9.2
The structures of some simple molecules.

The ebonite rubbed with fur has a negative charge. When the ebonite rod is replaced by a positively charged rod, the water is again deflected *towards* the rod. why is this?

The results in table 9.1 show what happens when water is replaced by other liquids. Those liquids which are affected are always deflected *towards* the charged rod.

- Why is trichloromethane deflected, but not tetrachloromethane?
- Why is nitrobenzene deflected, but not benzene?
- Why is cyclohexene deflected, but not cyclohexane?

Look at the structures of these six molecules in figure 9.2. Notice that those molecules which are unaffected by a charged rod are symmetrical. Those which are deflected are not symmetrical. This lack of symmetry, in $CHCl_3$ for example, means that its centre of positive charge does not coincide exactly with its centre of negative charge. This means that the molecule is affected by an electrostatic field from the charged rod.

When two different atoms are joined by a covalent bond, their attractions for the bonding electrons will not be the same. For example, in a molecule of HCl the bonding electrons will not be shared equally by the hydrogen and chlorine atoms. In fact, the chlorine atom has a greater attraction for the electrons in the covalent bond. In chemical language, we say that chlorine is more **electronegative** than hydrogen. Consequently, the centre of negative charge in the HCl molecule is drawn towards the chlorine atom and it is closer than the centre of positive charge to the nucleus of the chlorine atom (figure 9.3).

The overall distortion of charge in molecules such as HCl, which results from unequal sharing of electrons, is known as **polarisation**. The molecules are said to be **polar**. The separation of charge in the molecule is referred to as a **dipole**.

Molecules, such as tetrachloromethane, benzene and cyclohexane, with a symmetrical distribution of similar atoms and in which equal dipoles cancel each other exactly are **non-polar**. In these molecules, the centres of positive and negative charge are at the same point.

Molecules, such as water, trichloromethane, nitrobenzene and cyclohexene, in which dipoles do not cancel each other are polar. When these liquids stream

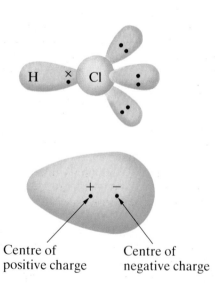

Centre of   Centre of
positive charge   negative charge

Figure 9.3
The centres of positive and negative charge in a molecule of HCl.

from a burette past a charged rod, molecules are attracted towards the charged rod and the jet is deflected. When a positive rod is used, it is the negative ends of the dipoles in the polar molecules that are attracted towards the rod. With a negative rod, positive ends of the dipoles are attracted.

9.2 Electronegativity, polarity and bonding

In non-symmetrical molecules such as HCl and H_2O, the centre of positive charge and the centre of negative charge do not coincide. The polarity in these molecules results from the presence of highly electronegative elements such as chlorine and oxygen.

Various attempts have been made to quantify the electronegativity or electron-attracting power of an atom within a molecule. The electron-attracting power of an atom will remain the same in different compounds. But, the effects of a particular atom's electron-attracting power will vary from one compound to another because it will depend on the other atoms to which the atom under consideration is attached. For this reason, it is difficult to estimate values of electronegativity with real accuracy. In spite of this, the concept of electronegativity is very useful.

One of the most widely used scales of electronegativity values is that devised by the American chemist, Linus Pauling (section 13.3). Pauling defined the electronegativity of an atom as *the power of that atom in a molecule to attract electrons*. He obtained values of electronegativity by considering the strengths of the bonds between atoms in molecules. Non-metals with a strong desire to gain electrons have the highest values of electronegativity. Metals have low values.

- Which element will have the highest electronegativity value?
- Which element will have the lowest electronegativity value?
- How will the electronegativities of the elements vary across a period?
- How will the electronegativities of the elements vary down a group as relative atomic mass increases?
- Will the trend in electronegativity values down Group I be the same as the trend down Group VII?

Check your answers to these questions by referring to table 13.3.

Notice that the concept of electronegativity is concerned with the attraction for electrons of *atoms in molecules*. Don't confuse electronegativity with electron affinity which is concerned with the attraction for electrons of *single gaseous atoms*. Unlike electronegativities, electron affinities can be measured directly.

Electronegativity values can be used to estimate the polarity of different bonds. The bonds between elements of widely differing electronegativities (i.e. between a metal and non-metal) will be ionic. The bonds between elements of similar electronegativity will be non-polar or slightly polar. If the two elements are non-metals the bonding will be covalent. If the elements are metals the bonding will be metallic. Thus, the bonding between sodium and fluorine like that between potassium and oxygen will be ionic. The bonding between fluorine and oxygen will be covalent and that between potassium and sodium will be metallic. Of course, two metals would not react with each other, but two liquid metals could mix intimately and there would be metallic bonding between different metal atoms in the mixture.

The existence of a dipole confers partial ionic character on a polar molecule. As the polarity of the molecule increases so does the extent of its ionic character.

As a result of differing electronegativities there is a distortion of the equal sharing which one would expect in a pure covalent bond. Similarly, chemists believe that there is also distortion of the charge in ionic substances. Cations attract the negative charge cloud of the anions with which they are associated (figure 9.4). This polarisation of the ions confers partial covalent character on the ionic bonding. Although it is convenient to regard bonds as either ionic or covalent, it is as well to remember that the wholly ionic and wholly covalent bonds described in chapter 8 are extreme types. All ionic bonds have some covalent character and most covalent bonds have some ionic character.

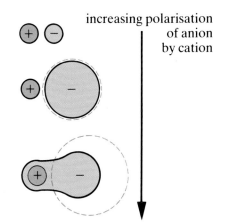

increasing polarisation of anion by cation

Figure 9.4
Polarisation of the negative charge cloud in anions of different sizes. (The dashed circles represent the shapes of the isolated anions before polarisation.)

9.3 Intermolecular forces

In the last chapter, we considered the bonding between ions and that between atoms. We must now consider the forces between molecules. We know, for example, that within a molecule of $CHCl_3$ the three Cl atoms and the H atom are joined to the central carbon atom by strong covalent bonds, but what kind of forces hold the separate $CHCl_3$ molecules together? What are the forces like between one $CHCl_3$ molecule and its neighbours?

(A) Dipole-dipole attractions

In section 9.1 we investigated the deflection of a jet of $CHCl_3$ by a charged ebonite rod. Our observations led us to conclude that $CHCl_3$ molecules were polar as a result of the non-symmetrical distribution of charge within each molecule. The interactions between permanent dipoles explain the attractions between neighbouring $CHCl_3$ molecules. These attractions are called **permanent dipole–permanent dipole attractions** (figure 9.5). In section 9.4, we shall meet another type of dipole called an *induced* dipole. The existence of dipole–dipole attractions will explain the forces holding together *polar* molecules in liquids

Figure 9.5
Permanent dipole–permanent dipole attractions in polar molecules.

such as trichloromethane ($CHCl_3$), propanone (acetone, CH_3COCH_3) and nitrobenzene ($C_6H_5NO_2$). But, what about *non-polar* molecules in liquids such as tetrachloromethane (CCl_4) and benzene (C_6H_6)? How can we explain the forces between molecules in these substances which have no permanent dipole? What holds CCl_4 molecules together in liquid CCl_4?

(B) Evidence for intermolecular forces between non-polar molecules–Van der Waals' forces

(i) The properties of noble gases

The noble gases are monatomic. They exist as single atoms in the gas phase at room temperature. These symmetrical, non-polar atoms have no permanent dipole and do not form any normal bonds. But all the noble gases will condense to liquids and ultimately form solids if the temperature is low enough. The possible liquefaction and solidification of noble gases suggests the existence of intermolecular forces in these non-polar substances. These intermolecular forces hold the molecules together in the solid and liquid state.

Furthermore, energy is required both to melt the solid and to boil the liquid noble gases, showing that cohesive forces are operating between molecules. For example, the energy of sublimation for solid xenon is $14.9\,kJ\,mole^{-1}$.

(ii) The non-ideal behaviour of gases

According to the kinetic theory, molecules of an ideal gas

1 occupy negligible volume, and

2 exert no forces on one another.

Using this model, the properties of ideal gas can be summarised in **the ideal gas equation**.

$$pV = nRT \quad \text{(see section 11.7)}$$

For one mole of gas $pV = RT$ and in this case $pV/RT = 1$.

This equation represents the behaviour of gases accurately at low pressures (i.e. $\leqslant$ atmospheric pressure) and high temperatures (i.e. well above their condensation point when the gas is least like a liquid). There are, however, increasing deviations at low temperatures and high pressures.

- What is the value of pV/RT for one mole of an ideal gas?
- Would pV/RT increase, remain constant or decrease as p increased for an ideal gas?

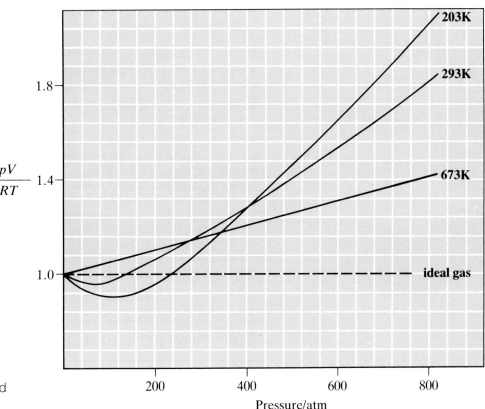

Figure 9.6
Variation of pV/RT with increasing pressure for N_2 at 673 K, 293 K and 203 K.

Look closely at figure 9.6.

- How does pV/RT vary as p increases for N_2 (a real gas) at 673 K?
- How does pV/RT deviate from ideal behaviour as temperature is reduced?

Figure 9.6 shows how the properties of nitrogen deviate from ideal behaviour at low temperature and high pressure. These deviations can be understood by reconsidering the two fundamental assumptions of the kinetic theory of gases stated at the beginning of this sub-section.

1 *In deriving the ideal gas equation, it is assumed that the molecules occupy a negligible volume.* Strictly speaking, the molecules do not have a negligible volume. Each one has a finite size which excludes a certain volume of the container from all the others. If we call this 'excluded' volume b, then the 'true' volume in which the molecules move is $(V - b)$.

Thus, the equation $pV = nRT$ should be written as $P(V - b) = nRT$.

The constant b has a value of about $30 \, \text{cm}^3 \, \text{mole}^{-1}$ for many gases. Since one mole of many gases occupies approximately $30 \, \text{cm}^3$ on liquefaction, this provides further evidence that b is roughly the same as the volume of the molecules.

2   *In deriving the ideal gas equation, it is assumed that the molecules exert no forces on each other.* Unfortunately, intermolecular forces cannot be neglected. This is particularly so at high pressure. The pressure of a gas results from particles of the gas bombarding the walls of their container. Within the bulk of the gas, intermolecular forces cancel each other out, but those molecules near the walls experience an overall net force tending to pull them back into the bulk. In effect, attractive forces cause molecules near the walls to transfer some of their momentum to other gas molecules rather than to the walls and this results in a measured pressure lower than the 'true' pressure. The magnitude of the 'pressure reduction' will be proportional to both the concentration of molecules near the wall ($\propto \frac{1}{V}$) and the concentration of molecules within the bulk (also $\propto \frac{1}{V}$). Thus, for one mole of gas the 'pressure reduction' can be written as a/V^2 where a is a constant. By adding the term a/V^2 to the actual pressure, we obtain the corrected pressure term $[p + (a/V^2)]$.

If we replace p and V in the ideal gas equation by their corrected expressions we obtain

$$[p + (\frac{a}{V^2})].(V - b) = nRT$$

This is known as the **Van de Waals' real gas equation**. This equation was first used by the Dutch physicist Van der Waals in 1873.

■ Why does this real gas equation reduce to $pV = nRT$ at low pressure and high temperature?

The properties of noble gases and the non-ideal behaviour of real gases provide evidence for the existence of cohesive forces between non-polar molecules. These weak, short-range forces of attraction between molecules are known as **Van der Waals' forces**. Van der Waals' bonds are, of course, much weaker than covalent and ionic bonds. For example, the energy of sublimation for solid chlorine (i.e. the energy required to overcome the Van der Waals' forces between one mole of Cl_2 molecules) is only $25\,kJ\,mole^{-1}$. In comparison, the bond energy of chlorine (i.e. the energy required to break one mole of Cl—Cl covalent bonds) is $244\,kJ\,mole^{-1}$. Roughly speaking, Van der Waals' bonds are between one-tenth and one-hundredth the strength of covalent bonds.

9.4   How do Van der Waals' forces arise?

The electrons in a molecule are in continual motion. At any particular moment, the electron charge cloud around the molecule will not be perfectly symmetrical. There is more negative charge on one side of the molecule than on the other. It possesses an instantaneous electric dipole. This dipole will induce dipoles in neighbouring molecules. If the positive end of the original dipole is pointing towards a neighbouring molecule, then the induced dipole will have its negative end pointing towards the positive of the original dipole. In this way, weak **induced dipole–induced dipole attractions** exist between molecules.

These induced dipoles will act first one way, then another way, continually forming and then disappearing as a result of electron movement. Notice that the force between the original dipole and the induced dipole is always an attraction. Consequently, even though the average dipole on every molecule over a period of time is zero, the resultant forces between molecules at any instant are not zero.

As the size of a molecule increases, the number of constituent electrons increases. As a result the induced dipole–induced dipole attractions become stronger.

The increase in boiling point for the elements in Group VII (F_2, Cl_2, Br_2 and I_2—figure 9.7) and the increase in boiling point for the homologous series of alkanes (CH_4, C_2H_6, C_3H_8, etc., in figure 27.3) result from stronger Van der Waals' attractions with increasing relative molecular mass.

Although the Van der Waals' forces between simple molecules such as CCl_4, Cl_2 and CH_4 are very small, the total Van der Waals' forces between the molecules of a large polymer with many contacts can be very large. The strength of non-polar polymers such as poly(ethene) and poly(propene) is dependent on significant Van der Waals' bonding between parallel molecules. Furthermore, experiments have shown that the tensile strength of high-density poly(ethene) which has tightly packed parallel molecules is three times as large as that of low-density poly(ethene) which is packed less tightly and therefore has weaker Van der Waals' attractions.

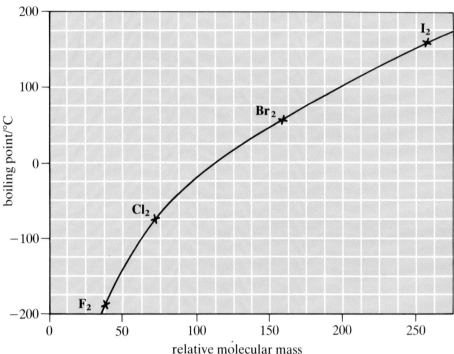

Figure 9.7
Boiling points for the elements in Group VII plotted against relative molecular mass.

Van der Waals' forces also account for the properties of graphite. Crystals of graphite are composed of parallel layers of hexagonally arranged carbon atoms (figure 9.8). Within each layer, carbon atoms are linked by strong covalent bonds. In comparison, the parallel layers are held together by Van der Waals' forces. The C—C distance within each layer is 0.14 nm but the distance between adjacent layers is 0.34 nm. The Van der Waals' bonding between the layers is strong enough to hold the layers together, but weak enough to allow them to slide over each other. Because of this, graphite is soft and acts as a solid lubricant.

9.5 Hydrogen bonding

Look closely at the graphs in figure 9.9. These show the boiling points of hydrides in Group IV, Group V, Group VI and Group VII. Notice that the boiling points of the Group IV hydrides decrease with decreasing relative molecular mass from SnH_4 to CH_4. Is there a similar decrease for the hydrides of Groups V, VI and VII?

Figure 9.9 shows the expected decrease in boiling point from H_2Te through H_2Se to H_2S, but H_2O has a much higher boiling point than one would expect. A similar pattern appears with the hydrides in Group V and those in Group VII. Here we find much higher boiling points for NH_3 and HF than extrapolation of the graphs would suggest.

How can we account for these unusually high boiling points for H_2O, NH_3 and HF?

In water, liquid ammonia and liquid hydrogen fluoride there must be unusually strong intermolecular forces. Why is this?

H_2O, NH_3 and HF are all very polar because they contain the three most electronegative elements (oxygen, nitrogen and fluorine) linked directly to

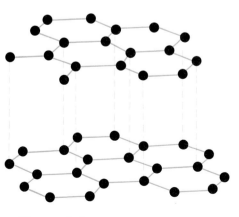

Figure 9.8
The structure of graphite.

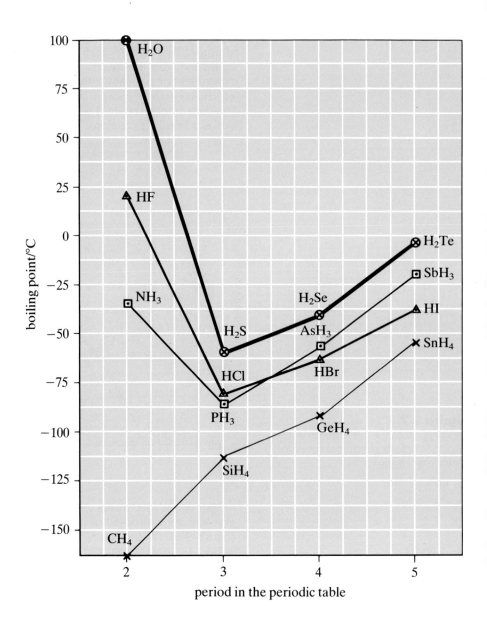

Figure 9.9
Variation of the boiling points of the
hydrides in Groups IV, V, VI and VII.

Table 9.2
Data for the hydrides of elements in Group VI

Compound	Melting point /K	Molar heat of fusion, $\Delta H_{fus}^{\ominus}$ /kJ mole^{-1}	Molar heat of vaporisation, $\Delta H_{vap}^{\ominus}$ /kJ mole^{-1}
H_2O	273	6.02	40.7
H_2S	188	2.39	18.7
H_2Se	207	2.51	19.3
H_2Te	225	4.18	23.2

hydrogen which is weakly electronegative. This results in exceptionally polar
molecules with stronger intermolecular forces than usual. These particularly
strong intermolecular forces are known as **hydrogen bonds.**

■ Look closely at the data in table 9.2. Do you think the melting point, the
molar heat of fusion and the molar heat of vaporisation of water are
influenced by hydrogen bonding?

■ Which other physical properties of a substance, besides those listed in table 9.2, will be influenced by the existence of hydrogen bonds?
Why do H atoms attached to N, O and F atoms form such strong intermolecular forces?

9.6   What is a hydrogen bond?

Nitrogen, oxygen and fluorine are the three most electronegative elements. When they are bonded to a hydrogen atom, the electrons in the covalent bond are drawn towards the electronegative atom. Now the H atom has no electrons other than its share of those in this covalent bond. These are being pulled away from it by the more electronegative N, O or F.

Since the H atom has no inner shell of electrons, the single proton in its nucleus is unusually 'bare' and readily available for any form of dipole–dipole attraction. Thus, H atoms attached to N, O or F are able to interpose themselves between two of these atoms exerting an attractive force on them and bonding them together (figure 9.10). The two larger atoms are drawn closer with an H atom effectively buried in their electron clouds. *H-bonds are therefore extra strong intermolecular, permanent dipole–permanent dipole attractions.*

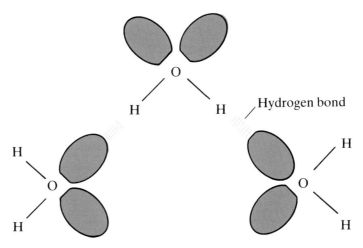

Figure 9.10
Hydrogen bonding in water.

The essential requirements for an H-bond are:

1   a hydrogen atom attached to a highly electronegative atom;

2   an unshared pair of electrons on the electronegative atom.

In the NH_3 molecule, there are three N—H bonds and one non-bonded electron pair. This means that there can be an average of one H-bond per molecule.

In water, however, there are two O—H bonds and two unshared electron pairs per molecule. This means that each H_2O molecule can form two hydrogen bonds and this helps to explain the three-dimensional lattice structure in ice (section 9.8(a)).

■ How many unshared electron pairs are there in each HF molecule?
■ What is the maximum number of H-bonds per HF molecule?

9.7   Estimating the strength of hydrogen bonds in water

Which one of the following values could be used to estimate the strength of an H-bond in water?

■ The strength of an O—H bond in H_2O.
■ The heat evolved when one mole of H_2O forms from its elements.

■ The heat of vaporisation of water.
■ The melting point of ice.

H-bonds in water are intermolecular forces. They are the forces which hold the molecules together. When water is vaporised the H_2O molecules, which are relatively close to each other in the liquid, are pulled much further apart in forming the vapour. The energy which vaporises the water is needed to overcome the forces holding the water molecules together in the liquid state. Thus, the heat of vaporisation of water will give a rough idea of the strength of H-bonds between H_2O molecules. But, the heat of vaporisation overcomes Van der Waals' bonds between H_2O molecules as well as the H-bonds. How can we estimate the strength of H-bonding alone?

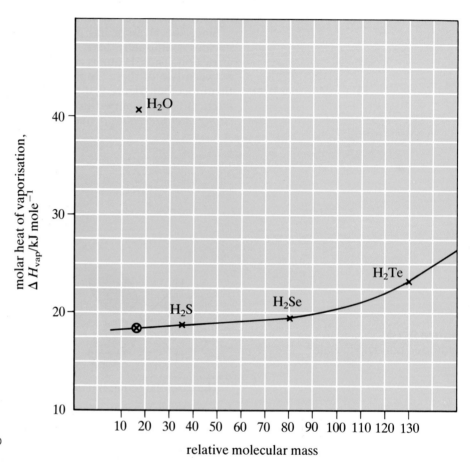

Figure 9.11
Variation of the molar heats of vaporisation for the hydrides of Group VI elements.

Figure 9.11 shows a graph of the molar heats of vaporisation for the hydrides of elements in Group VI plotted against relative molecular mass. Now, if we assume that H_2S, H_2Se and H_2Te have intermolecular forces due *only* to Van der Waals' bonds (negligible H-bonding) we can estimate a value for the strength of Van der Waals' forces in water. We do this by extrapolating the curve through the values of ΔH_{vap} for H_2S, H_2Se and H_2Te. The value we want is indicated by the circled cross in figure 9.11. This gives a predicted heat of vaporisation for water of $18.5 \, kJ \, mole^{-1}$ assuming that water has only Van der Waals' bonds.

Total strength of intermolecular forces (Van der Waals' bonds + H-bonds) in
water = $40.7 \, kJ \, mole^{-1}$.

Estimated strength of Van der Waals' forces in water = $18.5 \, kJ \, mole^{-1}$
$\therefore$ Approximate strength of H-bonds in water = $40.7 - 18.5 = 22.2 \, kJ \, mole^{-1}$

Usually, the strengths of H-bonds are in the range $5-40 \, kJ \, mole^{-1}$. Thus H-bonds are about one-tenth as strong as covalent bonds and roughly the same strength as Van der Waals' forces. Remember, however, that molecules which are H-bonded will also be attracted by Van der Waals' forces. In general, we can say that H-bonds are weak forces. They are strong enough to influence physical properties but they are not strong enough to change the chemical reactions of a substance.

Walking on water! A pond skater standing on the surface of pond water—it's all done by hydrogen bonding and surface tension.

9.8 The influence and importance of hydrogen bonding

(A) The structure and properties of water and ice

Water has unexpectedly high values for its melting point, boiling point, molar heat of fusion and molar heat of vaporisation. These unexpectedly high values result from extra attractions between water molecules due to the formation of H-bonds. This extra bonding between H_2O molecules also causes high surface tension and high viscosity. Because of the high surface tension of water, which provides a sort of 'skin effect', it is possible for small but relatively dense articles such as razor blades and beetles to float on an undisturbed water surface.

The presence of *two* hydrogen atoms and *two* lone electron pairs in each water molecule results in a three-dimensional tetrahdedral structure in ice. Each oxygen atom in ice is surrounded tetrahedrally by four others. Hydrogen bonds link each pair of oxygen atoms (figure 9.12).

The distance between adjacent oxygen atoms is 0.276 nm, more than twice the O–H distance of 0.096 nm in gaseous water molecules. This suggests that the hydrogen atom linking the two oxygen atoms in ice is not midway between them.

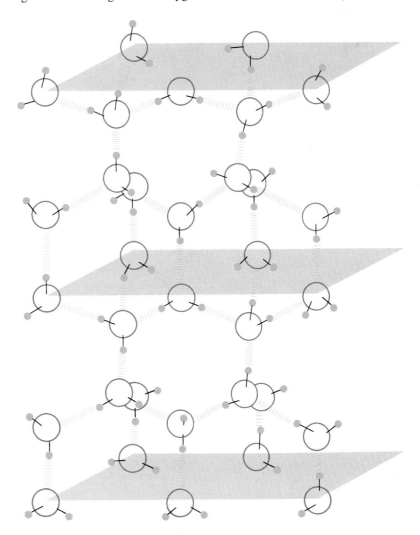

Figure 9.12
The structure of ice.

Ice floats on water. The arrangement of water molecules in ice creates a very open structure, making it less dense than water.

The arrangement of water molecules in ice creates a very open structure. This accounts for the fact that ice is less dense than water at 0°C. When ice melts, the regular lattice breaks up. The water molecules can then pack more closely, so the liquid has a higher density. In liquid water, the strong hydrogen bonding results in some ordered packing of water molecules over a short range. However, there is no long-range order as the H-bonds are continually being broken and formed. Also the regions of short-range order continually change.

The anomalous physical properties of ice and water which result from H-bonding have a great influence biologically and environmentally.

The fact that ice is less dense than water at 0°C means that ponds and lakes freeze from the surface downwards. The layer of ice insulates the water below, preventing complete solidification. If ice were the more dense, water would freeze from the bottom upwards. Ponds would freeze completely killing fish, aquatic plants and other water-living creatures.

Water has an unusually high boiling point owing to hydrogen bonds. Without these H-bonds, water would probably be a gas under normal atmospheric conditions. Oceans, lakes and rivers would never exist and it would never rain!

Earlier in this section, we mentioned the exceptionally high surface tension of water. Were it not for this, water would never rise through the capillary tubes in the roots and stems of plants.

The high polarity of water means that polar and ionic substances are usually soluble in water. Plants can obtain the salts which they require for growth by absorption of these materials into the sap through their roots.

In animals, essential ionic substances are absorbed into the bloodstream from the aqueous solution in their stomachs and intestines.

(B) The dimerisation of carboxylic acids

When carboxylic acids such as ethanoic acid (acetic acid) are dissolved in benzene, the solute particles appear to have a relative molecular mass twice as large as expected (table 9.3).

In order to explain these results, it is believed that the carboxylic acid forms hydrogen-bonded dimers in the benzene. The dimers are particularly stable because each pair of acid molecules is linked by two H-bonds, not just one (figure 9.13).

The presence of dimerised pairs of carboxylic acid molecules in benzene has been confirmed by electron diffraction measurements. Studies of the relative molecular mass of ethanoic acid in the vapour state also suggest the presence of dimers.

Figure 9.13
Hydrogen bonding in carboxylic acid dimers.

Table 9.3
The relative molecular masses of ethanoic acid and benzoic acid

Carboxylic acid	Formula	Relative molecular mass, M_r	Apparent relative molecular mass, M_r, in benzene
Ethanoic acid (acetic acid)	CH_3COOH	60	120
Benzoic acid	C_6H_5COOH	122	244

- One measurement of the relative molecular mass of ethanoic acid in the vapour state gave a value of 90. To what extent is the acid dimerised?
- Ethanoic acid could, in theory, form trimers, tetramers, etc., in benzene. Why are dimers formed in preference?
- When ethanoic acid is dissolved in water dimers do not form. Why is this?
- Name one other solvent, besides benzene, in which you would expect ethanoic acid to form H-bonded dimers.

The effect of dimerisation on the boiling points of carboxylic acids is discussed in section 33.1.

(C)   The hardness of ionic crystals

Crystals of anhydrous calcium sulphate (anhydrite, $CaSO_4$) are very hard and very difficult to cleave. On the other hand, crystals of hydrated calcium sulphate (gypsum, $CaSO_4.2H_2O$) are soft and easily cleaved. Why is there a big difference in hardness between such similar materials?

The structure of gypsum consists of layers containing both Ca^{2+} and SO_4^{2-} ions separated by layers of water molecules. Within the Ca^{2+}/SO_4^{2-} layers the ions are held together by strong electrovalent bonds. But these separated Ca^{2+}/SO_4^{2-} layers are linked by relatively weaker H-bonds. The weaker H-bonds link SO_4^{2-} ions in alternate layers to water molecules in the intermediate region. Consequently, the gypsum can be readily cleaved and scratched along the layer of water molecules.

In contrast, anhydrite has a completely ionic structure involving only Ca^{2+} and SO_4^{2-} ions. It is therefore much harder than gypsum and cannot be easily cleaved.

(D)   The structures and properties of biological molecules

H-bonds are present in the structures of proteins, carbohydrates and nucleic acids. The properties and functions of these biological molecules are dependent on their H-bonding.

Proteins

Proteins are composed of a long sequence of amino acids joined by the **peptide link**. The general formula of an amino acid is:

The $-C$ group of one amino acid molecule can react with the $-N$ group of a second amino acid to eliminate water and form a peptide bond:

$$
\begin{array}{c}
\text{H} \quad \text{R} \quad \text{O} \\
\diagdown \quad | \quad \diagup\!\!\diagup \\
\text{N}-\text{C}-\text{C} \\
\diagup \quad | \quad \diagdown \\
\text{H} \quad \text{H} \quad \text{OH}
\end{array}
\;+\;
\begin{array}{c}
\text{H} \quad \text{R} \quad \text{O} \\
\diagdown \quad | \quad \diagup\!\!\diagup \\
\text{N}-\text{C}-\text{C} \\
\diagup \quad | \quad \diagdown \\
\text{H} \quad \text{H} \quad \text{OH}
\end{array}
\;\longrightarrow\;
\begin{array}{c}
\text{H} \quad \text{R} \quad \text{O} \quad \text{H} \quad \text{R} \quad \text{O} \\
\diagdown \quad | \quad \| \quad | \quad | \quad \diagup\!\!\diagup \\
\text{N}-\text{C}-\text{C}-\text{N}-\text{C}-\text{C} \\
\diagup \quad | \qquad\qquad | \quad \diagdown \\
\text{H} \quad \text{H} \qquad\quad \text{H} \quad \text{OH}
\end{array}
\;+\;\text{H}_2\text{O}
$$

peptide link

Proteins are composed of anything from 10 to 10 000 amino acids linked by peptide bonds. They form a long-chain macromolecule which may be represented as:

$$
\begin{array}{c}
\text{H} \quad\; \text{R} \quad \text{O} \quad \text{H} \quad \text{R} \quad \text{O} \quad \text{H} \quad \text{R} \quad \text{O} \qquad\qquad \text{H} \quad \text{R} \qquad \text{O} \\
\diagdown \qquad | \quad \| \quad | \quad | \quad \| \quad | \quad | \quad \| \qquad\qquad\qquad | \quad | \quad\; \diagup\!\!\diagup \\
\quad \text{N}-\text{C}-\text{C}-\text{N}-\text{C}-\text{C}-\text{N}-\text{C}-\text{C}\ldots\ldots\ldots\text{N}-\text{C}-\text{C} \\
\diagup \qquad | \qquad\quad | \qquad\quad | \qquad\qquad\qquad\qquad | \qquad \diagdown \\
\text{H} \qquad \text{H} \qquad\quad \text{H} \qquad\quad \text{H} \qquad\qquad\qquad\qquad \text{H} \qquad \text{OH}
\end{array}
$$

The sequence of amino acids in the protein chain is referred to as **the primary structure** of the protein. The detailed configuration of the chain is called **the secondary structure**.

One of the commonest secondary structures in proteins is a spiral arrangement known as the α-helix. The amino acid units are arranged in a coiled helix. This is stabilised by H-bonds between the $>$N—H group of one amino acid unit and the $>$C=O group of the fourth unit further along the chain (figure 9.14).

H-bonding plays an important part in maintaining the structure of proteins. The precise position and sequence of amino acids in the protein structure means that proteins can carry coded information. This enables them to control growth and metabolism in plants and animals. In fact, most biological catalysts (which are usually known as **enzymes**) are proteins. So, life itself is very dependent on H-bonding.

The presence of polar $>$C=O and $>$N—H groups in proteins which can H-bond with water means that many proteins are water soluble. These soluble proteins are therefore present in the blood and in cellular fluids. They can take part in metabolism in plants and animals.

The properties of amino acids and the peptide link are discussed further in section 34.8.

Figure 9.14
Hydrogen bonding in the coiled helical structure of proteins.

Carbohydrates (polysaccharides)

Most common polysaccharides (e.g. starch, cellulose, glycogen) are built up from glucose. Glucose is a monosaccharide. The structure of glucose is:

Figure 9.16
The simplified structure of a single-stranded nucleic acid molecule.

An X-ray diffraction photograph of DNA. The arrangement of dots in the form of a cross is typical of a double helix structure.

Figure 9.15
The structure of cellulose.

In polysaccharides, the glucose units are linked by C—O—C bonds between the carbon atoms numbered 1, 4 and 6 in the structure above. Figure 9.15 shows the structure of cellulose in which the carbon atom labelled $_1$C in one glucose unit is linked to $_4$C in the next. The relative molecular masses of natural polysaccharides range from 10^5 to 10^7.

- What is the relative molecular mass of one glucose unit in a polysaccharide?
- Approximately how many glucose units are there in a polysaccharide of relative molecular mass 10^6?
- Cellulose is an important structural polysaccharide in the cell walls of plants. It is the best known example of a polysaccharide with considerable mechanical strength. What forces might operate between parallel chains of cellulose in order to hold them together?

Carbohydrates are non-ionic, but they contain a large number of OH groups. These give rise to the possibility of extensive hydrogen bonding with themselves or with water. In cellulose, the long chains of glucose units are packed very close. They form a strong bundle of parallel strands linked together by H-bonds.

Small carbohydrates, such as glucose, are very water soluble owing to their ability to hydrogen bond with the H_2O molecules. Carbohydrates also release a large amount of energy on reaction with oxygen. Because of this both animals and plants make use of polysaccharides as energy storage materials.

In plants, the principal energy storage carbohydrate is starch, whereas that in animals is glycogen. When energy is required the polysaccharides (starch or glycogen) break down releasing glucose. This dissolves easily in the blood for rapid transport to those parts of the body where energy supplies are needed. The chemistry of carbohydrates is discussed further in section 32.7.

Nucleic acids

The monomer units in nucleic acids are nucleotides. These consist of a complex organic base, B, a sugar, S, and a phosphate group as shown below:

In the nucleic acid molecule a chain of nucleotides are linked together through their sugar/phosphate groups (figure 9.16).

Essentially, there are two types of nucleic acids, deoxyribose nucleic acids (DNA) in which the sugar is deoxyribose, and ribose nucleic acids (RNA) in which the sugar is ribose. Both ribose and deoxyribose are monosaccharides containing five carbon atoms.

Figure 9.18
Above. Hydrogen-bonded base pairs in DNA.

Figure 9.17
The simplified structure of DNA showing the hydrogen bonds between bases in the parallel chains.

DNA is composed of two intertwining helices. This double helix is held together by hydrogen bonds between bases in the parallel helices (figure 9.17).

No attempt has been made to show the coiling of the helix in figure 9.17. The structure of DNA can be compared to a spiral staircase in which the sides of the staircase are composed of alternate sugar and phosphate groups and the steps of the staircase are composed of hydrogen-bonded base pairs.

There are four different bases in DNA: adenine, thymine, cytosine and guanine. Adenine can only form hydrogen bonds with thymine, and vice versa. In the same way, cytosine can only form hydrogen bonds with guanine and vice versa (figure 9.18). Thus, the sequence of bases in each strand of the double helix must be complementary. Adenine must be opposite thymine and cytosine opposite guanine. DNA is an essential constituent of the nuclei of cells. It makes up the genes which are responsible for the transmission of characteristics from one generation to the next. In a healthy plant or animal, cells are continually dividing and replacing those which die away. During cell division, the relatively weak hydrogen bonds between the two strands in the double helix break. New helices are then synthesised in the two 'daughter' cells using the separated single strands as templates—yet another biological process in which hydrogen bonding plays an essential part.

Summary

1 When two different atoms are joined by a covalent bond, their attractions for the bonding electrons will not be the same. The distortion of charge which results from the unequal sharing of electrons is known as polarisation. Molecules with an overall distortion of charge are said to be polar.

2 The electronegativity of an atom can be defined as the power of that atom in a molecule to attract electrons.

3 Do not confuse electronegativity with electron affinity. Electronegativity is concerned with the attraction for electrons of atoms in molecules. Electron affinity is concerned with the attraction for electrons of single gaseous atoms.

4 Van der Waals' forces are weak, short-range forces of attraction between molecules. Essentially, they are induced dipole–induced dipole attractions.

5 Hydrogen bonds are extra strong intermolecular forces. They are permanent dipole–permanent dipole attractions.

6 Hydrogen bonding can occur when
   (i) a hydrogen atom is attached to a highly electronegative atom (usually N, O or F) and
   (ii) this highly electronegative atom has an unshared pair of electrons.

7 Hydrogen bonding in water is responsible for the unexpectedly high values of its melting point, boiling point, latent heats, surface tension and viscosity.

8 Hydrogen bonds are present in the structures of proteins, carbohydrates and nucleic acids. The biological properties and functions of these molecules in living things are very dependent on their H-bonding.

Study questions

1 (a) Which of the following molecules would you expect to have a permanent dipole?
      (i) GeH_4 (ii) ICl (iii) SiF_4 (iv) CH_2Cl_2 (v) CO_2
   (b) The following molecules have no permanent dipole. What is their shape?
      (i) BCl_3 (ii) CS_2 (iii) C_2Cl_2 (iv) CBr_4
   (c) In which of the following compounds will hydrogen bonding occur?
      (i) $C_2H_5NH_2$ (ii) CH_3OH (iii) CH_3I (iv) CF_4
      (v) H_2SO_3 (vi) $CH_3.O.CH_3$
   (d) Which of the following molecules have a structure with a bond angle greater than 109° 28'?
      (i) SCl_2 (ii) CO_2 (iii) BF_3 (iv) NF_3 (v) CH_4

2 The solubility of iodine in tetrachloromethane at 10°C is 3 g per 100 g of solvent. The solubility of iodine in water at 10°C is 0.02 g per 100 g of solvent. The solution of iodine in tetrachloromethane is violet; the solution of iodine in water is yellow.
   (a) State whether the following are polar or non-polar;
      (i) iodine,   (ii) tetrachloromethane,   (iii) water.
   (b) Why is iodine very soluble in CCl_4?
   (c) Why is iodine only slightly soluble in water?
   (d) Would you expect CCl_4 and water to mix? Explain your answer.
   (e) A yellow solution of iodine in water is shaken with an equal volume of tetrachloromethane. Describe and explain what happens.

3 (a) The boiling point of *cis*-dichloroethene is 333 K, whereas that of *trans*-dichloroethene is 321 K. Draw the structural formulae of these two isomers and explain the difference in boiling point.

(b) The structural formulae, boiling points and densities of the isomers pentane and 2, 2-dimethylpropane are shown below.

	pentane	2, 2-dimethylpropane
Structural formula	$CH_3-CH_2-CH_2-CH_2-CH_3$	$CH_3-\overset{\overset{\textstyle CH_3}{\textstyle |}}{\underset{\underset{\textstyle CH_3}{\textstyle |}}{C}}-CH_3$
boiling point/°C	36	9
density/g cm^{-3}	0.626	0.591

(i)   Why does pentane have a higher boiling point than 2, 2dimethyl-propane?

(ii)   Why does pentane have a higher density than 2, 2-dimethylpropane?

(iii)   2-methylbutane is an isomer of pentane and 2,2-dimethylpropane. How do you think its boiling point and density will compare with these two substances? Explain your answer.

4   The relative molecular masses and molar heats of vaporisation for three of the hydrides of elements in Group V are given below.

Compound	Relative molecular mass	$\Delta H_{vap}^{\ominus}$/kJ mole^{-1}
NH_3	17	23.4
PH_3	34	14.6
AsH_3	78	17.5

(a)   Plot a graph of ΔH_{vap} against relative molecular mass for the three hydrides.

(b)   Why is the value of ΔH_{vap} for NH_3 unexpectedly high?

(c)   Use your graph to estimate a value for the ΔH_{vap} of NH_3 assuming that it has only Van der Waals' bonds.

(d)   Predict a value for the strength of H-bonds in NH_3.

(e)   The strength of H-bonding in water is approximately 22 kJ mole^{-1}. How do you explain the fact that H-bonding in NH_3 is only about half the strength of that in H_2O?

5   (a)   Water, in its state of lowest potential energy, exists as a regular, crystalline lattice, but it usually exists as a liquid at room temperature and pressure. Reconcile these facts with the statement that 'all systems tend to a state of minimum potential energy'.

(b)   The strength of the H-bond in ice is 22 kJ mole^{-1}. How much energy is required to break one mole of hydrogen bonds in ice?

(c)   What percentage of the hydrogen bonds are broken when ice melts, assuming that all the energy involved in the heat of fusion is used to break hydrogen bonds? (The heat of fusion of ice is 6 kJ mole^{-1}.)

(d)   What is the shape of the H_2O molecule? Explain why the H_2O molecule is polar.

(e)   What shape would you predict for the hypothetical molecule H_2X if it were non-polar? Explain your prediction.

6   Experiments were carried out with the three liquids, A, B and C (each with the empirical formula CHF), in order to find their polarity and relative molecular masses. The results of these experiments are given in the table below.

Compound	Empirical formula	Effect of a charged rod on a thin stream of the liquid issuing from a burette	Approx. mass of 1 dm^3 of gas at s.t.p./g
A	CHF	nil	3
B	CHF	liquid is attracted to charged rod	3
C	CHF	nil	6

(a) Which of the three liquids is (are) polar?
(b) What is the relative molecular mass of
   (i) A, (ii) B, (iii) C?
(c) Draw *one* possible structural formula for A. (H = 1, C = 12, F = 19)
(d) Draw *two* possible structural formulae for B.
(e) Draw *three* possible structural formulae for C.

7   Proteins are polypeptides with relative molecular masses in the range 10^3 to
10^6. They are composed of a sequence of amino acids joined by the peptide
link.
(a) Write a general formula for the monomer amino acid units which make
   up proteins.
(b) What is the approximate relative molecular mass for *one* amino acid unit
   in the protein?
(c) About how many amino acid units will there be in a protein of relative
   molecular mass 10^5?
(d) Draw a section of a protein structure.
(e) How does hydrogen bonding explain:
   (i)   the water solubility of certain proteins,
   (ii) the precise configurations of protein enzymes,
   (iii) the elasticity of natural protein fibres such as hair, wool and silk?

8   Suggest reasons for the following:
(a) The boiling points of water, ethanol and ethoxyethane (diethyl ether)
   are in the reverse order of their relative molecular masses unlike those of
   their analogous sulphur compounds H_2S, C_2H_5SH and $C_2H_5SC_2H_5$.
(b) BF_3 is non-polar, but NF_3 is polar.

10 Structure, Bonding and Properties —The Solid State

10.1 Introduction

One of the major achievements of chemistry has been the synthesis of new materials, such as plastics and alloys. Many of these new materials can be designed to have specific properties. In order to design these new substances, it is necessary to know how the structure of materials can affect their properties. Thus, it is not surprising that one of the most important aspects of chemistry is the investigation of the structure of materials and the bonds holding the particles together.

In this chapter, we shall begin by looking at the methods used to investigate the structures of solid materials. Then we shall consider the properties of these materials and the way in which properties are dictated by structure and bonding.

Strands of optical fibre specially designed to transmit light.

10.2 Evidence for the structure of materials

The physical properties of a material can often provide evidence for its structure. For example, the diffraction of X-rays by a solid provides evidence for the arrangement of particles within it. The amount of heat required to melt the solid

Professor Dorothy Hodgkin, who won a Nobel Prize for her work on X-ray crystallography.

An early version of the X-ray spectrometer designed by W. L. Bragg. X-rays generated at A pass through the slit at B and on to the inclined face of the crystal at C. The reflection of the X-rays from the crystal face is measured in the ionisation chamber D.

gives us information about the forces of attraction between these particles, whilst the effect of an electric current on the molten material can tell us something about the nature of the constituent particles.

Each of these phenomena gives different information about the material, but together they provide a detailed picture of its structure and bonding.

Evidence from X-ray diffraction

Look through a piece of thin stretched cloth (possibly your handkerchief) at a bright light. The pattern you see is caused by the deflection of light as it passes through the regularly spaced threads of the fabric. This deflection of light is called **diffraction** and the patterns produced are **diffraction patterns**.

When the cloth is rotated in front of the light, the diffraction pattern rotates also. If the cloth is stretched so that the strands of the fabric get closer, then the pattern spreads out further. If the strands are arranged differently the pattern changes. From the diffraction pattern which we can observe, it is possible to deduce how the strands are arranged in the fabric. The same idea is used to determine the arrangement of particles in a crystal.

In 1912, the German scientist, von Laue, obtained a photograph of the diffraction pattern produced by passing X-rays through a crystal. X-rays have a much smaller wavelength than ordinary light rays. Using a similar technique W.L. Bragg was able to determine the simple cubic structure of sodium chloride. Nowadays, the structures of extremely complex substances such as proteins and nucleic acids are investigated using X-ray diffraction.

When a beam of X-rays falls on a crystal composed of regularly spaced atoms or ions the X-rays are reflected. In most instances, waves from the reflected X-rays interfere with and destroy each other. However, it is possible for the X-rays to be reflected by the particles so that their waves coincide and reinforce each other (figure 10.1). Under what conditions will reinforcement occur?

2 original waves   resultant wave

(a) Waves in phase, coincide and reinforce each other

Figure 10.1
(a) Reinforcement of waves in phase.
(b) Interference of waves out of phase.

(b) Waves out of phase, interfere and destroy each other

no signal

Figure 10.2
Right. The diffraction of two waves of X-rays by the regularly spaced particles in a crystal.

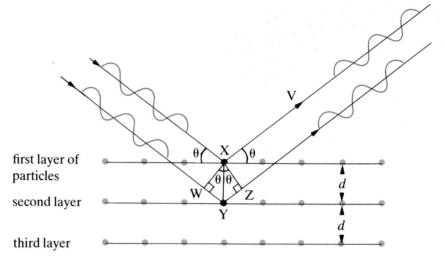

first layer of particles

second layer

third layer

Figure 10.2 shows a beam of X-rays being directed onto a crystal such as sodium chloride. Only three layers of the crystal are represented.

■ Figure 10.2 shows two waves in phase approaching the crystal. After reflection, the ray emerging along XV is ahead of the ray emerging along YZ. Why is this?
■ How much further does the ray reflected at the second layer travel?

If the waves are in phase after reflection, the difference in distance travelled by the two rays (i.e. the **path difference**) must equal a whole number of wavelengths, $n\lambda$.

$$
\begin{aligned}
\text{Path difference} &= WY + YZ \\
&= XY \sin\theta + XY \sin\theta \\
&= 2XY \sin\theta = 2d \sin\theta
\end{aligned}
$$

where d = distance between layers.

Thus, for reinforcement of the reflected X-rays

$$2d \sin\theta = n\lambda$$

Figure 10.3
Signals from a crystal of sodium chloride as the angle (θ) between the direction of the X-ray beam and the crystal face increases

This is known as the **Bragg equation**. For X-rays of a particular wavelength, λ, and for a given value of d, maximum reinforcement will occur only at certain values of θ. If θ is measured and λ is known, then the distance, d, between the layers in the crystal can be obtained.

By slowly increasing θ from 0° to 90°, an experimenter would obtain a series of reinforced, strong signals interspersed with regions of cancellation. Figure 10.3 shows some results for sodium chloride.

■ Why are there several values of θ at which reinforcement occurs?
■ How are the different values of θ explained?
■ What is the smallest value of θ for reinforcement?
■ What is the distance between layers in the sodium chloride crystal? (Assume the wavelength of the X-rays is 5.8×10^{-2} nm.)

The X-ray goniometer

Crystal structures can be determined by the methods described above using X-ray goniometers similar to the one shown in figure 10.5.

X-rays are directed through the narrow tube labelled X in figure 10.5 towards a single crystal . The crystal is mounted on a fine point on the goniometer head at the centre of the instrument. The goniometer head (figure 10.4) can be rotated in two arcs at right angles so that the crystal axis can be set in the correct orientation relative to the direction of the X-ray beam. The microscope at the bottom right-hand corner of the instrument in figure 10.5 is used for observing the crystal whilst setting its orientation on the goniometer head. In some instruments, X-ray film can be placed around the crystal in order to detect the diffracted X-rays. In figure 10.5, the diffracted X-rays are detected on a flat sensitive fibre optics screen. This would be placed immediately behind the

Figure 10.4
The goniometer head.

Figure 10.5
A goniometer.

crystal. The 'messages' received on the screen are intensified and converted for display on a TV monitor.

Figure 10.6 shows how a pattern of spots is produced during X-ray diffraction. As the crystal is rotated on the goniometer head, different planes of the crystal come into the correct angle for reinforcement. X-ray beams are then diffracted towards the photographic film where spots appear. If a flat film were used, the spots would occur along curves. Usually the film is bent into a cylinder with the crystal at its centre, so that the spots appear along straight lines. From the position of the spots and their relative intensities it is possible to obtain information about the structure of a crystal. Even non-crystalline materials such as raw cotton fibres (figure 10.7) produce diffraction patterns which can sometimes help in the elucidation of their structures.

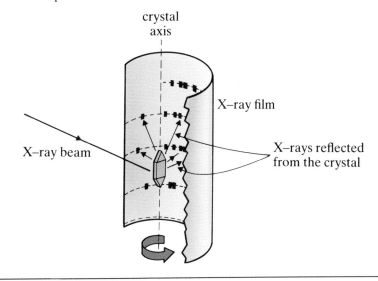

Figure 10.6
Right. X-rays reflected from a crystal produce a pattern of spots on X-ray film.

Figure 10.7
An X-ray diffraction photograph of raw cotton fibres.

10.3 Electron density maps

When X-rays strike a crystal, they are diffracted by the electrons in the atoms or ions. Consequently, the larger the atom, the more electrons it possesses and the brighter the spot will be on the diffraction pattern. By analysing both the positions and the intensities of the spots on a diffraction pattern, it is possible to determine the charge density of electrons within the crystal. The charge density is measured in terms of electrons per cubic nanometer (or electrons per cubic Ångström). Points of equal density in the crystal are joined by contours giving an electron density 'map' similar to the one in figure 10.8.

Look closely at figure 10.8.

■ How many atoms can be located from the electron density map?
■ Which atoms are they?
■ Why are the other atoms not evident from the electron density map?

Figure 10.8
(a) An electron density map of naphthalene (contours in electrons per nm^3). (b) The structural formula of naphthalene.

(a)   (b)

By mounting electron density maps on clear plastic or perspex it has been possible to assemble accurate three-dimensional structures of complex substances such as proteins and nucleic acids. It is then possible to see the positions of different atoms within them. There are many different arrangements in which atoms can be packed in repeating units to form crystals. A few of these arrangements which occur in natural crystals are considered in the following sections.

10.4 Giant metallic structures

Close-packed structures

The crystals of some substances can be imagined to consist of identical atoms packed together as close as possible. Most metals and many molecular substances have a close-packed structure. X-ray analysis shows that there are two possible close-packed arrangements: **hexagonal close-packed (h.c.p.)** and **cubic close-packed (c.c.p.)**.

In a close-packed layer, each atom is in contact with six others (figure 10.9). The atom marked 'a' has six other atoms (marked 'x') in the same layer in contact with it. Figure 10.9 also emphasises the hexagonal arrangement of atoms within each layer.

In the second layer, atoms pack as close as possible to those in the first layer by 'sitting' in the depressions between atoms in the first layer. Around each first-layer atom there are six depressions. These depressions are marked alternately 'b' and 'c' around the atom 'a'. If the depressions marked 'b' are used those marked 'c' cannot be. Figure 10.10 shows the second layer of atoms in place. Have the 'b' or 'c' depressions been used? Notice that three atoms in the second layer touch each atom in the first layer.

This means that any one atom touches 12 others in these close-packed arrangements—six in its own layer, three in the layer above and three in the layer below. This is summarised by saying that its **co-ordination number** is 12.

The third layer of atoms can now be added to figure 10.10 in two quite distinct ways. There are two types of depression available for atoms in the third layer. These two types of depression are not exactly the same. One type of depression, labelled 'a' in figure 10.10, lies directly above the centre of atom 'a' in the first layer. The other type of depression, denoted by 'c', lies directly over a hole in the first layer.

In figure 10.11, the darkest sphere is part of a third layer whose atoms are directly above the atoms of the first layer. In this case, the dark sphere occupies an 'a' type depression. Try to construct this arrangement using marbles, ping-pong balls or polystyrene spheres. Use books to hold the bottom layer in place. This type of arrangement in which alternate layers have a similar

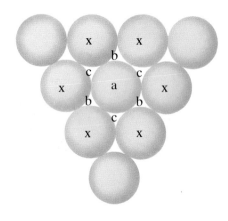

Figure 10.9
A layer of close-packed atoms.

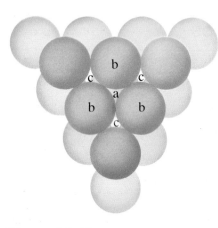

Figure 10.10
First and second layers of close-packed atoms.

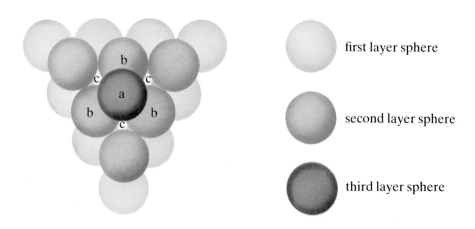

first layer sphere

second layer sphere

third layer sphere

Figure 10.11
First, second and third layers of close-packed atoms (ababab arrangement).

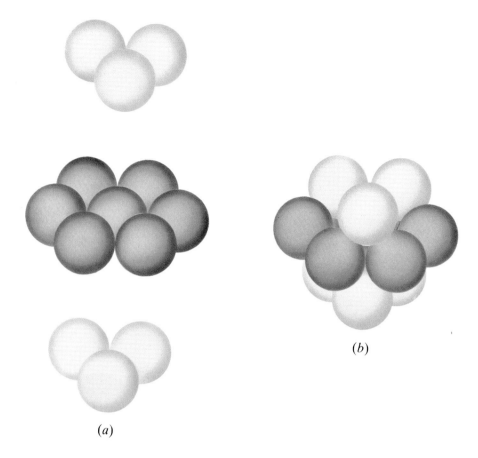

Figure 10.12
Hexagonal close-packed spheres:
(a) exploded view; (b) normal view.

positioning of atoms is denoted as 'ababab', etc., because alternate layers of atoms occupy either the 'a' depressions or the 'b' depressions. This arrangement is known as **hexagonal close-packing (h.c.p.)**.

The diagrams in figure 10.12 show the arrangement of atoms in the h.c.p. structure. (The spheres are shaded differently to show the packing of consecutive layers more clearly. There is, of course, no difference in the atoms in the different layers of a metal.) Zinc and magnesium are examples of metals with hexagonal close-packed structures.

An alternative close-packed arrangement occurs when atoms in the third layer occupy 'c' depressions, above neither the first layer nor the second layer atoms. This arrangement is shown in figure 10.13.

This type of arrangement in which the first layer of atoms occupy 'a' depressions, the second layer of atoms occupy 'b' depressions and the third layer of atoms occupy 'c' depressions can be repeated to give a sequence of layers denoted as 'abcabc', etc. This arrangement is known as **cubic close-packing (c.c.p.)**.

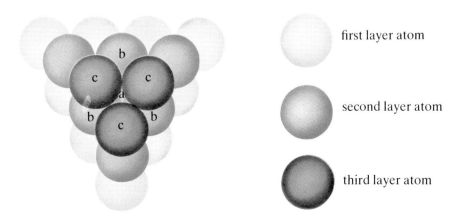

first layer atom

second layer atom

third layer atom

Figure 10.13
First, second and third layers of close-packed atoms (abcabc arrangement).

Figure 10.14
Cubic close-packed spheres:
(a) exploded view; (b) normal view;
(c) rotation of the normal view to show cubic symmetry more clearly.

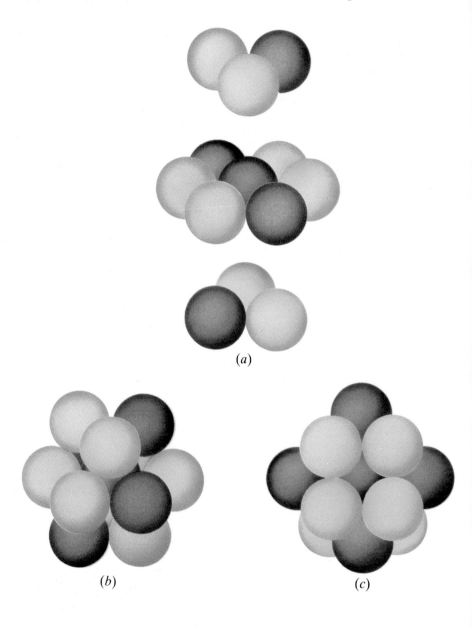

(a)

(b)   (c)

The atoms are arranged hexagonally within each layer, but this sequence of layers also has cubic symmetry. Hence, it is described as a *cubic* close-packed structure. The diagrams in figure 10.14 show the arrangment of three layers with respect to each other. Some of the spheres have been shaded so that the cubic symmetry is more obvious. Figure 10.14(c) which shows the cubic symmetry most clearly has been obtained by rotating figure 10.14 (b) through 60°.

Aluminium, copper, lead, silver, gold and platinum are examples of metals with a cubic close-packed structure.

The body-centred cubic structure

Some metals do not have their atoms arranged in a close-packed structure. These metals have a **body-centred cubic (b.c.c.)** structure.

This is shown in figure 10.15. As its name implies, this structure is basically cubic with an atom at the centre of each cube. In this case, each atom is surrounded by eight others. Thus, the co-ordination number is eight and the packing is more open than in the close-packed structures.

All the alkali metals, iron and manganese have a b.c.c. structure. There is no clear relationship between the structure of a metal and its position in the periodic table.

Figure 10.15
A model of the body-centred cubic structure of metal atoms.

Crystal grains on the surface of cast iron magnified 200 times. The black streaks show areas of carbon.

Figure 10.16
The arrangement of metal atoms (a) before, and (b) after 'slip'.

10.5 The properties of metals

If the surface of galvanised iron or unpolished tin plate is examined closely, it is possible to see small irregularly shaped areas clearly separated from each other by distinct boundaries. These irregularly shaped areas are known as **grains**. X-ray analysis shows that the metal atoms are packed regularly within the grains. Thus, these grains are small irregularly shaped crystals of the metal pushed tightly together and containing millions and millions of atoms in a giant structure.

The properties of a metal depend on its crystal structure (i.e. close-packed or body-centred cubic) and also on the size of its crystals (grains). Metals usually have high densities, high melting points, high boiling points and high molar heats of fusion and vaporisation. In addition, they are good conductors of heat and electricity, and they are malleable and ductile.

Look back at section 4.9.

- How do chemists visualise the strong forces between metal atoms?
- Why do most metals have a high density?
- Why do most metals have a high melting point and a high boiling point?
- Why are metals good conductors of electricity?

Chemists believe that the outermost electrons in metal atoms move about freely within the lattice. Thus, the metal consists of positive ions surrounded by a 'sea' of moving electrons. The outer electrons are said to be **delocalised** as they move from one place to another in the crystal, rather than staying in one locality. The negatively charged electrons attract *all* the positively charged ions and bind the nuclei together.

The relatively low affinity for electrons results in two other properties common to metals—low electronegativity and low ionisation energy.

There are no rigid, directed bonds in a metal, so that layers of atoms can slide over each other when a force is applied to it. This relative movement of layers in the metal lattice is called **'slip'**. After 'slipping' the atoms settle into new positions and the crystal structure is restored. Thus, a metal can be hammered into different shapes or drawn out into a wire, i.e. it is malleable and ductile. Figure 10.16 shows what happens when 'slip' occurs.

(a) before 'slip'   (b) after 'slip'

When a metal is placed under stress, it will behave elastically provided the stress is not too great. This is the way an engineer would like a metal to behave in machines and structures. However, a point is reached when the stress is so great that the metal behaves plastically and the changes in it are permanent and irreversible. This is 'slip'.

Can 'slip' be prevented so as to improve the elastic behaviour of the metal? Metallurgists can increase the strength of metals by inserting barriers in the metal lattice which prevent 'slip' occurring.

Two important methods of strengthening metals are:

(a) reducing the size of the crystal grains,
(b) alloying.

Slip does not readily take place across grain boundaries. Thus, metals with small grain size and more grain boundaries per unit volume are harder to deform, less malleable and less ductile than metals with larger grains. By careful choice of metallurgical techniques and casting processes the grain size can be reduced.

Metals will readily form alloys since the metallic bond is non-specific. The presence of small quantities of a second element in the metal frequently increases its strength. For example, brass—an alloy of zinc and copper—is much

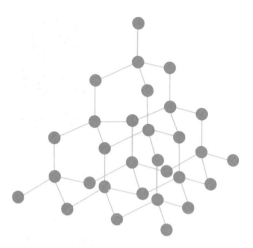

Figure 10.17
The structure of diamond.

stronger than either pure copper or pure zinc. Atoms of the second metal are different in size to those of the original metal. These differently sized atoms interrupt the orderly arrangement of atoms in the lattice and prevent them sliding over each other.

10.6 Giant molecular structures

Elements, such as carbon and silicon, which have medium electronegativity values and the ability to form four covalent bonds form giant molecular structures of covalently bonded atoms.

The structure of diamond

In diamond, every carbon atom can be imagined to be at the centre of a regular tetrahedron surrounded by four other carbon atoms whose centres are at the corners of the tetrahedron (figure 10.17). Every carbon atom forms four covalent bonds by sharing electrons with each of its four nearest neighbours.

■ What is the co-ordination number of carbon atoms in diamond?
■ Will the carbon atoms on the outside of the diamond form four covalent bonds?

Silicon (section 4.9) and silicon carbide (SiC, used in the abrasive 'carborundum') exist in a similar crystal structure to diamond. Silicon (IV) oxide (SiO_2) is another example of a compound with a giant molecular structure. The simplest way to appreciate the structure of SiO_2 is to picture the structure of pure silicon and imagine that each Si—Si bond is bridged by an oxygen atom. In each of these structures, atoms are linked by localised electrons by an oxygen atom. In each of these structures, atoms are linked by localised electrons in strong covalent bonds throughout the whole three-dimensional lattice.

It is therefore very difficult to distort a covalently bonded crystal, because this would involve breaking covalent bonds. Consequently, diamond, silicon carbide and silicon (IV) oxide are hard and brittle with very high melting points and very high boiling points. Furthermore, the localisation of the electrons within covalent bonds prevents their moving freely in an applied electric field and thus these materials do not conduct electricity.

A glass engraver using a diamond-tipped 'pencil' to produce a design on a glass goblet.

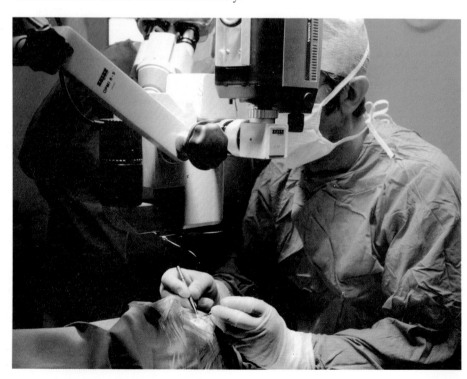

A special diamond-tipped cutting tool being used for eye surgery.

Above. Natural diamonds set in the surface of drill bits.

Above right. A workman uses a drill tipped with diamonds to extract granite samples from the Eleanor Cross monument at Charing Cross, London.

Almost all the industrial uses of diamond and silicon carbide depend on their hardness. Diamond is one of the hardest natural substances. Diamonds, unsuitable for gemstones, are used in glass cutters and in diamond-studded saws. Powdered diamond and carborundum (SiC) are used as abrasives for smoothing very hard materials.

The structure of graphite

In graphite, the carbon atoms are arranged in flat, parallel layers. Each layer contains millions of hexagonally arranged carbon atoms (figure 10.18). Each carbon atom is covalently bonded to three other atoms in its layer. Each layer can be viewed as a two-dimensional sheet polymer or layer lattice. The carbon–carbon bond length within a layer is 0.142 nm. This suggests some multiple bond character, because the value is intermediate in length between a single and a double carbon–carbon bond. The strong covalent bonds within the layers account for the very high melting point (3 730°C) of graphite. Owing to its high melting point, graphite is used to make crucibles for molten metals. A special form of heat-resistant graphite, pyrographite, is used for the exhaust cones of rockets.

The distance between the layers in graphite is 0.335 nm which is much longer than a single carbon–carbon bond. Thus, the bonding between the layers is restricted to relatively weak forces and the layers can slide over each other easily.

This accounts for the softness of graphite and its use as a lubricant.

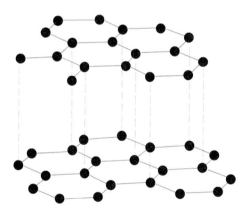

Figure 10.18
The structure of graphite.

- Why is it better to lubricate a car lock or a zip-fastener with a pencil rather than with oil?
- Why is graphite sometimes more suitable than oil for lubricating the moving parts of machinery?
- How do you think the proportion of graphite to clay affects the hardness of a pencil?

The bonding in graphite can be pictured as three trigonally arranged covalent bonds. These are formed by three of the four valence electrons of carbon, whilst the fourth electron is delocalised over the whole layer. This delocalisation of electrons similar to that in metals results in graphite conducting electricity and in its shiny appearance. The electrical conductivity of graphite enables it to be used as electrodes in electric furnaces and during electrolyses.

10.7 Giant ionic structures

Ionic structures are formed when atoms with large differences in electronegativity form compounds. Electrons are transferred from atoms of low electronegativity to those of high electronegativity. The oppositely charged ions which result are held together by strong electrostatic forces of attraction. The electrical force binding the ions together is described as an ionic or electrovalent bond.

The sodium chloride lattice

X-ray analysis shows that the particles in different ionic structures can be arranged in different patterns. One of the simplest structures for ionic compounds is the cubic arrangement of ions as in sodium chloride. Figure 10.19 shows a space-filling model of the structure of sodium chloride. The sodium and chloride ions are shown as solid spheres. The sizes of the ions are in the correct ratio. Notice that the ions are arranged in a cubic pattern. Although figure 10.19 shows only a few Na^+ and Cl^- ions, there will be millions and millions of ions in even the smallest visible crystal of sodium chloride.

The positions of Na^+ and Cl^- ions in the cubic lattice of sodium chloride are emphasised in figure 10.20. The small dots and circles in figure 10.20 represent the centres of Na^+ and Cl^- ions, respectively. The sodium lines in the diagram show the cubic geometry of the lattice.

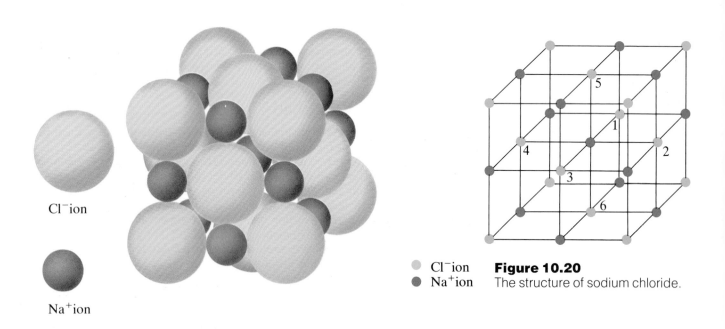

Cl⁻ion

Na⁺ion

○ Cl⁻ion
● Na⁺ion

Figure 10.20
The structure of sodium chloride.

Figure 10.19
A space-filling model of the structure of sodium chloride.

Notice in figure 10.20 that each positive sodium ion is surrounded by six Cl^- ions and each negative chloride ion is surrounded by six Na^+ ions. The six Cl^- ions around the central Na^+ ion in figure 10.20 have been numbered. Four of the Cl^- ions (numbered 1, 2, 3 and 4) are in the same horizontal layer of the crystal as the central Na^+ ion. One Cl^- ion (number 5) is in the layer above. The final Cl^- ion (number 6) is in the layer below. The structure of sodium chloride is said to have 6:6 co-ordination because the Na^+ ions have a co-ordination number of 6 and the Cl^- ions also have a co-ordination number of 6.

Measuring the size of ions

X-ray measurements on ionic solids can be presented in the form of electron density maps in order to determine the size of different ions. Figure 10.21 shows an electron density map for sodium chloride. The circular contours suggest that

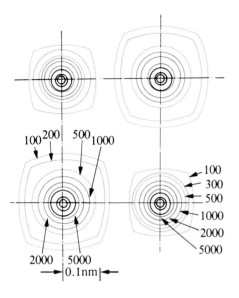

100 200 500 1000

100
300
500
1000
2000
5000

2000 5000
0.1nm

Figure 10.21
An electron density map for sodium
chloride. (Electron densities are
expressed as electrons per cubic
nanometre.)

the electron distribution in these ions is spherical. The spacing of the contour
lines enables us to distinguish particles with different numbers of electrons.

Look closely at figure 10.21.

■ Which are the sodium ions?
■ What is the interionic distance between neighbouring Na^+ and Cl^- ions?
■ What is the ionic radius of: (i) Na^+,   (ii) Cl^-?

Using electron density maps of this kind, it is possible to compile tables of ionic
radii. But remember, the size of a particular ion can vary slightly depending on
the size and charge of other ions in the crystal. So, values of ionic radii for the
same ion do not always agree. The trends in the size of ionic radii in relation to
the periodic table are discussed in section 13.3.

The caesium chloride lattice

Another relatively simple structure for ionic compounds is that shown by
caesium chloride. Figure 10.22 shows a space filling model of the structure of
caesium chloride. The caesium and chloride ions are shown as solid spheres. As
in figure 10.19, the sizes of the ions are in the correct ratio. Unlike Na^+, the Cs^+
ion is similar in size to the Cl^- ion. The general shape of the lattice is again
cubic, but it is different in detail from the sodium chloride lattice.

The positions of Cs^+ and Cl^- ions in the caesium chloride lattice are shown in
figure 10.23. The dots and circles in figure 10.23 represent the centres of Cs^+ and
Cl^- ions, respectively. The solid lines in the diagram show the overall cubic
shape of the crystal.

Cs$^+$ion

Cl$^-$ion

Figure 10.22
A space-filling model of the structure
of caesium chloride.

Cs$^+$ion   Cl$^-$ion

Figure 10.23
The structure of caesium chloride.

The dashed diagonal lines in figure 10.23 emphasise the arrangement of Cl^-
ions relative to Cs^+ ions and vice versa. This shows a single cube of Cl^- ions with
a Cs^+ ion at its centre. There are eight Cl^- ions around each Cs^+ ion, so the
co-ordination number of Cs^+ ions in the structure is eight. In the same way, the
co-ordination number of Cl^- ions is also eight. In fact, the lattice could be
redrawn with Cl^- ions exchanged for Cs^+ ions. The structure of caesium
chloride is therefore said to have 8:8 co-ordination.

10.8 The properties of ionic crystals

Crystals of ionic solids are:

1 Hard and brittle.

2 Involatile with high melting points and high boiling points.

3 Good conductors of electricity when molten, but non-conductors when solid.

4 Soluble in polar solvents such as water, but insoluble in non-polar solvents such as tetrachloromethane.

In an ionic solid, every ion is held in the crystal by strong attractions from the oppositely charged ions around it. Consequently, ionic solids like sodium chloride are hard and difficult to cut. However, they are also very brittle and may be split cleanly (**cleaved**) using a sharp-edged razor. When the crystal is tapped sharply along a particular plane it is possible to displace one layer of ions relative to the next. As a result of this displacement, ions of similar charge come together. Repulsion then occurs, forcing apart the two portions of the crystal (figure 10.24).

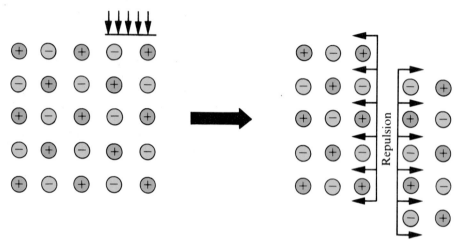

Figure 10.24
Crystal cleavage in ionic compounds.

(a) Arrangement of one layer of ions before displacement

(b) Arrangement of ions after displacement

The forces of attraction between oppositely charged ions in an ionic lattice are very strong. Large quantities of heat energy must be supplied to the crystal before the ions vibrate vigorously enough to overcome the forces of attraction between one another. The ions then move away from their relative positions. Thus, ionic solids have high melting points and high molar heats of fusion. Even in the molten ionic liquid there will still be strong forces between the oppositely charged ions. Consequently, ionic solids have high boiling points and high molar heats of vaporisation.

■ Why is it easier to cleave ionic solids in some directions rather than in others?
■ Why do molten ionic compounds conduct electricity?
■ Why are solid ionic compounds non-conductors?
■ Why are ionic compounds such as sodium chloride soluble in water, but insoluble in tetrachloromethane?

What happens when an ionic solid dissolves in water?

When sodium chloride dissolves in water, the crystal lattice is broken up forming separate Na^+ and Cl^- ions in aqueous solution. Where does the energy required to separate the oppositely charged ions come from? We have seen already that water contains highly polar molecules. The positive ends of polar water molecules are attracted to negative ions in the crystal and negative ends of the water molecules are attracted to positive ions. The formation of ion–solvent bonds results in a release of energy. This is sufficient to cause the detachment of ions from the lattice (figure 10.25). Thus, ionic crystals will often dissolve in polar solvents such as water, ethanol and propanone (acetone).

The attachment of polar solvent molecules to ions is known generally as **solvation**. Chemists say that the ions are **solvated**. Very often the solvent is water and in this specific case the ions are said to be **hydrated**. Figure 10.26 shows the structures of hydrated positive and negative ions. Frequently the attraction of

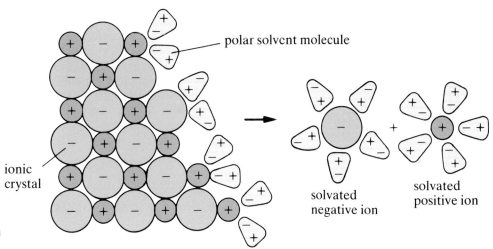

polar solvent molecule

ionic crystal

solvated negative ion

solvated positive ion

Figure 10.25
Polar solvent molecules dissolving an ionic solid.

the central ion extends beyond the first layer of polar solvent molecules to a whole sheath of solvent molecules. The sheath may be several layers thick. This envelope of attracted solvent molecules reduces the movement of solvated ions. The solvated ions diffuse much slower than expected and move less easily towards electrodes during electrolysis.

Non-polar liquids such as tetrachloromethane, benzene and hexane will not solvate ionic solids. The reasons for this are not difficult to see.

Non-polar molecules are held together by weak intermolecular (Van der Waals') forces. These are much smaller in magnitude than the forces between ions in an ionic crystal. The ion–ion attractions are in fact much stronger than either the solvent–solvent interactions or the ion–solvent interactions. So the non-polar solvent molecules cannot penetrate the ionic lattice. Thus, sodium chloride is virtually insoluble in tetrachloromethane and benzene.

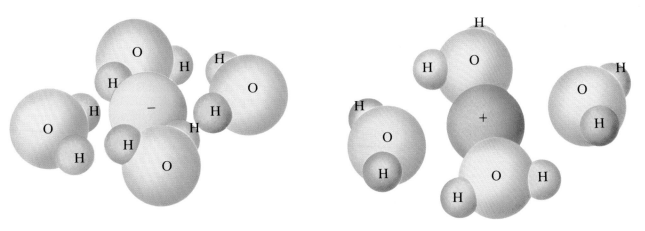

Figure 10.26
The structures of hydrated positive and negative ions. Only those water molecules in the horizontal plane are shown.

10.9 Determining the Avogadro constant from X-ray studies

One of the most valuable results of X-ray diffraction studies has been the accurate determination of the Avogadro constant, L. Using this method, L can be found to an accuracy of one in ten thousand.

1 The spacing of particles in a crystal is first determined.

2 Knowing the distance between atoms (or ions) in the crystal, it is then possible to find the volume occupied by one atom.

3 Next the volume of one mole of the substance is determined.

4 Finally the volume of one mole is divided by the volume of one atom to obtain the Avogadro constant.

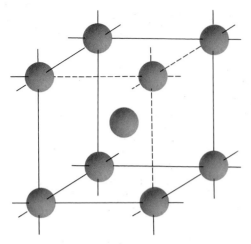

Figure 10.27
A unit cell of sodium.

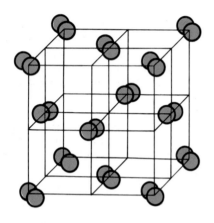

indicates the centre of an I_2 molecule

Figure 10.28
The crystal structure of iodine.

Figure 10.29
The distance between iodine atoms in a molecule and the distance between iodine atoms in adjacent molecules.

Consider the following example. Figure 10.27 shows a **unit cell** of sodium metal. This has a body-centred cubic structure. The unit cell is the simplest arrangement of atoms which when repeated will reproduce the whole structure. The central atom in figure 10.27 is wholly inside the unit cell. The eight atoms at the corners are shared equally between eight unit cells. Thus, the unit cell contains a total of $1 + 8 \times \frac{1}{8} (= 2)$ atoms. X-ray diffraction measurements show that the width of the unit cell is 0.429 nm (i.e. 0.429×10^{-7} cm).

Thus, the volume of the unit cell (i.e. two atoms) $= (0.429 \times 10^{-7})^3 \, cm^3$
$= 0.0790 \times 10^{-21} \, cm^3$

Therefore, the volume occupied by one sodium atom $= 0.0395 \times 10^{-21} \, cm^3$

The relative atomic mass of sodium $= 22.99$

The density of sodium $= 0.97 \, g \, cm^{-3}$

Using the equation, $volume = \dfrac{mass}{density}$,

the volume of one mole (L atoms) of sodium $= \dfrac{22.99}{0.97} \, cm^3$

$= 23.70 \, cm^3$

The Avogadro constant, $L = \dfrac{volume \ of \ one \ mole \ of \ atoms}{volume \ of \ one \ atom}$

$\therefore$ The Avogadro constant, $L = \dfrac{23.70}{0.0395 \times 10^{-21}} = 6.0 \times 10^{23} \, mole^{-1}$.

10.10 Simple molecular structures

Non-metal elements and compounds of non-metals such as iodine, carbon dioxide, water and naphthalene ($C_{10}H_8$) are usually composed of simple molecules. In these molecular substances, the atoms are joined together *within* the molecule by strong covalent bonds. But the separate molecules are attracted to each other by much weaker Van der Waals' bonds.

X-ray diffraction measurements on crystals of molecular substances have been used to determine both the distance between atoms within each molecule and the distance between separate molecules. Figure 10.28 shows the arrangement of I_2 molecules in solid iodine. The arrangement of molecules in the crystal lattice is described as **face-centred cubic**—the molecules are arranged in a *cube* with a molecule at each corner and a molecule at the *centre* of each *face*. Notice that the relative positions of molecules with 12 nearest neighbours in the face-centred cubic structure is the same as the relative positions of metal atoms in the cubic close-packed structure.

Analysis of the X-ray diffraction pattern for solid iodine suggests that the distance between two iodine atoms in a molecule is 0.256 nm. Thus, the I—I covalent bond length is 0.256 nm. The distance between the centres of iodine atoms in adjacent molecules is 0.43 nm, which gives a value of 0.215 nm for the Van der Waals' radius for iodine (figure 10.29).

The packing of molecules in crystals is often more complicated than the packing of simple atoms or ions. Even so, each substance is found to have a characteristic and uniform lattice arrangement.

The properties of simple molecular substances

Molecular substances such as iodine, naphthalene ($C_{10}H_8$), sucrose (sugar, $C_{12}H_{22}O_{11}$) and ice

1 are usually soft,

2 have low melting points and boiling points,

3 do not conduct electricity when molten or aqueous.

Non-polar molecular compounds such as iodine and naphthalene are almost insoluble in water, though they are usually very soluble in non-polar solvents such as benzene and tetrachloromethane.

■ How does the bonding in ice and sugar differ from that in non-polar simple molecules like naphthalene and iodine?
■ Why are ice and sugar harder than iodine and naphthalene?
■ Why is sugar soluble in water unlike iodine and naphthalene?

The properties of molecular compounds can be explained in terms of their structure which consists of simple molecules. Although the atoms within these molecules are joined by strong covalent forces, the separate molecules are held together by very weak Van der Waals' forces. These weak intermolecular bonds permit the molecules to be separated easily. Hence the crystals of simple molecular substances are usually soft with low melting points and low boiling points.

Molecular compounds contain neither delocalised electrons (like metals) nor ions (like ionic compounds). Thus, they cannot conduct electricity when molten or when dissolved in water.

Why are non-polar substances such as iodine insoluble in water, but soluble in benzene?

In liquids of high polarity such as water, there are strong water–water attractions. These are considerably stronger than either iodine–iodine attractions or iodine–water attractions. Consequently, iodine molecules cannot penetrate the water structure and there is little tendency for water molecules to solvate uncharged iodine molecules. Iodine is therefore almost insoluble in water.

In non-polar liquids, such as benzene and tetrachloromethane, there are weak intermolecular forces. The benzene–benzene attractions are similar in strength to iodine–benzene and iodine–iodine attractions. Thus, it is easy for benzene molecules to penetrate into the iodine crystal and solvate the iodine molecules. Consequently, iodine dissolves easily in benzene.

10.11   Comparing typical solid structures

The structure, bonding and properties of the four common solid structures (giant metallic, giant molecular, giant ionic and simple molecular) are summarised and compared in table 10.1 overleaf.

Use table 10.1 to consider each of the following in the solid state: Cu, Si, NH_3, NaI, Xe.

Which solid would

■ be a good electrical conductor?
■ be a poor electrical conductor, but conduct on melting?
■ be hard and brittle and insoluble in water?
■ be easy to cleave?
■ have strong hydrogen bonds between molecules?
■ have the lowest melting point?

Summary

1   When X-rays are directed at a crystalline solid they are diffracted by particles in the lattice. The diffracted X-rays can be detected using an X-ray film. The diffraction pattern which is obtained can be used to determine the arrangement of particles in the crystal.

Table 10.1
Comparing typical solid structures

		Giant metallic	Giant molecular (Giant covalent)	Giant ionic	Simple molecular
1.	**Structure**				
	(i) Examples	Na, Fe, Cu	Diamond, SiC, SiO_2	Na^+Cl^-, $Ca^{2+}O^{2-}$, $(K^+)_2SO_4^{2-}$	I_2, S_8, $C_{10}H_8$, HCl, CH_4
	(ii) Constituent particles	Atoms	Atoms	Ions	Molecules
	(iii) Type of substance	Metal element with low electro-negativity	Non-metal element in Group IV or its compound	Metal/non-metal compound (a compound of elements with a large difference in electronegativity)	Non-metal element or non-metal/non-metal compound (elements with high electronegativity)
2.	**Bonding** in the solid	Attraction of outer mobile electrons for positive nuclei binds atoms together by *strong* metallic bonds	Atoms are linked through the whole structure by *very strong* covalent bonds from one atom to the next	Attraction of positive ions for negative ions—*strong* ionic bonds	Strong covalent bonds hold atoms together within the separate molecules; separate molecules are held together by *weak* Van der Waals' bonds
3.	**Properties**				
	(i) Volatility	Non-volatile High m.pt., high b.pt., high latent heats	Non-volatile Very high m.pt., very high b.pt., very high latent heats	Non-volatile High m.pt., high b.pt., high latent heats	Volatile Low m.pt., low b.pt., low latent heats
	State at room temp.	Usually solid	Solid	Solid	Usually gases or volatile liquids
	(ii) Hardness/malleability	Hard, yet malleable	Very hard and brittle	Hard and brittle	Soft
	(iii) Conductivity	Good conductors when solid or liquid	Non-conductors (graphite is an exception)	Non-conductors when solid. Good conductors when molten or in aqueous solution—*electrolytes*	Non-conductors when solid, liquid and in aqueous solution. (A few (e.g. HCl) react with water to form electrolytes.)
	(iv) Solubility	Insoluble in polar and non-polar solvents, but soluble in liquid metals	Insoluble in all solvents	Soluble in polar solvents (e.g. H_2O), insoluble in non-polar solvents (e.g. CCl_4)	Insoluble in polar solvents (e.g. H_2O), soluble in non-polar solvents (e.g. CCl_4)

2 For reinforcement of reflected X-rays,

$$2d \sin \theta = n\lambda$$

This is known as the Bragg equation (d = distance between layers in the crystal, θ = angle between direction of X-rays and the crystal face, n = 1, 2, 3, etc., λ = wavelength of X-rays).

3 The structure and bonding of a substance determine its properties. Table 10.1 summarises the structure, bonding and properties of the four typical solid structures (giant metallic, giant molecular, giant ionic and simple molecular).

4 The co-ordination number of an atom or ion is the number of its nearest neighbours.

5 Metals form either close-packed structures (co-ordination number = 12) or body-centred structures (co-ordination number = 8).

6   Delocalised electrons are not held tightly between atoms in directed bonds. They can move freely from one atom to another.

7   In diamond, each carbon atom is joined by four tetrahedrally spaced covalent bonds to four other atoms (co-ordination number = 4).

8   In graphite, carbon atoms are arranged hexagonally in flat, parallel layers Graphite is described as a two-dimensional sheet polymer or layer lattice.

9   Sodium chloride forms a cubic structure in which the co-ordination numbers of both Na^+ and Cl^- ions are six.
   Caesium chloride also has a cubic arrangement of ions, but in this case the co-ordination numbers of both Cs^+ and Cl^- ions are eight.

10 The attachment of solvent molecules to the particles of a solute during the dissolving process is known as solvation. Very often the solvent is water and in this particular case the process is called hydration.

Study questions

1 (a) Draw a three-dimensional model of the sodium chloride crystal.
   (b) Name the constituent particles and explain their positions relative to each other.
   (c) What holds the crystal together?
   (d) Why does sodium chloride have a high melting point and a high boiling point?
   (e) Why is sodium chloride soluble in water?
   (f) Explain why sodium chloride will not conduct in the solid state, but is a good conductor when molten.

2 (a) What do you understand by the terms:
       (i) atom, (ii) molecule, (iii) ion?
   (b) Use one or more of these terms to describe the following structures.
       (i) copper, (ii) solid carbon dioxide, (iii) graphite.
   (c) How are the properties of copper and graphite related to their structure and bonding?

3 Explain in terms of the structure and bonding in diamond and graphite why:
   (a) diamond is more dense than graphite,
   (b) diamond is an insulator, whereas graphite is a conductor.
   (c) diamond has a larger heat of combustion than graphite,
   (d) diamond is used as an abrasive, whereas graphite is used as a lubricant.

4 Consider the following five types of crystalline solids:
   A Metallic,
   B Ionic,
   C Giant molecular (macromolecular),
   D Composed of monatomic molecules, and
   E Composed of molecules containing a small number of atoms.
   Select the letter (A–E) for the structure most likely to show the following properties.
   (a) An element which conducts electricity and boils at 1600°C to form gaseous monatomic atoms.
   (b) A solid which melts at −250°C.
   (c) A solid with a very high molar heat of vaporisation which does not conduct when liquid.
   (d) A hard, brittle solid which easily cleaves.
   (e) A substance which boils at −50°C and decomposes at high temperatures.

5   Consider the following five substances in the solid state:
A   sodium
B   silicon
C   tetrachloromethane
D   argon
E   potassium bromide

Select the letter (A–E) for the substance most likely to show the structure or property described in each of the following questions.
(a) A monatomic substance held together by Van der Waals' forces.
(b) A substance of low melting point composed of polyatomic molecules.
(c) A network solid of covalently bonded atoms.
(d) A non-conducting solid which melts to form a liquid which conducts electricity.
(e) A substance which exists as a liquid over a temperature range of only 3 K.
(f) A substance which is decomposed by an electric current in the liquid state.

6   Diamond is one of the hardest natural substances. Before the production of carborundum, powdered diamond was the most widely used abrasive. Carborundum (silicon carbide) is made by heating coke with sand (silicon(IV) oxide) at 2500°C.
(a) What is an abrasive?
(b) What are the main uses for abrasives?
(c) Write an equation for the manufacture of carborundum from coke and sand.
(d) Why do you think carborundum has superseded diamond as the most widely used abrasive?
(e) 'Diamonds are a girl's best friend'! Why? Carborundum has not superseded diamonds in this context. Why not?
(f) It has been said that the discovery of carborundum enabled the industrial revolution to occur swiftly during the late nineteenth and early twentieth centuries. Why was this?

7   Yellow phosphorus contains simple molecules of formula P_4 whereas red phosphorus forms a giant molecular layer lattice.
(a) Draw diagrams to illustrate the structures of yellow and red phosphorus.
(b) Explain the bonding in these structures by drawing simple electronic (dot/cross) diagrams.
(c) State two differences in properties between these allotropes.
(d) How are these differences in properties between yellow and red phosphorus explained by their different structures?

8   Solid iodine has a dark purple lustrous appearance. In addition, solid iodine displays a small electrical conductivity. When liquid iodine chloride (ICl) is electrolysed iodine is liberated at the cathode.
Discuss and explain each of these observations, explaining whether they are consistent with the expected structure of the substances involved and the position of iodine in the periodic table.

9   Discuss and explain the following:
(a) The ionic nature of $MgCl_2$ is greater than that of $AlCl_3$, which in turn is greater than that of $SiCl_4$.
(b) Silicon(IV) oxide is a solid at room temperature which does not melt until 1973 K, whereas carbon dioxide (m.pt. 217 K) is a gas at room temperature.
(c) Both calcium oxide and sodium chloride have a simple cubic structure and similar interionic distances yet calcium oxide melts at 2973 K whereas sodium chloride melts at 1074 K.
(d) Glucose ($C_6H_{12}O_6$) is much more soluble in water than in benzene, but cyclohexane (C_6H_{12}) is much more soluble in benzene than in water.

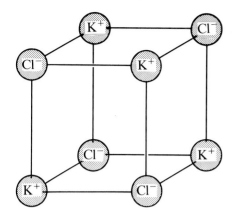

Figure 10.30
A unit cube of potassium chloride.

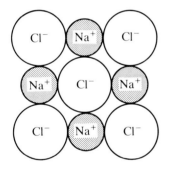

Figure 10.31
The arrangement of ions in sodium chloride.

Table 10.2
The ionic radii of some ions

Ion	Ionic radii/nm	Ion	Ionic radii/nm
F⁻	0.133	Li⁺	0.068
Cl⁻	0.181	Na⁺	0.098
Br⁻	0.196	K⁺	0.133
I⁻	0.219	Rb⁺	0.148
		Cs⁺	0.167

10 When X-rays of wavelength 0.1537 nm are directed towards a crystal of KCl, reinforcement of the X-rays first occurs when the angle between the direction of the rays and the crystal face is 14°.
   (a) Using the Bragg equation, calculate the distance between layers of ions in the crystal.
   (b) At what angle would the second reinforcement of X-rays occur with the crystal in the same orientation?
   (c) Figure 10.30 shows a unit cube for potassium chloride.
       (i)   Use your result in part (a) to calculate the volume of the unit cube.
       (ii) How many $K^+ Cl^-$ 'ion pairs' does the unit cube contain? (Remember that ions at the corners of a cube are shared equally amongst eight cubes.)
       (iii) What is the volume occupied by one $K^+ Cl^-$ 'ion pair'?
   (d) The density of KCl is $2.0\,g\,cm^{-3}$ and its relative formula mass is 74.6. Using these values, calculate the total volume of one mole of K^+Cl^-.
   (e) Using the volume of one mole of K^+Cl^- 'ion pairs' (obtained in part (d)) and the volume of one K^+Cl^- 'ion pair' (obtained in part (c)(iii)), calculate a value for the Avogadro constant, L.

11 (a) What is meant by the term co-ordination number?
   (b) Name two metals with a close-packed structure and two metals with a body-centred cubic structure.
   (c) What is the co-ordination number of atoms.
       (i)   in a close-packed structure,
       (ii) in a body-centred cubic structure?
   (d) Suppose one particular metal can have either a close-packed or a body-centred cubic structure. In which of these two forms would you expect it to have:
       (i)   the higher density,
       (ii) the higher melting point,
       (iii) the higher electrical conductivity,
       (iv) the greater malleability?
       Give reasons for your choice in each case.

12 The following data apply to the compounds XCl_x and YCl_y.

	Melting point /°C	Boiling point /°C	Solubility in water /g per 100g	Solubility in benzene /g per 100g
XCl_x	801	1 443	37	0.063
YCl_y	−22.6	76.8	0.08	miscible with benzene in all proportions.

   (a) What types of bond(s) are present in these two chlorides?
   (b) Explain clearly how the bonding in each chloride leads to the great differences in volatility and solubility shown in the data above.

13 Figure 10.31 shows the arrangement of ions in sodium chloride in which the smaller shaded Na^+ ions are in contact with the larger Cl^- ions.
   (a) Draw a diagram for an alkali metal halide in which the metal ions are in contact with larger halide ions and the halide ions also touch each other.
   (b) Show that the ratio between the radii of metal ions and halide ions in the structure you have drawn in part (a) is $\sqrt{2} - 1$.
   (c) Use the values in table 10.2 to find the alkali metal halide in which the situation described in part (a) is most likely to occur.

14 Aluminium crystallises with a cubic close-packed structure. The length of one side of its unit cell is 0.4041 nm.
   (a) Draw a diagram of the unit cell for aluminium.
   (b) How many Al atoms are there in the unit cell?
   (c) Calculate the density of aluminium (Al = 27.0).
   (d) Suggest a reason for the difference between your calculated value for the density of aluminium and the accepted value given in table 4.3.

11 The Gaseous State

11.1 Evidence for moving particles

The Greeks were probably the first people to believe that matter was made up from particles. In 60 B.C., Lucretius suggested that matter existed in the form of invisible particles. He wrote about this idea in his poem, *The Nature of the Universe*.

Unfortunately, the Greeks rarely performed experiments. Their theories remained nothing more than good ideas. They had, however, a vast range of everyday experience which supported their beliefs that matter was particulate. They knew, for instance, that a small amount of flavouring such as pepper or ginger could give a whole dish a strong, distinctive taste. This suggested that tiny particles in the pepper could spread throughout the stew to give it a particular flavour.

When dyes such as Tyrian purple were dissolved in water, a tiny amount of the dye could colour an enormous volume of solution. This supported the idea that there must be many particles of the purple pigment in only a little solid and that the particles must be very small.

Use the idea of particles to explain the following:

■ It is possible to smell the perfume a person is wearing from some distance.
■ Solid blocks of air freshener provide a pleasant smell in bathrooms. They disappear after a time without leaving any liquid.

Particles of vapour from the perfume and the air freshener which have a distinctive smell *mix* with air particles and *move* to other parts of the room. This mixing and movement of matter which results from the kinetic energy of moving particles is called **diffusion**.

The idea that particles in gases and liquids are moving has been confirmed by many other experiments involving diffusion.

One other phenomenon which provides strong evidence for moving particles is **Brownian Motion**.

During the 1820s, the botanist, Robert Brown, was carrying out a study of pollen grains. At first, Brown believed that he would be able to observe the pollen grains more effectively through his microscope if they were suspended in water. To his annoyance he found that the pollen continually jittered around in the water in a random manner. As a result of Brown's observation, the random movement of solid particles suspended in a liquid or in a gas is called Brownian Motion.

■ What particles are present in water? How are these particles moving?
■ What caused the pollen grains to jitter about in Brown's experiment? Why do they move so randomly and haphazardly?

Margaret Cavendish, Duchess of Newcastle. During the seventeenth century, Margaret Cavendish tried to interest people in science. However, she annoyed many scientists such as Robert Boyle because of her views about particles. Margaret Cavendish suggested that all atoms had the same mass, but differed in size and shape depending on the amount of earth, air, fire and water in them. She said atoms of earth were square, water atoms were round, air atoms were long and fire atoms were sharp.

11.2 Examining the Brownian Motion of smoke particles in air

Brown's observations of pollen grains was the first recorded example of Brownian motion. It is possible to watch Brownian Motion more conveniently using smoke particles in air. Figure 11.1 shows a suitable arrangement which you could use for this experiment.

Figure 11.1
An arrangement for examining the Brownian motion of smoke particles in air.

Use a teat pipette to inject smoke from a smouldering piece of string into the small glass smoke cell. Close the cell with a cover slip. When the illuminated cell is viewed through the microscope the smoke particles look like tiny jittering pin points of light. Why is this?

The smoke particles which you see through the microscope are under constant bombardment from even tinier and invisible molecules in the air. Each individual smoke particle is bombarded first on one side and then on another side. So, it moves first this way and then that way in a random jittery motion. The movement of the smoke particles which you can see provides strong evidence for the movement of air molecules which, of course, you cannot see. What do you think will happen if the heat from the small illuminating lamp causes the temperature of the air and smoke in the cell to rise significantly?

11.3 The Kinetic Theory of Matter

Our ideas about the movement of particles in solids, liquids and gases are summarised in the **Kinetic Theory of Matter**. The word kinetic is derived from the Greek word *kineo* which means 'I move'. The main points in the Kinetic Theory can be summarised as follows:

1   All matter is composed of tiny, invisible particles.

2   Particles of different substances are different in size.

3   In solids, the particles are relatively close together. They have smaller amounts of energy than the same particles in the liquid and gaseous states at higher temperatures. Consequently, solid particles cannot overcome the strong forces of attraction holding them together. They can only vibrate about fixed positions in the solid crystal. So, particles in a **solid** have vibrational and rotational motion, but no translational motion (figure 11.2).

4   In liquids, the particles are slightly further apart than in solids and possess larger amounts of energy. Thus, they are able to overcome the forces between

In translational motion, a particle moves from one place to another

In vibrational motion, a particle vibrates about the same point.

In rotational motion, a particle spins like a top.

Figure 11.2
The translational, vibrational and rotational motion of a particle.

each other to some extent. They can move freely around each other whilst in close proximity. The liquid particles have vibrational, rotational and translational motion.

5   In gases, the particles are much widely separated than those in solids and liquids and have much larger amounts of energy. The gas particles have sufficient energy to overcome the forces of attraction between each other almost completely. They move rapidly, randomly and haphazardly into any space available. According to the Kinetic Theory, an ideal gas can be imagined to consist of a collection of point mass particles in random motion.

6   An increase in temperature causes an increase in the average kinetic energy of particles. The kinetic energy is manifested as vibrational, rotational and translational energy in gases and liquids and as vibrational and rotational energy in solids.

11.4 Comparing solids, liquids and gases

In the last chapter, we looked at the structure, bonding and properties of different types of solids. This chapter concentrates on the properties and behaviour of gases, but before doing so it is instructive to compare solids with liquids and gases. Look closely at the information in table 11.1. This compares the bulk (macroscopic) characteristics of solids, liquids and gases.

Table 11.1
Comparing solids, liquids and gases

	Solids	Liquids	Gases
Volume	definite (fixed)	definite (fixed)	take the volume of their container
Shape	definite (fixed)	take the shape of their container, but do not necessarily occupy all of it	occupy the whole of their container
Realtive compressibility	nil	almost nil	large
Relative density	large	large	small

The particles in a gas are much more widely spaced than those in a liquid or solid. The forces between such widely spaced particles are weak. So the gas particles readily move away from each other and occupy the whole volume of their container.

Notice the similarities between solids and liquids, but the considerable differences between solids and gases. Solids and liquids have fixed volumes, they have similar densities and are almost non-compressible.

The particles in solids and liquids are very close, but in gases they are more widely spaced with much empty space in between. The closeness of the particles in solids and liquids means that these materials have relatively large densities and negligible compressibilities. There are strong forces of attraction between the particles and so the material is held together as a fixed volume.

In contrast, the widely spaced particles in a gas result in much lower densities and much larger compressibilities compared to solids and liquids. The forces between widely spaced gas molecules are very weak. So, the particles readily move away from each other and occupy the whole volume of their container.

In terms of their bulk characteristics, solids and liquids have certain features in common yet they are quite distinct. Liquids, unlike solids, are fluid, a property

Henry Cavendish (1731–1810). Cavendish was born in Nice of extremely wealthy English parents. He was one of the first chemists to investigate the reactions of gases. Cavendish determined the formula of water by exploding mixtures of hydrogen and oxygen. He was a very shy, retiring man, who died a millionaire.

which they share with gases. However, there are strong forces of attraction between the liquid particles. The particles are held tightly together in a fixed volume even though they have sufficient energy to move freely around each other. This freedom of movement of the particles enables a liquid to flow freely and smoothly on pouring and to take the shape of its container.

In a solid, particles are confined to fixed places in the lattice. It is possible to predict the positions of particles in the structure with some accuracy. The solid is said to have **long-range order**. In a liquid, particles are not confined to fixed positions. They can move freely around each other and it is difficult to predict the positions of particles around an initial reference molecule with any accuracy. The liquid is said to have **short-range order**.

11.5 Gay-Lussac's law and Avogadro's theory

During the later part of the eighteenth century, chemists were particularly interested in the reactions between different gases. In 1781, Henry Cavendish found that whenever he exploded hydrogen with oxygen, a given volume of hydrogen would always react with half of its own volume of oxygen. In 1805, Joseph Louis Gay-Lussac repeated this experiment. He confirmed that two volumes of hydrogen always reacted with one volume of oxygen provided that the volumes were measured under the same conditions of temperature and pressure.

Struck by the simplicity of the 2:1 ratio, Gay-Lussac examined the volumes in which other gases reacted together. Provided that all the volumes in each experiment were measured at the same temperature and pressure, he found that:

1 volume of hydrogen	+ 1 volume of chlorine	→ 2 volumes of hydrogen chloride
1 volume of nitrogen	+ 2 volumes of oxygen	→ 2 volumes of nitrogen dioxide
1 volume of hydrogen chloride	+ 1 volume of ammonia	→ solid ammonium chloride

In 1808, Gay-Lussac published his **law of combining volumes**, now known as **Gay-Lussac's law**.

The volumes of reacting gases and the volumes of any gaseous products are in the ratio of small whole numbers provided the volumes are measured at the same temperature and pressure.

As soon as Gay-Lussac had announced his law of combining volumes, chemists began to look for an interpretation of it in terms of Dalton's Atomic Theory of Matter published in 1803.

Chemists realised that since $100 \, cm^3$ of hydrogen just reacted with $100 \, cm^3$ of chlorine, the number of hydrogen and chlorine molecules in these volumes must be simply related. Otherwise some molecules of hydrogen or chlorine would be left over after the reaction.

In 1811, the Italian scientist Amedeo Avogadro suggested an important hypothesis which later became known as **Avogadro's theory**.

Equal volumes of all gases at the same temperature and pressure contain equal numbers of molecules.

This suggestion by Avogadro explained the results leading to Gay-Lussac's Law.

For example, experiments show that:

1 volume of hydrogen + 1 volume of chlorine → 2 volumes of hydrogen chloride

If we assume that one volume of hydrogen contains n molecules, then by Avogadro's theory, one volume of chlorine also contains n molecules and two volumes of hydrogen chloride contain $2n$ molecules.

Joseph Louis Gay-Lussac (1778–1850). Gay-Lussac was born near Limoges in France. In 1809 he became Professor of Chemistry at the Paris Polytechnic School. The Law of Combining Volumes, now known as Gay-Lussac's Law, was published in 1808.

$\Rightarrow n$ molecules of hydrogen $+ n$ molecules of chlorine $\rightarrow 2n$ molecules of hydrogen chloride

$\Rightarrow \frac{1}{2}$ molecule of hydrogen $+ \frac{1}{2}$ molecule of chlorine $\rightarrow 1$ molecule of hydrogen chloride

Notice that half of a molecule of hydrogen goes to form one molecule of hydrogen chloride. Now, since one molecule of hydrogen chloride must contain at least one hydrogen atom, half a molecule of hydrogen is at least one atom. So, the simplest formula for one hydrogen molecule is H_2.

By the same argument, the simplest formula for the chlorine molecule is Cl_2. This means that one atom of hydrogen and one atom of chlorine form one molecule of hydrogen chloride. Therefore, the formula of hydrogen chloride is HCl and the last equation can be rewritten using formulae as:

$$\frac{1}{2} \text{ molecule of } H_2 + \frac{1}{2} \text{ molecule of } Cl_2 \rightarrow 1 \text{ molecule of HCl}$$

Notice that by applying Avogadro's theory to the volumes of reacting gases it is possible to deduce the formulae of simple gases such as H_2, Cl_2 and HCl.

- What empirical formula would have been deduced for hydrogen chloride if we had assumed that the formula of the hydrogen molecule was H_4?
- Why can the formula of the hydrogen molecule *not* be H_3?
- When hydrogen chloride reacts with metals only one series of the salts can be formed. None of these salts contain hydrogen, i.e. there are no salts corresponding to $NaHSO_4$ from H_2SO_4. What does this suggest about the number of hydrogen atoms in one molecule of hydrogen chloride?

11.6 Obtaining relative molecular masses using Avogadro's theory

Avogadro's theory also introduced a means of obtaining the relative masses of molecules in gases and vapours.

Suppose, for example, that $1\,dm^3$ of a certain gas, X, is 20 times as heavy as $1\,dm^3$ of hydrogen at the same temperature and pressure. It follows from Avogadro's theory that molecules of X must be 20 times as heavy as hydrogen molecules and therefore 40 times as heavy as hydrogen atoms.

We can summarise this reasoning in the following way. By Avogadro's theory:

$$\frac{\text{mass of } x\,dm^3 \text{ of X}}{\text{mass of } x\,dm^3 \text{ of hydrogen}} = \frac{\text{mass of } n \text{ molecules of X}}{\text{mass of } n \text{ molecules of hydrogen}}$$

$$\therefore \quad \frac{\text{mass of } x\,dm^3 \text{ of X}}{\text{mass of } x\,dm^3 \text{ of hydrogen}} = \frac{\text{mass of 1 molecule of X}}{\text{mass of 1 molecule of hydrogen}} \quad \text{(dividing by } n\text{)}$$

Now, since one hydrogen molecule contains two atoms, we can write:

$$\Rightarrow \frac{\text{mass of } x\,dm^3 \text{ of X}}{\text{mass of } x\,dm^3 \text{ of hydrogen}} = \frac{\text{mass of 1 molecule of X}}{\text{mass of 2 atoms of hydrogen}}$$

$$= \frac{1}{2} \times \frac{\text{mass of 1 molecule of X}}{\text{mass of 1 atom of hydrogen}}$$

$$\Rightarrow \frac{\text{mass of } x\,dm^3 \text{ of X}}{\text{mass of } x\,dm^3 \text{ of hydrogen}} = \frac{1}{2} \times \text{relative molecular mass of X}$$

if we assume that the relative atomic mass of $H = 1$.

Thus, by assigning a relative mass of one unit to the hydrogen atom, it became possible for chemists to determine relative molecular masses. These could be obtained by simply comparing the mass of a given volume of a particular gas with the mass of the same volume of hydrogen. This important development led to increased understanding and experimentation in chemistry.

Table 11.2
Volumes occupied by 1 mole of various gases at 25°C and 1 atmosphere pressure

Gas	Mass of empty 1 dm³ flask/g	Mass of flask + gas/g	Mass of 1 dm³ of gas/g	Relative molecular mass of gas	Volume of 1 mole of gas/dm³
O_2	161.45	162.76	1.31	32	24.5
N_2	161.45	162.60	1.15	28	24.3
CO	161.45	162.60	1.15	28	24.3
CO_2	161.45	163.26	1.81	44	24.3

Having obtained relative molecular masses, chemists were able to find the volumes occupied by one mole of gas (table 11.2). The results of a large number of experiments show that one mole of any gas at 25°C and one atmosphere pressure occupies $24.4 \pm 0.1\,dm^3$ (or $22.4 \pm 0.1\,dm^3$ at s.t.p.).

The volume of 1 mole of a gas is known as the **Molar Volume**. It is not surprising that the molar volumes of all gases are the same, because they contain the same number of molecules. The molar volume is just a specific example of Avogadro's theory.

Gay-Lussaac's Law and Avogadro's Theory also enabled chemists to obtain the formulae of more complex gases than say H_2, Cl_2 and HCl. The following example illustrates this.

In order to find the formula of a hydrocarbon, $10\,cm^3$ of the gaseous hydrocarbon were mixed with $33\,cm^3$ of oxygen which was in excess. The mixture was exploded and after cooling to room temperature, the residual volume of gas occupied $28\,cm^3$. On adding concentrated potassium hydroxide the volume decreased to $8\,cm^3$. This remaining gas is excess oxygen.

Volume of hydrocarbon reacting $= 10\,cm^3$
Volume of oxygen reacting $= 33 - 8 = 25\,cm^3$
Volume of carbon dioxide (absorbed by potassium hydroxide) $= 28 - 8 = 20\,cm^3$

If we give the hydrocarbon a hypothetical formula of C_xH_y, we can write a general equation for combustion as:

$$C_xH_y + \left(x + \frac{y}{4} \right)O_2 \longrightarrow xCO_2 + \frac{y}{2} H_2O$$

$$1\ mole \quad \left(x + \frac{y}{4} \right) moles \quad x\ moles \quad \frac{y}{2} moles$$

Applying Avogadro's theory:

$$\Rightarrow 1\ volume\ C_xH_y + \left(x + \frac{y}{4} \right) volumes\ O_2 \longrightarrow x\ volumes\ CO_2 + negligible\ vol.\ H_2O \ldots (1)$$

(liquid at room temp.)

$$10\,cm^3\,C_xH_y \qquad 25\,cm^3\,O_2 \qquad\qquad 20\,cm^3\,CO_2$$
$$\Rightarrow 1\,cm^3\,C_xH_y \qquad 2.5\,cm^3\,O_2 \qquad\qquad 2\,cm^3\,CO_2$$
$$\Rightarrow 1\ volume\ C_xH_y + 2.5\ volumes\ O_2 \longrightarrow 2\ volumes\ CO_2 \ldots (2)$$

By comparing equations (1) and (2)

$$x = 2 \quad and \quad x + \frac{y}{4} = 2.5 \quad \therefore y = 2$$

$\therefore$ The molecular formula of the hydrocarbon is C_2H_2

By using Gay-Lussac's Law and Avogadro's Theory, it also became possible to analyse gas mixtures.

A sample of air was analysed by explosion with excess hydrogen. It was possible to determine its percentage by volume of oxygen. 25 cm³ of air was mixed with 100 cm³ of hydrogen and exploded. After cooling to room temperature, the final volume was 110 cm³.

- Write an equation to represent the reaction of hydrogen with oxygen. Put in state symbols and remember that the measurements are all made at room temperature.
- What are the relative volumes of hydrogen and oxygen which react?
- What is the decrease in volume on reaction?
- What proportion of the decrease in volume is due to oxygen?
- How many cm³ of oxygen are present in the original air sample?
- What is the percentage by volume of oxygen in the air sample?

11.7 Important Gas Laws and the Ideal Gas Equation

During the seventeenth, eighteenth and nineteenth centuries, a great deal of scientific research involved the physical and chemical properties of gases. Careful investigations were carried out in order to discover how the volume of gas varied with changes in temperature and pressure.

As early as 1662, Robert Boyle had discovered that *the volume of a fixed mass of gas is inversely proportional to its pressure, provided the temperature remains constant.*

This is known as **Boyle's Law**. It can be expressed mathematically as

$$V \propto \frac{1}{p}$$

About a century later in 1787, Charles showed that *the volume of a fixed mass of gas is directly proportional to its absolute temperature, provided the pressure remains constant.*

This is known as **Charles's law**. It can be expressed mathematically as

$$V \propto T$$

By combining Boyle's Law and Charles's Law we get

$$V \propto \frac{T}{p}$$

$$\therefore pV \propto T$$

$$\text{and } \frac{pV}{T} = \text{constant}$$

$$\text{so } \frac{p_1 V_1}{T_1} = \frac{p_2 V_2}{T_2}$$

This last equation enables the volume of a gas to be obtained under any conditions of temperature and pressure (say T_1 and p_1), provided its volume (V_2) is known under some other conditions of temperature (T_2) and pressure (p_2).

What is the value of the constant in the expression $pV/T = $ constant? Its value will, of course, depend on the amount of gas taken. Let us calculate the constant for one mole of gas. For one mole of gas, the constant is given the symbol R and called **the gas constant**.

We know that one mole of gas occupies a volume of 22.4 dm³ at one atmosphere pressure and 273 K.

$$\text{Using } R = \frac{pV}{T} = \frac{1 \times 22.4}{273}$$

Robert Boyle (1627–1691), one of the foremost scientists of the seventeenth century. Boyle carried out investigations into many areas of science including combustion, besides his research into the volume and pressure of gases.

* The internationally accepted unit of pressure is the pascal, Pa. One pascal is defined as a pressure of one newton per square metre, Nm^{-2}.

$$\Rightarrow R \;=\; 0.082 \text{ atm dm}^3 \text{ K}^{-1} \text{ mole}^{-1}$$

Operating in SI units, the volume of one mole of gas is 0.0224 m^3 at a pressure of $101\,325\,Nm^{-2}$ (i.e. $101\,325\,Pa^*$) and $273\,K$.

$$\therefore \text{ In SI units, } R = \frac{pV}{T} = \frac{101\,325 \times 0.0224}{273}$$

$$= 8.31 \text{ N m K}^{-1} \text{ mole}^{-1}$$

$$\Rightarrow R \;=\; 8.31 \text{ J K}^{-1} \text{ mole}^{-1}$$

So for one mole of gas, $pV = RT$
and for n moles of gas, $pV = nRT$ which is known as the **Ideal Gas Equation**.

A gas which obeys this equation exactly is called a 'Perfect' gas or an 'Ideal' gas. In practice, real gases obey the equation very closely at low pressure and high temperature. Under these conditions a gas is most like a gas and least like a liquid (section 9.3(b)(ii)).

Suppose the air in a room at one atmosphere pressure contains exactly one-fifth by volume of oxygen and four-fifths by volume of nitrogen.

■ What pressure is exerted by the oxygen?
■ What pressure is exerted by the nitrogen?

Imagine that you have two identical cylinders, joined by a tap, one containing pure oxygen at one atmosphere pressure, the other containing pure nitrogen at one atmosphere pressure. If all the oxygen in the first cylinder is forced through the tap into the second cylinder:

■ What is the total pressure in the second cylinder?
■ What is the pressure of nitrogen in the second cylinder?
■ What is the pressure of oxygen in the second cylinder?

In order to answer the last five questions you have assumed **Dalton's law of partial pressures**. This says that *the total pressure of a mixture of gases is equal to the sum of the pressures that each gas would exert if it occupied the space alone.* This can be expressed mathematically as

$$p_{\text{total}} = p_1 + p_2 + p_3 \ldots p_n$$

p_{total} is the total pressure of the mixture, whilst p_1 is the partial pressure of gas 1, p_2 is the partial pressure of gas 2, etc.

11.8 Investigating the distribution of molecular speeds in gases—the Zartmann experiment

All the molecules in a sample of gas will not have the same speed. At any particular moment, some will be almost stationary whilst others will be moving very rapidly. They will also be moving in very different directions as a result of their haphazard motion. During the early 1930s, Zartmann and Ko developed an ingenious technique for determining the distribution of molecular speeds. Their apparatus is shown in figure 11.3.

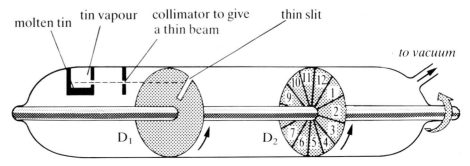

Figure 11.3
The Zartmann experiment.

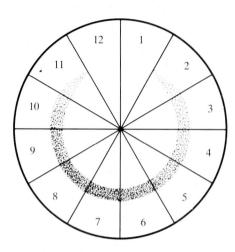

Figure 11.4
The distribution of tin over the pie slices of the target disc in the Zartmann experiment.

D_1 and D_2 are discs which rotate rapidly and at the same speed. The discs are on a common axle, within an evacuated container. In front of D_1 is an oven containing molten tin at a controlled temperature.

Tin vapour streams from a small hole in the oven and strikes D_1. Tin atoms pass through D_1 in small bursts each time the slit in D_1 is in line with the opening in the oven and the collimator. So, D_1 acts as a 'starting gate' for repeated 'molecular races' between D_1 and D_2.

No more molecules can get past D_1 until the disc has completed a full rotation and the slit is opposite the opening in the oven again. Those molecules passing D_1 travel towards the second rotating disc, D_2, and spread throughout its pie slices.

D_1 lets through burst after burst of tin atoms and a layer of tin builds up on D_2. The faster molecules hit the plate before the slower ones, since they all started from the slit simultaneously. Figure 11.4 shows how the deposited tin spreads over the pie slices. The distribution of tin on D_2 depends on the distribution of the speeds of the tin atoms. In order to find the distribution of tin on D_2, it is cut up into pie slices which are weighed to determine the amount of solid deposited on each slice.

- Why is no tin deposited on pie slice 1?
- Why is no tin deposited on pie slice 12?
- What happens to the distribution of tin on the pie slices if D_1 and D_2 rotate faster?
- Why must the space between D_1 and D_2 be continuously evacuated?

Figure 11.5 shows a distribution curve for the mass of tin deposited on each pie slice.

If we know the distance between D_1 and D_2 and the rate of rotation of the discs, we can calculate the velocity of those molecules hitting a particular point on D_2.

For example, suppose the distance from D_1 and D_2 is 1 metre and the discs rotate at 50 revolutions per second.

1 revolution of the discs takes $\frac{1}{50}$th second.

Now the pie slice 1–pie slice 2 boundary $= \frac{1}{12}$th of the disc.

$\therefore$ Atoms on the pie slice 1–pie slice 2 boundary travel 1 metre (from D_1 to D_2) in $\frac{1}{12}$th of a revolution (i.e. $\frac{1}{600}$th second).

$\therefore$ Speed of molecules on pie slice 1–pie slice 2 boundary $= 600 \, \mathrm{m \, s^{-1}}$.

What is the speed of those molecules on the pie slice 2–pie slice 3 boundary?

Proceeding in this way, it is possible to obtain a distribution curve for the molecular speeds.

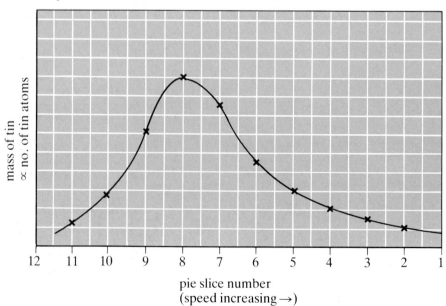

Figure 11.5
A distribution curve for the mass of tin deposited on the different pie slices in the Zartmann experiment.

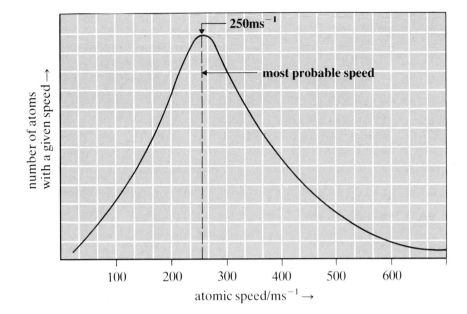

Figure 11.6
A distribution curve for the speeds of
tin atoms in the gas phase at 500 K.

Figure 11.6 shows a distribution curve for the speeds of tin atoms in the gas phase at 500 K. The most probable speed corresponds to the maximum of the curve. More molecules possess this speed than any other.

What is the effect of increased temperature on the distribution of molecular and atomic speeds?

Figure 11.7 shows the distribution curves for the speeds of tin atoms at 500 K and 1000 K. There are some important points to notice about the effect that increased temperature has had on the distribution and sizes of the atomic speeds.

1 The most probable speed increases as the temperature increases. At 500 K, the most probable speed for tin atoms is $250 \, \text{ms}^{-1}$. At 1000 K the most probable speed has increased to $360 \, \text{ms}^{-1}$.

2 The distribution curve becomes flatter as the temperature rises. This means that there are fewer molecules with the most probable speed, but there is a greater proportion of high-speed molecules.

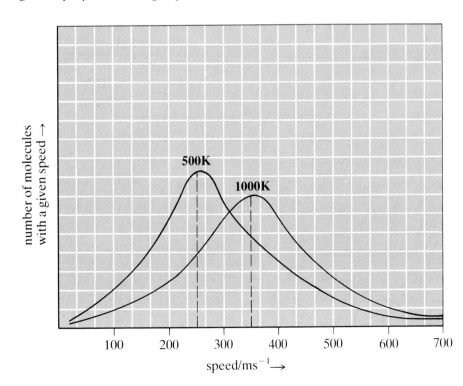

Figure 11.7
Distribution curves for the speeds of
tin atoms at 500 K and 1000 K.

It is important to remember that the graph in figure 11.7 is really a histogram. Thus, the area beneath each curve is proportional to the total number of molecules involved. As we are dealing with the *same total* number of molecules at 500 K and 1000 K, the areas beneath the two curves must be identical.

11.9 Determining the relative molecular mass of gases and volatile liquids

Nowadays, the most accurate and convenient method of measuring relative molecular masses is by mass spectrometry. Using this method, relative molecular masses can be obtained in a few minutes with an accuracy of one part in a million.

Originally, the relative molecular masses of gases and volatile substances were obtained by methods dependent on the gas laws discussed earlier in this chapter. Two of the more common methods employed were those designed by Regnault and Victor Meyer.

Finding the relative molecular mass of a gas by direct weighing—Regnault's method.

Using the Ideal Gas Equation, $pV = nRT$, and $n = \dfrac{m}{M}$

where m is the mass of gas whose molar mass is M.

We can write $pV = \dfrac{m}{M} . RT$

$\Rightarrow M = \dfrac{m}{pV} . RT$

By measuring m, p, V and T for a sample of gas and assuming R, it is possible to calculate M.

A container of known volume (V) is weighed full of the gas at pressure p and temperature T. The same container is then weighed after evacuation in order to obtain the mass of gas inside (m).

Alternatively, Regnault's method can be used to compare the mass of the gas under investigation with the mass of the same volume of a gas of known relative molecular mass. Although the method is limited to gaseous substances, it is very accurate provided the following precautions are taken.

1 The container must be large, so as to give an appreciable weight of gas. Gases are so light that any slight error in weighing produces a large percentage error if the mass of the substance weighed is very small.

2 The gas used must be perfectly pure and perfectly dry.

3 The container should be filled and emptied several times to make sure that all the air or the previous gas has been removed.

4 The container must be evacuated as completely as possible.

5 All weighings must be carried out at the same temperature and pressure so that the upthrust on the container remains constant.

Regnault's method has been used to determine relative molecular masses and relative atomic masses to an accuracy of more than one part in a thousand. For example, the relative molecular mass of H_2S ($M_r(H_2S)$) could be obtained accurately by this technique and then used to determine the relative atomic mass of sulphur ($A_r(S)$):

$$A_r(S) = M_r(H_2S) - 2 \times A_r(H)$$

Finding the relative molecular mass of a volatile liquid

This method can be used with volatile liquids. It is a development of an earlier technique used by Victor Meyer. Although it gives only an approximate value for the relative molecular mass, it enables an accurate value to be obtained if the empirical formula of the substance can be determined.

The principle of the experiment is the same as that described for gases earlier in this section.

$$pV = nRT$$

$$\text{and } n = \frac{m}{M}$$

$$\therefore pV = \frac{m}{M} . RT$$

$$\Rightarrow M = \frac{m}{pV} . RT$$

Thus, the molar mass (M) of the liquid can be calculated by finding the volume of vapour (V) formed from a known mass of liquid (m) at a measured temperature (T) and pressure (p), provided we assume the value of the gas constant (R).

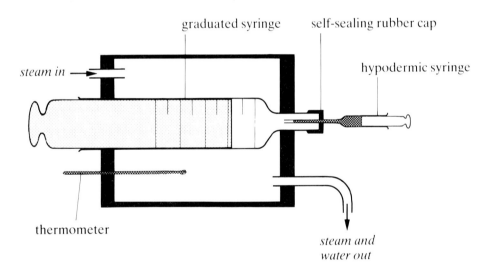

Figure 11.8
Determination of the relative molecular mass of a volatile liquid.

Figure 11.8 shows a suitable apparatus to use. Draw a few cm^3 of air into the graduated syringe and fit the self-sealing rubber cap over the nozzle. Pass steam through the outer jacket until the thermometer reading and the volume of air in the syringe become steady. Continue to pass steam through the jacket and record the temperature (T) and the volume of air in the syringe.

Now fill the hypodermic syringe with about $1 cm^3$ of the liquid under investigation, ensuring that all air is expelled from the needle. Weigh the hypodermic syringe and its contents and then push the needle through the self-sealing cap of the graduated syringe so that it is well clear of the nozzle. Inject about $0.2 cm^3$ of liquid into the larger syringe and withdraw the hypodermic syringe. Immediately, re-weigh the hypodermic syringe and its contents.

The liquid injected into the graduated syringe will evaporate. The final volume of air plus vapour in the graduated syringe should be recorded when the volume becomes steady. Finally record the atmospheric pressure. Here are some typical results:

Mass of liquid, X, vaporised, $m = 0.16 \text{g}$

Initial volume of air in graduated syringe $= 10 cm^3$

Final volume of air + vapour in syringe = 56cm^3

$\therefore$ Volume of X vaporised = 46cm^3

$$= 46 \times 10^{-6} \text{m}^3$$

Temperature of X = $100°C = 373 \text{K}$

Pressure of X = $1 \text{atm} = 101\,325 \text{N m}^{-2}$

Gas constant, $R = 8.31 \text{J K}^{-1} \text{mole}^{-1}$

Using $M = \dfrac{m}{pV} . RT$

$$M = \frac{0.16}{101\,325 \times 46 \times 10^{-6}} \times 8.31 \times 373 = 106.4 \text{g mole}^{-1}$$

i.e. relative molecular mass of X ≈ 106

- Why is this method unsuitable for liquids which boil above 80°C?
- Why should the hypodermic syringe be handled as little as possible between weighings?
- What are the main sources of error in this experiment?
- The substance X, for which the typical results were given, has the percentage composition by mass of 22.0% C, 4.6% H and 73.4% Br.

(i)   Calculate the empirical formula of X.
(ii)   What is the molecular formula of X?
(iii) What is the accurate relative molecular mass of X?

Summary

1 Diffusion is the mixing and movement of matter which results from the kinetic energy of moving particles.

2 Brownian motion is the random movement of particles of solid suspended in a liquid or in a gas.

3 All matter is composed of tiny, invisible particles.

4 In solids, the particles are relatively close and can only vibrate about fixed positions.
   In liquids, the particles are further apart than in solids. They have sufficient energy to overcome the forces of attraction between each other and they move freely around each other.
   In gases, the particles are much more widely separated. They have sufficient energy to overcome the forces of attraction between each other almost completely and move rapidly and randomly into any space available.

5 Solids are said to have long-range order. Liquids have short-range order.

6 Gay-Lussac's law summarises the relationship between volumes of reacting gases. The volumes of the reacting gases and the volumes of any gaseous products are in the ratio of small whole numbers provided all the volumes are measured at the same temperature and pressure.

7 Avogadro's theory says that equal volumes of all gases at the same temperature and pressure contain equal numbers of molecules.

8 At s.t.p., one mole of gas occupies 22.4dm^3.

9 The Ideal Gas Equation is $pV = nRT$. Under ordinary conditions (and at low pressure and high temperature) real gases obey the Ideal Gas Equation very closely.

Study questions

1   (a) Draw sketch graphs of (i) p against V,   (ii) p against $1/V$
            (iii) pV against p, (iv) pV against V.
        for a constant number of moles of a perfect gas at constant temperature.
    (b) Plot on separate sketch graphs, (i) V against T (°C),
            (ii) V against T (K)
        for a constant number of moles of a perfect gas at constant pressure.
    (c) Plot pV against T and pV/T against p for a constant number of moles of ideal gas.
    (d) Draw a sketch graph to show how the pressure of a constant mass of ideal gas will vary as the temperature rises from absolute zero in a container of constant volume. How will the graph change if the gas tends to dissociate ($X_2 \rightarrow 2X$) as temperature increases? (For each graph in sections (a), (b), (c) and (d), label the axes and show the zero for each axis.)

2   A balloon can hold $1000\,cm^3$ of air before bursting. The balloon contains $975\,cm^3$ of air at 5°C. Will it burst when it is taken into a house at 25°C? Assume that the pressure of the gas in the balloon remains constant.

3   What would you expect to be the effect of a change in
    (i) pressure,   (ii) volume,   (iii) temperature,
    on the rate of diffusion of a particular gas? Explain your answer in terms of the kinetic theory.

4   (a) A mixture of two gases in a container exerts a pressure of 800 mm Hg and occupies a volume of $400\,cm^3$. If one of these gases (A) occupies a volume of $300\,cm^3$ at the same temperature and pressure, what pressure does the other gas (B) exert in the mixture?
    (b) When gaseous argon is allowed to diffuse, it separates into a lighter and heavier fraction. What does this suggest about the nature of argon?

5   (a) Describe and explain in terms of molecular theory, what happens when
        (i) a small quantity of a volatile liquid is introduced into the space above mercury in a barometer.
        (ii) a solute in a solvent is shaken with a second solvent, which is immiscible with the first solvent.
    (b) 0.50 g of a volatile liquid was introduced into a globe of $1000\,cm^3$ capacity. The globe was heated to 91°C so that all the liquid vaporised. Under these conditions the vapour exerted a pressure of 190 mm mercury. What is the relative molecular mass of the liquid?

6   An industrial chemist who works for a firm which manufactures ethyne (acetylene) for use in oxyacetylene welding discovers a very cheap way of producing ethyne. Unfortunately, his ethyne is contaminated with ethene (ethylene). Unless the mixture contains at least 50% by volume of ethyne, it is useless. In order to determine the relative proportions of the two in the mixture, he exploded $10\,cm^3$ of the mixture with $30\,cm^3$ of oxygen. After absorbing the residual CO_2 in KOH, the uncombined O_2 occupied $2\,cm^3$. What was the composition of the mixture by volume?
    (*Hint*: let the volume of one gas be $x\,cm^3$. Write an equation for the combustion of that gas and hence work out, in terms of x, how many cm^3 of O_2 it reacts with. Repeat this with the other gas. Find the total volume O_2 used in terms of x. Find the actual volume of O_2 used from the data. Hence find x.)

7   The extent of combustion of petrol vapour in an internal combustion engine can be shown in terms of the percentage of CO_2 in the exhaust gases. $60\,cm^3$ of exhaust gas (assumed to contain only CO, CO_2 and N_2) were mixed with $20\,cm^3$ of O_2 (excess) and exploded. On cooling, there had been a decrease to $70\,cm^3$. On adding KOH, the volume decreased to $35\,cm^3$.

(a) What is the composition by volume of the mixture?

(b) What is the percentage by volume of CO_2 in the exhaust fumes?

8 On decomposition with electric sparks, two volumes of a gaseous compound containing only phosphorus and hydrogen, gave a deposit of phosphorus (solid) and three volumes of hydrogen at the same temperature and pressure.

(a) Apply Avogadro's theory and deduce the number of atoms of hydrogen in one molecule of the phosphorus hydride. (Assume hydrogen is diatomic.)

(b) Write an empirical formula for the phosphorus hydride in as much detail as you are able.

(c) What do you think its molecular formula is?

(d) What measurement would you make to try to confirm this?

(e) Predict two physical and two chemical properties of this compound using your knowledge of the periodic table.

9 It is required to find the composition by volume of a sample of Town gas which contains hydrogen, carbon monoxide and nitrogen. $40\,cm^3$ of the Town gas were carefully exploded with $40\,cm^3$ of oxygen (known to be in excess). Under the conditions used, only hydrogen and carbon monoxide would react with oxygen. On cooling to room temperature, the volume was $51\,cm^3$. On adding concentrated KOH, the volume decreased further to $41\,cm^3$.

(a) Write equations (including state symbols) for the reactions which occur on explosion. (Remember that water will be liquid, as the measurements are made at room temperature.)

(b) What are the relative volumes of gaseous reactants and products in these reactions?

(c) What volume of CO_2 is produced?

(d) What is the volume of CO in the original $40\,cm^3$ of Town gas?

(e) What is the total decrease in volume as a result of the explosion?

(f) What decrease in volume is caused by
(i) CO   (ii) H_2 on explosion?

(g) What are the volumes of
(i) H_2 (ii) N_2 in the original $40\,cm^3$ of Town gas?

10 $5.2\,cm^3$ of a gaseous hydrocarbon were exploded with an excess of oxygen. After cooling to the original room temperature and pressure, a contraction in volume of $7.8\,cm^3$ was observed. A further contraction of $10.4\,cm^3$ was noted after treatment with concentrated KOH.
What is the molecular formula of the hydrocarbon?

11 When $20\,cm^3$ of a gaseous hydrocarbon, A, are exploded with $150\,cm^3$ of oxygen, the residual gases occupied $110\,cm^3$. After shaking the residual gases with aqueous sodium hydroxide, the final volume was $30\,cm^3$. (All volumes were measured at the same temperature and pressure.)

(a) What is the molecular formula of A?

(b) Write six possible structural formulae for A and name each of these structures.

12 Energy Changes and Bonding

A miniature nuclear battery for a heart pace-maker, (a) before construction, (b) after construction. The nuclear battery (35 mm long × 15 mm across) uses heat from the radioactive decay of plutonium-238 to generate electricity in a miniature thermopile. The encapsulation of the plutonium ensures that there is no hazard from radiation. The batteries are designed to last 20 years.

12.1 Energy and energy changes

Energy is the most precious commodity we have. Without it there could be no life, no warmth, no movement. Energy gives us the power to do work. In every country, people's living standards are closely related to the availability of energy.

From the earliest times, people worshipped the sun. This is not surprising. Through the process of photosynthesis, the sun provides us with most of our food and, over millions of years, it has created our supplies of the fossil fuels (coal, oil and natural gas). More than ninety per cent of our present energy needs are supplied by these fossil fuels—valuable reserves of energy that cannot last forever.

Until recently, Western Europe was entirely dependent on the Middle East for its increasing demands for oil. After the outbreak of the Arab/Israeli war in 1973, the price of oil rose alarmingly and created serious inflation in the oil-importing countries such as Britain. Since 1975, however, the prospects for Europe have brightened with the discovery of first natural gas, and then vast oilfields, below the North Sea.

The transfer of energy to or from chemicals plays a crucial part in the chemical processes in industry and in living things. Consequently, the study of these energy changes are as important as the study of the changes in the materials themselves.

Our present-day living conditions rely heavily on the availability of energy in its various forms. Chemical energy is converted to heat energy when fuels such as gas, oil and coal are burnt in our homes and in industry. Just think of the vast quantities of chemical energy converted to mechanical energy each day from the petrol burnt in our vehicles!

Within our own bodies, energy changes are vital. Foods such as fats and carbohydrates are important biological fuels. During metabolism, the chemical energy in these foods is converted to heat energy to keep us warm, to mechanical energy in our muscles and to electrical energy in the signals within our nerve fibres. Chemical energy is also converted to electrical energy when the materials in cells and batteries are used to generate electricity.

All these important processes involve energy changes. This in itself would be a sufficient answer to the question 'Why study energy changes?' Apart, from their importance to industry and society, the study of energy changes can lead to a better understanding of the fundamental chemistry involved.

12.2 The ideas and language of thermochemistry

Very often chemical changes are accompanied by changes in the **heat content (enthalpy, H)** of the materials which are reacting. Usually, this change in the heat content is shown by a change in temperature. Indeed, the change in

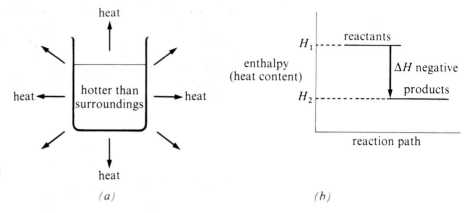

(a)   *(b)*

Figure 12.1
In an exothermic reaction, heat is lost to the surroundings and ΔH is negative.

temperature when substances react often provides evidence that a chemical change has taken place.

When an exothermic reaction occurs, heat is given out and the temperature of the products rises above room temperature. Eventually, the temperature of the products falls to room temperature again as the heat produced is lost to the surroundings (figure 12.1(*a*)). Thus, the heat content (enthalpy) of the products (H_2) is less than that of the reactants (H_1). Since the materials have *lost* heat, we can see that the enthalpy change for the reaction, ΔH (sometimes called the **heat of reaction**) is negative (figure 12.1(*b*)).

For example, when magnesium reacts with oxygen, heat is evolved:

$$2Mg(s) + O_2(g) \rightarrow 2MgO(s) \qquad \Delta H = -1204\,kJ$$

Figure 12.2
An energy level diagram for the exothermic reaction of magnesium with oxygen.

Chemical energy in the magnesium and in the oxygen is partly transferred to chemical energy in the magnesium oxide and partly evolved as heat. Thus, the magnesium oxide has less energy than the starting materials, magnesium and oxygen. The value of ΔH, the heat of reaction, relates to the amounts shown in the equation—2 moles of Mg atoms, 1 mole of O_2 molecules and 2 moles of MgO (figure 12.2).

When an endothermic reaction occurs, the heat required for the reaction is taken from the reacting materials. At first, the temperature of the products falls below the initial temperature. Eventually, the temperature of the products rises to room temperature again as heat is *absorbed* from the surroundings. In this case, the heat content of the products is greater than that of the reactants and the enthalpy change (heat of reaction), ΔH, is positive (figure 12.3).

We can summarise these ideas as:

$$\text{Enthalpy change} = \text{heat of reaction}$$
$$\Delta H = H_2 - H_1$$

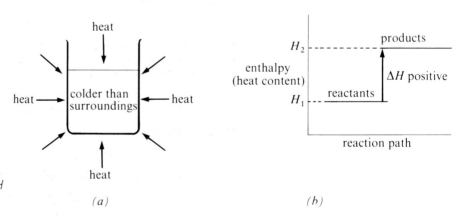

(a)   *(b)*

Figure 12.3
In an endothermic reaction, heat is gained from the surroundings and ΔH is positive.

Since the enthalpy change manifests itself as heat, the term 'heat of reaction' is often used in place of 'enthalpy change of reaction'.

Remember that ΔH refers only to the energy change *for the reacting materials*. The surroundings will necessarily gain whatever heat the reacting materials lose, and vice versa. Thus, the *total* energy is unchanged during a chemical reaction. *Energy may be exchanged between the materials and the surroundings but the total energy of the materials and the surroundings remains constant.* This important concept is known as the **Conservation of Energy.**

12.3 The standard conditions for thermochemical measurements

In order to compare energy changes, it is important to state the conditions under which a reaction is performed. Furthermore, we must ensure that the conditions of the system are the same before and after the reaction. In particular, the temperature and pressure should be stated. *The standard conditions of temperature and pressure for thermochemical measurements are 298 K (25°C) and 1 atmosphere*, respectively. Any enthalpy change measured under these conditions is described as a **standard heat of reaction** and given the symbol $\Delta H^{\ominus}$ or $\Delta H^{\ominus}_{298}$ with a special superscript and subscript. The symbol, $\Delta H^{\ominus}_{298}$, also implies that

—all the substances involved in the reaction are in their normal physical states at 298 K and 1 atmosphere and that
—any solutions involved have a concentration of unit activity.

This is effectively one mole per cubic decimetre (1.0 M) for our purposes.

Thus, $\Delta H^{\ominus}_{298}$ for the reaction

$$2H_2(g) + O_2(g) \rightarrow 2H_2O(l)$$

must relate to gaseous hydrogen, gaseous oxygen and liquid water (not steam).

In the case of elements which exist as different allotropes and compounds which exist as different polymorphs, the most stable form at 298 K and 1 atm is chosen as the standard. Consequently, $\Delta H^{\ominus}_{298}$ values for reactions involving carbon should relate to the allotrope graphite rather than diamond.

We can therefore define the standard heat of a reaction as: *the amount of heat absorbed or evolved when the molar quantities of reactants as stated in the equation react together under standard conditions.*

i.e., at a pressure of 1 atmosphere,
at a temperature of 298 K,
with substances in their normal physical states under these conditions and solutions having unit activity (effectively 1.0 M).

Consider the reaction

$$2H_2(g) + O_2(g) \rightarrow 2H_2O(l) \qquad \Delta H^{\ominus}_{298} = -575\,kJ$$

■ Is the reaction endothermic or exothermic?
■ Which have the greater enthalpy, the products or the reactants?
■ What is the value of $\Delta H^{\ominus}_{298}$ for

  (i) $2H_2O(l) \rightarrow 2H_2(g) + O_2(g)$
  (ii) $H_2(g) + \frac{1}{2}O_2(g) \rightarrow H_2O(l)$

■ Given also that

$$H_2O(l) \rightarrow H_2O(g) \qquad \Delta H^{\ominus}_{298} = +44\,kJ,$$

calculate the value of $\Delta H^{\ominus}_{298}$ for

$$2H_2(g) + O_2(g) \rightarrow 2H_2O(g)$$

12.4 Standard heats of formation and combustion

*The **standard heat of formation** of a substance is the heat evolved or absorbed when one mole of the substance is formed from its elements under standard conditions.* The standard heat of formation of a substance is given the symbol $\Delta H_f^\ominus$. The superscript $^\ominus$ indicates standard conditions and the subscript $_f$ refers to the formation reaction.

Thus, the statement $\Delta H_f^\ominus(MgO(s)) = -602\,kJ\,mole^{-1}$ relates to the formation of 1 mole of magnesium oxide from 1 mole of Mg atoms and $\frac{1}{2}$ mole of O_2 molecules, i.e.

$$Mg(s) + \tfrac{1}{2}O_2(g) \rightarrow MgO(s)$$

One important consequence of the definition of standard heat of formation is that the heat of formation of an element in its normal physical state under standard conditions is zero, since no heat change is involved when an element is formed from itself.

$$
\begin{array}{ll}
Cu\,(s) \rightarrow Cu(s) & \Delta H_f^\ominus(Cu(s)) = 0 \\
H_2(g) \rightarrow H_2(g) & \Delta H_f^\ominus(H_2(g)) = 0
\end{array}
$$

Obviously, $\Delta H_{298}^\ominus$ for the process $H_2(g) \rightarrow H_2(g)$ is zero, but this is not so for the process $H_2(g) \rightarrow 2H(g)$. This involves atomisation, i.e. the conversion of H_2 molecules to single H atoms. In fact, the **standard heat of atomisation** of an element is defined as *the enthalpy change when one mole of gaseous atoms is formed from the element under standard conditions*. Thus, the standard heat of atomisation of hydrogen ($\Delta H_{at}^\ominus (H_2(g))$) refers to the process

$$\tfrac{1}{2}H_2(g) \rightarrow H(g) \qquad \Delta H_{at}^\ominus (H_2(g)) = +218\,kJ\,mole^{-1}$$

A wide range of experimental techniques is available for determining the atomisation energies of elements. One technique widely used with gaseous diatomic elements is to find the minimum frequency of radiation (ν) required to dissociate the gaseous molecules into atoms. For example:

$$E = h\nu$$
$$I_2(g) \xrightarrow{} I(g) + I(g) \qquad \Delta H = h\nu$$

Since the energy (E) of the radiation is given by $h\nu$, where h is Planck's Constant,

$$\Rightarrow \Delta H_{at}^\ominus (I_2(g)) = \frac{\Delta H}{2} = \frac{h\nu}{2}$$

The heats of atomisation of solid and liquid elements are more difficult to measure. For many elements, they can be obtained using specific heat capacities and molar heats of fusion and vaporisation.

In discussing the heat changes in chemical reactions, it is useful to give each compound a definite heat content or enthalpy. For convenience, all elements are assigned a heat content of zero under standard conditions. The standard heat of formation of a compound then provides a measure of the heat content of the compound relative to its constituent elements. Remember, however, that these enthalpy values are *only* relative values. They give no information about the absolute energy content of a substance. This will depend on the potential energy in the electrical and nuclear interactions of the constituent particles plus the kinetic energy possessed by the atoms and sub-atomic particles.

The standard heat of formation of water ($\Delta H_f^\ominus(H_2O(l)) = -286\,kJ\,mole^{-1}$) tells us that, under standard conditions, water has a lower energy content than the hydrogen and oxygen from which it is formed (figure 12.4), On an atomic scale, we can imagine that some of the potential energy in the molecules of H_2 and O_2 has been converted to kinetic energy in the molecules of H_2O and this has resulted in a temperature rise. We can compare this with the conversion of potential energy to kinetic energy and heat when a stone rolls down a hill.

Figure 12.4
An energy level diagram for the standard heat of formation of water.

A bomb calorimeter can be used to determine the energy values of food and fuels.

One other enthalpy change of considerable importance is the standard heat of combustion. This is very relevant to the study of fuels and foods.

The **standard heat of combustion** of a substance, $\Delta H_c^\ominus$, is defined as *the enthalpy change when one mole of the substance is completely burnt in oxygen under standard conditions.* For example, the standard heat of combustion of ethanol ($\Delta H_c^\ominus(C_2H_5OH(l))$, is $-1368\,kJ\,mole^{-1}$, i.e.

$$C_2H_5OH(l) + 3\tfrac{1}{2}O_2(g) \rightarrow 2CO_2(g) + 3H_2O(l) \qquad \Delta H^\ominus = -1368\,kJ$$

■ Why is it necessary to say 'completely burnt in oxygen' in the definition of $\Delta H_c^\ominus$?

■ Draw the heats of combustion of both graphite and diamond on the same energy level diagram.

$$\Delta H_c(C\,(graphite)) = -393\,kJ\,mole^{-1}$$
$$\Delta H_c^\ominus(C\,(diamond)) = -395\,kJ\,mole^{-1}$$

■ Which allotrope has the larger enthalpy (energy content)?
■ Which allotrope is the more stable?
■ What is the enthalpy change for the process

$$C\,(graphite) \rightarrow C\,(diamond)?$$

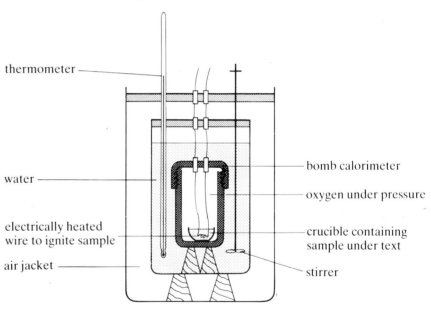

Figure 12.5
A bomb calorimeter.

Study question 1 at the end of this chapter shows a simple apparatus in which the heat of combustion of a liquid fuel could be determined. The apparatus in figure 12.5 shows an instrument, known as a bomb calorimeter, used for accurate thermochemical determinations. Scientists use equipment such as this to determine the energy values of foods and fuels. The apparatus is specially designed to avoid heat losses by completely surrounding the 'bomb' with water. Heat losses can be eliminated altogether if the thermochemical investigation is coupled with an electrical calibration. First of all, the chemical reaction is carried out in the calorimeter and the temperature is plotted against time before, during and after the reaction (figure 12.6). The experiment is now repeated, but this time an electrical heating coil replaces the reactants. The current in the coil is carefully adjusted so as to give a temperature/time curve identical to that obtained in the chemical reaction. By recording the current during the time of this electrical calibration, it is possible to calculate the electrical energy supplied with great accuracy. This electrical energy is exactly the same as the energy change in the chemical reaction. As it includes both the heat absorbed by the system and the heat lost from the system, it eliminates the need for a heat-loss correction.

The practical significance of heats of combustion is clear enough. Fuel engineers and dieticians call them energy values (calorific values). The prices of fuels are closely related to their energy values. Energy-providing foods, such as sugar, are classified by the amount of heat they liberate when they are metabolised in the body.

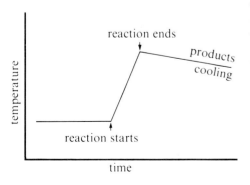

Figure 12.6
A graph of temperature against time for a reaction in a bomb calorimeter.

12.5 Measuring standard heats of formation

The standard heats of formation of carbon dioxide, magnesium oxide and many other oxides can be measured directly using a bomb calorimeter similar to that discussed in the last section.

However, there are many compounds for which $\Delta H_f^{\ominus}$ cannot be measured directly. Consider, for example, tetrachloromethane (CCl_4). Graphite and chlorine do not combine readily, nor does CCl_4 decompose easily into its constituent elements. In other words, neither the reaction

$$C \text{ (graphite)} + 2Cl_2(g) \rightarrow CCl_4(l)$$

nor the reaction

$$CCl_4(l) \rightarrow C \text{ (graphite)} + 2Cl_2(g)$$

can be carried out in a calorimeter.

Boron (III) oxide (B_2O_3) and aluminium oxide (Al_2O_3) provide a different problem in attempting to measure their standard heats of formation. In their case, it is difficult to burn the elements, boron and aluminium, completely in oxygen because a protective layer of oxide coats the unreacted element.

■ Write an equation for the standard heat of formation of carbon monoxide.
■ Why is it impossible to obtain $\Delta H_f^{\ominus}(CO(g))$ directly?

As a result of the problems involved in these measurements, chemists have had to obtain standard heats of formation indirectly. One indirect method involves measuring the heat of combustion of the compound and the heats of combustion of its constituent elements. These values can be linked in an energy cycle or an energy level diagram with the heat of formation of the compound. The energy cycle in figure 12.7 shows how this can be done for carbon monoxide. Notice that figure 12.7 shows two routes for converting graphite and oxygen to CO_2. One of these is the direct route straight from graphite and oxygen to CO_2. The alternative route goes via CO. It would seem reasonable that the overall enthalpy change for the conversion of graphite to carbon dioxide is independent of the route taken, so we can write

Figure 12.7
An energy cycle incorporating the heat of formation of carbon monoxide.

$$\Delta H_1 = \Delta H_2 + \Delta H_3 \ldots\ldots\ldots\ldots \text{Equation 1}$$

$$\text{now } \Delta H_1 = \Delta H_c^{\ominus}(C \text{ (graphite)}) = -393 \text{ kJ mole}^{-1},$$

$$\text{and } \Delta H_3 = \Delta H_c^{\ominus}(CO(g)) = -283 \text{ kJ mole}^{-1},$$

$$\text{so } \Delta H_2 = \Delta H_f^{\ominus}(CO(g)) = \Delta H_1 - \Delta H_3$$

$$= -393 - (-283)$$

$$= -110 \text{ kJ mole}^{-1}$$

The chemical processes and the different routes involved are represented in an energy level diagram in figure 12.8.

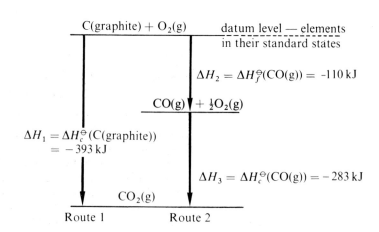

Figure 12.8
An energy level diagram incorporating the heat of formation of carbon monoxide.

The argument we used to obtain equation 1 above is a specific example of **Hess's Law of Constant Heat Summation**. This says that *the energy change in converting reactants, A + B, to products, X + Y, is the same, regardless of the route by which the chemical change occurs provided the initial and final conditions are the same.*

Of course, Hess's Law is simply an application of the more fundamental law of conservation of energy.

■ Why must Hess's Law hold if the law of conservation of energy is to be obeyed?

■ Suggested experiments which would enable you to verify Hess's Law.

We have used Hess's Law to find the standard heat of a reaction which cannot be measured directly. Before moving on, let us look at a more complex example involving butane (C_4H_{10}). Carbon and hydrogen will not react directly to produce butane. So, the formation of butane, represented by the equation

$$4C \text{ (graphite)} + 5H_2(g) \rightarrow C_4H_{10}(g)$$

cannot be carried out directly in a calorimeter. However, the standard heat of formation of butane can be obtained by measuring the standard heats of combustion of butane, carbon and hydrogen.

An energy level diagram for the reactions concerned is shown in figure 12.9.

Figure 12.9
An energy level diagram used to obtain the heat of formation of butane indirectly.

Notice that route 1 involves four times the standard heat of combustion of graphite and five times the standard heat of combustion of hydrogen.

By Hess's Law,

enthalpy change for route 1 = enthalpy change for route 2

$$\therefore -3002 = \Delta H_f^\ominus(C_4H_{10}(g)) - 2877$$

$$\Rightarrow \text{standard heat of formation of butane, } \Delta H_f^\ominus(C_4H_{10}(g)) = -125 \text{ kJ mole}^{-1}$$

12.6 Using standard heats of formation to calculate the energy changes in reactions

One of the most important uses of $\Delta H_f^\ominus$ values is in calculating the enthalpy changes in chemical reactions.

The *Apollo 11* project landed the first man on the moon on 21 July 1969. During this project, engines of the lunar module used methylhydrazine ($CH_3.NH.NH_2$) and dinitrogen tetraoxide (N_2O_4). These liquids were carefully chosen since they ignite spontaneously and very exothermically on contact. What is the enthalpy change for their rection?

$$4CH_3.NH.NH_2(l) + 5N_2O_4(l) \xrightarrow{\Delta H^{\ominus}_{reaction}} 4CO_2(g) + 12H_2O(l) + 9N_2(g)$$

$$4\Delta H^{\ominus}_f(CH_3.NH.NH_2(l) \qquad\qquad 4\Delta H^{\ominus}_f(CO_2(g))$$
$$+ 5\Delta H^{\ominus}_f(N_2O_4(l)) \qquad\qquad\qquad + 12\Delta H^{\ominus}_f(H_2O)l))$$

$$4C(graphite) + 12H_2(g) + 9N_2(g) + 10O_2(g)$$

Figure 12.10
An energy cycle incorporating the reaction of methylhydrazine with dinitrogen tetraoxide.

First we write the equation for the reaction:

$$4CH_3.NH.NH_2(l) + 5N_2O_4(l) \rightarrow 4CO_2(g) + 12H_2O(l) + 9N_2(g)$$

Now draw an energy cycle by adding the formation equations from the same elements to both sides of the equation (figure 12.10).

By Hess's Law, the total enthalpy change for the formation of carbon dioxide, water and nitrogen will be the same whether they are formed directly from their elements or via the intermediates, $CH_3.NH.NH_2$ and N_2O_4.

$$\therefore 4\Delta H^{\ominus}_f(CH_3.NH.NH_2(l)) + 5\Delta H^{\ominus}_f(N_2O_4(l)) + \Delta H^{\ominus}_{reaction}$$
$$= 4\Delta H^{\ominus}_f(CO_2(g)) + 12\Delta H^{\ominus}_f(H_2O(l))$$

The standard heats of formation are:

$$\Delta H^{\ominus}_f(CH_3.NH.NH_2(l)) = +53\,kJ\,mole^{-1}$$
$$\Delta H^{\ominus}_f(N_2O_4(l) = -20\,kJ\,mole^{-1}$$
$$\Delta H^{\ominus}_f(CO(g)) = -393\,kJ\,mole^{-1}$$
$$\Delta H^{\ominus}_f(H_2O(l)) = -286\,kJ\,mole^{-1}$$
$$\Rightarrow 4(+53) + 5(-20) + \Delta H^{\ominus}_{reaction} = 4(-393) + 12(-286)$$
$$so, \Delta H^{\ominus}_{reaction} = -5116\,kJ$$

Thus

$$4CH_3.NH.NH_2(l) + 5N_2O_4(l) \rightarrow 4CO_2(g) + 12H_2O(l) + 9N_2(g)$$
$$\Delta H^{\ominus} = -5116\,kJ$$

The booster stage for the Saturn V rocket. This was used to lift the American astronauts into orbit during the Apollo Project. This section is being hoisted into position at the Kennedy Space Centre, Florida. During its initial stage, Saturn V burns 15 tonnes of kerosene/oxygen mixture every second for the first $2\frac{1}{2}$ minutes.

12.7 Using standard heats of formation to predict the relative stabilities of compounds

Most compounds are formed exothermically from their elements. Thus, the standard heats of formation of water, carbon dioxide, aluminium oxide and many other compounds are negative. These compounds are therefore at a lower energy level than their constituent elements. This means that the compounds are energetically more stable than the elements from which they are formed. But, consider the following problem.

The standard heat of formation of hydrogen peroxide is $-188\,kJ\,mole^{-1}$. From this, we would expect H_2O_2 to be stable. But, H_2O_2 decomposes fairly readily into water and oxygen. How can this be explained?

The answer, lies in the fact that $\Delta H^{\ominus}_f(H_2O_2(l))$ only describes the stability of hydrogen peroxide *relative to its elements*:

$$H_2(g) + O_2(g) \rightarrow H_2O_2(l) \qquad \Delta H = -188\,kJ$$

H_2O_2 is obviously more stable than its elements, but on decomposition it produces not $H_2(g) + O_2(g)$ for which $\Delta H^{\ominus}$ is $+188\,kJ$, but $H_2O(l) + \frac{1}{2}O_2(g)$ for which $\Delta H^{\ominus}$ is $-98\,kJ$, i.e.

$$H_2O_2(l) \rightarrow H_2O(l) + \frac{1}{2}O_2(g) \qquad \Delta H = -98\,kJ$$

Thus, hydrogen peroxide is energetically stable with respect to its elements, but unstable with respect to water and oxygen. *This example shows how important it is to specify with respect to what substances a compound is stable or unstable.*

A few compounds, such as ethyne (C_2H_2) trioxygen (ozone, O_3), carbon disulphide (CS_2), and nitrogen oxide (NO), are formed endothermically from their elements:

$$2C(graphite) + H_2(g) \rightarrow C_2H_2(g) \qquad \Delta H^\ominus = +227\,kJ$$
$$\tfrac{1}{2}N_2(g) + \tfrac{1}{2}O_2(g) \rightarrow NO(g) \qquad \Delta H^\ominus = +90\,kJ$$

These compounds have positive standard heats of formation. They are therefore energetically unstable with respect to their elements. So, why don't these compounds decompose instantaneously into their constituent elements?

Ethyne (acetylene), trioxygen (ozone), carbon disulphide and nitrogen oxide can all be stored for long periods at room temperature and pressure in the absence of a catalyst. They do, however, begin to decompose at high temperatures or in the presence of a catalyst. In order to explain the unexpected stability of these compounds, we must distinguish between *energetic* stability and *kinetic* stability.

Thus, ethyne and nitrogen oxide are energetically unstable with respect to their elements. But at low temperatures and pressures the decomposition reactions are so slow that both ethyne and nitrogen oxide are kinetically stable. The kinetic stability of these energetically unstable compounds can be compared to the situation of a stone resting on a hillside. The stone is energetically unstable. Given the opportunity, it would roll to the bottom of the hill where it would come to rest in a position of lower energy. Resting on the hillside, stuck behind a tuft of grass, the stone's movement has been prevented. It is kinetically stable in spite of its energetic instability.

Diamond provides another example of an energetically unstable, yet kinetically stable, substance. At normal temperatures and pressures, diamond is unstable with respect to its allotrope, graphite:

$$C\,(diamond) \rightarrow C\,(graphite) \qquad \Delta H^\ominus = -2\,kJ$$

Fortunately, the rate of transformation of diamond to graphite is immeasurably slow at room temperature and so the diamond is kinetically stable. 'Diamonds are for ever', or are they?

The kinetic stability of nitrogen oxide with respect to its elements shows, yet again, how important it is to state clearly with respect to what a substance is stable. Obviously, nitrogen oxide is kinetically stable with respect to nitrogen and oxygen, but in the presence of air or oxygen it is energetically and kinetically unstable with respect to nitrogen dioxide. Hence, nitrogen oxide reacts rapidly with oxygen to form brown fumes of nitrogen dioxide:

$$2NO(g) + O_2(g) \rightarrow 2NO_2(g)$$

12.8   Predicting whether reactions will occur

The enthalpy change of a reaction is sometimes used as a rough guide to the likelihood that the reaction will occur. If $\Delta H^\ominus$ for a reaction is negative, energy is lost when the reaction occurs. The products are more stable than the reactants.

Thus, exothermic reactions are more likely to occur than endothermic reactions. Reactions which occur spontaneously are often very exothermic.

Although the value of $\Delta H^\ominus$ can provide some indication of the likelihood of a reaction, there are limitations to its use and it is important to bear these in mind.

1   $\Delta H^\ominus$ shows the relative energetic stabilities of the reactants and products for a reaction. It says nothing about the *kinetic* stability of the products relative to the reactants. Thus, $\Delta H^\ominus$ *is no guide to the rate of a reaction.* It cannot tell us whether the reaction is fast or slow. A reaction may be enormously exothermic, yet nothing happens. This is because the reaction rate is immeasurably slow and the reactants are kinetically stable with respect to the

An Atlas-Centaur rocket being launched. The rocket is carrying an intelligence satellite which will be released into orbit during the flight.

products. (For example, a mixture of hydrogen and oxygen at room temperature.) Reaction rates are studied in more detail in chapter 25.

2   In order to make accurate predictions concerning the relative energy levels of reactants and products, it is necessary to consider the energy lost or gained by the reacting system, and also any energy changes inside that system. For example, when a gas is produced in a reaction or a solid dissolves in a liquid, there is a marked increase in the disorder of the system itself and an increase in the number of ways in which the energy is distributed in the system. Consequently, in predicting the likelihood of any reaction we should take into account the change in order (or disorder) introduced into the system. At normal temperatures, this additional disorder is usually unimportant. Even so, it explains why certain endothermic reactions happen so spontaneously. For example, many solids dissolve in water readily and easily in spite of the fact that the process of dissolving is endothermic. The reason for this is that, although the enthalpy change is positive, there is an enormous increase in disorder (or entropy) as the solid dissolves, causing the reaction to occur spontaneously. Entropy is discussed further in chapter 24.

3   Predictions from $\Delta H_f^\ominus$ values relate only to standard conditions, i.e. 298 K and atmospheric pressure. The situation may be very different under different conditions or in the presence of a catalyst.

■ The heat of combustion of octane (in petrol) is $-5513 \text{ kJ mole}^{-1}$. Would you expect petrol to react with oxygen?
■ Why are petrol/oxygen mixtures stable at room temperature before a spark is applied?
■ How does the spark initiate a reaction between petrol and oxygen?

12.9   Heats of combustion and molecular structure

Look closely at the information in table 12.1. Notice the similarity in the heats of combustion for butane and methylpropane. Perhaps the similarity is not surprising. After all, they contain the same atoms and the same bonds.

Table 12.1
The formulae, bonds and heats of combustion of butane and methylpropane

	Butane	Methylpropane
Molecular formula	C_4H_{10}	C_4H_{10}
Structural formula		
Bonds	3 C—C 10 C—H	3 C—C 10 C—H
$\Delta H_c^\ominus/\text{kJ mole}^{-1}$	-2877	-2869

Table 12.2
The formulae and heats of combustion of the simpler alkanes

Alkane	Molecular formula	Structural formula	$\Delta H_c^{\ominus}$ /kJ mole^{-1}	Difference in $\Delta H_c^{\ominus}$ /kJ
Methane	CH_4	H | H—C—H | H	−890	
				670
Ethane	C_2H_6	H H | | H—C—C—H | | H H	−1560	
				660
Propane	C_3H_8	H H H | | | H—C—C—C—H | | | H H H	−2220	
				657
Butane	C_4H_{10}	H H H H | | | | H—C—C—C—C—H | | | | H H H H	−2877	
				643
Pentane	C_5H_{12}	H H H H H | | | | | H—C—C—C—C—C—H | | | | | H H H H H	−3520	

This suggests that each kind of bond makes a fixed contribution to the total enthalpy change. This idea is confirmed when we inspect the heats of combustion of the simpler alkanes (table 12.2). Two important points emerge from the data in table 12.2.

First, there is a regular difference in structure of a CH_2 group from one alkane to the next. Secondly, there is a similar difference of about 650 kJ in the value of $\Delta H_c^{\ominus}$ from one alkane to the next. This suggests that

—each additional CH_2 group is responsible for a fixed increment in the heat of combustion,
—each bond makes a characteristic contribution to the overall energy content of a substance.

What happens to the different bonds when an alkane burns? During combustion, the bonds between the atoms in the alkane molecule are broken. New bonds form between carbon and oxygen in CO_2 and between hydrogen and oxygen in H_2O. Notice that each alkane has one C—C bond and two C—H bonds more than the previous alkane. Check this for yourself.

Obviously, energy must be supplied in order to break a bond between two atoms. Consequently, energy is released when a bond forms. Thus, *bond-breaking is an endothermic process whereas bond-making is exothermic*.

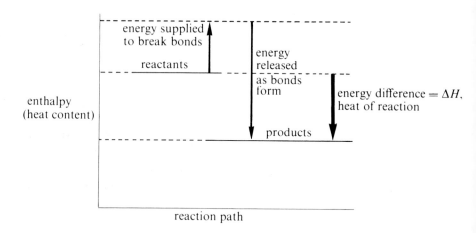

Figure 12.11
The heat of a reaction is the energy difference between bond-breaking and bond-making processes.

Chemical reactions involve bond-breaking followed by bond-making. This means that the heat of a reaction is the energy difference between bond-breaking and bond-making processes (figure 12.11). In section 12.11, we shall see how information concerning the strengths of bonds can be used to calculate the heats of reactions.

As the alkane molecules get larger in table 12.2, more energy is required to break the bonds between their carbon and hydrogen atoms. But, even larger amounts of energy are released as these atoms form carbon dioxide and water. On burning, each alkane molecule forms one CO_2 molecule and one H_2O molecule more than the previous alkane.

$$CH_4 + 2O_2 \rightarrow CO_2 + 2H_2O$$
$$C_2H_6 + 3\tfrac{1}{2}O_2 \rightarrow 2CO_2 + 3H_2O$$
$$C_3H_8 + 5O_2 \rightarrow 3CO_2 + 4H_2O$$

Thus, the heat of combustion gets larger by a constant amount as we progress along the homologous series of alkanes. The exothermic reactions show that the energy released on making the bonds in CO_2 and H_2O is greater than the energy required to break the bonds in the alkanes and O_2. An exothermic change is evidence for the formation of stronger bonds. An endothermic change is evidence for the formation of weaker bonds. Indeed, the energy we get from burning such fuels as coal, oil and natural gas results from the formation of strong C=O and H—O bonds in CO_2 and H_2O, respectively.

12.10   Finding the strength of the C—H bond and the C—C bond

From our discussions in section 12.9, it would seem that a definite quantity of energy (known as the **bond energy**) may be associated with each type of bond. This energy is absorbed when the bond is broken and evolved when the bond is formed. In order to find the strength of the C—H bond, it would seem appropriate to study methane. This contains C—H bonds only. Consider the following equation which relates to the heat of atomisation of methane:

$$
\begin{array}{c}
\text{H} \\
| \\
\text{H} - \text{C} - \text{H (g)} \rightarrow \text{C(g)} + 4\text{H(g)} \ldots \ldots \text{Equation 2} \\
| \\
\text{H}
\end{array}
$$

■ How many C—H bonds are broken down during the atomisation of 1 mole of methane?

■ Suppose ΔH for this process is x kJ mole^{-1}. What is the bond energy of one C—H bond in kJ mole^{-1}?

■ Why is equation 2 not simply the reverse of the heat of formation of methane?

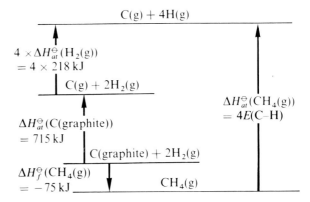

Figure 12.12
An energy level diagram which can be used in determining the C—H bond energy.

The standard heat of formation of methane does not relate directly to bond energies. We can, however, relate the C—H bond energy to the heat of formation of methane from its monatomic gaseous elements. This process is the reverse of equation 2 above.

Thus, in order to obtain the C—H bond energy we must first find the standard heats of atomisation for the elements present in methane as well as the standard heat of formation of methane. The C—H bond energy is usually written as $\overline{E}(C—H)$. An energy diagram connecting the relevant quantities for methane is shown in figure 12.12. From this diagram we can write:

$$4E(C—H) = -(-75) + 715 + 4 \times 218 \, kJ$$

$$\Rightarrow 4E(C—H) = 1662 \, kJ$$

∴ The average C—H bond energy in methane, $E(C—H) = 415.5 \, kJ \, mole^{-1}$

*Although average bond energies are usually referred to simply as bond energies, they are sometimes called **bond energy terms** and denoted by the symbol E.*

If we assume that the C—H bonds in ethane have the same strength as those in methane, we can now calculate the carbon–carbon bond energy, $E(C—C)$. Figure 12.13 shows the appropriate enthalpy changes in an energy diagram.

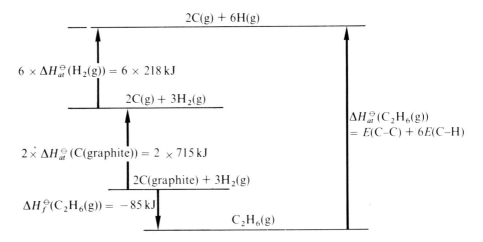

Figure 12.13
An energy level diagram used in determining the C—C bond energy.

Equating the two routes from $C_2H_6(g)$ to $2C(g) + 6H(g)$, we can write

$$E(C—C) + 6E(C—H) = -(-85) + 2 \times 715 + 6 \times 218 \, kJ$$

$$\Rightarrow E(C—C) + 6E(C—H) = 2823 \, kJ$$

Assuming that $E(C—H) = 415.5 \, kJ$;

$$E(C—C) + 6 \times 415.5 = 2823$$

$$\Rightarrow E(C—C) = 2823 - 2493 = 330 \, kJ$$

∴ The C—C bond energy in ethane, $E(C—C) = 330 \, kJ \, mole^{-1}$

Proceeding in this way, we could calculate the strength of other bonds. The idea that bond energies are additive and transferable from one molecule to another holds very well in most cases. However, results show small deviations in the strength of one particular type of bond in different molecules.

For example, the bond energy for C—C has been determined using several different compounds. The results range from 330 to 346 kJ mole⁻¹. This shows that a bond energy value depends to some small extent on the compound from which it was determined and upon the environment of the bond within the molecule. Thus, the strength of a C—C bond varies slightly due to the nature of the different atoms which are attached to the two carbon atoms.

12.11   What are the uses of bond energies?

The concept of bond energies is rather artificial because it does not relate to any everyday processes. It is nevertheless very useful. Four of its important uses are:

(a) comparing the strengths of bonds,
(b) understanding structure and bonding,
(c) estimating the enthalpy changes in reactions, and
(d) understanding the mechanisms of chemical reactions.

Table 12.3 shows a list of some bond energies.

The most important application of bond energies is probably their use in estimating the enthalpy changes in chemical reactions. These estimations are particularly useful when calorimetric or other experimental measurements cannot be made. Consider the following example.

Table 12.3
Average bond energies (bond energy terms)

Bond	Bond energy, $E(X{-}Y)$ /kJ per mole of bonds	Bond	Bond energy, $E(X{-}Y)$ /kJ per mole of bonds
C—H	413	O—O	146
C—C	346	O=O	497
C=C	610	N—N	163
C≡C	835	N≡N	945
C—F	495	N—H	390
C—Cl	339	O—H	463
C—Br	280	H—F	565
C—I	230	H—Cl	431
F—F	158	H—Br	365
Cl—Cl	242	C—O	360
Br—Br	193	C=O	740
I—I	151	Si—O	464
Si—Si	226		

Hydrazine is often used as a rocket fuel as it can be stored conveniently as a liquid and it reacts very exothermically with oxygen forming purely gaseous products.

$$N_2H_4(g) + O_2(g) \rightarrow N_2(g) + 2H_2O(g) \qquad \Delta H^\ominus = -622\,kJ$$

It has been suggested that hydrazine/fluorine mixtures might be more exothermic than hydrazine/oxygen mixtures. Using bond energies from table 12.3, we can find ΔH for the reaction of hydrazine with fluorine:

$$N_2H_4(g) + 2F_2(g) \rightarrow N_2(g) + 4HF(g)$$

The following equation shows clearly those bonds which are broken and those which form.

Bonds broken			Bonds made		
one	N — N		one	N≡N	
four	N — H		four	H — F	
two	F — F				

- Calculate the energy required to break the bonds in this reaction.
- Calculate the energy evolved when the bonds form in the products.
- What is the value of $\Delta H^\ominus$ for this reaction per mole of hydrazine?
- Is the reaction with fluorine more or less exothermic than the reaction with oxygen?
- If you had to decide whether to use oxygen or fluorine for the rocket flight what other factors besides the exothermicity of the reaction would influence your final decision?

For most reactions, the values of ΔH estimated from bond energies agree closely with experimentally determined values. This has further established the usefulness of bond energies. Occasionally, the estimated values for ΔH are in strong disagreement with the experimental results. This has resulted in a re-examination of the structure of compounds and an improvement in our understanding of bonding. A striking example of this is provided by benzene and other aromatic compounds. At one time, molecules of benzene were thought to contain hexagons of carbon atoms linked by alternate double and single bonds (figure 12.14).   Now, when cyclohexene, containing one (C=C) bond, undergoes an addition reaction with hydrogen, the enthalpy change is $-120\,kJ\,mole^{-1}$.

This would suggest that the hydrogen of 'Kekulé's' benzene, containing three (C=C) bonds should have an enthalpy change of $-360\,kJ$. When experiments are performed, the heat of hydrogenation of benzene is found to be $-208\,kJ\,mole^{-1}$ and not $-360\,kJ\,mole^{-1}$.

Figure 12.14
The Kekulé structure for benzene.

Figure 12.15
An energy level diagram to
emphasise the unexpected stability of
benzene.

It would seem that one mole of benzene is 152 kJ more stable than the Kekulé structure would suggest. This increased stability is emphasised in figure 12.15.

As a result of this discrepancy, chemists were forced to re-examine the structure of benzene. It is now known that all six C—C bonds in its molecule are equal in length and in strength. The molecule does not contain alternate double and single bonds as Kekulé had proposed. Furthermore, this new structure for benzene has improved our understanding of its properties and reactions and those of other aromatic compounds. The structure of benzene is discussed further in section 29.2.

12.12 Energy changes in forming ionic substances

In the last section, we found that bond energies were extremely useful in helping us to understand the properties and reactions of molecular substances. Bond energies for molecular substances can be compared with **lattice energies** for ionic compounds. The lattice energy of an ionic crystal is *the heat of formation for one mole of the ionic compound from gaseous ions under standard conditions*. Thus, the lattice energy of sodium chloride corresponds to the process:

$$Na^+(g) + Cl^-(g) \rightarrow Na^+Cl^-(s) \qquad \Delta H_{latt}(Na^+Cl^-(s))$$

As one might expect, lattice energies are very helpful in discussing the structure, bonding and properties of ionic compounds.

Unfortunately, lattice energies cannot be determined directly, but values can be obtained indirectly by means of an energy cycle. The energy cycle links the ionic solid, the gaseous ions and the elements in their standard states.

$$Na(s) + \tfrac{1}{2}Cl_2(g) \xrightarrow{\Delta H_x} Na^+(g) + Cl^-(g)$$

$$\Delta H_f^{\ominus}(Na^+Cl^-(s)) \searrow \qquad \swarrow \Delta H_{latt}(Na^+Cl^-(s))$$

$$Na^+Cl^-(s)$$

Since $\Delta H_f^{\ominus}$, the standard heat of formation of sodium chloride, can be measured conveniently in a calorimeter, ΔH_{latt} can be obtained if ΔH_x can be found.

In figure 12.16, the previous energy triangle has been extended to show the various stages involved in finding ΔH_x. The complete energy cycle, as shown in figure 12.16, is usually called a **Born–Haber Cycle**. Notice that

$$\Delta H_x = \Delta H_{at}(Na(s)) + \Delta H_i(Na(g)) + \Delta H_{at}(Cl_2(g)) + \Delta H_e(Cl)$$

The first two stages in this process involve atomising and then ionising sodium.

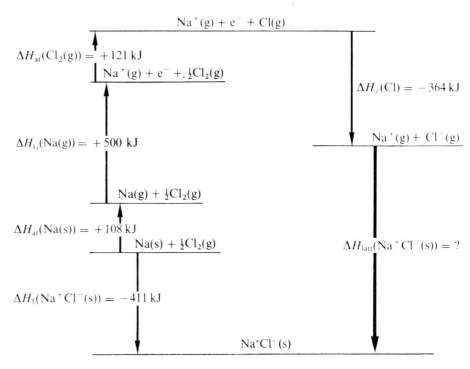

Figure 12.16
The Born–Haber cycle for sodium chloride.

The heat of atomisation of sodium

$$Na(s) \xrightarrow{\Delta H_{at}(Na(s))} Na(g)$$

can be obtained from its heat of fusion, heat of vaporisation and specific heat capacity.

The first ionisation energy of sodium,

$$Na(g) \xrightarrow{\Delta H_{i_1}(Na(g))} Na^+(g) + e^-$$

can be determined from spectra or by valve measurements (sections 6.2 and 6.3). The third and fourth stages, in the expression for ΔH_x, involve the atomisation of chlorine and the conversion of chlorine atoms to chloride ions respectively. The latter process is, of course, called the electron affinity for chlorine (section 8.3).

The heat of atomisation of chlorine,

$$\tfrac{1}{2}Cl_2(g) \xrightarrow{\Delta H_{at}(Cl_2(g))} Cl(g)$$

can be obtained from spectroscopic studies (section 12.4), whilst the electron affinity for chlorine,

$$Cl(g) + e^- \xrightarrow{\Delta H_e(Cl)} Cl^-(g)$$

can be found by methods similar to the valve measurements discussed in section 6.2.

The lattice energy for sodium chloride can now be obtained since all other values in the Born–Haber cycle can be determined experimentally.

Using the values in figure 12.16,

$$\Delta H_{latt}(Na^+Cl^-(s)) = (364 - 121 - 500 - 108 - 411) \text{kJ mole}^{-1}$$

$$= -776 \text{kJ mole}^{-1}$$

This lattice energy gives us some idea of the force of attraction between Na^+ and Cl^- ions in crystalline sodium chloride. The lattice energies of various other ionic crystals are given in table 12.4.

■ Why does the lattice energy become more exothermic along the series NaI, NaBr, NaCl, NaF?

■ The interionic distance between Na^+ and F^- ions in NaF is very similar to that

Table 12.4
Lattice energies of some ionic solids

Compound	Lattice energy /kJ mole^{-1}
NaF	−915
NaCl	−776
NaBr	−742
NaI	−699
MgCl$_2$	−2489
MgO	−3933

between Mg^{2+} and O^{2-} ions in MgO. Why, then, is the lattice energy of MgO about four times more exothermic than that of NaF?

Notice the similar size of the lattice energies in table 12.4 and the bond energies of covalent bonds in table 12.3. This would suggest that ionic bonds and covalent bonds are roughly similar in strength. Although discussion is often restricted to the two extreme types of bond, it is important to realise their underlying similarity. They both, of course, result from electrostatic forces, but involve different distributions of electrons.

12.13   Theoretical values for lattice energies— the ionic model

Essentially, the lattice energy of an ionic compound is the energy change which occurs when well-separated ions are brought together in forming the crystal. Consequently, it is possible to calculate a theoretical value for the lattice energy of a crystal. This is done by considering the interionic attractions and repulsions within the lattice.

The theoretical lattice energies of some ionic substances are compared with their corresponding experimental values in table 12.5. Notice the similarity between the theoretical and experimental lattice energies for the three alkali metal halides. The calculated values are only one or two per cent less than the experimental values. This close agreement provides strong evidence that the simple model of an ionic lattice is very satisfactory for alkali metal halides. This model is pictured as discrete spherical ions with evenly distributed charge.

Now look at the theoretical and experimental lattice energies for the silver halides in table 12.5. For these compounds the theoretical values are approximately $130 \, kJ \, mole^{-1}$ (i.e. about 15%) less than the experimental values. In this case, the simple ionic model is not completely satisfactory and requires some modification.

When the difference in electronegativity between the ions in the crystal is large, as in the case of alkali metal halides, the ionic model is satisfactory. However, when the difference in electronegativity gets smaller, as in the case of silver halides, there is a significant disagreement between experimental results and those calculated in terms of a simple ionic model. The experimental values suggest that the bonding in silver halides is stronger than the ionic model predicts. The explanation of this anomaly is that the bonding is not purely ionic but is intermediate in character between ionic and covalent. The partly covalent nature of the bonds can be interpreted by saying that the ionic bonds have been polarised (section 9.2) or by suggesting that the electrons are incompletely transferred in forming the ions. This further emphasises the fact that ionic and covalent bonds are simply extreme types. The bonding in most substances is intermediate in character between purely ionic and purely covalent.

Table 12.5
Theoretical and experimental lattice energies

Compound	Theoretical lattice energy/kJ mole^{-1}	Experimental lattice energy (obtained via Born–Haber cycle) /kJ mole^{-1}
NaCl	−766	−776
NaBr	−731	−742
NaI	−686	−699
AgCl	−768	−890
AgBr	−759	−877
AgI	−736	−867

12.14 Solution, hydration and lattice energy

When ionic solids dissolve in water, heat is usually evolved or absorbed. Why is this? Can we explain the enthalpy changes in terms of the processes taking place at a molecular and ionic level?

When sodium chloride is dissolved in water, the overall change can be represented as:

$$NaCl(s) + (aq) \rightarrow Na^+(aq) + Cl^-(aq)$$

For one mole of the solute and the formation of an infinitely dilute solution, this process is described as the **heat of solution** of sodium chloride, $\Delta H_{soln}(NaCl(s))$.

$$\Delta H_{soln}(NaCl(s)) = +5\,kJ$$

In order to understand this enthalpy change, the overall process of solution can be divided into two distinct stages which are shown in figure 12.17. First, the separation of the solid ionic crystal into monatomic gaseous ions. This is the reverse of the lattice energy process,

$$NaCl(s) \rightarrow Na^+(g) + Cl^-(g) \qquad -\Delta H_{latt} = +776\,kJ$$

Second, the solvation (hydration) of these gaseous ions by water molecules which is known as the **hydration energy**, ΔH_{hyd}.

$$Na^+(g) + Cl^-(g) + (aq) \rightarrow Na^+(aq) + Cl^-(aq)$$

The first of these stages is, of course, always endothermic since it involves separating the ions of the solute. The second stage, on the other hand, is always exothermic since it involves the attraction of ions in the solute for water molecules. The overall enthalpy change on solution will depend on whether the endothermic or the exothermic stage is the larger.

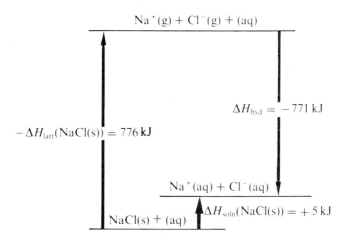

Figure 12.17
The relationship between lattice energy, hydration energy and heat of solution.

In the case of sodium chloride, the endothermic process is marginally greater than the exothermic process. So the heat of solution has a small positive value. The lattice energy and the hydration energy for ionic substances are nearly always large values. The heat of solution, which is the difference between these two values, is positive in some cases and negative in others. The relationship between these three quantities is:

$$\Delta H_{soln} = -\Delta H_{latt} + \Delta H_{hyd}$$

Check this for yourself in figure 12.17.

Notice that the hydration energy used in figure 12.17 is really the sum of the separate hydration energies of Na^+ and Cl^-. Clearly, the individual hydration energy for Na^+ cannot be measured directly because sodium ions always exist in combination with anions. Nevertheless, it is often very useful to know the individual hydration energies for particular ions. Various attempts have been made to estimate these values from the overall hydration energy of the ionic compound. The convention we shall follow in this book is to accept -1075 kJ as the hydration energy for H^+. All other enthalpies of hydration can then be obtained from the overall hydration energies of the compounds.

A few individual hydration energies are given in table 12.6. It does, of course, follow from what we have just said that the hydration energies of compounds can be obtained by addition of the hydration energies of their constituent ions. Look closely at table 12.6.

- Why are the hydration energies of *both* anions and cations negative?
- What is the hydration energy of $MgCl_2$?
- Why does the hydration energy get less exothermic along the series F^-, Cl^-, Br^-, I^-?
- Why does the hydration energy get more exothermic along the series Na^+, Mg^{2+}, Al^{3+}?

Table 12.6
Enthalpies of hydration of some ions

Ion	ΔH_{hyd} /kJ mole^{-1}	Ion	ΔH_{hyd} /kJ mole^{-1}
H^+	-1075	F^-	-457
Li^+	-499	Cl^-	-381
Na^+	-390	Br^-	-351
K^+	-305	I^-	-307
Mg^{2+}	-1891		
Ca^{2+}	-1562		
Al^{3+}	-4613		

12.15 Energy sources for the future

Until the Industrial Revolution, civilisations depended for their energy on the labour of people and animals or on the harnessing of wind and water.

Then, in the early nineteenth century, people began to use coal as a source of heat and power in their homes and industries. Civilisation progressed beyond the level of a simple agrarian society. At first, the use of coal was very small, but as the Industrial Revolution gained momentum towards the middle of the century, the consumption of coal began to double every fifteen years.

Before the Industrial Revolution, societies had to depend on the labour of people as their supply of energy. This engraving shows the construction of the pyramids.

Cutting peat, another form of fossil fuel, for domestic heating. Peat is still used as a source of fuel in some areas. It consists of plant fibres and mud and is probably the first stage in the formation of coal.

Early in the present century, oil began to make an impact. Discoveries multiplied fast and oil became plentiful, cheap and a source of many other products. Without oil, the internal combustion engine would have been impossible and the revolution in land, sea and air transport could never have taken place. Oil became the most convenient fuel for many industrial and domestic purposes and the basic raw material for the organic chemicals industry. Its consumption rose by leaps and bounds. By 1960, oil had outstripped coal as the major source of energy in most industrial countries.

A decade later, the first danger signals appeared as the rate of oil consumption increased faster than the discovery of new reserves.

Unfortunately, there is only a finite amount of oil on the Earth. From 1870 to 1970 we consumed only 15% of this oil, but in a single decade between 1970 and 1980 we consumed another 15%. If we continue at this rate, we shall exhaust our supplies of oil well before the year 2050.

The situation with coal is much less alarming since its rate of consumption is lower and reserves are possibly twenty times greater than those of oil. Nevertheless, it has been estimated that 80% of our coal reserves will have disappeared by the year 2500.

During the last decade we have begun to conserve our resources. Our thoughtless over-consumption has turned to a more intelligent use of fuel. We have begun to avoid frivolous wastes of energy like electrically operated tooth brushes. We have also reduced excessive heating and lighting in homes, offices, schools and factories. It has been shown that more than 20% of the energy used in heating buildings could be saved by better insulation of roofs and walls, and by double glazing windows. We have begun to accept slightly lower, but still comfortable, temperatures. It is vital, however, that we continue to conserve our resources of fuels.

One of the most wasteful users of fuel is the private car in which only 10–20% of the energy in the fuel ultimately goes towards moving the car. Hopefully, batteries will one day supersede the inefficient internal combustion engine. Vast quantities of precious oil are also being squandered on industrial and domestic heating and on the generation of electric power. Oil is much too valuable for this purpose. It should be conserved as feedstock for essential chemicals such as plastics, paints, pesticides and pharmaceuticals. A more suitable fuel for power generation is coal, but the increasing demands for electricity will also call for an increase in nuclear power generation. Furthermore, the electricity from nuclear power stations is likely to become cheaper than that from oil-powered stations over the next few years. By the year 2000, it is expected that the percentage of electricity generated from nuclear energy will have risen to about 70% from its present level of 18%.

Right. The huge turbine in the nuclear power station at Dounreay. Heat produced from nuclear reactions generates steam to drive the turbine.

Eventually, it is likely that there will be more 'fast-breeder' reactors. These can transform the energy in uranium 150 times more efficiently than present-day reactors. Unfortunately, the high capital cost of these new reactors and the problems of handling and storing large quantities of highly toxic waste with complete safety are retarding their construction and development.

Coal became important for home heating from the late nineteenth century onwards.

Water flowing from the turbine house below the dam at the hydroelectric power station at Genissiat on the River Rhône, France.

The tidal barrage on the River Rance near St. Malo, France. This uses energy from the tides to generate electricity.

Faced with these problems of dwindling energy resources and limited capital resources, it is essential to examine all possibilities for future energy provision. At the present time, hydro-electric schemes supply less than 1% of the world's electricity, yet there are possibilities for expansion in many areas. The production of hydro-electricity is essentially limited in its location. It has only been developed so far in those areas which are both mountainous and industrially advanced. It has been estimated that the full development of all suitable areas could increase the production of hydro-electricity by 20 times.

A less obvious source of water power lies in the rise and fall of the tides. The enormous movement of water between high and low tides results from the moon's gravitational pull on the seas. By damming tidal basins, it has been possible to generate electricity using low-speed turbines. One such installation is across the estuary of the Rance in Northern Brittany. There are many other places in the world, including the Bristol Channel, with potential for such development.

By far the biggest source of our energy is radiation from the sun. Scientists have calculated that one tenth of the solar energy falling on the Arizona Desert could provide enough energy for the whole of the USA. Unfortunately, we make very little direct use of solar energy. This is because technology has not yet provided an economic method of transforming solar radiation into useable energy. Solar cells are still very expensive. The most promising possibility for solar energy is the heating of buildings. Although sunlight is intermittent and unreliable in many areas, developments in the storage of solar energy may help to solve this problem. Alternatively, solar energy can be used in conjunction with conventional heating systems. This could save considerable quantities of our fossil fuels.

The sun's inexhaustible source of energy comes from nuclear fusion. The origin of this energy is exactly the same as that of the hydrogen bomb. Matter, below the sun's surface, is crushed together under a pressure millions upon millions of times greater than atmospheric pressure. Under this immense pressure atoms collide and undergo nuclear fusion. Only the smallest atoms can undergo fusion. The largest proportion of the sun's energy results from the fusion of heavy hydrogen (deuterium) atoms, forming helium.

$$\,^2_1H + \,^2_1H \rightarrow \,^3_2He + \,^1_0n + Energy$$

As a result of this fusion process, the energy radiated from the sun's surface in one year is equivalent to 250 thousand million million million tonnes of oil. If only we could harness the energy from such a fusion process on earth, we would have solved our energy problems for ever. Sea water, for example, contains 34 grams of deuterium per tonne, energy equivalent to 200 tonnes of oil. Another advantage of nuclear fusion is that the products are stable non-radioactive elements. This is very different from breeder reactors which pose considerable disposal problems and safety hazards. It is impossible to predict when, or indeed if ever, controlled fusion will be achieved. The major problem is that it requires temperatures of about 100000000°C before atomic collisions are vigorous enough to cause fusion. Although such elevated temperatures have not yet been reached, experiments have achieved temperatures close to those required. It is conceivable that full reaction conditions could be achieved within the next twenty years.

In the meantime, however, a world of limitless energy is still only a dream and we must learn to use and conserve our finite resources more wisely and more efficiently.

Summary

1 The transfer of energy to or from chemicals plays a crucial part in chemical processes in industry and in living things.

2 In an exothermic reaction, heat is lost from the reacting materials. In an endothermic reaction, heat is gained by the reacting materials.

Power from the wind—the Admiral's Cup inshore race at Cowes.

3 The standard heat of a reaction is the amount of heat absorbed or evolved when the molar quantities of reactants as stated in the equation react together under standard conditions.

4 Standard conditions for thermochemical measurements are:
a pressure of one atmosphere,
a temperature of 298 K,
substances in their normal physical states for these conditions,
concentrations of unit activity.

5 The standard heat of formation of a substance, $\Delta H_f^\ominus$, is the heat evolved or absorbed when one mole of the substance is formed from its elements under standard conditions.

6 The heat of atomisation of an element, $\Delta H_{at}^\ominus$, is the enthalpy change when one mole of gaseous atoms is formed from the element under standard conditions.

7 The standard heat of combustion of a substance, $\Delta H_c^\ominus$, is the enthalpy change when one mole of the substance is completely burnt in oxygen under standard conditions.

8 Hess's Law of Constant Heat Summation says that the energy change in converting reactants, A + B, to products, X + Y, is the same regardless of the route by which the chemical change occurs.

9 A substance is said to be stable if it tends neither to decompose nor to react spontaneously.

10 The heat of formation of a compound provides a measure of the energetic stability of that compound relative to its elements. If $\Delta H_f^\ominus$ is negative, the compound is more stable than its elements.

11 It is important to distinguish between energetic stability and kinetic stability. Compounds such as ethyne (C_2H_2) are energetically unstable with respect to their elements. But, they do not change at ordinary temperatures and pressures because their reaction rate is immeasurably slow. They are said to be kinetically stable.

12 Bond-breaking is an endothermic process whereas bond-making is exothermic.

13 A definite quantity of energy can be associated with each type of covalent bond. This energy is absorbed when the bond is broken and evolved when the bond is formed. Thus, the bond energy for the C—Cl bond, E(C—Cl), is defined as the energy required to break one mole of C—Cl bonds forming uncharged products under standard conditions.

14 Bond energies for molecular substances can be compared with lattice energies for ionic compounds. Lattice energies give the heat of formation of one mole of the ionic crystal from gaseous ions under standard conditions.

15 The close agreement between calculated and experimental lattice energies for many ionic compounds provides strong evidence that the simple model of an ionic lattice composed of discrete spherical ions with evenly distributed charge is generally very satisfactory.

Study questions

1 Figure 12.18 overleaf shows a diagram of the apparatus used by a student to determine the heat of combustion of ethanol. The heat produced by the burning fuel warms a known mass of water. By measuring the mass of fuel burnt and the temperature rise of the water, it is possible to obtain an approximate value for the heat of combustion of the fuel.

Figure 12.18
A simple apparatus to determine the
heat of combustion of a liquid fuel.

metal
calorimeter

thermometer

clamp

water

draught
shield

improvised
spirit lamp

ethanol

Volume of water in calorimeter	$= 400\,cm^3$
Initial temperature of water	$= 12°C$
Final temperature of water	$= 22°C$
Mass of ethanol burnt	$= 0.92\,g$
(Specific heat capacity of water	$= 4.2\,J\,g^{-1}\,K^{-1}$)

(a) How much heat is required to raise the temperature of the water from 12°C to 22°C?
(This is the amount of heat produced when 0.92 g of ethanol burn.)
(b) How much heat would be produced when 1 mole of ethanol burns? (Call this the heat of combustion of ethanol.)
(c) Why is the answer to (b) not described as the *standard* heat of combustion of ethanol?
(d) An accurate value for $\Delta H_c^{\ominus}(C_2H_5OH(l))$ is $-1368\,kJ\,mole^{-1}$. Mention four serious errors in the simple experiment which could be responsible for the poor result.

2   Ethanol (C_2H_5OH) cannot be prepared directly from its elements so the standard heat of formation of ethanol must be obtained indirectly.
(a) Define the terms: $\Delta H_f^{\ominus}$ and $\Delta H_c^{\ominus}$.
(b) Write an equation for the formation of ethanol from its elements in their standard states.
(c) Draw an energy cycle linking the heat of formation of ethanol with its heat of combustion and the heat of combustion of its constituent elements.
(d) Calculate $\Delta H_f^{\ominus}(C_2H_5OH(l))$.
$$\Delta H_c^{\ominus}(C(graphite)) = -393\,kJ\,mole^{-1}$$
$$\Delta H_c^{\ominus}(H_2(g)) = -286\,kJ\,mole^{-1}$$
$$\Delta H_c^{\ominus}(C_2H_5OH(l)) = -1368\,kJ\,mole^{-1}$$

3   Two campers are desperately short of calor gas, yet they badly need a hot drink. They estimate that they have the equivalent of $1.12\,dm^3$ of calor gas (measured at 0°C and 1 atm) in their 'gas bottle'.
(a) What is the maximum volume of water (at 20°C) which they could boil in order to make some hot coffee?
(Assume that calor gas is pure butane, $\Delta H_c(C_4H_{10}(g)) = -3000\,kJ$ $mole^{-1}$ and that 75% of the heat evolved in burning the calor gas is absorbed by the water.)
(b) State any other assumptions you make.

4   Table 12.7 shows the standard heats of formation of the oxides of elements in Group I and Group IV.
(a) Why does $\Delta H_f^{\ominus}$ for the oxides of Group I elements get less exothermic as the atomic numbers of the elements increase?
(b) Why is the $\Delta H_f^{\ominus}$ of SiO_2 much more exothermic than that for CO_2?
(c) Suggest reasons why there is a steady trend in $\Delta H_f^{\ominus}$ values for the oxides of Group I elements, but there is no similar trend in $\Delta H_f^{\ominus}$ values for the oxides of Group IV.

Table 12.7
Standard heats of formation of the oxides of elements in Group I and Group IV

Group I	$\Delta H_f^{\ominus}$ -kJ mole^{-1}	Group IV	$\Delta H_f^{\ominus}$ /kJ mole^{-1}
Li_2O	-596	CO_2	-393
Na_2O	-416	SiO_2	-909
K_2O	-361	GeO_2	-551
Rb_2O	-330	SnO_2	-581
Cs_2O	-318	PbO_2	-277

5 (a) Write equations, including state symbols, for the complete combustion of glucose ($C_6H_{12}O_6(s)$) and ethanol ($C_2H_5OH(l)$).
   (b) Calculate the enthalpy change during the fermentation of glucose.

$$C_6H_{12}O_6(s) \rightarrow 2C_2H_5OH(l) + 2CO_2(g)$$

Assume that the heats of combustion of glucose and ethanol are -2820 and $-1368\,kJ$ mole^{-1}, respectively.
   (c) In breweries, the vats in which fermentation is carried out are sometimes fitted with copper pipes to promote cooling during fermentation. Why is this?
   (d) With your answer to part (c) in mind, why is it that people who brew beer at home usually put their fermentation vessel in a warm place (such as an airing cupboard)?

6 A possible mechanism for the reaction of fluorine with methane is

$$CH_4 + Fe_2 \xrightarrow{\text{slow}} CH_3. + HF + F$$
$$CH_3. + F \xrightarrow{\text{fast}} CH_3F$$

   (a) Use the bond energies in table 12.3 to calculate the enthalpy change in each step of the reaction mechanism.
   (b) Is the suggested mechanism viable? Explain your answer.
   (c) Write the equations for a reaction between CH_4 and Cl_2 assuming a similar mechanism to that for CH_4 and F_2 and calculate the enthalpy change in each step.
   (d) Is such a mechanism viable for the CH_4/Cl_2 reaction? Explain your answer.
   (e) Why is that fluorine will react with methane in the dark, whereas chlorine only reacts appreciably with methane in sunlight?

7 (a) A simple rule regarding the solubility of solutes in solvents is 'Like dissolves like'. In general, non-polar solutes dissolve readily in non-polar solvents but not in polar ones. On the other hand, polar solutes dissolve readily in polar solvents but are insoluble in non-polar solvents. Discuss and explain this pattern in solubility, mentioning specific examples and referring to solute–solute, solvent–solvent and solute–solvent attractions.
   (b) Explain the terms lattice energy, hydration energy and heat of solution with reference to the hypothetical substance, X^+Y^-.
   (c) Draw an energy diagram relating the three terms in part (b).
   (d) Calculate the hydration energy of potassium iodide, assuming that its heat of solution is $+21\,kJ$ mole^{-1} and its lattice energy is $-642\,kJ$ mole^{-1}.

8 (a) Explain what is meant by the terms
  (i)   ionisation energy,
  (ii) atomisation energy,
  (iii) lattice energy.
 (b) Draw a complete, fully labelled Born–Haber Cycle for the formation of potassium bromide.
 (c) Using the information in the table below, calculate the lattice energy of potassium bromide.

Reaction	ΔH/kJ mole^{-1}
$K(s) + \frac{1}{2}Br_2(l) \rightarrow K^+Br^-(s)$	-392
$K(s) \rightarrow K(g)$	$+90$
$K(g) \rightarrow K^+(g) + e^-$	$+420$
$\frac{1}{2}Br_2(l) \rightarrow Br(g)$	$+112$
$Br(g) + e^- \rightarrow Br^-(g)$	-342

 (d) the values of the lattice energies of the other potassium halides are:

Compound	KF	KCl	KI
Lattice energy/kJ mole^{-1}	-813	-710	-643

What explanation can you give for the trend in these values?

9 The enthalpy changes involved in the synthesis of calcium oxide are represented in a Born–Haber cycle below. (The numerical values printed beside the cycle are in kJ mole^{-1}.)

$$Ca^{2+}(g) + O^{2-}(g)$$

$+790 =$	$\Delta H_e(O^-)$	
	$Ca^{2+}(g) + e^- + O^-(g)$	
$-141 =$	$\Delta H_e(O)$	
	$Ca^{2+}(g) + 2e^- + O(g)$	
$+249 =$	$\Delta H_{at}(O)$	
	$Ca^{2+}(g) + 2e^- + \frac{1}{2}O_2(g)$	
$+1\,100 =$	$\Delta H_{i_2}(Ca)$	
	$Ca^+(g) + e^- + \frac{1}{2}O_2(g)$	$\Delta H_{latt}(CaO(s))$
$+590 =$	$\Delta H_{i_1}(Ca)$	
	$Ca(g) + \frac{1}{2}O_2(g)$	
$+177 =$	$\Delta H_{at}(Ca)$	
	$Ca(s) + \frac{1}{2}O_2(g)$	
$-636 =$	$\Delta H_f(CaO(s))$	
	$Ca^{2+}O^{2-}(s)$	

 (a) Calculate the lattice energy for calcium oxide, $\Delta H_{latt}(CaO(s))$.
 (b) Why is $\Delta H_{i_2}(Ca)$ greater than $\Delta H_{i_1}(Ca)$?
 (c) why is $\Delta H_e(O)$ negative whereas $\Delta H_e(O^-)$ is positive?
 (d) State and explain how the value of
  (i) the first ionisation energy and
  (ii) the heat of atomisation
  of magnesium would compare with the corresponding values for calcium.

13 Patterns Across the Periodic Table

13.1 Periodic properties

In chapter 4 we noticed that a large number of chemical and physical properties of the elements vary periodically with atomic number. *The idea of periodicity embodied in the periodic table is one of the major unifying themes in chemistry.* It provides a structure for our knowledge and understanding of inorganic chemistry. It also reduces the need for memorising isolated pieces of factual information. In this chapter, we shall look more closely at the trends and gradations in properties of the elements in periods 2 and 3. We shall also look at their more important compounds. Several of the bulk physical properties of the elements, such as melting point and density, depend upon their structure and bonding. We discussed these relationships in chapter 4. However, the structure and bonding of the elements are in turn related to atomic properties such as electron structures, ionisation energies and atomic radii. It is therefore profitable to discuss the trends in these atomic properties and then relate them to physical and chemical properties. Table 13.1 provides a summary of the structure, bonding and properties of the elements in periods 2 and 3.

13.2 Atomic structure and the periodic table

Chemists now realise that the atomic number of an element dictates the number and arrangement of its electrons.

Table 13.2 shows the electronic (shell and sub-shell) structures for the elements in periods 2 and 3.

Elements in the same group of the periodic table have similar electron configurations. For example, Group I elements (Li to Fr) have one electron in their outer shells (ns^1). Group II elements (Be to Ra) have two electrons in their outer shell (ns^2) and Group VII elements (F to At) have seven outer shell electrons (ns^2np^5).

Since electron configurations recur in a periodic manner, it is not surprising that properties show a similar periodic recurrence. We shall investigate the similarities in the properties of elements within a *group* more thoroughly in chapters 15, 17, and 18. Meanwhile, we can investigate the variation of atomic properties across a period.

Table 13.1
A summary of the structure, bonding and properties of the elements in periods 2 and 3

Period 2	Li Be	B C	N O F Ne
Period 3	Na Mg Al	Si	P S Cl Ar
Type of element	metal	metalloid	non-metal
Structure	Metallic Close-packed or body-centred cubic arrangement of atoms	Giant molecular Infinite lattice structure	Simple molecular Discrete small molecules (e.g. P_4, Cl_2, O_2, Ne)
Bonding	Metallic: strong forces of attraction of positive ions for mobile outer electrons.	Covalent: very strong forces of attraction between atoms due to the attraction of nuclei for shared electrons.	Molecular (Van der Waals'): weak forces between molecules (strong covalent forces hold atoms together within the molecule).
Co-ordination number	8 or 12	4 or less	1 or 2 usually
Properties m.pt./b.pt.	High	Very high	Low
$\Delta H_{fus}/\Delta H_{vap}$	High	Very high	Low
Molar volume	Moderate	Low	High
Conductivity	Good	Poor	Nil

13.3 Atomic properties and the periodic table

Atomic and ionic size

The electron cloud of an atom has no definite limit. Because of this, the size of an atom cannot be defined simply and easily. However, *the radius of an atom is often defined as the distance of closest approach to another identical atom.* This means that the size of an atom is determined by the effective radius of the outer electrons. But, what do we mean when we say the atomic radius of iodine in the solid is 0.133 nm? Is it half the distance between the two iodine atoms in an I_2 molecule or half the distance between two iodine atoms in adjacent molecules which are not chemically bonded (figure 13.1)? The following terms clarify this situation. *Half the distance between two covalently bonded iodine atoms is defined as the covalent radius for iodine. Half the distance between two iodine atoms which are not chemically bonded is defined as the Van der Waals' radius.* The former gives a measure of the length of a covalent bond. The latter gives a measure of the length of a Van der Waals' bond. The covalent radius for iodine is 0.128 nm and the Van der Waals' radius is 0.215 nm. Clearly, the type of bonding greatly influences the size of an atom. When chemists refer to *atomic radii* they are usually thinking of *covalent radii*.

Table 13.2
The electronic structures of the elements in periods 2 and 3

Periodic 2	Li	Be	B	C	N	O	F	Ne
Electronic shell structure	2,1	2,2	2,3	2,4	2,5	2,6	2,7	2,8
Electronic sub-shell structure	$1s^2 2s^1$	$1s^2 2s^2$	$1s^2 2s^2 2p^1$	$1s^2 2s^2 2p^2$	$1s^2 2s^2 2p^3$	$1s^2 2s^2 2p^4$	$1s^2 2s^2 2p^5$	$1s^2 2s^2 2p^6$
Period 3	Na	Mg	Al	Si	P	S	Cl	Ar
Electronic shell structure	2,8,1	2,8,2	2,8,3	2,8,4	2,8,5	2,8,6	2,8,7	2,8,8
Electronic sub-shell structure	$1s^2 2s^2 2p^6$ $3s^1$	$1s^2 2s^2 2p^6$ $3s^2$	$1s^2 2s^2 2p^6$ $3s^2 3p^1$	$1s^2 2s^2 2p^6$ $3s^2 3p^2$	$1s^2 2s^2 2p^6$ $3s^2 3p^3$	$1s^2 2s^2 2p^6$ $3s^2 3p^4$	$1s^2 2s^2 2p^6$ $3s^2 3p^5$	$1s^2 2s^2 2p^6$ $3s^2 3p^6$

Figure 13.1
Covalent (atomic) and Van der Waals'
radii for iodine.

covalent radius
0.128 nm

Van der Waal's radius
0.215 nm

The size of atoms in metallic elements can be obtained by determining the inter-nuclear distance in the metallic crystal using X-ray diffraction and then dividing by two to give an atomic radius.

Figure 13.2 shows the atomic radii (covalent radii) of the elements Li–F and Na–Cl in the periodic table. Notice that along each period there is a gradual decrease in atomic size as the outer electron shell is being filled.

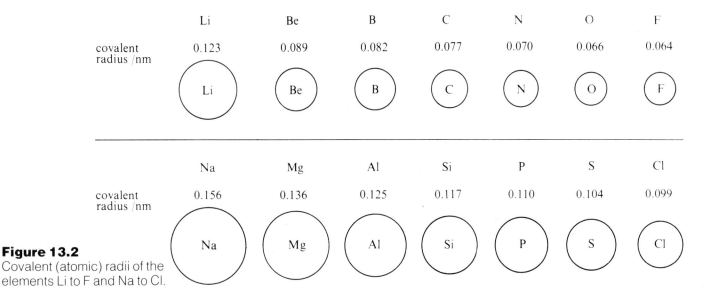

	Li	Be	B	C	N	O	F
covalent radius /nm	0.123	0.089	0.082	0.077	0.070	0.066	0.064

	Na	Mg	Al	Si	P	S	Cl
covalent radius /nm	0.156	0.136	0.125	0.117	0.110	0.104	0.099

Figure 13.2
Covalent (atomic) radii of the
elements Li to F and Na to Cl.

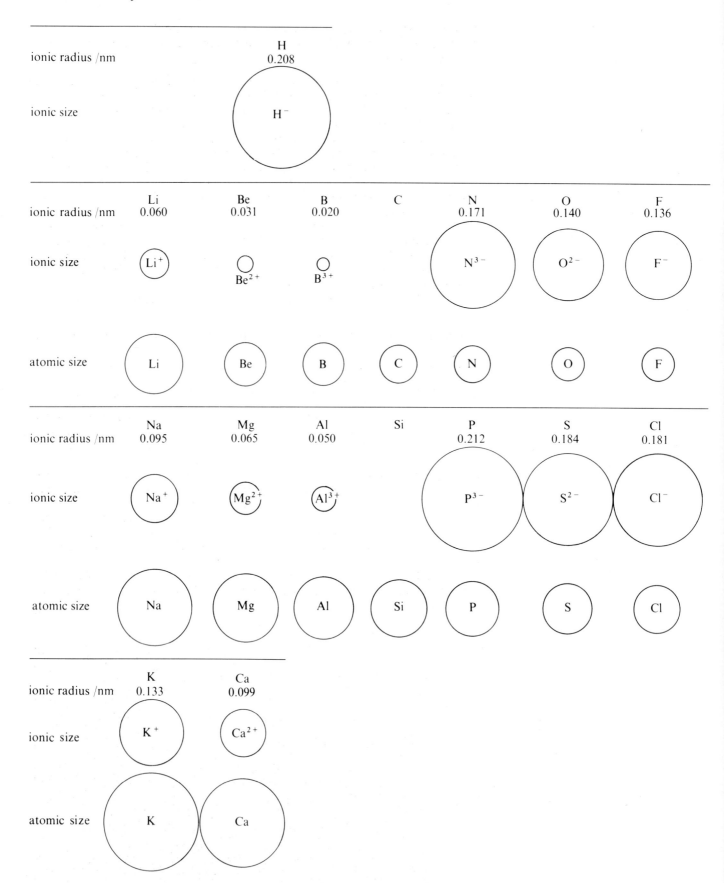

Figure 13.3
Radii of the most stable ions for some of the first 20 elements in the periodic table.

■ What causes the general decrease in size across a period?
■ How is the attraction of the positive nucleus for the outer electrons changing across the period?

Moving from one element to the next across a period, electrons are being added to the same shell at about the same distance from the nucleus. At the same time, protons are being added to the nucleus. Therefore, the electrons are attracted and pulled towards the nucleus with an increasing positive charge. So the radius of the atom decreases. However, the rate of decrease in the radius becomes smaller as the atoms get heavier. The addition of one more proton to the 11 already present in sodium causes a greater proportional change in nuclear attractive power than the addition of one more proton to the 16 already present in sulphur. Thus, the atomic radius falls by 0.020 nm from Na to Mg, but by only 0.005 nm from S to Cl.

So far we have considered only atomic sizes. Ionic sizes are also useful in explaining and understanding chemical properties. They provide a measure of the space occupied by an ion in the crystal lattice. Figure 13.3 shows the radii of the most stable ions for some of the first 20 elements in the periodic table.

Look for the following general patterns in figure 13.3.

(a) *The ionic radii of positive ions are smaller than their corresponding atomic radii.*

■ Which electrons are usually removed from metals in forming their positive ions?
■ How does the electronic structure of a positive ion (e.g. Na^+ or Mg^{2+}) compare with that of the corresponding metal atom?
■ Why is the radius of a positive ion less than that of the corresponding atom?

(b) *The ionic radii of negative ions are greater than their corresponding atomic radii.* When one or more electrons are added to the outer shell of an atom forming a negative ion, there is an increase in the repulsion between negative charge clouds. This results in an overall increase in size.

(c) *In a series of ions with the same number of electrons (an isoelectronic series), the ionic radius decreases as the atomic number increases.* In figure 13.4 the solid lines link those ions which are isoelectronic.

Why does ionic size decrease along these isoelectronic sequences? The nuclear charge increases along the isoelectronic series (e.g. from N^{3-} to Al^{3+}). This makes the electron cloud contract, because it is pulled in more effectively by an increasing positive charge.

Ionic radii provide a measure of ionic size in a *solid* crystal, but they give no accurate indication of the size of *aquated* ions in solution. Here the situation can

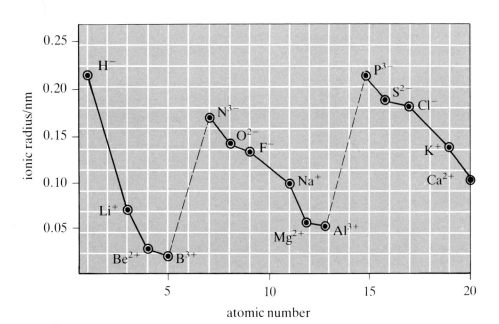

Figure 13.4
Variation in ionic size with atomic number.

be very different (section 12.14). The increasing ionic size from Li^+ through Na^+ to K^+ might lead us to expect that the electrical mobility would decrease in the order Li^+, Na^+, K^+. In practice, $K^+(aq)$ is the most mobile and $Li^+(aq)$ is the least mobile of the three aqueous ions. This is due to solvation by water molecules. The Li^+ ion has the largest charge density per unit of surface area and therefore attracts polar solvent (water) molecules around it more strongly than the Na^+ ion. Similarly, Na^+ exerts more attraction on polar solvent molecules than K^+. Thus, the effective size of the aqueous ions is $Li^+(aq) > Na^+(aq) > K^+(aq)$ and this makes $Li^+(aq)$ least mobile.

Ionisation energy

The striking periodic variation in the first ionisation energies of the elements was discussed in section 6.6. Figure 13.5 shows a graph of the first ionisation energy against atomic number for the first 20 elements. Notice that within a period the first ionisation energy tends to rise as atomic number increases. This increase in ionisation energy is associated to some extent with a decrease in metallic character from left to right in the periodic table. Metals with mobile outer-shell electrons have low ionisation energies. The ionisation energy gradually rises across the period as metallic character disappears and electrons are more tightly held in covalent bonds. But, how are ionisation energies related to atomic structure?

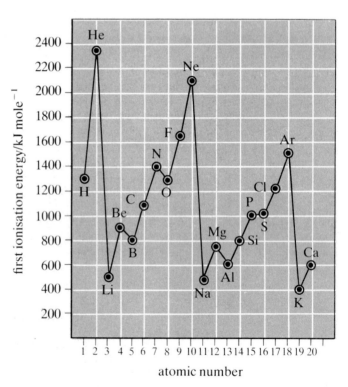

Figure 13.5
The first ionisation energies plotted against atomic number for the elements H to Ca.

The ionisation energy of an atom is strongly influenced by three atomic parameters.

1   *The distance of the outermost electron from the nucleus*
As this distance increases, the attraction of the positive nucleus for the negative electron will decrease. Consequently the ionisation energy will decrease.

2   *The size of the positive nuclear charge*
As the nuclear charge gets more and more positive, its attraction for the outermost electron increases and consequently the ionisation energy increases.

3   *The screening (shielding) effect of inner electrons*
The outermost electrons are repelled by all the other electrons in the atom besides being attracted by the positive nucleus. Chemists say that the outermost

electron is **screened** (shielded) from the attraction of the nucleus by the repelling effect of inner electrons. For example, the outermost electron in an atom with an atomic number of Z (i.e. Z protons in the nucleus) experiences an attraction from Z protons and a repulsion from $(Z - 1)$ electrons. At first sight, it might appear that the $(Z - 1)$ electrons could cancel all but one of the nuclear charges. In practice, the outermost electron occupies an orbital which is not completely outside the orbitals of other inner electrons. So the screening (shielding) effect is much less than perfect. This means that the electron experiences an overall effective nuclear attraction which is much greater than that from one proton. In general, however, *the screening effect by inner electrons is more effective the closer these inner electrons are to the nucleus*. Thus,

(a) electrons in shells of lower principal quantum number are more effective shields than electrons in shells of higher quantum number.
(b) electrons in the same shell exert a negligible shielding effect on each other.

This means that we need only consider inner shells of electrons when we discuss the screening effect on an outermost electron.

Moving from left to right across any period, there is a general increase in the first ionisation energy. The nuclear charge is increasing across the period from one element to the next. The increasing nuclear charge causes a decrease in the atomic radius (see earlier) and thus a decrease in the distance of the outermost electron from the nucleus. The screening effect remains almost the same from one element to the next across a period because electrons are added successively to the same shell and such electrons screen each other very little. Look closely at elements Li to Ne in figure 13.5. Notice that the ionisation energy of beryllium is higher than that of boron.

■ Which other element in period 2 has a higher first ionisation energy than the element immediately after it?
■ The electron structure of beryllium is $1s^2 2s^2$. Write out the electron structure of boron in the same notation.
■ Which of the two elements, beryllium or boron, has the more stable electron structure?
■ Why is the first ionisation energy of beryllium higher than that of boron?

The electron configuration of beryllium $1s^2 2s^2$, whereas that of boron is $1s^2 2s^2 2p^1$. All the sub-shells in beryllium are filled, but the outer sub-shell of boron contains only one electron. From our studies in chapter 6, we know that filled electron shells are associated with extra stability. *There is also some extra stability associated with filled sub-shells*. This means that the electron structure in beryllium is rather more stable than we might have expected and its first ionisation energy is greater than that of boron.

A similar situation arises with nitrogen which has a higher first ionisation energy than oxygen. The electron structures of nitrogen and oxygen are $1s^2 2s^2 2p^3$ and $1s^2 2s^2 2p^4$, respectively. The half-filled $2p$ sub-shell in nitrogen with one electron in each of the three $2p$ orbitals and its evenly distributed charge is more stable than the $2p$ sub-shell in oxygen which contains four electrons. This results in a higher first ionisation energy for nitrogen than oxygen.

Electronegativity

Elements that tend to acquire electrons in their chemical interactions are said to be **electronegative**. The electronegativity of an atom provides a numerical measure of the power of that atom in a molecule to attract electrons (section 9.2). The electronegativity values for main-group elements (based on Pauling's scale) are given in table 13.3. The values for the first 20 elements in the periodic table are plotted graphically against atomic number in figure 13.6.

Notice that electronegativities decrease down a group, but increase across a period. As expected, the most electronegative elements are the reactive non-metals in the top right-hand corner of the periodic table. In contrast, the least electronegative elements are the reactive metals in the bottom left-hand corner. Why does electronegativity increase from left to right across the periodic table? Moving from one element to the next across a period, the nuclear charge increases by one unit and one electron is added to the outer shell. As the positive charge on the nucleus rises, the atom has an increasing electron-attracting power and therefore an increasing electronegativity.

Table 13.3
Electronegativities for the main-group elements

H 2.1									He —
Li 1.0	Be 1.5			B 2.0	C 2.5	N 3.0	O 3.5	F 4.0	Ne —
Na 0.9	Mg 1.2			Al 1.5	Si 1.8	P 2.1	S 2.5	Cl 3.0	Ar —
K 0.8	Ca 1.0			Ga 1.6	Ge 1.8	As 2.0	Se 2.4	Br 2.8	Kr —
Rb 0.8	Sr 1.0			In 1.7	Sn 1.8	Sb 1.9	Te 2.1	I 2.5	Xe —
Cs 0.7	Ba 0.9			Tl 1.8	Pb 1.8	Bi 1.9	Po 2.0	At 2.2	Rn —
Fr 0.7	Ra 0.9								

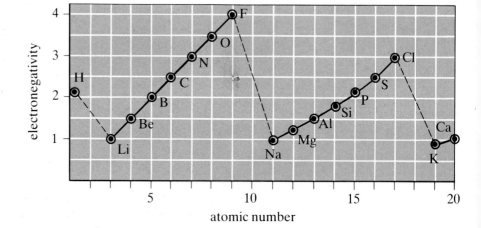

Figure 13.6
Variation in electronegativity with atomic number.

13.4 Physical properties and the periodic table

Melting point

The melting point of a substance is the temperature at which pure solid is in equilibrium with pure liquid at atmospheric pressure. Moving across a period from left to right, the melting point rises through the metals and metalloids and then drops abruptly to low values for the non-metals (see figure 4.8). Although there is some evidence of periodicity in the melting points of elements, it is not so obvious as the periodicity of other properties. This is partly because there are several factors which influence a melting point. Melting points depend on both the structure (packing) and on the bonding in a substance. Since there are abrupt changes in bonding and structure across a period there are also abrupt changes in the melting point. For example, in the period Na→Ar, the melting point rises sharply from the last metallic element, aluminium, to the giant molecule, silicon.

It then falls even more abruptly from silicon to the simple molecule phosphorus. In the period Li→Ne, the melting point rises steeply from the metallic element, beryllium, to the giant molecule, boron, and then falls sharply from the giant molecule, carbon, to the simple molecule, nitrogen.

Molar heat of fusion

The molar heat of fusion is the amount of energy required to melt one mole of the solid at its melting point. As expected, there is a close similarity in the periodicity of the heats of fusion and the periodicity of the melting points (figure 4.8). When a marked change of structure or bonding occurs, there is a sharp change in the ΔH_{fus} in the same way that we noticed sharp changes in the melting point in the last sub-section.

Consider the changes in ΔH_{fus} across the third period from Na to Ar. In metals (Na, Mg, Al) a good deal of the bonding remains when the metal liquefies. Only restricted bond-breaking has occurred, so the heat of fusion is only moderate. In giant molecules such as silicon, nearly all the bonds must be broken before the solid melts. Therefore the heat of fusion is very large. In simple molecules (P_4, S_8, Cl_2 and Ar), the heats of fusion are very small because melting in these elements involves only the breaking of molecular (Van der Waals') bonds.

Boiling point and molar heat of vaporisation

The periodic trends in this case can be related to those in the last two sub-sections. However, the structure (packing) has already been disrupted in forming the liquid. So, the boiling point and the ΔH_{vap} depend very much on the strength of the bonds which are broken in converting the liquid to a vapour. In the case of metals, most of the metallic bonding still exists in the liquid state. The atoms must be separated to a considerable distance in forming the vapour. This involves complete breakage of the metal bonds. Thus, the boiling points of metals are much higher than their melting points.

In the case of giant structures, such as carbon and silicon, most of the bonds have been broken during melting. So, the boiling point, though very high, is not much higher than the melting point. The boiling of elements with simple molecular structures, such as P_4, and S_8 and Cl_2, involves breaking weak Van der Waals' bonds. This is relatively easy for most of these elements. The boiling points occur at low temperatures, not much above the melting point.

- What happens to the particles of a substance during vaporisation?
- Why is the molar heat of vaporisation usually larger than the molar heat of fusion?
- Use the values in tables 4.2 and 4.3 to plot a graph of the molar heats of vaporisation for at least 10 elements against their boiling points. Is there any relationship between these two quantities?

Density

The density of a substance shows its mass per unit volume. Gaseous elements present a problem in comparing densities. With these elements, it is usual to take the density of the element as a liquid. What factors will influence the density of an element? The factors to consider are atomic mass, atomic radius (which determines the volume of each atom) and crystal structure (which determines how close to each other the atoms are packed). Although each of these factors varies from one element to the next across a period, there is a periodic trend in density (see figure 4.10).

Across a period there is a general increase in density until a maximum is reached at Group IV for periods 2 and 3 or at the end of the transition metals for the later periods. This increase in density can be related to the increasing atomic mass and decreasing atomic radius. A maximum is reached when the atomic radius reaches a low value and the strength of the metallic or giant covalent bonding is at a maximum. In spite of having higher atomic masses and lower atomic radii, elements at the right-hand side of each period have relatively low densities. This is because they form simple molecular structures in which molecules are only weakly bonded by Van der Waals' forces.

Molar volume

The molar volume of an element is the volume occupied by 1 mole $(6 \times 10^{23}$ *atoms) of the element.* It is very dependent on atomic radius and structure (packing). Suppose the relative atomic mass of the element is A_r and its density is ρ g cm^{-3}.

$$6 \times 10^{23} \text{ atoms of the element have a mass of } A_r \text{ g}$$

$$\therefore 6 \times 10^{23} \text{ atoms of the element have a volume of } \frac{A_r}{\rho} \text{ cm}^3$$

$$\therefore \text{molar volume} = \frac{A_r}{\rho} = \frac{\text{relative atomic mass in grams}}{\text{density}}$$

Since molar volume is *inversely related to density*, it is not surprising that trends in molar volume are the *reverse* of trends in density. Thus molar volumes gradually fall across a period to a minimum value and then rise again (figure 4.10). However, the major factor affecting volume is structure rather than atomic radius. So, there are abrupt changes in molar volume across a period where there are obvious changes in structure. For example, the atomic radius of phosphorus is less than that of silicon as expected, but phosphorus has the larger molar volume because it contains weakly bonded P_4 molecules whereas silicon is a strongly bonded giant molecule.

13.5 Patterns in the formulae of compounds

Table 13.4 shows the formulae of the oxides, chlorides and hydrides of the elements in periods 2 and 3.

Look at the formulae of the chlorides.

■ What is the oxidation number of each element in its chloride?
■ How do these oxidation numbers vary across periods 2 and 3?
■ Why do the oxidation numbers vary in this way?

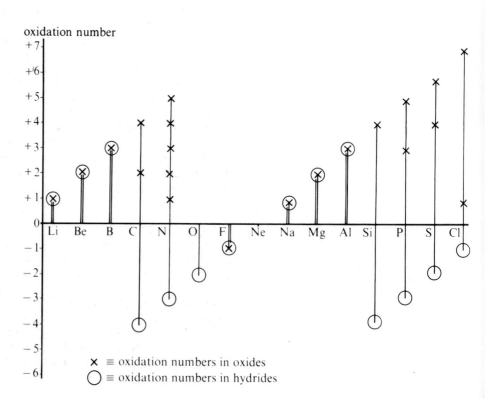

Figure 13.7
Oxidation numbers of the elements from Li to Cl.

Table 13.4
Formulae of the chlorides, oxides and hydrides of the elements in periods 2 and 3

Period 2	Li	Be	B	C	N	O	F	Ne
Formula of chloride	LiCl	BeCl$_2$	BCl$_3$	CCl$_4$	NCl$_3$	Cl$_2$O	ClF	—
Formula of oxide	Li$_2$O	BeO	B$_2$O$_3$	CO CO$_2$	N$_2$O NO NO$_2$ N$_2$O$_3$ N$_2$O$_4$ N$_2$O$_5$	O$_2$	OF$_2$	—
Formula of simplest hydride	LiH	BeH$_2$	BH$_3$	CH$_4$	NH$_3$	H$_2$O	HF	—
Period 3	Na	Mg	Al	Si	P	S	Cl	Ar
Formula of chloride	NaCl	MgCl$_2$	Al$_2$Cl$_6$	SiCl$_4$	PCl$_3$ PCl$_5$	SCl$_2$ S$_2$Cl$_2$	Cl$_2$	—
Formula of oxide	Na$_2$O Na$_2$O$_2$	MgO	Al$_2$O$_3$	SiO$_2$	P$_4$O$_6$ P$_4$O$_{10}$	SO$_2$ SO$_3$	Cl$_2$O Cl$_2$O$_7$	—
Formula of hydride	NaH	MgH$_2$	AlH$_3$	SiH$_4$	PH$_3$	H$_2$S	HCl	—

The oxidation numbers of the elements in their oxides are always positive (except for fluorine) because oxygen is the most electronegative element apart from fluorine. The oxidation numbers of the elements from Li to Ar are represented graphically in figure 13.7. Oxidation numbers of the elements in compounds with oxygen are indicated by a '×'. The pattern in these oxidation numbers provides further evidence of periodic properties. The maximum oxidation number of each element (apart from fluorine) is the same as its group number. For example, lithium in Group I has oxidation number +1, boron in Group III has oxidation number +3 and nitrogen in Group V has a maximum oxidation number of +5. Thus, the maximum oxidation number of each element corresponds to the number of electrons in the outermost shell of its atoms.

The oxidation numbers of the elements in their hydrides (indicated by a 'O') are also shown in figure 13.7. These have been obtained by assuming that the oxidation number of hydrogen in non-metal compounds is +1, whilst its oxidation number in metal hydrides is −1. Another interesting periodic pattern emerges this time. The oxidation numbers of the elements with hydrogen rise from +1 to +3 for the metals in Groups I, II, and III, plunge to −4 for the elements in Group IV and then rise through −3, −2, and −1 for the elements in Groups V, VI and VII.

Many of the oxidation numbers correspond to the loss or gain of enough electrons for the atom to obtain a completely filled outer shell of the type ns^2np^6. This tendency is certainly the case for elements in Groups I, II and III which contain 1, 2 and 3 electrons, respectively, in their outer shell. In forming compounds, the elements in these three groups lose 1, 2, or 3 electrons, respectively, forming ions such as Na$^+$, Mg^{2+} and Al^{3+} with oxidation numbers of +1, +2 and +3. In contrast, the elements in Groups V, VI and VII gain 3, 2 and 1 electrons, respectively, to achieve stable electron structures in ions such as P^{3-}, S^{2-} and Cl$^-$. Thus, the elements in these groups have oxidation numbers of −3, −2 and −1, respectively.

13.6 Patterns in the properties of chlorides

Table 13.5 shows various properties of the chlorides of the elements in the third period.

■ How do the states of the chlorides at 20°C vary across the third period?
■ How do the boiling points of the chlorides vary across the period?
■ How do the conductivities of the molten chlorides vary across the period?

Structure and bonding

The variations to which the above three questions point can be explained in terms of the structure and bonding in the chlorides.

NaCl and $MgCl_2$ are giant structures composed of oppositely charged ions attracted to each other by strong electrostatic forces in ionic bonds. This means

Table 13.5
Various properties of the chlorides of the elements in the third period

Formula of chloride	NaCl	$MgCl_2$	Al_2Cl_6	$SiCl_4$	PCl_3 (PCl_5)	S_2Cl_2	Cl_2
State of chloride (at 20°C)	s	s	s	l	l (s)	l	g
b.pt. of chloride/°C	1465	1418	423	57	74 (164)	136	−35
Conduction of electricity by molten or liquid chloride	good	good	v. poor	nil	nil	nil	nil
Structure of chloride	Giant structures		Simple molecular structures				
Heat of formation of chloride at 298 K/kJ mole^{-1}	−411	−642	−1408	−640	−320	−60	0
Heat of formation of chloride at 298 K per mole of Cl atoms/kJ	−411	−321	−235	−160	−107	−30	0
Effect of adding chloride to water	solid dissolves readily		chloride reacts with water, HCl fumes are produced.				some Cl_2 reacts with water
pH of aqueous solution of chloride	7	6.5	3	2	2	2	2

that the melting points and boiling points of these compounds are high. But, the molten substances will conduct electricity because the ions which they contain can move towards the electrode of opposite charge.

All the other chlorides are simple molecular structures composed of small discrete molecules attracted to each other by relatively weak Van der Waals' forces. So, the melting points and boiling points of these compounds are low and as liquids they will not conduct electricity.

Heats of formation

Elements on the left of the periodic table are highly electropositive metals keen to give up electrons. Elements on the right are highly electronegative non-metals (excluding the noble gases). Between these extremes, the electronegativity slowly increases from left to right. Consequently, there is a gradual decrease in the heat evolved when the elements react with 1 mole of chlorine (Cl). Sodium and chlorine react violently liberating $411\,kJ$ per mole of Cl.

$$2Na(s) + Cl(g) \rightarrow 2(Na^+Cl^-(s)) \qquad \Delta H = -822\,kJ$$

The strong ionic bonding in Na^+Cl^- results in a highly stable product and a reaction which is very exothermic. On the other hand, the reaction between sulphur and chlorine (two electronegative elements) is relatively feeble and much less exothermic.

$$2S(s) + Cl_2(g) \rightarrow S_2Cl_2 \qquad \Delta H = -60\,kJ$$

Reaction with water

When ionic chlorides are added to water, there is an immediate attraction of polar water molecules for ions in the chloride. The solid dissolves forming single aquated ions such as $Na^+(aq)$ and $Cl^-(aq)$. These are separate metal and non-metal ions surrounded by polar water molecules.

The solution of sodium chloride is neutral (pH = 7).

In contrast to this, the solution of aluminium chloride in water is very acidic (pH = 3). The acidic nature of the solution can be explained in terms of the relatively small but highly charged Al^{3+} ion. This draws electrons away from its surrounding water molecules and causes them to give up H^+ ions (see also section 16.5).

$$[Al(H_2O)_6]^{3+}(aq) \rightarrow [Al(H_2O)_5(OH)]^{2+}(aq) + H^+(aq)$$

The non-metal chlorides in period 3 all react with water forming acidic solutions.

$$SiCl_4(l) + \ 2H_2O(l) \longrightarrow SiO_2(s) + \underbrace{4H^+(aq) + 4Cl^-(aq)}_{\text{hydrochloric acid}}$$

$$PCl_3(l) + 3H_2O(l) \longrightarrow H_3PO_3(aq) + \underbrace{3H^+(aq) + 3Cl^-(aq)}_{\text{hydrochloric acid}}$$

phosphoric
(phosphorous)
acid

$$2S_2Cl_2(l) + 2H_2O(l) \longrightarrow 3S(s) + SO_2(aq) + \underbrace{4H^+ (aq) + 4Cl^- (aq)}_{\text{hydrochlocic acid}}$$

A similar pattern is found in the chlorides of period 2 (study question 2).

13.7 Patterns in the properties of oxides

Tables 13.6 and 13.7 show various properties of the oxides of the elements in period 2 and period 3, respectively. There are marked periodic patterns in the structure, bonding and properties of these oxides.

Structure and bonding

In each period, the oxides of metals and metalloids have giant structures, whereas the oxides of non-metals are composed of simple molecules. Thus, lithium oxide, Li_2O; beryllium oxide, BeO; sodium oxide, Na_2O; magnesium oxide, MgO; and aluminium oxide, Al_2O_3, may be regarded as ionic structures. Consequently, they are solids at room temperature, with high melting points and boiling points. These ionic oxides will conduct electricity in the molten state.

Table 13.6
Properties of the oxides of elements in period 2

Formula of oxide	Li_2O	BeO	B_2O_3	CO_2 (CO)	N_2O NO NO_2 N_2O_4 N_2O_5	O_2	OF_2
State of oxide (at 20°C)	solid	solid	solid	gases	gases (except N_2O_5, a solid)	gas	gas
Conduction of electricity by molten or liquid oxide	good	moderate	v. poor	nil	nil	nil	nil
Structure of oxide	**Giant structures**			**Simple molecular structures**			
Heat of formation of oxide at 298 K/kJ mole^{-1}	−596	−611	−1273	−394 (CO$_2$)	+33 (NO$_2$)		+22
Heat of formation of oxide at 298 K per mole of O/kJ	−596	−611	−424	−197	+17		+22
Effect of adding oxide to water	Reacts to form LiOH(aq) alkaline solution	BeO does not react with water but it is amphoteric	Reacts to form H_3BO_3, a very weak acid	CO_2 reacts to form H_2CO_3, a weak acid	NO_2 reacts to form an acid solution of HNO_3 and HNO_2		OF_2 reacts slowly forming O_2 and an acidic solution of HF
Nature of oxide	**Basic (alkaline)**	**Amphoteric**	**Acidic**	**Acidic**	**Acidic**		**Acidic**

Table 13.7
Properties of the oxides of elements in period 3

Formula of oxide	Na_2O	MgO	Al_2O_3	SiO_2	P_4O_{10} (P_4O_6)	SO_3 (SO_2)	Cl_2O_7 (Cl_2O)
State of oxide (at 20°C)	solid	solid	solid	solid	solid (solid)	liquid (gas)	liquid (gas)
Conduction of electricity by molten or liquid oxide	good	good	good	v. poor	nil	nil	nil
Structure of oxide	Giant structure				Simple molecular structure		
Heat of formation of oxide at 298 K/kJ mole^{-1}	−416	−602	−1676	−910	−2984 (P_4O_{10})	−395 (SO_3)	+80 (Cl_2O)
Heat of formation of oxide at 298 K per mole of O/kJ	−416	−602	−559	−455	−298	−132	+80
Effect of adding oxide to water	Reacts to form NaOH(aq) alkaline solution	Reacts to form $Mg(OH)_2$ weakly alkaline solution	Does not react with water but it is amphoteric	Does not react with water but it is acidic	P_4O_{10} reacts to form H_3PO_4 acid solution	SO_3 reacts to form H_2SO_4 acid solution	Cl_2O_7 reacts to form $HClO_4$ acid solution
Nature of oxide	**Basic (alkaline)**	**Basic (weakly alkaline)**	**Amphoteric**	**Acidic**	**Acidic**	**Acidic**	**Acidic**

The metalloids, boron and silicon, form oxides with giant molecular structures. In boron oxide, B_2O_3, boron and oxygen atoms are arranged in layers. Strong covalent bonds link one atom to the next in a giant sheet structure. The giant structures are therefore solids with high melting points and boiling points. The melting point of B_2O_3 is 577°C and that of SiO_2 is 1700°C. Unlike ionic solids, giant molecular solids do not conduct electricity in the molten state.

Carbon dioxide (CO_2), nitrogen dioxide (NO_2), oxygen difluoride (OF_2), phosphorus(v) oxide (P_4O_{10}), sulphur(vi) oxide (SO_3) and dichlorine heptoxide (Cl_2O_7) consist of discrete small molecules. These simple molecular oxides are much more volatile than the ionic metal oxides and the giant molecular metalloid oxides. They have low melting points and low boiling points and they do not conduct electricity in the liquid state.

Notice the gradations in structure and bond type across each period from ionic oxides and chlorides to simple molecular oxides and chlorides. These gradations can be correlated with changes in electronegativity across the period from low values on the left to high values on the right. Thus, atoms of low electronegativity, such as Na, Mg and Li, for compounds in which they have given up electrons to either chlorine atoms or oxygen atoms. In contrast to this, compounds formed between the more electronegative atoms (such as Si, P, S and F) and either chlorine or oxygen exist as discrete molecules (e.g. $SiCl_4$, ClF) or as giant covalent structures (e.g. SiO_2).

Thus, bonding in the oxides and chlorides become less ionic and more covalent as we move across a period from atoms of low to atoms of high electronegativity.

Heats of formation

Tables 13.6 and 13.7 show the standard heats of formation of some of the oxides in periods 2 and 3. Notice that most of the oxides have a negative heat of formation. This means that oxides are usually very stable. Indeed, only fluorides are generally more stable than oxides.

The standard heats of formation of the oxides vary from very large negative values to relatively small positive values. At first sight, there appears to be no distinct pattern in the standard heats of formation of the oxides. However, when we compare the heats of formation *per mole of oxygen atoms* an obvious pattern appears (tables 13.6 and 13.7 and figure 13.8). The heat of formation per mole of oxygen atoms gives the stability of bonds to one mole of oxygen atoms in the oxide. The heat of formation per mole of oxygen atoms becomes less negative as the atomic number increases across a period. This means that, in general, oxygen forms its most stable compounds with elements such as Li, Na, Mg and Al which are furthest removed from it in the periodic table. This is generally the case with other pairs of elements also.

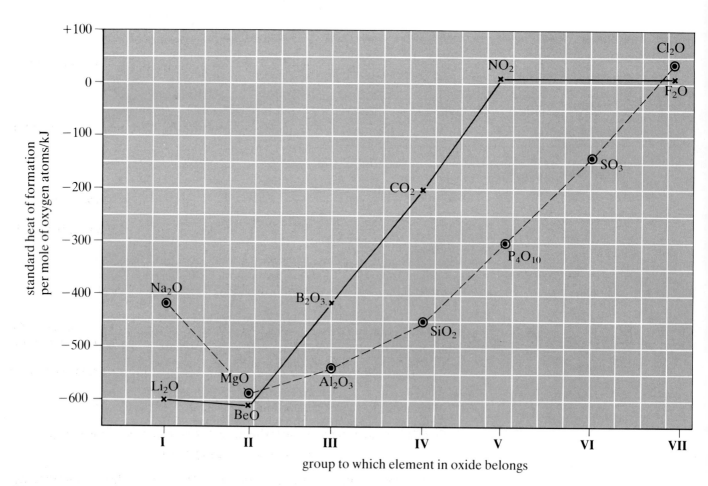

Figure 13.8
Variation in the heats of formation of oxides in periods 2 and 3 per mole of oxygen atoms/kJ per mole of O atoms.

Reaction with water: Acid–base character of oxides

■ Write equations for the reaction of each of the following oxides with water: Na_2O, MgO, P_4O_{10}, SO_3.
■ How does the acid–base character of the oxides vary across periods 2 and 3?

As we pass across a period from left to right there is a steady change in the structure of the oxide from ionic, through giant molecular to simple molecular. This change in structure leads to a profound difference in the way in which the oxides react with water, acids and alkalis.

The ionic oxides contain O^{2-} ions in the crystal lattice. Thus, O^{2-} ions in Li_2O and Na_2O, the oxides of Group I metals, react vigorously with water to form alkaline solutions.

$$O^{2-} + H_2O \longrightarrow 2OH^-$$

$$Li_2O(s) + H_2O(l) \longrightarrow 2Li^+(aq) + 2OH^-(aq)$$

$$Na_2O(s) + H_2O(l) \longrightarrow 2Na^+(aq) + 2OH^-(aq)$$

These oxides would react even more vigorously with acids forming a solution containing cations of the metal.

$$Na_2O(s) + 2H^+(aq) \rightarrow 2Na^+(aq) + H_2O(l)$$

The oxides of Group II metals do not react so readily with water or acids since the large charge density on the Group II cation holds the O^{2-} ions more firmly.

Thus, MgO is only slightly soluble in water, although it reacts readily with acids to form a solution of magnesium ions.

$$MgO(s) + H_2O(l) \rightleftharpoons Mg^{2+}(aq) + 2OH^-(aq)$$

$$MgO(s) + 2H^+(aq) \rightarrow Mg^{2+}(aq) + H_2O(l)$$

BeO is insoluble in water, but it shows basic properties by dissolving in acids to form Be^{2+} salts.

$$BeO(s) + 2H^+(aq) \rightarrow Be^{2+}(aq) + H_2O(l)$$

However, BeO also resembles acidic oxides by reacting with alkalis to form beryllates.

$$BeO(s) + 2OH^-(aq) + H_2O(l) \rightarrow Be(OH)_4{}^{2-}(aq)$$
beryllate

Oxides such as BeO, which show *both basic and acidic properties*, are called **amphoteric oxides.**

Aluminium oxide is also amphoteric. It does not react with water, but it will react with both H^+ and OH^- ions.

$$Al_2O_3(s) + 6H^+(aq) \rightarrow 2Al^{3+}(aq) + 3H_2O(l)$$

$$Al_2O_3(s) + 2OH^-(aq) + 3H_2O(l) \rightarrow 2Al(OH)_4(aq)$$
aluminate

The remaining oxides of the elements in periods 2 and 3 are all acidic except CO, OF_2, N_2O, NO and ClO_2 which are described as **neutral**. *Neutral oxides show neither acidic nor basic character.*

Boron(III) oxide, B_2O_3, reacts with water to form boric(III) acid, H_3BO_3, a very weak acid.

$$B_2O_3(s) + 3H_2O(l) \rightarrow 2H_3BO_3(aq)$$

CO_2 dissolves in water and reacts slightly to form weak carbonic acid, H_2CO_3.

$$CO_2(g) + H_2O(l) \rightleftharpoons H_2CO_3(aq)$$

CO_2 will react with the OH^- ions in alkalis to form first hydrogencarbonate, HCO_3^- and then carbonate, CO_3^{2-}.

$$CO_2(g) + OH^-(aq) \rightarrow HCO_3^-(aq)$$

$$CO_2(g) + 2OH^-(aq) \rightarrow CO_3^{2-}(aq) + H_2O(l)$$

Silicon(IV) oxide, SiO_2, does not react with water, but it reacts with concentrated alkalis forming silicate, SiO_3^{2-}.

$$SiO_2(s) + 2OH^-(aq) \rightarrow SiO_3^{2-}(aq) + H_2O(l)$$

NO_2 reacts with water to form a mixture of two acids, HNO_2 and HNO_3.

$$2NO_2(g) + H_2O(l) \rightarrow HNO_2(aq) + HNO_3(aq)$$

The oxides of P, S and Cl (except ClO_2) react readily with water to form strong acids.

$$P_4O_{10}(s) + 6H_2O(l) \rightarrow 4H_3PO_4(aq)$$
phosphorus (V) oxide (phosphorus pentoxide)   phosphoric (V) acid

$$SO_3(g) + H_2O(l) \rightarrow H_2SO_4(aq)$$
sulphur trioxide   sulphuric acid

$$Cl_2O_7(l) + H_2O(l) \rightarrow 2HClO_4(aq)$$
dichlorine heptoxide   chloric (VII) acid (perchloric acid)

$$Cl_2O(g) + H_2O(l) \rightarrow 2HClO(aq)$$
dichlorine oxide   chloric (I) acid (hypochlorous acid)

Notice that *as we cross the periodic table, we move from the ionic oxides of metals, which are basic, to the oxides of metalloids with giant molecular structure, which are weakly basic, weakly acidic or amphoteric, and finally to the simple molecular oxides of non-metals, which are acidic.*

Summary

1 Periodicity as summarised in the periodic table is one of the major unifying themes in chemistry.

2 Elements in the same group of the periodic table have similar electron configurations. This results in similar chemical properties.

3 Moving from left to right across a period there is a decrease in atomic radius since the increasing positive charge in the nucleus pulls the electrons closer.

4 Moving from left to right across a period, there is a general increase in the first ionisation energy. This results from an increasing nuclear charge and a decreasing atomic radius across the period causing the outermost electron to be held more tightly.

5 The electronegativity of an atom provides a measure of the power of that atom in a molecule to attract electrons.

6 Moving from left to right across a period there is an increase in the electronegativities of elements.

7 In forming compounds, the elements in Groups I, II and III form positive ions with charges of +1, +2 and +3, respectively. In contrast, the elements in Groups V, VI and VII gain electrons forming negative ions with charges of −3, −2 and −1, respectively.

8 Moving from left to right across a period:
   (i) The chlorides of the elements change from ionic, involatile metal chlorides to simple molecular, volatile non-metal chlorides.
   (ii) The hydrides of the elements change from ionic, involatile metal hydrides to simple molecular, non-metal hydrides. The hydrides also become more acidic across the period.

(iii) The oxides of the elements change from being ionic, involatile and basic in metal oxides, through giant molecular, involatile, amphoteric, metalloid oxides to simple molecular, volatile, and acidic non-metal oxides.

Study questions

1   Use the information in the following table to explain the statements below.

	Na	Mg	Al	Si	P	S	Cl
Atomic radius/nm	0.156	0.136	0.125	0.117	0.110	0.104	0.099
Ionic radius/nm	0.095	0.065	0.050			0.184	0.181
First ionisation energy/kJ mole^{-1}	492	743	579	791	1060	1003	1254

(a) The atomic radius decreases across the period from Na to Cl.
(b) The ionic radii of Na$^+$, Mg^{2+} and Al^{3+} are less than their respective atomic radii, whereas the ionic radii of Cl$^-$ and S^{2-} are greater than their respective atomic radii.
(c) The first ionisation energies show a general increase from Na to Cl.
(d) The first ionisation energy of Al is less than that for Mg.
(e) The first ionisation energy of S is less than that for P.

2   Table 13.8 shows various properties of the chlorides of the elements in period 2.
(a) Explain the variation in state and boiling point of the chlorides across period 2.
(b) Both LiCl and BeCl$_2$ are giant structures; LiCl(l) is a good conductor of electricity, but BeCl$_2$(l) is only a poor conductor of electricity.
   What type of structure does LiCl have?
   Why does LiCl(l) conduct electricity?
   Suggest a possible structure for BeCl$_2$(s).
(c) How would you expect LiCl and BCl$_3$ to behave when added to water?
(d) How is the type of bonding in the chlorides of the elements in period 2 related to their electronegativity?

Table 13.8
Various properties of the chlorides of the elements in period 2

Formula of chloride	LiCl	BeCl$_2$	BCl$_3$	CCl$_4$	NCl$_3$	Cl$_2$O	ClF
State at 20°C	s	s	g	l	l	g	g
B.pt. of chloride/°C	1350	487	12	77	71	2	−101
Conduction of electricity by liquid chloride	good	v. poor	nil	nil	nil	nil	nil
Structure	giant structures			simple molecular structures			

3   The behaviour of the hydrides of the elements Na–Cl with water is summarised below.

NaH	MgH$_2$	AlH$_3$	SiH$_4$	PH$_3$	H$_2$S	HCl
react forming H$_2$(g) and an alkaline soln.			no reaction	reacts to form a slightly alkaline solution	reacts to form a slightly acidic solution	reacts to form an acidic solution

(a) Write equations to summarise the reactions of NaH and MgH_2 with water.

(b) Suggest a reason why SiH_4 has no reaction with water.

(c) Write an equation to account for the formation of a slightly alkaline solution when PH_3 reacts with water.

(d) Write an equation to account for the formation of an acidic solution when HCl reacts with water.

(e) Explain the trends and differences in the reactions of these hydrides with water in terms of their structure and bonding.

4 Consider the elements Li, Be, B, N, F, Ne.

(a) Which exist as diatomic molecules in the gaseous state at room temperature?

(b) Which has the highest boiling point?

(c) Which form a chloride of formula XCl_3?

(d) Which has the largest first ionisation energy?

(e) Which has the smallest second ionisation energy?

(f) Which form hydrides which dissolve in water to give an alkaline solution?

5 (a) Draw dot/cross diagrams to show the electronic structures of
    (i) LiF, (ii) CF_4, (iii) NF_3.

(b) Predict the shapes with respect to atoms of molecules of
    (i) BF_3, (ii) CF_4, (iii) NF_3, (iv) OF_2.

(c) Why is it not meaningful to talk about the shape of LiF?

6 (a) What is the nature of the bonds in the oxides formed when Na, Mg, Al and S react with excess oxygen?

(b) How do these oxides react with
    (i) water, (ii) dilute acids, (iii) alkali?

(c) Magnesium chloride is a high melting point solid, aluminium chloride is a solid which sublimes readily at about 180°C and silicon tetrachloride is a volatile liquid. Explain the nature of the chemical bonding in these chlorides and show how this accounts for the above differences in volatility.

7 (a) Write down the symbols of the elements in the third period of the periodic table (ending with the noble gas, argon) in order of increasing atomic number.

(b) Which of these elements are
    (i) 's-block' elements,
    (ii) 'd-block' elements?

(c) (i) Write the empirical formula of the chloride formed by the element of atomic number 13.
    (ii) Describe briefly how you could prepare a sample of this chloride.

(d) The chloride of element X is represented as XCl_n. Suppose you are asked to find the value of n.
    (i)   Describe the experiments you would carry out.
    (ii) List the measurements you would take.
    (iii) Explain how you would calculate your result.

8 Element Y has an atomic number of 31. Use your knowledge of chemical periodicity to answer the following questions about Y.

(a) Write the electronic (s, p, d, f) configuration for Y.

(b) Is element Y in the 's-block', 'p-block' or 'd-block'?

(c) What is the principal oxidation number of Y?

(d) What is the probable formula of the oxide of Y?

(e) Is the oxide of Y likely to be acidic, amphoteric or basic?

(f) Write equations for any reactions which the oxide of Y would undergo with
    (i)   water,
    (ii) dilute nitric acid,
    (iii) dilute sodium hydroxide.

9 Some of the properties of elements P and Q are given below.

Element P

(i)   is soft and malleable,

(ii) floats on water,

(iii) has a melting point less than 100°C,

(iv) forms an ionic hydride of formula XH.

Element Q

(i)   has a density greater than $7\,\mathrm{g\,cm^{-3}}$,

(ii) has a melting point greater than 1500°C,

(iii) forms oxides of formulas QO, Q_2O_3 and QO_3,

(iv) does not react with concentrated nitric acid,

(v) forms some simple molecular compounds in its highest oxidation state.

    (a) State the deductions you can make about P and Q from each piece of evidence.

    (b) To which blocks of the periodic table do P and Q belong?

    (c) Identify P and Q as precisely as possible.

14 Competition Processes

14.1 Fundamental reactions in inorganic chemistry

The reactions of inorganic substances can be divided into four major classes. All these reactions involve competition in one way or another.

1 Redox

2 Acid–base

3 Precipitation

4 Complexing

Other processes such as decomposition and synthesis usually fit into one of these four categories.

Redox reactions

Redox processes have already been discussed in Chapter 3. It is worth recalling the following important types of redox reaction.

(a) The reaction of metals with non-metals

$$e.g. \; 2Na(s) + Cl_2(g) \rightarrow 2Na^+Cl^-(s)$$
$$2Mg(s) + O_2(g) \rightarrow 2Mg^{2+}O^{2-}(s)$$

(b) The reaction of metals with water and steam

$$e.g. \; Ca(s) + 2H_2O(l) \rightarrow Ca^{2+}(OH^-)_2(s) + H_2(g)$$

Rows of mercury cells used in the electrolytic manufacture of chlorine and sodium hydroxide from brine.

(c) The reaction of metals with acids

$$\text{e.g. } Mg(s) + 2H^+(aq) \rightarrow Mg^{2+}(aq) + H_2(g)$$

(d) Reactions at the electrodes during electrolysis

e.g. during the electrolysis of molten sodium chloride

$$\text{Anode } (+) \, 2Cl^-(l) \rightarrow Cl_2(g) + 2e^-$$
$$\text{Cathode } (-) \, 2Na^+(l) + 2e^- \rightarrow 2Na(s)$$

(e) Disproportionation reactions (see section 17.10)

Which one of the following processes involves redox? (Remember that the best check for redox is to consider the oxidation numbers of the elements involved in the reaction.)

- $Ag^+ + 2NH_3 \quad \rightarrow \quad [Ag(NH_3)_2]^+$
- $H^+ + NH_3 \quad \rightarrow \quad NH_4^+$
- $Ba^{2+} + CrO_4^{2-} \quad \rightarrow \quad BaCrO_4$
- $2Al + 3Cl_2 \quad \rightarrow \quad Al_2Cl_6$
- $CuCO_3 \quad \rightarrow \quad CuO + CO_2$

Acid–base reactions

You will have met many examples of acid–base reactions already. These include reactions of acids with insoluble bases (metal oxides and hydroxides),

$$CuO(s) + 2H^+(aq) \longrightarrow Cu^{2+}(aq) + H_2O(l)$$
$$Cu(OH)_2(s) + 2H^+(aq) \longrightarrow Cu^{2+}(aq) + 2H_2O(l)$$

reactions of acids with alkalis

$$Na^+(aq) + OH^-(aq) + H^+(aq) \rightarrow Na^+(aq) + H_2O(l)$$

and less obvious acid–base reactions such as that between aqueous ammonia and dilute acid

$$NH_3(aq) + H^+(aq) \rightarrow NH_4^+(aq)$$

In this chapter we shall extend our ideas of acid–base processes to include such reactions as acid + carbonate.

Precipitation reactions

Precipitation occurs when aqueous solutions are mixed and an insoluble substance forms. For example, a white precipitate of silver chloride forms when aqueous solutions of silver nitrate and sodium chloride are mixed.

$$Ag^+(aq) + NO_3^-(aq) + Na^+(aq) + Cl^-(aq) \rightarrow Ag^+Cl^-(s) + Na^+(aq) + NO_3^-(aq)$$

Reactions involving complexes

The formation of complexes is of great importance in the chemistry of transition metals (section 19.6), and indeed in the chemistry of most aqueous metal ions. In aqueous solution, most cations (M^{2+}) are surrounded by water molecules to form a hydrated ion of the form $[M(H_2O)_n]^{2+}$. This aquated ion can be regarded as a complex ion. Other important complexes form when cations are surrounded by anions or molecules other than water. For example, when aluminium hydroxide dissolves in excess NaOH(aq):

$$Al(OH)_3(s) + OH^-(aq) \rightarrow [Al(OH)_4]^-(aq)$$
$$\text{tetrahydroxoaluminate (III)}$$
$$\text{(aluminate)}$$

The $[Al(OH)_4]^-$ ion consists of an Al^{3+} ion surrounded by four OH^- ions.

14.2 Redox—competition for electrons

Most redox reactions involve a transfer of electrons from one reactant to another. When metals react with non-metals, electrons are transferred from the metal to the non-metal with the formation of ions.

$$2Na \rightarrow 2Na^+ + 2e^-$$
$$Cl_2 + 2e^- \rightarrow 2Cl^-$$

The metal, which loses electrons, is oxidised whilst the non-metal, which gains electrons, is reduced. The oxidising agent (oxidant) is the electron acceptor and the reducing agent (reductant) is the electron donor. In this case, the reactions occur because most non-metals are eager to gain electrons whilst metals will part with their electrons quite readily. In the competition for electrons, the non-metal wins 'hands down'. However, the competition for electrons in redox reactions is not always so clear cut. Consequently, chemists have developed the idea of electrode potentials. These can be used to check whether (and in which direction) a redox reaction is occurring between two systems.

In chapter 2, we discovered that electron transfer occurred when a $Zn(s)/Zn^{2+}$ (aq) half-cell was connected by a salt bridge to a $Cu(s)/Cu^{2+}$(aq) half-cell (figure 14.1). The bulb lights, showing that electrons are flowing through the wire.

Figure 14.1
Electron transfer between two half-cells connected by a salt bridge.

Zinc dissolves from the zinc rod which loses weight. Copper is deposited on the copper rod. The reactions at the terminals can be represented by the following half-equations.

$$Zn(s) \rightarrow Zn^{2+}(aq) + 2e^-$$
$$Cu^{2+}(aq) + 2e^- \rightarrow Cu(s)$$

Electrons flow from the negative zinc terminal through the external circuit to the positive copper terminal.

If the pea bulb is replaced by a high resistance voltmeter (e.g. a valve voltmeter) it is possible to measure the potential difference between the two half-cells.

By using a high-resistance voltmeter, the current in the external circuit is virtually zero and the cell registers its maximum potential difference. This maximum p.d. is called the **electromotive force (e.m.f.)**. This e.m.f. gives a quantitative measure of the likelihood of the redox reaction taking place in the cell.

When the pea bulb in figure 14.1 is replaced by a valve voltmeter, a voltage of 1.10 volts is recorded.

This combination of half-cells and the resulting cell e.m.f. can be summarised in a diagram as

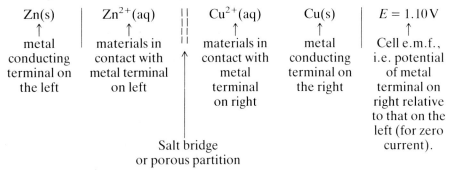

The voltage shown in the summary above relates to changes indicated by the order of material in the diagram, i.e.

$$Zn(s) \rightarrow Zn^{2+}(aq) + 2e^-$$

and

$$Cu^{2+}(aq) + 2e^- \rightarrow Cu(s)$$

So, for the change,

$$Zn(s) + Cu^{2+}(aq) \rightarrow Zn^{2+}(aq) + Cu(s) \qquad E = 1.10\,V$$

and for the reverse process,

$$Zn^{2+}(aq) + Cu(s) \rightarrow Zn(s) + Cu^{2+}(aq) \qquad E = -1.10\,V$$

Notice that this convention leads to a positive value of E if the reaction actually takes place in that direction when the cell is short-circuited. A negative value of E indicates that the cell reaction as written cannot take place.

When the $Zn(s)/Zn^{2+}(aq)$ half-cell in figure 14.1 was replaced by other metal/metal ion half-cells, the results shown in table 14.1 were obtained.

Table 14.1
The overall cell e.m.f. developed between various metal/metal ion half-cells and the $Cu(s)/Cu^{2+}(aq)$ half-cell

Metal/metal ion half-cell	Overall cell e.m.f./volts
$Zn(s)/Zn^{2+}(aq)$	+1.10
$Fe(s)/Fe^{2+}(aq)$	+0.78
$Pb(s)/Pb^{2+}(aq)$	+0.47
$Cu(s)/Cu^{2+}(aq)$	0.00
$Ag(s)/Ag^+(aq)$	−0.46

The first three metals (zinc, iron and lead) are all negative with respect to copper. These metals are stronger reducing agents than copper and go into solution as their ions when the cells are short-circuited. On the other hand, the $Ag(s)/Ag^+(aq)$ half-cell produces a negative e.m.f. relative to the copper half-cell. This means that the silver electrode is positive with respect to copper and in this case copper is acting as the reducing agent. The cell reaction can be summarised as

$$Ag(s)|Ag^+(aq)|\,|Cu^{2+}(aq)|Cu(s) \qquad E = -0.46\,V$$

As we might expect, the reaction

$$2Ag(s) + Cu^{2+}(aq) \rightarrow 2Ag^+(aq) + Cu(s)$$

has a negative e.m.f. of −0.46 volts. It is the reverse reaction which actually

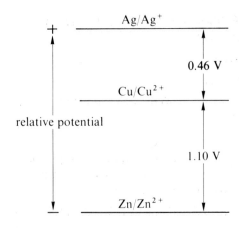

Figure 14.2
Relative cell potentials. What would be the e.m.f. of a
$Zn(s)|Zn^{2+}(aq)||Ag^+(aq)|Ag(s)$ cell?

occurs when the cell is short circuited. Figure 14.2 shows diagrammatically the relative potentials for the copper, zinc and silver half-cells.

■ Why is it important to measure the electrode potential for a single half-cell?
■ What other factors besides the particular metals and ions used will affect the size of an electrode potential?
■ Look at figure 14.2. What would be the e.m.f. of a $Zn(s)/Zn^{2+}(aq)||Ag^+(aq)/Ag(s)$ cell? Write half-equations for the reactions which take place in the two half-cells of this combination when it is short-circuited.

Notice that the e.m.f. values in table 14.1 can be used to compare the relative tendencies of the metals to release electrons and form ions. Thus zinc, at the head of the table is a strong reducing agent, releasing electrons readily, whilst silver at the bottom of the table is a very poor reducing agent. The e.m.f. values give a quantitative record of the position of the metal in the electrochemical series (ECS).

14.3   Standard electrode potentials

So far, we have seen that cell e.m.f.'s enable metals to be placed in order of their relative ability as reducing agents. Conversely, we can also draw up a list of metal ions showing their ability as oxidising agents. Thus, $Ag^+(aq)$ is a stronger oxidising agent than $Cu^{2+}(aq)$. i.e. it is a better competitor for electrons.

You will have realised already that it is impossible to obtain the electrode potential for a single half-cell. E.m.f.'s can only be measured for a complete circuit with two electrodes. In other words, *only differences in potential are measurable*. Nevertheless, it would be extremely useful if we could summarise all e.m.f. data by giving each half-cell a characteristic value. This can be done by arbitrarily assigning an electrode potential of zero to one particular half-cell. Other half-cells can then be compared with this standard.

The standard chosen for electrode potentials is not a metal but hydrogen. The so-called **standard hydrogen half-cell** (sometimes loosely referred to as the standard hydrogen electrode) is shown in figure 14.3.

This consists of H_2 gas at one atmosphere pressure and 25°C bubbling around a platinised platinum electrode. The electrode is immersed in a 1.0M solution of H^+ ions. It is alternately bathed in first $H^+(aq)$ and then $H_2(g)$. Hydrogen is

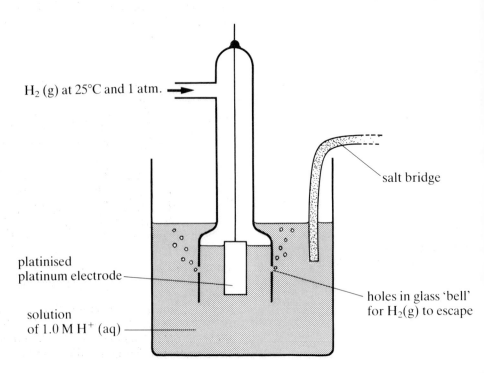

Figure 14.3
The standard hydrogen half-cell.

adsorbed on the platinum and an equilibrium is established between the adsorbed layer of H_2 and H^+ ions in the solution.

$$\tfrac{1}{2}H_2(g) \rightleftharpoons H^+(aq) + e^-$$

The platinised platinum electrode has three functions.

1 It acts as an inert metal connection to the H_2/H^+ system. (There is no tendency for Pt to form ions itself.)

2 It allows H_2 gas to be adsorbed on to its surface.

3 It is covered by a loosely deposited layer of finely divided platinum (i.e. it is *platinised*). This increases its surface area so that it can establish an equilibrium between $H_2(g)$ and $H^+(aq)$ as rapidly as possible.

The standard electrode potential ($E^\ominus$) for this reference half-cell is taken as zero, i.e.

$$\tfrac{1}{2}H_2(g) \rightarrow H^+(aq) + e^- \qquad E^\ominus = 0.00\,V$$

14.4 Measuring standard electrode potentials

The potential of the platinum electrode in a standard hydrogen half-cell will depend on the temperature, the concentration of H^+ ions and the pressure of the $H_2(g)$.

Consequently, *in measuring and comparing electrode potentials we must choose the same standard conditions for all measurements.* The standard conditions chosen are similar to those for thermochemical measurements.

1 All solutions have a concentration of 1 mole dm^{-3}.

2 Any gases involved have a pressure of 1 atmosphere.

3 The temperature is 25°C (298 K).

4 Platinum is used as the electrode when the half-cell system does not include a metal.

Figure 14.4 shows a standard $Cu^{2+}(aq)/Cu(s)$ half-cell and a standard $Fe^{3+}(aq)/Fe^{2+}(aq)$ half-cell. The latter system does not involve a metal so a platinum electrode must be used as the electrical connection. Notice also that the solution is 1.0 M with respect to both $Fe^{3+}(aq)$ and $Fe^{2+}(aq)$.

Figure 14.4
Two standard half-cells.

a standard $Cu^{2+}(aq)/Cu(s)$ half-cell   a standard $Fe^{3+}(aq)/Fe^{2+}(aq)$ half-cell

The standard potentials of all other half-cells are obtained by reference to that of the standard hydrogen half-cell. The standard electrode potential of the $Cu^{2+}(aq)/Cu(s)$ half-cell is thus the potential difference between the electrodes of a cell consisting of the standard hydrogen half-cell and the standard

Figure 14.5
Measuring the standard electrode potential of the $Cu^{2+}(aq)/Cu(s)$ half-cell.

$Cu^{2+}(aq)/Cu(s)$ half-cell (figure 14.5). When this cell is set up the potential is 0.34 volts and copper is the positive terminal. The reactions occurring at the electrodes are

$$H_2 \rightarrow 2H^+ + 2e^-$$

and

$$Cu^{2+} + 2e^- \rightarrow Cu.$$

Now, since the standard electrode potential for the system $H_2 \rightarrow 2H^+ + 2e^-$ is arbitrarily taken as zero, the standard electrode potential for the system $Cu^{2+} + 2e^- \rightarrow Cu$ is $+0.34$ volts. The plus sign is used because the copper electrode is positively charged with respect to the standard hydrogen electrode, i.e.

$$Cu^{2+} + 2e^- \rightarrow Cu \qquad\qquad E^{\ominus} = +0.34\,V$$
$$E^{\ominus} = +0.34\,V$$

$$Pt(s)|H_2(g), H^+(aq)\|Cu^{2+}(aq)|Cu(s)$$

By convention, the oxidised form is written first when a particular redox half-equation and its standard electrode potential are being referred to. Thus, $Cu^{2+}(aq)/Cu(s)$, $E^{\ominus} = +0.34$ volts means that the half-cell reaction

$$Cu^{2+}(aq) + 2e^- \rightarrow Cu(s)$$

has a standard electrode potential of $+0.34$ volts. Standard electrode potentials are sometimes called **standard reduction potentials** because they relate to the *reduction of the more oxidised species.*

When inert electrodes are present (as in the hydrogen half-cell), the least oxidised form of the components of the system should be written next to the electrode. Hence, the standard hydrogen electrode is written as $H^+(aq)$, $H_2(g)/Pt(s)$ or as $Pt(s)/H_2(g)$, $H^+(aq)$, but not as $Pt(s)/H^+(aq)$, $H_2(g)$.

We are now in a position to define the *standard electrode (reduction) potential,* $E^{\ominus}$, *of a standard half-cell. This is the potential of that half-cell relative to a standard hydrogen half-cell under standard conditions.*

When a standard zinc half-cell is connected to a standard hydrogen half-cell, the e.m.f. produced is 0.76 volts. In this case the zinc electrode is negative. Thus, the standard electrode potential for the $Zn^{2+}(aq)/Zn(s)$ half-cell is -0.76 volts, i.e.

$$Zn^{2+}(aq) + 2e^- \rightarrow Zn(s) \qquad E^{\ominus} = -0.76\,V$$

The reactions occurring at the electrodes are

$$Zn(s) \rightarrow Zn^{2+}(aq) + 2e^-$$

and

$$2H^+(aq) + 2e^- \rightarrow H_2(g)$$

Notice that the reaction at the hydrogen electrode is now in the opposite direction to that which occurred when a $Cu^{2+}(aq)/Cu(s)$ half-cell was connected to a standard hydrogen half-cell.

True or false: The statement $Cu^{2+}(aq)/Cu(s)$, $E^\ominus = +0.34$ volts means that:

- copper is more positive than Cu^{2+} ions.
- a solution of $Cu^{2+}(aq)$ is 0.34 volts more positive than a Cu electrode immersed in it.
- the reaction $Cu^{2+}(aq) + 2e^- \rightarrow Cu(s)$ has a standard electrode potential of +0.34 volts.
- the copper electrode of a standard $Cu^{2+}(aq)/Cu(s)$ half-cell is 0.34 volts more positive than the platinum electrode in a standard hydrogen half-cell to which it is connected.

14.5 Relative strengths of oxidising agents and reducing agents

Standard electrode potentials provide a direct measure of the relative oxidising and reducing power of different species.

The standard electrode potentials of the zinc, hydrogen and copper half-cells have been tabulated in table 14.2 in order to emphasise the relative strengths of the different oxidising and reducing agents. Cu^{2+} is a stronger oxidising agent than either H^+ or Zn^{2+}. On the other hand, Zn is a more powerful reducing agent than either H_2 or Cu since it will reduce H^+ to H_2 and Cu^{2+} to Cu. Notice that *the relative size of $E^\ominus$ gives a measure of the strengths of both oxidants and reductants*. Cu^{2+}, the most powerful oxidising agent in this table, has the largest (most positive) value for $E^\ominus$ Zn^{2+}, the least powerful oxidising agent, has the smallest (most negative) value for $E^\ominus$.

On the other hand, the value of $E^\ominus$ for the reaction

$$Zn(s) \rightarrow Zn^{2+}(aq) + 2e^-$$

is +0.76 volts. So, Zn, the most powerful reducing agent, has the most positive value for $E^\ominus$. It is important to remember that *the more positive the value of $E^\ominus$, the more energetically favourable is the reaction*.

The standard electrode potentials for a large number of redox half-reactions are shown in table 14.3. The table is arranged so that the strongest oxidising agent, F_2, which has the most positive value for $E^\ominus$, is at the top of the list. The weakest oxidising agent, K^+, with the most negative value for $E^\ominus$, is at the bottom. Thus, fluorine is a better competitor for electrons than any other oxidising agent in the list. Potassium ions have the weakest tendency to accept electrons. Conversely, F^- is the weakest reducing agent with the least tendency to denote electrons, whilst K is the most powerful reducing agent in the list. This illustrates the reciprocal or conjugate character of an oxidising agent and its corresponding reducing agent. *The stronger the oxidising agent, the weaker is its corresponding (conjugate) reducing agent and vice versa.*

Table 14.2
Standard electrode potentials of the zinc, hydrogen and copper half-cells

	Oxidising agent	Reducing agent	increasing strength of reducing agent	$E^\ominus$/volts
increasing strength of oxidising agent ↑	$Cu^{2+}(aq) + 2e^- \longrightarrow$	$Cu(s)$		+0.34
	$H^+(aq) + e^- \longrightarrow$	$\frac{1}{2}H_2(g)$		0.00
	$Zn^{2+}[aq] + 2e^- \longrightarrow$	$Zn(s)$	↓	−0.76

Table 14.3
Standard electrode potentials

				$E^{\ominus}$/volts
Strongest **oxidising agent**	$F_2(g) + 2e^-$	$\longrightarrow 2F^-(aq)$	**Weakest** **reducing agent**	+2.87
	$H_2O_2(aq) + 2H^+(aq) + 2e^-$	$\longrightarrow 2H_2O(l)$		+1.77
	$MnO_4^-(aq) + 4H^+(aq) + 3e^-$	$\longrightarrow MnO_2(s) + 2H_2O(l)$		+1.70
	$2HClO(aq) + 2H^+(aq) + 2e^-$	$\longrightarrow Cl_2(aq) + 2H_2O(l)$		+1.59
	$MnO_4^-(aq) + 8H^+(aq) + 5e^-$	$\longrightarrow Mn^{2+}(aq) + 4H_2O(l)$		+1.51
	$Cl_2(aq) + 2e^-$	$\longrightarrow 2Cl^-(aq)$		+1.36
	$MnO_2(s) + 4H^+(aq) + 2e^-$	$\longrightarrow Mn^{2+}(aq) + 2H_2O(l)$		+1.23
	$Br_2(aq) + 2e^-$	$\longrightarrow 2Br^-(aq)$		+1.09
	$NO_3^-(aq) + 2H^+(aq) + e^-$	$\longrightarrow NO_2(g) + H_2O(l)$		+0.80
	$Ag^+(aq) + e^-$	$\longrightarrow Ag(s)$		+0.80
	$Fe^{3+}(aq) + e^-$	$\longrightarrow Fe^{2+}(aq)$		+0.77
	$2H^+(aq) + O_2(g) + 2e^-$	$\longrightarrow H_2O_2(aq)$		+0.68
	$I_2(aq) + 2e^-$	$\longrightarrow 2I^-(aq)$		+0.54
	$Cu^{2+}(aq) + 2e^-$	$\longrightarrow Cu(s)$		+0.34
	$2H^+(aq) + 2e^-$	$\longrightarrow H_2(g)$		0.00
	$Pb^{2+}(aq) + 2e^-$	$\longrightarrow Pb(s)$		−0.13
	$Fe^{2+}(aq) + 2e^-$	$\longrightarrow Fe(s)$		−0.44
	$Zn^{2+}(aq) + 2e^-$	$\longrightarrow Zn(s)$		−0.76
	$Al^{3+}(aq) + 3e^-$	$\longrightarrow Al(s)$		−1.66
	$Mg^{2+}(aq) + 2e^-$	$\longrightarrow Mg(s)$		−2.37
	$Na^+(aq) + e^-$	$\longrightarrow Na(s)$		−2.71
Weakest **oxidising agent**	$K^+(aq) + e^-$	$\longrightarrow K(s)$	**Strongest** **reducing agent**	−2.92

(left vertical axis: increasing strength of oxidising agent; right vertical axis: increasing strength of reducing agent)

14.6 The uses of standard electrode potentials

Using standard electrode potentials to predict the possibility of reactions

Why can $Cu^{2+}(aq)$ oxidise $Zn(s)$, but $Zn^{2+}(aq)$ cannot oxidise $Cu(s)$?

As we have seen already, $E^{\ominus}$ values provide an indication of the relative strengths of oxidising agents and reducing agents.

The value of $E^{\ominus}$ for the reaction

$$Cu^{2+}(aq) + 2e^- \rightarrow Cu(s) \quad \text{is } +0.34 \text{ volts.}$$

That for the reaction

$$Zn(s) \rightarrow Zn^{2+}(aq) + 2e^- \quad \text{is } +0.76 \text{ volts.}$$

Consequently, the overall potential for the reaction

$$Cu^{2+}(aq) + Zn(s) \rightarrow Cu(s) + Zn^{2+}(aq) \quad \text{is } 1.10 \text{ volts,}$$

i.e.

$$
\begin{array}{ll}
Cu^{2+}(aq) + 2e^- \rightarrow Cu(s) & +\ 0.34 \text{ volts} \\
Zn(s) \rightarrow Zn^{2+}(aq) + 2e^- & +\ 0.76 \text{ volts} \\
\hline
Cu^{2+}(aq) + Zn(s) \rightarrow Cu(s) + Zn^{2+}(aq) & +\ 1.10 \text{ volts}
\end{array}
$$

Add:

The overall positive value for the reaction potential suggests that the process is energetically feasible.

Conversely, the overall potential for the reaction

$$Cu(s) + Zn^{2+}(aq) \rightarrow Cu^{2+}(aq) + Zn(s) \quad \text{is} - 1.10 \text{ volts.}$$

The negative value indicates that the reaction is unlikely to occur. In general, *reactions with an overall positive potential are energetically feasible whereas those with an overall negative value are not so.* Relating this to table 14.3, it means that any oxidising agent on the left will oxidise any reducing agent below it on the right. In other words,

stronger + stronger $\longrightarrow$ weaker + weaker
oxidising   reducing   reducing   oxidising
agent   agent   agent   agent

Use table 14.3 to predict whether a reaction will occur between

- F_2 and Na
- F^- and Na^+
- F_2 and Na^+
- Cl_2 and Fe^{2+}
- I_2 and Br^-

In predicting reactions, it is important to remember that $E^\ominus$ relates to the *probability of a reaction occurring, not to the quantity of materials reacting.* The reaction is just as likely (or unlikely) to occur whether we have a milligram or a kilogram, a thimbleful or a bucketful, one mole or two moles. The electrode e.m.f. is independent of the number of electrons being transferred.

Thus, if $E^\ominus$ for the reaction $Fe^{3+}(aq) + e^- \rightarrow Fe^{2+}(aq)$ is +0.77 volts, $E^\ominus$ for the reaction $2Fe^{3+}(aq) + 2e^- \rightarrow 2Fe^{2+}(aq)$ is also +0.77 volts, and *not* +1.54 volts.

Although on overall positive value of $E^\ominus$ for a redox pair suggests that the reaction should take place, in practice the reaction may be too slow to occur. The important point is that $E^\ominus$ relates to the relative stabilities of reactants and products. Therefore, it indicates the feasibility of the reaction from an energetic standpoint. However, $E^\ominus$ gives no information whatsoever about *the rate of a reaction or its kinetic feasibility* (sections 12.7 and 12.8).

Thus, $E^\ominus$ values predict that $Cu^{2+}(aq)$ should oxidise $H_2(g)$ to $H^+(aq)$, i.e.

$Cu^{2+}(aq) + 2e^- \rightarrow Cu(s)$	+0.34
$H_2(g) \rightarrow 2H^+(aq) + 2e^-$	0.00

Add: $\overline{Cu^{2+}(aq) + H_2(g) \rightarrow Cu(s) + 2H^+(aq)}$ \qquad +0.34

But, nothing happens when hydrogen is bubbled into copper (II) sulphate solution because the reaction rate is effectively zero.

Another important point to remember in using $E^\ominus$ values in that they *relate only to standard conditions.* Changes in temperature, pressure and concentration will affect the values of electrode potentials. In particular, all electrode potentials for metal ion/metal systems become more positive when the concentration of metal ion is increased. This is because the reduction of these ions is then more likely to occur. Conversely, electrode potentials become less positive when the concentration of reactant ions is reduced.

For example, the standard electrode potential of copper is +0.34 volts, but the electrode potential of copper in contact with 0.1 M $Cu^{2+}(aq)$ is only +0.31 volts. Similarly, the standard electrode potential of zinc is -0.76 volts, but the electrode potential of zinc in contact with 0.1 M $Zn^{2+}(aq)$ is reduced to -0.79 volts.

The standard electrode potential for the half-reaction

$$Fe^{3+}(aq) + e^- \rightarrow Fe^{2+}(aq) \quad \text{is} +0.77 \text{ volts.}$$

What happens to the electrode potential when the concentration of

- $Fe^{3+}(aq)$ rises,
- $Fe^{3+}(aq)$ falls,
- $Fe^{2+}(aq)$ rises,
- $Fe^{2+}(aq)$ falls,
- both $Fe^{3+}(aq)$ and $Fe^{2+}(aq)$ are doubled?

The effect of non-standard conditions on electrode potentials is nicely illustrated by the reaction between $MnO_2(s)$ and $HCl(aq)$. Under standard conditions, MnO_2 will not oxidise 1.0M $HCl(aq)$ to Cl_2.

$$MnO_2(s) + 4H^+(aq) + 2e^- \rightarrow Mn^{2+}(aq) + 2H_2O(l) \qquad +1.23 \text{ volts}$$
$$2Cl^-(aq) \rightarrow Cl_2(g) + 2e^- \qquad -1.36 \text{ volts}$$
$$MnO_2(s) + 4H^+(aq) + 2Cl^-(aq) \rightarrow Mn^{2+}(aq) + 2H_2O(l) + Cl_2(g) \quad -0.13 \text{ volts}$$

However, when manganese (IV) oxide is warmed with *concentrated* hydrochloric acid the electrode potentials for each half-equation become more positive, and the overall electrode potential becomes positive. Hence, chlorine is produced.

Using standard electrode potentials to calculate the e.m.f.'s of cells

The calculation of cell e.m.f.'s can be done in a similar manner to the prediction of likely reactions. One of the first practical cells to be used was the **Daniell Cell**, invented in 1836 (figure 14.6). Notice how closely the Daniell cell resembles the arrangement in figure 14.1 by using $Zn^{2+}(aq)/Zn(s)$ and $Cu^{2+}(aq)/Cu(s)$ half-cells. The only difference is that the Daniell cell uses a rigid porous pot to allow movement of ions between the two solutions. The arrangement in figure 14.1 uses a soggy filter paper soaked in potassium nitrate solution.

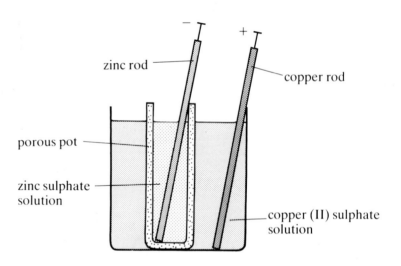

zinc rod

copper rod

porous pot

zinc sulphate solution

copper (II) sulphate solution

Figure 14.6
The Daniell Cell.

When the Daniell cell is operating, zinc goes into solution as zinc ions and copper deposits on the copper electrode. Under standard conditions, the e.m.f. of the cell is 1.10 volts, i.e.

$$Zn(s) \rightarrow Zn^{2+}(aq) + 2e^- \qquad +0.76 \text{ volts}$$
$$Cu^{2+}(aq) + 2e^- \rightarrow Cu(s) \qquad +0.34 \text{ volts}$$
Add: $$Cu^{2+}(aq) + Zn(s) \rightarrow Zn^{2+}(aq) + Cu(s) \qquad +1.10 \text{ volts}$$

The arrangement in the Daniell cell can be summarised as:

$$Zn(s)|Zn^{2+}(aq)| \; |Cu^{2+}(aq)|Cu(s) \qquad E = 1.10 \text{ volts}$$

Using standard electrode potentials to show the relative oxidising and reducing power of different species

The relative oxidising and reducing power of different species and the relative reactivity of metals and non-metals can be shown using standard electrode potentials (see sections 14.2 and 14.5).

14.7 Commercial cells

Cells and batteries provide a useful and economic way of obtaining energy from chemical reactions. The Daniell cell, discussed in the last section, has now been replaced by cheaper and more convenient cells.

The Leclanché dry cell

Probably the commonest, cheapest and most convenient cell in use at the present time is the Leclanché dry cell (figure 14.7). It is used in a wide range of small electrical appliances such as torches, bicycle lamps, radios and electric bells.

Figure 14.7
The Leclanché Dry Cell.

- Why is the Leclanché dry cell cheaper than the Daniell cell?
- Why is the Leclanché dry cell more convenient and more portable than the Daniell cell?
- Why is the ammonium chloride used as a paste rather than as a dry solid?

The Leclanché cell can be summarised as:

$$Zn(s)|Zn^{2+}(aq)||NH_4^+(aq), [2NH_3(g) + H_2(g)]|C \text{ (graphite)} \qquad E = 1.5 \text{ volts}$$

The negative terminal in the cell is zinc, as in the Daniell cell.

$$Zn(s) \rightarrow Zn^{2+}(aq) + 2e^-$$

The carbon rod is the positive terminal at which ammonium ions are converted to ammonia and hydrogen.

$$2NH_4^+(aq) + 2e^- \rightarrow 2NH_3(g) + H_2(g)$$

The surface area of the positive terminal is increased by surrounding the carbon rod with a mixture of powdered charcoal and manganese(IV) oxide. The purpose of the manganese(IV) oxide is to oxidise hydrogen produced at the electrode to water. This prevents bubbles of the gas from coating the carbon terminal which would reduce its efficiency. The ammonia gas causes no such problems because it dissolves rapidly in the water of the paste. It is necessary to use a paste because dry solid ammonium chloride would not conduct electricity.

A single dry cell can produce a potential difference of 1.5 volts, but batteries of these cells giving 100 volts and more are frequently used.

When the cell is in use, the casing is eaten away as the zinc is converted to Zn^{2+} ions. In time, the zinc case disintegrates and the paste oozes through the outer carboard.

Accumulators (secondary cells)

Obviously, the dry cell cannot be used to provide continuous supplies of electrical energy indefinitely. Nor is there any means of restoring or *recharging*

This chin-controlled robot has been designed to aid severely handicapped people.

the cell so that it can be used again. Cells such as this which can be used once only and cannot be recharged are called **primary cells**. In contrast, **secondary cells (accumulators)** can be recharged and used again.

When an accumulator is recharged, an electric current is passed through it in the opposite direction to that which the cell generates. Chemical reactions occur at the terminals as the original substances reform. This recharging is an example of electrolysis and is the reverse of discharge.

The lead–acid accumulator

This is certainly one of the most commonly used types of accumulator. Virtually all car batteries are composed of either three or six lead–acid cells in series. This gives a total potential difference of approximately 6 or 12 volts, respectively.

Most milk floats are powered by lead–acid accumulators. They have only a limited mileage and can be conveniently recharged for each day's work. In recent years, the escalating cost of petrol has made the electrically powered motor car an increasingly viable proposition. Unfortunately, attempts to develop and market an economic vehicle of this type are still unsuccessful. Problems are caused by high battery costs, the smaller power/weight ratio of the 'battery' engine compared to a petrol engine and the limited mileage between recharging. Battery cars will not become a reality until a cheap, low-weight, rechargeable battery is developed.

The negative terminal in the lead–acid accumulator is a lead plate. This gives up electrons, forming lead (II) ions during discharge.

$$Pb(s) \rightarrow Pb^{2+}(aq) + 2e^-$$

The positive terminal is lead(IV) oxide. This reacts with H^+ ions in the sulphuric acid electrolyte, also forming lead(II) ions.

$$PbO_2(s) + 4H^+(aq) + 2e^- \rightarrow Pb^{2+}(aq) + 2H_2O(l)$$

The Pb^{2+} ions formed during discharge react with SO_4^{2-} ions in the electrolyte forming insoluble lead(II) sulphate.

$$Pb^{2+}(aq) + SO_4^{2-}(aq) \rightarrow PbSO_4(s)$$

It is important that discharge does not continue for too long. Otherwise the precipitate of fine lead sulphate changes to a coarser, inactive and non-reversible form. If this happens the accumulator becomes much less efficient.

14.8 Acids and bases

During the nineteenth century, the Swedish chemist Arrhenius put forward ideas about acids and bases. He suggested that *acids were substances which dissociated in water to produce hydrogen ions, H⁺.*

$$HCl \rightarrow H^+ + Cl^-$$

Arrhenius suggested that strong acids were completely dissociated in aqueous solution. Weak acids were only partially dissociated with a high proportion of the acid remaining in the undissociated form. Thus, a solution of ethanoic (acetic) acid could be represented by the following equilibrium.

$$CH_3COOH \rightleftharpoons CH_3COO^- + H^+$$

With the development of knowledge concerning atomic structure, chemists realised that the H^+ ions was simply a proton. It was highly improbable that such a small ion could exist independently in aqueous solution.

Thus, H^+ ions with their high charge density were believed to associate with polar water molecules in aqueous solution as H_3O^+ ions. H_3O^+ is now called the oxonium ion. At one time it was known as the hydronium ion. Experimental evidence for the hydration of H^+ ions was obtained by observing the movement of H^+ ions during electrolysis.

A life-saving acid–base reaction. During the Apollo 13 space project, the astronauts discovered that carbon dioxide was building up in their crippled spacecraft as they travelled towards the Earth. By an ingenious use of lithium hydroxide, they were able to repair the air purification equipment in their spacecraft. The photograph shows test pilot Scott MacLeod holding one of the lithium hydroxide containers.

Consequently, the dissociation of acids in water can be represented more accurately as

$$HCl + H_2O \rightarrow H_3O^+ + Cl^-$$

$$CH_3COOH + H_2O \rightleftharpoons H_3O^+ + CH_3COO^-$$

Two important advantages of this refinement to Arrhenius's theory are immediately apparent.

1 It recognised the essential role of the 'solvent' water molecules in the dissociation of acids.

2 It explains why substances such as hydrogen chloride and ethanoic acid only show their acidic properties in the presence of water.

Pure dry ethanoic acid and solutions of hydrogen chloride or ethanoic acid in organic solvents such as toluene contain no H^+ or H_3O^+ ions. Thus, they are non-electrolytes. They do not affect dry litmus paper and they do not react with carbonates to produce CO_2.

Although H^+ ions exist in aqueous solutions as H_3O^+, for the sake of simplicity we usually write H^+ rather than H_3O^+ in equations.

According to the Arrhenius theory, *bases were substances which reacted with H^+ ions to form water*. For example,

$$CuO + 2H^+ \rightarrow Cu^{2+} + H_2O$$

$$Cu(OH)_2 + 2H^+ \rightarrow Cu^{2+} + 2H_2O$$

Alkalis, soluble bases, were regarded as substances which dissociated in water producing OH^- ions.

$$NaOH(s) \xrightarrow{\text{water}} Na^+(aq) + OH^-(aq)$$

■ Write an equation for the reaction of sodium oxide with dilute hydrochloric acid.

■ Write an equation for the reaction of sodium sulphide with dilute hydrochloric acid.

■ Compare the two equations you have just written. Do you regard the second equation as an acid–base reaction? Explain your answer.

14.9 The Brønsted–Lowry theory of acids and bases

The definitions of acids and bases in the last section are somewhat unsatisfactory because they only apply to aqueous solutions.

Furthermore, the Arrhenius definition of a base is restricted to those substances which react with H^+ ions *to form water*.

In order to widen the scope of acid–base reactions and include non-aqueous systems and a wider range of bases, Brønsted and Lowry independently suggested the following definitions in 1923.

An acid is a proton donor.
A base is a proton acceptor.

Thus, the relationship between an acid and its corresponding base is

$$\underset{\substack{\text{acid} \\ \text{(proton donor)}}}{HB} \rightleftharpoons \underset{\text{proton}}{H^+} + \underset{\substack{\text{base} \\ \text{(proton acceptor)}}}{B^-}$$

HB and B^- are said to be *conjugate* and to form a **conjugate acid–base pair**. HB is the conjugate acid of B^- and B^- is the conjugate base of HB.

■ What is the conjugate acid of
(i) CH_3COO^-, (ii) HSO_4^-, (iii) NH_3, (iv) OH^-?

■ What is the conjugate base of
(i) HCl, (ii) H_3O^+, (iii) HSO_4^-, (iv) NH_3?

According to the Brønsted–Lowry theory, acid salts (such as $NaHSO_4$) and ammonium ions are recognised as acids. What is more, the definition of a base now includes all anions, water and ammonia as well as oxide and hydroxide ions.

Acid		Base		Base		Acid
HSO_4^-	+	OH^-	→	SO_4^{2-}	+	H_2O
NH_4^+	+	OH^-	→	NH_3	+	H_2O
$2H_3O^+$	+	S^{2-}	→	$2H_2O$	+	H_2S
H_3O^+	+	NH_3	→	H_2O	+	NH_4^+

Notice in the equations above that *water can act as a base and as an acid*. It does this simultaneously during its dissociation.

$$\underset{\text{acid}}{H_2O} + \underset{\text{base}}{H_2O} \rightleftharpoons \underset{\text{base}}{OH^-} + \underset{\text{acid}}{H_3O^+}$$

14.10 Relative strengths of acids and bases

When the acid HB dissociates in water the following equilibrium is established.

$$HB + H_2O \rightleftharpoons B^- + H_3O^+$$

Acids differ in the extent to which they donate protons to water in aqueous solution. Those which donate all their protons to water molecules are known as strong acids. The equilibrium in the equation above is well over to the right. Thus, the conjugate base of a strong acid is weak, and vice versa.

If HB is a weak acid, relatively few molecules will donate H^+ ions to water. The equilibrium will be well over to the left.

Using the idea of an acid as a proton donor and a base as a proton acceptor, we can arrange all acids and bases in a 'league table'. This shows their order of relative strengths (table 14.4).

If an acid is weak (i.e. has little tendency to donate protons) it follows that its conjugate base is strong. The base will have a strong affinity for protons. For example, hydrogen sulphide is a weak acid, but the sulphide ion is a strong base. Thus, in table 14.4 the acids increase in strength down the page, but their conjugate bases gradually become weaker. The relative strengths of acids are discussed further in sections 24.5 and 24.9.

Table 14.4
Relative strengths of acids and bases

	Name of acid	Acid	H^+ + Base	Name of base	
acid strength increases ↓	ethanol	C_2H_5OH	$\rightleftharpoons H^+ + C_2H_5O^-$	ethoxide	base strength decreases ↓
	water	H_2O	$\rightleftharpoons H^+ + OH^-$	hydroxide	
	ammonium	NH_4^+	$\rightleftharpoons H^+ + NH_3$	ammonia	
	hydrogen sulphide	H_2S	$\rightleftharpoons 2H^+ + S^{2-}$	sulphide	
	ethanoic acid	CH_3COOH	$\rightleftharpoons H^+ + CH_3COO^-$	ethanoate	
	sulphurous acid	H_2SO_3	$\rightleftharpoons 2H^+ + SO_3^{2-}$	sulphite	
	oxonium	H_3O^+	$\rightleftharpoons H^+ + H_2O$	water	
	sulphuric acid	H_2SO_4	$\rightleftharpoons 2H^+ + SO_4^{2-}$	sulphate	
	hydrochloric acid	HCl	$\rightleftharpoons H^+ + Cl^-$	chloride	
	chloric (VII) acid	$HClO_4$	$\rightleftharpoons H^+ + ClO_4^-$	chlorate (VII)	

14.11 Acid–base reactions: competition for protons

When dilute hydrochloric acid is added to a solution of sodium sulphide, hydrogen sulphide is produced.

$$S^{2-} + \underbrace{2H^+ + 2Cl^-}_{\text{hydrochloric acid}} \rightarrow H_2S + 2Cl^-$$

This reaction emphasises that hydrochloric acid is a stronger acid than hydrogen sulphide. It donates protons to sulphide ions in forming H_2S. Alternatively, we might say that sulphide is a stronger base than chloride. In competition for protons the sulphide wins convincingly.

Thus, *acid–base reactions involve competitions between bases for protons. In this respect, they can be compared to redox reactions which involve competitions between oxidising agents for electrons.*

When dilute hydrochloric acid or sulphurous acid is added to a solution of sodium benzoate, a white precipitate of benzoic acid appears.

$$\underset{\text{benzoate ion}}{C_6H_5COO^-(aq)} + \underbrace{H^+(aq) + Cl^-(aq)}_{\text{hydrochloric acid}} \rightarrow \underset{\text{benzoic acid}}{C_6H_5COOH(s)} + Cl^-(aq)$$

$$\underset{\text{benzoate ion}}{C_6H_5COO^-(aq)} + \underset{\text{sulphurous acid}}{H_2SO_3(aq)} \rightarrow \underset{\text{benzoic acid}}{C_6H_5COOH(s)} + HSO_3^-(aq)$$

Hence, both hydrochloric acid and sulphurous acid are stronger than benzoic acid.

However, when ethanoic acid is added to a solution of sodium benzoate no apparent reaction occurs. Ethanoic acid does not protonate benzoate ions, because ethanoic acid is a weaker acid than benzoic acid. Thus, we could place benzoic acid between ethanoic acid and sulphurous acid in the 'league' showing relative strengths of acids in table 14.4.

■ When Universal Indicator solution was added to separate portions of
(i)   ammonium chloride solution,
(ii) a dilute solution of phenol (C_6H_5OH) in water and
(iii) water,
solutions (i) and (ii) produced an orange colour, whereas (iii) was yellow–green. What can you conclude from this about the relative acidities of the three solutions?
■ When a few magnesium turnings were added to the three solutions mentioned in the 1st paragraph, solution (i) evolved hydrogen most vigorously and (iii) least vigorously. What can you conclude from this about the relative acidities of (i), (ii) and (iii)?
■ Where would you place phenol (C_6H_5OH) in the 'league' of relative acidity in table 14.4?

14.12 Complex ions

In the last few sections, we have been discussing the reactions of acids with bases. Acids donate protons (H^+ ions). In aqueous solution these H^+ ions are strongly attracted to polar water molecules by co-ordinate bonds forming H_3O^+ ions (figure 14.8).

In the same way as H^+, other cations (e.g. M^{2+}) will also exist in aqueous solution as hydrated ions (e.g. $[M(H_2O)_n]^{2+}$). Since the H^+ ion (a bare proton) is more than 50 000 times smaller than any other cation, the bonding between H_2O and H^+ is much stronger than that between H_2O molecules and other cations. However, the larger size of other cations enables them to associate with two, four or even six water molecules. H^+ associates with only one.

Figure 14.8
Co-ordinate bonding in an aqueous H^+ ion (oxonium ion).

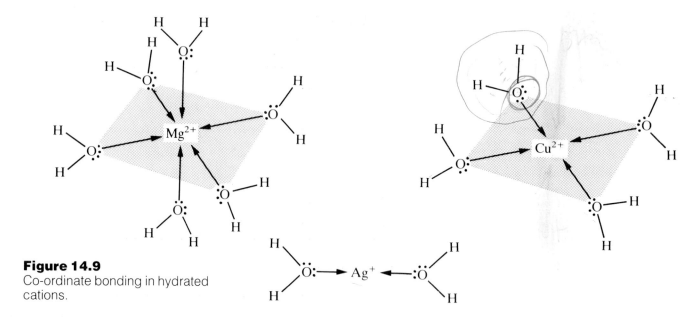

Figure 14.9
Co-ordinate bonding in hydrated cations.

The number of water molecules associated with a given cation has been firmly established in some cases. In other cases it has been established more tentatively through indirect evidence. Thus, $Mg^{2+}(aq)$ is represented as $[Mg(H_2O)_6]^{2+}$, $Cu^{2+}(aq)$ is represented as $[Cu(H_2O)_4]^{2+}$ and $Ag^+(aq)$ is represented as $[Ag(H_2O)_2]^+$.

In these ions, the water molecules are bound to the central cation by co-ordinate bonds (figure 14.9).

Other polar molecules, besides water, can also co-ordinate to metal cations. For example, in ammonia solution Cu^{2+} exists principally as $[Cu(NH_3)_4]^{2+}$ whilst Ag^+ exists as $[Ag(NH_3)_2]^+$.

It is not only polar molecules but also anions that can associate with cations in this way. Thus, when anhydrous copper (II) sulphate is added to concentrated hydrochloric acid, the solution becomes yellow–green. This is due to the formation of $[CuCl_4]^{2-}$ ions.

Ions such as $[Cu(H_2O)_4]^{2+}$, $[Cu(NH_3)_4]^{2+}$ and $[CuCl_4]^{2-}$, in which a metal ion is associated with a number of anions or neutral molecules, are known as **complex ions.** The anions and molecules firmly bonded to the central cation are called **ligands**. Each ligand contains at least one atom with a lone pair of electrons. These can be donated to the central cation forming a co-ordinate (dative) bond. The ligand is said to be co-ordinated to the central ion.

Water is by far the most common ligand. The strengths of the co-ordinate bonds between different cations and water molecules can be compared by determining their **hydration energies** (sections 10.8 and 12.14). These relate to the process

$$M^{x+}(g) + nH_2O(l) \rightarrow [M(H_2O)_n]^{x+}(aq)$$

Look closely at table 14.5.

■ How does the hydration energy change as ionic radius increases?
■ How does the hydration energy change as the charge on an ion increases?
■ In general, how will the strengths of co-ordinate bonds from ligands to a central cation vary with size of the cation and with the charge on the cation?

14.13 Naming complex ions

The systematic naming of complex ions is based on four simple rules.

1 *State the number of ligands around the central cation* using Greek prefixes: mono-, di-, tri-, etc.

Table 14.5
Hydration energies of some cations

Ion	Ionic radius /nm	Hydration energy /kJ mole^{-1}
Li$^+$	0.068	-499
Na$^+$	0.098	-390
K$^+$	0.133	-305
Rb$^+$	0.148	-281
Cs$^+$	0.167	-248
Mg^{2+}	0.065	-1891
Ca^{2+}	0.094	-1562
Sr^{2+}	0.110	-1413
Ba^{2+}	0.134	-1273
Al^{3+}	0.045	-4613
Ga^{3+}	0.062	-4650

2 *Identify the ligands* using names ending in -o for anions, e.g. F$^-$ -fluoro, CN$^-$ -cyano, Cl$^-$, -chloro, OH$^-$ -hydroxo, H$_2$O aqua, NH$_3$ -ammine.

3 *Name the cation* using the English name for the cation in a positively charged complex, but the Latinised name ending with the suffix -*ate* for the central metal atom in a negatively charged complex, i.e. aluminate for aluminium, cuprate for copper, ferrate for iron, plumbate for lead, zincate for zinc, etc.

4 *Indicate the oxidation number of the central cation* using Roman numerals (I, II, III, etc.).

The following examples show you how to apply these rules.

Formula of complex ion	1 State the number of ligands	2 Name the ligand	3 Name the central cation (suffix -*ate* for anions)	4 Indicate oxidation number of central cation
[Cu(NH$_3$)$_4$]$^{2+}$	tetra	..........ammine	........copper	...........(II)
[CuCl$_4$]$^{2-}$	tetra	..........chloro	........cuprate	...........(II)
[Fe(CN)$_6$]$^{3-}$	hexa	..........cyano	........ferrate	...........(III)
[Cu(NH$_3$)$_2$]$^+$	di	..........ammine	........copper	...........(I)

Try to name the following ions:

■ [Al(OH)$_4$]$^-$, [Zn(NH$_3$)$_4$]$^{2+}$, [Fe(CN)$_6$]$^{4-}$, [CrCl$_2$(H$_2$O)$_4$]$^+$.

14.14 Polydentate ligands and chelation

Most ligands can form only one co-ordinate bond with a cation. They are said to be **unidentate** because they have only 'one tooth' with which to attach themselves to the central cation. (The word *dens* in Latin means tooth.) In some cases, each ligand molecule or anion can form more than one link with the metal ion. These are said to be **polydentate** ('many teeth'). Examples of bidentate ligands are: ethanedioate (oxalate), 1, 2-diaminoethane and 2-hydroxybenzoate (salicylate).

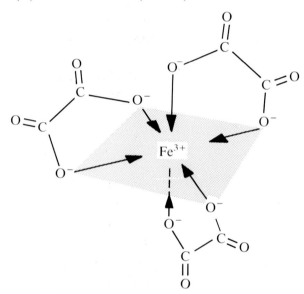

Some ligands such as edta (ethylenediaminetetraacetate) can form as many six dative bonds with the central ion. They are known as hexadentate ligands.

The complex ions which form between polydentate ligands and cations are known as **chelates** or **chelated complexes**. These names come from the Greek word *chelos* meaning 'a crab's claw'. In these complexes the ligand forms a clawing pincer-like grip on the metal ion. Figure 14.10 shows the chelate formed between iron(III) and ethanedioate (oxalate) ions.

Figure 14.10
The structure of the iron(III)/
ethanedioate chelate, $[Fe(C_2O_3)_3]^{3-}$.

In general, polydentate ligands are more powerful and more versatile than simple unidentate ligands. The stability of a complex is much enhanced by chelation as the pincer-like grip of the polydentate ligand can hold the central cation more securely.

14.15 Complexing—competition for cations

In section 14.11 we noticed that *acid–base reactions involve competitions between different bases for H⁺ ions*. In a similar fashion, *complexing reactions involve competitions between different ligands for metal cations*. Thus, when excess ammonia solution is added to aqueous copper(II) sulphate, ammonia molecules displace water molecules from hydrated copper(II) ions forming $[Cu(NH_3)_4]^{2+}$. The colour changes from pale blue to a much deeper blue.

$$[Cu(H_2O)_4]^{2+}(aq) + 4NH_3(aq) \rightarrow [Cu(NH_3)_4]^{2+}(aq) + 4H_2O(l)$$

In this case, ammonia acts as a stronger ligand than water.

In general, it can be shown that a more powerful ligand will displace a less powerful ligand from a cation complex. Thus, when 2-hydroxybenzoate ions (salicylate, $HOC_6H_4COO^-$) are added to aqueous iron(III) chloride the colour changes from yellow to a deep purple. 2-Hydroxybenzoate ions displace water molecules from hydrated iron(III) ions forming purple $[Fe(HOC_6H_4COO)_3]$.

$$[Fe(H_2O)_6]^{3+}(aq) + 3HOC_6H_4COO^-(aq) \rightarrow [Fe(HOC_6H_4COO_3](s) + 6H_2O(l)$$

When edta (an even stronger ligand than 2-hydroxybenzoate) is added to the purple suspension, the edta displaces 2-hydroxybenzoate ions. The colour changes from purple to pale yellow which is the colour of the iron(III)/edta complex.

One of the most fascinating demonstrations of the relative strengths of different ligands for a particular cation is outlined in the sequence of reactions shown in table 14.6. Going down the table, the ligands are increasing in strength and becoming stronger competitors for Ag^+ ions, i.e.

$$edta^{2-} > S^{2-} > CN^- > I^- > S_2O_3^{2-} > Br^- > NH_3 > Cl^- > H_2O$$

Table 14.6
The relative strengths of different ligands for Ag^+

	Ligand	Colour and state of complex
$[Ag(H_2O)_2]^+$	H_2O	clear solution
add ↓ NaCl(aq)		
AgCl(s)	Cl^-	white precipitate
add ↓ NH_3(aq)		
$[Ag(NH_3)_2]^+$	NH_3	clear solution
add ↓ KBr(aq)		
AgBr(s)	Br^-	cream precipitate
add ↓ Na_2S_2O_3(aq)		
$[Ag(S_2O_3)_2]^{3-}$	$S_2O_3^{2-}$	clear solution
add ↓ KI(aq)		
AgI(s)	I^-	yellow precipitate
add ↓ KCN(aq)		
$[Ag(CN)_2]^-$	CN^-	clear solution
add ↓ Na_2S(aq)		
$Ag_2S(s)$	S^{2-}	black precipitate
add ↓ sodium edta(aq)		
$[Ag(edta)]^-$	$edta^{2-}$	clear solution

Notice that some reactions in the sequence result in the precipitation of insoluble solids. These precipitates can be regarded in some ways as neutral complexes. Being uncharged, they are less readily hydrated by polar water molecules than the charged complexes and so they are less likely to dissolve in water.

14.16 Investigating the stoichiometry of complex ions

The determination of the stoichiometry of a complex ion involves measuring the number of ligands complexing with one metal cation. This can be investigated by various methods. The most important methods are:

1 titration methods involving competitive complexing (described below), and

2 colorimetric methods which measure the colour intensity of the mixture as the proportion of metal cation to ligand is varied (see section 19.6).

Determination of the stoichiometry of the iron(III)/edta complex by competitive complexing

When sodium 2-hydroxbenzoate (sodium salicylate) is added to a solution of Fe^{3+} ions a deep purple complex ion is formed. As edta is added to this purple solution, the colour slowly fades to a pale yellow iron(III)/edta complex. Of course, the purple colour remains as long as there are more than enough Fe^{3+} ions for the edta. Any excess Fe^{3+} ions will form a purple complex with the 2-hydroxybenzoate. When the purple colour just disappears, the quantities of Fe^{3+} and edta just react and neither is in excess. Thus, 2-hydroxybenzoate ions can act as indicators. They will show when just sufficient edta has been added to react with all the Fe^{3+}.

In a particular experiment, $20\,cm^3$ of 0.1M iron(III) chloride was pipetted into a flask. A few drops of sodium 2-hydroxybenzoate were added. 0.1M edta was then added from a burette and the solution became a clear yellow colour after the addition of $20.1\,cm^3$.

- How many moles of Fe^{3+} were taken?
- How many moles of edta reacted with the Fe^{3+} taken?
- How many moles of edta react with one mole of Fe^{3+}?
- What is the formula of the Fe^{3+}/edta complex?

14.17 The use and importance of complex ions

Complex ions are important in industry, in the laboratory and in biology.

Complex ions of biological importance

Two essential biological macromolecules composed of complex ions are **chlorophyll** and **haemoglobin**. Chlorophyll is the green pigment in plant cells. It is responsible for absorbing the radiant energy of sunlight. It converts sunlight into chemical energy in the bonds of carbohydrate molecules synthesised by the plant.

The chlorophyll molecule is composed of a complicated cyclic tetradentate ligand. This contains four nitrogen atoms surrounding magnesium in a square planar arrangement (figure 14.11).

Dark parallel lines of chlorophyll in two chloroplasts from a leaf of flax. The magnification is 46 000 times.

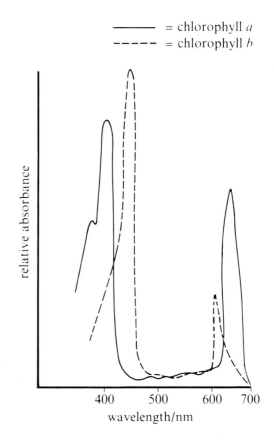

Figure 14.11
Complex ions of biological importance.

Chlorophyll a
(In chlorophyll b, the methyl group marked by an asterisk is replaced by a —CHO group)

Haemoglobin

Sunlight reaching the Earth's surface has a maximum intensity in the blue–green region of the spectrum in the wavelength range 450–550 nm. Curiously, the chlorophyll molecule has its weakest absorption in this portion of the visible spectrum. The maximum absorption peaks for chlorophyll are, in fact, at 680 nm in the red region of the spectrum and at 440 nm in the violet (figure 14.12). In spite of this inefficiency in absorbing the radiation of greatest

——— = chlorophyll *a*
- - - - - = chlorophyll *b*

Figure 14.12
The absorption spectrum of chlorophyll.

relative absorbance

wavelength/nm

intensity, chlorophyll has other properties that make it particularly suitable as a photosynthetic pigment. It can receive energy both directly from light and indirectly from other pigments in plants such as carotenoids. It can store radiant energy and transfer it to other molecules. Its ring structure of alternating double and single bonds enables it to function in the transfer of hydrogen atoms from one molecule to another.

Figure 14.11 also shows the structure of haemoglobin, the red pigment present in red blood cells. Haemoglobin acts as the transport of oxygen in the blood. It is composed of the complex cyclic tetradentate haem ligand and the protein, globin. Notice the striking similarity of the haem portion of the complex to the chlorophyll molecule. This suggests that the two may have been adapted from the same original substance during the course of evolution. In haemoglobin, the central metal ion is iron(II), not magnesium. The active part of the haemoglobin complex is the Fe^{2+} ion. This is co-ordinated to four nitrogen atoms in the haem ligand and also to two nitrogen atoms in the globin. One of the latter two positions can be occupied weakly and reversibly by oxygen. It is this property which enables haemoglobin to transport oxygen in the bloodstream from the lungs to other parts of the body. Unfortunately, much stronger ligands than oxygen, such as cyanide and carbon monoxide, can also occupy this position. Unlike oxygen, they attach themselves to the haem group irreversibly. So, they act as acute poisons by eliminating the oxygen-transporting ability of haemoglobin.

Complex ions of industrial importance

Complex ions play an important part in the methods used to soften hard water and in the extraction of metals from their ores.

Soap contains the sodium salts of long-chain carboxylic acids such as sodium hexadecanoate (sodium palmitate) and sodium octadecanoate (sodium stearate) (section 33.3), When hard water, containing calcium and magnesium ions, is mixed with soap, it reacts with the anions in these salts. The products are insoluble compounds which we see as 'scum' on the surface of the water.

$$Ca^{2+}(aq) + 2CH_3(CH_2)_{16}COO^- \rightarrow Ca(CH_3(CH_2)_{16}COO)_2(s)$$

| Ca²⁺(aq) ions in hard water | octadecanoate (stearate) ions in soap | insoluble ppte of calcium octadecanoate (stearate) in 'scum' |

Various ligands will react with the calcium and magnesium ions to form soluble and stable complex ions. As a result of this, the calcium and magnesium ions are unable to react with the anions in soap and the water is softened. Thus, sodium polyphosphate ($Na_6P_6O_{18}$), which contains the powerful polyphosphate ligand ($P_6O_{18}^{6-}$), is sold for domestic and industrial water softening under the trade name Calgon. The name Calgon is derived from the expression 'calcium gone'. Edta and its sodium salt are also used in water softening. However, these two substances are poisonous. So, they cannot be used to soften drinking water. Consequently, their use is restricted to the softening of water required for industrial processes such as the dyeing of textiles.

Complex formation frequently plays an essential part in the purification of metal ores and the subsequent extraction of the metal. An excellent example of this is provided by the purification of aluminium oxide from bauxite (section 16.6). The extraction of gold and silver involve the formation of complex cyanides. Impure silver ores such as argentite (containing Ag_2S) and horn silver (containing AgCl) are first mixed with a solution of sodium cyanide. This forms the soluble complex cyanide.

$$Ag_2S(s) + 4CN^-(aq) \rightarrow 2[Ag(CN)_2]^-(aq) + S^{2-}(aq)$$
$$AgCl(s) + 2CN^-(aq) \rightarrow [Ag(CN)_2]^-(aq) + Cl^-(aq)$$

Silver is then precipitated by adding zinc dust to the solution. Any excess zinc is removed by adding dilute acid.

$$2[Ag(CN)_2]^-(aq) + Zn(s) \rightarrow [Zn(CN)_4]^{2-}(aq) + 2Ag(s)$$

Summary

1 The major classes of reaction involving inorganic substances are redox, acid–base, precipitation and complexing.

2 Redox reactions involve competition for electrons. Oxidising agents (oxidants) accept electrons. The stronger the oxidising agent, the greater is its competitive desire for electrons. Reducing agents (reductants) are electron donors.

3 The electromotive force of a cell is the maximum potential difference which the cell can generate.

4 The standard conditions for electrochemical ($E^\ominus$) measurements are similar to those for the thermochemical measurements: 25°C, gases at a pressure of 1 atm, aqueous solutions 1.0M, platinum as the electrode when the half-cell system does not include a metal.

5 The standard electrode (reduction) potential, $E^\ominus$, of a standard half-cell is the potential of that half-cell relative to a standard hydrogen half-cell under standard conditions. $E^\ominus$ which relates to the reduction half-equation has the same sign as the particular half-cell relative to the standard hydrogen half-cell.

6 The size and sign of $E^\ominus$ values provide a measure of the relative strength of oxidising agents and their conjugate reducing agents.

7 An overall positive $E^\ominus$ for a redox pair suggests that a reaction should take place. In practice:

   (i) some likely reactions may be too slow to occur,

   (ii) reactions which are unlikely under standard conditions (to which $E^\ominus$ values refer) may take place readily under different conditions of temperature, pressure and concentration.

8 Arrangements which generate electric currents from chemical reactions are called cells. A battery is composed of several cells.

9 Cells which can be used only once and which cannot be recharged are called primary cells. Cells which can be recharged and used again and again are called secondary cells or accumulators.

10 According to the Brønsted–Lowry theory of acids and bases, an acid is a proton donor, whereas a base is a proton acceptor.

11 Acid–base reactions involve competitions between bases for protons (H^+ ions). The stronger the base the greater is its competitive desire for protons.

12 Water can act as a base and as an acid. It does so simultaneously during its dissociation.

$$H_2O + H_2O \rightleftharpoons OH^- + H_3O^+$$
$$\text{acid} \quad \text{base} \quad \text{base} \quad \text{acid}$$

13 A complex ion is an ion in which a central metal cation is associated with a number of anions or neutral molecules.

14 The anions and molecules bound to the central cation in a complex ion are called ligands.

15 Chelates or chelated complexes are complex ions in which each ligand forms more than one link with the central cation.

16 Complexing reactions involve competitions between ligands for metal cations. The stronger the ligand the greater is its competitive desire for cations.

Study questions

1 (a) How did Arrhenius define (i) an acid, (ii) a base?

(b) How did Brønsted and Lowry define (i) an acid, (ii) a base?

(c) Show, by writing appropriate equations how:
  (i) the HSO_4^- ion can act as an Arrhenius acid,
  (ii) the HSO_4^- ion can act as a Brønsted–Lowry base.

(d) Explain, using examples, how the Brønsted-Lowry definition allows a much wider range of substances to be classified as bases.

2 2.5 g of a hydrate of sodium carbonate, $Na_2CO_3.xH_2O$, were dissolved in 250 cm^3 of water. 25 cm^3 of the resulting solution required 17.5 cm^3 of 0.1 M hydrochloric acid to neutralise it completely.

(a) How many moles of sodium carbonate react with one mole of hydrochloric acid?

(b) How many moles of hydrochloric acid are there in 17.5 cm^3 of a 0.1 M solution?

(c) How many moles of sodium carbonate react with 17.5 cm^3 of 0.1 M hydrochloric acid?

(d) How many moles of sodium carbonate were there in the original 2.5 g?

(e) How many moles of water of crystallisation must there have been in the original 2.5 g?

(f) What is the value of x? [Na = 23, C = 12, O = 16, H = 1]

3 (a) Explain what you understand by the terms oxidation and reduction.

(b) Illustrate your answer to part (a) by writing separate half-equations involving electrons for the oxidation and reduction processes when the following substances react:
  (i) Mg with Cl_2,
  (ii) I_2 with $S_2O_3^{2-}$,
  (iii) MnO_2 with conc. HCl,
  (iv) acidified $KMnO_4$ with SO_2,
  (v) H_2O_2 with I^-,
  (vi) Fe^{3+} with I^-,
  (vii) acidified BrO_3^- with Br^-.

(c) Suppose you were required to determine the standard electrode potential of a $Zn^{2+}(aq)/Zn(s)$ half-cell.
  (i) List all the essential materials and equipment required for this determination.
  (ii) Draw a diagram showing how the apparatus would be assembled for the determination.

4 Use oxidation numbers to decide which elements are oxidised or reduced in the following conversions.

(a) Cr^{3+} $\rightarrow Cr_2O_7^{2-}$

(b) CrO_4^{2-} $\rightarrow Cr_2O_7^{2-}$

(c) $S_2O_3^{2-}$ $\rightarrow S_4O_6^{2-}$

(d) $C_2O_4^{2-}$ $\rightarrow CO_2$

(e) IO_3^- $\rightarrow I_2$

5 (a) Draw a fully labelled diagram of a Daniell cell and show the direction in which electrons flow in the external circuit when it is used to generate electricity.

(b) Write equations for the reactions at each electrode when the cell produces an electric current.

(c) Deduce a value for the e.m.f. of the cell operating under standard conditions.
($Cu^{2+}(aq)/Cu(s) = +0.34$ V, $Zn^{2+}(aq)/Zn(s) = 0.76$ V.)

(d) How would the e.m.f. of the cell be affected if:
  (i) the concentration of Cu^{2+} ions was increased,
  (ii) the concentration of Zn^{2+} ions was increased?
  Give an explanation of your reasoning in each case.

6 The rusting of iron is an oxidation process.

(a) Explain, with appropriate equations, what happens when iron rusts.

(b) Iron may be protected from rusting by coating with zinc or tin. By reference to the following data, explain why zinc protects iron more effectively than tin, once the protective metal coating has been scratched to expose the iron below.

$$E^\ominus/\text{volts}$$

$$\text{Zn} \rightarrow \text{Zn}^{2+} + 2e^- \qquad +0.76$$

$$\text{Fe} \rightarrow \text{Fe}^{2+} + 2e^- \qquad +0.44$$

$$\text{Sn} \rightarrow \text{Sn}^{2+} + 2e^- \qquad +0.14$$

(c) Why do you think tin is used rather than zinc to coat and protect the inside of 'tin' cans, whereas zinc is used rather than tin to galvanise and protect buckets?

7 (a) The element nitrogen can form compounds in various oxidation states from -3 to $+5$.

    (i) Construct an oxidation number chart for nitrogen in the following compounds:

$$\text{NH}_3, \text{NH}_4\text{Cl}, \text{HNO}_3, \text{HNO}_2, \text{NO}_2$$

    (ii) What are the principal oxidation states of nitrogen?

    (iii) What are the oxidation numbers of nitrogen in the following compounds?
hydrazine, N_2H_4;
dinitrogen oxide, N_2O;
nitrogen oxide, NO;
dinitrogen tetraoxide, N_2O_4;
ammonium nitrate, NH_4NO_3.

(b) Nitrogen forms a variety of bonds in different compounds–ionic, covalent, co-ordinate and hydrogen bonds.
Draw dot/cross diagrams to show the electronic structures of atoms and/or ions in the following compounds and indicate the type of bonds involved:

    (i)   magnesium nitride, Mg_3N_2;

    (ii) ammonia, NH_3;

    (iii) nitrogen trichloride, NCl_3;

    (iv) ammonium chloride, NH_4Cl;

    (v) sodium nitrate, NaNO_3.

8 The following table gives the standard electrode potentials for a number of half-reactions.

$$E^\ominus/\text{volts}$$

$$\text{Zn}^{2+} + 2e^- \rightarrow \text{Zn} \qquad -0.76$$

$$\text{Fe}^{2+} + 2e^- \rightarrow \text{Fe} \qquad -0.44$$

$$\text{I}_2 \quad + 2e^- \rightarrow 2\text{I}^- \qquad +0.54$$

$$\text{Fe}^{3+} + e^- \rightarrow \text{Fe}^{2+} \qquad +0.77$$

$$\text{Ce}^{4+} + e^- \rightarrow \text{Ce}^{3+} \qquad +1.61$$

(a) Which half-equation is used as the standard for these electrode potentials?

(b) Which of the substances in the table above is:

    (i)   the strongest oxidising agent,

    (ii) the strongest reducing agent?

(c) Which substance(s) in the table could be used to convert iodide ions to iodine? Write balanced equations for any possible conversions.

(d) A half-cell is constructed by putting a platinum electrode in a solution which is 1.0M with respect to both Fe^{2+} and Fe^{3+} ions. This half-cell is then connected by means of a 'salt bridge' to another half-cell containing an iron electrode in a 1.0M solution of Fe^{2+} ions.

    (i)   What is the e.m.f. of this cell?

    (ii) If the two electrodes are connected externally, what reactions take place in each half-cell?

    (iii) In which direction do electrons flow in the external circuit?

(e) Write an equation for the reaction you would expect to occur when an iron nail is placed in a solution of iron(III) sulphate.

9 When edta (ethylenediaminetetraacetic acid) is titrated against calcium salts, experiments show that one mole of edta reacts with one mole of Ca^{2+}.

    This technique is used in clinical analysis to estimate the concentration of calcium in blood serum. Concentrations between 7 and 13 mg of calcium per 100 cm^3 of blood serum are normal.

    One cubic centimetre of blood serum was taken from a patient and found to require 2.75 cm^3 of 0.0015 M edta solution.

(a) Calculate the concentration of calcium in the patient's blood serum in mg per 100 cm^3 (Ca = 40).

(b) What assumption(s) have you made in your calculation?

(c) Is there an abnormal level of calcium concentration in the patient's blood?

15 Groups I and II— The Alkali Metals and the Alkaline-Earth Metals

---

15.1 Introduction

The alkali metals and the alkaline-earth metals are members of Groups I and II, respectively, in the periodic table (table 15.1). Whilst you are studying these elements, look for the following important features:

1 The similarities between the elements in Group I.

2 The similarities between the elements in Group II.

3 The differences between the elements in Group I and those in Group II.

4 The gradual trends in properties of the elements and their compounds in Group I and the similar trends in Group II.

In all these metals, the only electrons in their outermost shell occupy an s-orbital. Hence these elements are sometimes called the **s-block elements.**

■ Write the electronic structures of Na and K in terms of s, p, d, f notation. (The electronic structure of Li is $1s^2 2s^1$.)

■ Write the electronic structures of Be, Mg and Ca in terms of s, p, d, f notation.

Table 15.1
The members of Group I and Group II in the periodic table

Group I	Group II
Li	Be
Na	Mg
K	Ca
Rb	Sr
Cs	Ba
The alkali metals	**The alkaline-earth metals**

You will now realise that *all Group I metals have one outermost s-electron, whereas Group II metals have two outermost s-electrons*. These *s*-electrons are situated much further from the nucleus than all other electrons in the metals. They are only weakly held by the positive nucleus. Consquently, the atoms of Group I metals readily lose their single outermost s-electron to form stable ions. These ions have a noble gas electron structure, and one positive charge, e.g.

$$Na \rightarrow Na^+ + e^-$$

Similarly, the atoms of Group II metals readily lose two electrons to form stable ions, with a noble gas structure, carrying two positive charges, e.g.

$$Ca \rightarrow Ca^{2+} + 2e^-$$

15.2 Oxidation numbers and s-block elements

Tables 15.2 and 15.3 summarise various physical properties of the elements in Groups I and II, respectively.

Look closely at the first and second ionisation energies for Group I elements. The first ionisation energy is much lower than the second. In the case of Na, it is nine times easier to remove the first electron than the second. The first electron, in an s-orbital, can be removed from the element easily. The second electron must be removed from a noble gas core. This is closer to the nucleus and therefore requires much more energy. Thus, sodium readily forms Na^+ ions, but never forms Na^{2+} ions. This means that sodium and the other elements in Group I have only the one oxidation state of $+1$ in their compounds.

Look closely at the first, second and third ionisation energies of Group II elements in table 15.3.

- What is the ratio of the first ionisation energy: second ionisation energy: third ionisation energy for Mg?
- Why is there a greater increase from the second to the third ionisation energy than from the first to the second?
- What is the most stable oxidation state for Group II elements?

Table 15.2
Physical properties of the elements in Group I

Element	Li	Na	K	Rb	Cs
Electron structure	$(He)2s^1$	$(Ne)3s^1$	$(Ar)4s^1$	$(Kr)5s^1$	$(Xe)6s^1$
First ionisation energy/kJ mole^{-1}	520	500	420	400	380
Second ionisation energy/kJ mole^{-1}	7300	4600	3100	2700	2400
Atomic radius/nm (metallic radius)	0.15	0.19	0.23	0.25	0.26
Melting point/°C	180	98	64	39	29
Boiling point/°C	1330	892	760	688	690
Density/g cm^{-3}	0.53	0.97	0.86	1.53	1.90
Standard electrode potential, M$^+$(aq) /M(s)/volts	−3.05	−2.71	−2.93	−2.92	−2.92

Table 15.3
Physical properties of the elements in Group II

Element	Be	Mg	Ca	Sr	Ba
Electron structure	$(He)2s^2$	$(Ne)3s^2$	$(Ar)4s^2$	$(Kr)5s^2$	$(Xe)6s^2$
First ionisation energy/kJ mole^{-1}	900	740	590	550	500
Second ionisation energy/kJ mole^{-1}	1800	1450	1150	1060	970
Third ionisation energy/kJ mole^{-1}	14 800	7700	4900	4200	—
Atomic radius/nm (metallic radius)	0.11	0.16	0.20	0.21	0.22
Melting point/°C	1280	650	838	768	714
Boiling point/°C	2770	1110	1440	1380	1640
Density/g cm^{-3}	1.85	1.74	1.55	2.6	3.5
Standard electrode potential, M^{2+}(aq) /M(s)/volts	−1.85	−2.37	−2.87	−2.89	−2.91

15.3 Physical properties

The atomic radii of alkali metals are included in table 15.2 and those of the alkaline-earth metals are given in table 15.3. As one might expect, the atomic radii increase with atomic number down the groups. Each succeeding element has electrons in one more shell than the previous element.

The outermost s-electron in these metals are held very weakly by the nucleus. Thus, the outer electrons can drift further from the nucleus than in most other atoms. So, the elements in Groups I and II have larger atomic radii than those elements which follow them in their respective periods. The large atomic size results in weaker forces between neighbouring atoms because there is a reduced attraction of the nuclear charge for the shared mobile outer electrons.

Consequently, the elements in Groups I and II have melting points and lower boiling points than we normally associate with metals. Apart from beryllium, all their melting points are less than 840°C and all the Group I metals melt below 200°C. In contrast, most of the transition metals melt at temperatures above 1000°C.

The reduced interatomic forces in these metals makes them relatively soft. All the Group I metals can be cut with a pen-knife. Although the Group II metals are harder than those in Group I, they are noticeably softer than transition metals.

As we have seen, s-block elements have larger atomic radii than transition metals of approximately the same relative atomic mass. Thus, the s-block elements will be less dense and have larger molar volumes. Of the 10 elements in tables 15.2 and 15.3, only barium is more dense than aluminium, a metal normally associated with low density. Their densities vary from approximately

$1 \, g \, cm^{-3}$ to $3.5 \, g \, cm^{-3}$ for barium. Most transition metals have a density greater than $7 \, g \, cm^{-3}$.

15.4 Chemical properties

All the metals in Groups I and II are high in the activity (electrochemical) series. The standard electrode potential (tables 15.2 and 15.3) for the conversion of each metal *to its ions* is positive and greater than 2.0 volts for all metals except beryllium.

Hence, *these metals are very good reducing agents.* They all react vigorously with water, reducing it to hydrogen.

$$M(s) \rightarrow M^+(aq) + e^-$$

$$H_2O(l) + e^- \rightarrow \tfrac{1}{2}H_2(g) + OH^-(aq)$$

Excluding lithium, which reacts slower than all the other elements of Group I, the reactivity of the elements with water closely follows the value of $E^\ominus$.

In Group I, for example, sodium reacts vigorously, fizzing and skating about on the water surface. Potassium reacts even more vigorously. It cracks and pops as the hydrogen explodes and produces a lilac flame as the hydrogen burns. Rubidium and caesium explode violently in contact with water.

■ Do $E^\ominus$ values relate to reaction rates or to equilibrium? Do they indicate how fast a reaction will go, or simply the direction in which it should go?

■ Why do you think lithium reacts more slowly with water than other Group I metals, even though it has the largest $E^\ominus$?

Although electrode potentials relate only to aqueous solutions, they also provide a guide to the general reactivity of a substance. All the *s*-block elements are good reducing agents. They react with chlorine, bromine, sulphur, hydrogen and oxygen on heating. They all, of course, tarnish rapidly in air, forming a layer of oxide. The reactivity of Group I metals is such that lithium, sodium and potassium are usually stored under oil.

15.5 Thermal stability of salts

The stability of the salt $M^+X^-(s)$ is dependent on the size of its ions and the charge on these ions. The greater the charge, the stronger is the attraction between the ions and the more stable is $M^+X^-(s)$. The smaller the ions become, the closer they can approach each other in the solid crystal and the more stable is $M^+X^-(s)$.

Thus, we might expect M^+X^- to become more stable as ionic charge increases and as ionic radius decreases.

There is, however, one other important factor to bear in mind during any discussion of thermal stability. When large anions such as CO_3^{2-} decompose on heating to form smaller anions such as O^{2-}, the crystal containing the latter will generally be more stable. This is because the charge density on a small O^{2-} ion will be greater than that on a larger CO_3^{2-} ion, and the former will be more strongly attracted to the cations in the crystal. Furthermore, the smaller O^{2-} ions can get closer to the cations than the larger CO_3^{2-} ions.

This point is crucial in considering the thermal stability of nitrates, carbonates and hydroxides of the *s*-block elements.

Nitrates

Group I nitrates (except $LiNO_3$) decompose on heating in the bunsen flame to form their corresponding nitrites which are stable:

$$KNO_3(s) \rightarrow KNO_2(s) + \tfrac{1}{2}O_2(g)$$

However, Group II nitrates and $LiNO_3$ decompose on heating to form their corresponding oxide:

$$Mg(NO_3)_2(s) \rightarrow MgO(s) + 2NO_2(g) + \tfrac{1}{2}O_2(g)$$

$$2LiNO_3(s) \rightarrow Li_2O(s) + 2NO_2(g) + \tfrac{1}{2}O_2(g)$$

This suggests that the decrease in size from NO_3^- to NO_2^- stabilises the Group I nitrates to heat. But the Group II nitrates can only achieve thermal stability by decomposing from their nitrate to form the much smaller oxide ion.

Carbonates

The pattern of thermal stability amongst the carbonates is very similar to that of the nitrates. The carbonates of Group I (except Li_2CO_3) are stable at the temperature of the bunsen flame, but those of Group II decompose at this temperature to form the corresponding oxide:

$$MgCO_3(s) \rightarrow MgO(s) + CO_2(g)$$

Here again the Group II compounds are achieving thermal stability by forming their oxide.

Hydroxides

Yet again, the pattern is repeated. The hydroxides of Group I (except LiOH) are stable, but those of Group II decompose on heating.

$$Mg(OH)_2(s) \rightarrow MgO(s) + H_2O(g)$$

LiOH starts to decompose at about 650°C, but NaOH does not decompose until temperatures well above this and above the highest temperature obtainable with a bunsen.

15.6   The solubility of salts

Table 15.4 shows the results obtained when 0.1 M solutions of Mg^{2+}, Ca^{2+}, Sr^{2+} and Ba^{2+} are treated with various solutions.

Notice that the sulphates, carbonates, ethanedioates (oxalates) and chromates(VI) of the Group II metals become more insoluble as atomic number increases, whereas their hydroxides become more soluble. These results are confirmed by the values of solubilities quoted for the sulphates, carbonates chromates(VI), nitrates and hydroxides in table 15.5.

Table 15.4
Results obtained when 0.1 M solutions of Mg^{2+}, Ca^{2+}, Sr^{2+} and Ba^{2+} are treated with various solutions

0.1 M solution of Group II cation	Reaction with 1M Na_2SO_4	(Group II sulphates become more insoluble)	Reaction with sat'd. $CaSO_4$	Reaction with 1M Na_2CO_3	(Group II carbonates become more insoluble)	Reaction with 1M $Na_2C_2O_4$	(Group II ethanedioates (oxalates) become more insoluble)	Reaction with 1M K_2CrO_4	(Group II chromates(VI) become more insoluble)	Reaction with 1M NaOH	(Group II hydroxides become more soluble)
Mg^{2+}	no ppte		no ppte	white ppte		no ppte		no ppte		thick white ppte	
Ca^{2+}	thin white ppte		no ppte	white ppte		thin white ppte		no ppte		white ppte	
Sr^{2+}	white ppte		white ppte	thick white ppte		white ppte		pale yellow ppte		white ppte	
Ba^{2+}	thick white ppte		white ppte	very thick white ppte		thick white ppte		thick pale yellow ppte		thin white ppte	
General equations for precipitations	$M^{2+} + SO_4^{2-}$ $\rightarrow MSO_4(s)$		$M^{2+} + SO_4^{2-}$ $\rightarrow MSO_4(s)$	$M^{2+} + CO_3^{2-}$ $\rightarrow MCO_3(s)$		$M^{2+} + C_2O_4^{2-}$ $\rightarrow MC_2O_4(s)$		$M^{2+} + CrO_4^{2-}$ $\rightarrow MCrO_4(s)$		$M^{2+} + 2OH^-$ $\rightarrow M(OH)_2(s)$	

Table 15.5
The solubilities (at 25°C) of some compounds of Group II metals

Compound	Solubility/moles per 100g water
$MgSO_4$	3600×10^{-4}
$CaSO_4$	11×10^{-4}
$SrSO_4$	0.62×10^{-4}
$BaSO_4$	0.009×10^{-4}
$MgCO_3$	1.3×10^{-4}
$CaCO_3$	0.13×10^{-4}
$SrCO_3$	0.07×10^{-4}
$BaCO_3$	0.09×10^{-4}
$MgCrO_4$	8500×10^{-4}
$CaCrO_4$	870×10^{-4}
$SrCrO_4$	5.9×10^{-4}
$BaCrO_4$	0.011×10^{-4}
$Mg(OH)_2$	0.2×10^{-4}
$Ca(OH)_2$	16×10^{-4}
$Sr(OH)_2$	330×10^{-4}
$Ba(OH)_2$	240×10^{-4}
$Mg(NO_3)_2$	4.9×10^{-1}
$Ca(NO_3)_2$	7.7×10^{-1}
$Sr(NO_3)_2$	3.4×10^{-1}
$Ba(NO_3)_2$	0.35×10^{-1}

Table 15.6
The solubilities (at 25°C) of various magnesium salts containing singly charged anions

Compound	Solubility/moles per 100g water
$MgCl_2$	5.6×10^{-1}
$MgBr_2$	5.5×10^{-1}
$Mg(OH)_2$	2.0×10^{-5}
$Mg(CH_3COO)_2$	5.6×10^{-1}
$Mg(NO_3)_2$	4.9×10^{-1}

When an ionic compound dissolves in water, the following process occurs:

$$M^+X^-(s) + (aq) \rightarrow M^+(aq) + X^-(aq)$$

We can picture this process as the result of two stages. First, the separation of ions in the solid, i.e.

$$M^+X^-(s) \rightarrow M^+(g) + X^-(g)$$

This is the reverse of the lattice formation process. It requires the *input* of the lattice energy. The second stage involves hydration of these separated ions by water, i.e.

$$M^+(g) + X^-(g) + (aq) \rightarrow M^+(aq) + X^-(aq)$$

As the ionic radii of M^+ and X^- increase, the heat change in both of these processes decreases. The reverse lattice energy process becomes less endothermic and the hydration process becomes less exothermic. Consequently, a decrease in the endothermic process may be cancelled by a decrease in the exothermic process. It is therefore difficult to make predictions about changes in solubility from changes in ionic size (section 12.14),

There are, however, one or two general patterns and trends in the solubility of Group I and Group II salts.

1   All common Group I salts are soluble.

2   Salts of Group II cations containing anions with a charge of −1 (e.g. chlorides, nitrates) are generally soluble, except for the hyroxides.

Look at the solubiliites of the various magnesium salts (all of which contain an anion with charge −1) in table 15.6.

■ Approximately how many times more soluble are the other compounds in the table than $Mg(OH)_2$?
■ Why are the solubilities given in moles per 100g of water rather than g/100g of water?

3   Salts of Group II cations containing anions with a charge of −2 (e.g. sulphates, carbonates) are generally insoluble (except for some magnesium salts and a few calcium salts).

Furthermore, there is a distinct decrease in solubility as the atomic number of the metal rises. Thus, beryllium and magnesium sulphates are very soluble in water, calcium sulphate is sparingly soluble, but strontium and barium sulphates are virtually insoluble. The solubilities of some compounds of the Group II metals are shown in table 15.5.

15.7 Occurrence of the *s*-block elements

The alkali metals occur in nature only as +1 ions. Sodium and potassium are, of course, the commonest Group I metal ions. They are the sixth and seventh most abundant elements in the earth's crust (figure 15.1). Lithium is found in trace amounts in virtually all rocks and clays, but rubidium and caesium are very rare. Francium, the last element in Group I, does not exist naturally. It has only been obtained in tiny amounts during nuclear reactions. Even these amounts disappear rapidly as the radioactive francium disintegrates and forms more stable elements.

Since many of the compounds of alkali metals are soluble in water, it is not surprising that they are found in relatively large quantities in sea water. Compounds of the alkali metals also occur in deposits which have formed by the evaporation of brine. However, many insoluble clays also contain alkali metal ions combined as complex silicates and aluminates.

The alkaline-earth metals occur in nature only as +2 ions. These +2 ions react with −2 ions to form insoluble compounds, so it is not surprising that Group II elements exist in nature principally as their carbonates, silicates and sulphates.

Figure 15.1
The relative abundance of elements on the earth. (The figures given indicate the % by mass of the particular element in the earth's crust to a depth of 40 kilometres, including the atmosphere.)

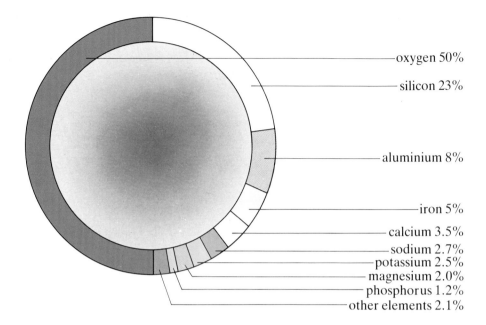

oxygen 50%
silicon 23%
aluminium 8%
iron 5%
calcium 3.5%
sodium 2.7%
potassium 2.5%
magnesium 2.0%
phosphorus 1.2%
other elements 2.1%

The only beryllium mineral found in any quantity is the silicate, beryl ($3BeSiO_3.Al_2(SiO_3)_3)_3$). Emeralds are, in fact, crystals of beryl, coloured green by traces of chromium.

Magnesium is the eighth most abundant element in the earth's crust. It is found extensively as the carbonate in magnesite ($MgCO_3$) and dolomite ($MgCO_3.CaCO_3$). It also occurs as the silicate in minerals such as asbestos ($3MgSiO_3.CaSiO_3$) and as Mg^{2+} ions in sea water.

Calcium is the fifth most abundant element in the earth's crust and the most abundant s-block element. Vast quantities of calcium occur as the carbonate in chalk, limestone and marble. These deposits of calcium carbonate have been formed by the effect of high temperature and high pressure on the accumulated skeletons of dead marine animals.

Smaller quantities of calcium occur as the sulphate in anhydrite ($CaSO_4$) and gypsum ($CaSO_4.2H_2O$). These sulphate ores are thought to have resulted from the action of sulphuric acid (produced from the oxidation of sulphide minerals) on limestone.

Strontium and barium are much rarer elements. The commonest ore of strontium is strontianite ($SrCO_3$) and that of barium is barite ($BaSO_4$).

The average abundance of radium (the last element in Group II) in the earth's crust has been estimated at less than 1 part per million million. Radium exists naturally as a radioactive element. It is formed during the nuclear disintegration of heavier elements such as uranium.

Natural crystals of rock salt growing in a brine spring.

Right. An Ethiopian working slabs of salt into conveniently sized bars.

Far right. Harvesting salt from lagoons in Tanzania.

The white horse carved in the chalk downland at Uffington, Berkshire.

15.8 Manufacture of the *s*-block elements

In order to extract the *s*-block elements from their naturally occurring compounds, it is necessary to reduce their positive ions, i.e.

$$M^+(aq) + e^- \rightarrow M(s)$$

or

$$M^{2+}(aq) + 2e^- \rightarrow M(s)$$

This reduction can be done either chemically or electrolytically. But the large negative electrode potentials of the *s*-block elements mean that chemical methods are virtually impossible. The processes would require a reducing agent stronger than the alkali or alkaline-earth metals themselves.

Consequently, these elements are usually extracted by electrolysis. The electrolytes are always fused compounds rather than aqueous solutions, since water in the solution would be discharged at the cathode in preference to the *s*-block ion. For example,

$$Na^+ + e^- \rightarrow Na \qquad E^\ominus = -2.71 \text{ volts}$$
$$H_2O + e^- \rightarrow \tfrac{1}{2}H_2 + OH^- \qquad E^\ominus = -0.42 \text{ volts}$$

The more positive (less negative) electrode potential for the lower reaction means that it will occur in preference to the one above.

Sodium and magnesium are in much higher demand than the other metals in Groups I and II. Each of these metals is obtained by electrolysis of its molten chloride.

Look closely at figure 15.2. This shows a simplified diagram of the Downs cell in which sodium is manufactured.

- The electrolyte contains $Na^+(l)$ and $Cl^-(l)$. Write equations for the process at each electrode during the manufacture of sodium.
- In the Downs cell, the cathode is made of steel, but the anode is made of graphite. Why is the anode made of graphite in spite of the fact that steel would be a better conductor?
- Why is it necessary to incorporate hoods and a steel gauze cylinder in order to prevent the product at the anode from mixing with the product at the cathode?

  The extraction of these *s*-block elements involves fused compounds rather than aqueous solutions. Large quantities of energy are therefore required not only for the electrolytic process itself, but also to melt the electrolyte.

Jane Marcet (1769–1858)

In the early nineteenth century, Michael Faraday carried out many experiments involving electrolysis. It is said he was first introduced to chemistry by the books of Jane Marcet when he was an apprentice bookbinder. A chemist herself, Jane was a well-known science writer.

Michael Faraday (1791–1867)

Figure 15.2
The extraction of sodium in the Downs Cell.

Liquid sodium is used as coolant in the Prototype Fast Reactor at Dounreay. The photograph shows the bottom end of one of the sodium pumps with the central inlet pipe and two outlet pipes.

■ Why do you think that many electrolytic manufacturing plants are located near supplies of hydro-electricity?
■ In these electrolyses, the naturally occurring raw material is usually mixed with some other compound. For example, in the extraction of sodium, the electrolyte of NaCl(l) contains some calcium chloride. How does the addition of calcium chloride reduce the energy expenditure?
■ In view of the calcium chloride added to the electrolyte, what substance will contaminate the sodium produced at the cathode?

15.9 Uses of the *s*-block elements and their compounds

In spite of being so reactive, sodium, magnesium and the other *s*-block elements have many important uses. Probably the most important use of sodium is as a liquid coolant in nuclear reactors.

Fast nuclear reactors, such as that at Dounreay, operate at a temperature of about 600°C. Water, which boils at 100°C, is therefore unsuitable as a cooling liquid in this case. But sodium, which melts at 98°C and boils at 892°C, is ideal for this purpose. Furthermore, sodium is a much better thermal conductor than water and almost as mobile a liquid. It can therefore be circulated around the hot reactor core and will transfer the heat away more effectively than water.

Assuming that the sodium can be prevented from coming into contact with air or moisture, the major problem in its use is its action as a solvent for various metals. These metals might otherwise be used for the reactor core and the cooling system. Fortunately, iron and other transition metals are virtually insoluble in the liquid. So, they can be used in the construction of the reactor core and the cooling pipes.

Sodium's high electrical conductivity and low density have led to extensive testing to see whether sodium could be used in electricity cables. Of course, sodium wire could never be used because of its reaction with air and water and its poor mechanical strength. However, a method has been devised to overcome these problems. This involves extruding a tube of poly(ethene)(polythene) and simultaneously filling it with sodium.

This large tank contains a solution of sodium cyanide which is used in extracting gold from crushed rock.

Another important use of sodium is as an alloy with lead in the manufacture of tetraethyllead(IV) from chloroethane.

$$\underbrace{Pb + 4Na}_{\text{lead–sodium alloy}} \quad + \quad 4C_2H_5Cl \quad \rightarrow \quad \underbrace{(C_2H_5)_4Pb}_{\text{tetraethyllead(IV)}} \quad + \quad 4NaCl$$

Tetraethyllead(IV) (TEL) is used as an essential 'anti-knock' additive for petrol. Every gallon of 'leaded' petrol contains about $2\,cm^3$ of TEL (section 27.6).

Sodium is also used in the reduction of titanium(IV) chloride to titanium by the Goldschmidt process and in the familiar yellow sodium vapour street lamps. In the metals industry, sodium cyanide is used in the extraction of gold and in the hardening of steel.

Sodium hydroxide is the cheapest industrial alkali. Thousands of tonnes are used annually in the manufacture of rayon (cellulose acetate), paper and soap. Sodium polyphosphate ($Na_6P_6O_{18}$) is used as a water softener and sodium nitrite, $NaNO_2$, is used as a preservative in food.

The other metals in Group I are used much less extensively than sodium and its compounds. Potassium is used to make potassium superoxide, KO_2, which reacts with water to give oxygen.

$$4KO_2 + 2H_2O \rightarrow 4KOH + 3O_2$$

KO_2 can therefore be used as an emergency source of oxygen in mines and in submarines. Potassium chloride, KCl, is used as a fertiliser and potassium nitrate is an essential constituent of gunpowder.

Owing to its very low first ionisation energy, caesium is used as the light-sensitive surface in photocells. These convert light signals into electrical signals. Figure 15.3 shows a simplified diagram of a photocell circuit. The evacuated tube contains two electrodes, one of which is coated with caesium or an alloy of caesium, antimony and silver. In the absence of light, the photocell does not conduct electricity. However, when the caesium-coated electrode is struck by light, it emits electrons. These are attracted to the positive electrode and an electric current flows in the circuit. Television pickup devices and photographers' lightmeters incorporate the use of photocells.

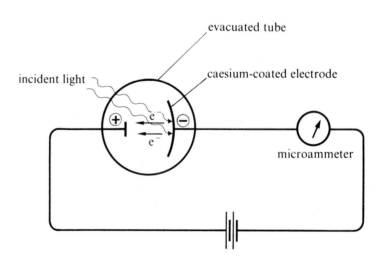

Figure 15.3
A simplified photocell circuit.

Magnesium is by far the most commonly used metal in Group II. It is used extensively in the preparation of lightweight alloys with a high tensile strength and particularly in the construction of aircraft parts and household goods. Although pure magnesium has poor structural strength, this can be increased by alloying it with alluminium, zinc and manganese. The aluminium increases the tensile strength of the alloy, the zinc improves its machining properties and the manganese reduces corrosion.

When magnesium burns in air it emits a bright, intense light. Because of this, magnesium powders and ribbons are used as light sources in photography. Flash bulbs contain magnesium (or aluminium) foil in an atmosphere of oxygen. When

The use of magnesium flares in photography being demonstrated at an early meeting of the British Association in Birmingham.

the bulb is fired, an electric current passes through the metal. The metal is heated up and then suddenly flares into reaction with oxygen.

The most commonly used compounds of magnesium are magnesium oxide (MgO), magnesium hydroxide (Mg(OH)$_2$) and hydrated magnesium sulphate (MgSO$_4$.7H$_2$O). Magnesium oxide is used as a lining for high-temperature furnaces because of its exceptionally high melting point and low reactivity. Magnesium hydroxide is a very weak alkali. Because of this, it is an important constituent of toothpastes. It neutralises acids that would otherwise cause decay. A suspension of Mg(OH)$_2$ in water (commonly called milk of magnesia) is used to treat acid indigestion.

Beryllium is too rare and costly for any large-scale uses. But it is used for hardening alloys and other metals such as copper and nickel. Beryllium–copper alloys are particularly resistant to sea-water corrosion and are therefore used for small marine parts.

Calcium, strontium and barium have very few uses owing to their high reactivity. The affinity of these elements for oxygen results in their use as deoxidisers in steel production. They are also used as 'getters' (i.e. removers of traces of oxygen) inside electron tubes and radio valves. By far the most

Quarrying limestone.

Volatile compounds of calcium, strontium and barium are often used in fireworks.

commercially important compounds of calcium are lime (CaO) and limestone ($CaCO_3$), Limestone, of course, is the source of lime. Many thousands of tonnes of lime are used annually in counteracting soil acidity. Enormous quantities of lime are also used in the manufacture of plaster, mortar and cement.

The salts of calcium, strontium, and barium produce beautiful flame colours when heated to high temperatures: calcium salts—red; strontium salts—crimson; and barium salts—pale green. As a result of this, volatile compounds of these elements are used in the production of coloured military signals and in fireworks.

Summary

1　In the alkali and alkaline-earth metals, the electrons in the outermost shell occupy an *s*-orbital.

2　The atoms of these *s*-block metals readily lose their outermost *s*-electrons to form stable ions: M^+ if the elements are in Group I, M^{2+} if the elements are in Group II.

3　The metals in Groups I and II show only one oxidation number in their compounds: +1 for Group I elements, +2 for Group II elements.

4　The *s*-block metals have lower melting points, lower boiling points, lower densities and they are softer than transition metals.

5　The *s*-block metals are high in the activity series. They will all reduce water to hydrogen and they all tarnish rapidly in air forming a layer of oxide.

6　The compounds of Group I metals are thermally more stable than those of Group II metals. The carbonates and hydroxides of Group I metals (except Li_2CO_3 and LiOH) do not decompose at 1000°C, but those of Group II metals decompose to their oxides on heating.

7　All common Group I salts are soluble.

8　Salts of Group II cations containing anions with a charge of −1 are generally soluble except for the hydroxides. Salts of Group II cations containing anions with a charge of −2 are generally insoluble.

9　The *s*-block elements are manufactured by electrolysis of their molten compounds.

10　Sodium and magnesium are the most widely used *s*-block elements. Sodium is used as a liquid coolant in nuclear reactors and as an alloy with lead in the manufacture of tetraethyllead(IV) ('anti-knock' in petrol). Magnesium is used extensively in the manufacture of low-density, high-strength alloys.

Study questions

1　(a)　How do each of the following properties of the elements in Group II change with increasing atomic number?
　　　(i)　Atomic radius
　　　(ii)　Ionisation energy
　　　(iii)　Strength as reducing agents
　　　(iv)　Vigour of reaction with chlorine
　　　(v)　Electropositivity
　　(b)　In each case, explain why the five properties change in the way you have suggested.

2 Group II consists of the elements Be, Mg, Ca, Sr, Ba.
   (a) Are these elements metals or non-metals? Give one reason for your answer.
   (b) What ion or ions are formed by calcium?
   (c) Describe briefly five properties that show a general gradation as the group is descended from Be to Ba.
   (d) Briefly explain how ionisation energies and redox potentials are related to the reactivity of these elements.
   (e) It is easier to form Mg^+ ions than Mg^{2+} ions from magnesium. In spite of this, Mg^+ is not found in magnesium compounds. Why is this?
   (f) Beryllium shows certain properties that are not typical of the rest of the group. Mention two of these properties. Suggest reasons why the difference should occur.
   (g) Group II elements frequently form hydrated salts, while the corresponding compounds of Group I elements are anhydrous. Suggest reasons for this difference.

3 The elements in Group II of the periodic table are barium (Ba), beryllium (Be), calcium (Ca), magnesium (Mg), radium (Ra) and strontium (Sr).
   (a) Arrange these elements in order of increasing relative atomic mass.
   (b) Write the electronic structures of any two of these elements except beryllium (e.g. Be would be $1s^2 2s^2$).
   (c) Draw a sketch graph of ionisation energy against number of electrons removed for the successive ionisation energies of magnesium.
   (d) Why do all the metals in Group II have an oxidation number of +2?
   (e) Why do the elements in Group II not have an oxidation number of +1 or +3?
   (f) How and why does the first ionisation energy of the metals vary as the group is descended?
   (g) When solutions of strontium nitrate and potassium chromate(VI) are mixed, a yellow precipitate is formed. This dissolves in dilute hydrochloric acid. Explain these observations.

4 Group I consists of the elements Li, Na, K, Rb, Cs.
   (a) The first member of a group often shows anomalous properties. Give two respects in which the behaviour of Li is anomalous.
   (b) Describe briefly five properties that show a general gradation as the group is descended from Li to Cs.
   (c) Explain the meaning of the term 'ionisation energy'.
   (d) How will successive ionisation energies of Na vary?
   (e) Why is the ion Na^+ formed in normal chemical reactions rather than the ion Na^{2+}?
   (f) How are ionisation energies related to the reactivity of these elements?

5 The isotope, $^{42}_{19}K$ decays by beta-particle emission to a stable nuclide. The rate of emission of β-particles from $^{42}_{19}K$ was followed by a suitable method.
   (a) Draw a sketch graph showing the rate of emission of β-particles (vertically) against time.
   (b) 0.02 g of $^{42}_{19}K$ were allowed to decay.
      (i) Write a nuclear equation (balanced for mass number and atomic number) for the decay process.
      (ii) What is the stable nuclide produced in the decay process?
      (iii) If the half-life of the potassium isotope is 12.5 h, what mass of the stable nuclide will have been formed after 25 h?

6 Make a table to compare the first three elements in Group II of the periodic table: beryllium, magnesium and calcium. Your table should include
   (a) electronic structures,
   (b) relative values of ionisation energies,
   (c) methods of extraction of elements,
   (d) action of water, acids and alkalis on the elements,
   (e) the basic strength of oxides and hydroxides,
   (f) the volatility and hydrolysis of chlorides.
   (If necessary, use your knowledge of the periodic table to predict properties for beryllium and its compounds.)

7 Magnesium and calcium occur naturally in the mineral dolomite, $MgCO_3.CaCO_3$, a mixture of insoluble magnesium and calcium carbonates. This can be used to produce calcium sulphate and magnesium sulphate. Calcium sulphate is used in the manufacture of building materials such as plaster-board. Magnesium sulphate is used in fireproofing fabrics and as a purgative (Epsom salts).

(a) Describe carefully how you would prepare pure samples of $MgSO_4.7H_2O$ and $CaSO_4$ from dolomite. You may find the following information useful.

Compound	Solubility/g per 100g water at 20°C
$MgCO_3$	0.01
$CaCO_3$	0.0014
$MgSO_4$	33.0
$CaSO_4$	0.21

(b) How would you obtain pure $MgSO_4$ from crystals of $MgSO_4.7H_2O$?

8 Sodium sulphite (Na_2SO_3) is sometimes added to sausage meat to act as a preservative. The amount of Na_2SO_3 present can be determined by boiling a sample of the meat with acid. The quantity of sulphur dioxide which forms is then determined by titration against iodine.

100g of sausage meat was boiled with $500\,cm^3$ of M HCl. The sulphur dioxide evolved was dissolved in water. It was found to require $12.00\,cm^3$ of $0.025\,M\ I_2$ solution in order to oxidise the SO_2 as in the equation below.

$$SO_2 + 2H_2O + I_2 \rightarrow 4H^+ + SO_4^{2-} + 2I^-$$

In order to check the results of the titration, excess barium chloride is added to the final solution after the titration. The resulting precipitate is collected and weighed.

(Na = 23, S = 32, O = 16, Ba = 137)

Tunnels in an anhydrite mine.

(a) How many moles of SO_2 are evolved from $100\,g$ of the sausage meat?
(b) How many grams of Na_2SO_3 are present in $100\,g$ of the sausage meat?
(c) Government scientists often express the amount of Na_2SO_3 in meat as parts per million (p.p.m.).

$$(1\,p.p.m. = 1\,g\ of\ Na_2SO_3\ in\ 10^6\,g\ of\ meat)$$

Express the amount of Na_2SO_3 in the sausage meat in p.p.m.
(d) Write an equation for the reaction which occurs when $BaCl_2(aq)$ is added to the solution at the end of the titration.
(e) Calculate the mass of precipitate formed when excess $BaCl_2(aq)$ is added to the solution at the end of the titration.

9 Calcium sulphate is found naturally in two forms: anhydrous calcium sulphate ($CaSO_4$—anhydrite) and hydrated calcium sulphate ($CaSO_4.2H_2O$—gypsum). When anhydrite is heated with coke, sulphur dioxide is obtained. This can then be used to manufacture sulphuric acid.
   If gypsum is heated at about 125°C it dehydrates partially forming plaster of Paris, $(CaSO_4)_2.H_2O$. When this is mixed with water it changes back to $CaSO_4.2H_2O$. The paste expands slightly as it hardens and sets quickly to a firm solid.
   When gypsum is heated to 200°C, it loses all its water of crystallisation. The anhydrous salt which forms sets only very slowly when mixed with water.
(a) Explain what is meant by the terms:
    (i) anhydrous,   (ii) hydrated,   (iii) water of crystallisation.
(b) Write equations for the reactions which occur when:
    (i)   anhydrite is heated with coke,
    (ii)   plaster of Paris is mixed with water,
    (iii) gypsum is heated at 200°C.
(c) Draw a diagram of the apparatus *you* would use to make plaster of Paris from gypsum.
(d) Explain why plaster of Paris is so suitable for:
    (i)   immobilising broken limbs,
    (ii)   making models from moulds.
(e) Why is anhydrite not a suitable alternative for the uses mentioned in (d)?

A disused lime kiln in which lime (calcium oxide) was once made by heating limestone (calcium carbonate).

16 Aluminium

16.1 Introduction

Aluminium is by far the most important element in Group III (table 16.1). Although the metal itself is not found free in nature, its compounds are widespread. Aluminium is the third most abundant element (after oxygen and silicon) in the Earth's crust. Economically, the most important ore is bauxite (hydrated aluminium oxide, $Al_2O_3.2H_2O$) from which the metal is obtained. There are also considerable and widespread deposits of aluminosilicates in clays, mica and other minerals and deposits of cryolite (sodium hexafluoroaluminate(III), Na_3AlF_6) are found in Greenland.

Table 16.1
The symbols, atomic numbers and electron structures of the elements in Group III

Element	Symbol	Atomic number	Electron structure
Boron	B	5	$(He)2s^22p^1$
Aluminium	Al	13	$(Ne)\,3s^2\,3p^1$
Gallium	Ga	31	$(Ar)3d^{10}\,4s^2\,4p^1$
Indium	In	49	$(Kr)4d^{10}\,5s^2\,5p^1$
Thallium	Tl	81	$(Xe)4f^{14}\,5d^{10}\,6s^2\,6p^1$

Photograph of mica showing its giant layer-lattice structure. Mica contains silicates of aluminium.

The brown piles in the background are heaps of bauxite (impure aluminium oxide). The crates in the foreground contain purified white aluminium oxide.

254

Table 16.2
Values of charge, ionic radius and charge/radius ratio for some common cations

Cation	Unit charge	Ionic radius/nm	Charge/radius ratio/unit charge nm^{-1}
Na$^+$	+1	0.098	10
Mg^{2+}	+2	0.065	31
Al^{3+}	+3	0.048	63
Zn^{2+}	+2	0.074	27
Cu^{2+}	+2	0.069	29

Much of the chemistry of aluminium compounds is dictated by *the high charge and small radius of the Al^{3+} ion*. These result in an unusually large charge density. The high charge density of Al^{3+} is reflected in its large charge/radius ratio which is compared with the values for some other common cations in table 16.2.

16.2 Bonding in aluminium compounds

Aluminium and the other elements in Group III are, of course, predominantly trivalent showing an oxidation state of +3. They form ions with a charge of 3+ by losing all three electrons from their outermost shell:

$$Al \rightarrow Al^{3+} + 3e^-$$

However, the high charge density of the Al^{3+} ion will cause some distortion (polarisation) of the electron cloud around any ion in contact with it. If the polarisation is sufficiently large, there is an effective electron density in the region between the aluminium ion and its neighbouring anion. This comprises partial covalent bonding. With large anions such as Br$^-$ and I$^-$, in which polarisation occurs more easily, the distortion can be so great that the bonding is best described as covalent.

- Why are large anions polarised more easily than smaller anions?
- Which anion do you think will be least polarised by cations?
- Will Ga^{3+} have a greater or smaller polarising effect than Al^{3+} on anions? Explain.

The influence of polarisation on bonding, structure and properties is beautifully illustrated by the aluminium halides (figure 16.1).

16.3 The bonding and properties of aluminium oxide

Aluminium oxide is normally regarded as ionic containing Al^{3+} and O^{2-} ions. However, the oxide ions are polarised by the Al^{3+} ions and the bonds have a marked degree of covalent character. This factor and the small size of the Al^{3+} and O^{2-} ions combine to make the bonding in aluminium oxide very strong indeed.

Figure 16.1
The influence of polarisation on the bonding, structure and properties of aluminium halides.

	Aluminium fluoride	Aluminium chloride	Aluminium bromide
Bonding:	Ionic	Intermediate between ionic and covalent	Covalent
	F^- ion is too small to be polarised	Cl^- ion is intermediate in size and partly polarised.	Br^- ion is easily polarised forming a covalent bond to Al.
Structure:		Solid shows intermediate bonding, but this sublimes to yield Al_2Cl_6 dimers with the same structure as Al_2Br_6. At high temps. $AlCl_3$ mols. form.	
	Symmetrical ionic crystal.		Simple molecular (dimer) structure. (The outer shell of each Al atom is completed by a co-ordinate bond from a Br atom in the associated $AlBr_3$ molecule—hence the dimer.)
Melting point:	1265°C.	Sublimes at 180°C.	97°C.

Aluminium oxide may be prepared by heating powdered aluminium in oxygen or by heating aluminium hydroxide:

$$2Al + \tfrac{3}{2}O_2 \rightarrow Al_2O_3$$

$$2Al(OH)_3 \rightarrow Al_2O_3 + 3H_2O$$

The strongly bound structure of Al_2O_3 results in its insolubility in water and very high melting point (2050°C). Its intermediate (ionic/covalent) bonding is

A collection of sapphires around the impure aluminium oxide (known as corundum) from which they are obtained. Sapphires get their blue colour from impurities of titanium(IV) oxide or cobalt(II) oxide. Artificial sapphires have been made by heating aluminium oxide with a mixture of Fe_3O_4 and TiO_2.

reflected in its amphoteric character. It dissolves slowly in both dilute acids and dilute alkalis:

$$Al_2O_3(s) + \underset{\text{acid}}{6H^+(aq)} \rightarrow 2Al^{3+}(aq) + 3H_2O$$

$$Al_2O_3(s) + \underset{\text{alkali}}{2OH^-(aq)} + 3H_2O(l) \rightarrow \underset{\substack{\text{tetrahydroxoaluminate(III)}\\ \text{(aluminate)}}}{2Al(OH)_4{}^-(aq)}$$

Aluminium oxide occurs naturally in both hydrated and anhydrous forms. The hydrated form (bauxite) has already been mentioned. The anhydrous forms, emery and corundum, are less well known. Corundum is a hard, crystalline form. It may be coloured by traces of red impurity to form *ruby* or by traces of blue impurity to form *sapphire*. The extremely hard, stable structure of Al_2O_3 explains its use in cement, in refractory-furnace bricks and as an inert surface catalyst in cracking petroleum fractions.

Concorde in flight; much of the aircraft's structure is aluminium.

16.4 Properties and uses of aluminium

The position of aluminium in the electrochemical series and its electrode potential suggest that it should react readily with oxygen. It should also dissolve rapidly in dilute acids, liberating hydrogen:

$$Al(s) \rightarrow Al^{3+}(aq) + 3e^- \qquad E^{\ominus} = +1.66\,V$$

$$3H^+(aq) + 3e^- \rightarrow \tfrac{3}{2}H_2(g) \qquad E^{\ominus} = 0.00\,V$$

However, the rapid formation of a thin layer of oxide, no more than $10^{-8}\,m$ thick, prevents further attack by oxygen. This layer also retards reaction of the aluminium with dilute acids. Furthermore, this extremely thin oxide layer is virtually non-porous to water. So, the aluminium is completely protected from further oxidation. This is very different from iron, which forms a porous oxide layer (rust). This is readily penetrated by water, allowing the process of corrosion to continue beneath the superficial layer of rust.

Most of the uses of aluminium are, of course, only possible because of the protective oxide coating—saucepans, aircraft and vehicle bodywork, window frames, etc.

In order to protect the aluminium even more, it is possible to increase the thickness of the neutral oxide layer to about $10^{-5}\,m$ by a process known as **anodising**. The aluminium is anodised by making it the anode during the electrolysis of sulphuric acid. Oxygen, released at the anode, combines with the aluminium and thickens the oxide layer. By carrying out the electrolytic anodising process in the presence of dyes which are adsorbed by the oxide layer, the anodised material can be coloured attractively for use in domestic and personal articles.

■ What properties of aluminium (besides its resistance to corrosion) make it suitable for use as: (i) milk bottle tops, (ii) baking foil, (iii) aircraft bodywork, (iv) electricity cables, (v) kettles and pans?
■ Why does aluminium not corrode away like iron?
■ Aluminium and iron are the two most widely used metals. What advantages are there in using iron rather than aluminium?

Once the oxide layer is removed or penetrated, aluminium reacts readily with many reagents. When strongly heated in air, it will burn to form the oxide and a little nitride:

$$2Al + \tfrac{3}{2}O_2 \rightarrow Al_2O_3$$

$$2Al + N_2 \rightarrow 2AlN$$

Aluminium dissolves in hot dilute hydrochloric and sulphuric acids, evolving hydrogen:

$$Al(s) + 3H^+(aq) \rightarrow Al^{3+}(aq) + \tfrac{3}{2}H_2(g)$$

The bodyworks of cars entering an electrolytic bath for anodising.

With concentrated sulphuric acid and both concentrated and dilute nitric acid, however, the metal remains unreactive. This so-called 'passivity' of the aluminium is caused by the oxidising agents reacting with the metal to form an inert, impervious oxide layer.

Aluminium dissolves in hot dilute sodium hydroxide, forming hydrogen and sodium tetrahydroxoaluminate(III):

$$Al(s) + OH^-(aq) + 3H_2O(l) \rightarrow Al(OH)_4^-(aq) + \tfrac{3}{2}H_2(g)$$

The reactions with dilute acids and alkalis are usually very slow at first. They become increasingly vigorous as the oxide layer is removed and the solutions become hotter. Owing to the reaction of aluminium with alkali, aluminium pans should never be cleaned with washing soda which forms an alkaline solution.

16.5 Reactions of aqueous Al^{3+} ions

When aluminium salts are added to water, Al^{3+} ions are immediately attracted to the negative end of polar water molecules. They form hexaaquaaluminium(III) ions, $[Al(H_2O)_6]^{3+}$. This is often written simply as $Al^{3+}(aq)$.

However, the electric field associated with a small, highly charged Al^{3+} ion is so intense that it draws electrons in the O—H bonds of water molecules towards itself. This enables the water molecules to become proton donors. In aqueous solution, free water molecules act as bases and the following equilibrium is established:

$$[Al(H_2O)_6]^{3+} + H_2O \rightleftharpoons [Al(H_2O)_5(OH)]^{2+} + H_3O^+$$

Thus, solutions of Al^{3+} salts are acidic—in fact, as acidic as vinegar. The salts of other divalent and trivalent ions (e.g. Mg^{2+}, Cu^{2+}, Fe^{3+}) behave in a similar fashion to Al^{3+}. When bases stronger than water, such as S^{2-} or CO_3^{2-}, are added to aqueous aluminium salts further H^+ ions are removed from $[Al(H_2O)_6]^{3+}$ and insoluble aluminium hydroxide precipitates:

$$2[Al(H_2O)_6]^{3+}(aq) + 3S^2(aq) \rightarrow 2[Al(H_2O)_3(OH)_3](s) + 3H_2S(g)$$

A similar reaction occurs when NaOH(aq), an even stronger base, is added to aqueous aluminium salts:

$$[Al(H_2O)_6]^{3+}(aq) + 3OH^-(aq) \rightarrow [Al(H_2O)_3(OH)_3](s) + 3H_2O(l)$$

With excess NaOH(aq), further protons are removed from the insoluble precipitate and an anion forms which is soluble in water:

$$\underset{\text{white ppte}}{[Al(H_2O)_3(OH)_3](s)} + OH^-(aq) \rightarrow \underset{\text{colourless soln}}{[Al(H_2O)_2(OH)_4]^-} + H_2O(l)$$

These reactions of the hydrated Al^{3+} ion involve the progressive removal of H^+ ions from $[Al(H_2O)_6]^{3+}$ by bases of increasing strength. The effects of different bases are summarised in the following:

$$[Al(H_2O)_6]^{3+}(aq) \rightleftharpoons \underbrace{[Al(H_2O)_5(OH)]^{2+}(aq) + H^+(aq)}_{} \rightleftharpoons$$

Even $H_2O(\rightarrow H_3O^+)$ is
a strong enough
base to remove one
H^+. Thus, aqueous Al^{3+}
salts are acidic

$$\underbrace{[Al(H_2O)_3(OH)_3](s) + 3H^+(aq)}_{} \rightleftharpoons \underbrace{[Al(H_2O)_2(OH)_4]^-(aq) + 4H^+(aq)}_{}$$

$NH_3(\rightarrow NH_4^+)$, S^{2-} $(\rightarrow H_2S)$ and CO_3^{2-} $(\rightarrow H_2CO_3)$ are strong enough bases to remove $3H^+$ ions, thus precipitating $[Al(H_2O)_3(OH)_3](s)$

Only very strong bases such as $OH^-(\rightarrow H_2O)$ can remove $4H^+$ ions

The reactions of aqueous aluminium ions with NaOH(aq) are frequently represented more simply but less correctly as:

$$Al^{3+}(aq) + 3OH^-(aq) \rightarrow Al(OH)_3(s)$$
white ppte

$$Al(OH)_3(s) + OH^-(aq) \rightarrow Al(OH)_4{}^-(aq)$$
clear soln

Aluminium hydroxide, like aluminium oxide, is amphoteric. It dissolves in alkalis, as just described, to form aluminates and in acids to give aluminium salts.

When aluminium hydroxide is freshly prepared, the occluded water molecules make the precipitate very gelatinous. Because of this, aluminium hydroxide is used as a *mordant* in dyeing (Latin *mordere*—to bite). A fine precipitate of aluminium hydroxide is deposited in the fibres of the cloth. The gelatinous nature of the precipitate enables the dye to be easily absorbed and held by the Al^{3+} ions. In effect, the aluminium hydroxide acts as a *mordant* by 'biting' both the dye and the cloth.

■ Aluminium hydroxide is precipitated in the fibres of the cloth by soaking the material in aluminium sulphate and then adding alkali. Why is this preferable to adding freshly prepared aluminium hydroxide?

■ Aluminium sulphate can be used for coagulating organic material in water and for coagulating blood. How do Al^{3+} ions in this substance act as coagulating agents?

16.6 The extraction of aluminium

Aluminium is obtained industrially by the electrolysis of molten aluminium oxide. This is obtained from bauxite, impure hydrated aluminium oxide. Bauxite contains impurities such as iron(III) oxide and silicon(IV) oxide. Clearly, the first stage in the production of aluminium must be to obtain pure aluminium oxide from the bauxite (figure 16.2).

The amphoteric nature of Al_2O_3 is an essential feature on which its purification is based.

When the impure bauxite is treated with concentrated sodium hydroxide, Al_2O_3 and SiO_2 dissolve, but Fe_2O_3 and other basic materials remain insoluble:

$$Al_2O_3(s) + 2OH^-(aq) + 3H_2O(l) \rightarrow 2[Al(OH)_4]^-(aq)$$

$$SiO_2(s) + 2OH^-(aq) \rightarrow SiO_3{}^{2-}(aq) + H_2O(l)$$

The basic insoluble materials are removed by filtrations. The solution is then diluted with water or treated with carbon dioxide to precipitate aluminium hydroxide:

$$2[Al(OH)_4]^- + CO_2 \rightarrow 2Al(OH)_3(s) + CO_3{}^{2-}(aq) + H_2O(l)$$

Figure 16.2
The production of pure aluminium oxide from bauxite.

Rows of electrolytic cells at Alcan's aluminium smelter at Kitimat, British Columbia. What is the white powder on the floor? What are the different workmen doing?

Finally, the aluminium hydroxide is heated to obtain pure Al_2O_3:

$$2Al(OH)_3 \rightarrow Al_2O_3 + 3H_2O$$

Owing to the vast, ubiquitous deposits of clay throughout the world, many attempts have been made to obtain aluminium oxide from clays. However, its separation and purification from large quantities of impurity in the clay is very difficult and, at present, uneconomical.

During the nineteenth century, attempts were made to reduce aluminium oxide chemically. However, the oxide is so stable that it required such unconventional and undesirable reducing agents as sodium or potassium. At one time, aluminium was so precious and so novel that Napoleon III used aluminium knives and forks at his most prestigious banquets.

The answer to the problem of aluminium production was obviously electrolysis, but this also presented problems. Aluminium is so reactive that the electrolytic process would have to use molten solids rather than aqueous solutions. This in turn would require very high temperatures in order to maintain the molten state.

Fortunately, the electrolytic process (figure 16.3) can be performed by dissolving the oxide in molten cryolite (Na_3AlF_6). This has a much lower melting

Figure 16.3
The electrolytic manufacture of aluminium.

The original samples of aluminium obtained by Michael Faraday using electrolysis and exhibited at the Royal Institute in 1855.

temperature than aluminium oxide. Even so, the temperature of the molten electrolyte must be raised to 850°C and maintained at this temperature by the current through the electrolyte. Obviously, the amount of electrical energy required for this process is enormous, and it is usually carried out where electrical energy is cheap and plentiful.

Aluminium is discharged at the graphite lining of the cell which acts as the cathode:

$$Al^{3+} + 3e^- \rightarrow Al$$

The molten aluminium collects at the bottom of the cell and is tapped off periodically. Oxygen is evolved at the anodes:

$$2O^{2-} \rightarrow O_2 + 4e^-$$

The oxygen reacts with the carbon of the anode to form oxides of carbon at the high temperatures involved. As a result, the anodes are gradually burnt away and must be replaced from time to time.

Summary

1 Aluminium is obtained industrially by the electrolysis of aluminium oxide in molten cryolite. The pure oxide is obtained from bauxite.

2 Aluminium is a reactive metal but its reactivity is suppressed by a thin coating of inert oxide.

3 The high charge density of Al^{3+} is responsible for:
   (i) the intermediate ionic/covalent bonding of aluminium oxide and its amphoteric character;
   (ii) the polarisation of anions with which it is associated resulting in the partially covalent nature of aluminium chloride and other aluminium salts;
   (iii) the hydration of Al^{3+} ions in aqueous aluminium salts.

4 Aluminium oxide and aluminium hydroxide are amphoteric, dissolving in acids to form aluminium salts and in alkalis to form aluminates.

5 Although pure aluminium is a reactive metal, its compounds have little resemblance to the corresponding compounds of the reactive metals in Groups I and II. The compounds of aluminium are more like those of such 'poor' metals as tin and lead.

Study questions

1 The superficial oxide coating on aluminium is effectively removed by washing the metal in mercury(II) chloride solution which is slightly acidic. The solution reacts with the oxide layer and then with the aluminium below. Mercury is displaced by aluminium and then forms an amalgam (alloy) with more aluminium. Oxide does not adhere to this amalgam. If the metal is left in moist air 'feather-like' growths appear on the aluminium as it reacts with oxygen and water vapour in the atmosphere.
   Write equations for the reaction of:
   (a) H^+ ions in the acidic solution with the aluminium oxide layer.
   (b) the $HgCl_2(aq)$ with aluminium.
   (c) aluminium with oxygen and with water vapour in producing 'feather-like' growths.

Large carbon anode blocks being prepared for use in the electrolytic manufacture of aluminium.

2 This question concerns a group of compounds called the *alums*. The alums are *double salts* with the general formula $X_2SO_4.Y_2(SO_4)_3.24H_2O$.
Two of the commonest alums are: potash alum $(K_2SO_4.Al_2(SO_4)_3.24H_2O)$ and chrome alum $(K_2SO_4.Cr_2(SO_4)_3.24H_2O)$.
All alums can be prepared by allowing equimolar proportions of the two sulphates to crystallise together. The alums are *isomorphous*. Thus, if a crystal of chrome alum is suspended in a solution of potash alum, a colourless *overgrowth* forms on the dark violet chrome alum crystal and the similarity in structure is very apparent.
(a) Explain what is meant by the terms:
   (i)   double salt,
   (ii) isomorphous,
   (iii) overgrowth.
(b) What is the empirical formula of alums?
(c) What can you conclude about the charges on X and Y and the groups in the periodic table to which X and Y may belong from the general formula of alums given above?
(d) Name two elements, other than potassium, which could take the place of X in an alum.
(e) Describe how you would obtain pure crystals of potash alum if you were provided with KOH(aq), $Al_2O_3(s)$ and $H_2SO_4(aq)$.

3 The density of anhydrous aluminium chloride vapour was measured at 200, 600 and 800°C at atmospheric pressure and the results are given below.

Temp/°C	200	600	800
density/g dm^{-3}	6.9	2.7	1.5

(a) Calculate the relative molecular mass of anhydrous aluminium chloride vapour at each temperature. (The gas constant, $R = 8.31$ J K^{-1} mole^{-1} (0.082 litre atm K^{-1} mole^{-1}).)
(b) What are the probable molecular formulae of aluminium chloride vapour at 200° and 800°C? (Al = 27.0, Cl = 35.5.)
(c) Comment on the state of aluminium chloride vapour at 600°C.
(d) Outline the laboratory preparation of solid anhydrous aluminium chloride from aluminium.

4 Describe and explain with equations what happens in each of the following.
(a) Anhydrous aluminium chloride is added to water and aqueous sodium hydroxide is then added slowly until it is present in excess.
(b) Lithium is heated in dry hydrogen and excess of the product is added to a suspension of aluminium chloride in ethoxyethane (diethyl ether).
(c) A mixture of aluminium fluoride, ammonium fluoride and sodium nitrate is heated.

5 When 10 g of a compound, A, were analysed they contained 5.0 g of carbon, 1.25 g of hydrogen and 3.75 g of aluminium.
(i)   0.120 g of A reacts with excess water at s.t.p. to evolve 0.112 dm^3 of a gas, B and form a white precipitate, C.
(ii) C dissolves in dilute sodium hydroxide solution and in dilute hydrochloric acid.
(iii) 10 cm^3 of B require 20 cm^3 of oxygen for complete combustion, producing only carbon dioxide and water.
(a) Identify A, B and C. (H = 1.0; C = 12.0; O = 16.0; Al = 27.0.)
(b) Write equations for the reactions in (i), (ii) and (iii) above.
(c) Write a structural formula for A.

17 Group VII—The Halogens

F
Cl
Br
I
At

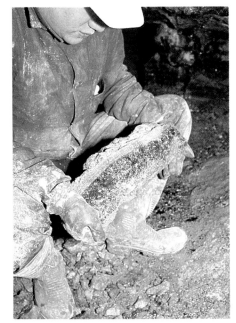

A miner examining a large piece of Blue John stone at the Treak Cliff Cavern in Derbyshire. Blue John is still mined in Treak Cliff Cavern and made into jewellery and ornaments. Blue John is calcium fluoride (fluorspar) containing streaks of oil. Over millions of years, the oil has been transformed into beautiful colours in the stone. What do you think caused the changes in the oil?

17.1 Introduction

The halogens are the elements in Group VII of the periodic table—fluorine (F), chlorine (Cl), bromine (Br), iodine (I) and astatine (At) (table 17.1). They are known as the *halogens*—a name derived from Greek meaning 'salt formers'—because they combine readily with metals to form salts.

Generally speaking, the halogens comprise the most reactive group of non-metals. In this chapter we shall contrast their properties with those of the reactive metals in Groups I and II of the periodic table. We shall notice that *the halogens are strong oxidising agents* whilst the alkali and alkaline-earth metals in Groups I and II are strong reducing agents. Furthermore, we shall see that the halogens exhibit various oxidation numbers in their compounds. In contrast, the *s*-block elements have only one oxidation number in their compounds.

Although the halogens show distinct trends in behaviour down their group, they also show remarkable similarities. These similarities in their properties and reactions result very largely from their similar electron structures. Each of the halogens has an outer shell containing seven electrons (i.e. ns^2np^5).

■ Write the electron shell structure for fluorine and chlorine. (For example; that for beryllium would be $1s^2, 2s^2$.)
■ Why are these elements strong oxidising agents?
■ What is the most likely oxidation number of these elements in their compounds?

17.2 Sources of the halogens

The halogens are so reactive that they cannot exist free in nature. Indeed, fluorine is reactive enough to combine directly with almost all the known elements including some of the noble gases. Chlorine reacts directly with all elements except carbon, nitrogen, oxygen and the noble gases.

Consequently, the halogens always occur naturally as compounds with metals. The halogens are present as negative ions: fluoride (F^-), chloride (Cl^-), bromide (Br^-) and iodide (I^-). The last element in the Group, astatine (At), does not occur naturally at all. It is a very unstable, radioactive element which was first synthesised by chemists in the USA in 1940. It has never been obtained in anything other than minute amounts and the most stable isotope, $^{210}_{85}At$, has a half-life of only 8.3 hours. The name astatine is derived from the Greek word *astatos* meaning 'unstable'.

Fluorine and chlorine are by far the most abundant halogens. The most widespread compounds of fluorine are fluorspar (fluorite), CaF_2, such as Derbyshire 'Blue John' and cryolite, Na_3AlF_6. Unfortunately these extensive deposits of fluoride are dispersed very thinly over the earth's surface. Only a few sources can be worked economically.

The commonest chlorine compound is, of course, sodium chloride (NaCl) which occurs in sea water and in rock salt. Each kilogram of sea water contains about 30 g of sodium chloride.

Bromides and iodides occur in much smaller amounts than either fluorides or chlorides. Sea water contains only small concentrations of bromide, about 70 parts per million by weight. But the extraction of bromine from brine is still economically feasible.

Iodides are even scarcer than bromides. Sea water contains traces of iodide (0.05 parts per million by weight). However, the laminarian seaweeds concentrate iodide from the sea to such an extent that fresh wet weed can contain up to 800 parts per million of iodine. Biologists are still perplexed by this capacity of laminaria to hoard such large concentrations of iodide. Curiously, this is not the only problem posed by naturally occurring iodine compounds. The main source of iodine is sodium iodate(v) (NaIO$_3$). This is found only in Chile, mixed with larger proportions of sodium nitrate. Why does iodine occur as iodate(v) in such high concentration in only one part of the world, and why does the iodine exist in such a high oxidation state? These two questions still remain unanswered.

Chilean workmen drilling holes in the layers of caliche (impure sodium iodate (v)) prior to blasting.

17.3 Obtaining the halogens

The halogens are usually obtained by oxidation of halide ions:

$$2Hal^- \rightarrow Hal_2 + 2e^-$$

Fluorine is a very powerful oxidising agent. Fluoride ions cannot be oxidised by any of the common chemical oxidising agents such as conc. H$_2$SO$_4$, MnO$_2$ and KMnO$_4$. Consequently, fluorine is obtained commercially by electrolysis. Furthermore, electrolysis cannot be carried out in aqueous solution, since water would be discharged in preference to F$^-$. Even if fluorine were produced it would react with water. The electrolyte is usually potassium fluoride dissolved in liquid anhydrous hydrogen fluoride and the electrodes consist of a graphite anode and a steel cathode.

- Write the symbols, with charges, of all ions present in the electrolyte.
- Write an equation to summarise the formation of fluorine at the anode.
- Write an equation to summarise the process at the cathode.

The other halogens are much less reactive than fluorine. They can be obtained in the laboratory by oxidising the appropriate halide ions using MnO$_2$ or KMnO$_4$.

Figure 17.1
Preparing the halogens.

Figure 17.1 shows how chlorine, bromine and iodine can be prepared on a small scale. These preparations should be carried out in a fume cupboard. All halogens are poisonous and should be handled with great care. Notice that each preparation is essentially the oxidation of halide ions by MnO_2 in the presence of conc.H_2SO_4. The apparatus varies from one preparation to the next because chlorine is a gas, bromine is a liquid and iodine is a solid at room temperature.

$$2Hal^- \rightarrow Hal_2 + 2e^-$$
$$MnO_2 + 4H^+ + 2e^- \rightarrow Mn^{2+} + 2H_2O$$

■ Why is the acid necessary?
■ Why is conc.H_2SO_4 used rather than dilute H_2SO_4?

Some hydrogen halide is always evolved from the interaction of the halide and conc.H_2SO_4. The chlorine, for example, will be contaminated with hydrogen chloride fumes. Pure, dry chlorine can be obtained by passing the gas through water to remove the HCl fumes and then through conc.H_2SO_4 to dry it. Finally, the gas is collected by downward delivery.

■ Why is the white paper placed behind the gas jar in which Cl_2 collects?
■ The bromine which is collected will be contaminated with fumes of HBr. How could the liquid bromine be purified? (Hint: how are liquids normally purified?)
■ Why is cold water passed through the flask on which solid iodine is deposited?

Chlorine is less reactive than fluorine. It can be obtained industrially by the electrolysis of *aqueous* sodium chloride (brine) (figure 17.2).

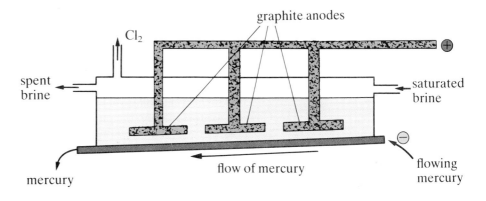

Figure 17.2
A flowing mercury cell in which chlorine is manufactured by electrolysis of aqueous sodium chloride.

In the electrolyte process, chlorine is liberated at the graphite anodes. Sodium is liberated at the mercury cathode.

anode (+) (graphite)

$$2Cl^-(aq) \rightarrow Cl_2(g) + 2e^-$$

cathode (−) (mercury)

$$2Na^+(aq) + 2e^- \rightarrow 2Na(l)$$

The sodium dissolves in the mercury forming an amalgam. This then flows into a second cell, called the soda cell. Here it reacts with water to form hydrogen and a solution of sodium hydroxide (caustic soda):

$$\underbrace{Na/Hg(l)}_{amalgam} + H_2O(l) \rightarrow \underbrace{Na^+(aq) + OH^-(aq)}_{\text{sodium hydroxide solution}} + \tfrac{1}{2}H_2(g) + Hg(l)$$

All three products (chlorine, sodium hydroxide and hydrogen) are invaluable materials for the chemical industry. The mercury is, of course, recovered from the soda cell and recycled. Why is mercury an ideal cathode for use in this process?

In 1975, the estimated worldwide production of chlorine was 25 million tonnes. Since it is a strong, inexpensive oxidising agent it is used in manufacturing bromine by oxidation of bromide ions in sea water:

$$Cl_2 + 2Br^- \rightarrow 2Cl^- + Br_2$$

Unlike chlorine and bromine which normally exist in the −1 oxidation state, iodine occurs naturally in the +5 oxidation state as sodium iodate(v). Consequently, iodine is normally obtained by reduction, with sodium hydrogensulphite.

$$(IO_3^- + 6H^+ + 5e^- \rightarrow \tfrac{1}{2}I_2 + 3H_2O) \times 2$$

$$(HSO_3^- + H_2O \rightarrow HSO_4^- + 2H^+ + 2e^-) \times 5$$

A second, somewhat obsolete method of obtaining iodine is to extract the soluble iodides from dried seaweed with water. The iodine is then displaced from the solution by treatment with chlorine.

17.4 Mercury poisoning—the tragedy of Minamata

In 1970, about one quarter of the total mercury produced industrially was used in flowing mercury cells. These cells are used for the manufacture of chlorine (figure 17.2). In theory, no mercury is consumed during the process. In practice losses of mercury do occur. Sad to say, it was the heavy losses of mercury from flowing mercury cells which led to the tragedy at Minamata during the 1950s. The major losses of mercury are in the effluent liquids and in the products. Since 1965, government legislation and law court actions in various parts of the world have required manufacturers to improve their recovery plants and reduce losses to the environment.

During 1953, the inhabitants of Minamata, a heavily industrialised town in Japan's southern-most island of Kyushu, noticed a bizarre illness affecting their cats. One by one, the cats owned by the families of fishermen in the area were affected. At first, the cats would stagger around unable to support themselves as their muscles weakened. Later, they became paralysed, eventually going into a state of coma followed by death.

Unfortunately, the public health authorities were slow to realise the symptoms of heavy-metal poisoning in the animals owing to the widespread distribution of the fishing community.

In December of the same year, the first human case of this mysterious illness was recorded. During the next decade, 43 other humans died of the same disease. More than 60 others were permanently disabled.

The appalling effects of mercury poisoning. Fifteen-year-old Tadsuko was deformed at birth. His mother ate fish from Minamata Bay during her pregnancy.

In 1969, millions of fish were killed when the River Rhine became heavily polluted.

Unable to find the cause of this mysterious disease, the health authorities sought the help of the medical department at Kumamoto University. By November 1956, the Kumamoto investigators had eliminated any naturally occurring brain disease and had turned to the possibility of heavy-metal poisoning. By this time, other domestic animals, seabirds and the fish in Minamata Bay were dying of a similar disease to that in humans and cats.

Fish and shellfish were examined for toxic matter and found to contain 20 to 60 times as much mercury as would normally accumulate from the minute traces naturally present in sea water. All fishing in Minamata Bay was banned as it became obvious that fish, the main source of food in the area, were responsible for passing the toxic mercury on to humans.

The search for the source of the mercury began immediately. It was traced to a large industrial plant which produced plastics including poly(chloroethene) (PVC, polyvinyl chloride) and industrial organic chemicals.

Chlorine for making PVC was produced by the electrolysis of concentrated brine using a flowing mercury cathode. Several of the organic chemicals were prepared from ethanol (acetaldehyde, CH_3CHO) which was synthesised using a mercury(II) sulphate catalyst. A further significant fact was that large-scale production of PVC commenced in 1952 shortly before the outbreak of the disease.

The effluents from these plants flowed into settling ponds and then via a canal into Minamata Bay. The mud from these settling ponds was found to contain as much as 700 parts per million (p.p.m.) wet weight of inorganic mercury compounds. These were mainly mercury(II) chloride and mercury(II) sulphide. Small concentrations of organic mercury compounds were present in the water effluent.

As a result of these findings the industrial plant started to build a new effluent treatment system to remove heavy metals, but a return to fishing resulted in a new outbreak of poisoning. Figure 17.3 shows a correlation of the production of chloroethene (vinyl chloride) at Minamata with the number of cases of Minamata disease.

Look at figure 17.3 and then answer the following questions.

- Was the fishing ban in 1956 effective?
- What happened after a partial return to fishing in 1958?
- Did the new effluent system which came into use in early 1961 prove effective?

Figure 17.3
The annual production of chloroethene and the number of cases of Minamata disease between 1952 and 1960.

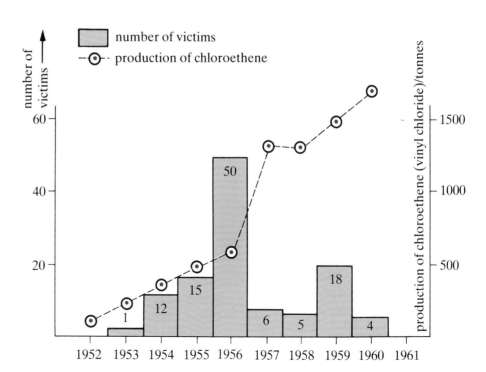

Large amounts of foam polluting a Japanese river.

At one time, it was thought that Napoleon suffered from mercury poisoning. This would explain his strange moods. Since then, tests carried out on a lock of Napoleon's hair have shown that it contained an unusually high proportion of arsenic. This suggests that he may have suffered from arsenic poisoning, which has similar symptoms to mercury poisoning.

In 1960, methyl mercury ethanoate was identified in the effluent water from the industrial plant at Minamata. This compound was also identified in fish. The appalling tragedy has been solved, but this was not the end of Minamata disease. In their summary, the Kumamoto study group stressed that the control of industrial contamination was insufficient. Further outbreaks of mercury poisoning were inevitable in the area. Less than four years later a similar outbreak of mercury poisoning occurred at Niigata in Japan. This time 28 people were critically ill and 5 of them died.

17.5 The effects of mercury and mercury compounds on the body

Mercury and its compounds are not the only poisons amongst metals. Barium, cadmium, lead, arsenic and other heavy metals and their compounds are also highly toxic to both animals and plants. They all act in a similar way to mercury. There is some evidence that Napoleon, Ivan the Terrible and Charles II may each have died from heavy metal poisoning.

Traces of mercury and its inorganic compounds containing Hg^{2+} ions have always been present in nature. Therefore, all plants and animals will carry traces of mercury. Over millions of years, humans and other living creatures have become tolerant to these small concentrations of mercury. It is likely that some forms of life may now be dependent on mercury and other poisonous metals as trace elements.

Mithridates VI Eupator, King of Pontus (132–63 BC) is reputed to have put together 54 substances including mercury compounds as a universal antidote against poisoning. Mithridates was determined not to suffer a similar fate to other kings. Consequently, he protected himself against would-be poisoners by taking small, but steadily increasing doses of toxic materials.

By the nineteenth century, the toxic nature of mercury compounds was well known. It had been noticed by practitioners in almost every trade in which mercury was used—felt-hatters, alchemists, physicians, etc. For example, it was the use of mercury(II) nitrate in felt-making that led to the condition called 'hatters' shakes' or 'hatters' madness' in the felt trade. Perhaps mercury(II) nitrate was responsible for the Hatter's madness in *Alice in Wonderland*.

A biochemical interpretation of mercury poisoning

Mercury has a strong affinity for sulphur and for the sulphydryl group (SH) in the cysteine units in proteins (figure 17.4),

cysteine

a cysteine unit
in a protein

Figure 17.4
The structural formula of cysteine and the cysteine unit in a protein.

Alkyl-mercury compounds, such as methyl mercury ethanoate, are particularly dangerous because they can form strong covalent bonds to the sulphur atoms in these SH groups. The alkyl-mercury compounds replace hydrogen atoms in the SH groups with alkyl-mercury groups. The attachment of a mercury atom to the protein can seriously alter its properties. Proteins play an important part in cells as enzymes which catalyse biological processes and as constituents of cell membranes. Thus, the binding or attraction of mercury to sulphur in proteins can inhibit the activity of an enzyme and also interfere with the movement of materials across a cell membrane.

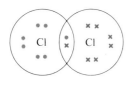

The Mad Hatter's Tea Party from *Alice in Wonderland*. Mercury (II) nitrate may have caused the Hatter's madness.

Electron structure   2,8,8   2,8,8

Figure 17.5
Chlorine is a diatomic molecule. The two chlorine atoms are linked by a covalent bond.

17.6 Structure and properties of the halogens

All the halogens exist as diatomic molecules. The two atoms are linked by a covalent bond (figure 17.5). These molecules persist in the gaseous, liquid and solid states. Table 17.2 shows some of the physical properties of the halogens.

They are all coloured—the depth of colour increasing with increase in atomic number. Fluorine is pale yellow, chlorine is pale green, bromine is red-brown and iodine is shiny black.

Notice also the change in volatility down the group. Fluorine and chlorine are gases, bromine is a liquid and iodine is a solid. This decreasing volatility is, of course, related to the increasing strength of Van der Waals' forces with increasing relative molecular mass. This results in increasing melting points, boiling points and molar heats of vaporisation.

All the halogens except fluorine dissolve slightly in water and colour it. Fluorine is such a powerful oxidising agent that it converts water to oxygen. The halogens, being non-polar simple molecules, are more soluble in organic solvents, the colour of the solution depending upon the particular halogen and the solvent. In non-polar solvents such as tetrachloromethane and cyclohexane, chlorine is colourless, bromine is red and iodine is violet. In these solvents the elements exist as relatively free molecules as in the gas phase. In polar (electron-donating) solvents such as water, ethanol and propanone (acetone), bromine and particularly iodine tend to give brownish solutions. The change in colour is due to the formation of complexes involving charge transfer. This involves a donation of electrons from the polar solvent molecules to the iodine molecule. It is thought that an electron normally located in an orbital of the solvent molecule can absorb energy from visible light and jump into an orbital of the iodine molecule. Thus, the complex removes visible light of a particular wavelength and appears coloured.

The blue compound which iodine forms with starch is also a complex. The iodine molecules are absorbed reversibly within the helical chains of glucose molecules which make up starch.

Table 17.2
Physical properties of the halogens

Element	Fluorine	Chlorine	Bromine	Iodine
Atomic number	9	17	35	53
Electron configuration	2, 7	2, 8, 7	2, 8, 18, 7	2, 8, 18, 18, 7
Outer shell electron configuration	$2s^2\,2p^5$	$3s^2\,3p^5$	$4s^2\,4p^5$	$5s^2\,5p^5$
Relative atomic mass	19.0	35.5	79.9	126.9
State at 20°C	gas	gas	liquid	solid
Colour	pale yellow	pale green	red-brown	black
Melting point/°C	-220	-101	-7	113
Boiling point/°C	-188	-35	59	183
Molar heat of vaporisation of liquid /kJ mole^{-1}	3.3	10.2	15	30
Solubility/g per 100g of water at 20°C	reacts readily with water	0.59 (reacts slightly)	3.6	0.018

You will need to refer to table 17.2 in order to answer the following questions. Astatine is the halogen element immediately below iodine in the periodic table. Predict the following properties of astatine:

- Its colour.
- Its state at room temperature.
- Its relative atomic mass.
- Its melting point.
- Its molar heat of vaporisation.
- The abundance of its compounds.

17.7 Chemical properties of the halogens

All the halogens have one electron less than the noble gas which follows them in the periodic table. As we might expect, their chemistry is dominated by a tendency to gain a completely filled outermost electron shell.

Consequently, *the halogens react with metals to form ionic compounds containing the Hal$^-$ ion. With non-metals and with some metals in high oxidation states, they tend to form simple molecular compounds. In these, the halogen is linked by a single covalent bond (—Hal) to the other element.*

Reactions with metals

The halogens are very reactive with metals. The vigour of reaction depends on

1 the position of the metal in the activity series, and

2 the particular halogen which is reacting.

The reactivity of the halogen decreases with increase in atomic number. Fluorine is, in fact, the most reactive of all non-metals. Because of its great reactivity it was not isolated until 1886. Fluorine combines readily and directly with all metals, whereas iodine reacts very slowly even at high temperatures with metals low in the activity series such as silver and gold. Gold and platinum are readily attacked by fluorine on heating but some metals, such as copper and nickel alloys, become coated with a superficial layer of fluoride which prevents further

Table 17.3
Properties of the halogens and their compounds

Element	Fluorine	Chlorine	Bromine	Iodine
Standard heat of formation of NaHal/kJ mole^{-1}	−573	−414	−361	−288
Heat of atomisation of Hal$_2$/kJ per mole of atoms formed	79	121	112	107
Bond energy E(Hal—Hal) /kJ per mole of bonds	158	242	193	151
Electrode potential Hal$_2$(aq), 2 Hal$^-$(aq)/Pt/volts	2.87	1.36	1.09	0.54

reaction. Hence, vessels of these alloys are used for preparing and storing the gas.

The standard heats of formation of the sodium halides (table 17.3) provide quantitative evidence for the relative reactivity of the halogens. The heat evolved is greatest for the reaction between sodium and fluorine and least for the reaction between sodium and iodine.

The most reactive halogen is fluorine. Chlorine is the next most reactive, then bromine, then iodine. Thus, reactivity decreases as the relative atomic mass of the halogen increases. A similar decrease in reactivity with increasing atomic mass is found in other groups of non-metals. This is the reverse of the trend in Group I and other groups of metals where reactivity increases as the relative atomic mass increases.

■ List three factors which influence the reactivity of the halogen elements.
■ How will each of these three factors influence the reactivity of fluorine compared to chlorine?
■ Why is fluorine (F = 19.0) more reactive than chlorine (Cl = 35.5)?
■ Why is sodium (Na = 23.0) less reactive than potassium (K = 39.1)?

Reactions with non-metals

A similar pattern of reactivity emerges here also. Fluorine reacts directly with all non-metals except nitrogen, helium, neon and argon. It will even react with diamond and xenon on heating.

$$C(diamond) + 2F_2 \rightarrow CF_4$$

$$Xe + 2F_2 \rightarrow XeF_4$$

Fluorine will also attack glass, quartz and silica, displacing oxygen.

$$SiO_2 + 2F_2 \rightarrow SiF_4 + O_2$$

If these substances are carefully dried, the reaction is very slow. Thus, fluorine can be studied in dry glass equipment. Any taps must be lubricated with 'fluorocarbon grease' containing fluorinated hydrocarbons rather than grease containing hydrocarbons since fluorine readily attacks organic compounds, displacing hydrogen.

$$—C—H + F_2 \rightarrow —C—F + HF$$

Chlorine and bromine are much less reactive than fluorine. They will not react directly with any of the noble gases, carbon, nitrogen or oxygen. Iodine does not combine with these elements nor with sulphur, but it reacts readily with phosphorus forming the triiodide.

$$\tfrac{1}{4}P_4 + \tfrac{3}{2}I_2 \rightarrow PI_3$$

The relative reactivity of the halogens with non-metals is well illustrated by their reaction with hydrogen. Fluorine explodes with hydrogen even in the dark at −200°C. Chlorine and hydrogen explode in bright sunlight but react slowly in the dark. Bromine reacts with hydrogen only on heating and in the presence of a platinum catalyst. Iodine combines only partially and slowly with hydrogen even on heating.

$$H_2(g) + Cl_2(g) \longrightarrow 2HCl(g)$$

$$H_2(g) + I_2(g) \rightleftharpoons 2HI(g)$$

The great reactivity of fluorine in these reactions with non-metals is explained partly in terms of its low atomisation energy (table 17.3), This means that the initial stage (in which the F—F bond breaks) requires little energy. Hence the activation energy for reactions involving fluorine is low. The low atomisation energy of fluorine is reflected in a low F—F bond energy (table 17.3). Fluorine's reactivity is further explained by its ability to form very stable bonds with other elements. Look at table 12.3 and compare the C—F bond energy with the energies of other C—Hal bonds. How does the strength of the H—F bond compare with other H—Hal bonds?

17.8   The halogens as oxidising agents

When the halogens combine with metals or non-metals they normally act as oxidising agents. The elements with which they react have positive oxidation numbers in the resultant compounds.

When the halogens combine with metals to form ionic compounds, they gain electrons from the metals to form negative halide ions:

$$2Na(s) + Cl_2(g) \rightarrow 2Na^+Cl^-(s)$$

The halogens accept electrons during these reactions and act as oxidising agents:

$$Hal_2 + 2e^- \rightarrow 2Hal^-$$

Fluorine is the most reactive halogen and the most powerful oxidising agent. *The order of decreasing power as oxidising agents is $F_2 > Cl_2 > Br_2 > I_2$.* The electrode potentials (Hal_2/Hal^-) for the halogens shown in table 17.3 become less positive from fluorine to iodine. This reflects the decreasing oxidising power.

We have already noticed the different oxidising powers of the halogens with metals, non-metals and hydrogen. The following reactions further emphasise their relative oxidising ability.

Fluorine, chlorine and bromine will all oxidise $Fe^{2+}(aq)$ to $Fe^{3+}(aq)$:

$$2Fe^{2+}(aq) \rightarrow 2Fe^{3+}(aq) + 2e^-$$

$$Hal_2(aq) + 2e^- \rightarrow 2Hal^-(aq)$$

Iodine, however, is such a weak oxidising agent that it cannot remove electrons from iron(II) ions to form iron(III) ions. These observations are exactly as we might have predicted since $E^{\ominus}$ for $Fe^{3+}(aq)$, $Fe^{2+}(aq)/Pt$ is 0.77 volts.

Fluorine and chlorine are such powerful oxidising agents that they can oxidise various coloured dyes to colourless substances. Thus, indicators such as litmus and Universal Indicator are decolorised when exposed to fluorine and chlorine. Chlorine acts as an oxidising agent when it is used for bleaching.

When chlorine water ($Cl_2(aq)$) is added to aqueous KI, the solution becomes brown. This is due to the formation of iodine. When bromine water ($Br_2(aq)$) is used in place of chlorine water, iodine is again liberated from KI(aq).

In these two reactions, iodide ions are oxidised to iodine:

$$2I^- \rightarrow I_2 + 2e^-$$

Chlorine acts as an oxidising agent when it bleaches. The piece of printed material on the right was bleached by immersion in chlorine water for a few hours.

Antoine (1743–1794) and Marie Lavoisier (1758–1836). The Lavoisiers were the first chemists to identify the role of oxygen as an oxidising agent during burning.

The chlorine and bromine act as oxidising agents. They accept electrons from iodide to form chloride and bromide, respectively.

$$Cl_2 + 2e^- \rightarrow 2Cl^-$$

$$Br_2 + 2e^- \rightarrow 2Br^-$$

Chlorine and bromine can oxidise iodide to iodine, but iodine cannot oxidise chloride or bromide ions. This is because chlorine and bromine are more reactive than iodine, i.e. chloride will not release electrons to iodine.

Since chlorine is a stronger oxidising agent than bromine, chlorine can oxidise bromide ions to bromine:

$$Cl_2 \rightarrow 2Br^- \rightarrow 2Cl^- + Br_2$$

Thus, when chlorine water is added to KBr(aq), yellow-orange bromine is produced.

■ Which halide ions can F_2 oxidise?
■ Which of the following species is the strongest reducing agent?
  I^-, I_2, Cl^-, F_2, F^-.
■ The standard electrode potential for the half-reaction
  $MnO_4^-(aq) + 8H^+(aq) + 5e^- \rightarrow Mn^{2+}(aq) + 4H_2O$ is 1.51 volts.
  Which halide ions could this half-cell oxidise under standard conditions? (Refer to the electrode potentials for halogens in table 17.3.)

Explaining the relative oxidising power of the halogens

When halogens act as oxidising agents they undergo the reaction:

$$\tfrac{1}{2}Hal_2 + e^- \rightarrow Hal^-$$

We can picture the reaction in terms of three stages as shown in figure 17.6.

The enthalpy change for each of these stages for F, Cl, Br and I is also shown in figure 17.6. The order of oxidising power, $F_2 > Cl_2 > Br_2 > I_2$, is due mainly to the relative hydration energies and to a lesser degree to the small atomisation energy of fluorine.

Figure 17.6
Stages in the conversion of diatomic halogen elements to aqueous halide ions.

	F	Cl	Br	I
ΔH_{at} /kJ mole^{-1}	79	121	112	107
ΔH_e / kJ mole^{-1}	−333	−364	−342	−295
ΔH_{hyd} /kJ mole^{-1}	−457	−381	−351	−307
ΔH_r /kJ mole^{-1} for $\tfrac{1}{2}Hal_2 \longrightarrow Hal^-(aq)$	−711	−624	−581	−495

Since the lattice energies of ionic halides also get less exothermic from fluorides to iodides, the order of oxidising power is the same for solid-state reactions as for reactions in aqueous solution.

Estimating the concentration and molar amounts of oxidising agents

Iodine is by far the weakest oxidising agent of the four common halogens. Even mild oxidising agents, such as $Fe^{3+}(aq)$, can oxidise aqueous I^- to I_2. Because of this, iodine/thiosulphate titrations are used in the quantitative estimation of oxidising agents.

The oxidising agent to be estimated is first added to excess potassium iodide. This liberates iodine:

$$2I^-(aq) \rightarrow I_2(aq) + 2e^-$$

The amount of iodine liberated is then determined by titration against a solution of sodium thiosulphate of known concentration.

$$I_2(aq) + 2e^- \rightarrow 2I^-(aq)$$

$$\underset{\text{thiosulphate}}{2S_2O_3{}^{2-}(aq)} \rightarrow \underset{\text{tetrathionate}}{S_4O_6{}^{2-}(aq)} + 2e^-$$

The end point of the titration can be detected by adding starch solution. The starch gives a blue colour as long as any iodine is present. By calculating the amount of thiosulphate used in the titration, we can determine the amount of iodine liberated and hence the amount of oxidising agent taken.

17.9 The reactions of halogens with water

The electrode potential for the oxidation of water to oxygen,

$$2H_2O(l) \rightarrow O_2(g) + 4H^+(aq) + 4e^-$$

is -1.23 volts. The electrode potential data for the halogens in table 17.3 show that iodine and bromine cannot oxidise water to oxygen. But fluorine and chlorine can do so.

Fluorine reacts with water vapour to form oxygen and ozone:

$$2F_2(g) + 2H_2O(g) \rightarrow 4HF(g) + O_2(g)$$
$$3F_2(g) + 3H_2O(g) \rightarrow 6HF(g) + O_3(g)$$

The reaction of water with chlorine is very slow. In fact, chloric(I) acid (hypochlorous acid) is first formed which then decomposes to oxygen and hydrochloric acid.

$$Cl_2(g) + H_2O(l) \rightarrow H^+(aq) + Cl^-(aq) + HClO(aq)$$

$$2HClO(aq) \rightarrow 2H^+(aq) + 2Cl^-(aq) + O_2(g)$$

The second reaction is accelerated by sunlight and by catalysts such as platinum and metallic oxides.

17.10 The reactions of halogens with alkalis

Chlorine, bromine and iodine undergo very similar reactions with alkalis. The products depend upon the temperature at which the reaction occurs.

With cold, dilute alkali at 15°C, the halogen reacts to form a mixture of halide (Hal^-) and halate(I) (hypohalite, $HalO^-$):

$$Hal_2 + 2OH^-(aq) \rightarrow Hal^-(aq) + HalO^-(aq) + H_2O(l)$$

The $HalO^-$ which is produced in the first reaction then decomposes to form halide and halate(v) ($HalO_3^-$):

$$3 HalO^- \rightarrow 2 Hal^- + HalO_3^-$$

For chlorine, this second reaction is very slow at 15°C, but rapid at 70°C. Therefore, sodium chlorate(I) (sodium hypochlorite) can be obtained by passing chlorine into sodium hydroxide at 15°C. Sodium chlorate(v) ($NaClO_3$) is obtained by carrying out the same reaction at 70°C.

With bromine, both reactions are rapid at 15°C, but decomposition of BrO^- is slow at 0°C.

With iodine, decomposition of IO^- occurs rapidly even at 0°C, so it is difficult to prepare $NaIO$ free from $NaIO_3$.

These two reactions of halogens with alkali involve **disproportionation**. *This is a change in which one particular molecule, atom or ion is simultaneously oxidised and reduced.*

When Cl_2 reacts with alkali to form Cl^- and ClO^-, chlorine atoms are both oxidised and reduced. The oxidation number of one Cl atom in Cl_2 changes from 0 to −1 in Cl^- (reduction). The oxidation number of the other Cl atom changes from 0 to +1 in ClO^- (oxidation). Oxidation numbers are shown above each atom in the following equation:

$$\overset{0}{Cl_2}(g) + 2\overset{-2\ +1}{OH^-}(aq) \longrightarrow \overset{-1}{Cl^-}(aq) + \overset{+1\ -2}{ClO^-}(aq) + \overset{+1\ -2}{H_2O}(l)$$

The decomposition of ClO^- to Cl^- and ClO_3^- is also disproportionation.

■ Write an equation for this change.
■ What is the oxidation number of chlorine in each of ClO^-, Cl^- and ClO_3^-?
■ What is the commonest oxidation number for chlorine in its compounds?
■ Write the formula of a compound containing chlorine in a different oxidation state to that in Cl^-, ClO^- or ClO_3^-.

The halogens, other than fluorine, form compounds in which they have positive oxidation numbers up to +7. Figure 17.7 shows an oxidation number chart for chlorine. The exhibition of several oxidation numbers by each halogen is, of course, very different from the metals in Groups I and II. These metals exhibit only one oxidation number in their compounds. As we might expect, the most stable oxidation state for halogens is −1. The relative stability of the −1 state decreases as the group is descended and the electronegativity of the halogen decreases.

Fluorine, the most electronegative element, never exhibits a positive oxidation number. This is because it can never form a compound in which it is the less electronegative element. In this respect, it is significant that fluorine reacts with alkalis differently to the other halogens. Fluorine forms a mixture of fluorine and oxygen difluoride in both of which its oxidation number is −1.

$$2 F_2(g) + 2 OH^-(aq) \rightarrow OF_2(g) + 2 F^-(aq) + H_2O(l)$$

A further consequence of the decreasing electronegativity down Group VII is that the relative stability of the positive oxidation states increases from Cl to I.

Thus, the ease of decomposition of $HalO^-$ into Hal^- and $HalO_3^-$ (halogen in the +5 state) is in the order $IO^- > BrO^- > ClO^-$. This suggests that IO_3^- is relatively more stable than BrO_3^- which in turn is more stable than ClO_3^-. In fact, the positive oxidation state has become sufficiently stable in iodine to permit the existence of iodine cations. Thus, iodine(I) nitrate ($I^+NO_3^-$) and iodine(I) chlorate(VII) ($I^+ClO_4^-$) have been prepared in the presence of pyridine which forms a complex with the I^+ ion. Not surprisingly, these salts react with potassium iodide to give iodine,

$$I^+ + I^- \rightarrow I_2$$

and when electrolysed they give iodine at the cathode. I^+ ions are probably also produced when iodine chloride (ICl) is dissolved in conc. H_2SO_4.

The existence of I^+ ions is evidence for the slight metallic character of iodine. Other features which show this are the metallic lustre of solid iodine and the small but definite conductivity of liquid iodine.

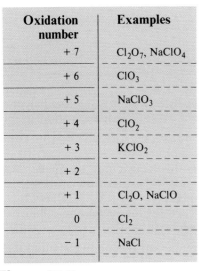

Oxidation number	Examples
+7	Cl_2O_7, $NaClO_4$
+6	ClO_3
+5	$NaClO_3$
+4	ClO_2
+3	$KClO_2$
+2	
+1	Cl_2O, $NaClO$
0	Cl_2
−1	$NaCl$

Figure 17.7
The range of oxidation numbers for chlorine.

Table 17.4
The electrode potentials of the halogens and some of their oxyanions

Electrode process	$E^{\ominus}$/volts
$F_2(g) + 2e^- \rightarrow 2F^-(aq)$	+2.87
$2HClO(aq) + 2H^+(aq) + 2e^- \rightarrow Cl_2(aq) + 2H_2O(l)$	+1.59
$2HBrO(aq) + 2H^+(aq) + 2e^- \rightarrow Br_2(aq) + 2H_2O(l)$	+1.57
$2BrO_3^-(aq) + 12H^+(aq) + 10e^- \rightarrow Br_2(aq) + 6H_2O(l)$	+1.52
$Cl_2(aq) + 2e^- \rightarrow 2Cl^-(aq)$	+1.36
$2IO_3^-(aq) + 12H^+(aq) + 10e^- \rightarrow I_2(aq) + 6H_2O(l)$	+1.19
$Br_2(aq) + 2e^- \rightarrow 2Br^-(aq)$	+1.09
$HIO(aq) + H^+(aq) + 2e^- \rightarrow I^-(aq) + H_2O(l)$	+0.99
$ClO^-(aq) + H_2O(l) + 2e^- \rightarrow Cl^-(aq) + 2OH^-(aq)$	+0.89
$I_2(aq) + 2e^- \rightarrow 2I^-(aq)$	+0.54
$IO^-(aq) + H_2O(l) + 2e^- \rightarrow I^-(aq) + 2OH^-(aq)$	+0.49

One sensible deduction that we might make from all this is that the halogens are likely to act as oxidising agents in all their positive oxidation states (table 17.4).

Notice two points from table 17.4.

1 The redox potentials for the various systems involving iodine are smaller than the corresponding systems involving bromine. These in turn are less than those for chlorine (i.e. $F_2 > Br_2 > Cl_2 > I_2$; $HClO > HBrO > HIO$; $BrO_3^- > IO_3^-$; $ClO^- > IO^-$). This is further evidence for the increasing stability of positive oxidation states as Group VII is descended.

2 The $HalO_3^-$ ions are stable in alkali and require acid conditions before they become effective oxidising agents. This is not surprising in view of the fact that each $HalO_3^-$ ion requires six H^+ ions during the reaction.

17.11   Reactions of halide ions

Some reactions of fluorides, chlorides, bromides and iodides in aqueous solution are summarised in table 17.5. All common halides are soluble except all lead halides, AgCl, AgBr and AgI. Notice in table 17.5 that precipitates of these halides are produced when aqueous solutions of halides are treated with either $Pb^{2+}(aq)$ or $Ag^+(aq)$.

$$Ag^+(aq) + Cl^-(aq) \rightarrow AgCl(s)$$

$$Pb^{2+}(aq) + 2Cl^-(aq) \rightarrow PbCl_2(s)$$

Silver nitrate solution followed by sunlight or ammonia solution can be used as a test for halide ions.

$F^-(aq)$ gives no precipitate of AgF.
$Cl^-(aq)$ gives a white precipitate of AgCl which becomes purple-grey in sunlight and dissolves in conc. NH_3.
$Br^-(aq)$ gives a cream precipitate of AgBr which becomes green-yellow in sunlight and dissolves in conc. NH_3.
$I^-(aq)$ gives a yellow precipitate of AgI which is unaffected by sunlight and which is insoluble in conc. NH_3.

The colour changes which occur when AgCl and AgBr are exposed to sunlight result from the superficial conversion of these silver halides to silver and halogen.

$$Ag^+Br^-(s) \rightarrow Ag(s) + \tfrac{1}{2}Br_2(g)$$

Table 17.5
Reactions of aqueous halide ions

Solution added	F^-(aq)	Cl^-(aq)	Br^-(aq)	I^-(aq)
$Pb(NO_3)_2$(aq)	white ppte of PbF_2	white ppte of $PbCl_2$	yellow ppte of $PbBr_2$	yellow ppte of PbI_2
$AgNO_3$(aq)	no reaction	white ppte of AgCl	cream ppte of AgBr	yellow ppte of AgI
Solubility of AgHal in (a) dil. NH_3(aq) (b) conc. NH_3(aq)	soluble soluble	soluble soluble	insoluble soluble	insoluble insoluble
Effect of sunlight on AgHal	no effect	white AgCl turns purple-grey	cream AgBr turns green-yellow	no effect

This photochemical change involving AgBr plays an essential part in black and white photography. Photographic plates and films contain silver bromide. This decomposes to silver on exposure to light. When the photograph is developed, the plate/film is treated with 'hypo' (sodium thiosulphate solution). This removes excess AgBr as a soluble complex ion $[Ag(S_2O_3)_2]^{3-}$, and the silver remains on the plate/film as an opaque shadow.

The reactions of solid halides with concentrated sulphuric acid, with concentated phosphoric(v) acid and with a mixture of manganese(iv) oxide plus concentrated sulphuric acid are summarised in table 17.6.

When conc.H_2SO_4 is added to solid halides the first product is the hydrogen halide. Being relatively volatile, these hydrogen halides are evolved as gases.

$$Hal^-(s) + H_2SO_4(l) \rightarrow HHal(g) + HSO_4^-(s)$$

However, concentrated sulphuric acid is also an oxidising agent and it is powerful enough to oxidise HBr to Br_2 and HI to I_2. But, it cannot oxidise HF and HCl.

$$2HBr + H_2SO_4 \rightarrow Br_2 + 2H_2O + SO_2$$
$$2HI + H_2SO_4 \rightarrow I_2 + 2H_2O + SO_2$$

Table 17.6
Reactions of solid halides

Reagent added	Fluoride	Chloride	Bromide	Iodide
Conc. H_2SO_4	HF(g) produced	HCl(g) produced	HBr(g) + a little Br_2(g) produced	a little HI(g) + I_2(g) produced
Conc. H_3PO_4	HF(g) produced	HCl(g) produced	HBr(g) produced	HI(g) produced
Conc. H_2SO_4 + MnO_2	HF(g) produced	Cl_2(g) produced	Br_2(g) produced	I_2(g) produced

When concentrated H_2SO_4 is used in the presence of an even stronger oxidising agent, such as MnO_2, the oxidising conditions are sufficient to oxidise HCl to Cl_2, but HF is still not oxidised to F_2.

$$4HCl + MnO_2 \rightarrow Cl_2 + MnCl_2 + 2H_2O$$

- Write an equation for the reaction of concentrated phosphoric(v) acid (H_3PO_4) with NaBr(s).
- Why does the conc. H_3PO_4 cause HBr(g) to be evolved during the reaction?
- Why is no Br_2 produced in this reaction unlike that between conc. H_2SO_4 and NaBr(s)?

Figure 17.8
The structure of poly(tetrafluoroethene).

17.12 Uses of the halogens

Although the halogen elements have very few direct uses themselves, their compounds are used extensively in industry, in agriculture, in medicine and in the home. Chlorine is used as a cheap industrial oxidant in the manufacture of bromine (section 17.3). It is also used as a bleach and as a germicide in the treatment of household and swimming bath water supplies.

Iodine dissolved in alcohol, commonly known as 'tincture of iodine', is used as a mild antiseptic for cuts and scratches. Iodine is also mixed with the detergents used in cleaning dairy equipment.

Small quantities of fluorine are used in rocket propulsion. Much larger quantities are used to make uranium(vi) fluoride for the separation of $^{238}_{92}U$ and $^{235}_{92}U$:

$$UF_4(s) + F_2(g) \rightarrow UF_6(s)$$

By allowing gaseous UF_6 to diffuse slowly it is possible to separate $^{238}_{92}UF_6$ from $^{235}_{92}UF_6$. This allows the separation of isotopes of uranium for use in atomic power stations.

Fluorine is also used to make a wide range of fluorocarbon compounds for use as refrigerants, aerosol propellants, anaesthetics and fire-extinguisher fluids. Many of these compounds are discussed in section 30.3. One of the most important fluorcarbons is poly(tetrafluoroethene), PTFE, frequently sold under the trade name 'Fluon' or 'Teflon' (figure 17.8).

PTFE, like other fully fluorinated hydrocarbons, is very unreactive and resists almost all corrosive chemicals. For this reason, it is used for valves, seals and gaskets in chemical plants and laboratories. It is also an excellent electrical insulator and is used for wire coverings. Furthermore, PTFE has an extremely low coefficient of friction and has anti-stick properties. Because of this, thin layers of it are coated on the cooking surface of non-stick sauce-pans and on the running surface of skis.

Chlorine is used in the manufacture of many familiar materials. Hydrogen chloride is produced by the reaction between chlorine and hydrogen, itself a by-product in the electrolytic manufacture of chlorine. Hydrogen chloride made in this way is dissolved in water to produce hydrochloric acid:

$$H_2(g) + Cl_2(g) \rightarrow 2HCl(g) \xrightarrow{aq} 2H^+(aq) + 2Cl^-(aq)$$

Hydrochloric acid is the cheapest industrial acid. It is used industrially for removing (de-scaling) rust from steel sheets before galvanising.

The other inevitable by-product of the electrolysis of brine is sodium hydroxide. The sodium hydroxide solution is treated with gaseous chlorine to obtain sodium chlorate(I) (sodium hypochlorite, NaClO). This is used as a bleach in laundries, as a disinfectant and in sewage treatment.

$$Cl_2(g) + 2NaOH(aq) \rightarrow NaCl(aq) + NaClO(aq) + H_2O(l)$$

On warming, the NaClO decomposes to sodium chlorate(v) ($NaClO_3$) which is used as a weed killer.

$$3NaClO(aq) \xrightarrow{heat} 2NaCl(aq) + NaClO_3(aq)$$

PTFE has an extremely low coefficient of friction and anti-stick properties. Because of this, thin layers of it are coated on the running surface of skis.

In recent years, chlorine compounds have been developed for use as degreasing solvents such as tetrachloromethane and trichloroethene, as plastics such as PVC (section 28.7), as disinfectants and antiseptics such as 'Dettol' and TCP (section 31.5) and as pesticides.

During the 1940s and 1950s a range of highly chlorinated aromatic compounds was developed and used as pesticides. Probably the best known of these compounds are DDT (dichlorodiphenyltrichloroethane), BHC (benzene hexachloride), aldrin, dieldrin and heptachlor (figure 17.9).

dichlorodiphenyltrichloroethane
('DDT')

benzene hexachloride
('BHC')

aldrin

dieldrin

Figure 17.9
The structures of some chlorinated pesticides.

The spraying of large areas of land with these chlorine-containing pesticides has eliminated many insect-borne diseases, such as malaria. This had led to enormous improvement in the quality and yield of crops. Unfortunately, however, these chlorinated compounds are so stable that they remain unchanged on the crops or accumulate in the soil. Furthermore, these compounds are fat-soluble but not water-soluble. This means that they concentrate, after ingestion, in the fatty tissues of birds and animals, possibly reaching hazardous levels.

In the spring of 1956, large numbers of seed-eating birds were found dead in cereal-growing areas in the UK. Analysis showed that the corpses of these birds contained dieldrin and aldrin which had been sprayed on the spring-sown wheat. In 1964, the International Advisory Committee on Poisonous Substances Used in Agriculture and Food Storage recommended that the use of dieldrin and aldrin should be reserved for the treatment of heavily infected areas. At that time, no restrictions were placed on the use of DDT and BHC. But, concern over their use continued to grow. It appears that insects can develop a tolerance to small, non-lethal doses of these chemicals. These insects are eaten by small carnivorous animals and birds which concentrate the DDT in their own fatty tissue. Larger predators who eat these small carnivores concentrate the DDT even further. Ultimately, the DDT may reach toxic levels in animals or birds several stages along the food chain. At one time, for instance, there was some concern that humans might be affected by drinking the milk of cows which had eaten grasses sprayed with DDT. The use of DDT was also thought to be responsible for the decreasing populations of birds of prey and frogs, whose bodies were found to contain the chemical. As a result of concern over its use, the world consumption of DDT fell from 400 000 tonnes in 1963 to only about half that quantity in 1971.

Since 1972, both the British and American Governments have imposed restrictions on the use of DDT, but the need for an effective substitute is most urgent. Nevertheless, it is sensible to use less toxic insecticides for general purposes and reserve DDT for special control schemes.

Birds killed by insecticide poisoning.

Summary

1 The halogens are a group of reactive non-metals. They contrast strongly with the alkali and alkaline-earth elements which are reactive metals.

2 The halogens show remarkable similarities in their properties and reactions. There is also an obvious trend in their behaviour and reactivity with increase in atomic number. As atomic number increases, the halogens get less reactive.

3 The halogens are so reactive that they occur naturally only in compounds.

4 The halogens can be obtained by oxidation of halide ions.

5 The halogens exist as diatomic molecules. Their atoms have seven electrons in the outermost shell. Consequently, the chemistry of the halogens is dictated by a tendency to act as oxidising agents in forming negative halide ions (Hal^-). The most stable oxidation number of the halogens is -1.

6 The order of oxidising power for the halogens is $F_2 > Cl_2 > Br_2 > I_2$.

7 A disproportionation reaction is one in which a particular molecule, atom or ion is simultaneously oxidised and reduced. When halogens react with dilute alkali to form a mixture of Hal^- and $HalO^-$, the halogen is both oxidised and reduced.

8 The action of $AgNO_3(aq)$ followed by sunlight or $NH_3(aq)$ can be used as a test for halide ions.

Study questions

1 (a) Chlorine and sodium hydroxide are manufactured by the electrolysis of brine. Write equation to summarise what happens during the electrolysis when
    (i) precautions are taken to prevent the products mixing with each other,
    (ii) the products are deliberately mixed with one another.
   (b) Iodine is obtained from iodate(v) by treatment with sodium hydrogensulphite.
    (i) Write an equation (or half-equations) for the reaction of iodate(v) with hydrogen sulphite ions in acid solution.
    (ii) Why must excess hydrogensulphite *not* be used?
   (c) Fluorine can be obtained by the electrolysis of KF dissolved in liquid HF, but not by the electrolysis of KF dissolved in water. Why not?
   (d) Astatine ($^{211}_{85}At$) has been made by bombarding bismuth with high-energy alpha particles.
    (i) Write a nuclear equation for this reaction.
    (ii) The name 'astatine' is derived from a Greek word meaning unstable. How do you think astatine might decompose? Write a nuclear equation for the decay process.

2 From your knowledge of the halogens, predict what happens in the following situations. Write equations for any reactions which take place. (Ignore the radioactive nature of astatine.)
   (a) Astatine vapour is mixed with hydrogen at 100°C.
   (b) Astatine is added to aqueous sodium hydroxide.
   (c) Concentrated sulphuric acid is added to solid sodium astatide.
   (d) Aqueous silver nitrate is added to aqueous sodium astatide.
   (e) Astatine is added to sodium thiosulphate solution.

3   The halogens (F, Cl, Br and I) form a well-defined group of elements.
   (a) Explain how the following support this statement:
      (i)   electron structure,
      (ii) redox behaviour,
      (iii) physical properties of the elements,
      (iv) usual oxidation state,
   (b) Describe four specific properties that show a regular gradation as the group is descended from F to I.
   (c) Explain the meaning of the term 'electron affinity'. How does this vary among the halogens, and how is it related to their reactivity?
   (d) Fluorine and fluorides show some properties which are not typical of the rest of the group. Mention three of these properties and suggest a reason or reasons for the difference.

4   Explain the following observations:
   (a) A mass spectrograph of chlorine shows five particles with relative masses of 35, 37, 70, 72 and 74.
   (b) As Group VII of the periodic table is descended, the halogens become weaker oxidising agents.
   (c) In its compounds, fluorine shows only one oxidation state, whereas chlorine shows several.
   (d) Hydrogen fluoride has a higher boiling point than hydrogen chloride and hydrogen iodide.

5   The percentage of copper in a sample of brass was determined as follows, 2.0 g of the brass was converted to 200 cm^3 of a solution of copper(II) nitrate free from nitric or nitrous acid and acidified with ethanoic (acetic) acid. 20.0 cm^3 of this solution liberated sufficient iodine from potassium iodide solution to react with 25.0 cm^3 of 0.1 M sodium thiosulphate solution. (Cu = 64, the reaction between $Cu^{2+}(aq)$ and $I^-(aq)$ can be represented as
   $2Cu^{2+}(aq) + 4I^-(aq) \rightarrow 2CuI(s) + I_2(aq)$.)
   (a) Write an equation or half-equations for the reaction between iodine and sodium thiosulphate.
   (b) How many moles of I_2 were liberated by 20 cm^3 of aqueous Cu^{2+} solution?
   (c) How many moles of copper are there in 200 cm^3 of aqueous Cu^{2+} solution?
   (d) What is the percentage by mass of copper in the brass?

6   Read the second half of section 17.12 concerning pesticides.
   (a) What is the correct systematic name for BHC?
   (b) Why are chlorinated pesticides such as BHC and DDT soluble in fat, but not in water?
   (c) Seed-eating birds are extremely vulnerable to chlorinated pesticides sprayed on spring crops. Why is there much less danger to these birds when autumn-sown crops are treated with pesticides?
   (d) Suggest an explanation for the pesticide action of DDT.
   (e) Why can insects develop a tolerance to small doses of chemicals such as DDT?
   (f) Suppose you were given the task of synthesising the ideal insecticide for spraying on spring-sown wheat. Make a list of the properties that you would look for in your product.

18 Group IV— Carbon to Lead, Non-Metal to Metal

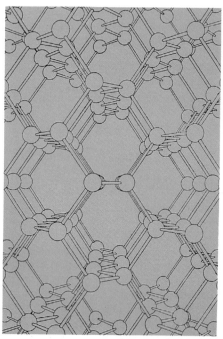

An unusual view through the structure of diamond.

Right. A rough diamond being cut to form a gem. The cast-iron surface below the diamond rotates 2500 turns per minute. It is coated with a paste of diamond powder in olive oil.

Far right. A diamond-impregnated saw being used to cut glass.

18.1 Introduction

The similarity between elements in the same family which was so obvious in Groups I, II and VII is much less apparent in Group IV. Here there is a considerable change in the character of the elements as atomic number rises. Carbon is unquestionably a non-metal. Silicon and germanium are metalloids. Tin and lead show typical metallic properties (table 18.1).

Although carbon occurs naturally as diamonds (Brazil and South Africa) and as graphite (Sri Lanka, Germany and USA), it occurs much more abundantly as carbon compounds. These are present in coal, oil and natural gas, in limestone and other carbonates and in living things.

Small synthetic diamonds produced by the General Electric Company, USA, in 1971. These were the first gem-quality diamonds ever created. No price could be put on their value if research and other intangible costs were included.

A view of the planes of atoms in the structure of graphite.

Table 18.1
The elements in Group IV

| C |
| Si |
| Ge |
| Sn |
| Pb |

Group IV

The properties and uses of diamond and graphite were discussed in section 10.6. The high refractive index and dispersive power of diamonds led to their use as jewellery in the very earliest civilisations. The modern industrial uses of diamond almost all result from its hardness—drilling, cutting, grinding and for bearings in precision instruments such as watches.

The uses of graphite are related to its conducting and lubricating qualities. It is used in 'lead' pencils, as a high-temperature lubricant and as inert electrodes in various electrolytic processes. It is also used as a moderator for slowing down neutrons in nuclear reactors.

After oxygen, silicon is the most abundant element on the Earth. It occurs as silicon(IV) oxide (SiO_2) in sand and sandstone and as many forms of silicate in rocks and clays. In comparison with carbon and silicon, germanium, tin and lead are rare elements. Traces of germanium are present in coal and accumulate in flue dust as germanium(IV) oxide, GeO_2. Silicon and germanium are used as semi-conductors in transistorised electronic equipment. These two elements are obtained from their oxides, SiO_2 and GeO_2, by very similar processes.

The flow scheme in figure 18.1 shows how pure silicon is obtained industrially. The final zone-refining process allows the production of ultra-pure silicon or germanium. These are necessary for use in transistors. The material to be purified is packed in a cylindrical tube and suspended vertically (figure 18.2). At the top of the tube, a short length of the cylinder is surrounded by an electrical heating coil. This melts a narrow band of material within the tube. During the zone-refining operation, the tube is raised through the heating element and the zone of molten material moves slowly down the tube. As the sample moves away from the heating element, it begins to recrystallise. Impurities are left in the molten zone. In this way, the impurities collect in the molten phase and end up concentrated at one end of the tube.

Figure 18.1
A flow scheme showing the industrial manufacture of pure silicon.

mechanical device lifts tube
slowly through the heating coil

sample packed in cylindrical
tube

purified crystals

molten zone

heating coil

impure crystals

Figure 18.2
The production of ultrapure silicon by zone refining.

- Write equations for
  (i)   the reduction of SiO_2 with carbon,
  (ii) the reaction of Cl_2 with Si.
- Why is the impure silicon converted to $SiCl_4$?
- Why can the impure silicon not be zone-refined?

Tin occurs as tin(IV) oxide (SnO_2) in cassiterite (tinstone). Lead occurs as the sulphide ore, galena (PbS). Lead is obtained by roasting the sulphide in air to obtain PbO, which is then reduced with carbon:

$$2\,PbS + 3\,O_2 \rightarrow 2\,PbO + 2\,SO_2$$

$$PbO + C \rightarrow Pb + CO$$

Tin can be obtained by direct reduction of SnO_2 with carbon:

$$SnO_2 + 2C \rightarrow Sn + 2CO$$

Tin and lead have important uses as relatively inert metals. Tin is used to tin-plate steel. This is then used to make the so called 'tins' for canning meats, soup, fruit, etc. Lead has been used as an inert material for gas and water pipes, for cable sheathing and for chemical vessels. Lead is also used for the plates of lead–acid accumulators (batteries) and as a screen from radioactivity.

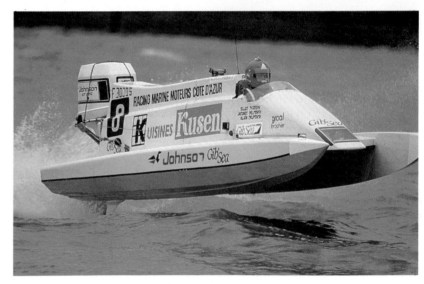

This speed boat has a hull reinforced with carbon fibre.

Another important use of both tin and lead is in alloying. Indeed, there are several important alloys containing both of these metals, for example:

solder (50% Sn 50% Pb),

pewter (80% Sn, 20% Pb) and

type metal (10% Sn, 75% Pb, 15% Sb).

Carbon-fibre cloth.

18.2 Variation in the physical properties of the elements

Table 18.2 shows a list of physical properties for the elements in Group IV. Notice how many of these properties vary more from one element to the next than with Group I or Group VII. These changes in property are, of course, related to the *increasing metallic (electropositive) character and the decreasing non-metallic (electronegative) character as atomic number rises*.

The structural changes from giant molecular lattices in carbon and silicon (see section 10.6) to giant metallic structures in tin and lead help to explain the changes in physical properties. Silicon and germanium crystallise in the same structure as diamond. Tin and lead have distorted close-packed metal structures. As the atoms get larger and the atomic radius increases, the interatomic bonding

Table 18.2
Various atomic and physical properties of the elements in Group IV

Element	C	Si	Ge	Sn	Pb
Atomic number	6	14	32	50	82
Electron configuration (outer shell only)	$2s^2\,2p^2$	$3s^2\,3p^2$	$4s^2\,4p^2$	$5s^2\,5p^2$	$6s^2\,6p^2$
Atomic radius/nm	0.077	0.117	0.122	0.141	0.154
Melting point/°C	3730^d	1410	937	232	327
Boiling point/°C	4830^d	2680	2830	2270	1730
Density/g cm^{-3}	2.26$^{gr.}$ 3.51$^{d.}$	2.33	5.32	7.3	11.4
Thermal conductivity/J cm^{-1} s^{-1} K^{-1}	0.24$^{gr.}$	0.84	0.59	0.63	0.35
Conductivity	fairly good$^{gr.}$ non-cond.$^{d.}$	semi-cond.	semi-cond.	good	good
Electrical conductivity/ohm^{-1} m^{-1}		1×10^6	2×10^6	8×10^6	5×10^6
Heat of atomisation/kJ mole^{-1}	716^{gr}	456	376	302	195
First ionisation energy/kJ mole^{-1}	1086	787	760	707	715
Type of structure	Giant molecular	Giant molecular similar to diamond	Giant molecular similar to diamond	Giant metallic	Giant metallic

gr. = graphite   d. = diamond

becomes weaker. As a result the attraction of neighbouring nucleii for intervening electrons gets less.

The weaker interatomic forces result in a change in bonding from covalent to metallic down the Group. Hence there is a decrease in melting point, boiling point, heat of atomisation and first ionisation energy. At the same time, the increasing metallic character causes a general increase in density and conductivity.

The first ionisation energy decreases considerably from carbon to silicon. After that, it falls relatively little. The reason for this is that, after silicon, there is a larger increase in nuclear charge. This is associated with the filling of 'd' and 'f' sub-shells and it counterbalances the increase in atomic radius.

18.3 Variation in the chemical properties of the elements

Information about the chemical properties of the Group IV elements is given in table 18.3. *The group trends further emphasise the increase in metallic character down the group.* Notice how carbon, the first member of the group, is much more electronegative than the remainder.

In general, chemical reactivity increases from carbon to lead. Electrode potentials show that only tin and lead are strong enough reducing agents to liberate hydrogen from dilute acids.

$$Pb(s) \rightarrow Pb^{2+}(aq) + 2e^- \qquad E^{\ominus} = 0.13 \text{ volts}$$

The twelfth-century leaded windows in the cathedral at Le Mans, France. Lead is an ideal material for holding the different-coloured pieces of glass in place on account of its malleability and inertness.

Table 18.3
Various atomic and chemical properties of the elements in Group IV

Element	C	Si	Ge	Sn	Pb
Atomic number	6	14	32	50	82
Electron configuration (outer shell only)	$2s^2\,2p^2$	$3s^2\,3p^2$	$4s^2\,4p^2$	$5s^2\,5p^2$	$6s^2\,6p^2$
Electronegativity	2.5	1.8	1.8	1.8	1.8
Electrode potential $M^{2+}(aq)/M(s)$/volts			0.23	−0.14	−0.13
$\Delta H_f^{\ominus}(XO_2)$/kJ mole^{-1}	−394	−910	−551	−581	−277
$\Delta H_f^{\ominus}(XO)$/kJ mole^{-1}	−111		−212	−286	−217
$\Delta H_f^{\ominus}(XH_4)$/kJ mole^{-1}	−75	+34	+90	+163	
Bond energy, $E(X{-}H)$/kJ mole^{-1}	435	318	285	251	
$\Delta H_f^{\ominus}(XCl_4)$/kJ mole^{-1}	−136	−640	−544	−511	−320
Bond energy, $E(X{-}Cl)$/kJ mole^{-1}	327	402	339	314	235

Lead will react very slowly with soft water containing dissolved oxygen to form $Pb(OH)_2$. This reaction has resulted in cases of lead poisoning in areas where householders have drunk hot water directly from lead pipes. The same problem does not arise in hard-water areas. Here the lead piping develops a protective layer of insoluble lead sulphate or lead carbonate.

As the heats of formation would suggest, all Group IV elements except lead react with oxygen on heating to form the dioxide.

$$C + O_2 \rightarrow CO_2$$
$$Sn + O_2 \rightarrow SnO_2$$

Lead, however, reacts to form PbO.

The heats of formation of the hydrides, XH_4, would suggest that only carbon could react directly with hydrogen. In practice, carbon does not react with hydrogen, even at very high temperatures, because of the large activation energy involved.

All the elements in Group IV react directly on heating with chlorine. They form the tetrachloride, except lead which forms the dichloride (see $\Delta H_f^{\ominus}(XCl_4)$ values in table 18.3). However, the reaction between carbon and chlorine is so slow that CCl_4 (tetrachloromethane) is manufactured by the reaction between carbon disulphide and chlorine:

$$CS_2 + 3\,Cl_2 \rightarrow CCl_4 + S_2Cl_2$$

18.4   General features of the compounds

The most striking feature of the compounds of these elements is the existence of two oxidation states, +2 and +4. The formation of compounds in which the elements show more than one stable oxidation number is typical of the p-block elements.

It is interesting to consider the relative stabilities of the +2 and +4 oxidation states for the different elements. In carbon and silicon compounds, the +4 state is very stable relative to +2. The +2 state is rare and easily oxidised to +4. Thus, CO reacts very exothermically to form CO_2. SiO is too unstable to exist under normal conditions, although it has been obtained at $2000°C$.

Germanium forms oxides in both +4 and +2 states. However, GeO_2 is rather more stable than GeO. GeO_2 does not act as an oxidising agent and GeO is readily converted to GeO_2.

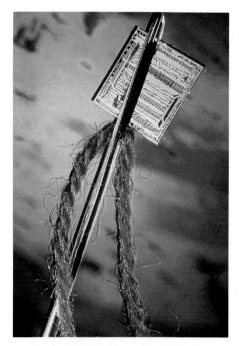

Through the eye of a needle!
This photograph shows the extremely small size of some 'silicon chips'. The 'chip' shown carries a very complex electrical circuit with 120 components. 'Silicon chips' such as this are used extensively in transistors and other electrical equipment. The material which looks like 'rope' is ordinary sewing cotton.

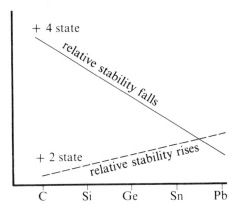

Figure 18.3
Relative stabilities of the +4 and +2 oxidation states for the elements in Group IV.

In tin compounds, the +4 state is only slightly more stable than the +2 state. Thus, aqueous tin(II) ions are mild reducing agents. They will convert mercury(II) ions to mercury and iodine to iodide:

$$Sn^{2+} + Hg^{2+} \rightarrow Sn^{4+} + Hg$$

$$Sn^{2+} + I_2 \rightarrow Sn^{4+} + 2I^-$$

In lead compounds, however, the +2 state is unquestionably more stable. PbO_2 is a strong oxidising agent, whilst PbO is relatively stable. Thus, PbO_2 can oxidise hydrochloric acid to chlorine and hydrogen sulphide to sulphur.

$$PbO_2 + 4HCl \rightarrow PbCl_2 + Cl_2 + 2H_2O$$

Notice the steady increase in the stability of the lower oxidation state relative to the higher oxidation state on moving down the group from carbon to lead (figure 18.3).

The greater stability of the +2 oxidation state with respect to the +4 state as the atomic number rises is nicely illustrated by the standard electrode potentials of the $M^{4+}(aq)/M^{2+}(aq)$ systems for germanium, tin and lead.

$$Ge^{4+} + 2e^- \rightarrow Ge^{2+} \qquad E^{\ominus} = -1.6 \text{ volts}$$

$$Sn^{4+} + 2e^- \rightarrow Sn^{2+} \qquad E^{\ominus} = +0.15 \text{ volts}$$

$$Pb^{4+} + 2e^- \rightarrow Pb^{2+} \qquad E^{\ominus} = +1.8 \text{ volts}$$

As the electrode potentials get more positive from Ge^{4+} to Pb^{4+}, the oxidised form is more readily reduced to the +2 state.

All the Group IV elements have four electrons in their outermost shell (ns^2np^2). So, it is not surprising that they show a well-defined oxidation state of +4 (−4 in the hydrides). However, none of the elements forms an M^{4+} cation in its solid compounds. This is due to the high ionisation energies involved in removing four successive electrons from an atom. Consequently, the bonding in the tetravalent compounds is predominantly covalent.

Compounds of tin and lead in which the Group IV element has an oxidation number of +2 (e.g. PbF_2, $PbCl_2$, PbO) are normally regarded as ionic. In these compounds, the Sn^{2+} and Pb^{2+} ions are believed to form by loss of the two '*p*' electrons in the outer shell. The two '*s*' electrons remain relatively stable and unreactive in their filled sub-shell. This is sometimes referred to as the **'inert pair' effect**.

One of the most curious features regarding Group IV is the unique ability of carbon to form stable compounds containing long chains and rings of carbon atoms. This property is called **catenation**. It results in carbon forming an enormous range of compounds. Indeed, there are many thousands of compounds containing only carbon and hydrogen of which methane (CH_4), ethane (C_2H_6), ethene (C_2H_4) and ethyne (C_2H_2) are four of the simplest.

This ability of carbon to catenate results from the fact that the C—C bond is almost as strong as the C—O bond (table 18.4). This makes the oxidation of carbon compounds to such products as carbon dioxide and water less energetically favourable than in the case of silicon. With silicon, the Si—Si bond is much weaker than the Si—O bond. Consequently, catenated compounds of silicon are energetically unstable with respect to SiO_2 and therefore do not occur naturally. Nevertheless, chemists have succeeded in synthesising a whole series of silicon hydrides, called silanes. Some of these have as many as 11 silicon atoms linked together. In a similar fashion, three hydrides have been synthesised for germanium. But, tin and lead form only one hydride each, SnH_4 and PbH_4. The ability of carbon to form long chains is considered further in section 26.1.

Figure 18.4
The molecular shape of the simple hydrides of Group IV elements.

- What structure and shape do you predict for the hydrides CH_4, SiH_4, GeH_4, SnH_4 and PbH_4?
- How will the volatility of these hydrides vary from CH_4 to PbH_4?
- The hydrides become less stable from CH_4 to PbH_4. Why is this?
- Suggest a reason why there are no compounds of Si, Ge, Sn and Pb analogous to C_2H_4 and C_2H_2.

Some of the important properties of the hydrides, chlorides and oxides of the Group IV elements are summarised in tables 18.5, 18.6 and 18.7 respectively.

Table 18.4
Average energies for corresponding carbon and silicon bonds

Bond	Average bond energy $E(X—Y)/kJ\ mole^{-1}$
C—C	346
C—O	360
Si—Si	226
Si—O	464

Table 18.5
The preparation and properties of the simple hydrides of Group IV elements

Preparation

1. Reaction of magnesium compounds (e.g. magnesium silicide) or magnesium alloy with dilute acid

$$Mg_2Si + 4H^+ \rightarrow 2Mg^{2+} + SiH_4$$

2. Reduction of tetrachloride with lithium aluminium hydride at 0°C in ether

$$SnCl_4 + LiAlH_4 \rightarrow SnH_4 + LiCl + AlCl_3$$

Properties	
Structure:	Simple molecular.
Molecular shape:	Tetrahedral (figure 18.4).
Volatility:	Low melting points and boiling points, all are gases at room temperature.
Thermal stability:	The X—H bonds become longer and weaker down the Group (table 18.3). Thus, the hydrides become less stable. CH_4, SiH_4 and GeH_4 are stable up to high temperatures. SnH_4 is unstable and decomposes slowly at room temperature. PbH_4 is so unstable that it cannot be isolated at room temperature.
Reducing powers:	As their ease of decomposition into the element plus hydrogen increases, they become stronger reducing agents. Thermodynamically, the hydrides should all react with oxygen since their heats of combustion are all negative. However, CH_4 is kinetically stable due to the high activation energy for its reaction with oxygen.

Table 18.6
The preparation and properties of the tetrachlorides of Group IV elements

Preparation

1. Direct synthesis for $SiCl_4$, $GeCl_4$ and $SnCl_4$

$$Si + 2Cl_2 \rightarrow SiCl_4$$

2. CCl_4 is prepared more easily by the action of Cl_2 on CS_2

$$CS_2 + 3Cl_2 \rightarrow CCl_4 + S_2Cl_2$$

3. $PbCl_4$ is prepared by the reaction of cold conc. HCl with PbO_2.

Properties	
Structure:	Simple molecular.
Molecular shape:	Tetrahedral (similar to XH_4 hydrides).
Volatility:	Low melting points and boiling points. All are volatile liquids at room temperature.
Thermal stability:	As the X—Cl bonds become longer and weaker down the group, the tetrachlorides become less stable. Thus, CCl_4, $SiCl_4$ and $GeCl_4$ are stable even at high temperatures. $SnCl_4$ decomposes on heating to form $SnCl_2 + Cl_2$. $PbCl_4$ decomposes readily to form $PbCl_2 + Cl_2$.
Hydrolysis:	All the chlorides (except CCl_4) are readily hydrolysed to form hydroxy compounds + HCl

$$SnCl_4 + 4H_2O \rightarrow Sn(OH)_4 + 4HCl$$

Table 18.7
The preparation and properties of the oxides of the Group IV elements

The dioxides   Group IV element in the +4 oxidation state

Oxide	CO_2	SiO_2	GeO_2	SnO_2	PbO_2
Preparation	Direct combination of element and oxygen				Electrolytic oxidation of Pb^{2+} in acid solution
Properties Boiling point/°C	−78	2590	1200	1900	decomposes on heating
Structure:	Simple molecular	Giant molecular	Intermediate between giant molecular and ionic		
Nature:	Acidic		Amphoteric		

react with aq. alkalis giving XO_3^{2-} salts
$$CO_2 + 2OH^- \rightarrow CO_3^{2-} + H_2O$$
$$\text{carbonate}$$
$$SiO_2 + 2OH^- \rightarrow SiO_3^{2-} + H_2O$$
$$\text{silicate}$$

(i) react with fused alkalis giving XO_3^{2-} salts
$$SnO_2 + 2OH^- \rightarrow SnO_3^{2-} + H_2O$$
$$\text{stannate(IV)}$$
$$PbO_2 + 2OH^- \rightarrow PbO_3^{2-} + H_2O$$
$$\text{plumbate(IV)}$$
(ii) react with conc. acids forming +4 salts
$$SnO_2 + 4H^+ \rightarrow Sn^{4+}(aq) + 2H_2O$$
$$PbO_2 + 4HCl \rightarrow PbCl_4 + 2H_2O$$
$$\text{conc.}$$

Thermal stability:	stable even at high temps.				Decomposes to PbO on warming $PbO_2 \rightarrow$ $PbO + \frac{1}{2}O_2$

The monoxides   Group IV elements in the +2 oxidation state

Oxide	CO	SiO	GeO	SnO	PbO
Preparation	Reduction of CO_2 with carbon $CO_2 + C \rightarrow 2CO$	Only exists at very high temps.	Reduction of GeO_2 with Ge	Heat appropriate hydroxide or nitrate $Pb(OH)_2 \rightarrow PbO + H_2O$ $Pb(NO_3)_2 \rightarrow PbO + 2NO_2 + \frac{1}{2}O_2$	
Properties Boiling point/°C	−191			1470	
Structure:	Simple molecular		Predominantly ionic		
Nature:	Neutral oxides— react with neither acids nor alkalis.		Amphoteric oxides		

(i) react with acids to form salts
$$PbO + 2H^+ \rightarrow Pb^{2+} + H_2O$$
$$SnO + 2H^+ \rightarrow Sn^{2+} + H_2O$$
(ii) react with alkalis to form salts
$$PbO + OH^- + H_2O \rightarrow Pb(OH)_3^-$$
$$\text{trihydroxoplumbate(II)}$$
$$SnO + OH^- + H_2O \rightarrow Sn(OH)_3^-$$
$$\text{trihydroxostannate(II)}$$

Thermal stability:	Readily oxidised to dioxide. (SiO, GeO and SnO revert to dioxide on standing in air.)			Stable.	

Look closely for the following points in these tables:

1 The simple molecular structures of the hydrides and the tetrachlorides.

2 The decreasing thermal stability from C to Pb of the hydrides, tetrachlorides and dioxides.

3 The change in the nature of the dioxides down the group from acidic to amphoteric.

4 The change in the nature of the monoxides down the group from neutral to amphoteric.

Drikold (solid carbon dioxide) has been sprayed onto the base of this huge axle. The Drikold causes the axle to contract so that it can be fitted onto the wheel on which the workers are standing. As the axle warms up, it will expand and fit the wheel tightly.

Summary

1 The elements in Group IV emphasise the *differences* between one element and the next in a particular group of the periodic table.

2 Catenation is the ability of one element to form chains or rings in which its atoms are bonded to one another.

3 As we go down the group from carbon to lead:
   (a) the elements change from non-metallic to metallic. This change in the elements is reflected in their properties and uses.
   (b) the +2 oxidation state becomes more stable relative to the +4 oxidation state.
   (c) the bonding in compounds changes from covalent to ionic. Covalent compounds (e.g. hydrides and tetrachlorides) become less stable down the group.
       Ionic compounds (e.g. monoxides) become more stable down the group.
   (d) the nature of the dioxides changes from acidic to amphoteric.
   (e) the nature of monoxides changes from neutral to amphoteric.

Study questions

1 Discuss the following points with respect to the elements in Group IV of the periodic table. Write equations for any reactions which you mention.
   (a) The reactivity of the elements.
   (b) The state, structure, and thermal stability of the hydrides.
   (c) The state, structure, thermal stability and hydrolysis of the tetrachlorides.
   (d) The acidic/basic character of the dioxides.
   (e) The relative stabilities of the +2 and +4 oxidation states.

Watercolour paintings containing lead (II) compounds as constituents of their pigments darken over many decades. This is caused by the reaction of traces of hydrogen sulphide in the atmosphere with the lead (II) salts.

The 'white lead' pigment which contains lead (II) carbonate becomes darker owing to the formation of black lead (II) sulphide. The original colour of the pigment can be restored by oxidation of the black lead (II) sulphide to white lead (II) sulphate using hydrogen peroxide. The photographs above show a watercolour painting by Jacques Le Moyne, c.1586, before and after restoration. In this case the petals of the flowers, which had turned black, were restored to their original white appearance.

2 (a) What is the principal ore of tin?
(b) Explain briefly, with an equation, how tin is obtained from its principal ore.
(c) What are the main uses of tin and its alloys?
(d) What are the relative advantages and disadvantages of tin-plating and galvanising (zinc-plating) iron in order to prevent corrosion?

3 (a) How and under what conditions does lead react with:
(i)   air (oxygen),
(ii) water (soft and hard),
(iii) hydrochloric acid,
(iv) nitric acid?
(b) Describe briefly the preparation of tin(II) oxide and tin(IV) oxide from tin.
(c) How (if at all) and under what conditions do tin(II) oxide and tin(IV) oxide react with:
(i)   oxygen,
(ii) hydrochloric acid,
(iii) sodium hydroxide?

4 The following passage describes the preparation of tin(IV) iodide.
Add 4.0 g of powdered tin to a solution of 12.7 g of iodine in 100 cm³ of tetrachloromethane. Reflux the mixture gently until the reaction is complete. Now filter the mixture through a pre-heated funnel. Wash the residue with hot tetrachloromethane, adding the washings to the filtrate. Cool the filtrate in ice until orange crystals of tin(IV) iodide form. Filter and dry the crystals.
(a) Write an equation for the formation of tin(IV) iodide in the above preparation.
(b) Give two reasons for using CCl_4 as the solvent in this preparation.
(c) Calculate the maximum possible yield of tin(IV) iodide. Explain your calculation.
(d) How would you know when the reaction was complete?
(e) Why was the mixture filtered through a pre-heated funnel?
(f) Why was the residue washed with hot CCl_4?
(g) What would you predict for
(i)   the structure,
(ii) the thermal stability
of tin(IV) iodide?

5 (a) Draw electron dot/cross diagrams for carbon monoxide and carbon dioxide. (You need only show the electrons in the outer shells of the constituent atoms.)
(b) 'Carbon monoxide is iso-electronic with nitrogen'. Explain what is meant by this statement.
(c) Use the standard heats of formation of CO_2 and CO in table 18.3 to find the enthalpy changes of the following reactions.

$$CO_2(g) + C(s) \rightarrow 2CO(g)$$
$$CO(g) + \tfrac{1}{2}O_2(g) \rightarrow CO_2(g)$$

Comment on the relative stabilities of CO_2 and CO.

6 (a) State four important similarities in the chemistry of tin and lead or their corresponding compounds.
(b) Explain the following in terms of atomic or electronic properties.
(i)   Tin(IV) compounds are more stable than tin(II) compounds.
(ii) Lead(II) compounds are more stable than lead(IV) compounds.
(iii) $PbCl_4$ has a simple molecular structure, whereas $PbCl_2$ is ionic.
(c) Tin and lead have variable oxidation states and form complex ions. Why then are they not classed as transition metals?
(d) Oil paintings containing lead(II) compounds in their pigments darken over many decades. This is due to the reaction of traces of hydrogen sulphide in the atmosphere with the lead(II) salts.
(i)   Write an equation for the darkening process.
(ii) Explain why dilute solutions of hydrogen peroxide can be used to restore the oil paintings.

19

The Transition Metals

19.1 Introduction

The elements from scandium (atomic number 21) to zinc (atomic number 30) in the periodic table form what is generally regarded as the first sequence of transition elements. But, what exactly is a transition element? And why is it that these elements have similarities to each other across their period, unlike other parts of the periodic table?

The answer to these questions concerning the transition metals lies in their electronic structures shown in table 19.1. Indeed, *virtually all the properties of transition elements are related to their electronic structures and the relative energy levels of the orbitals available for their electrons.*

Table 19.1
The electron structures of atoms and ions of the elements K to Zn ((Ar) $\equiv$ electron structure of argon)

Element	Symbol	Electronic structure of atom	Common ion	Electronic structure of ion
potassium	K	$(Ar)4s^1$	K^+	(Ar)
calcium	Ca	$(Ar)4s^2$	Ca^{2+}	(Ar)
scandium	Sc	$(Ar)3d^1 4s^2$	Sc^{3+}	(Ar)
titanium	Ti	$(Ar)3d^2 4s^2$	Ti^{4+}	(Ar)
vanadium	V	$(Ar)3d^3 4s^2$	V^{3+}	$(Ar)3d^2$
chromium	Cr	$(Ar)3d^5 4s^1$	Cr^{3+}	$(Ar)3d^3$
manganese	Mn	$(Ar)3d^5 4s^2$	Mn^{2+}	$(Ar)3d^5$
iron	Fe	$(Ar)3d^6 4s^2$	Fe^{2+}	$(Ar)3d^6$
			Fe^{3+}	$(Ar)3d^5$
cobalt	Co	$(Ar)3d^7 4s^2$	Co^{2+}	$(Ar)3d^7$
nickel	Ni	$(Ar)3d^8 4s^2$	Ni^{2+}	$(Ar)3d^8$
copper	Cu	$(Ar)3d^{10} 4s^1$	Cu^+	$(Ar)3d^{10}$
			Cu^{2+}	$(Ar)3d^9$
zinc	Zn	$(Ar)3d^{10} 4s^2$	Zn^{2+}	$(Ar)3d^{10}$

As the shells of electrons get further and further from the nucleus, successive shells become closer in energy. Thus, the difference in energy between the second and third shells is less than that between the first and second. By the time the fourth shell is reached, there is, in fact, an overlap between the third and fourth shells. In other words, from scandium onwards, the orbitals of highest energy in the third shell (i.e. the 3d orbitals) have higher energy than those of lowest energy in the fourth shell (the 4s orbital) (figure 19.1).

Orbitals in the 4th shell

Orbitals in the 3rd shell

4p (max.: 6 electrons in 3 orbitals)

3d (max.: 10 electrons in 5 orbitals)
4s (max.: 2 electrons in 1 orbital)

3p (max.: 6 electrons in 3 orbitals)

3s (max.: 2 electrons in 1 orbital)

Figure 19.1
Relative energy levels of the 3s, 3p, 3d, 4s and 4p orbitals.

The 3d sub-shell is 'on average' nearer the nucleus than the 4s sub-shell, but at a higher energy level. So, once the 3s and 3p sub-shells are filled at argon, further electrons enter the 4s sub-shell because it is at a lower energy level than the 3d sub-shell. Hence, potassium and calcium have, respectively, one and two electrons in the 4s sub-shell (table 19.1). Once the 4s sub-shell is filled at calcium, electrons enter the 3d level. Hence scandium has the electron structure $(Ar)3d^14s^2$, titanium has the electron structure $(Ar)3d^24s^2$ and so on (table 19.1 and figure 19.2).

Notice the unexpected electron structures for chromium and copper. The arrangement of electrons in chromium is $(Ar)3d^54s^1$ (figure 19.2) not $(Ar)3d^44s^2$ which we might have expected. The explanation of this anomaly is that the electron structure $(Ar)3d^54s^1$ with half-filled 3d and 4s sub-shells has a lower energy level than $(Ar)3d^44s^2$. The extra stability of a half-filled sub-shell is thought to result from the occupation of each orbital by one electron. This produces an equal distribution of charge around an atom.

Copper atoms have an electron structure $(Ar)3d^{10}4s^1$ (figure 19.2) rather than $(Ar)3d^94s^2$. In this case, it appears that $(Ar)3d^{10}4s^1$ with a filled 3d sub-shell and a half-filled 4s sub-shell is more stable than $(Ar)3d^94s^2$.

Some transition metals are used for jewellery. This photograph shows earrings made from anodised titanium. The anodised layer absorbs and reflects the various colours of light slightly differently. This produces different bands of colour on the earrings.

Figure 19.2
The 'electrons-in-boxes' representation of the electronic structures of certain transition metals.

Element		Electronic structure					
		3d					4s
Scandium	(Ar)	↑					↑↓
Titanium	(Ar)	↑	↑				↑↓
Chromium	(Ar)	↑	↑	↑	↑	↑	↑
Iron	(Ar)	↑↓	↑	↑	↑	↑	↑↓
Copper	(Ar)	↑↓	↑↓	↑↓	↑↓	↑↓	↑

19.2 Ions of the transition metals

Look closely at table 19.1.

■ Which electrons does a calcium atom lose in forming a Ca^{2+} ion?
■ Which electrons would you expect an iron atom to lose in forming an Fe^{2+} ion? Which electrons are lost in practice by the Fe atom?
■ Why do you think Fe^{3+} ions are more stable than Fe^{2+} ions?

When transition metals form their ions, electrons are lost first from the $4s$ sub-shell rather than the $3d$ sub-shell. Thus, Fe^{2+} ions have the electron structure $(Ar)3d^6$ (figure 19.3) rather than $(Ar)3d^44s^2$. V^{3+} ions have the electron structure $(Ar)3d^2$ not $(Ar)4s^2$. At first sight, this seems very odd because prior to occupation by electrons, the $4s$ level is energetically more stable than the $3d$ level. However, once the $3d$ level is occupied by electrons, these repel the $4s$ electrons even further from the nucleus. The $4s$ electrons are therefore pushed to a higher energy level, higher in fact than the $3d$ level now occupied. Consequently, when transition metal atoms form ions, they lose electrons from the $4s$ level before the $3d$ level. This means that *all transition metals will have similar chemical properties* which are dictated by the behaviour of the $4s$ electrons in the outermost shell.

This horizontal similarity contrasts sharply with the marked changes across a period of the s- and p- block elements, say from sodium to argon.

Figure 19.3
The 'electrons-in-boxes' representation of the electronic structures of Fe, Fe^{2+} and Fe^{3+}.

19.3 What is a transition element?

The simplest answer to this question is to say that *transition elements are those in the 'd-block' of the periodic table*. The neatness of this definition lies in the fact that it emphasises the four blocks of elements in the periodic table. This division is useful for the s-block elements in Groups I and II which are so alike in their properties. It is much less satisfactory for the elements in the p-block which include metals such as aluminium, reactive non-metals within Group VII and the noble gases. Furthermore, the simple definition of transition metals as 'd-block elements' is rather unsatisfactory because it leads to the inclusion of scandium and zinc as transition elements. As you may have already realised, scandium and zinc show some fairly obvious differences to the elements in the sequence titanium to copper.

They have only one oxidation state in their compounds (scandium 3+, zinc 2+), whereas the others have two or more.

Their compounds are usually white, unlike the compounds of transition metals which are usually coloured.

They show little catalytic activity.

Owing to the divergence of scandium and zinc from typical transition metal properties, it would seem sensible to choose a more satisfactory definition for transition metals. This definition would exclude these elements, but include the elements titanium to copper. In order to achieve this, we can define a transition metal as *one which forms at least one ion with a partially filled d sub-shell*. Look closely at table 19.1 and you will see that neither Sc^{3+} nor Zn^{2+} has a partly filled *d* sub-shell.

19.4 Trends across the period of transition metals

In building up the elements from sodium to argon, electrons are being added to the outer shell and the nuclear charge is increasing by the addition of protons. The added electrons shield each other only weakly from the extra nuclear charge, so atomic radii decrease sharply from sodium to argon. At the same time the electronegativities and ionisation energies steadily rise. The increasing number of outer-shell electrons also results in major differences in structure and chemical properties from one element to the next.

Across the series of metals from scandium to zinc, however, the nuclear charge is also increasing, but electrons are being added to an *inner d* sub-shell. These inner *d* electrons shield the outer 4*s* electrons from the increasing nuclear charge much more effectively than outer shell electrons can shield each other. Consequently, the atomic radii decrease much less rapidly.

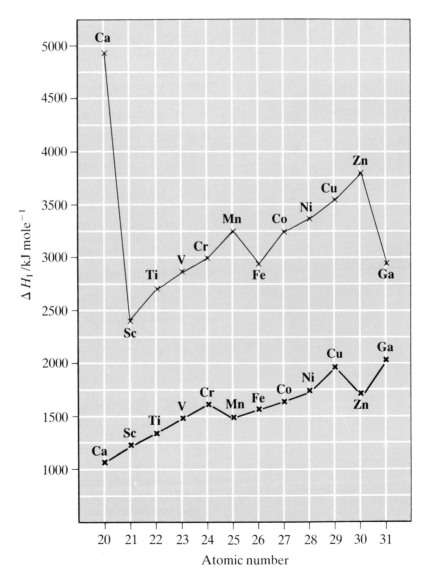

Figure 19.4
Graphs of the second and third ionisation energies of the elements from calcium to gallium.

Table 19.2
Electronegativities, ionisation energies and electrode potentials for the elements Sc to Zn

	Sc	Ti	V	Cr	Mn	Fe	Co	Ni	Cu	Zn
Metallic (atomic) radius/nm	0.16	0.15	0.14	0.13	0.14	0.13	0.13	0.13	0.13	0.13
Electronegativity	1.2	1.3	1.45	1.55	1.6	1.65	1.7	1.75	1.75	1.6
First ionisation energy/kJ mole^{-1}	630	660	650	650	720	760	760	740	750	910
Second ionisation energy/kJ mole^{-1}	1240	1310	1410	1590	1510	1560	1640	1750	1960	1700
Third ionisation energy/kJ mole^{-1}	2390	2650	2870	2990	3260	2960	3230	3390	3560	3800
Electrode potential for $M^{2+}(aq) + 2e^- \rightarrow M(s)$/V			−1.20	−0.91	−1.19	−0.44	−0.28	−0.25	+0.34	−0.76
Electrode potential for $M^{3+}(aq) + 3e^- \rightarrow M(s)$/V	−2.1	−1.2	−0.86	−0.74	−0.28	−0.04	+0.40			

Similarly, electronegativities and ionisation energies increase from scandium to zinc, but only marginally compared with the increase across period 3 from sodium to argon (table 19.2).

The increasing electronegativity from scandium to copper means that the elements become slightly less metallic. This is reflected in the increasingly positive electrode potentials (sections 14.3 and 14.4) of their M^{2+} and M^{3+} ions. Look closely at figure 19.4 which shows graphs of the second and the third ionisation energies of the elements from calcium to gallium.

■ Write the electronic structure for the following:
(a) Cu (b) Cu$^+$ (c) Cu^{2+} (d) Cr (e) Cr$^+$ (f) Mn^{2+} (g) Zn^{2+}.
(Use (Ar) to represent the electron structure of argon as we have done before.)

■ By referring to electronic structures explain why:
(a) the second ionisation energies of both Cr and Cu are higher than those of the next element.
(b) the third ionisation energies of both Mn and Zn are higher than those of the next element.

Remember that the second ionisation energy refers to the process

$$M^+(g) \rightarrow M^{2+}(g) + e^-$$

and the third ionisation to

$$M^{2+}(g) \rightarrow M^{3+}(g) + e^-$$

19.5 General properties of the first transition series (Sc to Zn)

Most of the transition metals have a close-packed structure in which each atom has twelve nearest neighbours. Furthermore, transition metals have a relatively low atomic radius because the electrons being added to the $3d$ sub-shell are nearer the nucleus than the electrons in the outermost $4s$ orbital. Consequently, the double effect of close packing and small atomic size results in strong metallic bonds between atoms. Hence, *the transition metals have higher melting points, higher boiling points, higher densities and higher heats of fusion and vaporisation than metals such as potassium and calcium in the s-block of the periodic table*

(table 19.3). The strong interatomic bonding in transition metals is also reflected in high tensile strengths and good mechanical properties.

The transition metals are less electropositive than the s-block metals. However, their electrode potentials indicate that all of them (except copper) should react with dilute solutions of strong acids, such as 1 M HCl. In practice, however, many of the metals react only slowly with dilute acids. This is due to protection of the metal from chemical attack by a thin impervious and unreactive layer of oxide. Chromium provides an excellent example of this. Despite its electrode potential, it can be used as a protective, non-oxidising, non-corroding metal owing to the presence of an unreactive layer of Cr_2O_3. This can be compared with the protection of aluminium by a layer of Al_2O_3.

The ions of transition metals are smaller than those of the s-block metals in the same period (table 19.3). Owing to their smaller ionic radii and also to a larger charge in many of their ions, the charge/radius ratios for transition metals are greater than the values for s-block metals. The small, highly charged cations also cause greater polarisation of associated anions. These factors result in the following properties of transition metal compounds compared to those of the s-block metals:

(a) their oxides and hydroxides in oxidation states 2+ and 3+ are less basic and less soluble;
(b) their salts are less ionic and less thermally stable;
(c) their salts and aqueous ions are more hydrated and more readily hydrolysed forming acidic solutions;
(d) their ions are more easily reduced.

It is interesting to compare these features of transition metal ions and their compounds with those of aluminium discussed in chapter 16.

Although the compounds of transition elements in oxidation states +2 and +3 are usually regarded as ionic, polarisation of anions by small, highly charged cations is undoubtedly apparent. Some of the oxides begin to show acidic features and the compounds begin to show covalent bonding. Thus, Cr_2O_3 and Mn_2O_3 are amphoteric and Fe_2Cl_6 is regarded as a molecular solid. Of course, these features begin to predominate in the higher oxidation states. Here the oxides become increasingly acidic and the compounds increasingly molecular. Hence, CrO_3 and Mn_2O_7 are regarded as simple molecular, acidic oxides.

There are relatively small changes in ionic radii from scandium to copper. Because of this, compounds of the simple hydrated +2 and +3 ions have very similar crystalline structures, hydration and solubility. Thus, all the M^{3+} ions form an alum of the type $K_2SO_4.M_2(SO_4)_3.24H_2O$ and all the M^{2+} ions form isomorphous double sulphates of formula $(NH_4)_2SO_4.MSO_4.6H_2O$.

Table 19.3
Physical properties of the elements K to Zn

| Element | s-block metals | | transition metals | | | | | | | | | |
	K	Ca	Sc	Ti	V	Cr	Mn	Fe	Co	Ni	Cu	Zn
Atomic radius/nm	0.24	0.20	0.16	0.15	0.14	0.13	0.14	0.13	0.13	0.13	0.13	0.13
Melting point/°C	64	850	1540	1680	1900	1890	1240	1540	1500	1450	1080	420
Boiling point/°C	770	1490	2730	3260	3400	2480	2100	3000	2900	2730	2600	910
Density/g cm^{-3}	0.86	1.54	3.0	4.5	6.1	7.2	7.4	7.9	8.9	8.9	8.9	7.1
Ionic radius/nm												
M$^+$	0.130											
M^{2+}		0.094		0.090	0.088	0.084	0.080	0.076	0.074	0.072	0.070	0.074
M^{3+}			0.081	0.076	0.074	0.069	0.066	0.064	0.063	0.062		

19.6 Characteristic properties of transition metals and their compounds

Variable oxidation states

Transition metals have electrons of similar energy in both the $3d$ and $4s$ levels. This means that one particular element can form ions of roughly the same stability by losing different numbers of electrons. Thus, all the transition metals from titanium to copper exhibit two or more oxidation states in their compounds.

The formulae of the common oxides and chlorides are shown in figure 19.5. All the oxidation numbers of the elements are given below these formulae. The more important oxidation states are emphasised by bold print. Notice that both scandium and zinc have only one common oxide, one common chloride, and one oxidation state in their compounds. Notice also, how closely the oxidation states of the elements in their common oxides and chlorides compare with the more important oxidation states of the elements.

	Sc	Ti	V	Cr	Mn	Fe	Co	Ni	Cu	Zn
Common oxides	Sc_2O_3	Ti_2O_3 TiO_2	V_2O_3 V_2O_5	Cr_2O_3 CrO_3	MnO MnO_2 Mn_2O_7	FeO Fe_2O_3	CoO Co_2O_3	NiO	Cu_2O CuO	ZnO
Common chlorides	$ScCl_3$	$TiCl_3$ $TiCl_4$	VCl_3	$CrCl_2$ $CrCl_3$	$MnCl_2$ $MnCl_3$	$FeCl_2$ Fe_2Cl_6	$CoCl_2$	$NiCl_2$	$CuCl$ $CuCl_2$	$ZnCl_2$
Oxidation numbers that occur in compounds	**3**	**4** **3** 2 1	**5** **3** 2 1	**6** 5 4 **3** 2 1	**7** 6 5 **4** 3 **2** 1	6 5 4 **3** **2** 1	5 4 **3** **2** 1	5 4 3 **2** 1	3 **2** **1**	**2**

Figure 19.5
Oxidation states of the elements Sc to Zn. (Common oxidation numbers are in bold print.)

The following generalisations emerge from a study of the oxidation states.

(a) *The common oxidation states for each element include +2 or +3 or both.* +3 states are relatively more common at the beginning of the series. +2 states are more common towards the end.

(b) The highest oxidation states up to manganese correspond to the involvement of all the electrons outside the argon core; (4 for Ti, 5 for V, 6 for Cr and 7 for Mn). After this, the increasing nuclear charge binds the d electrons more strongly. So, one of the more important oxidation states is that which involves the weakly held electrons in the outer $4s$ shell only (2 for Fe, 2 for Co, 2 for Ni and 1 for Cu).

(c) The transition metals usually exhibit their highest oxidation states in compounds with oxygen or fluorine. These are two of the most electronegative elements.

(d) Ti, V, Cr and Mn never form simple ions in their highest oxidation state since this would result in ions of extremely high charge density. Hence the compounds of these elements in their highest oxidation state are either covalently bonded (e.g. TiO_2, V_2O_5, CrO_3, Mn_2O_7) or contain complex ions (e.g. VO_3^-, CrO_4^{2-}, MnO_4^-).

One of the most beautiful and effective demonstrations of the range of oxidation states can be shown by shaking a solution containing a compound of vanadium(v) with zinc and dilute acid. The solution of vanadium(v) can be made by dissolving about 3 g of ammonium vanadate(v) (NH_4VO_3) in 40 cm^3 of 2 M NaOH and then adding 80 cm^3 of M H_2SO_4. This solution is a yellow colour. It contains dioxovanadium(v) ions, VO_2^+, in acid solution.

$$VO_3^-(aq) + 2H^+(aq) \rightarrow VO_2^+(aq) + H_2O(l)$$

When the yellow solution is shaken with granulated zinc or zinc amalgam, it changes gradually through green to blue oxovanadium(iv) ions, $VO^{2+}(aq)$. It then changes to green vanadium(iii) ions, $V^{3+}(aq)$, and finally to violet vanadium(ii) ions, $V^{2+}(aq)$ (figure 19.6).

Oxidation state	+5	+4	+3	+2
Colour in aqueous solution	yellow	blue	green	violet
Ion	VO_2^+	VO^{2+}	V^{3+}	V^{2+}
Name	dioxovanadium(v) ion	oxovanadium(iv) ion	vanadium(iii) ion	vanadium(ii) ion

Figure 19.6
The oxidation states of vanadium.

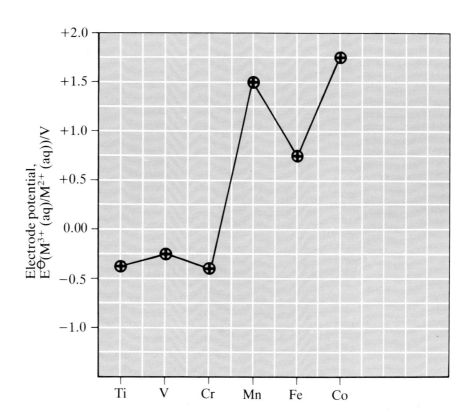

Figure 19.7
Redox potentials of the M^{3+}/M^{2+} systems for transition metals.

All the transition metals from titanium to copper exhibit oxidation numbers of +3 and +2 in their compounds. But what are the relative stabilities of the +3 and +2 states for the different elements? Why, for example, is manganese more stable in the +2 state than the +3 state, whilst the reverse is true for iron? Look closely at figure 19.7 which shows the electrode potentials of the $M^{3+}(aq)/M^{2+}(aq)$ systems for the elements from titanium to cobalt. There is a fairly steady rise in $E^{\ominus}$ values across the series. This is interrupted by what appears to be an abnormally high value for manganese and a low value for iron.

The more positive the value for $E^{\ominus}$, the more likely is the aqueous M^{3+} ion to become reduced to M^{2+}.

$$M^{3+}(aq) + e^- \rightarrow M^{2+}(aq)$$

Thus, low values of $E^{\ominus}$ for Ti, V, Cr and Fe indicate that the 3+ oxidation state is relatively more stable in these elements than in Mn and Co which have higher values of $E^{\ominus}$.

The relative stabilities of the +3 and +2 states in manganese and iron can be interpreted using an 'electrons-in-boxes' representation of their electronic structures (figure 19.8).

Mn^{2+} and Fe^{3+} each have half-filled $3d$ orbitals. This makes them more stable than Mn^{3+} and Fe^{2+}, respectively. Hence, in manganese the +2 state is more stable than +3, whereas in iron the +3 state is more stable than +2.

- Write the electronic structures of Cu, Cu^+ and Cu^{2+} using the 'electrons-in-boxes' notation.
- In which oxidation state would you expect copper to be more stable, +1 or +2?
- From experience you will know that copper compounds usually exist in the +2 state. What factors might increase the relative stability of the +2 state over the +1 state for copper? We shall return to this anomalous stability of copper(II) over copper(I) in the next chapter.

Figure 19.8
Electronic structures of manganese and iron and some of their respective ions.

Atom/Ion		Electronic structure					
				$3d$			$4s$
Mn	(Ar)	↑	↑	↑	↑	↑	↑↓
Mn^{2+}	(Ar)	↑	↑	↑	↑	↑	
Mn^{3+}	(Ar)	↑	↑	↑	↑		
Fe	(Ar)	↑↓	↑	↑	↑	↑	↑↓
Fe^{2+}	(Ar)	↑↓	↑	↑	↑	↑	
Fe^{3+}	(Ar)	↑	↑	↑	↑	↑	

Formation of complex ions

Complex ions are composed of a central metal ion surrounded by anions or molecules, called **ligands**. In transition metal complexes, non-bonded pairs of electrons on the ligand form co-ordinate bonds to the central ion. The unshared electron pairs are donated into vacant orbitals of the transition metal ion. The number of co-ordinate bonds from ligands to the central ion is known as the **co-ordination number** of the central ion. Generally, the co-ordination number of a particular ion is the same whatever the ligand. Thus, Cu^{2+} ions have a co-ordination number of four in $[Cu(H_2O)_4]^{2+}$, in $[Cu(NH_3)_4]^{2+}$, in $[CuCl_4]^{2-}$ and in $[Cu(H_2NCH_2CH_2NH_2)_2]^{2+}$. On the other hand, Fe^{3+} ions usually have a co-ordination number of six. This is the case in $[Fe(H_2O)_6]^{3+}$, in $[FeF_6]^{3-}$, in $[Fe(CN)_6]^{3-}$ and in $[Fe(EDTA)]^-$. Ag^+ ions normally have a co-ordination number of two ($[Ag(NH_3)_2]^+$, $[Ag(CN)_2]^-$).

In the case of the transition metals, the most common co-ordination number is six. But, examples of four and two are not uncommon. In aqueous solution, transition metal ions exist as hydrated complexes with water molecules. Owing to the high charge density on the central metal ion, these hydrated complexes dissociate like the hydrated Al^{3+} ion (section 16.5).

$$[Fe(H_2O)_6]^{3+}(aq) \rightleftarrows [Fe(H_2O)_5OH]^{2+}(aq) + H^+(aq)$$

Consequently, the aqueous solutions of most transition metal compounds such as $CuSO_4(aq)$, $FeCl_3(aq)$ and $Co(NO_3)_2(aq)$ are acidic. In oxidation states higher than $+3$, the polarising power of the central ion is even greater. This causes release of protons and loss of water molecules which results in the formation of oxy-anions. For example, neither $[Cr(H_2O)_6]^{6+}$ nor $[Mn(H_2O)_6]^{7+}$ exist in aqueous solution. The loss of two water molecules and eight protons from each of these ions results in the formation of CrO_4^{2-} and MnO_4^-, respectively:

$$[Cr(H_2O)_6]^{6+} \rightarrow 2H_2O + 8H^+ + CrO_4^{2-}$$

$$[Mn(H_2O)_6]^{7+} \rightarrow 2H_2O + 8H^+ + MnO_4^-$$

In complexes with a **co-ordination number of six**, the ligands usually occupy **octahedral positions**. The six electron pairs around the central metal ion are repelled as far as possible from each other (figure 19.9).

In complexes with a **co-ordination number of four**, the ligands usually occupy **tetrahedral positions**. A few four-co-ordinated complexes (such as $[Cu(H_2O)_4]^{2+}$ and $[Cu(NH_3)_4]^{2+}$) have a square-planar structure (figure 19.9).

Complexes with a **co-ordination number of two** usually have a **linear arrangement** of ligands (figure 19.9).

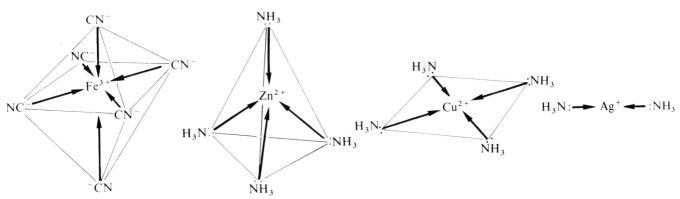

Complexes with a co-ordination number of 6 have an octahedral arrangement of ligands.

Complexes with a co-ordination number of 4 usually have a tetrahedral arrangement of ligands.

A few complexes with a co-ordination number of 4 have a square-planar arrangement of ligands.

Complexes with a co-ordination number of 2 have a linear arrangement of ligands.

Figure 19.9
The stereochemistry of complex ions.

Owing to the stereochemical positions of the ligands isomerism can occur in four-co-ordinated and six-co-ordinated complexes. Isomers are compounds with the same molecular formula, but different arrangements of their constituent atoms. (See section 26.6.)

The neutral complex $PtCl_2(NH_3)_2$ has two isomers.

■ Does $PtCl_2(NH_3)_2$ have a tetrahedral or a square-planar structure? Explain your answer.
■ The isomers of $PtCl_2(NH_3)_2$ are described as *cis* and *trans*. Draw these isomers. Indicate which is the *cis* form and which is the *trans* form.

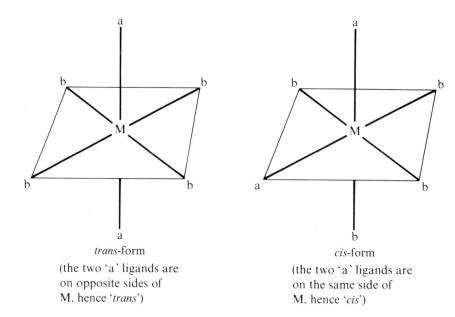

Figure 19.10
Isomers of complexes of the type Ma₂b₄.

trans-form
(the two 'a' ligands are
on opposite sides of
M, hence 'trans')

cis-form
(the two 'a' ligands are
on the same side of
M, hence 'cis')

Octahedral complexes of the type Ma_2b_4 also exhibit *cis/trans* (geometric) isomerism. Figure 19.10 shows the two isomers for Ma_2b_4.

One of the most striking examples of this type of isomerism occurs with tetra-amminedichlorocobalt(III) chloride ($[Co(NH_3)_4Cl_2]^+Cl^-$). In this compound, the two isomers have different-coloured crystals. The *cis* form is blue-violet in colour, whereas the *trans* form is green.

Another interesting situation arises with salts of the formula $CrCl_3(H_2O)_6$. Three differently coloured isomers have been isolated with this formula (table 19.4). These isomers show different conductivities in aqueous solution since they produce different numbers of ions. They also precipitate different amounts of silver chloride with silver nitrate solution because they contain different numbers of free Cl^- ions.

Table 19.4
Isomers of $CrCl_3(H_2O)_6$

Molecular formula	Total number of moles of ions per mole of $CrCl_3(H_2O)_6$ (deduced from conductivity)	Number of moles of chloride ions (Cl^-) per mole of $CrCl_3(H_2O)_6$ (deduced from amount of AgCl pptd)	Structural formula	Colour of isomer
$CrCl_3(H_2O)_6$	4	3	$[Cr(H_2O)_6]^{3+}.3Cl^-$	violet
$CrCl_3(H_2O)_6$	3	2	$[Cr(H_2O)_5Cl]^{2+}.2Cl^-.H_2O$	light green
$CrCl_3(H_2O)_6$	2	1	$[Cr(H_2O)_4Cl_2]^+.Cl^-.2H_2O$	dark green

Coloured compounds

Most of the compounds of transition elements are coloured. The colour of these compounds can often be related to incompletely filled *d*-orbitals in the transition metal ion.

In general, when light hits a substance, part is absorbed, part is transmitted (if the substance is transparent) and part may be reflected.

If all the incident radiation is absorbed then the substance looks '*black*'. If all the incident radiation is reflected, then the substance looks '*white*'. On the other hand, if only a very small proportion of the incident white light is absorbed and if all the radiations in the visible region of the spectrum are transmitted equally, then the substance will appear '*colourless*'.

However, many substances only absorb the light photons in certain areas of the visible spectrum. In this case, the transmitted or reflected light is relatively richer in the radiations of the remaining regions and the substance has a characteristic colour. Suppose a material absorbs all radiations in the yellow–orange–red region of the spectrum. It will appear blue in white light because only the radiations in the blue region are reflected towards your eyes.

When light energy is absorbed by a substance, an electron in the substance is promoted from an orbital of lower to one of higher energy. The atom or ion absorbing the radiation changes from its **ground state** (i.e. stable state) to an **excited state** (section 4.3). The different electron transitions involve the absorption of radiation of different frequencies (i.e. different quanta of energy). If the absorbed frequencies are in the visible region of the spectrum, then the material appears coloured. Why are solutions of transition metal compounds usually coloured, whereas solutions of compounds of non-transition metals are usually colourless?

Consider first a solution of a non-transition metal compound such as aqueous sodium chloride. The sodium ion has the electronic structure $1s^2 2s^2 2p^6$. The energy difference between the $2p$ orbitals and the next available orbital, $3s$, is very large. To undergo the electron transition $1s^2 2s^2 2p^6$ (ground state) $\rightarrow$ $1s^2 2s^2 2p^5 3s^1$ (excited state), a sodium ion must absorb a photon of high-energy radiation well beyond the range of visible light. Consequently, none of the photons of visible light has sufficient energy to promote electrons in Na^+ ions to even the lowest possible excited state. Hence, $Na^+(aq)$ does not absorb any photons of visible light and so solutions of sodium salts are usually colourless. The $Cl^-(aq)$ ion is also colourless for the same reason.

Now let us consider an aqueous solution of titanium(III) chloride. This contains the octahedral complex ions, $[Ti(H_2O)_6]^{3+}$, and $Cl^-(aq)$ ions. A solution of titanium(III) chloride is violet. An absorption spectrum shows that the solution absorbs most effectively in the green–yellow region of the spectrum (figure 19.11). Red, blue and violet radiations are absorbed less efficiently. So, the solution looks violet. Both water and $Cl^-(aq)$ are colourless, so the $Ti^{3+}(aq)$ ion must be responsible for the colour of the solution.

The Ti^{3+} ion has the electron structure $(Ar)3d^1$. When a solution of titanium(III) chloride absorbs photons of visible light, electrons in the $Ti^{3+}(aq)$ ion are promoted from $(Ar)3d^1$ (ground state) to $(Ar)4s^1$ (excited state). However, calculations show that this excitation of Ti^{3+}, like that for Na^+, requires photons of radiation well beyond the visible region of the spectrum.

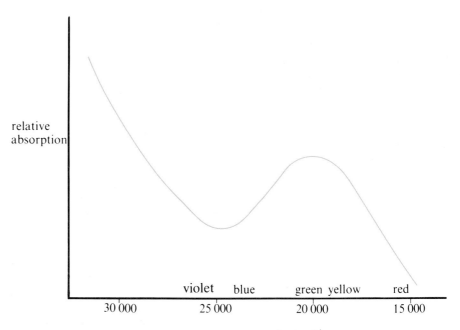

Figure 19.11
The absorption spectrum of $Ti(H_2O)_6^{3+}$.

How then do we explain the absorption of green light by Ti^{3+}(aq)? Which electron transition is responsible for the absorption of green light? The answer lies in electron transitions *within* the set of five 3d orbitals.

All five 3d orbitals have exactly the same energy level in the isolated gaseous Ti^{3+} ion (figure 19.12(a)). However, when the Ti^{3+} ion is surrounded by ligands, the 3d orbitals are no longer symmetrically arranged. Orbitals closer to the ligands are pushed to a slightly higher energy level than those orbitals further away. This splitting of the 3d orbitals in the octahedral $[Ti(H_2O)_6]^{3+}$ ion is shown in figure 19.12(b).

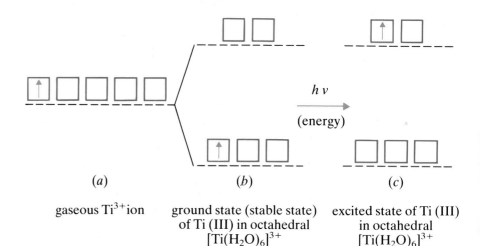

Figure 19.12
Relative energy levels for the five 3d orbitals of the gaseous and hydrated Ti^{3+} ion.

(a)
gaseous Ti^{3+}ion

(b)
ground state (stable state) of Ti (III) in octahedral $[Ti(H_2O)_6]^{3+}$

(c)
excited state of Ti (III) in octahedral $[Ti(H_2O)_6]^{3+}$

In the ground (stable) state, the single d electron occupies one of the d orbitals of slightly lower energy as shown in figure 19.12(b). It is now possible for the Ti^{3+} ion in $[Ti(H_2O)_6]^{3+}$ to absorb sufficient energy for the 3d electron to be promoted from its lower to one of the higher 3d orbitals (figure 19.12(c)). The difference in energy between the two sets of 3d orbitals in $[Ti(H_2O)_6]^{3+}$ is relatively small. It coincides with the wavelength for green light. Thus, when aqueous $TiCl_3$ is exposed to white light, it absorbs photons of green light but transmits those of red and blue light. Hence $TiCl_3$(aq) appears purple or violet.

In considering the colour of aqueous $TiCl_3$, the situation is relatively simple because the $[Ti(H_2O)_6]^{3+}$ ion has only one d electron. Ths situation is much more complex for most transition metal ions since many more d–d transitions are possible. In principle, however, the explanation of the colour of other transition metal compounds is similar to that for $TiCl_3$(aq). The five d orbitals are split into two or more slightly different energy levels. Thus, the promotion of an electron from the lower to the higher of these d orbitals just happens to require energies within the range of visible light.

Assume that the 3d orbitals of Cu(II) in the aqueous Cu^{2+} ion are split into two levels as in figure 19.12.

■ The electron structure for a copper atom is $(Ar)3d^{10}4s^1$. Write the electron structure for a Cu^{2+} ion.
■ Draw a diagram similar to figure 19.12(b) to show how electrons occupy the higher and lower 3d orbitals in the ground state for Cu^{2+}(aq),
■ Draw a diagram similar to figure 19.12(c) to show how electrons occupy the higher and lower 3d orbitals in the excited state for Cu^{2+}(aq).
■ What colour of light do you think the Cu^{2+}(aq) ion absorbs when the d–d transition occurs?

The colour of a transition metal complex depends mainly on the central cation. It is influenced much less by the co-ordinating ligand. For example the colours of most Cu^{2+} complexes are blue or violet as in the reaction scheme below. The complexes of Cu^+ are usually colourless.

$[Cu(H_2O)_4]^{2+}$ $\xrightarrow{\text{NH}_3\text{(aq)}}$ $[Cu(NH_3)_4]^{2+}$ $\xrightarrow{\text{H}_2\text{NCH}_2\text{CH}_2\text{NH}_2\text{(aq)}}$ $[Cu(H_2NCH_2CH_2NH_2)_2]^{2+}$
pale blue   dark blue   violet

Since different oxidation states have different numbers of d electrons, the colours of complexes often change when the oxidation state of the transition metal changes. For example, when potassium manganate(VII) is heated with 50% KOH it is reduced through the following colour changes.

substance present in alkaline soln	MnO_4^-	MnO_4^{2-}	MnO_3^-	MnO_2	$Mn(OH)_3$	$Mn(OH)_2$
oxidation state	7+	6+	5+	4+	3+	2+
colour		purple	green	blue	dark brown	brown	pink

Virgin desert scrub with reclaimed land in the background. This barren land in Israel was made productive simply by the application of minute quantities of transition metals to the soil. This made good the mineral deficiencies that had prevented previous settlement of the area.

Catalytic properties

Transition metals and their compounds are important **catalysts** in industry and in biological systems. Many transition metal ions are required by ourselves and other living things in minute, but definite quantities. Elements required in such small amounts are referred to as 'trace elements' These trace elements include copper, manganese, iron, cobalt, nickel and chromium. They are essential for the effective catalytic activity of various enzymes (section 9.8). One of the most important enzymes containing copper is cytochrome oxidase. This enzyme is involved when energy is obtained from the oxidation of food. In the absence of copper, cytochrome oxidase is completely inhibited and the animal or plant is unable to metabolise food effectively.

A large number of industrial catalysts are either transition metals or their compounds. Table 19.5 shows a list of some of the more important examples.

Table 19.5
Transition metals and their compounds used as catalysts in industry

Transition element	Substance used as catalyst	Reaction catalysed
Ti	$TiCl_3/Al_2(C_2H_5)_6$	$nC_2H_4 \longrightarrow \left(-\overset{\mid}{\underset{\mid}{C}}-\overset{\mid}{\underset{\mid}{C}}-\right)_n$ Polymerisation of ethene → poly(ethene)
V	V_2O_5 or vanadate (VO_3^-)	$2SO_2 + O_2 \rightarrow 2SO_3$ Contact process
Fe	Fe or Fe_2O_3	$N_2 + 3H_2 \rightarrow 2NH_3$ Haber process
Ni	Ni	$RCH{=}CH_2 + H_2 \rightarrow RCH_2CH_3$ Hardening of vegetable oils (e.g. manufacture of margarine)
Cu	Cu or CuO	$CH_3CH_2OH + \frac{1}{2}O_2 \rightarrow CH_3CHO + H_2O$ Oxidation of ethanol to ethanal (acetaldehyde)
Pt	Pt	$2SO_2 + O_2 \rightarrow 2SO_3$ Contact process
Pt	Pt	$4NH_3 + 5O_2 \rightarrow 4NO + 6H_2O$ then $NO \rightarrow NO_2 \rightarrow HNO_3$ Manufacture of nitric acid from ammonia

Chemists believe that the catalytic activity of transition metals depends on their ability to exist in various states of oxidation or co-ordination. For example, in the **Contact Process**, vanadium compounds in the +5 state (either V_2O_5 or a vanadate) are used to oxidise sulphur dioxide to sulphur(VI) oxide:

$$SO_2 + \tfrac{1}{2}O_2 \xrightarrow{V_2O_5} SO_3$$

It is thought that the actual oxidation process involves two stages. In the first of these, V^{5+} in the presence of oxide ions converts SO_2 to SO_3. At the same time, V^{5+} is reduced to V^{4+}:

$$2V^{5+} + O^{2-} + SO_2 \rightarrow 2V^{4+} + SO_3$$

In the second stage, V^{5+} is regenerated from V^{4+} by oxygen:

$$2V^{4+} + \tfrac{1}{2}O_2 \rightarrow 2V^{5+} + O^{2-}$$

The overall process is, of course, the sum of these two stages:

$$2\cancel{V}^{5+} + \cancel{O}^{2-} + SO_2 \rightarrow 2\cancel{V}^{4+} + SO_3$$
$$2\cancel{V}^{4+} + \tfrac{1}{2}O_2 \rightarrow 2\cancel{V}^{5+} + \cancel{O}^{2-}$$
$$\text{Sum:} \quad SO_2 + \tfrac{1}{2}O_2 \rightarrow SO_3$$

Transition metals and their compounds can catalyse reactions because they are able to introduce an entirely new reaction mechanism. This new reaction mechanism has a lower activation energy (section 25.10) than the uncatalysed reaction. Since the activation energy of the catalysed reaction is lower, the reaction rate is faster. This is shown diagrammatically in figure 19.13.

During the course of the reaction, the catalyst undergoes changes in oxidation state. It is, however, regenerated in its original form when the reactants form the products.

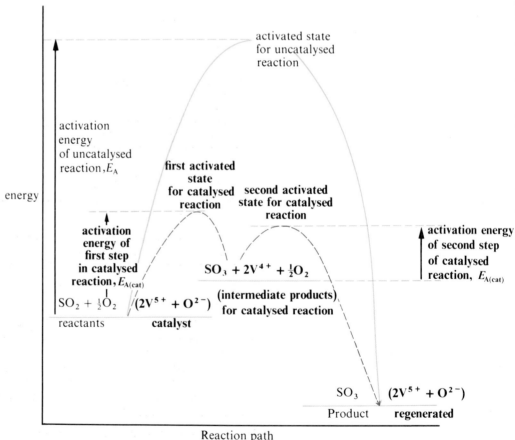

Figure 19.13
The activation energies of catalysed and uncatalysed reactions.

—— reaction path of uncatalysed reaction
--- reaction path of catalysed reaction

An excellent demonstration of the change in oxidation state of a transition metal during catalysis can be shown using cobalt(II) ions with a mixture of 2,3-dihydroxybutanedioate (tartrate) ions and hydrogen peroxide.

When aqueous solutions of 2,3-dihydroxybutanedioate (tartrate) ions and hydrogen peroxide are mixed, there is little or no evolution of gas from the decomposition of hydrogen perioxide. If cobalt(II) chloride solution is added, the solution becomes pink and there is still little evolution of gas. After an induction period, the solution begins to darken, going brown and then green due to the presence of cobalt(III). At the same time, effervescence from the solution gradually increases and the maximum evolution of gas occurs when the solution is green. As the reaction subsides, the green colour of the cobalt(III) complex fades to be replaced by the original pink of cobalt(II) ions.

Summary

1   A transition metal is an element which forms at least one ion with a partially filled *d* sub-shell.

2   Unlike elements in other parts of the periodic table, the elements in a transition series have very similar properties to one another.

3   Many of the properties of transition elements can be attributed to their electronic structures and the relative energy levels of the orbitals available for their electrons.

4   Prior to occupation by electrons, the 4*s* level is energetically more stable than the 3*d* level. Once the 3*d* level is occupied by electrons, the latter repel the 4*s* electrons further from the nucleus to an energy level higher than the 3*d* level. Thus, transition metals lose electrons from the 4*s* level before the 3*d* level in forming ions.

5   Transition metals have higher melting points, higher boiling points and higher densities than other metals.

6   Transition metals show the following characteristic properties:
    (a) variable oxidation states,
    (b) formation of complex ions,
    (c) coloured compounds,
    (d) catalytic properties.

7   The more important oxidation states for each transition element include +2 or +3 or both.

8   Complex ions are composed of a central cation surrounded by a cluster of anions or molecules known as ligands. Non-bonded pairs of electrons on the ligand form co-ordinate bonds to the central metal cation.

9   The number of co-ordinate bonds from ligands to the central cation is known as the co-ordination number of the central ion. Generally, the co-ordination number of a particular ion is the same whatever the ligand.

10 In complexes with a co-ordination number of six, the ligands usually occupy octahedral positions.
    In complexes with a co-ordination number of four the ligands occupy tetrahedral or square-planar positions.
    Complexes with a co-ordination number of two usually have a linear arrangement of ligands.

11 Transition metals and their compounds are important catalysts in industry and in biological systems.

12 Many enzymes rely on transition metals for their catalytic activity.

Study questions

1 (a) By reference to copper or its compounds, give three properties that are characteristic of transition metals.
   (b) Explain the following:
       (i)   Aqueous copper(II) sulphate turns blue litmus paper red.
       (ii) The addition of ammonia solution to aqueous copper(II) sulphate gives a pale blue precipitate at first and then a deep blue solution when more ammonia solution is added.
       (iii) Copper(I) compounds are sometimes white and rarely show catalytic properties. Copper(II) compounds are usually coloured and can act as catalysts for many reactions.

2 (a) Give the name and formula of:
       (i)   one complex cation containing a transition metal,
       (ii) one complex anion containing a transition metal.
   (b) Describe the overall shape (octehedral, tetrahedral, etc.) of the two ions in part (a).
   (c) Outline the electronic structures of the ions that you chose in (a), restricting yourself to a consideration of the outer shell of each atom.

3 (a) How do the electronic structures of transition elements differ from those of the elements in the main groups of the periodic table?
   (b) Describe the electronic structure of either vanadium or chromium. How are the more important oxidation states of the element determined by this electronic structure?
   (c) Give three characteristic features of transition metals or their compounds (other than that of variable oxidation states). Illustrate your answer with reference to vanadium or chromium.

4 Complex ions can sometimes exhibit isomerism.
   (a) The compound $[Co(NH_3)_5Br]^{2+}SO_4^{2-}$ is isomeric with the compound $[Co(NH_3)_5SO_4]^+Br^-$.
       (i)   What ions will these two isomers yield in solution?
       (ii) How would you confirm which isomer was which? (You are required to describe a simple positive test for each isomer.)
       (iii) What is (*i*) the oxidation state,
                      (*ii*) the co-ordination number,
             of cobalt in each complex ion?
       (iv) Draw the structure of the $[Co(NH_3)_5Br]^{2+}$ ion indicating its shape and the co-ordinate bonds involved.
   (b) The compound $NiCl_2(NH_3)_2$ has *cis-trans* isomers. These have a complex non-ionic structure.
       (i)   Does $NiCl_2(NH_3)_2$ have a tetrahedral or a square-planar structure? Explain your answer.
       (ii) Draw the *cis* and *trans* isomers for $NiCl_2(NH_3)_2$.

5 Ascorbate oxidase is a metallo-protein enzyme in plants. The enzyme consists of a protein associated with a copper(II) ion.
       (i)   The enzyme protein alone has no catalytic activity.
       (ii) Copper(II) ions alone can act as a catalyst for the reaction, but much less efficiently than the metal–protein combination.
       (iii) When egg albumen is added to aqueous copper(II) ions, the mixture shows greater catalytic activity than copper(II) ions alone. But, this activity is not so good as with the specific metal–protein in ascorbate oxidase.
   (a) Write the electronic structure of copper(II) ions.
   (b) Explain how copper(II) ions might act as catalysts.
   (c) What explanation can you offer for observations (i), (ii) and (iii) above?

6 Early this century, the German scientist, Werner, succeeded in clarifying the situation concerning the five compounds of $PtCl_4$ and ammonia. The properties of these compounds are listed in the table below.

Compound	Empirical Formula	Total no. of ions in the empirical formula	No. of Cl^- ions in the empirical formula
A	$PtCl_4.6NH_3$	5	4
B	$PtCl_4.5NH_3$	4	3
C	$PtCl_4.4NH_3$	3	2
D	$PtCl_4.3NH_3$	2	1
E	$PtCl_4.2NH_3$	0	0

(a) What is the oxidation state of Pt in each of the compounds, A–E?

(b) The co-ordination number of Pt in each compound is six. Write a formula for each of the five compounds. Show the complex ion and the other ions and/or molecules present.

(c) Each of the compounds forms an octahedral complex ion. Draw structures for the complex ions in A, B, C and D.

(d) Which of the complex ions in (c) have isomers?

(e) Draw structures to show the various structural isomers of A, B, C and D.

20 Metals and the Activity Series

A sample of 'kidney' iron ore.

20.1 Iron—a typical transition metal

Iron typifies transition metals in many respects.

(a) It has **two** relatively stable **oxidation states** in its compounds (+2 and +3).
(b) It forms a variety of **complex ions** (such as $[Fe(H_2O)_6]^{2+}$, $[Fe(H_2O)_6]^{3+}$, $[Fe(CN)_6]^{4-}$ and $[Fe(CN)_6]^{3-}$).
(c) It has characteristically **coloured compounds**: pale green for iron(II) salts and yellow or brown for iron(III) salts.
(d) It has important **catalytic properties**. For example, it is used to catalyse the synthesis of ammonia in the Haber process.

The typical properties of transition metals were considered in detail in the last chapter.

How do transition metals compare and contrast with the typical metals of Group I and Group II in their:

- melting points and boiling points,
- density,
- reactivity with water,
- reactivity with dilute HCl,
- number of oxidation states,
- colour of compounds,
- formation of complex ions?

20.2 The occurrence of iron

Roman iron objects. Successful methods for smelting and casting iron existed even in Roman times.

Apart from its presence in certain meteorites, iron does not occur as the element in the earth's crust. Nevertheless, its compounds are abundant. The commonest ores are iron pyrites (FeS_2), haematite (Fe_2O_3), magnetite (Fe_3O_4) and limonite or brown iron ore ($Fe_2O_3 . H_2O$).

Iron is also present in silicates, sands and clays, and in all living matter. It is essential for the production of haemoglobin in blood and chlorophyll in plants. Many pregnant women must take iron tablets (usually in the form of iron(II) sulphate). These tablets enable them to produce sufficient haemoglobin for themselves and their growing babies without becoming anaemic.

Iron has considerable economic and industrial importance because:
(a) its ores are abundant in the Earth's crust,
(b) it is easily manufactured from its ores,
(c) it has very useful mechanical properties.

310

Mining iron ore. Notice the crushing plant and railway terminus in the foreground.

Blast furnaces at Ravenscraig.

Figure 20.1
Right. A blast furnace.

20.3 The manufacture of iron

One hundred years ago, the total production of iron in the UK was only about 20 000 tonnes. Nowadays, its production in the UK is about 20 000 000 tonnes.

The most important raw materials from which iron is obtained are haematite (Fe_2O_3) and brown iron ore ($Fe_2O_3.H_2O$). Other minerals containing iron, such as sulphides and carbonates, must be roasted in air to obtain the oxide which is then reduced to the metal. In most parts of the world, there are still large quantities of high quality iron ores from which iron can be obtained.

The first stage in the manufacture of iron involves preparing the ore for reduction in a blast furnace. This involves crushing the ore to produce lumps about the size of one's fist. These lumps are then pre-heated using hot gases from the furnace. This drives off water and other volatile impurities.

The prepared iron ore contains between 30% and 95% iron oxides. The main impurities are silicon(IV) oxide (silica, SiO_2) and aluminium oxide (alumina, Al_2O_3).

The second stage in the manufacture of iron involves reducing the iron oxides. This is done using carbon (coke) in the presence of limestone (calcium carbonate) in a blast furnace (figure 20.1).

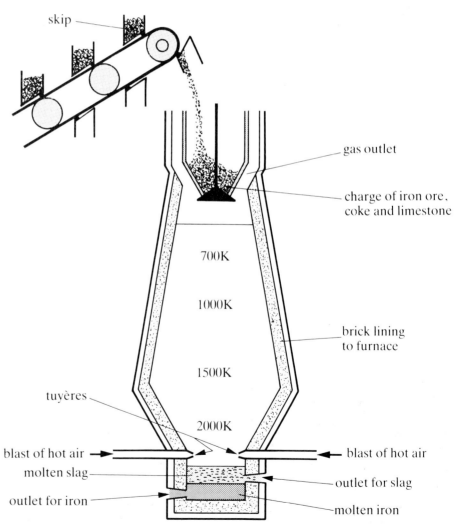

The furnace itself is a tapered cylindrical tower, 30–40 metres in height. It is made of steel and lined with refractory bricks. The furnace is fed mechanically so

Molten iron being tapped off from No 3 blast furnace at Llanwern, Newport.

that no gas escapes during the charging. A mixture of iron ore, coke and limestone is added at the top of the furnace. At the same time, blasts of hot air at about 1 000 K are driven into the furnace through small holes or 'tuyères' near its base. These blasts of air give the furnace its name.

As the air enters the furnace, it reacts with coke. A highly exothermic reaction occurs forming carbon dioxide:

$$C + O_2 \rightarrow CO_2 \qquad \Delta H = -394 \, kJ$$

This raises the temperature at the bottom of the furnace to about 2 000 K. As the CO_2 rises up the furnace, it reacts with more coke forming carbon monoxide:

$$CO_2 + C \rightarrow 2CO \qquad \Delta H = +173 \, kJ$$

The carbon monoxide then reduces the iron oxides to iron in the upper parts of the furnace. Here the temperature is between 750 K and 1 000 K.

$$Fe_2O_3 + 3CO \rightarrow 2Fe + 3CO_2$$

As the iron falls towards the base of the furnace, it eventually melts (m.pt. = 1812 K). It flows to the hearth where it collects and is tapped off every few hours. A typical blast furnace produces about 1 000 tonnes of iron every 24 hours.

So far, we have said nothing about the use of limestone in the blast furnace. The limestone plays an essential part in the extraction of impurities from the iron.

At the high temperatures inside the furnace, the limestone decomposes into calcium oxide and carbon dioxide:

$$CaCO_3 \rightarrow CaO + CO_2$$

The calcium oxide then reacts with impurities in the iron such as sand (silicon(IV) oxide) and alumina (aluminium oxide). It forms calcium silicate and calcium aluminate, respectively:

$$CaO + SiO_2 \rightarrow CaSiO_3$$
$$CaO + Al_2O_3 \rightarrow CaAl_2O_4$$

The mixture of calcium silicate and calcium aluminate which remains molten at the furnace temperature is known as **'slag'**. This flows to the bottom of the furnace where it floats on the molten iron. It is tapped off at a different level, separately from the iron. The 'slag' is not wasted. It is used for road-making materials, for cement manufacture and for lightweight building materials.

The iron obtained from the blast furnace is far from pure. It is known generally as **'pig iron'**. It is hard, but brittle, and melts at about 1 500 K.

It is re-melted, mixed with scrap steel and cooled in moulds to form **'cast iron'**. This has much the same properties as 'pig iron' and can be used for articles such as gates, pipes and lamp posts where cheapness is more important than strength.

Above. Workmen constructing a cast-iron gate.

Right. When steel was first produced it was an expensive alloy. Thus, the first metal bridge which was erected at Coalbrookdale in 1799 was made of 400 tonnes of cast iron not steel.

A large cast-iron gun.

20.4   The conversion of iron into steel

Most of the pig iron manufactured nowadays is used for the production of **steel**. This is an alloy of iron with carbon and other elements such as manganese, nickel, chromium, tungsten and vanadium.

The main impurities in pig iron and in mild steel are shown in table 20.1.

Table 20.1
The main impurities in pig iron and in mild steel

Impurity	% impurity in pig iron	% impurity in mild steel
Carbon	3–5	0.15
Silicon	1–2	0.03
Sulphur	0.05–0.10	0.05
Phosphorus	0.05–1.5	0.05
Manganese	0.5–1.0	0.50

The carbon impurity comes from coke used during the reduction of iron ore. The silicon, sulphur and phosphorus are present as a result of the reduction of silicates, sulphates and phosphates by carbon and carbon monoxide during the blast furnace processes. The manganese arises from the reduction of manganese compounds in the iron ore.

In order to obtain steel from pig iron, a large proportion of these impurities must be removed.

During the last 50 years, the techniques of **steel production** have undergone vast changes in scale. New processes have also been developed to keep pace with the demands of quantity and quality. At present, there are only three major steel-making processes; **Open Hearth, Electric Arc**, and **Basic Oxygen**. The Basic Oxygen Process is a development of the old Bessemer process.

All of these processes share the same general principle. This entails removing the impurities of carbon, silicon, sulphur and phosphorus from molten 'pig iron' by oxidation. Known quantities of other elements are then added to obtain steel of the desired composition and properties.

During steel production, carbon and sulphur are oxidised to CO_2 and SO_2 which then escape as gases. On the other hand, silicon and phosphorus are oxidised to less volatile oxides, phosphorus(v) oxide and silicon(iv) oxide. These combine with the lime added to the furnace to form 'slag'.

$$6CaO + P_4O_{10} \rightarrow 2Ca_3(PO_4)_2$$
calcium phosphate(v)

$$CaO + SiO_2 \rightarrow CaSiO_3$$
calcium silicate

} 'slag'

The Basic Oxygen process

In the Basic Oxygen Process, no external heating is required because the reactions inside the furnace are very exothermic. Initially the furnace is charged with hot, molten pig iron and lime (figure 20.2).

Oxygen is blown onto the surface of the metal at great speed through water-cooled pipes. The oxygen penetrates into the melt and oxidises the impurities rapidly. The heat evolved as the impurities are oxidised keeps the contents of the furnace in a molten state, despite a rise in the melting point as impurities are removed.

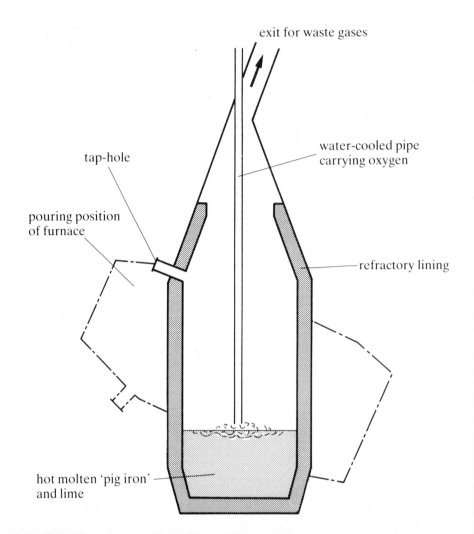

Figure 20.2
The Basic Oxygen furnace.

Charging one of the Basic Oxygen furnaces with hot molten pig iron at the Frodingham Works, Scunthorpe.

Nowadays, over 80% of the steel produced in the UK is obtained by the Basic Oxygen Process. This process is increasingly used because it is so fast. A large furnace can convert 300 tonnes of pig iron into steel in only 40 minutes.

The Open Hearth Process

In the Open Hearth Process (figure 20.3), heat is provided by burning gas or oil in a mixture of air and oxygen.

Oxygen is injected into the furnace through water-cooled pipes. This oxidises the impurities in the iron.

About 10% of the steel produced in the UK is obtained by the Open Hearth Process. The process is rather slow. A large furnace takes 10 hours to convert 300 tonnes of pig iron to steel. This process is being replaced by the very much faster Basic Oxygen Process.

By adding different metals to the steel it is possible to impart specific properties, such as hardness, resistance to corrosion or high tensile strength. 'Manganese steel' containing as much as 13% manganese can be made incredibly tough by heating to 1000°C and 'quenching' (rapid cooling) in water. It is used for parts of rock-breaking machinery and for railway cross-overs.

Stainless steel, containing about 20% chromium and 10% nickel, is used for cutlery, surgical instruments and car bumpers.

'High-speed' steels used as the cutting edges on lathes for metalwork, contain about 18% tungsten and 5% chromium. The tungsten hardens the steel and makes it less brittle. Using 'high-speed' steels, metal parts can be machined up to twenty times faster than ordinary steel.

'Manganese steel' is used for military steel helmets. The dents in the steel are caused by bullets striking the helmet at the speeds shown. 912 F.S. means 912 feet per second.

Above. Wrought-iron beetle. Wrought iron is very pure iron. It is much more malleable than steel or cast iron. Everything on this baroque-style Volkswagen works.

Figure 20.3
Right. The Open Hearth furnace.

20.5 The reactions of iron and its aqueous ions

Several reactions of iron and the aqueous Fe^{2+} and Fe^{3+} ions are summarised in figure 20.4.

Iron combines with most non-metals on heating, forming compounds such as Fe_3O_4, FeS, $FeCl_3$ and FeI_2. Those non-metals which are strong oxidising agents, such as chlorine and oxygen, form products containing iron in the more oxidised +3 state.

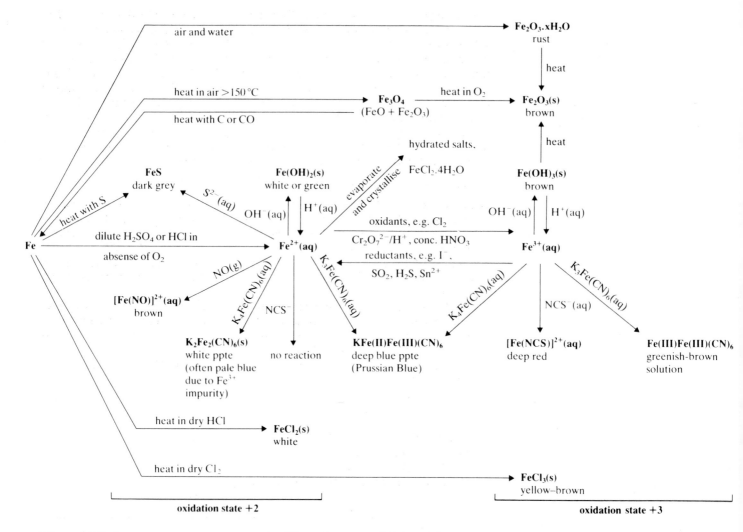

Figure 20.4
The reactions of Fe, Fe^{2+}(aq) and Fe^{3+}(aq).

The redox potentials involving iron it its $+2$ and $+3$ states are given below.

In 1.0 M acid solution

$$Fe(s) \xrightarrow{+0.44} Fe^{2+}(aq) \xrightarrow{-0.77} Fe^{3+}(aq)$$

In 1.0 M alkaline solution

$$Fe(s) \xrightarrow{+0.89} Fe(OH)_2(s) \xrightarrow{+0.56} Fe(OH)_3(s)$$

The figures show that in acid solution the more stable state of iron is $+2$ and in alkaline solution the more stable state of iron is $+3$.

In alkaline solution, oxygen in the air or oxygen dissolved in the solution will oxidise $Fe(OH)_2$ to $Fe(OH)_3$:

$$2Fe(OH)_2(s) + 2OH^-(aq) \rightarrow 2Fe(OH)_3 + 2e^- \qquad +0.56\,V$$

$$\tfrac{1}{2}O_2(g) + H_2O(l) + 2e^- \rightarrow 2OH^-(aq) \qquad +0.40\,V$$

So, freshly prepared pale green $Fe(OH)_2$ rapidly turns dirty green and finally red-brown on contact with the air as it is oxidised to $Fe(OH)_3$.

In acid solution and under standard conditions, Fe^{2+}(aq) is not oxidised to Fe^{3+}(aq) by oxygen in the air or by oxygen dissolved in the solution.

$$2Fe^{2+}(aq) \rightarrow 2Fe^{3+}(aq) + 2e^- \qquad -0.77\,V$$

$$2H^+(aq) + O_2(g) + 2e^- \rightarrow H_2O_2(aq) \qquad +0.68\,V$$

Thus, Fe^{2+}(aq) ions are stable in acid solution under standard conditions.

Look again at the redox potentials given at the beginning of this section. The figures show that iron will react with dilute acids (in the absence of air) to yield solutions of iron(II) salts and hydrogen. However, dilute nitric acid and concentrated sulphuric acid yield a mixture of Fe^{2+} and Fe^{3+} ions together with the reduction products of the acids (mainly NH_4^+ and SO_2, respectively). This is because these two acids are oxidising agents. On the other hand, concentrated nitric acid and concentrated chromic(VI) acid have only a momentary effect on iron. After this, the metal becomes 'passive'. It is thought that the initial attack of the acid coats the metal with an exceptionally tight, impervious and unreactive layer of oxide. This layer prevents further attack. Not surprisingly, iron is sometimes treated in this way to prevent corrosion.

Try to answer these questions about the chemistry of iron.

■ Aqueous iron(II) salts are only slightly acidic whereas iron(III) salts are strongly acidic. Why is this? (Hint: your explanation should involve a consideration of the charge density on Fe^{2+} and Fe^{3+} and the possible hydrolysis of $[Fe(H_2O)_6]^{2+}$ and $[Fe(H_2O)_6]^{3+}$ in aqueous solution.)
■ Draw structural formulae to show the octahedral stereo-structures of the hexacyanoferrate(III) ion ($[Fe(CN)_6]^{3-}$) and the hexacyanoferrate(II) ion ($[Fe(CN)_6]^{4-}$).
■ Iron(III) compounds are more covalent than iron(II) compounds. Why is this? (For example, anhydrous iron(III) chloride is very similar to anhydrous aluminium chloride. It sublimes readily. It is soluble in alcohol and ether. It has a co-ordinately bonded dimer structure, $Fe_2.Cl_6$.)
■ Use figure 20.4 to tabulate the reactions of $Fe^{2+}(aq)$ and $Fe^{3+}(aq)$ with:
  (a) $OH^-(aq)$
  (b) $NCS^-(aq)$
  (c) $[Fe(CN)_6]^{4-}(aq)$
  (d) $[Fe(CN)_6]^{3-}(aq)$
  Which of the above ions would be most effective in:
  (i)   testing for $Fe^{2+}(aq)$
  (ii) testing for $Fe^{3+}(aq)$
  (iii) distinguishing $Fe^{2+}(aq)$ from $Fe^{3+}(aq)$?

20.6 The corrosion of iron and its prevention

Rusting costs us millions of pounds each year. The costs arise because we need to protect iron and steel objects as well as replace rusted articles.

It is well known that *rusting requires the presence of both oxygen and water*. Other factors can influence the rate of rusting considerably. These factors include impurities in the iron, availability of dissolved oxygen and electrolytes in the solution in contact with the iron.

When rusting occurs a hydrated form of iron(III) oxide with variable composition ($Fe_2O_3.xH_2O$) is produced. This oxide is very permeable to both air and water. It cannot protect the metal from further corrosion, which continues unhindered below the rusted surface. The process is represented schematically in figure 20.5.

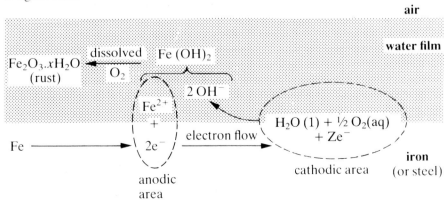

Figure 20.5
The chemical reactions involved in rusting.

A wrecked and rusting tanker.

In our climate, any iron or steel object open to the atmosphere is often wet. The layer of water on the surface of the metal dissolves oxygen and carbon dioxide from the air. In addition to these, the water will also contain dissolved sulphur dioxide in any industrialised area and sodium chloride from salt spray near the sea.

Look closely at figure 20.5.

■ Why are oxygen and water both required for rusting?
■ Why do electrolytes, particularly acids, accelerate rusting?
■ Which substances, normally present in the atmosphere, can provide the H^+ ions to accelerate rusting?
■ Why is rusting accelerated in polluted areas?

In the initial stages of rusting, iron(II) ions pass into solution at the anodic area. At the same time, reduction of atmospheric oxygen to hydroxide ions takes place at the cathodic area. Here the concentration of dissolved oxygen is higher.

anode (+) $Fe(s) \rightarrow Fe^{2+}(aq) + 2e^-$

cathode (−) $H_2O(l) + \frac{1}{2}O_2(aq) + 2e^- \rightarrow 2OH^-(aq)$

The Fe^{2+} and OH^- ions diffuse away from the 'electrodes' and precipitate as iron(II) hydroxide. This is then oxidised by dissolved oxygen to form rust.

Notice that the rust is formed by a secondary process *in the solution* as Fe^{2+} and OH^- ions diffuse away from the metal surface. Thus, the rust does not form as a protective layer in contact with the iron surface. However, if the solution contains a relatively high concentration of dissolved oxygen, Fe^{2+} ions are converted into rust more rapidly. In this case a protective layer may be formed on the iron surface which retards further rusting.

Acid conditions (low pH) accelerates rusting by promoting the dissolution of iron. Electrolytes in the water and impurities in the iron also assist the process. They increase the conductivity of the solution and initiate cell action, respectively.

Two general methods are used to protect iron and steel from corrosion: the **application of a protective layer** and the **application of a sacrificial metal**.

Application of a protective layer

Many different substances are used to protect iron and steel objects from rusting.

In the motor car industry, the chassis of vehicles are painted with phosphoric(V) acid. This reacts with the iron to form an insoluble, tenacious film of iron(III) phosphate(V). Engine parts, like other machinery and tools, are often protected by a film of oil or grease. However, the most common method of rust prevention is painting. This often incorporates the use of dilead(II) lead(IV) oxide (red lead) or zinc chromate(VI) as protective priming coats.

Other methods are also used in the fight against corrosion. Iron can be alloyed with other metals, such as chromium, nickel and manganese. This produces corrosion-resistant steels. Alternatively, iron can be coated with non-rusting metals. Buckets and watering cans are coated with a layer of zinc ('galvanised'). This is done by dipping them in molten zinc at 450°C or by depositing the zinc by electrolysis. 'Tin plate' is formed by immersing steel sheeting in molten tin. Other articles, such as car bumpers, are chromium-plated by making them the cathode in an electrolytic cell.

Application of a sacrificial metal

In this case, the iron or steel must be connected to a more electropositive metal. This is oxidised in preference to iron, thus preventing the formation of Fe^{2+} ions and hence rust. 'Galvanising' provides a good example of a sacrificial method of rust prevention. When the 'galvanised' surface is undamaged, zinc is protected from corrosion by a firmly adhering layer of zinc oxide. When the galvanised surface is scratched and the iron is exposed, zinc ions pass into solution rather

'Tin plate' being rolled into thin sheets at the British Steel Corporation's tin plate works in South Wales.

than iron(II) ions. This occurs because zinc is more electropositive. Electrode potentials confirm this fact.

$$E^{\ominus}/\text{volts}$$
$$Fe(s) \rightarrow Fe^{2}(aq) + 2e^{-} \quad +0.44$$
$$Zn(s) \rightarrow Zn^{2+}(aq) + 2e^{-} \quad +0.76$$

Thus, zinc is sacrificed in the protection of iron. 'Tin plating', however, is not so effective as 'galvanising'. Provided the coating is undamaged, the tin surface protects the iron below. But when the tin plate is scratched rusting occurs. Iron is more electropositive than tin, so Fe^{2+} ions pass into solution and then form rust.

$$E^{\ominus}/\text{volts}$$
$$Fe(s) \rightarrow Fe^{2+}(aq) + 2e^{-} \quad +0.44$$
$$Sn(s) \rightarrow Sn^{2+}(aq) + 2e^{-} \quad +0.14$$

Cathodic protection is used to prevent corrosion in underground pipelines made of steel. Pieces of zinc or manganese alloy are connected to the buried pipeline at intervals. In oil refineries, magnesium alloys are bolted inside distillation and cracking plant. These metals will then corrode in preference to steel between regular maintenance inspections.

20.7 Copper—extraction and manufacture

Metallic copper does occur naturally in a few places. However, most copper is obtained from its naturally occurring compounds. The most important ores are copper pyrites ($CuFeS_2$) and copper glance (Cu_2S). These two sulphides provide 80% of the element.

Most copper ores contain only a few per cent of the metal. So, the first stage in obtaining copper is to crush the ore and concentrate it by froth flotation.

The second stage involves reduction of the sulphide ore to the metal. Initially, the concentrate is roasted in air to form copper(I) sulphide, iron(II) oxide and gaseous sulphur dioxide:

$$2CuFeS_2 + 4O_2 \rightarrow Cu_2S + 2FeO + 3SO_2(g)$$

The product is then heated with silica in a closed furnace. Most of the iron(II) oxide reacts with the silica to form a molten slag which floats on the molten copper(I) sulphide and can be tapped off separately:

$$FeO + SiO_2 \rightarrow FeSiO_3$$
$$\text{iron(II) silicate (slag)}$$

The open-cast copper mine at Chuquicamata, in the Atacama Desert, Chile. This is the largest copper-producing mine in the world.

Purifying crushed copper ore by froth flotation.

The impure copper(I) sulphide is then heated in air so that part of it reacts, forming copper(I) oxide:

$$2Cu_2S + 3O_2 \rightarrow 2Cu_2O + 2SO_2$$

The copper(I) oxide, mixed with unchanged copper(I) sulphide, is now heated strongly in the absence of air. This forms copper and sulphur dioxide:

$$2Cu_2O + Cu_2S \rightarrow 6Cu + SO_2$$

The product is known as 'blister copper'. It releases bubbles of SO_2 as it solidifies and therefore gets a blistered appearance. It still contains 2 or 3% of impurities, mainly iron and sulphur.

The final stage in the manufacture of copper involves purification of the blister copper. This is usually achieved by electrolysis of copper(II) sulphate solution. A thin sheet of pure copper is used as the cathode and the impure blister copper as the anode. During the electrolytic purification, copper dissolves away from the anode, whilst a thickening deposit of pure copper appears on the cathode.

■ Write equations for the processes at the anode and the cathode during this electrolytic purification of blister copper.
■ If the impure blister copper contains metals above copper in the electrochemical series, these metals will dissolve from the anode in preference to copper. Why do these metals, which get into solution, not deposit on the pure copper cathode?
■ Suppose the impure blister copper contains metals below copper in the electrochemical series such as silver. Why do these metals not contaminate the copper which deposits on the cathode?

In practice, silver and gold can be obtained as valuably by-products from the impure 'blister copper'. They help to pay for the copper-refining process.

Electrolytic refining of copper.

20.8 The properties and uses of copper

The uses of copper result from its malleability and ductility, its high electrical and thermal conductivity, its resistance to corrosion and its ability to form alloys. The pure metal is used extensively as electrical conducting wires and cables. It is also used as pipes and radiators in central heating systems.

Copper provides a wide range of different alloys with other metals. Alloying copper with zinc produces various types of brass which are harder and stronger than pure copper. Alloying copper with tin produces bronze, an alloy which has greater tensile strength and is more readily cast into moulds than pure copper. Copper also alloys readily with nickel, aluminium and gold. For example, 75% copper and 25% nickel are used in our present 'silver coins', whilst 9 carat gold is about two-thirds copper and one-third gold.

20.9 Copper—Dr Jekyll or Mr Hyde

Copper has a rather complex, two-faced character with two sets of very different compounds. *In the +2 state, copper shows typical transition metal properties. In the +1 state its compounds are more like those of silver.*

The reactions of copper and the more important copper compounds are shown in figure 20.6. Various atomic properties of copper are tabulated in table 20.2.

The chemical properties of copper can be related to its relatively high ionisation energies, high heat of atomisation and relatively low heats of hydration. These result in its negative electrode potentials and generally low reactivity (see table 20.2).

The electrode potentials for both $Cu(s)/Cu^{2+}(aq)$ and $Cu(s)/Cu^{+}(aq)$ are negative. Consequently, copper will not react with dilute strong acids to produce hydrogen (figure 20.6).

$$2H^{+}(aq) + 2e^{-} \rightarrow H_2(g) \qquad E^{\ominus} = 0.00 \text{ volts}$$

However, copper can be oxidised very slowly to the copper(II) state $(CuCO_3.Cu(OH)_2)$ by oxygen and water vapour in the air, and much more rapidly by concentrated nitric acid (figure 20.6). The following potentials further confirm this:

	$E^{\ominus}$/volts
$Cu(s) \rightarrow Cu^{2+}(aq) + 2e^{-}$	-0.34
$\frac{1}{2}O_2(g) + H_2O(g) + 2e^{-} \rightarrow 2OH^{-}(aq)$	$+0.40$
$NO_3^{-}(aq) + 2H^{+}(aq) + e^{-} \rightarrow NO_2(g) + H_2O$	$+0.80$

Dr Jekyll . .

. . and Mr Hyde.
Copper is like Jekyll and Hyde. It has a complex, two-faced character.

Table 20.2
Atomic properties of copper

Electron structure	$(Ar)3d^{10}4s^{1}$
Heat of atomisation/kJ mole^{-1}	339
First ionisation energy/kJ mole^{-1}	744
Second ionisation energy/kJ mole^{-1}	1956
Hydration energy of $Cu^{2+}(g)$/kJ mole^{-1}	-2244
Hydration energy of $Cu^{+}(g)$/kJ mole^{-1}	-481
Standard electrode potential, $Cu(s)/Cu^{2+}(aq)$/volts	-0.34
Standard electrode potential, $Cu(s)/Cu^{+}(aq)$/volts	-0.52
Standard electrode potential, $Pt/Cu^{+}(aq), Cu^{2+}(aq)$/volts	-0.15

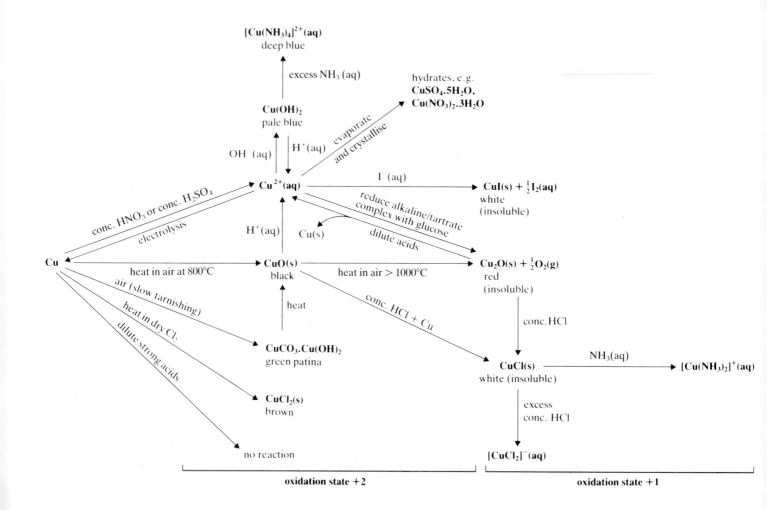

Figure 20.6
The reactions of copper and copper compounds.

Furthermore, copper will react directly on heating with chlorine and oxygen (both strong oxidising agents) forming copper(II) chloride and copper(II) oxide, respectively (figure 20.6).

Notice that all these reactions of copper result in the formation of copper(II), rather than copper(I) compounds. This emphasises another enigma concerning copper.

- Write the electronic structure for Cu, Cu^+ and Cu^{2+}.
- Which ion would you expect to be more stable, Cu^+ or Cu^{2+}? Explain your answer.

In terms of electron structure we would expect Cu^+ to be more stable than Cu^{2+}. Yet most common copper compounds are those of the element in the +2 state. Why is this? In the next section we must try to unravel this problem.

20.10 Exploring the relative stabilities of copper(I) and copper(II) compounds

Figure 20.7 shows two energy cycles relating to the formation of $Cu^+(aq)$ and $Cu^{2+}(aq)$ from copper. The cycles show that both $Cu^+(aq)$ and $Cu^{2+}(aq)$ are unstable relative to copper. This is not surprising, when one considers their negative electrode potentials. But, in this section, we are concerned not so much

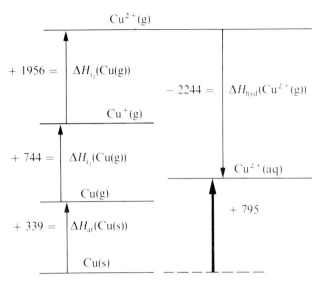

(a) The formation of $Cu^+(aq)$ from Cu

(b) The formation of $Cu^{2+}(aq)$ from Cu

Figure 20.7
The formation of $Cu^+(aq)$ and $Cu^{2+}(aq)$ from Cu (all values in kJ $mole^{-1}$).

with the stability of these ions with respect to copper, but their stability with respect to each other.

From figure 20.7, we can obtain the following values.

$$Cu(s) \rightarrow Cu^+(aq) \qquad \Delta H = +602\,kJ$$
$$Cu(s) \rightarrow Cu^{2+}(aq) \qquad \Delta H = +795\,kJ$$

These values can be used to calculate the energy change for the process

$$2Cu^+(aq) \rightarrow Cu(s) + Cu^{2+}(aq)$$

which involves the disproportionation of copper(I).

This is shown schematically in figure 20.8. The exothermic value for this change suggests that aqueous copper(I) ions are thermodynamically unstable with respect to aqueous copper(II) ions and copper. So, as we might expect, $Cu^+(aq)$ does not exist in aqueous solution.

Thus, when solid copper(I) oxide (Cu_2O) is treated with dilute acids, it forms a precipitate of copper and a solution of the copper(II) salt. It does not form a solution of the copper(I) salt (figure 20.6).

$$Cu_2O(s) + H_2SO_4(aq) \rightarrow Cu(s) + CuSO_4(aq) + H_2O$$

This means that the only copper(I) compounds which are stable in the presence of water must either be
(a) *insoluble* (e.g. Cu_2O, CuI and CuCl in figure 20.6) *or*
(b) *complexed with ligands other than water* (e.g. diamminecopper(I) ($[Cu(NH_3)_2]^+$) and dichlorocuprate(I)($[CuCl_2]^-$) in figure 20.6.)

By Hess's Law
$$2 \times 602 + \Delta H = 795$$
$$\therefore \Delta H = -409\,kJ$$

Figure 20.8
Enthalpy change for the disproportionation of Cu^+.

20.11 The reactions of metals and the activity series

Many of the reactions of metals can be summarised using either the periodic table or the activity series. The reactions of some common metals with oxygen, chlorine, water, steam and dilute acids are summarised in table 20.3. The metals are written in order of their position in the activity (electrochemical) series.

Table 20.3

Summary of the reactions of metals in the Activity Series (M is used as the general symbol for a metal and assumed to be divalent)

Metal	Reaction with $O_2(g)$ on heating	Heat evolved when metal reacts with 1 mole of O_2 to form oxide shown /kJ		Reaction with cold water	Reaction with steam	Reaction with dilute strong acids
K	form oxides (e.g. Na_2O) in limited supplies of O_2, but peroxides (e.g Na_2O_2) with excess O_2	K_2O	723	displace $H_2(g)$ from cold water with decreasing reactivity (K, violently; Mg, very slowly)	displace $H_2(g)$ from steam with decreasing vigour (K, very violently; Fe, very slowly)	displace $H_2(g)$ from dilute strong acids with decreasing vigour (K, explosively; Mg, very vigorously; Fe, steadily; Pb, very slowly)
Na		Na_2O	832			
Ca		CaO	1272			
Mg		MgO	1204			
Al	burn with decreasing vigour to form oxides	Al_2O_3	1114			
Zn		ZnO	697			
Fe		Fe_2O_3	548			
Pb	do *not* burn, but only form a surface layer of oxide	PbO	436	do *not* displace $H_2(g)$ from cold water	do *not* displace $H_2(g)$ from steam	do *not* displace $H_2(g)$ from dilute strong acids
Cu		CuO	311			
Hg		HgO	182			
Ag	do *not* burn or oxidise on surface	Ag_2O	61			
Pt						
Au		Au_2O_3	54			
General Eqtn	$2M + O_2 \rightarrow 2M^{2+}O^{2-}$			$M + H_2O \rightarrow$ $\rightarrow M^{2+}O^{2-} + H_2$	$M + H_2O \rightarrow$ $\rightarrow M^{2+}O^{2-} + H_2$	
Half-Eqtn	$2M \rightarrow 2M^{2+} + 4e^-$ $O_2 + 4e^- \rightarrow 2O^{2-}$			$M \rightarrow M^{2+} + 2e^-$ $H_2O + 2e^- \rightarrow$ $\rightarrow H_2 + O^{2-}$	$M \rightarrow M^{2+} + 2e^-$ $H_2O + 2e^- \rightarrow$ $\rightarrow H_2 + O^{2-}$	
Comment				the oxides which form react with water to form hydroxides	oxides of the top four metals react with water to form their hydroxides	

Reaction with Cl_2 on heating	Heat evolved when metal reacts with 1 mole of $Cl_2(g)$ to form the chloride shown/kJ		Strength as reducing agents	Reactions with other aqueous cations	Standard electrode potential, M(s) /M^{n+}(aq)/volts	Metal
	KCl	873	decreasing strength as reducing agents (i.e. electrons are donated less readily by the metal)		2.92	**K**
	NaCl	826			2.71	**Na**
	$CaCl_2$	795			2.87	**Ca**
	$MgCl_2$	642			2.37	**Mg**
all the metals react with $Cl_2(g)$ on heating to form their chloride with decreasing vigour	Al_2Cl_6	464		any metal will displace from solution ions of a metal below it in the activity series	1.66	**Al**
	$ZnCl_2$	416			0.76	**Zn**
	$FeCl_3$	270			0.44 ($Fe(s)/Fe^{2+}$(aq))	**Fe**
	$PbCl_2$	359			0.13	**Pb**
	$CuCl_2$	206			-0.34	**Cu**
	$HgCl_2$	230			-0.79 ($Hg(l)/Hg_2^{2+}$(aq))	**Hg**
	AgCl	255			-0.80	**Ag**
	$PtCl_3$	142			-1.20	**Pt**
	$AuCl_3$	79				**Au**

$M + Cl_2 \rightarrow M^{2+}(Cl^-)_2$			**General Eqtn**
$M \rightarrow M^{2+} + 2e^-$ $Cl_2 + 2e^- \rightarrow 2Cl^-$		$M(s) \rightarrow M^{2+}(aq) + 2e^-$	**Half-Eqtn**
metals with 2 oxidation states form the chloride of the metal in the highest state (e.g. $FeCl_3$ *not* $FeCl_2$)			**Comment**

The heat evolved when each metal reacts with one mole of oxygen and with one mole of chlorine is also included in the table. The electrode potentials of the metals are also given. Notice that there is a similar order of reactivity of these metals for each series of reactions.

Metals act as reducing agents in their reactions. They donate electrons to non-metals, water or H^+ ions.

$$M \xrightarrow{\text{oxidation}} M^{2+} + 2e^-$$
$$\text{reducing agent}$$

Of the metals listed in table 20.3, potassium is the strongest reducing agent whereas gold is the weakest. Potassium donates its electrons most readily, gold least readily. The ease with which a metal loses electrons is, of course, directly linked to the value of its electrode potential and hence its position in the electrochemical series.

20.12   The extraction of metals from their ores

Table 20.4 gives some information concerning the methods of extraction, the major sources, the annual productions and the uses of nine common metals.

Some of these extraction processes have already been considered in previous chapters. References to these earlier sections are given in the final column of table 20.4.

■ Why is iron used in much larger quantities than any other metal?
■ Which compounds do metals occur as most frequently in their ores?
■ As a general rule, the most electropositive metals occur in nature in combination with the most electronegative anions such as oxide and chloride. The less electropositive metals are found in combination with the less electronegative anions such as sulphide. Do you think this is a good generalisation? Illustrate your answer with several examples.
■ How do very poor metals such as gold and platinum occur in nature?
■ Why is there a clear division between those metals discovered in ancient times and those discovered in the eighteenth and nineteenth centuries?

The extraction of all metals from their naturally occurring ores involves three stages.

(A) Purification and concentration of the ore

Very few metal ores are sufficiently concentrated to be used without purification. The main impurities are earthy materials such as rocks, clays and sand. In many cases, the ore is separated from useless impurities before transportation. This reduces costs.

Usually the material is first crushed and ground into small pieces. Magnetic material can then be separated by using strong electromagnets. Alternatively, differences in density between the ore and its impurities may be sufficient to allow separation of crushed material on an agitated sloping table. Jets of water play across the line of inclination of the table. This causes the denser material to accumulate at the lower end.

One of the most common methods of purifying metal ores is **froth flotation**. In this process, the finely ground ore is added to water which contains special oils known as 'frothing agents'. These include pine oil, creosote and xanthates. Separation is possible because of the different densities of the materials and their different wetting characteristics with the frothing agent and water.

Table 20.4
The extraction and uses of metals

Metal	Date of discovery	Main ore from which metal is obtained	Main method of extraction	Annual World Production /1000 tonnes (1986)	Price per tonne (1987)	Main uses	Reference to other sections in the book
Sodium	1807	Rock salt, NaCl	Electrolysis of molten NaCl	400	—	Manufacture of tetraethyllead(IV), sodium vapour lamps, titanium	15.7–15.9, 20.14
Magnesium	1808	Magnesite, $MgCO_3$ and Mg^{2+} ions in sea water	Electrolysis of molten $MgCl_2$	340	£1410	Low-density alloys, e.g. in aircraft and nuclear reactors	15.7–15.9, 20.14
Aluminium	1827	Bauxite Al_2O_3	Electrolysis of Al_2O_3 in molten cryolite (Na_3AlF_6)	12000	£785	Kitchen utensils, packaging foil, structural alloys in aircraft, automobiles	16.4–16.6, 20.14
Zinc	1746	Zinc blende, ZnS	Heat sulphide in air → oxide. Dissolve oxide in H_2SO_4, electrolyse	5000	£475	Galvanising iron, die castings, alloys (brass)	20.14
Iron	Ancient	Haematite, Fe_2O_3	Reduce Fe_2O_3 with carbon monoxide	710000 (raw steel)	£130	Most important structural metal (as steel), vehicles, engines, tools	20.1–20.6, 20.14
Tin	Ancient	Tinstone, SnO_2	Reduce SnO_2 with carbon	133	£4500	Tin plate (coating iron), alloys (e.g. solder, pewter, bronze, etc.)	18.1–18.2
Lead	Ancient	Galena, PbS	Heat sulphide in air → oxide. Reduce oxide with carbon	4000	£300	Roof and cable covering, battery plates, alloys (e.g. solder, printing metals)	18.1–18.2
Copper	Ancient	Copper pyrites, $CuFeS_2$ (CuS + FeS)	Controlled heating with correct amount of air → Cu + SO_2	7500	£940	Electrical wires, cables, etc., water pipes, alloys (e.g. brass, bronze)	20.7–20.10, 20.14
Mercury	Ancient	Cinnabar, HgS	Heat in air → Hg + SO_2	6	£6500	Scientific equipment (e.g. barometers, thermometers, etc.), mercury vapour lamps, chlor-alkali cells	20.14

When air is blown through the mixture, air bubbles adhere to certain minerals, but not to others. The 'frothing agent' is carefully chosen so that particles of the ore become attached to air bubbles. The particles then float to the surface while earthy particles sink to the bottom. The ore particles collect as a froth and are removed periodically. The process of froth flotation is particularly suitable for concentrating dense ores such as galena (lead sulphide) and zinc blende (zinc sulphide).

Above. An Egyptian worker smelting metal ore. Why is he using a blow pipe?

Right. Egyptian workmen casting molten metal.

(B) Reduction of the purified ore to the metal

The various methods of reducing metal ores are discussed in section 20.14. The method used for a particular metal depends mainly on the position of the metal in the activity series.

(C) Purification of the metal

The metal obtained by reduction of the ore is usually contaminated with impurities. These include the unchanged ore, other metals present in the ore and non-metals from the anions in the ore. For example, the 'blister copper' obtained by reduction of copper ores contains small amounts of copper sulphide, iron and sulphur. The purification of 'blister copper' is described in section 20.7

Impure 'pig iron' obtained from the blast furnace contains about 8% impurities. These include carbon, silicon, sulphur, phosphorus and manganese. The removal of these impurities from 'pig iron' is described in section 20.4.

A derelict Cornish tin mine.

20.13 Factors influencing the choice of method used to reduce a metal ore

Chemical factors

Here, the most important question to ask is 'How easily can the ore be reduced to the metal?' This will dictate the cost of the method used. It is directly related to the position of the metal in the activity series.

Other factors which must be considered are:

—the accessibility of the ore;
—the ease with which the ore can be purified;
—the scale on which the metal is required, and
—the most suitable type of furnace or electrolytic technique to be used.

The enormous quantities of iron required for the production of steels could not be extracted by any method other than carbon reduction. This is the cheapest method available. In contrast, a much wider choice of techniques is possible for metals which command higher prices and those required in smaller quantities. Thus, titanium can be obtained economically by reduction of titanium(IV) chloride using sodium in an inert atmosphere.

$$TiCl_4 + 4Na \rightarrow Ti + 4NaCl$$

Economic factors

The final choice of extraction method will usually depend on cost. In this respect, quality of the ore is obviously important. Low-grade ores may not be economically worth exploitation. In this respect, it is interesting that Cornish tin mines, which closed down some decades ago, have now re-opened. The escalating prices of tin have made their operation economically profitable once more.

The demand for a particular metal affects its scale of production. Large-scale operations require plentiful and cheap materials. These may include a cheap reducing agent, such as coke, and possibly a cheap supply of electricity.

Consequently, almost all iron works are located near coalfields and aluminium smelters are usually situated near a source of cheap hydro-electric power.

Another crucial economic factor in the choice of extraction method relates to the value of any by-products. During the 1950s, a world surplus of zinc arose. Prices sunk to an almost uneconomic level. Nevertheless, extraction of zinc continued because the process yielded sulphur dioxide. This was needed for the manufacture of sulphuric acid which was in short supply at the time.

20.14 Metal extractions and the electrochemical series

Electrolysis of fused compounds for metals at the top of the electrochemical series

This method is used to extract potassium, sodium, calcium, magnesium, and aluminium. Chemical reduction of the oxides of these metals by carbon or carbon monoxide is not feasible. The temperature required for reduction is too high to make the process economical or practical.

Furthermore, these metals cannot be obtained by electrolysis of their aqueous solutions. Water in the solution would be discharged at the cathode in preference to the metal ions.

As neither chemical reduction nor aqueous electrolysis is possible, the only viable method of extraction is electrolysis of their fused compounds. This usually involves the chlorides.

In practice, the cathode at which the metal ions are discharged is usually made of steel. The anode is made of graphite.

| **cathode (−) (steel)** | $M^{2+}(l) + 2e^- \rightarrow M(l)$ |
| **anode (+) (graphite)** | $2Cl^-(l) \rightarrow Cl_2(g) + 2e^-$ |

Large quantities of electricity are required not only to electrolyse the chloride but also to maintain the molten state. Suitable impurities are normally added to the chloride. This reduces its melting point and less electrical energy is then needed to keep the electrolyte molten.

■ Why is the anode made of graphite rather than steel, even though steel is a better conductor than graphite?
■ Why must the metal liberated at the cathode be protected from the chlorine produced at the anode?
■ In the electrolytic manufacture of sodium, calcium chloride is added to sodium chloride electrolyte. This reduces the melting point of the sodium chloride. Will this result in any impurity in the sodium produced? Explain.

Chemical reduction of compounds for metals in the middle of the electrochemical series

Metals in the middle of the activity series (such as zinc, iron, tin, lead and copper) can be extracted from their oxides and sulphides by chemical reduction. Very often, the sulphides are converted to oxides before reduction to the metal.

$$2PbS + 3O_2 \rightarrow 2PbO + 2SO_2$$
galena

(a) Reduction of oxides by carbon

The oxides are reduced by coke in a closed furnace. This forms the metal and carbon monoxide. This method is used to obtain lead by heating a mixture of powdered lead oxide and coke at 1400°C.

$$PbO + C \rightarrow Pb + Co$$

(b) Reduction of oxides by carbon monoxide

Reduction of iron ore (Fe_2O_3) by carbon monoxide takes place in a blast furnace.

$$Fe_2O_3 + 3CO \rightarrow 2Fe + 3CO_2$$

(c) Self-reduction of sulphide ores

Self-reduction plays an essential part in the extraction of copper from its sulphide ores.

Part of the sulphide is first converted to oxide by roasting in air.

$$2Cu_2S + 3O_2 \rightarrow 2Cu_2O + 2SO_2$$

The supply of air is then cut off and the temperature is raised. The rest of the sulphide then reacts with oxide forming the metal and sulphur dioxide.

$$2Cu_2O + Cu_2S \rightarrow 6Cu + SO_2$$

(d) Reduction of compounds by more reactive metals

This method is particularly useful for those metals such as chromium and titanium which are expensive. Usually they are required in only small quantities.

For example titanium is obtained by heating titanium(IV) chloride to 850°C with magnesium or sodium in an atmosphere of argon.

$$TiCl_4 + 2Mg \rightarrow Ti + 2MgCl_2$$
$$TiCl_4 + 4Na \rightarrow Ti + 4NaCl$$

■ Why is it necessary to use an atmosphere of argon?
■ Chromium is obtained from chromium(III) oxide by heating with aluminium powder. Write an equation for this process.

Heating the ore alone or displacement from aqueous solution for metals at the bottom of the electrochemical series

(a) Heating the ore alone

The compounds of mercury and silver are so unstable that they decompose to the metal on heating. Consequently, mercury can be extracted from the ore, cinnabar (HgS), by heating in air.

$$HgS + O_2 \rightarrow Hg + SO_2$$

The mercury distils over at the temperature of the furnace. It is condensed in water-cooled receivers.

(b) Displacement from aqueous solution

In this case, the metal is precipitated from a solution of its ions. This requires a metal higher in the electrochemical series.

Silver can be extracted from very low-grade ore by this method. The insoluble silver sulphide ore is first dissolved in a solution of cyanide ions. It reacts with this to form dicyanoargentate(I) ions.

$$Ag_2S(s) + 4CN^-(aq) \rightarrow 2[Ag(CN)_2]^-(aq) + S^{2-}(aq)$$

Silver is then precipitated from the solution of dicyanoargentate(I) ions by the addition of powdered zinc dust.

$$2[Ag(CN)_2]^-(aq) + Zn(s) \rightarrow [Zn(CN)_4]^{2-}(aq) + 2Ag(s)$$

Summary

1 The world production of iron far exceeds that of any other metal. The reasons for this are that iron ores are abundant, rich, readily accessible and easily reduced to iron.

2 Iron can be alloyed easily with other elements to produce different steels with a great variety of properties and uses.

3 In many respects, iron is a typical transition metal. It produces coloured ions. It forms stable complexes. It shows variable oxidation state. The element and its compounds show catalytic activity.

4 Rusting requires the presence of both oxygen and water. During rusting, iron is first oxidised to Fe^{2+} ions by dissolved oxygen in the water. Iron(II) hydroxide then precipitates. This is further oxidised by dissolved oxygen to rust (hydrated iron(III) oxide, $Fe_2O_3.xH_2O$).

5 Two general methods are used to protect iron and steel from corrosion: the application of a protective layer or the application of a sacrificial metal.

6 The manufacture of copper from its ores involves three stages.
   (a) Conversion of the sulphide ore to Cu_2O by roasting in air.
   (b) Reduction of the Cu_2O to copper by heating with more sulphide ore in the absence of air.
   (c) Purification of the impure copper by electrolysis.

7 Although copper(II) compounds show typical transition metal properties, copper(I) compounds do not do so. Several copper(I) compounds are white. They show little or no catalytic activity and there are relatively few complexes containing Cu^+ ions.

8 $Cu^+(aq)$ is unstable in aqueous solution. It disproportionates spontaneously into $Cu^{2+}(aq)$ and Cu. Thus, copper(I) compounds are stable in contact with water only if they are insoluble or when they are complexed with ligands other than water.

9 Metals at the top of the electrochemical series (e.g. K, Na, Ca, Mg, Al) are usually obtained by electrolysis of their fused compounds.
   Metals in the middle of the electrochemical series (e.g. Zn, Fe, Sn, Pb, Cu) are usually obtained by chemical reduction of their oxides with carbon or carbon monoxide or by self-reduction of sulphide ores.
   Metals at the bottom of the electrochemical series (e.g. Cu, Hg, Ag, Au) are usually obtained by heating the ore alone or by displacement from aqueous solution.

Study questions

1 (a) What do you understand by the term 'transition metal'?
   (b) Which of the following do you regard as transition metals? Explain your answer.
        (i)   Scandium
        (ii) Iron
        (iii) Zinc
   (c) Although the salts of transition elements are usually coloured, there are several copper(I) compounds which are white. Suggest an explanation for this.

Table 20.5
The densities of four metals in the first transition series

Metal	Relative atomic mass	Density/g cm^{-3}
Vanadium	50.9	6.1
Manganese	54.9	7.4
Cobalt	58.9	8.9
Copper	63.5	9.0

   (d) The densities of transition elements in the same period gradually increase with relative atomic mass (table 20.5). Why is this?

2 Comment upon and explain the following observations.
   (a) An iron pipe, buried in the earth is protected from rusting by joining it to a bar of magnesium.
   (b) A transition metal, M, forms only two isomers of formula $[M(NH_3)_4Cl_2]^+Cl^-$.
   (c) Copper does not react appreciably with hydrochloric acid. But, in the presence of strong complexing agents, it will do so, evolving hydrogen steadily.
   (d) Iron immersed in copper(II) sulphate solution is rapidly coated with copper. This does not occur when the iron has previously been dipped in concentrated nitric acid.
   (e) Copper(II) sulphate does not react with potassium chloride solution. It does, however, react with potassium iodide solution to form a precipitate of copper(I) iodide and a solution of iodine.

3 Metals are usually isolated from their ores by reduction. Describe briefly the methods of reduction for the following metals. Suggest reasons for the method chosen for each metal.
   (a) Zinc from zinc blende (ZnS)
   (b) Magnesium from magnesium chloride extracted from sea water
   (c) Iron from haematite (Fe$_2$O$_3$)
   (d) Silver from silver glance (Ag$_2$S)

4 10 g of an impure iron(II) salt were dissolved in water and made up to 200 cm^3 of solution. 20 cm^3 of this solution acidified with dilute H$_2$SO$_4$ required 25 cm^3 of 0.03 M KMnO$_4$(aq) before a faint pink colour appeared.
   (a) Write a balanced equation (or half-equations) for the reaction of acidified manganate(VII) ions with iron(II) ions.
   (b) How many moles of iron(II) ions react with 1 mole of MnO$_4^-$ ions?
   (c) How many moles of Fe^{2+} react with 25 cm^3 of 0.03 M KMnO$_4$(aq)?
   (d) How many grams of Fe^{2+} are there in the 200 cm^3 of original solution? (Fe = 56)
   (e) What is the percentage by mass of iron in the impure iron(II) salt?

5 (a) Most metals exist naturally in a combined state. They are too reactive to occur native (uncombined). However, a few metals do occur native.
       Give examples of:
       (i)   two metals which never occur native,
       (ii) two metals which occur both native and combined,
       (iii) two metals which almost always occur naturally in an uncombined form.
   (b) Before extraction of the metal can begin, many ores must be purified and concentrated. Mention two different processes by which this is done.
   (c) Why do metal extractions often produce slag?
   (d) In many cases, the primary reduction of an ore does not produce metal of sufficient purity. The metal requires further purification. Outline two different processes by which further purification is carried out.

6 Economic considerations are important in deciding the method by which a metal is extracted from its ore.
       How may economic considerations affect:
   (a) the nature of the ore used,
   (b) the nature of the reducing agent used,
   (c) the location of the industrial plant at which extraction is carried out?
       Give examples wherever possible to illustrate the point you make.

7 The method by which any metal is extracted is dependent upon the position of that metal in the electrochemical series.
   (a) Highly reactive metals must be extracted by electrolysis.
       (i) Give a brief outline of the extraction process for one such metal. Mention the electrolyte used, the materials of which the electrodes are made and the reactions taking place at each electrode.
       (ii) Why are such reactive metals not obtained by chemical methods of reduction?
   (b) A few metals are obtained by reduction of their purified ores with a second more reactive metal.
       (i) Outline the chemical principles of one such extraction process.
       (ii) Why is this type of process fairly uncommon?
   (c) Less reactive metals can be extracted by reduction of their ores with carbon.
       (i) Outline the chemical principles of the extraction of two different metals using carbon.
       (ii) What are the disadvantages of reduction with carbon?

21 Equilibria

21.1 Introduction

Have you ever been in the difficult position of having your loyalty and responsibility pulled in two directions at the same time? If you have, you will know that very often you must settle for a compromise (or balance) between the alternatives.

In the same way, many physical and chemical processes also exist in a position of balance. In chapter 14, we studied the competition processes taking place in acid–base, redox and complexing processes. In these and other reactions, the starting materials are not always completely converted to the products. Sometimes, they reach an intermediate position or **equilibrium** in which both reactants and products are present.

In real life, compromise in difficult circumstances is usually dictated by many factors—responsibility, loyalty, honesty, age, parents, finance, etc. Fortunately, the factors which dictate the balance position in chemical reactions are fewer and less complicated than those in real life.

The tendency for materials to exist in a state of maximum disorder is a key factor in dictating the position of equilibrium. Disorder is, in fact, a more natural state than order. Even so, it may be rather difficult to convince your parents of this when you are next told to tidy your room.

One of the key factors which dictates the position of equilibrium in any physical or chemical process is the tendency for materials to exist in the lowest energy state. Will equilibrium ever be achieved in the waterfall?

21.2 Equilibria in physical processes

Liquid–vapour equilibrium: vapour pressure

When liquid bromine is shaken in a stoppered flask, some of the liquid evaporates forming an orange gas. Gradually, the gas becomes thicker. Eventually, the intensity of the gas does not change any more. However much we shake the flask, its colour remains constant. Provided we took more than a tiny volume of liquid bromine, some liquid remains in the flask. The constant colour of the gas in the flask suggests that a position of balance has been reached. Some of the bromine has formed a vapour and some of the bromine remains as a liquid. A position of equilibrium has been reached between bromine liquid and bromine gas. This equilibrium can be summarised as:

$$Br_2(l) \rightleftharpoons Br_2(g)$$

The equilibrium sign ($\rightleftharpoons$) is used rather than an arrow to indicate that both bromine liquid and bromine gas are present in the flask.

What is happening to the molecules in the flask at equilibrium? Do all those in the gas remain as gas while all those in the liquid remain as liquid (**a static equilibrium**)? Or, are some gas molecules becoming liquid molecules while an equal number of liquid molecules become gas molecules (**a dynamic equilibrium**)?

Experiments show that liquid and gas molecules are moving rapidly and randomly in all directions. So, it would seem likely that molecules in the flask are in a dynamic rather than a static equilibrium. Thus, the rate at which molecules leave the liquid surface and enter the vapour is equal to the rate at which other molecules in the vapour return to the liquid. Random molecular activity occurs even after all the external signs of change have disappeared. But, we cannot *see* the molecules moving. Nor can we measure the rate at which they enter or leave the vapour.

The difference between a static and a dynamic equilibrium is illustrated in figure 21.1 on the next page using two familiar situations.

Consider the equilibrium between bromine liquid and bromine vapour once again. Figure 21.2(*a*) shows a perfect balance between the rate of evaporation and the rate of condensation at equilibrium.

Now, suppose that some of the vapour is suddenly removed without affecting the system in any other way (figure 21.2(*b*)).

■ Will the rate of condensation of gas molecules still equal the rate of evaporation of liquid molecules? Explain.
■ What happens to the concentration of molecules in the gas phase as time goes on? Explain.
■ How and when is equilibrium attained once more?

Suppose, now, the equilibrium is disturbed by injecting an excess of bromine vapour into the flask. The concentration of gas molecules suddenly rises and the rate of condensation increases. Condensation occurs faster than evaporation. Thus the concentration of gas molecules in the vapour phase decreases. Gradually, the rate of condensation falls until finally the rate of condensation equals the rate of evaporation. The system is in equilibrium once more and the vapour pressure of the bromine is the same as it was before the equilibrium was disturbed.

Notice that equilibrium can be reached from either direction. We can start with either bromine liquid or with excess bromine vapour. *Provided we use a closed container*, an equilibrium will eventually be obtained.

Solute–solution equilibrium: solubility

When one teaspoon of sugar is added to a cup of tea, all the sugar dissolves forming a solution. If more sugar is added, this also dissolves. If you continue to add sugar, a stage is eventually reached when no more will dissolve. The solution

Figure 21.1
The difference between a static and a
dynamic equilibrium.

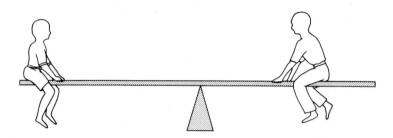

(*a*) Static equilibrium: children on a see-saw
At the balance point (i.e. the equilibrium
position) no movement of the children or
the see-saw occurs.

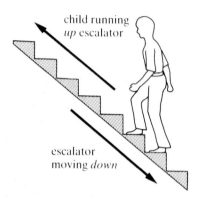

child running
up escalator

escalator
moving *down*

(*b*) Dynamic equilibrium: child ascending
escalator at the same rate as the escalator
descends. At the balance point (i.e. the
equilibrium position) the child and the
escalator are moving at the same rate in
opposite directions. Thus, the equilibrium
position is maintained.

(tea) is now saturated with solute (sugar) at the temperature involved. Solute
particles in the undissolved sugar are in equilibrium with solute particles in the
solution.

$$\text{sugar(s)} \rightleftharpoons \text{sugar(aq)}$$

Provided the system is closed, no evaporation of solvent occurs and the amounts
of dissolved and undissolved solute remain constant. As in the liquid–vapour
system, macroscopic properties become constant at equilibrium. In this case, the
constant macroscopic properties are the concentration of the solution and the
amount of undissolved solute.

*Investigating the nature of the equilibrium between a solute and its
saturated solution*

What are the molecules doing when equilibrium is established between a solute
and its saturated solution? No changes are apparent at a macroscopic level, but
what is happening at a molecular (microscopic) level?

It would seem reasonable to predict that the equilibrium in this case is also
dynamic. This would involve particles leaving the undissolved solute and
entering the solution at the same rate as dissolved particles rejoin the solid.

In order to check these ideas about a dynamic equilibrium, we can 'label'
some of the particles in either the saturated solution or in the undissolved solute.
This is done by using a radioactive material.

When the atmosphere is saturated
with water vapour, water is in equilibrium with its vapour.

Figure 21.2
Liquid–vapour equilibria.

(*a*) **Equilibrium**
Rate of evaporation
= rate of condensation

(*b*) **Imbalance**
Gas removed
from vapour phase

(*c*) **Imbalance**
Gas added to vapour
phase
Rate of condensation
> rate of evaporation

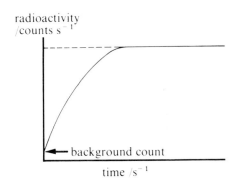

Figure 21.3
Following the radioactivity of a
saturated solution in contact with its
radioactive solute.

If solid radioactive $^{212}PbCl_2$ is added to a saturated solution of non-radioactive $PbCl_2(aq)$, no increase in the amount of dissolved lead chloride can occur. Nevertheless, if our ideas about a dynamic equilibrium are correct, there will be an interchange of Pb^{2+} between the undissolved solid and the saturated solution. As a result, radioactive $^{212}PbCl_2$ will appear in the solution.

The background radiation of the saturated aqueous lead chloride is first measured. It is then shaken with radioactive solid $^{212}PbCl_2$ for 10 minutes. After this, the saturated solution and undissolved solute are separated by centrifuging and the radioactivity of the saturated solution is measured a second time. The mixing and centrifuging are repeated at intervals and the radioactivity of the saturated solution is measured at each stage. Figure 21.3 shows how the radioactivity of the solution changes with time.

The increasing radioactivity of the saturated solution suggests that solid radioactive $^{212}PbCl_2$ must have dissolved. But, the solution was already saturated before any $^{212}PbCl_2$ dissolved. So, the radioactive particles dissolving from the solid must have been replaced by non-radioactive particles crystallising onto the solid from the solution. This means that a dynamic equilibrium is taking place betwen the undissolved solute and the saturated solution:

$$PbCl_2(s) \rightleftharpoons PbCl_2(aq)$$

Figure 21.3 shows that the radioactivity of the saturated solution eventually reaches a constant value. Why is this?

21.3 Characteristic features of a dynamic equilibrium

The last section has highlighted four important features of dynamic equilibria.

1 *At equilibrium, macroscopic properties are constant* under the given conditions of temperature, pressure and initial amounts of substances.

2 *At equilibrium, microscopic (molecular-scale) processes continue but these are in balance. This means that no overall macroscopic (large-scale) changes occur.* The particles participate in both forward and reverse processes. The rate of the forward process is equal to the rate of the reverse process so that no net change results.

3 *The equilibrium can be attained from either direction* beginning with only the materials on one side of the change. Changes of this kind are described as **reversible**.

4 *Equilibrium can only be achieved in a closed system.* A closed system is one in which there is absolutely no loss or gain of materials to or from the surroundings. An open system may allow matter to escape or to enter. This cannot reach equilibrium.

21.4 Equilibria in chemical reactions

The decomposition of calcium carbonate

When calcium carbonate is heated strongly, it decomposes forming calcium oxide and carbon dioxide:

$$CaCO_3(s) \xrightarrow{\text{heat}} CaO(s) + CO_2(g)$$

Normally, CO_2 escapes well away from the solid CaO. So, their recombination to form $CaCO_3$ never occurs in an open container. The system can never reach equilibrium.

If, however, a few grams of $CaCO_3(s)$ are heated at 800°C in a *closed* evacuated container, only part of the solid is decomposed no matter how long it is heated. The pressure of CO_2 inside the container rises, and then remains steady at a constant value. As long as the temperature stays at 800°C, the pressure remains constant at 25 kPa (0.25 atm). The reaction appears to have reached an equilibrium. Constant macroscopic properties have been achieved in a closed system.

We can check that the system is at equilibrium, by showing that it reaches the same macroscopic composition when approached from the opposite direction.

A rotatory lime kiln at the Westbury Works, near Trowbridge, Wiltshire. The kiln is viewed from the lower heated end. Raw limestone slurry is pumped in at the higher end. As the slurry flows down the kiln, it is first dried and then decomposed to lime at the lower heated end. The kiln is inclined at 1 in 30. It makes one revolution per minute.

Solid CaO is first placed in the reaction vessel. This is then evacuated and refilled with CO_2 at a pressure well above 25 kPa. Finally, the container is closed and maintained at 800°C. The pressure begins to fall and becomes steady at 25 kPa (0.25 atm). The same equilibrium pressure of CO_2 results whether we start with $CaCO_3$ or with CaO and CO_2. This is further evidence that the materials are in equilibrium. This equilibrium can be summarised as

$$CaCO_3(s) \rightleftharpoons CaO(s) + CO_2(g)$$

In this case, $CaCO_3$ is decomposing to form CaO + CO_2, at the same rate as CaO + CO_2 are reforming $CaCO_3$.

Although this reaction will come to equilibrium in a closed vessel, the industrial process is never allowed to do so. Millions of tonnes of calcium oxide are produced annually by heating limestone in large open kilns. The CaO is used in liming the soil and as a constituent of plaster and mortar.

The iodine chloride–iodine trichloride equilibrium

When dry chlorine is passed over brown liquid iodine chloride (ICl), yellow crystals of iodine trichloride (ICl_3) form and the brown liquid disappears:

$$\underset{\text{pale green}}{Cl_2(g)} + \underset{\text{brown}}{ICl(l)} \rightarrow \underset{\text{yellow}}{ICl_3(s)}$$

If the supply of chlorine is stopped, the ICl_3 decomposes if it is left in an open container. The ICl_3 decomposes to a brown liquid (ICl) and a pale green gas (Cl_2):

$$\underset{\text{yellow}}{ICl_3(s)} \rightarrow \underset{\text{brown}}{ICl(l)} + \underset{\text{pale green}}{Cl_2(g)}$$

This is another example of a reversible reaction. But what happens when a mixture of $Cl_2(g)$ and ICl(l) is left in a closed vessel from which the air has been removed? Yellow crystals of ICl_3 form, but not all the reactants are used up. As the reaction proceeds, the amounts of ICl and Cl_2 fall and the amount of ICl_3 rises (figure 21.4).

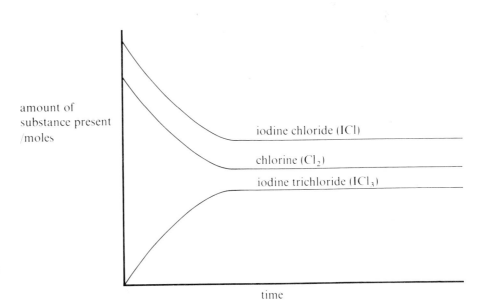

Figure 21.4
A sketch graph showing the amount of substances present as equilibrium is approached in the reaction between ICl and Cl_2.

After some time, the composition of the mixture becomes constant. The amount of each reactant remains steady. The reaction seems to have reached an equilibrium because the macroscopic properties of the closed system have become constant.

- What is happening to the molecules of ICl_3, ICl and Cl_2 inside the container at equilibrium?
- How is the rate of formation of ICl_3 related to the rate of decomposition of ICl_3 at equilibrium?
- How could you show experimentally that the equilibrium is a dynamic one?

Now look closely at figure 21.4.

- What is the relationship between the decrease in amount of ICl, the decrease in amount of Cl_2 and the increase in amount of ICl_3?
- Why is there always a constant difference between the amounts of ICl and Cl_2?
- Why do the three curves become horizontal at the same time?

21.5 The equilibrium of a solute between two immiscible solvents—the partition coefficient

The following experiment investigates the equilibrium of a solute between two immiscible solvents. Dissolve a small crystal of iodine in $5 cm^3$ of aqueous potassium iodide. A brown solution forms. Then carefully add $5 cm^3$ of tetrachloromethane to the mixture. The tetrachloromethane is clear at first, but when the mixture is shaken gently, iodine dissolves from the KI solution into the tetrachloromethane. A purple solution of iodine forms in the latter. Now shake vigorously. The density of the purple colour in the tetrachloromethane darkens whilst the brown colour of the KI(aq) becomes paler. Eventually, the colours of the two solutions remain constant. The iodine has distributed itself between the two solvents and an equilibrium has been attained. No matter how much the mixture is shaken, no further changes in colour intensity occur. The concentrations of iodine in the two solutions remain constant.

In the same way, other solutes which are soluble in two immiscible solvents will distribute themselves between both solvents when the three substances are shaken together. These are further examples of equilibrium systems. But, how do the concentrations of the solute in the different solvents depend on the initial amount of solute taken? Is there a relationship between the equilibrium concentrations of the solute in the two solvents?

Investigating the partition of butanedioic acid (succinic acid) between water and ethoxyethane (diethyl ether)

One gram of butanedioic acid was added to a separating funnel with $25 cm^3$ of water and $25 cm^3$ of ethoxyethane (ether). This was shaken until all the solid had dissolved and equilibrium had been established. The mixture was then left to stand for some time. This allows the organic and aqueous layers to separate as fully as possible.

The two layers were then separated. The concentration of butanedioic acid in each layer was determined by titration against sodium hydroxide solution of known molarity using phenolphthalein indicator.

Further experiments were carried out using different initial amounts of butanedioic acid between 0.5 and 1.5 grams. The results obtained are shown in table 21.1. When these equilibrium concentrations in table 21.1 are plotted graphically, a straight line is obtained (figure 21.5),

- Would you expect the graph to go through the origin? Explain.
- What does the straight line graph suggest about the ratio of the equilibrium concentrations of butanedioic acid in the two solvents?
- How does the ratio of concentrations depend upon the amount of butanedioic acid taken?

Table 21.1
Equilibrium concentrations of butanedioic acid in ether and in water

Experiment Number	Equilibrium concentration of butanedioic acid in ether layer/mole dm^{-3}	Equilibrium concentration of butanedioic acid in water layer/mole dm^{-3}
1	0.023	0.152
2	0.028	0.182
3	0.036	0.242
4	0.044	0.300
5	0.052	0.358
6	0.055	0.381

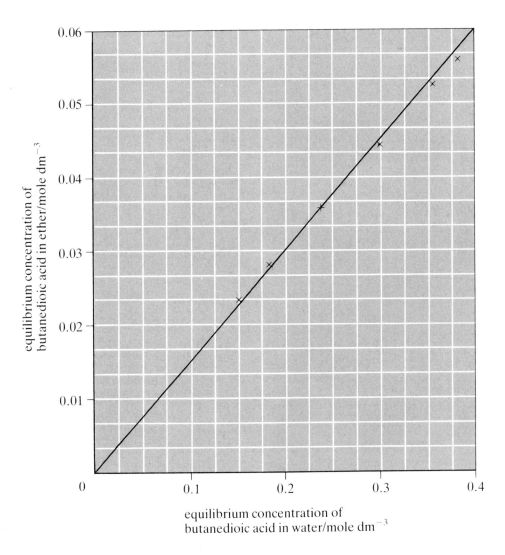

Figure 21.5
A graph of the equilibrium concentration of butanedioic acid in ether against that in water.

The results in figure 21.5 show that the ratio:

$$\frac{\text{concentration of butanedioic acid in ether}}{\text{concentration of butanedioic acid in water}} \text{ is constant.}$$

Using square brackets, [ ], to denote the concentrations, we can deduce from the graph in figure 21.5 that

$$\frac{[\text{butanedioic acid (ether)}]_{\text{eqm}}}{[\text{butanedioic acid (water)}]_{\text{eqm}}} = 0.15$$

This ratio is known as the **partition coefficient** (or the **distribution ratio**) for butanedioic acid between ether and water. *The partition coefficient is independent of the amount of solute taken and also independent of the volumes of the solvents used.* Notice that it is *the ratio of concentrations and not the ratio of masses* of solute which matter. Other investigations of the distribution of solutes between two immiscible liquids at equilibrium also give constant partition coefficients. In each case, the partition coefficient is constant provided:

(i)   the temperature is constant,
(ii) the solvents are immiscible and do not react with each other,
(iii) the solute does not react, associate or dissociate in the solvents.

21.6 Solvent extraction

The most important application of partition is in solvent extraction. Organic compounds are generally more soluble in non-polar solvents such as methylbenzene (toluene), tetrachloromethane and ethoxyethane (ether) than in water. These solvents are themselves immiscible with water. Thus, organic compounds can be extracted from aqueous solutions or suspensions by shaking with a non-polar organic solvent and then separating the two layers. The pure organic compound can then be obtained by distilling off the solvent. For example, penicillin is extracted from a dilute aqueous solution using trichloromethane. Phenylamine (aniline) can be reclaimed from a mixture with water by using ether.

During solvent extraction, it is more efficient to use a given volume of organic solvent in small portions rather than all at once. The following calculation shows why this is so.

Suppose that $100\,\text{cm}^3$ of ether is available for extracting the solute, X, from $100\,\text{cm}^3$ of water. Suppose also that the partition coefficient of X between ether and water is 4.

Fraction of X extracted using $100\,\text{cm}^3$ of ether all at once

After shaking the aqueous solution of X with ether, we know that

$$\frac{\text{concentration of X in ether}}{\text{concentration of X in water}} = 4$$

$$\Rightarrow \frac{\text{mass of X in ether}}{\text{volume of ether}} \times \frac{\text{volume of water}}{\text{mass of X in water}} = 4$$

Let us suppose that m_e is the mass of X in ether and m_w is the mass of X in water. Then we can write

$$\frac{m_e}{100} \cdot \frac{100}{m_w} = 4$$

$$\therefore \frac{m_e}{m_w} = \frac{4}{1}$$

Penicillium mould growing on filter paper. The penicillin can be washed into aqueous solution and then extracted using trichloromethane.

i.e. four parts of X are in the ether and one part is in the water. Thus, using $100\,cm^3$ of ether in one extraction, four-fifths of X are extracted into the organic solvent.

Fraction of X extracted using two $50\,cm^3$ portions of ether

When $50\,cm^3$ of ether is used to extract X from $100\,cm^3$ of water, we can write

$$\frac{m_c}{50} \cdot \frac{100}{m_w} = 4$$

$$\therefore \frac{m_c}{m_w} = 4 \times \frac{50}{100} = \frac{2}{1}$$

i.e. two parts of X are in the ether and one part in the water. This means that two-thirds of X are extracted by the first $50\,cm^3$ of ether. This leaves one-third in the water.

The second $50\,cm^3$ portion of ether extracts two-thirds of the remaining one-third, i.e. $\frac{2}{3} \times \frac{1}{3} = \frac{2}{9}$ths of the original amount of X.

Thus, the total fraction of X extracted by two $50\,cm^3$ portions of ether is $\frac{2}{3} + \frac{2}{9} = \frac{8}{9}$. This is more than the four-fifths extracted by one $100\,cm^3$ portion of ether. The total fraction of X extracted would, of course, be even greater than eight-ninths if three portions of $33\frac{1}{2}\,cm^3$ were used, and so on.

21.7 The equilibrium constant

Earlier in this chapter we obtained a constant value for the ratio of concentrations of a solute between two immiscible solvents at equilibrium. Thus, it would seem sensible to investigate whether there is a similar relationship between the concentrations of reactants and products at equilibrium in chemical reactions.

Table 21.2 shows information concerning the following reaction at 731 K:

$$H_2(g) + I_2(g) \rightleftharpoons 2HI(g)$$

In experiments 1, 2, 3 and 4 the sealed reaction vessel contains gaseous hydrogen and gaseous iodine initially. After a time the mixture in the flask remains unchanged and equilibrium is attained. In experiments 5 and 6, equilibrium is approached from the opposite direction. The reaction vessel contains only gaseous hydrogen iodide initially. Table 21.2 shows the initial and equilibrium concentrations of hydrogen, iodine and hydrogen iodide.

Table 21.2
Initial and equilibrium concentrations of H_2, I_2 and HI

Experiment number	Initial concentrations/mole dm^{-3}			Equilibrium concentrations/mole dm^{-3}		
	$[H_2(g)]$	$[I_2(g)]$	$[HI(g)]$	$[H_2(g)]_{eqm}$	$[I_2(g)]_{eqm}$	$[HI(g)]_{eqm}$
1	2.40×10^{-2}	1.38×10^{-2}	0	1.14×10^{-2}	0.12×10^{-2}	2.52×10^{-2}
2	2.40×10^{-2}	1.68×10^{-2}	0	0.92×10^{-2}	0.20×10^{-2}	2.96×10^{-2}
3	2.44×10^{-2}	1.98×10^{-2}	0	0.77×10^{-2}	0.31×10^{-2}	3.34×10^{-2}
4	2.46×10^{-2}	1.76×10^{-2}	0	0.92×10^{-2}	0.22×10^{-2}	3.08×10^{-2}
5	0	0	3.04×10^{-2}	0.345×10^{-2}	0.345×10^{-2}	2.35×10^{-2}
6	0	0	7.58×10^{-2}	0.86×10^{-2}	0.86×10^{-2}	5.86×10^{-2}

■ According to the equation

$$H_2(g) + I_2(g) \rightleftharpoons 2HI(g)$$

one mole of H_2 reacts with one mole of I_2 forming two moles of HI.
Check the data from experiments 1 and 2 to see whether:
no. of moles H_2 reacted = no. of moles I_2 reacted = $\frac{1}{2}$ no. of moles HI formed.
■ Why is $[H_2(g)]_{eqm} = [I_2(g)]_{eqm}$ in experiments 5 and 6?
■ How could you show that the system involving H_2, I_2 and HI is in dynamic equilibrium?

By comparing the reaction $H_2(g) + I_2(g) \rightleftharpoons 2HI(g)$ with a partition equilibrium, it would seem reasonable to calculate the ratio of product and reactant concentrations as

$$\frac{[HI(g)]_{eqm}}{[H_2(g)]_{eqm}\,[I_2(g)]_{eqm}}$$

For each set of equilibrium concentrations, we want to see if the expression is constant. The computed values are shown in the second column of table 21.3. The ratios are far from constant.
   If, however, we evaluate the values for the ratio

$$\frac{[HI(g)]^2_{eqm}}{[H_2(g)]_{eqm}\,[I_2(g)]_{eqm}}$$

we obtain the results in column three of table 21.3. The values obtained this time are constant, within experimental limits. Therefore, we can write

$$\frac{[HI(g)]^2_{eqm}}{[H_2(g)]_{eqm}\,[I_2(g)]_{eqm}} = \text{a constant} = 46.8 \text{ at } 731\,K$$

Notice that the ratio of concentrations which gives a constant value is related to the stoichiometry of the reaction.

$$H_2(g) + I_2(g) \rightleftharpoons 2HI(g)$$

The ratio uses the *second* power of the [HI] in the equilibrium expression and the first power for both [H₂] and [I₂]. Thus, *the power to which we raise the concentration of a substance in the equilibrium expression is the same as its coefficient in the stoichiometric equation.*

Table 21.3
Equilibrium constants for the reaction $H_2 + I_2$ 2HI

Experiment number	$\dfrac{[HI(g)]_{eqm}}{[H_2(g)]_{eqm}\,[I_2(g)]_{eqm}}$	$\dfrac{[HI(g)]^2_{eqm}}{[H_2(g)]_{eqm}\,[I_2(g)]_{eqm}}$
1	1840	46.4
2	1610	47.6
3	1400	46.7
4	1520	46.9
5	1970	46.4
6	790	46.4

The value of this ratio of concentrations at equilibrium is known as the **equilibrium constant**. The equilibrium constant is represented by the symbol K_c. Thus, for the reaction $H_2(g) + I_2(g) \rightleftharpoons 2\,HI(g)$

$$K_c = \frac{[HI(g)]^2_{eqm}}{[H_2(g)]_{eqm}\,[I_2(g)]_{eqm}}$$

Very often, the 'eqm' subscripts are omitted from the concentration terms because it is assumed that the concentrations in the expression for K_c are necessarily equilibrium ones. Hence, the expression for K_c is shortened to

$$K_c = \frac{[HI(g)]^2}{[H_2(g)]\,[I_2(g)]}$$

The subscript 'c' indicates that K_c is expressed in concentrations.

21.8 The equilibrium law

The equilibria in many other chemical reactions have also been studied. In each case an equilibrium constant has been obtained. This relates to the stoichiometric equation in a similar manner to that for the hydrogen, iodine and hydrogen iodide system.

For example, the reaction

$$Fe^{3+}(aq) + NCS^-(aq) \rightleftharpoons Fe(NCS)^{2+}(aq)$$

has been studied by colorimetry. The reaction has an equilibrium constant in which

$$K_c = \frac{[Fe(NCS)^{2+}(aq)]}{[Fe^{3+}(aq)]\,[NCS^-(aq)]}$$

These observations lead to the general statement known as the **Equilibrium Law** or the Law of Chemical Equilibrium.

If an equilibrium mixture contains substances A, B, C and D related by the equation

$$aA + bB \rightleftharpoons cC + dD$$

it is found experimentally that

$$\frac{[C]^c[D]^d}{[A]^a[B]^b} = K_c$$

K_c, *the equilibrium constant, is constant at a given temperature.*

In writing expressions for the equilibrium constants of reactions, there are important conventions. Concentrations of substances on the right-hand side of the equation are written in the numerator. Concentrations of substances on the left-hand side are written in the denominator.

Thus, it is essential to relate any numerical value for an equilibrium constant to the particular equation concerned. Suppose the equilibrium constant for the reaction $H_2 + I_2 \rightleftharpoons 2\,HI$

$$K_c = \frac{[HI]^2}{[H_2][I_2]} = x$$

The equilibrium constant for the reverse reaction $2\,HI \rightleftharpoons H_2 + I_2$ at the same temperature is

$$K_c' = \frac{[H_2][I_2]}{[HI]^2} = \frac{1}{x}$$

As you would expect, $K_c' = \dfrac{1}{K_c}$

On the other hand, the equilibrium constant for the reaction

$$\tfrac{1}{2}H_2 + \tfrac{1}{2}I_2 \rightleftharpoons HI$$

is

$$K_c'' = \frac{[HI]}{[H_2]^{\frac{1}{2}}[I_2]^{\frac{1}{2}}} = \sqrt{x}$$

■ What is the relation between K_c and K_c''?

K_c has no units in reactions with equal numbers of particles on both sides of the equation. This is because the concentration units cancel out in the expression for K_c, i.e. for the reaction

$$H_2(g) + I_2(g) \rightleftharpoons 2HI(g)$$

$$K_c = \frac{[HI(g)]^2}{[H_2(g)][I(g)]} \qquad units = \frac{(\text{mole dm}^{-3})^2}{(\text{mole dm}^{-3})(\text{mole dm}^{-3})}$$

For reactions in which the numbers of reactant and product particles are not equal, K_c will, of course, have units.

What are the units of K_c for the following reactions?

■ $N_2(g) + 3H_2(g) \rightleftharpoons 2NH_3(g)$,
■ $2NH_3(g) \rightleftharpoons N_2(g) + 3H_2(g)$,
■ $2NO(g) + O_2(g) \rightleftharpoons 2NO_2(g)$,
■ $NO(g) + \tfrac{1}{2}O_2(g) \rightleftharpoons NO_2(g)$.

Determination of equilibrium constants

The essential stages in determining an equilibrium constant are listed below.

(i)   Write the stoichiometric equation.
(ii)   Mix known molar amounts of either the reactants or the products.
(iii)   Allow the mixture to reach equilibrium.
(iv)   Determine the equilibrium concentration of at least one substance in the equilibrium mixture. This analysis might be carried out using colorimetry, pressure measurements, titration or some other method.
(v)   Deduce the equilibrium concentrations of the other materials in the mixture.
(vi)   Substitute the equilibrium concentrations in the expression for K_c.
(vii)   Repeat the determination of K_c using different initial concentrations.

In determining and using equilibrium constants it is important to realise the following points:

1   *The Equilibrium Law only applies to systems at equilibrium.*

2   *K_c is constant only so long as the temperature remains constant.* If the temperature changes, the value of K_c will change.

3   *The numerical value of K_c is unaffected by changes in concentration of either reactants or products.* Obviously, when more reactant is suddenly added to a system in equilibrium, more of the products will tend to form. Eventually the system will adjust itself at a new equilibrium position in which the concentrations of reactants and products give the same numerical value for K_c.

4   *The magnitude of K_c provides a useful indication of the extent of a chemical reaction.* A large value for K_c indicates a high proportion of products to reactants (i.e. an almost complete reaction). A low value for K_c indicates that only a small fraction of reactants has been converted to products.

5   *The equilibrium constant for a reaction indicates the extent of a reaction. It gives no information about the rate of reaction. K_c tells us how far, but not how fast the reaction goes.* In fact, the extent and the rate of a reaction are quite independent. For example, the conversion of sulphur trioxide and oxygen to sulphur trioxide at 450°C occurs very slowly but almost completely. In

Absorber towers for sulphur trioxide (SO_3). The sulphur trioxide passes up the tower down which concentrated H_2SO_4 trickles.

contrast, the conversion of nitrogen oxide and oxygen to nitrogen dioxide at the same temperature occurs rapidly but only partially.

The following worked example will help you to understand the ideas we have covered so far. It will show you how equilibrium constants and equilibrium concentrations can be calculated.

When 1 mole of hydrogen iodide is allowed to dissociate in a $1.0 \, dm^3$ vessel at $440°C$, only 0.78 moles of HI are present at equilibrium. What is the equilibrium constant at this temperature for the reaction, $2HI \rightleftharpoons H_2 + I_2$?

	$2HI$	$\rightleftharpoons$	H_2	$+$	I_2	
No. of moles initially	1		0		0	

No. of moles at equilibrium	0.78	0.11	0.11	Since 0.22 moles of HI have decomposed, 0.11 moles of H_2 and 0.11 moles of I_2 have formed at equilibrium

Concentration at equilibrium/moles dm^{-3}	$\dfrac{0.78}{1}$	$\dfrac{0.11}{1}$	$\dfrac{0.11}{1}$

$$K_c = \frac{[H_2][I_2]}{[HI]^2} = \frac{0.11 \times 0.11}{(0.78)^2} = \frac{1}{50} = 0.02$$

$$\Rightarrow K_c \text{ for the reaction} = 0.02$$

Suppose 2 moles of hydrogen and 1 mole of iodine are mixed together in a $1.0 \, dm^3$ vessel at $440°C$. How many moles of HI, H_2 and I_2 will be present at equilibrium?

	$2HI$	$\rightleftharpoons$	H_2	$+$	I_2	
No. of moles initially	0		2		1	

No. of moles at equilibrium	$2x$	$2-x$	$1-x$	We have assumed that $2x$ moles of HI have formed at equilibrium. Hence, x moles of H_2 and x moles of I_2 must have disappeared.

Concentration at equilibrium/moles dm^{-3}	$\dfrac{2x}{1}$	$\dfrac{2-x}{1}$	$\dfrac{1-x}{1}$

$$\Rightarrow K_c = \frac{(2-x)(1-x)}{(2x)^2} = 0.02$$

$$\therefore \frac{2 - 3x + x^2}{4x^2} = 0.02$$

$$\Rightarrow 0.92x^2 - 3x + 2 = 0$$

Solving this quadric equation using the formula,

$$x = \frac{-b \pm \sqrt{b^2 - 4ac}}{2a}$$

we get

$$x = 0.935$$

Hence, at equilibrium:
number of moles of HI $= 2x = 1.870$
number of moles of $H_2 = (2 - x) = 1.065$
number of moles of $I_2 = (1 - x) = 0.065$

See if you can now answer the following questions.

■ The equilibrium constant for the reaction

$$2NO_2(g) \rightleftharpoons N_2O_4(g)$$

at $298 \, K$ is $200 \, mole^{-1} \, dm^3$.

(a) Write an expression for the equilibrium constant for the reaction.
(b) If the concentration of $N_2O_4(g)$ in the equilibrium mixture at 298 K is 2×10^{-2} mole dm^{-3}, what is the concentration of $NO_2(g)$?
(c) Calculate the equilibrium constant at 298 K for the reaction

$$\tfrac{1}{2}N_2O_4(g) \rightleftharpoons NO_2(g)$$

■ The equilibrium constants for the synthesis of hydrogen chloride, hydrogen bromide and hydrogen iodide are given below.

			K_c
$H_2(g) + Cl_2(g)$	$\rightleftharpoons$	$2\,HCl(g)$	10^{17}
$H_2(g) + Br_2(g)$	$\rightleftharpoons$	$2\,HBr(g)$	10^9
$H_2(g) + I_2(g)$	$\rightleftharpoons$	$2\,HI(g)$	10

(a) What do the values of K_c tell you about the extent of each reaction?
(b) Which of these reactions would you regard as virtually complete conversions?

21.9 Equilibrium constants in gaseous systems

Equilibrium constants are normally expressed in terms of concentrations using the symbol, K_c. For reactions involving gases, however, it is usually more convenient to express the amount of gas present in terms of its partial pressure rather than its molar concentration.

Using the ideal gas equation

$$pV = nRT$$

$$\Rightarrow p = \frac{n}{V}RT$$

In this equation, p is the pressure in atmospheres, n is the number of moles of gas, V is the volume in cubic decimetres and T is the temperature in kelvins. In this case, R, the gas constant, has units of atm dm^3 K^{-1} mole^{-1}.

$$\therefore p = [\text{gas}]RT$$

where [gas] is the concentration of the gas in moles dm^{-3}. Thus, at a constant temperature, the pressure of a gas is proportional to its concentration, i.e.

$$p \propto [\text{gas}]$$

This means that for the equilibrium

$$H_2(g) + I_2(g) \rightleftharpoons 2\,HI(g)$$

we can write either

$$K_c = \frac{[HI(g)]^2}{[H_2(g)][I_2(g)]}$$

or

$$K_p = \frac{(P_{HI})^2}{(P_{H_2})(P_{I_2})}$$

Now, since

$$P_{HI} = [HI(g)]RT$$
$$P_{H_2} = [H_2(g)]RT$$

and

$$P_{I_2} = [I_2(g)]RT$$

it follows that

$$K_p = \frac{(P_{HI})^2}{(P_{H_2}) \cdot (P_{I_2})} = \frac{[HI(g)]^2(\cancel{RT})^2}{[H_2(g)]\cancel{RT}[I_2(g)]\cancel{RT}} = \frac{[HI(g)]^2}{[H_2(g)][I_2(g)]} = K_c$$

In this particular example, $K_p = K_c$. Neither K_p nor K_c has any units, but this is not always the case. Just consider the reaction

$$N_2(g) + 3H_2(g) \rightleftharpoons 2NH_3(g)$$

$$K_p = \frac{(P_{NH_3})^2}{(P_{N_2}).(P_{H_2})^3} = \frac{[NH_3(g)]^2(RT)^2}{[N_2(g)]RT.[H_2(g)]^3(RT)^3}$$

$$= \frac{[NH_3(g)]^2}{[N_2(g)][H_2(g)]^3}.(RT)^{-2} = K_c(RT)^{-2}$$

In this case, $K_p = K_c(RT)^{-2}$.

Can you see that the numerical value of K_p is the same as that of K_c only when there are the same number of moles on each side of the stoichiometric equation?

In general,

$$K_p = K_c(RT)^{\Delta n}$$

Where Δn = number of moles on the right of the equation − number of moles on the left. In calculating values for K_p, pressure should be expressed in pascals and *not* in atmospheres.

$$1 \text{ pascal (Pa)} = 1 Nm^{-2}$$

$$1 \text{ atm} \approx 10^5 Pa$$

The equilibrium constant, K_c, for the reaction

$$N_2(g) + 3H_2(g) \rightleftharpoons 2NH_3(g)$$

at 620 K is 2 $(\text{mole dm}^{-3})^{-2}$.

■ What is the value of K_p for this reaction at 620 K? ($R = 8.31 J K^{-1} \text{mole}^{-1}$)
■ What are the units of K_p for this reaction?

21.10 Heterogeneous equilibria

Most of the equilibria that we have discussed so far may be described as **homogeneous**. In these systems, all the reactants and products co-exist in the *same phase* at equilibrium. Either they are all gases or they are all mixed together in aqueous solution.

In this section, we shall be looking at **heterogeneous equilibria** in more detail. In these systems, *two or more phases are present* at equilibrium. Examples of these include a solid in equilibrium with its aqueous solution or a solid in equilibrium with a gas. We have already discussed three types of heterogenous equilibria:

(i)   liquid in equilibrium with its saturated vapour;
(ii) solid solute in equilibrium with its saturated solution;
(iii) solute partitioned between two immiscible solvents.

Data concerning the equilibrium of water with its vapour at various temperatures are given in table 21.4.

The important feature in these results is that the vapour pressure exerted by the water at a particular temperature is independent of the mass of water present. How can this be explained?

For equilibrium between water and its vapour

$$H_2O(l) \rightleftharpoons H_2O(g)$$

we can write an equilibrium constant as

$$K_c = \frac{[H_2O(g)]}{[H_2O(l)]} \quad \dots\dots\dots\dots\dots\dots\dots\dots (1)$$

Table 21.4
Saturation vapour pressures of water at various temperatures

Mass of water taken/g	Pressure/Pa		
	20°C	**40°C**	**60°C**
1	23.4×10^2	73.8×10^2	199×10^2
2	23.4×10^2	73.8×10^2	199×10^2
10	23.4×10^2	73.8×10^2	199×10^2
50	23.4×10^2	73.8×10^2	199×10^2

Now, the value of $[H_2O(l)]$ is effectively constant, whatever the amount of water taken. We can, in fact, calculate its value from the density of water:

$$\text{density of water} = 1\,g\,cm^{-3}$$
$$= 1000\,g\,dm^{-3}$$
$$\Rightarrow [H_2O(l)] = \frac{1000}{18} = 55.56\ \text{moles}\ dm^{-3}$$

Using equation (1) above we can write

$$K_c.[H_2O(l)] = K_c' = [H_2O(g)]$$

where K_c' may be described as a modified equilibrium constant. Instead of writing $K_c' = [H_2O(g)]$, we can write

$$K_p' = P_{H_2O}$$

Thus, it is not surprising that the saturation vapour pressure of water is constant at a particular temperature.

Similar results are obtained when other liquids or solids are used in place of water. In each case, the results suggest that *the effective reacting concentration of a pure liquid or a pure solid is constant. The reacting concentration is also independent of the amount of liquid or solid present.*

In other words, $[X(s)]$ and $[X(l)]$ are constant whatever the amount of X taken. But, $[X(g)]$ and $[X(aq)]$ will vary as the amount of X in a given volume varies.

An interesting example of heterogeneous chemical equilibrium involves the thermal dissociation of calcium carbonate:

$$CaCO_3(s) \rightleftharpoons CaO(s) + CO_2(g)$$

From this equation, we can write $K_c = \dfrac{[CaO(s)][CO_2(g)]}{[CaCO_3(s)]}$

but $[CaCO_3(s)]$ and $[CaO(s)]$ are both constant, so the modified equilibrium constant for the system becomes

$$K_c' = [CO_2(g)]$$

or

$$K_p' = P_{CO_2}$$

This shows that at a particular temperature, there is a constant pressure (or concentration) of CO_2 in equilibrium with $CaO(s)$ and $CaCO_3(s)$. It does not matter what masses of these two solids are present. This prediction is confirmed by experiment.

Thus, at 800°C, the pressure of CO_2 in equilibrium with $CaCO_3(s)$ and $CaO(s)$ is $2.5 \times 10^4\,Pa$. Hence, the equilibrium constant at 800°C is:

$$K_p' = P_{CO_2} = 2.5 \times 10^4\,Pa$$

When solid NH_4HS is allowed to dissociate in an evacuated vessel it forms ammonia and hydrogen sulphide:

$$NH_4HS(s) \rightleftharpoons NH_3(g) + H_2S(g)$$

If excess NH_4HS is used at 25°C, the total pressure at equilibrium is $6 \times 10^4 \, Pa$.

■ Write an expression for the modified equilibrium constant for the system, K_p'.
■ What are the partial pressures of ammonia and hydrogen sulphide at equilibrium?
■ What is the value of K_p' at 25°C?

Summary

1 A dynamic equilibrium is characterised by the following features:
(a) Constant macroscopic properties.
(b) Continuing microscopic processes. Particles participate in both forward and reverse processes, but the rate of the forward process is equal to the rate of the reverse process. Thus, no overall change occurs.
(c) Equilibrium attained from either direction.
(d) Equilibrium achieved only in a closed system.

2 When a dissolved solute, X distributes itself between two immiscible solvents A and B, at equilibrium

$$\frac{\text{concentration of X in solvent A}}{\text{concentration of X in solvent B}} = \text{a constant}$$

This constant is called the distribution ratio or the partition coefficient. It remains constant provided
(a) the temperature is constant,
(b) the solvents are immiscible and do not react with each other.
(c) the solute neither reacts, nor associates, nor dissociates in the solvents.

3 The most important application of partition is in solvent extraction both industrially and in the laboratory.

4 If an equilibrium mixture contains substances A, B, C and D related by the equation

$$aA + bB \rightleftharpoons cC + dD$$

it is found experimentally that

$$\frac{[C]^c[D]^d}{[A]^a[B]^b} = K_c$$

K_c is called the Equilibrium Constant. It is constant at a given temperature. This experimental result is known as the Equilibrium Law.

5 The partial pressure of a gas is proportional to its concentration. So, we can express the equilibrium constant of a gaseous reaction either in terms of partial pressures or in terms of concentrations.
   This means that for the equilibrium, $H_2(g) + I_2(g) \rightleftharpoons 2HI(g)$ we can write either

$$K_c = \frac{[HI]^2}{[H_2][I_2]} \quad \text{or} \quad K_p = \frac{(P_{HI})^2}{(P_{H_2}) \cdot (P_{I_2})}$$

6 The concentration of a solid or a pure liquid is constant. So, we can write modified equilibrium constants for heterogeneous equilibria excluding these constant concentrations. Thus, for the equilibrium

$$AgCl(s) + 2NH_3(aq) \rightleftharpoons Ag(NH_3)_2^+(aq) + Cl^-(aq)$$

$[AgCl(s)]$ is constant, so we can write a modified equilibrium constant:

$$K_c' = \frac{[Ag(NH_3)_2^+(aq)][Cl^-(aq)]}{[NH_3(aq)]^2}$$

Study questions

1 The partition coefficient of solute X between the solvent Y and the solvent Z is 9. Suppose you are given a solution containing 10 g of X in 1 dm³ of Z. You are asked to extract the X using 1 dm³ of Y.
Calculate the % of X left in Z:
(a) using one extraction with the whole 1 dm³ of Y,
(b) using two extractions each with 500 cm³ of Y.
(c) using three extractions each with $333\frac{1}{3}$ cm³ of Y.

2 The Mogul Oil Company is worried about the impurity, M in its four-star petrol. One dm³ of petrol contains 5 g of M. In an effort to reduce the concentration of M in the petrol, Mogul have discovered the secret solvent, S. The partition coefficient of M between petrol and S is 0.01.
(a) What is meant by the term 'partition coefficient'?
(b) Explain the principles of solvent extraction.
(c) Calculate the total mass of M removed from one dm³ of petrol using
(i)   one portion of 100 cm³ of solvent, S,
(ii) two 50 cm³ portions of solvent, S.

3 5 moles of ethanol, 6 moles of ethanoic acid, 6 moles of ethyl ethanoate and 4 moles of water were mixed together in a stoppered bottle at 15°C.
After equilibrium had been attained the bottle was found to contain only 4 moles of ethanoic acid.
(a) Write an equation for the reaction between ethanol and ethanoic acid to form ethyl ethanoate and water.
(b) Write an expression for the equilibrium constant, K_c, for this reaction.
(c) How many moles of ethanol, ethyl ethanoate and water are present in the equilibrium mixture?
(d) What is the value of K_c for this reaction?
(e) Suppose 1 mole of ethanol, 1 mole of ethanoic acid, 3 moles of ethyl ethanoate and 3 moles of water are mixed together in a stoppered flask at 15°C.
How many moles of:
(i) ethanol,   (ii) ethanoic acid,   (iii) ethyl ethanoate,   (iv) water are present at equilibrium?

4 At a certain temperature and a total pressure of 10^5 Pa, iodine vapour contains 40% by volume of I atoms:

$$I_2(g) \rightleftharpoons 2I(g)$$

(a) Calculate K_p for the equilibrium.
(b) At what total pressure (without temperature change) would the percentage of I atoms be reduced to 20%?

5 (a) Deduce the relationship between K_p and K_c for the gaseous equilibria:
(i) $2NO(g) + O_2(g) \rightleftharpoons 2NO_2(g)$,
(ii) $NO(g) + \frac{1}{2}O_2(g) \rightleftharpoons NO_2(g)$.
(b) What are the units of K_p and K_c for the two equilibria referred to in (a)?

6   Consider the following reaction:

$$H_2(g) + I_2(g) \rightleftharpoons 2HI(g) \qquad \Delta H = -10\,kJ.$$

(a) Write an expression for the equilibrium constant in terms of partial pressures.

At a certain temperature, analysis of an equilibrium mixture of the gases yielded the following results:

$P_{H_2} = 2.5 \times 10^4\,Pa$
$P_{I_2} = 1.6 \times 10^4\,Pa$
$P_{HI} = 4.0 \times 10^4\,Pa$

(b) Calculate the equilibrium constant for the reaction. What are its units?

(c) In a second experiment at the same temperature, iodine and hydrogen iodide were mixed together. Each gas had at a partial pressure of $3 \times 10^4\,Pa$. What are the partial pressures of hydrogen, iodine and hydrogen iodide at equilibrium?

(d) In a third experiment at the same temperature, pure hydrogen iodide was injected into the flask at a pressure of $6 \times 10^4\,Pa$. What are the partial pressures of hydrogen, iodine and hydrogen iodide at equilibrium?

(e) What effect, if any, will decreasing the temperature have on the value of K_p? Explain your answer.

22 Factors Affecting Equilibria

22.1 The effect of concentration changes on equilibria

When a system in equilibrium is suddenly disturbed, it will respond until the equilibrium is eventually restored.

Consider the equilibrium

$$Fe^{3+}(aq) + NCS^-(aq) \rightleftharpoons FeNCS^{2+}(aq)$$
$$\text{pale yellow} \quad \text{colourless} \quad \text{deep red}$$

When 10^{-3}M iron(III) nitrate solution is added to an equal volume of 10^{-3}M potassium thiocyanate, a red solution is produced. The red colour is due to the formation of thiocyanatoiron(III) ions. The system forms an equilibrium mixture containing unreacted Fe^{3+}, unreacted NCS^- and the product $FeNCS^{2+}$. But what happens to the equilibrium when one of the concentrations is suddenly changed?

If a soluble iron(III) salt is added to an equilibrium solution containing $Fe^{3+}(aq)$, $NCS^-(aq)$ and $FeNCS^{2+}(aq)$, the colour of the solution becomes darker (figure 22.1).

A new state of equilibrium is quickly attained. In this equilibrium the concentration of $FeNCS^{2+}(aq)$ is obviously greater than before the addition of Fe^{3+}. Increasing the concentration of Fe^{3+} has increased the concentration of $FeNCS^{2+}(aq)$. In the same way, the concentration of $FeNCS^{2+}(aq)$ also rises when a soluble thiocyanate is added to the system. On the other hand, removal of Fe^{3+} or NCS^- from the equilibrium mixture causes the solution to become paler. A decrease in the concentration of Fe^{3+} or NCS^- results in the conversion of some $FeNCS^{2+}$ into Fe^{3+} and NCS^- in an attempt to replace the substance removed.

The results of these experiments can be summarised by the following statement:

If the concentration of one of the substances in a reversible equilibrium is altered, the equilibrium will shift to oppose the change in concentration.

Figure 22.1
The effect of Fe^{3+} and NCS^- on the equilibrium, $Fe^{3+}(aq) + NCS^-(aq) \rightleftharpoons$ $[Fe(NCS)]^{2+}(aq)$.

add soluble Fe^{3+} salt

add soluble salt containing NCS^-

10^{-3} M Fe^{3+} (100 cm³)
$+10^{-3}$ M NCS^- (100 cm³)

354

Thus, if a reactant is added to a system in equilibrium, that reaction will occur which uses up the added reactant. Conversely, if a reactant is removed, that reaction will occur which replenishes the removed reactant. The statement in italics above is a specific application of an important generalisation known as **Le Chatelier's Principle**. The Frenchman, Henri Louis Le Chatelier, was one of the first chemists to investigate the effects of different factors such as temperature, pressure and concentration on equilibria. After studying data concerning equilibria, Le Chatelier proposed the following generalisation:

If a system in equilibrium is subjected to a change, processes will occur which tend to counteract the change imposed.

Although the concentration of individual substances in an equilibrium may vary, the equilibrium constant is always the same at one particular temperature. This, of course, is the crucial point of the equilibrium law. We can emphasise this further by considering the effect of suddenly increasing the concentration of hydrogen in an equilibrium mixture of $H_2(g)$, $I_2(g)$ and $HI(g)$. In the initial equilibrium mixture (figure 22.2),

$$[HI(g)] = 0.07\,M, [H_2(g)] = 0.01\,M \quad \text{and} \ [I_2(g)] = 0.01\,M$$

$$\therefore K_c = \frac{[HI(g)]^2}{[H_2(g)][I_2(g)]} = \frac{(0.07)^2}{(0.01) \times (0.01)} = 49$$

When the concentration of $H_2(g)$ is suddenly doubled,

$$\Rightarrow \frac{[HI(g)]^2}{[H_2(g)][I_2(g)]} = \frac{(0.07)^2}{(0.02) \times (0.01)} = 24.5 < K_c$$

The system is no longer in equilibrium. In order to restore the equilibrium, the concentration of $HI(g)$ must rise, whilst those of $H_2(g)$ and $I_2(g)$ must fall. This is achieved by an *overall* conversion of *some* of the hydrogen and iodine in the mixture to hydrogen iodide:

$$H_2(g) + I_2(g) \rightarrow 2\,HI(g)$$

When equilibrium is restored once more (figure 22.2), we find that

$$[HI(g)] = 0.076\,M, \quad [H_2(g)] = 0.017\,M \quad \text{and} \ [I_2(g)] = 0.007\,M$$

$$\Rightarrow \frac{[HI(g)]^2}{[H_2(g)][I_2(g)]} = \frac{(0.076)^2}{(0.017) \times (0.007)} \ 49 = K_c$$

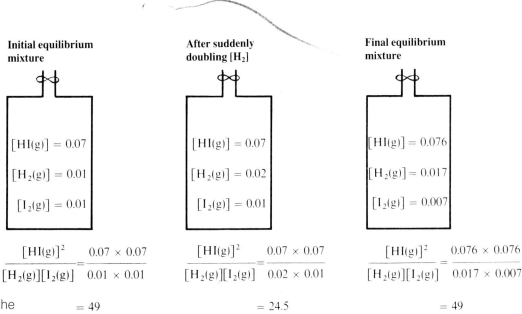

Figure 22.2
Effect of suddenly increasing the concentration of one species in a mixture at equilibrium.

Notice that only part of the added hydrogen is used up in restoring equilibrium. The concentration of $H_2(g)$ was suddenly doubled from 0.01 M in the initial equilibrium to 0.02 M. When equilibrium is achieved once more the final concentration of hydrogen is not 0.01 M but 0.017 M. Obviously, [HI(g)] in the final equilibrium is greater than that in the initial equilibrium whilst [$I_2(g)$] in the final equilibrium is less than that initially.

22.2 The effect of pressure changes on equilibria

If the partial pressure of *only one* of the gases in an equilibrium mixture is changed, the overall effect can be predicted like those involving changes in concentration in the last section. But what happens when the *total* pressure of a gaseous system at equilibrium is suddenly increased or decreased? In this case, the partial pressures of *all* the gases increase or decrease. Consider, first, the reaction

$$N_2(g) + 3H_2(g) \rightleftharpoons 2NH_3(g)$$

for which

$$K_p = \frac{(P_{NH_3})^2}{P_{N_2} \cdot (P_{H_2})^3}$$

Now, suppose that the equilibrium partial pressures of nitrogen, hydrogen and ammonia are a, b, and c atm, respectively. Thus:

$$K_p = \frac{(P_{NH_3})^2}{P_{N_2} \cdot (P_{H_2})^3} = \frac{c^2}{ab^3} \text{ atm}^{-2}$$

What happens when the total pressure is suddenly doubled? Let us follow the arguments in the last section. What do the Equilibrium Law and Le Chatelier's Principle predict will happen after a change in pressure?

Using the equilibrium law to predict the results of a change in pressure

When the total pressure is suddenly doubled, all of the partial pressures are doubled. Hence

$$P_{N_2} = 2a \text{ atm,}$$
$$P_{H_2} = 2b \text{ atm,}$$
$$P_{NH_3} = 2c \text{ atm.}$$

$$\Rightarrow \frac{(P_{NH_3})^2}{P_{N_2} \cdot (P_{H_2})^3} = \frac{(2c)^2}{2a.(2b)^3} = \frac{4c^2}{2a.8b^3} = \frac{c^2}{ab^3} \cdot \frac{1}{4} \text{ atm}^{-2}$$

Momentarily, the equilibrium constant ratio is reduced to one quarter of its value at equilibrium. So, nitrogen and hydrogen react to form ammonia until equilibrium is restored once more. Table 22.1 shows how the percentage of ammonia in the equilibrium mixture rises as the total pressure on the system increases.

Table 22.1

The effect of pressure on the equilibrium percentage of ammonia in the system, $N_2 + 3H_2 \rightleftharpoons 2NH_3$

Total pressure/atm	1	50	100	200
Equilibrium percentage of NH$_3$ at 723 K	0.24	9.5	16.2	25.3

Using Le Chatelier's principle to predict the results of a change in pressure

When the total pressure is suddenly increased, the molecules are crowded closer together. The additional pressure can be relieved if the molecules are able to react and reduce the number of molecules present.

In the reaction we are considering, one molecule of nitrogen reacts with three molecules of hydrogen to form two molecules of ammonia. In other words, four molecules of gas are reacting to form only two molecules of gas. This reduction in the total number of gas molecules results in a reduction in the total pressure. Hence, any increase in pressure in the $N_2/H_2/NH_3$ system at equilibrium can be relieved by a conversion of nitrogen and hydrogen to ammonia. Conversely, a decrease in pressure will favour the formation of nitrogen and hydrogen. This results in an increase in the number of molecules present thereby counteracting the pressure reduction.

In general, for gaseous reactions increase in pressure favours the reaction which produces fewer molecules. Decrease in pressure favours the gaseous reaction which produces more molecules.

On the other hand, pressure has no effect on a gaseous reaction if there is no change in the number of molecules. Consider the reaction

$$H_2(g) + I_2(g) \rightleftharpoons 2HI(g)$$

Suppose the equilibrium partial pressures of hydrogen, iodine and hydrogen iodide are x, y and z atm, respectively, at a particular temperature.

- Write an expression for the equilibrium constant, K_p, in terms of x, y and z.
- Suppose the overall pressure is halved. What are the partial pressures of hydrogen, iodine and hydrogen iodide now?
- What is the value of the equilibrium constant expression

$$\frac{(P_{HI})^2}{(P_{H_2}).(P_{I_2})}$$

when the overall pressure is suddenly halved?
- Use the equilibrium law to explain why reducing the pressure has no effect on this system.
- How does Le Chatelier's principle explain why pressure changes have no effect on this system?

22.3 The effect of catalysts on equilibria

The equilibria constant expression includes *only* those substances shown in the overall stoichiometric equation. Catalysts do not appear in the overall equation for a reaction. Therefore, it is not surprising that they have no effect on the equilibrium position.

Experiments show that catalysts can increase the *rates* of both forward and backward reactions in an equilibrium. They also enable equilibrium to be achieved much more rapidly, but they do not alter the concentrations of reacting substances at equilibrium.

22.4 The effect of temperature changes on equilibria

Although the equilibrium concentrations of reactants and products can vary over a wide range, the numerical value of the equilibrium constant remains constant at one particular temperature. This means that K_c and K_p are unaffected by catalysts or by changes in pressure and concentration. The equilibrium constant

Table 22.2
Values of K_p for three different reactions at various temperatures

$N_2(g) + 3H_2(g) \rightleftharpoons 2NH_3(g)$ $\Delta H^\ominus = -92kJ$		$N_2O_4(g) \rightleftharpoons 2NO_2(g)$ $\Delta H^\ominus = +57kJ$		$2SO_2(g) + O_2(g) \rightleftharpoons 2SO_3(g)$ $\Delta H^\ominus = -197kJ$	
T/K	$K_p = \dfrac{(p_{NH_3})^2}{p_{N_2} \cdot (p_{H_2})^3}$ atm^{-2}	T/K	$K_p = \dfrac{(p_{NO_2})^2}{p_{N_2O_4}}$ atm	T/K	$K_p = \dfrac{(p_{SO_3})^2}{(p_{SO_2})^2 \cdot p_{O_2}}$ atm^{-1}
400	1.0×10^2	200	1.9×10^{-6}	600	3.2×10^3
500	1.6×10^{-1}	300	1.7×10^{-1}	700	2.0×10^2
600	3.1×10^{-3}	400	5.1×10	800	3.2×10
700	6.3×10^{-5}	500	1.5×10^3	900	6.3
800	7.9×10^{-6}	600	1.4×10^4	1000	2.0

does, however, vary with temperature. Table 22.2 shows the values of K_p at different temperatures for three important reactions. The corresponding enthalpy changes for the complete conversion of reactants to products are also shown.

Notice that the two exothermic reactions have K_p values which decrease with increase in temperature. In contrast, the endothermic reaction has K_p values which increase as the temperature rises. Evidence from other investigations also fits this pattern of results. In general, it is found that:

(i)   equilibria in which the **forward reaction is exothermic (i.e. $\Delta H^\ominus$ negative)** have equilibrium constants that decrease as temperature rises;
(ii) equilibria in which the **forward reaction is endothermic (i.e. $\Delta H^\ominus$ positive)** have equilibrium constants that increase as temperature rises.

Look at the information in table 22.2.

■ How does the proportion of ammonia in the $N_2/H_2/NH_3$ system change as temperature increases?
■ What is the value of $\Delta H^\ominus$ for the reaction

$$2NH_3(g) \rightleftharpoons N_2(g) + 3H_2(g)?$$

What is the value of K_p for this reaction at 400 K?
■ Predict the effect of increasing temperature on K_p for the reaction

$$2NH_3(g) \rightleftharpoons N_2(g) + 3H_2(g)$$

Le Chatelier's principle can be used once again to predict the effect of temperature on chemical systems in equilibrium. Changing the temperature of a system in equilibrium provides a constraint. The system will try to remove the constraint. Hence, increase in temperature favours the endothermic process which will absorb the additional heat. Decrease in temperature favours the exothermic process. This may be summarised as:

$$A+B \underset{\substack{\text{endothermic process favoured} \\ \text{by increase in temperature}}}{\overset{\substack{\text{exothermic process favoured} \\ \text{by decrease in temperature}}}{\rightleftharpoons}} C+D_{; \Delta H = -x\,kJ}$$

The following demonstration illustrates the effect of temperature on the position of an equilibrium. Three identical sealed tubes are prepared containing dark brown nitrogen dioxide (NO_2) in equilibrium with pale yellow dinitrogen tetraoxide (N_2O_4).

$$N_2O_4(g) \quad \rightleftharpoons \quad 2NO_2(g)$$
$$\text{pale yellow} \qquad\qquad \text{dark brown}$$

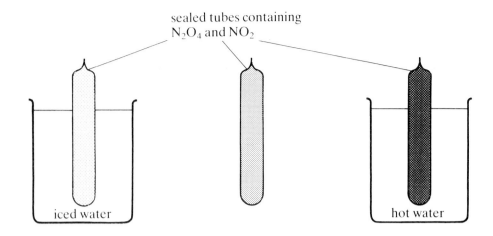

sealed tubes containing
N_2O_4 and NO_2

iced water

hot water

Figure 22.3
The effect of temperature on the
equilibrium, $N_2O_4 \rightleftharpoons 2NO_2$

All three tubes contain the same amounts of NO_2 and N_2O_4. Initially they have the same brown appearance.

One tube is now placed in iced water. A second tube is left at room temperature and the third tube is placed in hot water (figure 22.3).

The tube in cold water becomes much paler whilst that in hot water turns dark brown. This shows that the equilibrium in the reaction is displaced towards the formation of darker NO_2 at higher temperatures. This is the endothermic direction. The equilibrium moves towards the formation of paler N_2O_4 at lower temperatures.

22.5 Applying the principles of reaction kinetics and equilibria to industrial processes

The principles of reaction kinetics and chemical equilibria play an important part in the design and conditions of industrial processes. The economic and commercial competitiveness of any process rests on the speed, efficiency and economy with which products can be obtained from starting materials. Many people (including managers, economists and engineers) will be involved in decisions about the methods and materials in any industrial process, but the major problems confronting the chemist are to convert reactants into products:

(i)   as quickly as possible,
(ii) as completely as possible.

The first of these problems is clearly a kinetic one. It involves obtaining a maximum rate of reaction and rate of product formation.

The second problem is one of equilibrium. It involves the choice of those conditions which favour an optimum proportion of the product in the equilibrium mixture.

The solution to each of these problems requires a careful choice of reaction conditions such as temperature, pressure and catalyst. In this respect, Le Chatelier's principle is invaluable in predicting those conditions for maximum yield of a particular product at equilibrium.

The manufacture of sulphuric acid: the contact process

The essential stages in the manufacture of sulphuric acid are shown diagrammatically in figure 22.4. Sulphur dioxide is first obtained by burning sulphur or by roasting sulphide ores in air. SO_2 is then mixed with excess air and thoroughly purified. This prevents 'poisoning' of the catalyst by dust and other impurities. The SO_2 is now ready for further oxidation to sulphur trioxide using the Contact Process. Finally, the SO_3 is combined with water to form sulphuric acid.

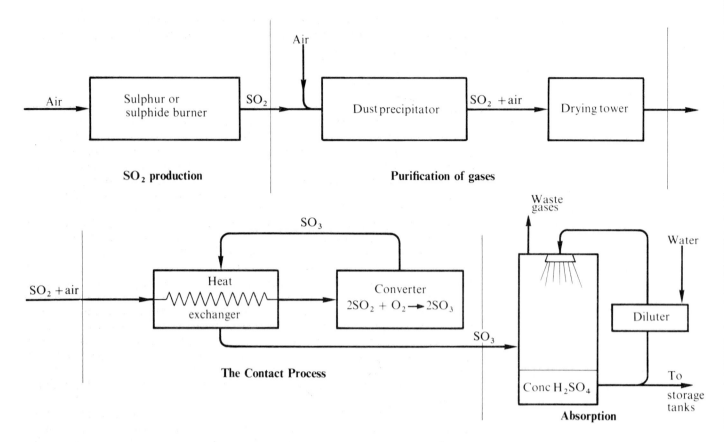

Figure 22.4
Essential stages in the manufacture of
sulphuric acid.

The bottleneck in the production of sulphuric acid is the slow third stage. This is the conversion of SO_2 to SO_3 in the Contact Process:

$$2SO_2(g) + O_2(g) \rightleftharpoons 2SO_3(g) \qquad \Delta H = -197 \, kJ$$

However, the rate of SO_3 production can be improved by

(i)   increasing the concentration (pressure) of O_2 or SO_2,
(ii) increasing the temperature,
(iii) employing a catalyst.

Both vanadium compounds and platinum have been used as catalysts for the Contact Process (section 19.6). The vanadium catalysts (incorporating either VO_3^- or V_2O_5) are less efficient than platinum. They are, however, cheaper and less susceptible to poisoning. Very few platinum catalyst plants have been built since 1945.

A sulphuric acid plant.

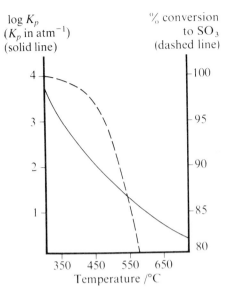

log K_p
(K_p in atm^{-1})
(solid line)

% conversion
to SO_3
(dashed line)

Temperature /°C

Figure 22.5
The effect of temperature on the equilibrium constant and the percentage conversion to SO_3 at equilibrium for a typical mixture of SO_2 and O_2 at one atmosphere pressure.

As the reaction is exothermic and three moles of reactants form two moles of products, Le Chatelier's principle predicts that the maximum yield of SO_3 at equilibrium will be obtained:

(i) at high pressure,
(ii) at low temperature.

Notice how kinetic and equilibrium considerations conflict in the choice of reaction temperature. The greatest yield of sulphur dioxide would be obtained at low temperatures, but under these conditions the reaction rate would be very slow indeed. In practice, a compromise temperature of 450°C is chosen. This is the lowest that can be used without reducing the reaction rate to an unacceptable level. There are two other important reasons for keeping the temperature as low as possible. Fuel costs and corrosion of reaction chambers increase rapidly with rise in temperature. Figure 22.5 shows the effect of temperature on the equilibrium constant and the percentage conversion to SO_3 at equilibrium for a typical gas mixture. Notice how the percentage conversion to SO_3 falls rapidly above about 450°C. At 450°C, conversion to SO_3 is 97%, but at 550°C conversion to SO_3 falls to 86%. The high conversion to SO_3 at 450°, even at atmospheric pressure, makes it unnecessary to carry out the process at increased pressure.

There is one other aspect of the Contact Process worth consideration. As the reaction proceeds, the heat evolved in the exothermic reaction moves the system to a higher temperature. At this higher temperature the percentage conversion to SO_3 is much reduced (figure 22.5). Thus, it is necessary to cool gases between successive beds of catalyst. By clever use of heat exchangers to heat the incoming gases, the operating temperature can be maintained at 450°C without external heating.

After passing through the heat exchange system, the product gases pass into an absorption tower. Here SO_3 dissolves in concentrated H_2SO_4. Direct absorption in water is unsatisfactory because the heat evolved vaporises the H_2SO_4. The H_2SO_4 vapour would condense as a fog of tiny droplets which is slow to settle out.

The uses and importance of sulphuric acid

Sulphuric acid is one of the most widely used chemicals. About $2\frac{1}{2}$ million tonnes are manufactured each year in the UK. The annual world production of

An unusual use for sulphur: an early advertisement for Frazer's sulphur tablets.

Figure 22.6
The uses of sulphuric acid in the UK (percentages relate to 1985).

Pie chart labels:

other uses 28% oil/petrol, explosives, batteries, pharmaceuticals, insecticides

fertilizers 30% phosphate fertilizer, ammonium sulphate, mixed fertilizer

dyestuffs 2%

metallurgy 2% cleaning steel and metals

plastics 5% synthetic rubber, cellophane

detergents and soaps 14%

fibres 8% viscose rayon, acetate rayon, paper, textile finishing

paints and pigments 11% paints, enamels, linoleum, coated fabrics, lithopone (BaSO$_4$)

sulphuric acid is about 110 million tonnes. Figure 22.6 shows the main uses of sulphuric acid. Notice that sulphuric acid is required for the manufacture of many basic materials, including fertilisers, paints, fibres, detergents, plastics, dyes and steel. Because of this, its level of production can be used as a reliable guide to a country's industrial activity. What is more, problems of bulk storage of the acid mean that production must respond quickly to any changes in consumption. Hence, the economic and technical development of a country can be estimated from its sulphuric acid production.

The manufacture of ammonia: the Haber process

During the last century, the agricultural industry began to require rapidly increasing quantities of nitrogenous fertilisers. Larger and larger supplies of food were being needed to feed growing populations, particularly in Europe and North America.

At the same time, the chemical industry also required increasing quantities of nitrogen compounds to make nitric acid for dyes and explosives such as TNT and dynamite.

Consequently, agriculture and industry were competing with each other for dwindling supplies of nitrogenous raw material. By 1900, Peruvian guano, a valuable fertiliser from the droppings of sea-birds, had already been worked out. It was also clear that supplies of sodium nitrate from Chile would soon become exhausted.

An alternative supply of nitrogen in the form of ammonia or nitrate had to be found, or the chemical industry would stagnate and the world's growing population would starve. Ironically, it was the war preparations in Germany between 1909 and 1914 which solved the problem. Military leaders in Germany realised that once war was declared, their country would be subjected to a strict blockade. The importation of goods and raw materials from the rest of unoccupied Europe and America would cease. German industry, therefore, had to be capable of meeting its country's requirements for nitrogenous fertilisers, and the tremendous demand for explosives and hence for nitric acid that a war would create.

In 1909, the leading German chemical company, Badische Anilin und Soda Fabric (BASF), turned its research expertise and financial resources towards the problem. They started to investigate the possible manufacture of ammonia from atmospheric nitrogen. In the previous year, a young German research chemist,

Fritz Haber (1868–1934). Haber was the son of a merchant of Breslau. After studying chemistry, Haber obtained a post as lecturer at the technical college in Karlsruhe. It was here that he discovered his method of synthesising ammonia, a discovery for which he was awarded the Nobel Prize for Chemistry in 1918.

Haber's apparatus for the synthesis ammonia.

Fritz Haber, had discovered that nitrogen and hydrogen would form an equilibrium mixture containing ammonia. The reaction needed a suitable catalyst at 600°C and 200 atmospheres.

$$N_2(g) + 3H_2(g) \rightleftharpoons 2NH_3(g); \qquad \Delta H^{\ominus} = -92\,kJ$$

■ What conditions did Haber employ to increase the reaction rate?
■ Haber's experiments yielded an equilibrium mixture containing only 8% by volume of ammonia. What conditions of temperature and pressure does Le Chatelier's principle predict for maximum yield of ammonia at equilibrium?
■ Why do you think Haber employed a heat exchanger in his equipment?

BASF bought from Haber the rights to his ammonia process. Then, they spent more than £1m in transforming Haber's simple pilot process into a giant industrial plant capable of producing 10000 tonnes of ammonia per year. By 1913, German production of nitrogen compounds had reached 120000 tonnes per year. Without this effort, Germany would have run out of food and explosives and the war would certainly have ended before 1918.

Figure 22.7 shows a flow diagram for the modern Haber process. Nitrogen can be obtained by the fractional distillation of liquid air, whilst hydrogen is obtained from naphtha or natural gas. Naphtha is a mixture of hydrocarbons

The world's first ammonia manufacturing plant. This was built by BASF at Oppau in Germany and opened in 1912.

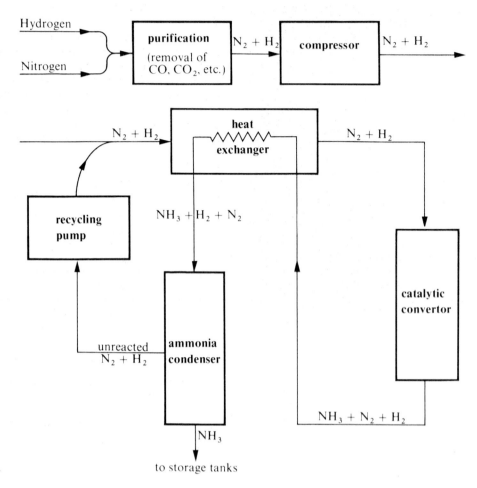

Figure 22.7
A flow diagram for the Haber process.

containing 5–9 carbon atoms. The production of hydrogen involves either catalysis with steam:

$$C_6H_{14}(g) + 6H_2O(g) \rightarrow 6CO(g) + 13H_2(g)$$
in naphtha

$$CH_4(g) + H_2O(g) \rightarrow CO(g) + 3H_2(g)$$
in natural gas

or partial oxidation with oxygen:

$$C_6H_{14}(g) + 3O_2(g) \rightarrow 6CO(g) + 7H_2(g)$$
in naphtha

$$CH_4(g) + \tfrac{1}{2}O_2(g) \rightarrow CO(g) + 2H_2(g)$$
in natural gas

Thorough purification of both the nitrogen and hydrogen is necessary. This removes carbon monoxide from the hydrogen and also sulphur compounds, water vapour and carbon dioxide, because these would poison the catalyst in the converter. Hot product gases are used to warm up the purified nitrogen and hydrogen in a heat exchanger before they enter the converter.

Le Chatelier's principle suggests that increase in pressure and decrease in temperature will increase the proportion of ammonia at equilibrium. These predictions are borne out by the results in figure 22.8.

High pressure obviously gives a higher yield of ammonia, but the higher the pressure the greater the cost and maintenance of equipment. Although pressures up to 600 atm have been used, the favoured pressure nowadays is 250 atm.

In contrast to pressure, the temperature must be low to give a high yield of ammonia. But at low temperature the rate of reaction is so slow that it makes the

Figure 22.8
The percentage of ammonia in the equilibrium mixture obtained from a 1:1 mixture of N_2 and H_2 at different temperatures and pressures.

Carl Bosch (1874–1940). Bosch was the son of a plumber. After studying chemistry at university, Bosch joined BASF in 1899 and quickly gained a reputation as a brilliant chemical engineer. Bosch was responsible for developing an industrial plant to manufacture ammonia from Haber's laboratory process. Bosch was awarded the Nobel Prize for Chemistry in 1931 for his work on high-pressure reactions.

process uneconomical. In practice, the operating temperature is usually about 450°C.

In addition to temperature and pressure, the catalyst is a vitally important variable in any industrial process. A more efficient catalyst permits lower operating temperatures. Experience has shown that the best catalyst is iron mixed with small amounts of promoters such as potassium oxide and aluminium oxide. These improve its catalytic activity (table 22.3).

The hot gases leaving the converter pass through the heat exchange system and are then cooled to −50°C. The ammonia liquefies (b.pt. −33°C) and collects in the storage vessels. The unconverted nitrogen and hydrogen are recycled.

Table 22.3
The effect of promoters on the efficiency of iron as a catalyst for the Haber process (at 200 atm and 400°C)

Catalyst	Promoter	% ammonia in exit gases
Fe	nil	3–5
Fe	K_2O	8–9
Fe	$K_2O + Al_2O_3$	13–14

The uses and importance of ammonia

Ammonia forms the basis of the nitrogen industry. It will react with acids to give ammonium salts and it can also be oxidised to nitric acid which in turn can give nitrates. Both ammonium salts and nitrates are used as fertilisers which provide the outlet for about 85% of ammonia. Of the remaining 15%, one third is used in the production of nylon.

22.6 From ammonia to nitric acid

The manufacture of nitric acid from ammonia involves three stages.

1 Catalytic oxidation of ammonia to nitrogen oxide (NO).

$$4NH_3(g) + 5O_2(g) \rightleftharpoons 4NO(g) + 6H_2O(g) \qquad \Delta H^\ominus = -950\,kJ$$

The top sheet of platinum-alloy catalyst used in the manufacture of nitric acid from ammonia.

2 Oxidation of nitrogen oxide (NO) to nitrogen dioxide (NO_2).

$$2NO(g) + O_2(g) \rightarrow 2NO_2(g) \qquad \Delta H^{\ominus} = -114\,kJ$$

3 Reaction of nitrogen dioxide with water to form nitric acid.

$$3NO_2(g) + H_2O(l) \rightarrow 2HNO_3(aq) + NO(g) \qquad \Delta H^{\ominus} = -117\,kJ$$

Look closely at stage 1 above.

- What conditions of temperature and pressure would give the maximum yield of nitrogen oxide (NO) at equilibrium?
- In practice, the first stage is carried out by passing dry ammonia and air at 7 atm over a platinum gauze catalyst at 900°C. Explain why the conditions for the process differ from those predicted in your answer to the last question.

After stage 1, the product gases are cooled to 25°C. They are then mixed with more air so that nitrogen oxide is immediately oxidised to red–brown nitrogen dioxide:

$$2NO(g) + O_2(g) \rightarrow 2NO_2(g)$$

In the third stage, nitrogen dioxide reacts with water in large absorption towers. These are designed to ensure thorough mixing of the ascending gases and the descending solution. The final product contains about 60% nitric acid. More concentrated acid can be obtained by distilling the 60% solution with concentrated sulphuric acid.

22.7 Fertilisers and explosives from nitric acid

Nitric acid plays an important part in the production of fertilisers and explosives which consume 75% and 15% of its production, respectively.

Fertilisers

Nitrogen is an essential element for plant growth. It is required for the formation of proteins, chlorophyll and nucleic acids. Plants suffering from nitrogen deficiency become stunted with yellow leaves.

A range of ICI fertilisers. The numbers across the top of each bag represent the percentage of nitrogen(N), phosphorus oxides (P_2O_5 and P_2O_3) and potassium oxide (K_2O) in each fertiliser. What do the terms 'Low-P', 'Hi-P', 'NP', 'Nitram', 'Nitro-chalk' and 'Kaynitro' stand for?

Above. 'Nitram' fertiliser.

Above right. A huge silo in which the fertiliser, ammonium sulphate, is stored.

■ Why does a deficiency of nitrogen cause plants to become stunted?
■ Why does a deficiency of nitrogen cause yellowing of the leaves?

The intensive cropping of agricultural land means that nitrogen removed from the soil must be replaced by the application of fertilisers. Otherwise, the soil will become barren and infertile. Nitrogenous fertilisers are usually nitrates or ammonium salts. Indeed, ammonium nitrate ('Nitram') is the most widely used fertiliser in many countries, because of its relatively high percentage of nitrogen.

■ What is the percentage by mass of nitrogen in ammonium nitrate, NH_4NO_3? (N = 14, H = 1, O = 16)
■ How do you think ammonium nitrate is obtained industrially?

'Nitro-chalk' contains about 20% nitrogen. It consists of ammonium nitrate crystals coated with chalk (calcium carbonate). It is used less frequently than 'Nitram', but it is non-deliquescent (unlike ammonium nitrate crystals). It is also a convenient way of liming the soil at the same time.

Explosives

Although 90% of ammonium nitrate is used in fertilisers, a large proportion of the remaining 10% goes towards the production of explosives. Nowadays, virtually all explosives are manufactured by processes which use concentrated nitric acid. The formulae of some of these explosives are shown in figure 22.9. All of these substances undergo rapid chemical reactions on explosion. The explosions produce large amounts of gas and heat and a sudden increase in pressure. Carbon and hydrogen in the explosives undergo combustion with oxygen from the nitro-groups forming carbon dioxide and water. For example, the explosion of 'nitroglycerine' made by the action of concentrated nitric acid on glycerol (propane-1, 2, 3-triol), can be summarised by the equation:

$$4C_3H_5(NO_3)_3(l) \rightarrow 12CO_2(g) + 10H_2O(g) + 6N_2(g) + O_2(g)$$

Notice that in this case, there is more than enough oxygen within the 'nitroglycerine' molecule to ensure complete combustion to CO_2 and water. This is not the case with aromatic nitro compounds such as TNT and picric acid. These explosives are frequently blended with compounds containing a high percentage of oxygen, such as chlorates and nitrates, to ensure complete oxidation during explosion.

Figure 22.9
Explosives from nitric acid.

cellulose nitrate
(nitrocellulose, gun cotton)

propane-1,2,3-triyl trinitrate
(nitroglycerine)

methyl-2,4,6-trinitrobenzene
(trinitrotoluene, T.N.T.)

2,4,6-trinitrophenol
(picric acid)

Dynamite is a general term for explosives which contain both nitroglycerine and nitrocellulose. Cordite and gelignite are each different forms of dynamite.

The manufacture of explosives requires elaborate and very special safety precautions. You *should never* attempt to make any explosive yourself.

The explosion of a French ammunitions depot in January 1871 during the Franco-Prussian War.

Summary

1 If the concentration of one of the substances in a reversible equilibrium is altered, the equilibrium will shift in such a way as to oppose the change in concentration.

2 The influence of different factors, such as temperature, pressure and concentration, on a system in equilibrium can be predicted using Le Chatelier's Principle.

  If a system in equilibrium is subjected to a change, processes occur which tend to counteract the change imposed.

3 The values of K_c and K_p are unaffected by changes in pressure or concentration.

4 Temperature is the only factor which influences the values of K_c and K_p.

5 Changes in concentration or pressure may, however, result in changes in the concentration or partial pressure of substances in the equilibrium mixture.

6 Catalysts affect neither the position of equilibrium nor the values of K_c and K_p.

7 Equilibria for which the forward reaction is exothermic (i.e. $\Delta H^\ominus$ negative) have equilibrium constants that decrease as temperature rises.
Equilibria for which the forward reaction is endothermic (i.e. $\Delta H^\ominus$ positive) have equilibrium constants that increase as temperature rises.

Study questions

1   Nowadays, hydrogen can be obtained from natural gas by partial oxidation with steam. This involves the following endothermic reaction:

$$CH_4(g) + H_2O(g) \rightleftharpoons CO(g) + 3H_2(g)$$

(a) Write an expression for K_p for this reaction.
(b) How will the value of K_p be affected by
    (i)   increasing the pressure,
    (ii) increasing the temperature,
    (iii) using a catalyst?
(c) How will the composition of the equilibrium mixture be affected by:
    (i)   increasing the pressure,
    (ii) increasing the temperature,
    (iii) using a catalyst?

2   The first step in the manufacture of nitric acid from ammonia involves the exothermic oxidation of ammonia to nitrogen oxide (NO) and steam.
(a) Write an equation for the reaction of ammonia with oxygen to form nitrogen oxide and steam.
(b) Predict, qualitatively, the conditions of temperature and pressure for maximum yield of nitrogen oxide in the equilibrium mixture.
(c) The industrial manufacture of nitrogen oxide from ammonia employs high temperature and a pressure of 7 atm. How and why are these industrial conditions different to those you predicted in (b) for maximum yield of nitrogen oxide at equilibrium?
(d) Describe, with equations, how nitrogen oxide produced by this process is converted to nitric acid.

3   At 25°C, the value of K_c for the following system is 10^{10}:

$$Sn^{2}(aq) + 2Fe^{3+}(aq) \rightarrow Sn^{4+}(aq) + 2Fe^{2+}(aq)$$

(a) Write an expression for K_c for this reaction.
(b) Explain why K_c has no units.
(c) What is the value of K_c for:
    (i) $Sn^{4+}(aq) + 2Fe^{2+}(aq) \rightarrow Sn^{2+}(aq) + 2Fe^{3+}(aq)$,
    (ii) $2Sn^{2+}(aq) + 4Fe^{3+}(aq) \rightarrow 2Sn^{4+}(aq) + 4Fe^{2+}(aq)$?

4   At 200°C, K_c for the reaction

$$PCl_5(g) \rightleftharpoons PCl_3(g) + Cl_2(g) \qquad \Delta H^\ominus = +124\,kJ$$

has a numerical value of 8×10^{-3}.
(a) Write an expression for K_c for this reaction.
(b) What are the units of K_c?
(c) What is the value of K_c for the reverse reaction at 200°C and what are its units?

(d) How will the amounts of PCl_5, PCl_3 and Cl_2 in the equilibrium mixture change if (i) more PCl_5 is added, (ii) the pressure is increased, (iii) the temperature is increased?

(e) What would be the effect on K_c if (i) more PCl_5 is added, (ii) the pressure is increased, (iii) the temperature is increased?

(f) A sample of pure PCl_5 was introduced into an evacuated vessel at 200°C. When equilibrium was obtained, the concentration of PCl_5 was 0.5×10^{-1} mole dm^{-3}.
What are the concentrations of PCl_3 and Cl_2 at equilibrium?

5   At 488 K the equilibrium constant for the reaction

$$COCl_2(g) \rightleftharpoons CO(g) + Cl_2(g)$$

is 0.2 Pa (2×10^{-6} atm).

(a) Assuming that the total pressure at equilibrium is P and the degree of dissociation of $COCl_2$ is α, deduce a relationship between K_p, α and P. (Degree of dissociation (α) is the fraction dissociated.)

(b) Calculate the degree of dissociation
   (i)   at 10^5 Pa (1 atm) pressure assuming the temperature remains constant,
   (ii) at 2×10^5 Pa (2 atm) pressure assuming the temperature remains constant.
   (Hint: when α is very small, $(1 - \alpha) \simeq 1$; $(1 + \alpha) \simeq 1$.)

6   State and explain what happens to the concentrations of hydrogen, carbon monoxide and methanol which are in equilibrium according to the reaction

$$CO(g) + 2H_2(g) \rightleftharpoons CH_3OH(g) \qquad \Delta H = -92 \, kJ$$

when
(a) the volume in which they are contained is suddenly reduced to half,
(b) the temperature is increased,
(c) the partial pressure of hydrogen is suddenly doubled,
(d) a catalyst is added,
(e) an inert gas is added to the system.

7   The densities of diamond and graphite are 3.5 and 2.3 g cm^{-3}, respectively. The change from graphite to diamond is represented by the equation

$$C(graphite) \rightleftharpoons C(diamond) \qquad \Delta H = +2 \, kJ$$

Is the formation of diamond from graphite favoured by
(a) high or low temperature,
(b) high or low pressure?
   Explain your answers.

8   (a) Describe in outline the manufacture of *one* of the following industrial chemicals:
   (i) sulphuric acid,   (ii) nitric acid,   (iii) ammonia.

(b) Discuss the chemical principles which determine the optimum operating conditions for the process which you describe.

(c) For the substances you have chosen, mention:
   (i)   two large-scale uses,
   (ii) two important features of its chemistry. Give reactions and equations to illustrate the points you make.

23 Ionic Equilibria in Aqueous Solution

23.1 Introduction

Equilibria involving ions in aqueous solution are important in industrial, analytical and biological processes. The principles and characteristics of these ionic equilibria are very similar to those in other systems in chemical equilibrium.

In this chapter, we shall concentrate on two fundamental types of ionic equilibria:

(i) the equilibrium between an undissolved solid solute and its dissolved products in solution, i.e. **solubility equilibria**;

(ii) the equilibrium between a dissolved undissociated molecule and its dissociated ions, i.e. **dissociation equilibria**.

In most cases, these dissociation equilibria involve either acids or bases.

23.2 The solubility of sparingly soluble ionic solids in water

---

When increasing quantities of a sparingly soluble ionic solid are added to water, a saturated solution is eventually formed. Ions in the saturated solution are in equilibrium with the excess undissolved solute:

$$MX(s) \rightleftharpoons M^+(aq) + X^-(aq)$$

What is the relationship between the concentrations of the aqueous ions and the undissolved solute at equilibrium?

Table 23.1 shows the equilibrium concentrations of $Ag^+(aq)$ and $BrO_3^-(aq)$ in contact with undissolved $AgBrO_3$ when different initial volumes of $0.1\,M$ $AgNO_3$ and $0.1\,M$ $KBrO_3$ are added to $200\,cm^3$ of distilled water at $16°C$.

The concentrations of BrO_3^- were obtained as follows. A measured volume of the aqueous solution was pipetted. Acid was then added followed by excess potassium iodide. The liberated iodine was then titrated against sodium thiosulphate solution of known concentration.

Having determined the concentration of $BrO_3^-(aq)$, the concentration of $Ag^+(aq)$ can be calculated.

Notice that the products of the concentration of $Ag^+(aq)$ and $BrO_3^-(aq)$ are constant (column 5 in table 23.1). Thus, for this equilibrium we can write

$$[Ag^+(aq)][BrO_3^-(aq)] = \text{a constant at a given temperature}$$

$$= 3.39 \times 10^{-5}\,mole^2\,dm^{-6} \text{ at } 16°C.$$

Table 23.1

Concentrations of $Ag^+(aq)$ and $BrO_3^-(aq)$ ions in contact with undissolved $AgBrO_3$ when different initial volumes of 0.1 M $AgNO_3$ and 0.1 M $KBrO_3$ are added to 200 cm^3 of distilled water

Initial volume of 0.1 M $AgNO_3$/cm^3	Initial volume of 0.1 M $KBrO_3$/cm^3	Concentration of $Ag^+(aq)$ at equilibrium /mole dm^{-3}	Concentration of $BrO_3^-(aq)$ at equilibrium /mole dm^{-3}	$[Ag^+]_{eqm} \times [BrO_3^-]_{eqm}$ /$mole^2$ dm^{-6}
40	10	0.0144	0.0024	3.45×10^{-5}
30	20	0.0081	0.0041	3.32×10^{-5}
25	25	0.0058	0.0058	3.36×10^{-5}
20	30	0.0042	0.0082	3.44×10^{-5}
10	40	0.0033	0.0102	3.37×10^{-5}

In other words, the product of the concentrations of Ag^+ and BrO_3^- is independent of the amount of $AgBrO_3$ present, provided there is some undissolved $AgBrO_3$ in contact with the solution. This situation is comparable to other heterogeneous systems involving solid–liquid or solid–gas equilibria (section 21.10).

How can we explain the constant value for the product $[Ag^+(aq)][BrO_3^-(aq)]$?

When equilibrium between pure $AgBrO_3$ and its solution is reached, we have

$$AgBrO_3(s) \rightleftharpoons Ag^+(aq) + BrO_3^-(aq)$$

Hence we can write an equilibrium constant expression as

$$K_c = \frac{[Ag^+(aq)][BrO_3^-(aq)]}{[AgBrO_3(s)]}$$

But, $[AgBrO_3(s)]$, which represents the concentration of a pure solid, is constant (see section 21.10). So,

$$[Ag^+(aq)][BrO_3^-(aq)] = K_c.[AgBrO_3(s)] = \text{a new constant}$$

This new constant is known as the **solubility product** and is given the symbol, $K_{s.p.}$.

Using the general formula A_xB_y for a sparingly soluble salt, we can deduce a general expression for the solubility product as follows.

At equilibrium, let us suppose

$$A_xB_y(s) \rightleftharpoons xA^{y+}(aq) + yB^{x-}(aq)$$

Hence

$$K_c = \frac{[A^{y+}(aq)]^x[B^{x-}(aq)]^y}{[A_xB_y(s)]}$$

But $[A_xB_y(s)]$ is constant. Therefore

$$[A^{y+}(aq)]^x[B^{x-}(aq)]^y = K_{s.p.}, \text{ the solubility product of } A_xB_y$$

Write an expression for the solubility product of:

- Bi_2S_3,
- $AgCl$,
- PbI_2.

The solubility products of some common compounds are given in table 23.2.

Table 23.2
The solubility products of some common compounds at 25°C

Compound	Solubility product	Compound	Solubility product
Barium sulphate	1.0×10^{-10}	Lead(II) sulphate	1.6×10^{-8}
Calcium carbonate	5.0×10^{-9}	Lead(II) sulphide	1.3×10^{-28}
Calcium fluoride	4.0×10^{-11}	Nickel sulphide	4.0×10^{-21}
Calcium sulphate	2.0×10^{-5}	Silver bromide	5.0×10^{-13}
Copper(II) sulphide	6.3×10^{-36}	Silver chloride	2.0×10^{-10}
Lead(II) bromide	3.9×10^{-5}	Silver iodide	8.0×10^{-17}
Lead(II) chloride	2.0×10^{-5}	Zinc sulphide	1.6×10^{-24}
Lead(II) iodide	7.1×10^{-9}		

The solubility product of a salt can also be obtained from its solubility. The following example shows how this is done.

A saturated solution of silver chloride contains 1.46×10^{-3} g dm^{-3} at 18°C. What is the solubility product of silver chloride at this temperature?

The solubility of silver chloride at 18°C $= 1.46 \times 10^{-3}$ g dm^{-3}

$$= \frac{1.46 \times 10^{-3}}{143.5} \text{ mole dm}^{-3}$$

$$= 1 \times 10^{-5} \text{ mole dm}^{-3}$$

Now at equilibrium,

$$AgCl(s) \rightleftharpoons Ag^+(aq) + Cl^-(aq)$$

$$\therefore [Ag^+(aq)] = 1 \times 10^{-5} M$$

and

$$[Cl^-(aq)] = 1 \times 10^{-5} M$$

$$\Rightarrow K_{s.p.} (AgCl) = [Ag^+(aq)][Cl^-(aq)] = 1 \times 10^{-5} \times 1 \times 10^{-5}$$

$$= 10^{-10} \text{ mole}^2 \text{ dm}^{-6}$$

so, the solubility product of silver chloride at 18°C is 10^{-10} mole^2 dm^{-6}.

Conversely, the solubility of a salt can be obtained from its solubility product. The following calculation shows how this is done.

The solubility product of silver carbonate at 20°C is 8×10^{-12} mole^3 dm^{-9}. What is its solubility at this temperature?

Let us suppose the solubility is s mole dm^{-3}. At equilibrium,

$$Ag_2CO_3(s) \rightleftharpoons 2Ag^+(aq) + CO_3^{2-}(aq)$$

$\therefore$ if the solubility of Ag_2CO_3 is s mole dm^{-3}

$$[Ag^+(aq)] = 2s \text{ and } [CO_3^{2-}(aq)] = s$$

$$\therefore K_{s.p.} (Ag_2CO_3) = [Ag^+(aq)]^2[CO_3^{2-}(aq)] = 8 \times 10^{-12} \text{ mole}^3 \text{ dm}^{-9}$$

$$= (2s)^2 s = 8 \times 10^{-12}$$

$$\therefore 4s^3 = 8 \times 10^{-12}$$

$$\Rightarrow s^3 = 2 \times 10^{-12}$$

$$\therefore s = 1.25 \times 10^{-4} \text{ mole dm}^{-3}$$

i.e. the solubility of silver carbonate at 20°C is 1.25×10^{-4} mole dm^{-3}.

Notice that the concentration of Ag^+ is both doubled and squared in the solubility product expression relative to the solubility of Ag_2CO_3. This is because the stoichiometry of the equation means that the concentration of Ag^+ will be double the solubility (i.e. $2s$ not s) and the expression for $K_{s.p.}$ requires this concentration of Ag^+ to be squared (i.e. $(2s)^2$ not $2s$).

23.3   Limitations to the solubility product concept

The solubility product concept is valid only for saturated solutions in which the total concentration of ions is no more than about 0.01 M. For concentrations greater than this, the value of $K_{s.p.}$ is no longer constant. This means that it is quite inappropriate to use the solubility product concept for soluble compounds such as NaCl, $CuSO_4$ and $AgNO_3$. As a consequence of this, the numerical value of solubility products is always very small, rarely exceeding 10^{-4}. For substances of extremely low solubility, $K_{s.p.}$ may be less than 10^{-40}.

The solubility product of a sparingly soluble salt is essentially a modified equilibrium constant. Like other equilibrium constants, its value will change with temperature. Consequently, the temperature at which a solubility product is measured should always be specified unless it relates to the selected standard temperature of 298 K.

23.4   Using the solubility product concept

The common ion effect

Although the solubility product of a salt is constant at constant temperature, the concentrations of the individual ions may vary over a very wide range. When a saturated solution is obtained by dissolving the pure salt in water the concentrations of the ions are in the ratio determined by the stoichiometry of the compound. For example, the concentrations of Ca^{2+} and F^- ions in pure saturated calcium fluoride solution must be in the ratio 1:2. However, when a saturated solution is obtained by mixing two solutions containing a common ion, there may be a big difference in the concentration of the ions of any sparingly soluble electrolyte. In these cases, solubility products can be used to determine the concentration of ions in the solution. We can illustrate this by considering the solubility of $BaSO_4$, first in water and then in 0.1 M sodium sulphate solution $(K_{s.p.}(BaSO_4) = 1 \times 10^{-10}$ mole^2 dm$^{-6})$.

Solubility of $BaSO_4$ in water

Suppose the solubility of $BaSO_4$ in water $= s$ mole dm^{-3}

$$BaSO_4(s) \rightleftharpoons Ba^{2+}(aq) + SO_4^{2-}(aq)$$

$$\therefore K_{s.p.}(BaSO_4) = [Ba^{2+}][SO_4^{2-}]$$

$$\Rightarrow 1 \times 10^{-10} = s \times s = s^2$$

$$\Rightarrow s = 10^{-5}\text{ mole dm}^{-3}$$

$\therefore$ solubility of $BaSO_4$ in water $= 10^{-5}$ mole dm^{-3}.

Solubility of $BaSO_4$ in 0.1 M Na_2SO_4

Suppose the solubility of $BaSO_4$ in 0.1 M $Na_2SO_4 = s'$ mole dm^{-3}

$$BaSO_4(s) \rightleftharpoons Ba^{2+}(aq) + SO_4^{2-}(aq)$$

$$Na_2SO_4(s) \rightarrow 2Na^+(aq) + SO_4^{2-}(aq)$$

In this case, $[Ba^{2+}(aq)] = s'$ mole dm^{-3}

but, $[SO_4^{2-}(aq)] = (s' + 0.1)$ mole dm^{-3}

$$\Rightarrow K_{s.p.} = [Ba^{2+}][SO_4^{2-}] = s'(s' + 0.1)$$

$$\Rightarrow s'(s' + 0.1) = 1 \times 10^{-10}$$

Now since $s' \ll 0.1$; $(s' + 0.1) \simeq 0.1$

$$\Rightarrow s' \times 0.1 = 1 \times 10^{-10}$$

$$\Rightarrow s' = 10^{-9} \text{ mole dm}^{-3}$$

$\therefore$ solubility of $BaSO_4$ in $0.1\,M\,Na_2SO_4 = 10^{-9}$ mole dm^{-3}

This calculation illustrates the important generalisation known as the **common ion effect**. This says that *in the presence of either A^+ or B^- from a second source, the solubility of the salt AB is reduced.*

Calculate the solubility of silver chloride

■ in water
■ in 0.1 M NaCl.
  ($K_{s.p.}$ (AgCl) $= 2.0 \times 10^{-10}$ mole^2 dm^{-6})

Predicting precipitation

The Ogof Rhyd Sych cave, near Merthyr Tydfil, Wales. The formation of stalagmites and stalactites in caves results from the precipitation of calcium carbonate from saturated solution.

Another important application of solubility products is that they enable chemists to predict the maximum concentrations of ions in a solution. Hence, it is possible to tell whether or not precipitation will occur.

Suppose we mix a 10^{-3}M solution of Ca^{2+} ions with an equal volume of a 10^{-3}M solution of SO_4^{2-} ions at 25°C. Will a precipitate of $CaSO_4$ form?

The solubility product for calcium sulphate is 2×10^{-5} mole^2 dm^{-6} at 25°C. So,

$$K_{s.p.}(CaSO_4) = [Ca^{2+}][SO_4^{2-}] = 2 \times 10^{-5} \text{ mole}^2 \text{ dm}^{-6}$$

Immediately after mixing equal volumes of the two solutions and before any precipitation has occurred,

$$[Ca^{2+}] = [SO_4^{2-}] = \frac{10^{-3}}{2} \text{ M} = 5 \times 10^{-4} \text{ M}$$

(The concentration of each ion is halved since each solution is diluted by mixing with the other.) Hence, the ionic product for $CaSO_4$ immediately after mixing,

$$[Ca^{2+}][SO_4^{2-}] = 5 \times 10^{-4} \times 5 \times 10^{-4}$$
$$= 25 \times 10^{-8} = 2.5 \times 10^{-7} \text{ mole}^2\text{dm}^{-6}$$

This ionic product is less than the value of $K_{s.p.}$ for $CaSO_4$ and so no precipitate will form.

Let us now suppose that we mix equal volumes of 10^{-2}M solutions.

Immediately after mixing,

$$[Ca^{2+}] = [SO_4^{2-}] = 5 \times 10^{-3}\text{M}$$

and the ionic product
$$= [Ca^{2+}][SO_4^{2-}]$$
$$= 5 \times 10^{-3} \times 5 \times 10^{-3}$$
$$= 25 \times 10^{-6}$$
$$= 2.5 \times 10^{-5} \text{ mole}^2 \text{ dm}^{-6}$$

In this case, the ionic product is greater than the solubility product. Therefore precipitation of $CaSO_4$ occurs. The concentration of aqueous Ca^{2+} and SO_4^{2-} ions are lowered by the reaction

$$Ca^{2+}(aq) + SO_4^{2-}(aq) \rightarrow CaSO_4(s)$$

The product $[Ca^{2+}][SO_4^{2-}]$ falls from 2.5×10^{-5} to 2.0×10^{-5}.

The precipitation of solids from aqueous solution is of great importance in nature and industry. Stalagmites and stalactites precipitate slowly from water in which the concentrations of $Ca^{2+}(aq)$ and $CO_3^{2-}(aq)$ have an ionic product greater than the solubility product of calcium carbonate. Coral reefs grow in a

Stag's horn coral growing off the shore of the Seychelle Islands in the Indian Ocean.

The oyster, in growing its shell, must adjust conditions so that the concentration of carbonate ions and calcium ions is large enough to precipitate calcium carbonate from sea water.

similar fashion. In this case, the concentration of Ca^{2+} and CO_3^{2-} ions around the coral must be large enough to precipitate calcium carbonate from the surrounding sea water.

Selective precipitation

The different solubilities of salts can be used as a means of separating different substances from each other by carefully selected precipitation reactions. Just suppose that we have a solution containing magnesium chloride, calcium chloride and barium chloride. How can a separation of the metal ions be achieved? Both magnesium and calcium chromates(VI) are soluble but barium chromate(VI) is insoluble. So addition of a solution of $K_2CrO_4(aq)$ will precipitate $BaCrO_4(s)$, which can then be removed by filtration.

The remaining solution now contains Mg^{2+} and Ca^{2+} ions. However, $MgSO_4$ is soluble, while $CaSO_4$ is insoluble. Thus, addition of $Na_2SO_4(aq)$ to the mixture will precipitate $CaSO_4$ and leave $Mg^{2+}(aq)$ in solution. Finally, the $Mg^{2+}(aq)$ can be removed as solid $MgCO_3$ by adding Na_2CO_3 solution.

Notice that the order in which reagents are added is important. If we added Na_2SO_4 solution before adding K_2CrO_4 solution, a mixture of $BaSO_4$ and $CaSO_4$ would be precipitated. On the other hand, if Na_2CO_3 solution was added to the mixture of the three cations, $MgCO_3$, $CaCO_3$ and $BaCO_3$ would all be precipitated. Thus, both the precipitating reagents and their order of addition must be chosen carefully.

The results in table 23.3 show how three similar cations (Cu^{2+}, Zn^{2+} and Ni^{2+}) can be separated by careful, selective precipitation.

When hydrogen sulphide is bubbled through a solution containing these cations, the following equilibrium is established:

$$H_2S(aq) \rightleftharpoons 2H^+(aq) + S^{2-}(aq)$$

In acid solution, H^+ ions will displace the equilibrium to the left. This reduces the concentration of $S^{2-}(aq)$ to very low values. Hence, only those metal sulphides with very low solubility products (such as CuS, $K_{s.p.}$ (CuS) $= 6.3 \times 10^{-36}$) will be precipitated. So, in acid solution;

$$[Cu^+][S^{2-}] > K_{s.p.}(CuS) = 6.3 \times 10^{-36}$$

but

$$[Zn^{2+}][S^{2-}] < K_{s.p.} (ZnS) = 1.6 \times 10^{-24}$$

and

$$[Ni^{2+}][S^{2-}] < K_{s.p.} (NiS) = 4 \times 10^{-21}$$

Thus CuS is precipitated, but ZnS and NiS do not precipitate.

In neutral solution, the concentration of $S^{2-}(aq)$ from H_2S will be higher and this enables ZnS with a higher solubility product ($K_{s.p.}(ZnS) = 1.6 \times 10^{-24}$) to precipitate as well as CuS.

Table 23.3
Selective precipitation of CuS, ZnS and NiS

Cation under test	H₂S bubbled into acidic solution of cation	H₂S bubbled into neutral solution of cation	H₂S bubbled into alkaline solution of cation
Cu^{2+}	Black ppte of CuS $Cu^{2+} + S^{2-} \rightarrow CuS(s)$	Black ppte of CuS	Black ppte of CuS
Zn^{2+}	No ppte	White ppte of ZnS $Zn^{2+} + S^{2-} \rightarrow ZnS(s)$	White ppte of ZnS
Ni^{2+}	No ppte	No ppte	Black ppte of NiS $Ni^{2+} + S^{2-} \rightarrow NiS(s)$

When the solution is made slightly alkaline with ammonia, the H_2S equilibrium is displaced further to the right and the concentration of $S^{2-}(aq)$ is even higher. Under these conditions, NiS is precipitated as well as CuS and ZnS.

Thus, by careful adjustment of the pH before passing in H_2S, it is possible to separate a mixture of Cu^{2+}, Zn^{2+} and Ni^{2+}.

23.5 The strengths of acids and bases

The strengths of different acids and bases can be roughly compared using conductivity measurements.

Strong electrolytes, such as hydrochloric acid and sodium hydroxide, are virtually completely dissociated into ions in aqueous solutions. Therefore, they are better conductors than weak electrolytes, such as ethanoic (acetic) acid, which are only partially dissociated:

$$HCl(aq) \rightleftharpoons H^+(aq) + Cl^-(aq) \quad \text{almost complete dissociation}$$

$$CH_3COOH(aq) \rightleftharpoons H^+(aq) + CH_3COO^-(aq) \quad \text{only partial dissociation}$$

The simple descriptive terms 'strong' and 'weak' are much too limited and inaccurate as a method of comparing the strengths of electrolytes. So, chemists looked for a more accurate and quantitative comparison. In the case of acids, relative strengths can be compared by measuring the concentration of H^+ ions or by measuring their **pH**. The 'p' in pH comes from the German word 'potenz' meaning power and the 'H' from $[H^+]$.

The pH of a solution is the negative logarithm to base ten of the molar hydrogen ion concentration, i.e.

$$pH = -lg[H^+(aq)]$$

Notice that the accepted abbreviation for logarithm to base 10 is lg not log_{10}. Hydrogen ion concentrations in aqueous solution range from about 10^{-15} to 10 mole dm^{-3}, so it is convenient to have a scale that is both negative and logarithmic. The negative sign produces positive pH values for almost all solutions encountered in practice. The logarithmic scale reduces the extremely wide variation in $[H^+(aq)]$ to a narrow range of pH from about 15 to -1.

The following examples show how pH can be obtained from $[H^+(aq)]$ and vice versa.

(a) What is the pH of 10^{-1} M HCl?
Since HCl is fully dissociated and monobasic,

$$[H^+(aq)] \text{ in } 10^{-1}\text{M HCl} = 10^{-1}\text{M}$$

$$\therefore pH = -lg[H^+(aq)] = -lg[10^{-1}] = -1(-1) = +1$$

(b) What is the pH of 10^{-3} M H_2SO_4?
Since H_2SO_4 is fully dissociated and dibasic,

$$[H^+(aq)] \text{ in } 10^{-3}\text{M } H_2SO_4 = 2 \times 10^{-3}\text{M}$$

$$\therefore pH = -lg[H^+(aq)] = -lg(2 \times 10^{-3})$$
$$= -(+0.30 - 3.00)$$
$$= -(-2.70)$$
$$= +2.70$$

(c) The pH of pure water at 25°C is 7. What is its hydrogen ion concentration?

$$pH = -lg[H^+(aq)]$$
$$\Rightarrow 7 = -lg[H^+(aq)]$$
$$\therefore lg[H^+(aq)] = -7$$
$$\therefore [H^+(aq)] = 10^{-7}$$

What are the pH values of the following solutions:

- 10^{-3} M HCl,
- 1.0 M HCl,
- 3 M HX which is only 50% dissociated?

23.6 The dissociation of water

When water is purified by repeated distillation, its conductance falls to a constant, low value. Even the purest water has a tiny electrical conductivity. This is further evidence that water dissociates to form ions, i.e.

$$H_2O(l) \rightleftharpoons H^+(aq) + OH^-(aq)$$

Obviously, the concentration of ions is very small, as the pH of pure water shows. The equilibrium in this reaction lies far to the left. Nevertheless, we can write an equilibrium constant for the dissociation of water as

$$K_c = \frac{[H^+][OH^-]}{[H_2O]}$$

Now, as only a minute trace of the water is ionised, each cubic decimetre of water will contain virtually 1000g of H_2O. Furthermore, as 1 mole of water weighs 18g, we can say,

$$[H_2O] \text{ in water} = \frac{1000}{18} = 55.56 \text{ M, which is constant}$$

Thus, we can incorporate this constant for $[H_2O]$ in the value of K_c, just as we did with the concentration of undissolved solute in considering the solubility product of a sparingly soluble salt, i.e.

$$\underset{\text{constant}}{\nearrow} K_c \underset{\text{constant}}{\overset{[H_2O]}{\nwarrow}} = [H^+][OH^-] = \text{a new constant, } K_w$$

Table 23.4
Values of K_w at various temperatures

Temperature/°C	K_w/mole^2 dm^{-6}
0	0.11×10^{-14}
10	0.30×10^{-14}
20	0.68×10^{-14}
25	1.00×10^{-14}
50	5.47×10^{-14}
100	51.3×10^{-14}

This overall constant, K_w, is called the **ionic product for water**. At 25°C,

$$[H^+] = [OH^-] = 10^{-7} \text{M}$$

hence

$$K_w = [H^+][OH^-] = 10^{-7} \times 10^{-7} = 10^{-14} \text{ mole}^2 \text{ dm}^{-6}$$

- Why is $[H^+] = [OH^-]$ in pure water?
- At 25°C, $K_w = 10^{-14}$. What is the value of K_c for the reaction

$$H_2O(l) \rightleftharpoons H^+(aq) + OH^-(aq) \text{ at 25°C?}$$

- Look at the information in table 23.4. How does the value of K_w change with increase in temperature?
- Explain the effect of temperature on K_w using Le Chatelier's principle ($H_2O(l) \rightleftharpoons H^+(aq) + OH^-(aq); \quad \Delta H = +58 \text{kJ}$).

23.7 The pH scale

In pure water and in neutral solutions such as sodium chloride, H^+ and OH^- ions arise only from the ionisation of water.

Hence in water and in neutral solutions,

$$[H^+] = [OH^-]$$

At 25°C, $[H^+][OH^-] = 10^{-14}$

$$\therefore [H^+] = [OH^-] = 10^{-7} \therefore pH = 7$$

Notice, however, in table 23.4 that K_w which equals $[H^+][OH^-]$ rises with temperature. At 50°C, for example,

$$K_w = [H^+][OH^-] = 5.47 \times 10^{-14}$$

So, at 50°C,

$$[H^+] \text{ in neutral solution} = (5.47 \times 10^{-14})^{1/2}$$
$$= 2.34 \times 10^{-7}$$
$$\therefore \text{ pH of neutral solutions at 50°C}$$
$$= -\lg[2.34 \times 10^{-7}]$$
$$= 6.6$$

This shows that the pH of neutral solutions varies with temperature. It is only 7.0 at 25°C. At 0°C, the pH of a neutral solution is 7.5, at 50°C it is 6.6 and at 100°C it is 6.1.

So far, we have been considering neutral solutions in which H^+ and OH^- ions can only come from water. In acidic and alkaline solutions H^+ and OH^- ions may arise from sources other than water. Nevertheless, the system

$$H_2O(l) \rightleftharpoons H^+(aq) + OH^-(aq)$$

is still in equilibrium. So, the product $[H^+][OH^-]$ remains constant for all solutions at the same temperature. Thus, it is possible to determine both the $[H^+]$ and the $[OH^-]$ in any solution. For example, in 10^{-2} M HCl,

$$[H^+] = 10^{-2} \text{ and } \therefore \text{ pH} = 2$$

but since $[H^+][OH^-] = 10^{-14}$ for this solution,

$$[OH^-] \text{ in } 10^{-2} \text{M HCl} = 10^{-12}$$

Likewise, in 10^{-1} M NaOH,

$$[OH^-] = 10^{-1}$$

but since $[H^+][OH^-] = 10^{-14}$ for this solution,

$$[H^+] \text{ in } 10^{-1} \text{M NaOH} = 10^{-13} \text{ and } \therefore \text{ pH} = 13$$

These results will help you to appreciate the following important generalisations.

For neutral solutions,
$[H^+] = [OH^-] = 10^{-7}$ and pH = 7 at 25°C.

For acidic solutions,
$[H^+] > [OH^-]$ and pH < 7.

For basic (alkaline) solutions,
$[H^+] < [OH^-]$ and pH > 7.

Figure 23.1 relates the pH scale to the hydrogen ion concentration and to changing acidity and alkalinity.

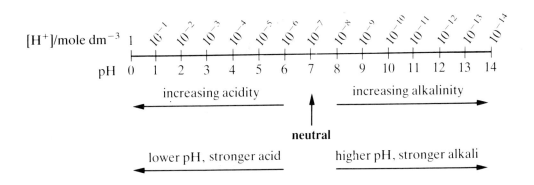

Figure 23.1
The pH scale.

23.8 The measurement of hydrogen ion concentration and pH

The most obvious method of measuring the hydrogen ion concentration of a solution is by using the hydrogen half-cell (hydrogen electrode) described in section 14.3. Under standard conditions, the hydrogen half-cell is assigned an electrode potential of 0.00 volts. Thus, when the standard hydrogen electrode is combined with another electrode to form a complete cell, the numerical value of the overall e.m.f. will depend on the second electrode. However, if the concentration of H^+ ions in the hydrogen half-cell changes, then the electrode potential of this cell is no longer 0.00 volts. So the overall e.m.f. will also change. A calibration curve can be constructed showing the overall e.m.f. against the $[H^+]$ in the hydrogen half-cell using various solutions of known $[H^+]$. It is then possible to place the hydrogen electrode in any solution and obtain its pH from the resulting e.m.f.

Unfortunately, the hydrogen half-cell is awkward and inconvenient to use. It requires the use of a bulky hydrogen cylinder, and it is difficult to adjust the gas pressure so as to provide a steady and satisfactory flow rate. It takes some time to reach equilibrium and the platinum electrode is easily 'poisoned' by impurities in the gas or in the solution.

Consequently, more convenient electrodes were sought, of which the so-called '**glass electrode**' is the most widely used.

The glass electrode consists of a silver/silver chloride electrode (a silver wire coated with AgCl) in a solution of fixed acidic pH. The electrode is placed inside a thin glass membrane permeable only to H^+ ions. The acid solution inside the glass electrode is 0.1 M HCl. This half-cell is combined with a reference half-cell which is usually a calomel electrode (mercury in contact with mercury(I) chloride in saturated potassium chloride solution). This arrangement is attached to a sensitive voltmeter to form a **pH meter** (figure 23.2).

The electrode potential of the glass electrode is dependent on the $[H^+]$ of the solution in which it is placed. In practice, the voltmeter is calibrated to give a direct reading of the pH of the solution rather than the e.m.f. of the cell. The calibration is done using solutions of known pH. In commercial pH meters, the glass electrode and the reference half-cell are normally combined in a single unit. This unit can be dipped into the solution under test.

The whole cell (glass electrode + calomel electrode) can be written as

$$\underbrace{\text{Ag(s)/AgCl(s)/HCl (0.1 M)/glass/test solution}}_{\text{glass electrode}} \, \| \, \underbrace{\text{KCl(sat)/Hg}_2\text{Cl}_2\text{(s)/Hg(l)}}_{\text{calomel electrode}}$$

Figure 23.2
A pH meter consisting of a glass electrode and a reference calomel half-cell.

The overall cell e.m.f. is determined by

— the electrode potential of the Ag(s)/AgCl(s) electrode, E_1, which is constant;
— the potential set up at the thin glass membrane which separates two solutions of different $[H^+]$, E_2,
— the electrode potential of the calomel electrode which is constant, E_3.

Now since E_1 and E_3 are constant and since the $[H^+]$ *inside* the glass bulb is constant, the overall e.m.f. depends only on the $[H^+]$ of the test solution in which the glass electrode is placed.

23.9 Dissociation constants of acids and bases

The pH of a solution is sometimes used to indicate the strength of a constituent acid or base. The use of pH is, however, very limited in this context since its value will change as the concentration changes. Consequently, chemists looked for a more useful means of representing the strengths of acids and bases. They found this by considering the dissociation of these substances in aqueous solution.

When the weak acid HA is dissolved in water, we can write

$$HA(aq) \rightleftharpoons H^+(aq) + A^-(aq)$$

Hence,

$$K_c = \frac{[H^+(aq)][A^-(aq)]}{[HA(aq)]}$$

In dealing with acids, K_c is usually replaced by the symbol, K_a, which is known as the **dissociation constant** of the acid. By eliminating the (aq) state symbols, we can simplify the last equation to

$$K_a = \frac{[H^+][A^-]}{[HA]}$$

Since dissociation constants are effectively equilibrium constants, they are unaffected by concentration changes. They are influenced only by changes in temperature. Hence, the numerical value of K_a provides an accurate measure of the extent to which an acid is dissociated (i.e. the strength of the acid).

The stronger the acid, the greater the extent of dissociation, the greater are $[H^+]$ and $[A^-]$ and the larger is K_a. The values of K_a for some acids are given in table 23.5.

The value of K_a can be calculated using the equation above if we know the concentration of the acid and its pH. The following example shows how this is done.

The pH of 0.01 M ethanoic(acetic) acid (CH_3COOH) is 3.40 at 25°C. What is the dissociation constant of ethanoic acid at this temperature?

$$CH_3COOH(aq) \rightleftharpoons H^+(aq) + CH_3COO^-(aq)$$

$$\Rightarrow K_a = \frac{[H^+][CH_3COO^-]}{[CH_3COOH]}$$

Since the concentration of H^+ ions arising from the water is much smaller than the concentration of those from the acid, we can say that

$$[H^+] \simeq [CH_3COO^-]$$

and

$$[CH_3COOH[= 0.01 - [H^+] \simeq 0.01$$

if $[H^+] \ll 0.01$, which it will be since ethanoic acid is a weak acid.

$$pH = -lg[H^+] = 3.40$$

Table 23.5
The values of K_a for some acids

Acid	Equilibrium in aqueous solution	K_a at 25°C /mole dm^{-3}
Sulphuric acid	$H_2SO_4 \rightleftharpoons H^+ + HSO_4^-$	very large
Nitric acid	$HNO_3 \rightleftharpoons H^+ + NO_3^-$	40
Trichloroethanoic acid	$CCl_3COOH \rightleftharpoons H^+ + CCl_3COO^-$	2.3×10^{-1}
Dichloroethanoic acid	$CHCl_2COOH \rightleftharpoons H^+ + CHCl_2COO^-$	5.0×10^{-2}
Sulphurous acid	$H_2SO_3 \rightleftharpoons H^+ + HSO_3^-$	1.6×10^{-2}
Chloroethanoic acid	$CH_2ClCOOH \rightleftharpoons H^+ + CH_2ClCOO^-$	1.3×10^{-3}
Nitrous acid	$HNO_2 \rightleftharpoons H^+ + NO_2^-$	4.7×10^{-4}
Methanoic acid	$HCOOH \rightleftharpoons H^+ + HCOO^-$	1.6×10^{-4}
Benzoic acid	$C_6H_5COOH \rightleftharpoons H^+ + C_6H_5COO^-$	6.4×10^{-5}
Ethanoic acid	$CH_3COOH \rightleftharpoons H^+ + CH_3COO^-$	1.7×10^{-5}
Hydrated aluminium ion	$[Al(H_2O)_6]^{3+} \rightleftharpoons H^+ + [Al(H_2O)_5OH]^{2+}$	1.0×10^{-5}
Carbonic acid	$H_2CO_3 \rightleftharpoons H^+ + HCO_3^-$	4.5×10^{-7}
Hydrogen sulphide	$H_2S \rightleftharpoons H^+ + HS^-$	8.9×10^{-8}
Boric acid	$H_3BO_3 \rightleftharpoons H^+ + H_2BO_3^-$	5.8×10^{-10}
Hydrogen peroxide	$H_2O_2 \rightleftharpoons H^+ + HO_2^-$	2.4×10^{-12}
Water	$H_2O \rightleftharpoons H^+ + OH^-$	1.8×10^{-16}

(decreasing acid strength)

Hence,

$$\lg[H^+] = -3.40 = (-4.00 + 0.60)$$
$$\therefore [H^+] = 4.0 \times 10^{-4} \text{ mole dm}^{-3}$$

and

$$[CH_3COO^-] = 4.0 \times 10^{-4} \text{ mole dm}^{-3}$$

Now, assuming $[CH_3COOH] = 0.01$ mole dm^{-3},

$$K_a = \frac{[H^+][CH_3COO^-]}{[CH_3COOH]} = \frac{4 \times 10^{-4} \times 4 \times 10^{-4}}{0.01}$$

$$K_a = 1.60 \times 10^{-5} \text{ mole dm}^{-3}$$

By reversing this calculation, it is possible to predict the pH of a solution of a weak acid. We must, however, know the molarity of the acid and its dissociation constant.

The following questions will help you to calculate the pH of 1.0 M benzoic acid (C_6H_5COOH).

$$K_a(C_6H_5COOH) = 6.4 \times 10^{-5} \text{ mole dm}^{-3}$$

■ Write an equation for the dissociation of C_6H_5COOH.
■ Using this equation, write an expression for the dissociation constant of benzoic acid.

■ Assuming $[H^+] = [C_6H_5COO^-]$ and $[C_6H_5COOH] \gg [H^+]$, calculate $[H^+]$ for 1.0M benzoic acid.

■ What is the pH of 1.0M benzoic acid?

Just as acid dissociation constants can be used to compare the strengths of different acids, we can also use base dissociation constants for bases.

If the base BOH is in equilibrium with water as

$$BOH \rightleftharpoons B^+ + OH^-$$

then

$$K_b = \frac{[B^+][OH^-]}{[BOH]}$$

where K_b is known as the dissociation constant of the base.

23.10 Acid–base indicators

Acid-base indicators such as methyl orange, phenolphthalein and bromothymol blue *are substances which change colour according to the hydrogen ion concentration of the solution to which they are added* (figure 23.3). Consequently, they are used to test for acidity and alkalinity. They are also used to detect the end point in acid–base titrations.

Most indicators can be regarded as weak acids of which either the undissociated molecule or the dissociated anion, or both, are coloured. If we take methyl orange as our example and write the undissociated molecule as HMe,

$$\underset{\text{red}}{\text{HMe}} \quad \rightleftharpoons \quad \underset{\text{colourless}}{H^+} \quad + \quad \underset{\text{yellow}}{\text{Me}^-}$$

Addition of acid (i.e. H^+ ions) displaces this equilibrium to the left. When this happens $[\text{HMe}] \gg [\text{Me}^-]$ and the solution becomes red.

On the other hand, when alkali (containing OH^- ions) is added to methyl orange it removes H^+ ions forming water. The equilibrium in the above system moves to the right in order to replace some of the H^+ ions. In this case, $[\text{Me}^-] \gg [\text{HMe}]$, and the methyl orange becomes yellow.

The dissociation of phenolphthalein in aqueous solution can be represented as

$$HPh \rightleftharpoons H^+ + Ph^-$$

Methyl orange (yellow form)

Phenolphthalein (colourless form)

Figure 23.3
The formulae of two common indicators.

Table 23.6
The values of K_a for some indicators

Indicator	K_a at 25°C /mole dm^{-3}
Phenolphthalein	7×10^{-10}
Bromothymol blue	1×10^{-7}
Litmus	3×10^{-7}
Methyl orange	2×10^{-4}

■ What colour is phenolphthalein in
   (a) strongly acid solution,
   (b) strongly alkaline solution?
■ What are the colours of
   (a) HPh   (b) Ph$^-$ for phenolphthalein?

Indicators can be regarded as weak acids, so it is possible to determine their dissociation constants. Using HIn for the undissociated form of the indicator, we can write

$$HIn(aq) \rightleftharpoons H^+(aq) + In^-(aq)$$

$$\Rightarrow K_a(HIn) = \frac{[H^+][In^-]}{[HIn]}$$

The numerical values of these dissociation constants can, of course, be obtained by measuring the pH of a solution of known molarity for each indicator (see section 23.8). The dissociation constants for some indicators are shown in table 23.6.

The end point of an indicator

The aim of any titration is to determine the volumes of two solutions which just react with each other. The **end point** is the point at which the titration is stopped. This must coincide with the **equivalence point** for the two reacting solutions. In order to achieve this, the indicator should change colour sharply at the equivalence point on adding a single drop of either acid or alkali. At the exact equivalence point of the titration, the colour of the indicator will be mid-way between the acid colour of HIn and the alkaline colour of In$^-$, since

$$[HIn] = [In^-]$$

Now,

$$K_a(HIn) = \frac{[H^+][In^-]}{[HIn]}$$

But, at the end point, $[HIn] = [In^-]$

$\therefore$ At the end point, $K_a(HIn) = [H^+]$

$\Rightarrow$ pH at end point $= -\lg[H^+] = -\lg K_a(HIn)$.

The pHs at the end point for the four indicators previously mentioned are shown in table 23.7.

Table 23.7
The pH at the end point for some indicators

Indicator	pH at end point $= -\log K_a(HIn)$
Phenolphthalein	9.1
Bromothymol blue	7.0
Litmus	6.5
Methyl orange	3.7

The range of an indicator

The colour change of an indicator is due to the change from one coloured form to another. Near the end point, both coloured forms will be present in appreciable quantities. It is not possible to say precisely when the two forms are

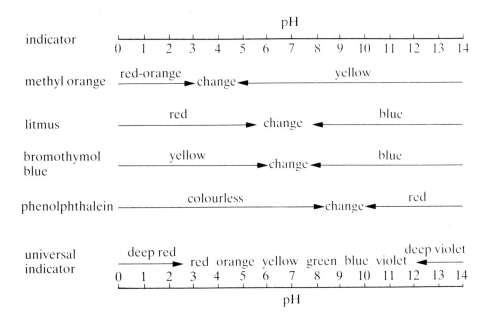

Figure 23.4
The pH ranges of some indicators.

at equal concentrations. Experience shows that our eyes cannot judge the exact end point, and indicators effectively change colour over a range of about 2 pH units.

The range of an indicator is the pH range over which it changes colour. The values in tables 23.6 and 23.7 show that the dissociation constants of indicators differ widely. Consequently, they change colour over widely differing ranges of pH. This last point is illustrated more clearly in figure 23.4.

Look closely at figure 23.4.

■ What is the pH of pure water?
■ What colours will the following indicators give in pure water:
  (a) methyl orange,
  (b) phenolphthalein,
  (c) bromothymol blue?
■ When too much phenolphthalein is added to water, it forms a cloudy solution. Why is this? (*Hint*: Phenolphthalein is an organic compound of high relative molecular mass.)

Many indicators, like phenolphthalein, are only slightly soluble in water. They are therefore often used as solutions in alcohol or in a mixture of alcohol and water. Because of this, the colour of an indicator may be confusing before it is added to an aqueous solution.

Notice that the end point of each indicator is in the centre of its pH range. Figure 23.4 also shows that many indicators change colour over a range of pH well away from 7. This is an important point. Many students misunderstand the use of indicators and expect them all to change colour at pH 7. In fact, it is because different indicators change colour at different pHs that acid–base titrations have such a wide application in industry and in the laboratory.

23.11 pH changes during titrations

During a titration there is a change in pH as alkali is added to acid or vice versa. At the equivalence point, the pH must change sharply by several units for it to be identifiable using an indicator.

The change in pH during the course of a titration depends largely upon the strength of the acid and alkali used.

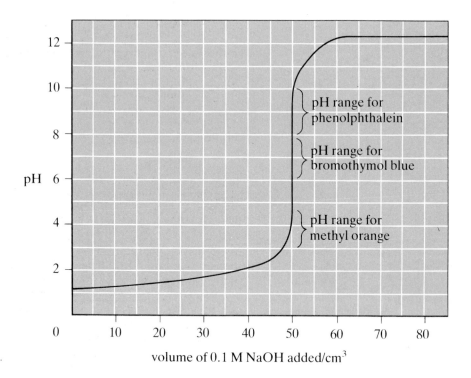

Figure 23.5
pH changes during the titration of
50 cm³ of 0.1 M HCl with 0.1 M NaOH.

Titrating strong acid against strong alkali

The graph in figure 23.5 shows how the pH changes during the titration of 50 cm³ of 0.1 M HCl with 0.1 M NaOH. As the alkali is added, the pH changes slowly at first. It does, however, change very rapidly from about 3.5 to 9.5 at the equivalence point. Thus, any indicator which changes colour between pH 3.5 and 9.5 will identify the equivalence point. This means that any one of methyl orange, bromothymol blue or phenolphthalein could be used.

Titrating strong acid against weak alkali

The graph in figure 23.6 shows how the pH changes when 0.1 M NH₃ solution (weak alkali) is added to 50 cm³ of 0.1 M HCl. As before, there is little variation in pH when the alkali is first added. But, at the equivalence point the pH

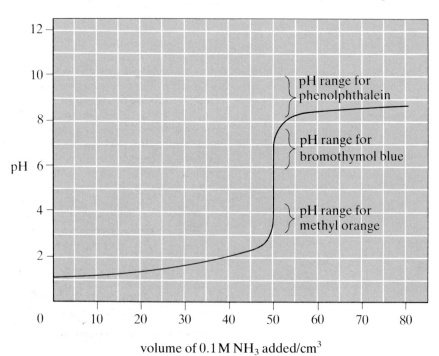

Figure 23.6
pH changes during the titration of
50 cm³ of 0.1 M HCl with 0.1 M NH₃.

changes rapidly from about 3.5 to 7.0. Thus, any indicator which changes colour between 3.5 and 7.0 will identify the equivalence point accurately. In this case, methyl orange is a very suitable indicator. Bromothymol blue may also be used, but phenolphthalein is useless because it does not begin to change colour until about pH 8.

The pH change during the titration of a weak acid with a strong alkali and a weak acid with a weak alkali are discussed in study question 10 at the end of this chapter.

23.12 Buffer solutions

When $0.1\,cm^3$ of $1.0M$ HCl is added to $1\,dm^3$ of water or sodium chloride solution, the pH changes sharply from 7.0 to 4.0 (i.e. by 3 units of pH). Clearly, the pH's of water and sodium chloride solution are extremely sensitive to even small additions of acid or alkali. If this happened when small amounts of acid or alkali were added to biological systems, living organisms would be killed instantly. Fortunately, animals and plants are protected against sharp changes in pH by the presence of **buffers**. *Buffers are solutions which resist changes in pH on addition of acid or alkali.*

Buffers are also important in many industrial processes where the pH must not deviate very much from an optimum value. Furthermore, many synthetic and processed foods must be prepared in a buffered form so that they may be eaten and digested in our bodies without undue change in pH. The pH of blood is normally 7.4. Under most circumstances, a change of only 0.5 units in the pH of blood would be fatal.

How does a buffer act? How does it resist changes in pH when acid or alkali is added?

Buffer solutions usually consist of:

either a solution of a weak acid in the presence of one of its salts (e.g. ethanoic acid and sodium ethanoate, carbonic acid and sodium hydrogencarbonate);
or a solution of a weak base in the presence of one of its salts (e.g. ammonia solution and ammonium chloride).

In order to understand how a buffer works, we can consider the hypothetical weak acid, HA, in a solution with its salt MA. In this solution, HA will be slightly dissociated whilst MA is fully dissociated:

$$HA \rightleftharpoons H^+ + A^-$$

$$MA \rightarrow M^+ + A^-$$

Hence the mixture contains a relatively high concentration of un-ionised HA (an acid) and a relatively high concentration of A^- (a base).

If acid is suddenly added to this system, the H^+ ions in the acid will combine with A^- ions to form un-ionised HA. Provided there is a large reservoir of A^- ions in the buffer, nearly all the added H^+ ions are removed. Thus, $[H^+]$ changes very little and the pH is only slightly altered.

What happens when alkali, such as sodium hydroxide, is added to the system? In this case, the OH^- ions combine with H^+ ions to form water. This reduces the concentration of H^+ ions in the buffer. So, more HA dissociates to restore the equilibrium and the $[H^+]$ rises almost to its original value. Provided there is a large reservoir of HA in the buffer, the $[H^+]$ changes very little and again the pH is only slightly altered. By having these reserves of both HA and A^- in the buffer mixture, changes in the pH resulting from the addition of acid or alkali can be minimised.

Essentially, the stable pH of the buffer is due to

(a) a high $[A^-]$ which traps added H^+ ions, and
(b) a high $[HA]$ which can supply H^+ ions to trap added OH^- ions.

Solutions of this kind can provide a 'buffer' against the effects of adding acid or alkali.

Calculating the pH of buffer solutions

In a buffer composed of the weak acid HA and its salt MA,

$$HA \rightleftharpoons H^+ + A^-$$

and

$$MA \rightarrow M^+ + A^-$$

We can write an expression for the dissociation constant of HA as

$$K_a = \frac{[H^+][A^-]}{[HA]}$$

$$\therefore [H^+] = K_a . \frac{[HA]}{[A^-]}$$

In the buffer mixture, [HA] is effectively the concentration of acid taken ([acid]) because the acid will be only very slightly dissociated. $[A^-]$ is effectively the concentration of salt taken ([salt]) because the salt is fully dissociated into ions.
Thus we can write

$$[H^+] \simeq K_a \times \frac{[acid]}{[salt]}$$

This last equation explains why the $[H^+]$, and therefore the pH of a buffer, is affected very little by dilution. This is because the ratio [acid]/[salt] will remain constant on dilution.
Using the last equation, notice the special case which applies when [acid] = [salt], i.e.

$$[H^+] \simeq K_a$$

The following example will help you to understand the points which have just been discussed.
A buffer solution was made by adding 3.28 g of sodium ethanoate to 1 dm³ of 0.01 M ethanoic acid. What is the pH of this buffer?

$$(K_a(CH_3COOH) = 1.7 \times 10^{-5} \text{ mole dm}^{-3})$$

$$K_a = \frac{[H^+][CH_3COO^-]}{[CH_3COOH]}$$

$$\therefore [H^+] = K_a \frac{[CH_3COOH]}{[CH_3COO^-]}$$

But,

$$[CH_3COOH] = [acid] = 0.01 \text{ M}$$

and

$$[CH_3COO^-] = [salt] = \frac{3.28}{82} = 0.04 \text{ M}$$

$$\therefore [H^+] = 1.7 \times 10^{-5} \times \frac{0.01}{0.04} = 4.25 \times 10^{-6} \text{ M}$$

$$\therefore \text{pH of the buffer} = -lg[H^+] = -lg(4.25 \times 10^{-6})$$
$$= -(+0.63 - 6.00) = -(-5.37)$$
$$\Rightarrow \underline{\text{pH of the buffer} = 5.37}$$

In order to appreciate the action of a buffer let us first consider the change in pH when 1 cm³ of M NaOH is added to 1 dm³ of the buffer in the last example. Then, we should consider the change in pH when 1 cm³ of M NaOH is added to 1 dm³ of 0.01 M ethanoic acid.

1 *The change in pH when 1 cm³ of M NaOH is added to 1 dm³ of buffer in the last example*

$$\text{pH of buffer initially} = 5.37$$

When NaOH is added to the buffer it reacts with CH_3COOH and forms CH_3COONa (i.e.$[CH_3COOH]$ falls and $[CH_3COO^-]$ rises). $1 cm^3$ of $1.0 M$ NaOH is 0.001 moles of OH^-. This removes 0.001 moles of CH_3COOH and forms 0.001 moles of CH_3COONa.

∴ After adding 0.001 moles of NaOH,

$$[CH_3COOH] = 0.010 - 0.001 = 0.009$$

$$[CH_3COO^-] = 0.040 + 0.001 = 0.041$$

$$\Rightarrow [H^+] = K_a \times \frac{[\text{acid}]}{[\text{salt}]} = 1.7 \times 10^{-5} \times \frac{0.009}{0.041}$$

$$[H^+] = 3.73 \times 10^{-6}$$

$$\therefore \text{pH after adding NaOH} = 5.43$$

Therefore, the pH of the buffer changes by only 0.06 units.

2 *The change in pH when 1 cm³ of M NaOH is added to 1 dm³ of 0.01 M ethanoic acid.* Before the addition of alkali,

$$[H^+] = [CH_3COO^-] \ll [CH_3COOH]$$

$$\therefore \text{using } K_a = \frac{[H^+][CH_3COO^-]}{[CH_3COOH]}$$

$$[H^+]^2 = K_a \times [CH_3COOH] = 1.7 \times 10^{-5} \times 10^{-2} = 17 \times 10^{-8}$$

$$\therefore [H^+] = 4.12 \times 10^{-4}$$

$$\therefore \text{pH of } 0.01 M \text{ ethanoic acid} = 3.39$$

On adding 0.001 moles of NaOH,

$$[CH_3COOH] = 0.010 - 0.001 = 0.009$$

$$\text{and } [CH_3COO^-] = 0.001$$

$$\therefore [H^+] = K_a \times \frac{[\text{acid}]}{[\text{salt}]} = 1.7 \times 10^{-5} \times \frac{0.009}{0.001} = 1.53 \times 10^{-4}$$

$$\Rightarrow \text{pH after adding NaOH} = 3.82$$

Therefore, the pH of the 0.01 M ethanoic acid changes by 0.43 units on adding alkali.

This is more than seven times the pH change of the buffer. It illustrates the pH-stabilising nature of the buffer very nicely.

The main use of buffers in the laboratory is in preparing solutions of known and constant pH. Buffer solutions cannot be made by simply preparing acid or alkaline solutions of a given concentration. The pH of such solutions will vary slightly as gases from the atmosphere, such as CO_2, dissolve in them, or as traces of alkali dissolve from the glass vessel.

Buffer solutions are also important in medicine and in agriculture because the pH's of living systems must be maintained at certain critical values. Because of this, intravenous injections must be carefully buffered so as not to change the pH of the blood from its normal value of 7.4. In the same way, most fermentation processes must also be buffered. Otherwise relatively small changes in the pH would cause the death of the fermenting organisms. In living systems, the buffering action is usually provided by H_2CO_3 and HCO_3^-, by $H_2PO_4^-$ and HPO_4^{2-} and by various proteins which can both accept and donate H^+ ions.

Summary

1 For the sparingly soluble strong electrolyte, A_xB_y, in contact with excess undissolved solute,

$$[A^{y+}]^x[B^{x-}]^y = K_{s.p.}$$

$K_{s.p.}$, known as the solubility product of A_xB_y, is constant at constant temperature.

2 The solubility product concept is valid only for saturated solutions in which the total concentration of ions is no more than about 0.01 M. It is therefore inappropriate for soluble salts such as NaCl, KNO_3 and $CuSO_4$.

3 In the presence of either A^{y+} or B^{x-} from a second source, the solubility of the salt A_xB_y is reduced. This generalisation is known as the common ion effect.

4 Precipitation of an insoluble salt occurs when

ionic product > solubility product

In this case, precipitation occurs until

ionic product = solubility product

5 The pH of a solution provides a quantitative measure of its acidity or alkalinity.

$$pH = -lg[H^+]$$

6 At 25°C, the product, $[H^+][OH^-] = 10^{-14}$ for all aqueous solutions
For neutral solutions, $[H^+] = [OH^-] = 10^{-7}$ and pH = 7.
For acidic solutions, $[H^+] > [OH^-]$ and pH < 7.
For alkaline solutions, $[H^+] < [OH^-]$ and pH > 7.

7 For the weak acid, HA, dissolved in water

$$K_a = \frac{[H^+][A^-]}{[HA]}$$

K_a, known as the dissociation constant of the acid, is constant at constant temperature.
   The numerical value of K_a provides a quantitative indication of the extent to which an acid is dissociated (i.e. the strength of the acid). The greater the extent of dissociation, the greater are $[H^+]$ and $[A^-]$, the larger is K_a and the stronger the acid.

8 K_a, like $K_{s.p.}$, is a modified equilibrium constant.

9 Acid–base indicators change colour according to the $[H^+]$ of the solution to which they are added. They are used to test for acidity and alkalinity and to detect the end point in acid–base titrations.

10 The range of an indicator is the pH range over which it changes colour.

11 Buffers are solutions which resist changes in pH on dilution or on addition of acid or alkali. Animals and plants are protected against changes in pH by the presence of buffers.

Study questions

1 The solubility of silver bromide in water is 7×10^{-7} mole dm^{-3} at 25°C. Calculate its solubility product.

2 (a) Calculate the solubility of silver ethanedioate (oxalate) ($Ag_2C_2O_4$) in water. ($K_{s.p.}$ ($Ag_2C_2O_4$) = 5×10^{-12} mole^3 dm^{-9}.)
  (b) How would you expect the value of a solubility product to vary with temperature? Explain your answer.
  (c) Explain why $Ag_2C_2O_4$ is very soluble in dilute nitric acid, but only sparingly soluble in water.

3 The solubility product of lead(II) sulphate, $PbSO_4$, in water, is 1.6×10^{-8} mole^2 dm^{-6}.
  (a) Calculate the solubility of lead(II) sulphate in
    (i)   pure water,
    (ii) 0.1 M $Pb(NO_3)_2$ solution,
    (iii) 0.01 M Na_2SO_4 solution.
  (b) Why is lead(II) sulphate more soluble in water than in any solution containing either Pb^{2+} or SO_4^{2-} ions?
  (c) Use your understanding of solubility product to explain the so-called 'common ion' effect. Illustrate your answer with an example.

4 Suppose the acid, HX, is a weak electrolyte.
  (a) What happens to the degree of dissociation of HX in aqueous solution if:
    (i)   water is added,
    (ii) gaseous HCl is added,
    (iii) solid NaX is added?
  (b) Assume that K_a(HA) = 10^{-6}, K_a(HB) = 10^{-8} and K_a(HC) = 10^{-10} mole dm^{-3}, respectively.
    (i)   Which solution has the highest [H^+], 1.0 M HA, 1.0 M HB or 1.0 M HC?
    (ii) What is the [H^+] in the solution with the highest value of [H^+]?

5 Chloric(I) (hypochlorous) acid, HClO, is a weak acid. K_a(HClO) = 3.2×10^{-8} mole dm^{-3}
  (a) Calculate the [H^+] and [OH^-] in 1.25×10^{-2} M HClO.
  (b) What is the pH of 1.25×10^{-2} M HClO?

6 Explain the following observations:
  (a) When 1.0 M hydrochloric acid is diluted, the pH rises. Eventually, the pH reaches a static value and does not change on further dilution.
  (b) A solution of iron(III) chloride is acid to litmus.
  (c) Phenolphthalein can be used to determine the equivalence point in the titration of ethanoic acid with sodium hydroxide, but methyl orange is no use in this case.
  (d) Zinc sulphide dissolves in dilute hydrochloric acid, but not in ethanoic acid or water.
  (e) The pH of 10^{-8} M HCl is not 8.

7 Benzoic acid, C_6H_5COOH, is a weak monobasic acid (K_a = 6.4×10^{-5} mole dm^{-3}).
  (a) Explain how a mixture of benzoic acid and sodium benzoate can act as a buffer on the addition of small amounts of either HCl(aq) or NaOH(aq).
  (b) What is the [H^+] in 0.02 M benzoic acid?
  (c) What is the pH of 0.02 M benzoic acid?
  (d) What is the pH of a solution containing 7.2 g of sodium benzoate in 1 dm^3 of 0.02 M benzoic acid?
  (e) By how much will the pH change if 1 cm^3 of 1.0 M NaOH is added to the buffer in part (d)?

8 Assuming that the pH of blood is maintained at 7.4 by the acid, $H_2PO_4^-$, and its salt, HPO_4^{2-}, calculate the ratio of the concentration of $H_2PO_4^-$ to that of HPO_4^{2-} in blood (K_a($H_2PO_4^-$) = 6.4×10^{-8} mole dm^{-3}).

9 Calculate the pH of
  (a) 10^{-4} M HCl,
  (b) 10^{-4} M Ba(OH)$_2$,
  (c) 1.0 M H$_2$X which is only 50% dissociated,
  (d) 0.01 M propanoic acid ($K_a = 1.45 \times 10^{-5}$ mole dm^{-3}).
  (e) 1.0 M NH$_4$OH ($K_b = 1.7 \times 10^{-5}$ mole dm^{-3}).

10 Figure 23.7 shows how the pH changes when 0.1 M CH$_3$COOH is titrated against 0.1 M NaOH (curve a). Curve b shows what happens when 0.1 M CH$_3$COOH is titrated against 0.1 M NH$_3$.

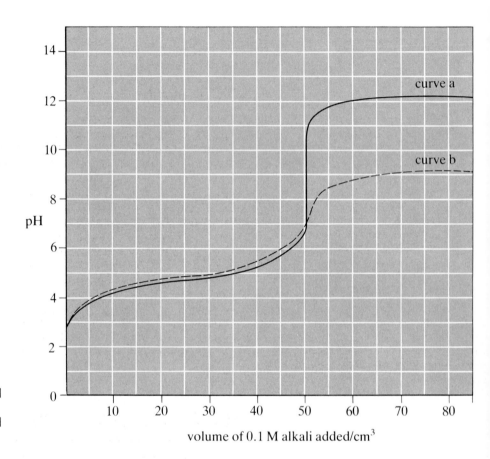

Figure 23.7
pH changes during the titration of 50 cm^3 of 0.1 M CH$_3$COOH with 0.1 M NaOH (curve a) and during the titration of 50 cm^3 of 0.1 M CH$_3$COOH with 0.1 M NH$_3$ (curve b).

volume of 0.1 M alkali added/cm^3

  (a) The pH of 0.1 M HCl is 1.0. Why is the pH of 0.1 M CH$_3$COOH about 2.8 rather than 1.0?
  (b) Is methyl orange a suitable indicator to use when titrating 0.1 M CH$_3$COOH against 0.1 M NaOH? Explain.
  (c) Is phenolphthalein a suitable indicator to use when titrating 0.1 M CH$_3$COOH against 0.1 M NaOH? Explain.
  (d) No indicator will detect the equivalence point accurately when titrating 0.1 M CH$_3$COOH against 0.1 M NH$_3$. Why is this?
  (e) In view of the statement in (d), how could you determine the equivalence point in titrating a weak acid against a weak alkali?

24 Entropy and Free Energy

Jar L
contains H₂S

Jar R
contains air

cover
slip

remove
cover
slip

both jars contain H₂S and air

Figure 24.1
Mixing hydrogen sulphide and air.

Jar L   Jar R

Figure 24.2
Two gas jars—there are six molecules of hydrogen sulphide in jar L, and jar R is empty.

Jar L   Jar R

Figure 24.3
Two gas jars—the cover slip has been removed and some of the molecules can now pass from jar L into jar R.

24.1 Introduction

Why do things happen the way they do? If you put a block of ice over the flame of a bunsen burner, it melts. Why? Why does the water not refreeze over the bunsen? Take the bunsen burner itself for that matter, and the methane gas burning in it. Why do methane and oxygen readily turn to carbon dioxide and water, giving out heat? Why do carbon dioxide and water never form methane and oxygen of their own accord?

Chemists are always interested in the direction of change. The answers to the questions above may seem obvious, but there are many changes for which it is not at all easy to predict the direction in which a reaction will go. Consider, for example, the Haber process (section 21.5). Will nitrogen and hydrogen form ammonia of their own accord or vice versa? Here it is very important to choose the right conditions—temperature and pressure in particular. Much of industrial chemistry involves choosing the right conditions in order to get as much of the product as possible. By understanding the factors that make a reaction go in a particular direction, it is possible to predict the conditions which might allow a reluctant reaction to go in the way you want.

24.2 A simple example—the diffusion of gases

Hydrogen sulphide, H_2S, is a very smelly gas. If you opened a gas jar of H_2S in one corner of a room, it would not be long before everyone in the room could smell it. But suppose you wanted to remove the smell from the room. It would be pointless to open a new, empty gas jar and hope the H_2S would go back in. It just would not. Why? The answer, of course, is that all gases diffuse naturally to fill any available space. Why do they do this? To answer this question, let us simplify the problem a little and reduce the room to the size of a second gas jar (figure 24.1).

Look at figure 24.1. Why will the gases never unmix?

Simplify things still further and imagine there are only six molecules of H_2S in jar L, and that jar R is completely empty (figure 24.2). We can represent this arrangement of the six molecules in jar L as LLLLLL. The molecules move around in jar L completely at random, colliding haphazardly with each other and with the walls of the container and not caring where they go. They continue like this after the cover slip has been removed, but now some of them can pass into jar R. To start with, perhaps only one molecule might go into jar R (figure 24.3). This arrangement could be represented as LLLLLR. Of course, there are many other possible arrangements of the six molecules. In fact, each of the six molecules can be arranged in one of two ways, L or R. There are two arrangements for each molecule, and we can calculate the total number of possible arrangements by multiplying the number of individual arrangements together.

For the six molecules, this is

$$2 \times 2 \times 2 \times 2 \times 2 \times 2 = 2^6 = 64$$

Only one of these 64 arrangements is

LLLLLL

—the arrangement at the beginning. So there is a 1 in 64 chance that the molecules will all end up in jar L. However, it is much more likely (a 63 in 64 chance) that the six molecules will be distributed in some other way.

So far we have considered only six molecules. For 100 molecules, the number of arrangements will be 2^{100}—about 10^{30}. Only one of these arrangements will have all 100 molecules in jar L—a 1 in 10^{30} chance. It is extremely likely that the gas molecules will arrange themselves between the two jars.

In real life, the numbers of molecules involved are much, much bigger. The volume of a gas jar might be $0.5\,dm^3$, and at room temperature and pressure this will hold $0.5/24$ moles—about $1/50$ moles of gas, containing approximately 10^{22} molecules. The number of ways these molecules can be arranged between jar L and jar R is therefore $2^{10^{22}}$—an unimaginably huge number. Only one of these $2^{10^{22}}$ arrangements has all the molecules in jar L. Therefore it is overwhelmingly likely that the molecules will spread out and occupy both jars—there are so many different ways of doing it.

Chance, probability and the direction of change

We have just seen that gases diffuse into any available space *because they are overwhelmingly likely to do so*. They never 'undiffuse' because this is overwhelmingly unlikely. For a single molecule, the chance of 'undiffusion' is quite good, but the numbers of molecules are so huge that the probability of all of them undiffusing is zero—and we can be certain that they will stay diffused.

Diffusion is not the only process dictated by chance.

All changes, physical and chemical, are governed by the laws of chance and probability. Quite simply, the most likely thing will always happen. If you appreciate this, it will be very helpful in predicting changes.

The delicious smell of baking bread soon diffuses to fill the kitchen. The laws of chance and probability say it must.

24.3 Entropy

The examples in the last section show that probability is vital in deciding whether a physical or chemical change will occur. The more ways a change can happen, the higher the chance of its happening. Clearly, the 'number of ways' in which a

Below 0°C, ice does not melt spontaneously, even though this would increase its entropy. Why not?

change can take place is of vital importance in deciding whether the change occurs.

When you are dealing with just a few molecules, the number of ways in which the change can occur, W, is quite small and the figures are quite easy to cope with. But in real situations the numbers of molecules are huge. We have already seen that when 1/50 mole of gas (about 10^{22} molecules) diffuses from one gas jar to fill two jars, the increase in the number of ways of arranging these molecules is

$$W = 2^{10^{22}}$$

This is an enormous number and very difficult to handle. Fortunately, W itself is not usually used. Instead, a quantity called **entropy**, S, is used. Entropy is closely related to the number of ways, W.

Entropy, $S = k \ln W$

(ln means the natural logarithm, $\log_e$) k is a constant, called the **Boltzmann Constant**. Its value is 1.38×10^{-23} J K^{-1}.

Using this expression, we can calculate the value of S corresponding to $W = 2^{10^{22}}$

$$S = k \ln W = 1.38 \times 10^{-23} \ln 2^{10^{22}}$$
$$= 0.096 \text{ or about } 0.1 \text{ JK}^{-1}$$

This is a much easier number to handle.

$k \ln W$ is an extremely important quantity. It shows, in a 'scaled-down' form, the number of ways of achieving a particular physical or chemical situation, and therefore indicates the probability of that situation arising.

We have already seen that changes that happen of their own accord are those with the greatest probability, i.e. the greatest number of ways, W, of happening. So, for all changes that happen spontaneously,

W must increase

But entropy $S = k \ln W$. So for all spontaneous changes

entropy, S must increase

or

entropy change, ΔS must be positive

This is one way of stating the *Second Law of Thermodynamics*.

At the beginning of this section we asked the question 'why does ice melt spontaneously?' Using the ideas that have been developed, we can begin to see an answer. Look at figure 24.4.

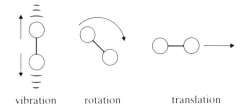

water molecules in an ice crystal water molecules in the liquid state

Figure 24.4
Changes in the arrangement of water molecules when ice melts.

■ Are there more ways of arranging water molecules in an ordered, regular ice crystal or in the disordered liquid state?
■ Which is the more probable—solid water (ice) or liquid water?
■ Which has the greater entropy—ice or liquid water?

This simple argument suggests that ice will tend to melt spontaneously because there are more arrangements for liquid water than for solid water, and ΔS for this process is positive.

But if you were reading this at the South Pole, you would disagree—ice does not melt spontaneously below 0°C—indeed, at that temperature liquid water *freezes* spontaneously. Clearly we have not yet considered the whole picture.

Entropy and energy

For spontaneous changes, the number of arrangements, W, must always increase. Furthermore, because entropy $S = k \ln W$, entropy must always increase too. So far, we have considered only the number of ways, W, that molecules can be arranged in space. But there are other kinds of arrangement to consider—the ways *energy* can be distributed among the molecules.

You should already be familiar with the idea of energy levels. Electrons in atoms can exist in certain energy levels, but not between these levels. Their energy comes in packets, or **quanta**, and the number of quanta they possess defines the energy level they occupy.

vibration rotation translation

Figure 24.5
Distribution of energy in a diatomic molecule.

Figure 24.6
Vibrational energy levels of hydrogen. The molecule represented here possesses two energy quanta. Its vibrational energy is 2 E.

Figure 24.7
Transfer of one quantum of vibrational energy from one H₂ molecule to another.

When the electrons in an atom or molecule move between energy levels, energy changes occur. But atoms and molecules can change their energy states in other ways. Molecules can vibrate, rotate, and translate (move about) as well. All these movements involve energy (figure 24.5). *Energy comes in quanta for each of these changes too.*

A diatomic molecule such as hydrogen has definite vibrational energy levels spaced rather like the rungs of a ladder. At any instant, the molecule will have a definite amount of vibrational energy, depending on how many quanta it possesses. The molecule represented by the energy level diagram in figure 24.6 has two quanta, each of energy E. Now suppose there are two hydrogen molecules, each with energy $2E$, i.e. two quanta (figure 24.7a). These molecules might exchange one quantum, giving the arrangement in figure 24.7b.

There are, of course, other ways in which these two molecules could share their four energy quanta. The full list of arrangements is

Molecule 1	Molecule 2
2 quanta	2 quanta
1	3
3	1
0	4
4	0

—five different ways. If the molecules were given one extra quantum to share, there would be six ways. Try to list these six arrangements in a table similar to the one above. The more quanta there are to share, the greater the number of ways of sharing them. And the greater the number of molecules involved, the greater still the number of ways of sharing the quanta (table 24.1).

Table 24.1
Number of ways of sharing quanta between molecules

Number of molecules	Number of quanta	Number of different ways of sharing the quanta
10	100	$\sim 10^{12}$
100	10	$\sim 10^{13}$
100	100	$\sim 10^{60}$
200	110	$\sim 10^{86}$

Consider two small pieces of metal in contact, each containing 100 atoms. One piece of metal is hot and has 100 energy quanta, the other is cold and has only 10 quanta (figure 24.8).

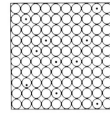

hot metal
100 quanta

cold metal
10 quanta

Figure 24.8
Hot and cold metal in contact.

Use table 24.1 to answer these questions.

■ How many ways can the quanta be distributed in (a) the hot piece of metal, and (b) the cold piece?
■ What is the total number of arrangements of quanta for the hot piece and the cold piece?
■ Suppose all the quanta are now distributed between the two pieces of metal. What is the number of ways this could be done?
■ Which is the more probable arrangement—to have 100 quanta in one piece and 10 in the other, or to distribute the quanta between both pieces?
■ What will happen to the temperatures of the pieces of metal when they are placed in contact like this?

You should now see that *sharing energy quanta increases the number of energy states and therefore the number of ways in which atoms and molecules can arrange themselves and their energy.* Thus, when entropy changes are worked out, using $S = k \ln W$, we must include in W the number of ways of distributing the available energy among the particles, as well as the number of ways of arranging the particles in space.

It must also be remembered that increasing the temperature of a substance will increase the number of energy quanta available, and so increase the number of ways and thus the entropy.

Some entropy values

Using a variety of experimental methods, it is possible to work out the total entropy of elements and compounds under different conditions. Table 24.2 lists the standard entropies of some substances. Examine the table carefully.

■ Try to formulate a general rule concerning which types of substance have low standard entropies, and which have high ones.
■ What will happen to all these entropy values when the temperature is increased?

In general, gases have higher standard entropies than solids. This is reasonable because there are more ways of arranging molecules and energy quanta in a disordered gas like carbon dioxide than in an ordered solid like diamond. Indeed, entropy is sometimes loosely described as 'disorder'. This is valid to an extent, but remember that in calculating entropy it is not enough to think simply of the arrangement of the particles in space—the distribution of energy among them must be considered as well.

Table 24.2
Standard entropies, $S^{\ominus}$ of various substances at 298 K

Substance	State	$S^{\ominus}/\text{J K}^{-1}\text{ mole}^{-1}$
diamond	s	2.4
argon	g	154.7
carbon dioxide	g	213.6
aluminium	s	28.3
iron	s	27.2
water (solid)	s	48.0
water (liquid)	l	70.0
water (gas)	g	188.7
sodium chloride	s	72.4
quartz (SiO_2)	s	41.8

Table 24.3
Standard entropies of some alkanes at 298 K

Substance	State	$S^{\ominus}/\text{J K}^{-1}\text{ mole}^{-1}$
methane	g	186.2
ethane	g	229.5
propane	g	269.9
butane	g	310.1
pentane	l	261.1

Table 24.3 gives some more standard entropy values.

■ What conclusions can you draw concerning the effect of molecular complexity on the entropy of a substance?
■ Try to explain this effect.
■ Why is the trend of steadily rising $S^{\ominus}$ values broken by pentane?

24.4 Entropy changes

The entropy change, ΔS, for a reaction can be worked out using tabulated standard entropy values. However, a qualitative idea of the entropy change can be obtained by inspecting the equation for the reaction and then using the following simple rules.

1 Gases generally have higher entropies than solids or liquids, so if the number of molecules of gas increases during a reaction, it is likely the overall entropy will also increase.

2 Ions and molecules in solution generally have higher entropies than solids.

3 If a large molecule breaks down into smaller ones, entropy increases since there are more ways of arranging several small molecules than one large one.

■ For each of the following reactions, say whether you would expect the entropy change of the chemicals to be

A positive
B negative
C approximately zero

(i)　$2\text{Mg(s)} + O_2(g) \rightarrow 2\text{MgO(s)}$
(ii)　$\text{CaCO}_3(s) \rightarrow \text{CaO(s)} + \text{CO}_2(g)$
(iii)　$N_2(g) + 3H_2(g) \rightarrow 2\text{NH}_3(g)$
(iv)　$\text{Zn(s)} + \text{Cu}^{2+}(aq) \rightarrow \text{Zn}^{2+}(aq) + \text{Cu(s)}$

Can entropy ever decrease?

For spontaneous changes it appears that *entropy, S, must always increase*.

We do not have to look far, however, to find spontaneous changes which seem to involve an entropy *decrease*. For example, at room temperature, steam spontaneously condenses to water:

$$\text{H}_2\text{O(g)} \longrightarrow \text{H}_2\text{O(l)}$$
$$\text{steam} \qquad\qquad\qquad \text{water}$$
$$S^{\ominus} = 188.7\text{ J K}^{-1}\text{ mole}^{-1} \qquad S^{\ominus} = 70.0\text{ J K}^{-1}\text{ mole}^{-1}$$

For this change, ΔS is negative—the entropy has *decreased* by $(188.7 - 70) = 118.7\text{ J K}^{-1}\text{ mole}^{-1}$. How can this be reconciled with the rule that entropy must always *increase*?

To find the answer, we must look a little further than the water alone, and remember the importance of *energy* in deciding entropy changes.

When steam condenses, energy is given out.

$$H_2O(g) \rightarrow H_2O(l); \qquad \Delta H^\ominus = -44.1\,\text{kJ mole}^{-1}$$

Where does this energy go? If the water condenses on a window, it goes to the glass. If it condenses as droplets in the air, it goes to the air molecules. The atoms in the glass, or the air molecules, or whatever—call them the **surroundings**—gain energy quanta and this increases the number of ways in which they can share energy (figure 24.9). Thus the entropy of the surroundings *increases*, whereas the entropy of the water *decreases*. There are two entropy changes to consider—the entropy change of the water, ΔS_{water}, and the entropy change of the surroundings, $\Delta S_{surroundings}$.

$$\Delta S_{total} = \Delta S_{water} + \Delta S_{surroundings}$$

As we shall see shortly, the positive $\Delta S_{surroundings}$ outweighs the negative ΔS_{water}, so ΔS_{total} is positive and the *total* entropy increases. The Second Law of Thermodynamics has not been broken after all.

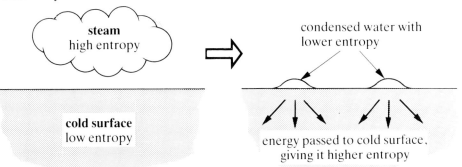

Figure 24.9
Entropy changes when water condenses.

Calculating the entropy change in the surroundings

In any change in which energy is passed to or taken from the surroundings, it is vital to consider the entropy change in the surroundings, $\Delta S_{surroundings}$, as well as the entropy change in the chemicals or **system**, ΔS_{system}. ΔS_{system} is easily worked out from tabulated values of S. But $\Delta S_{surroundings}$ cannot be calculated in the same way, because the surroundings cannot be accurately defined.

Fortunately, there is a simple way of calculating $\Delta S_{surroundings}$. When a quantity of energy Q is passed to a body whose temperature is T, it can be shown that the increase in entropy, ΔS, of the body is just

$$\Delta S = Q/T$$

In a chemical reaction for which the enthalpy change is ΔH, the amount of energy Q passed to the surroundings is $-\Delta H$. (A minus sign is needed because if the surroundings gain energy, the chemicals must lose it.) Therefore

$$\Delta S_{surroundings} = -\Delta H/T$$

This enables us to calculate $\Delta S_{surroundings}$ very easily. Going back to the condensation of steam.

$$H_2O(g) \rightarrow H_2O(l); \qquad \Delta H^\ominus = -44.1\,\text{kJ mole}^{-1}$$

$$\Delta S_{system}^\ominus = -118.7\,\text{J K}^{-1}\,\text{mole}^{-1}$$

At 298 K,

$$\Delta S_{surroundings}^\ominus = \frac{-\Delta H^\ominus}{T} = \frac{44100}{298} = 148.0\,\text{J K}^{-1}\,\text{mole}^{-1}$$

(ΔH must be in joules, T in kelvins). Hence

$$\Delta S_{total}^\ominus = \Delta S_{system}^\ominus + \Delta S_{surroundings}^\ominus$$

$$\Delta S_{total}^\ominus = -118.7 + 148.0 = +29.3\,\text{J K}^{-1}\,\text{mole}^{-1}$$

When the surroundings as well as the system are taken into account, ΔS is positive.

Table 24.4
Possible combinations of ΔS_{system} and $\Delta S_{surroundings}$

ΔS_{system}	$\Delta S_{surroundings}$	ΔS_{total}	Is the reaction thermodynamically feasible? Will it go?	Example
+	+	must be +	yes	$C + O_2 \rightarrow CO_2$
+	−	may be +	maybe	liquids evaporating; solids melting; solids dissolving
−	+	may be +	maybe	gases condensing; solutions crystallising; solids freezing
−	−	must be −	no	$CO_2 \rightarrow C + O_2$

This change is an example of a situation where ΔS_{system} is negative and unfavourable, but more than compensated for by a positive $\Delta S_{surroundings}$.

For many changes, however, ΔS_{system} is positive. For example,

$$C(s) + O_2(g) \rightarrow CO_2(g), \qquad \Delta S_{system}^{\ominus} = 6.31 \, \text{J K}^{-1} \, \text{mole}^{-1}$$

In such cases, $\Delta S_{surroundings}$ need not be positive, indeed it could be negative, and the change will be thermodynamically favourable so long as ΔS_{system} is bigger and ΔS_{total} remains positive. Table 24.4 sums up the various possibilities.

24.5 Entropy changes and free energy changes

Entropy changes are very useful to the chemist, because they help us to predict whether a particular reaction will 'go', and if not, what can be done to make it go. However, the *total* entropy change, including both system and surroundings, must always be considered. This is rather a nuisance, because chemists are always more interested in what is going on *inside* a test tube or *inside* a reaction vessel than what is happening to the vessel itself and its surroundings. Fortunately there is a convenient way round this problem.

Since

$$\Delta S_{total} = \Delta S_{surroundings} + \Delta S_{system}$$

and

$$\Delta S_{surroundings} = \frac{-\Delta H}{T}$$

then

$$\Delta S_{total} = \frac{-\Delta H}{T} + \Delta S_{system}$$

Multiplying through by T

$$T\Delta S_{total} = -\Delta H + T\Delta S_{system}$$

or

$$-T\Delta S_{total} = \Delta H - T\Delta S_{system}$$

The quantity $(-T\Delta S_{total})$ is called the **free energy change, ΔG**

$$\Delta G = \Delta H - T\Delta S_{system}$$

(The symbol G for free energy is in honour of the American chemist Willard Gibbs.)

■ What are the units of ΔG?

For changes that happen spontaneously, ΔS_{total} must be positive—the total entropy must increase.

Now, since $\Delta G = -T\Delta S_{total}$, for ΔS_{total} to be positive, ΔG must be negative. *For all spontaneous changes, ΔG must be negative.* This is another way of stating the Second Law of Thermodynamics.

Free energy changes, ΔG, are extremely helpful to chemists, because their use enables the surroundings to be ignored. ΔG values are tabulated in data books in much the same way as ΔH values. In particular, the free energy of formation, ΔG_f, is known and tabulated for a wide variety of compounds. Look up a few examples in your own data book.

ΔG and ΔH

In many chemical reactions, $T\Delta S_{system}$ is a lot smaller than ΔH, at least at room temperatures, and in these cases it is fairly safe to say that if ΔH is negative, ΔG will also be negative. This is why ΔH values can often be used to predict whether or not reactions could go spontaneously. But while it is true that for *all* spontaneous changes ΔG must be negative, a negative ΔH does not *always* mean a spontaneous reaction.

If ΔS_{system} is large and negative, $T\Delta S_{system}$ may exceed ΔH and ΔG may be positive. This is particularly likely when T is large, i.e. at high temperatures. Table 24.5 gives some values of $\Delta G^{\ominus}$ and $\Delta H^{\ominus}$ for comparison. Look at the figures in table 24.5, then answer these questions.

■ For which two reactions is the discrepancy between ΔG and ΔH greatest? Suggest reasons for these discrepancies.
■ Why does ammonium chloride dissolve spontaneously under standard conditions even though ΔH for the process is positive?
■ Why does liquid water not freeze under standard conditions even though ΔH for the process is negative?
■ For the reaction of zinc with copper ions, why are the values of ΔG and ΔH so close?

Table 24.6 on the next page summarises the different possible combinations of ΔH and ΔS. Compare the table with table 24.4.

Table 24.5
Standard enthalpy changes and free energy changes (note the units of $\Delta G^{\ominus}$ and $\Delta H^{\ominus}$ are the same, kJ mole^{-1}. Note too that these values refer to standard conditions: 298 K and 1 atm)

Reaction	Equation	$\Delta G^{\ominus}$/kJ mole^{-1}	$\Delta H^{\ominus}$/kJ mole^{-1}
burning carbon	$C(s) + O_2(g) \rightarrow CO_2(g)$	−394.4	−393.5
condensing steam	$H_2O(g) \rightarrow H_2O(l)$	−8.6	−44.1
freezing water	$H_2O(l) \rightarrow H_2O(s)$	+0.6	−6.01
burning magnesium	$Mg(s) + \frac{1}{2}O_2(g) \rightarrow MgO(s)$	−1138.8	−1204
reaction of zinc with copper ions	$Zn(s) + Cu^{2+}(aq) \rightarrow Zn^{2+}(aq) + Cu(s)$	−212.1	−216.7
making ammonia by the Haber process	$\frac{1}{2}N_2(g) + \frac{3}{2}H_2(g) \rightarrow NH_3(g)$	−33.4	−92.0
combining hydrogen and oxygen to make water	$H_2(g) + \frac{1}{2}O_2(g) \rightarrow H_2O(g)$	−457.2	−483.6
dissolving ammonium chloride in water	$NH_4Cl(s) \rightarrow NH^+_4(aq) + Cl^-(aq)$	−6.7	+16

Table 24.6
Possible combinations of ΔH and ΔS_{system}

ΔH	ΔS_{system}	ΔG	Result	Example
0	+	−	change occurs spontaneously (if rate is reasonable)	mixing of gases
0	−	+	change does not occur spontaneously	unmixing of gases
−	+	−	change occurs spontaneously (if rate is reasonable)	exothermic reactions producing gases, e.g. burning most fuels
−	−	+ or −	change may occur spontaneously depending on conditions. Most likely at low temperatures.	exothermic reactions in which number of moles of gas decreases, e.g. $N_2 + 3H_2 \rightarrow 2NH_3$, or condensing of a gas
+	+	+ or −	change may occur spontaneously depending on conditions. Most likely at high temperatures.	endothermic reactions in which number of moles of gas increases, e.g. liquids evaporating
+	−	+	change does not occur spontaneously	endothermic reactions in which number of moles of gas decreases, e.g. $6C(s) + 3H_2(g) \rightarrow C_6H_6(g)$, or photosynthesis

How to make reluctant reactions go

A reaction for which ΔG is positive will not happen spontaneously, because its total entropy change will be negative. This does not mean it can *never* happen—if this were the case, photosynthesis would never occur, and it would be impossible to make ice cubes in warm climates.

To make a reluctant reaction take place we need to alter the conditions, and this usually means altering the temperature. Consider the problem of making ice cubes in a warm climate (figure 24.10). To make the water freeze we put it in the freezer compartment of a refrigerator. At room temperature (298 K), ΔG for the change

$$H_2O(l) \rightarrow H_2O(s)$$

is $+546 J$ mole^{-1} and freezing does not occur. At the temperature of the freezer (263 K), ΔG is $-224 J$ mole^{-1} and the water freezes.

Why does the water freeze at 263 K but not at 298 K? When freezing occurs, heat equivalent to 6010 J $(-\Delta H)$ passes to the surroundings, and this increases the entropy of these surroundings by $6010/T$ JK^{-1} mole^{-1}, where T is the temperature of the surroundings. The amount of heat passed to the surroundings is the same at both temperatures, but at 263 K this heat has a bigger effect on the

At 298 K (25 °C)
$\triangle G = \triangle H - T \triangle S$
$\quad = -6010 - 298(-22)$
$\quad = +546 J$ mole^{-1}
water does not freeze

At 263 K (-10 °C)
$\triangle G = \triangle H - T \triangle S$
$\quad = -6010 - 263(-22)$
$\quad = -224 J$ mole^{-1}
water freezes

Figure 24.10
Making ice cubes.

colder surroundings than at 298 K when the surroundings are already quite warm.

At 263 K, $\Delta S_{\text{surroundings}} = 6010/263$, but at 298 K it is only 6010/298.

Another reluctant reaction which can be made to occur by changing the conditions is the production of quicklime, CaO, from limestone, $CaCO_3$:

$$CaCO_3(s) \rightarrow CaO(s) + CO_2(g); \qquad \Delta H^{\ominus} = 178\,\text{kJ}$$
$$\Delta S^{\ominus} = 165\,\text{J K}^{-1}$$

Answer these questions.

- Use $\Delta G^{\ominus} = \Delta H^{\ominus} - T\Delta S^{\ominus}$ to calculate $\Delta G^{\ominus}$ (in J mole^{-1}) for this reaction at 298 K.
- Will the reaction happen spontaneously at 298 K?
- Will (a) raising the temperature, (b) lowering the temperature, help to make ΔG more negative, and hence help the reaction to go?
- At what temperature will ΔG for this reaction be zero?
- What is the minimum temperature to which $CaCO_3$ must be heated to make it decompose?

Using ΔG you should have worked out that, to decompose $CaCO_3$, the temperature must be raised to 805°C (1078 K), assuming the pressure remains constant at 1 atm. Even though ΔG shows that under standard conditions at 298 K decomposition will not occur, by altering the conditions of temperature it is possible to make ΔG negative and ensure decomposition.

- What other change in the conditions would encourage decomposition?

You should now be able to see how useful ΔG can be, not only in telling whether a reaction will go, but also in deciding what can be done to make it go if it is reluctant to do so.

Limitations of ΔG

There are many reactions for which ΔG is negative, and which should therefore go spontaneously at room temperature. Surprisingly, some of these reactions do not happen spontaneously.

Fuels provide an example. ΔG values suggest that fuels should burn spontaneously at room temperature, yet they do not. This is because they have a high *activation energy*, and until they are heated (i.e. ignited) the reaction is infinitely slow. So beware: ΔG *provides information only about the thermodynamic or energetic feasibility* of a reaction. In practice, a reaction may not occur because of *kinetic stability*. To find out about kinetic factors, we need to know the activation energy for the reaction, and this is quite independent of ΔG. The kinetics or *rates* of reaction are considered in chapter 25.

ΔG values suggest that matches should burn spontaneously at room temperature. Fortunately, a high activation energy stops them burning before they are struck.

What's so 'free' about free energy?

Reactions which give out heat can be made to do useful work. For example, the energy from burning coal can be used in a power station to drive a generator. At first sight, we might expect that the energy available to do useful work would be given by ΔH. It turns out, in fact, that ΔG, not ΔH, gives the maximum amount of energy available. ΔG, the free energy change, tells us the amount of chemical energy *free* to do useful work, and that is how it gets its name. In reality, the actual amount of work obtained is always less than ΔG, so the free energy change represents the *maximum* amount of useful work available.

24.6 Free energy, electrode potentials and equilibrium constants

The free energy change, ΔG, indicates whether a reaction is likely to proceed spontaneously. There are, however, other criteria that can be used to answer this question.

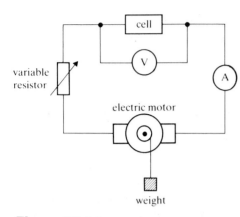

variable
resistor

electric motor

weight

Figure 24.11
Obtaining work from a cell.

Electrode potentials

A useful way of deciding whether a redox reaction will take place is to look at standard electrode potentials. In general, if $E^\ominus$ for a reaction is positive, it will go, provided the reaction is kinetically feasible. We might expect $\Delta G^\ominus$ and $E^\ominus$ to be related, and this relationship is quite easy to derive.

Figure 24.11 shows an electric cell driving an electric motor which is doing some useful work—raising a weight.

If the resistance in the circuit is low, the current will be quite high, and work will be done quickly. However, the voltage V of the cell will be quite low because of the cell's internal resistance. To get a higher voltage, the variable resistance must be increased, which naturally reduces the current. As the resistance approaches infinity, the voltage tends to a maximum, which is the e.m.f. of the cell, E_{cell}. With the voltage of the cell at this maximum, the maximum amount of work can now be done by the cell. But, unfortunately, since this maximum voltage corresponds to an infinitesimal current, the work will be obtained infinitely slowly.

We can now calculate the maximum amount of work obtained. Suppose a mole of electrons are driven round the circuit by a potential difference E. Since 1 volt = 1 joule per coulomb,

$$\text{maximum work done} = FE \text{ joules}$$

(F = charge on one mole of electrons—the Faraday).

For z moles of electrons,

$$\text{maximum work done} = zFE$$

This is the *absolute maximum* amount of work available. In reality, much less than this will be obtained because of inefficiencies such as friction in the motor. Furthermore, to obtain the work in a finite time, the voltage must be less than E.

Now, we already know the maximum amount of energy available = $-\Delta G$. (The minus sign arises because the energy is *lost* by the chemicals.) Hence

$$\Delta G = -zFE$$

Under standard conditions,

$$\Delta G^\ominus = -zFE^\ominus$$

(z is the number of electrons transferred during the cell reaction).

This equation applies quite generally and shows the close relationship between $E^\ominus$ and $\Delta G^\ominus$.

■ Consider the Daniell Cell

$$Zn|Zn^{2+}|Cu^{2+}|Cu; \quad E^\ominus = 1.1\,V$$

Calculate the value of $\Delta G^\ominus$ for the cell reaction.

$$Zn(s) + Cu^{2+}(aq) \rightarrow Zn^{2+}(aq) + Cu(s)$$

Equilibrium constants

The extent of a reaction can be judged directly by looking at its equilibrium constant. There is a simple relationship between K and $\Delta G^\ominus$.

$$\Delta G^\ominus = -RT \ln K$$

If K is large and positive the reaction is expected to go, providing it is kinetically feasible. This equation is quite general and applies to all reactions, whether or not they involve redox cells. It shows the close relationship between $\Delta G^\ominus$ and K.

■ Use the value of $\Delta G^\ominus$ for the Zn/Cu^{2+} reaction which you derived above to calculate the equilibrium constant for the reaction at 298 K.

Your result should be of the order of 10^{37}, indicating that the reaction should certainly go under these conditions. However, even with this big number, a tiny amount of Zn and Cu^{2+} will be left unreacted at the end. If it was not, K would be infinite.

■ What is the value of the equilibrium constant for the reverse reaction

$$Cu(s) + Zn^{2+}(aq) \rightleftharpoons Cu^{2+}(aq) + Zn(s)$$

What does your answer suggest about this reaction?

Comparison of $\Delta G^{\ominus}$, $E^{\ominus}$ and K

Table 24.7 gives the corresponding values of $\Delta G^{\ominus}$, $E^{\ominus}$ and K for certain types of reaction. Note that a 'complete' reaction has been defined arbitrarily as one for which $K>10^{10}$, and a reaction which 'doesn't go' as one with $K<10^{-10}$. These figures correspond to $\Delta G^{\ominus}$ values of $-60\,kJ\,mole^{-1}$ and $+60\,kJ\,mole^{-1}$, respectively. Up to now we have simply said that any reaction with a negative ΔG will be spontaneous, but table 24.7 shows how ΔG can be extended to see just *how far* a reaction goes. But remember, ΔG only provides information about *thermodynamic* or *energetic* feasibility. Many reactions appear to be feasible on thermodynamic grounds, but in fact do not go because of kinetic stability: the reaction *rate* is too slow. Reaction rates are covered in the next chapter.

Table 24.7
Values of K, $\Delta G^{\ominus}$ and $E^{\ominus}$ for certain types of reaction

K	$\Delta G^{\ominus}/kJ\,mole^{-1}$	$E^{\ominus}/volt$	**Extent of reaction**
$>10^{10}$	<-60	$>+0.6$	reaction 'complete'
1	0	0	reaction balanced between reactants and products
$<10^{10}$	$>+60$	<-0.6	reaction 'doesn't go'

24.7 A final point

In this chapter we have asked the question 'Why do things happen the way they do?' The answer has been 'Because that is the most probable way'. There are, however, other, quite valid answers. For example, we have asked the question 'Why does ice melt when heated?' The answer given is 'Because there are more ways of melting than unmelting.' Nevertheless, it would have been equally correct to say 'Because heat provides the energy needed to break bonds between water molecules in the solid state.'

Entropy and thermodynamics provide one way of explaining natural phenomena, but there are other approaches. Nevertheless, the Second Law of Thermodynamics has proved to be one of the most valuable and universally applicable scientific laws.

Summary

1 Spontaneous changes occur if they result in a more probable arrangement of molecules and energy, i.e. an increase in the number of ways, W, in which molecules and energy can be distributed.

2 Entropy, S, is equal to $k\ln W$. In spontaneous changes entropy always increases.

3 In general, the entropy values for gases and aqueous solutions are greater than those for solids.

4 When considering entropy changes, it is important to consider the entropy change of the surroundings as well as that of the chemical system:

$$\Delta S_{surroundings} = -\Delta H/T$$

5 Free energy changes, ΔG, provide a way of accounting for the surroundings as well as the system:

$$\Delta G = \Delta H - T\Delta S_{system}$$

For all spontaneous changes, ΔG must be negative.

6 Reactions for which ΔG is positive can sometimes be made to go by altering conditions, particularly temperature.

7 ΔG indicates the *thermodynamic* feasibility of a reaction. It says nothing about its *kinetic* feasibility.

8 $\Delta G^{\ominus}$ is closely related to K and $E^{\ominus}$:

$$\Delta G^{\ominus} = -RT\ln K$$

$$\Delta G^{\ominus} = zFE^{\ominus}$$

Study questions

1 (a) Consider the equilibrium between solid iodine and iodine dissolved in CCl_4:

$$I_2(s) \rightleftharpoons I_2(CCl_4)$$

On which side does this equilibrium normally lie?
Does iodine normally dissolve in CCl_4, or not?

(a)

(b)

Figure 24.12
(a) A lattice containing eight iodine molecules, surrounded by ten CCl_4 molecules. (b) One iodine molecule has dissolved.

$I_2(s)$ crystal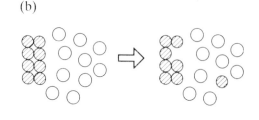

CCl_4 solvent

(b) Consider the simple situation of a lattice containing eight iodine molecules, surrounded by ten CCl_4 molecules (figure 24.12a). Suppose one iodine molecule dissolves, thereby occupying the space previously occupied by a CCl_4 molecule (figure 24.12b). How many different CCl_4 molecules could this I_2 molecule displace? In how many different ways could one I_2 molecule arrange itself among the CCl_4 molecules?

(c) Suppose now that the I_2 molecule returned to its vacant place in the $I_2(s)$ crystal. How many ways are there of doing this?

(d) Which is the more probable arrangement for the I_2 molecule—in the crystal or among the CCl_4 molecules?

(e) Does this agree with your answer to (a)?

(f) Suppose this example is now scaled up and we have one mole of $I_2(s)$ surrounded by one mole of CCl_4. If just one I_2 molecule dissolves in the CCl_4, how many different arrangements are possible?

(g) Suppose the molecule now returns to its original place in the lattice. In how many ways can it do this? Is it likely to?

(h) (Hard) If iodine is steadily added to CCl_4, the solution eventually reaches saturation, and no more iodine dissolves. Suggest a reason why.

2 Consider the entropy changes involved in the burning of magnesium.

$$2\,Mg(s) + O_2(g) \rightarrow 2\,MgO; \qquad \Delta H^{\ominus} = -1204\,kJ$$
$$S^{\ominus}(Mg) = 32.7\,J\,K^{-1}\,mole^{-1}$$
$$S^{\ominus}(O_2) = 204.9\,J\,K^{-1}\,mole^{-1}$$
$$S^{\ominus}(MgO) = 26.8\,J\,K^{-1}\,mole^{-1}$$

(a) What is the standard entropy of the reactants, two moles of magnesium and one mole of oxygen?
(b) What is the standard entropy of the products, two moles of magnesium oxide?
(c) What is the standard entropy change for the reaction as written in the equation?
(d) Will this entropy change alone favour the reaction?
(e) Use the value of $\Delta H^{\ominus}$ to calculate $\Delta S^{\ominus}_{surroundings}$ at 298 K. (Remember to work in joules.)
(f) What is the overall entropy change, $\Delta S^{\ominus}_{total}$?
(g) Is this a favourable entropy change?

3 For each of the following reactions, say whether you think the entropy change of the *chemicals* (the system) will be

A positive
B negative
C approximately zero

(a) $NH_4Cl(s) \rightarrow NH_3(g) + HCl(g)$
(b) $CH_4(g) + 2O_2(g) \rightarrow CO_2(g) + 2H_2O(g)$
(c) $C_3H_8(g) + 5O_2(g) \rightarrow 3CO_2(g) + 4H_2O(g)$
(d) $H_2(g) + I_2(g) \rightarrow 2HI(g)$

(e) $NaCl(s) \xrightarrow{aq} Na^+(aq) + Cl^-(aq)$

(f) $H^+(aq) + OH^-(Aq) \rightarrow H_2O(l)$
(g) amino acids $\rightarrow$ proteins
(h) starch $\rightarrow$ glucose

4 Some standard entropy values are given in the table below.

Substance	Formula	$S^{\ominus}/J\,K^{-1}\,mole^{-1}$
helium	He	126
neon	Ne	146
argon	Ar	155
krypton	Kr	164
xenon	Xe	167
hydrogen	H_2	65
water (gaseous)	$H_2O(g)$	189
water (liquid)	$H_2O(l)$	70
water (solid)	$H_2O(s)$	48

(a) What pattern do you notice in the standard entropies of the noble gases? Suggest a reason for the pattern.
(b) Compare the standard entropy values of hydrogen and helium. Suggest a reason for any difference.
(c) Compare the standard entropy values of steam, water and ice. What do the values suggest about the three states of matter?

5 (a) If you have not already done so, answer the questions on p. 403 relating to the decomposition of calcium carbonate.
(b) Repeat the calculations for magnesium carbonate instead of calcium carbonate, using these data:

$$MgCO_3(s) \rightarrow MgO(s) + CO_2(g); \qquad \Delta H^{\ominus} = 100.3\,kJ$$
$$\Delta S^{\ominus} = 174.8\,J\,K^{-1}\,mole^{-1}$$

(c) Use your results to compare the thermal stability of magnesium carbonate with that of calcium carbonate.
(d) Suggest a reason for the difference in thermal stability between the two compounds.

25 Reaction Rates

25.1 Introduction

The rates of chemical reactions are just as important to you as they are to industrialists and chemical engineers.

At home you might be interested in the rate at which you can boil an egg or bake a cake. Out-of-doors you might be interested in the rate at which your car is rusting, the rate at which your lettuces are growing and possibly the rate at which the stonework of buildings is being weathered by acidic gases in the atmosphere.

Why does a pressure cooker reduce the time needed to cook food?

Stonework from the Houses of Parliament. The left-hand specimen has been slowly weathered by acidic gases in the atmosphere.

In industry, engineers and other workers are concerned with the rates of chemical reactions in industrial processes and in constructional engineering. These might include the rate at which ammonia can be obtained from nitrogen and hydrogen, the rate at which concrete sets or the rate of growth of fruit and vegetable crops.

Industrialists and chemical engineers are not satisfied with merely turning one substance into another. In most cases, they want to obtain products rapidly, easily and as cheaply as possible. Time and money are important in industry. It is often necessary to accelerate reactions so that they are economically worthwhile.

At normal temperatures and pressures and in the absence of a catalyst, ammonia cannot be obtained from nitrogen and hydrogen. Fortunately, chemical engineers have found that a reasonable reaction rate results when the process is carried out at 250 atmospheres pressure and 450°C in the presence of an iron catalyst (section 22.5).

$$N_2(g) + 3H_2(g) \rightarrow 2NH_3(g)$$

Biological reactions also rely on the presence of catalysts. Almost every chemical reaction in your body is controlled by one or more catalysts. Catalysts are involved in simple reactions like the hydrolysis of starch to sugars. They are also involved in highly complex reactions like the replication of DNA which forms the genes in the nuclei of your cells. These biological catalysts, called **enzymes**, are proteins.

Why are these carcases stored in a refrigerator? What other foods must be stored in refrigerated containers?

Reaction rates are also important in archaeology. Archaeologists can estimate the age of rocks, fossils or prehistoric remains by a process known as radioactive dating. This entails measuring the concentration of a decaying radioactive isotope such as $^{14}_{6}C$ in the object under scrutiny.

■ Why does a pressure cooker enable vegetables to be cooked more rapidly?
■ What conditions and processes are used to slow down the rate at which perishable foods deteriorate?
■ How do gardeners accelerate the growth of their crops?

25.2 The concept of reaction rate

During a chemical reaction, reactants are being converted to products. The reaction rate tells us how fast the reaction is taking place by indicating how much of a reactant is consumed or how much of a product forms in a given time. Hence,

$$\text{reaction rate} = \frac{\text{change in amount (or concentration) of a substance}}{\text{time taken}}$$

Thus, *we can define reaction rate as the rate of change of amount or concentration of a particular reactant or product.*

When acidified hydrogen perioxide is added to a solution of potassium iodide, iodine is formed.

$$H_2O_2 + 2I^- + 2H^+ \rightarrow 2H_2O + I_2$$

The concentration of iodine rises from 0 to 10^{-5} mole dm^{-3} in 10 seconds.

$$\therefore \text{reaction rate} = \frac{\text{change in concentration of iodine}}{\text{time taken}}$$

Using the symbol Δ to represent the change in a particular quantity, we can write

$$\frac{\Delta[I_2]}{\Delta t} = \frac{10^{-5}\,\text{mole dm}^{-3}}{10\,\text{s}}$$

$$= 10^{-6}\,\text{mole dm}^{-3}\,\text{s}^{-1}$$

Strictly speaking, this result gives the average reaction rate during the 10 seconds that it took for the concentration of iodine to become 10^{-5} mole dm^{-3}. By measuring the change in concentration (or amount) over shorter and shorter time intervals we obtain an increasingly accurate estimate of the reaction rate at any moment. The disadvantages of 'clock' techniques such as this one are illustrated more effectively in figure 25.1. Using 'clock' techniques, the rate is obtained as the inverse of the time for a certain proportion of the reaction to occur. Provided the reaction has gone only a little way towards completion, very little error is introduced, but serious errors result if the 'end point' is, say, half-way to completion.

Ideally we should make the time interval almost zero, and then we obtain what is effectively the reaction rate at a particular instant, i.e.

$$\frac{\Delta[I_2]}{\Delta t} = \frac{d[I_2]}{dt}$$

$$\Delta t \rightarrow 0$$

In practice, it is usual to plot a graph of the concentration or amount of a particular substance against time. Then, the reaction rate can be obtained at particular times by drawing tangents to the resulting curve. This technique is illustrated in section 25.5.

Normally it is convenient to express reaction rates in mole dm^{-3} s^{-1} or in mole s^{-1}, but occasionally it is more convenient to use minutes or even hours as the unit of time.

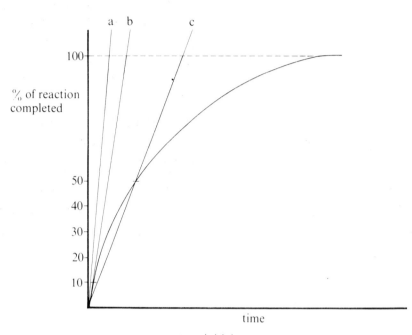

Figure 25.1
Errors in the 'clock technique' for measuring reaction rates.

a = true initial rate
b = average reaction rate for 10% completion
c = average reaction rate for 50% completion

In the reaction of acidified hydrogen peroxide with potassium iodide

$$H_2O_2 + 2I^- + 2H^+ \rightarrow 2H_2O + I_2$$

we measured the reaction rate in terms of the rate of formation of iodine.

$$\text{Rate of formation of iodine} = 10^{-6} \text{ mole dm}^{-3} \text{ s}^{-1}$$

■ What is the rate of consumption of H_2O_2 in this reaction?
■ What is the rate of consumption of I^- in this reaction?

25.3 Factors affecting the rate of a reaction

Our studies have already indicated several factors which can influence the rate of a reaction.

The availability of reactants and their surface area

Anyone who has camped knows that it is easier to start a fire using sticks rather than tree trunks. Similarly, magnesium powder will react much more rapidly than magnesium ribbon with dilute sulphuric acid. In general, the smaller the size of reacting particles, the greater is the total surface area exposed for reaction and, consequently, the faster the reaction. In the case of heterogeneous systems, in which the reactants are in different states, the area of contact between the reacting substances will influence the reaction rate considerably. In homogeneous systems, reacting substances normally occur in their *maximum* state of subdivision. They are usually present as individual particles in the gaseous or aqueous phase. In this case, the idea of surface area becomes meaningless.

The concentration of reactants

Increasing the concentration of a reactant normally causes an increase in the rate of a reaction, but this is not always the case. Furthermore, the different reactants can affect the rate of a particular reaction in different ways. For example, when nitrogen oxide reacts with oxygen

$$2NO(g) + O_2(g) \rightarrow 2NO_2(g)$$

the reaction rate doubles when the oxygen concentration doubles. But doubling the concentration of NO quadruples the rate of reaction. The effect of concentration on reaction rates is considered further in sections 25.5 and 25.6.

The temperature of the reactants

Perishable foods like milk 'go bad' much more rapidly in summer than in winter. In summer, the chemical reactions in the deterioration processes occur more rapidly at the higher temperatures. In general, increasing the temperature increases the rate of chemical reactions.

■ Why do you think reactions speed up at higher temperatures?

Look at the graphs of reaction rate against temperature in figure 25.2.

Figure 25.2
Graphs of reaction rate against temperature.

(a) most reactions

(b) explosive reactions

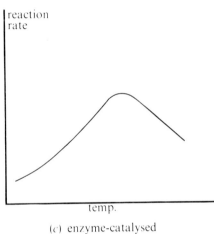

(c) enzyme-catalysed reactions

■ Why is there a sudden change in the rate of an explosive reaction at one particular temperature?
■ The rate of an enzyme-catalysed reaction increases at first, but then decreases as the temperature increases. Why is this? (How are enzymes susceptible to increase in temperature?)

The effect of temperature on reaction rates is considered further in sections 25.8 and 25.9.

Catalysts

Catalysts are substances which alter the rate of chemical reactions without undergoing any overall chemical change themselves. Although catalysts are not used up in reactions, they participate by forming intermediate compounds in the conversion of the reactants to products (section 19.6).

Normally, catalysts are used to accelerate reactions. Certain catalysts can, however, slow reactions down. For example, propane-1, 2, 3-triol (glycerine) is sometimes added to hydrogen peroxide as a negative catalyst in order to slow down its rate of decomposition.

Catalysts play an important part in biological and industrial processes by enabling reactions to take place which would never occur in their absence. Many industrial processes including the manufacture of ammonia, sulphuric acid, nitric acid, ethene, polythene and polystyrene rely heavily on the use of catalysts. Living things rely even more heavily on catalysts because every chemical reaction in animals, plants and micro-organisms requires its own specific enzyme (catalyst).

The last of six blocks of solid CO_2 (Drikold) being loaded into the refrigeration system of a Walls' ice cream van.

Light

Photosynthesis and photography both involve light-sensitive reactions. The leaves of plants contain a green pigment called chlorophyll. This can absorb radiation in the visible region of the electromagnetic spectrum and use this energy to synthesise chemicals and provide food for the plant.

During photosynthesis, plants transform carbon dioxide and water into oxygen and sugars such as glucose:

$$6CO_2(g) + 6H_2O(l) \xrightarrow{hv} C_6H_{12}O_6(aq) + 6O_2(g) \quad \Delta H = +2820\,kJ$$

In the absence of sunlight, energy is no longer provided and photosynthesis ceases.

White silver chloride turns purple and finally dark grey when it is exposed to sunlight. Sunlight provides the energy required to decompose the silver chloride.

$$AgCl(s) \rightarrow Ag(s) + \tfrac{1}{2}Cl_2(g)$$

The use of silver salts in photography depends on photosensitivity of this kind.

The reactions of halogens with hydrogen and with alkanes are further examples of photochemical reactions.

Thus, chlorine reacts slowly with hydrogen or methane in diffused daylight. When the reaction is exposed to intense ultraviolet radiation it becomes explosive. The effect of sunlight is believed to result from its ability to split chlorine molecules into highly reactive single atoms. These are known as **radicals**. They contain an unpaired electron (figure 25.3). See also section 27.5.

Figure 25.3
The reaction of chlorine with hydrogen.

25.4 Measuring reaction rates

The rate of a chemical reaction can be obtained by following some property which alters as the reaction occurs. By analysing the reaction mixture at suitable intervals, it is possible to determine the concentration of both reactants and products. Hence, we can obtain a measure of the reaction rate (i.e. the rate at which the concentration of a particular substance changes with time).

The method used to analyse the reaction mixture depends on the reaction under consideration. The following techniques illustrate four possible approaches.

Titrimetric analysis

This is particularly suitable for reactions in solution such as that between iodine and propanone catalysed by acid:

$$CH_3.CO.CH_3(aq) + I_2(aq) \xrightarrow{H^+} CH_2I.CO.CH_3(aq) + H^+(aq) + I^-(aq)$$

Titration techniques can also be used to analyse the alkaline hydrolysis of an ester (such as methyl methanoate):

$$HCOOCH_3 + OH^- \rightarrow HCOO^- + CH_3OH$$

In these cases, the reaction can be followed by removing and analysing small portions of the reaction mixture at intervals. Very often, the removed portion must be added to some reagent which will stop the reaction (i.e. 'quench' it).

This prevents further changes in concentration before the analysis is carried out. For example, in studying the reaction between iodine and propanone, portions of the reaction mixture can be pipetted into sodium hydrogencarbonate solution. This 'quenches' the reaction by neutralising the acid catalyst. The quenched mixture can then be analysed by titrating the unreacted iodine against a standard solution of sodium thiosulphate.

Colorimetric analysis

This method is especially convenient for systems in which one of the substances is coloured (e.g. the reaction of iodine with propanone or the reaction of bromine with methanoic (formic) acid).

The intensity of colour can be followed during the reaction using a photoelectric colorimeter (figure 25.4). From these measurements the concentration of the coloured substance can be obtained at different times.

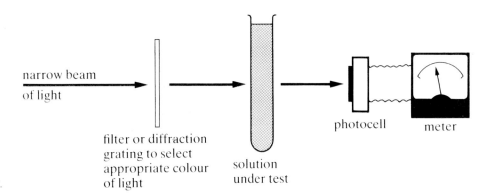

Figure 25.4
A simplified diagram of a colorimeter.

narrow beam
of light

filter or diffraction
grating to select
appropriate colour
of light

solution
under test

photocell   meter

In a colorimeter, a narrow beam of light passes through the solution under test towards a sensitive photocell. In many colorimeters, it is possible to select the most appropriate colour of light by choosing a particular filter or by adjusting a diffraction grating. The current generated in the photocell is, of course, proportional to the amount of light transmitted by the solution. This, in turn, depends upon the depth of colour of the substance under test. Thus, the current from the photocell will be greatest when the light transmitted by the solution is the greatest. This occurs when the coloured substance is most dilute. However, the meter is usually calibrated to show not the fraction of light transmitted but the fraction of light absorbed. This will be proportional to the concentration of the coloured substance in the test solution.

Pressure measurements

This technique is particularly suitable for gaseous reactions which involve changes in pressure. For example, the gaseous decomposition of 2-methyl-2-iodopropane can be followed conveniently by measuring the pressure at suitable time intervals.

Conductimetric analysis

Many reactions in aqueous solution involve changes in the ions present as the reaction proceeds. Consequently the electrical conductivity of the solution will change during the reaction. This can be used to determine the changing concentrations of reactants and products with time. Essentially, this method involves immersing two inert electrodes in the reaction mixture and then following the change in electrical conductivity of the solution with time (figure 25.5).

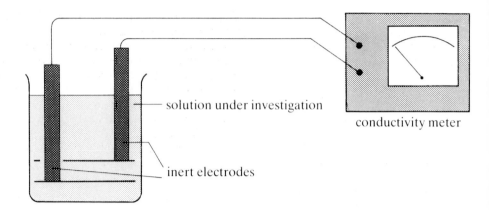

Figure 25.5
Following the change in electrical
conductivity of a solution with time.

The last three methods have one advantage over titrimetric analysis. They do not require the removal of samples from the reacting mixture. In these three cases, the extent of the reaction is determined at regular intervals without disturbing the reaction mixture.

It is important to realise that measurements on the reacting system do not give the rate of reaction directly. They simply give the concentration of a particular reactant or product, X, at a given time, t. By plotting a graph of the concentration of X against time, it is possible to determine the reaction rate. This involves measuring the change in concentration of X with time ($d[X]/dt$) from the gradient of the tangent at a given point (figure 25.6).

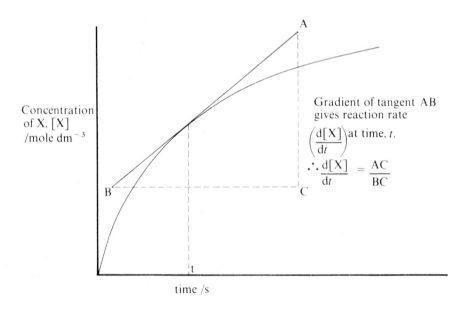

Figure 25.6
Obtaining the rate of reaction at a
given time from a graph of
concentration against time.

25.5 Investigating the effect of concentration on the rate of a reaction

We must now consider the influence of concentration on reaction rates in more detail. A convenient reaction to study is that between bromine and methanoic acid (formic acid) in aqueous solution.

The reaction is catalysed by acid:

$$Br_2(aq) + HCOOH(aq) \xrightarrow{H^+} 2Br^-(aq) + 2H^+(aq) + CO_2(g)$$

Table 25.1

Results for the kinetic study of the reaction between bromine and methanoic acid

Time/s	$[Br_2]$/mole dm^{-3}
0	0.0100
30	0.0090
60	0.0081
90	0.0073
120	0.0066
180	0.0053
240	0.0044
360	0.0028
480	0.0020
600	0.0013

The reaction can be followed colorimetrically by measuring the intensity of the red-brown bromine at suitable time intervals. By plotting a calibration curve of known bromine concentrations against colorimeter readings, it is possible to deduce the concentrations of bromine from the colorimeter readings obtained in the experiment. Some typical results are shown in table 25.1. The concentration of methanoic acid was virtually constant throughout the experiment because it was present in large excess.

The concentrations of bromine in table 25.1 are plotted graphically against time in figure 25.7.

The concentration of bromine ($[Br_2]$) falls during the course of the reaction. So, the rate of the reaction can be expressed in terms of the rate at which the bromine concentration changes.

$$\text{Reaction rate} = - \text{ rate of change of bromine concentration}$$

$$= - \frac{d[Br_2]}{dt}$$

Notice the negative sign in the last expression. $d[Br_2]$ is negative because the bromine is being used up. So, the negative sign is necessary to give the rate of reaction, $-d[Br_2]/dt$, a positive value.

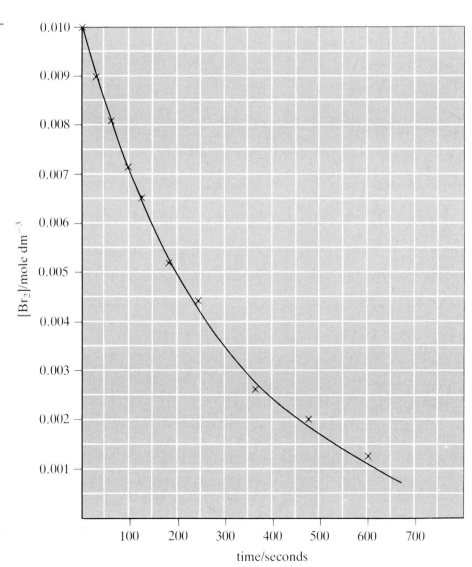

Figure 25.7

The variation of bromine concentration with time in the reaction between methanoic acid and bromine.

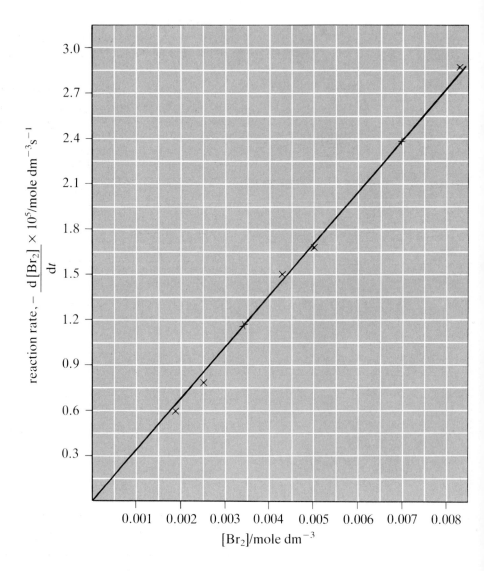

Figure 25.8
Variation of reaction rate with bromine
concentration.

In order to obtain the reaction rate at any given time, we must draw a tangent
to the curve at this particular time and measure its gradient. Values of the
reaction rate corresponding to different bromine concentrations at different
times are shown in table 25.2. These values of reaction rate are plotted vertically
against bromine concentration in figure 25.8.

■ How does the bromine concentration change with time?

■ How does the reaction rate change with time?

■ Is the rate of reaction affected by the bromine concentration?

■ How does the rate of reaction depend on the bromine concentration?

■ Write a mathematical expression relating reaction rate to bromine concentration.

Table 25.2
Values of the reaction rate corresponding to different bromine concentrations

Time/s	$[Br_2]$ obtained from figure 25.7/mole dm^{-3}	Reaction rate $(-d[Br_2]/dt)$ obtained from gradients in figure 25.7/mole $dm^{-3} s^{-1}$
50	0.0083	2.9×10^{-5}
100	0.0070	2.4×10^{-5}
200	0.0050	1.7×10^{-5}
250	0.0042	1.5×10^{-5}
300	0.0035	1.2×10^{-5}
400	0.0025	0.8×10^{-5}
500	0.00185	0.6×10^{-5}

The graph in figure 25.8 shows that the reaction rate is directly proportional to the bromine concentration, i.e.

$$\text{reaction rate} \propto [Br_2]$$
$$\Rightarrow \text{reaction rate} = k [Br_2]$$

where k is a constant, known as the **rate constant** or the **velocity constant** for the reaction.

25.6 Order of reaction and rate equations

Experiments show that the rates of most reactions can be related to the concentrations of individual reactants by an equation of the form

$$\text{Rate} = k[X]^n$$

This expression, in which X is the reactant under consideration and n is usually 0, 1 or 2, is known as a **rate equation** or **rate law**. The value of n gives the **order of the reaction**.
When $n = 0$, the reaction rate is said to be **zero order** with respect to X, i.e.

$$\text{Rate} = k[X]^0$$
$$\text{but, since } [X]^0 = 1,$$
$$\text{Reaction rate} = k$$

In other words, the reaction rate is independent of the concentration of X. This means that changing the concentration of X will not affect the rate of the reaction (figure 25.9).

When $n = 1$, the reaction rate is proportional to $[X]^1$ and the reaction is said to be **first order** with respect to X (figure 25.9).

When $n = 2$, the reaction rate is proportional to $[X]^2$ and the reaction is said to be **second order** with respect to X (figure 25.9).

■ What is the order with respect to Br_2 for the reaction we studied in the last section?

Figure 25.9
The variation of reaction rate with concentration for reactions which are zero, first and second order.

Look at the information in table 25.3. This relates to the reaction between hydrogen and nitrogen oxide at 800°C:

$$2\,H_2(g) + 2\,NO(g) \rightarrow 2\,H_2O(g) + N_2(g)$$

In experiments 1, 2 and 3 the initial concentration of nitrogen oxide is the same.

- What happens to the initial rate, when the initial concentration of hydrogen is doubled in experiment 2 compared to experiment 1?
- What happens to the initial rate, when the initial concentration of hydrogen is trebled in experiment 3 compared to experiment 1?
- How does the reaction rate depend on $[H_2]$?
- What is the order of the reaction with respect to hydrogen?

In experiments 4, 5 and 6 the initial concentration of hydrogen is the same. When [NO] is doubled in experiment 5 compared to experiment 4, the reaction rate is *not* doubled, *but* quadrupled (i.e. 2^2).

Table 25.3
Information concerning the rate of reaction between hydrogen and nitrogen oxide at 800°C

Experiment number	Initial concentration of nitrogen oxide /mole dm^{-3}	Initial concentration of hydrogen /mole dm^{-3}	Initial rate of production of nitrogen /mole dm^{-3} s^{-1}
1	6×10^{-3}	1×10^{-3}	3×10^{-3}
2	6×10^{-3}	2×10^{-3}	6×10^{-3}
3	6×10^{-3}	3×10^{-3}	9×10^{-3}
4	1×10^{-3}	6×10^{-3}	0.5×10^{-3}
5	2×10^{-3}	6×10^{-3}	2.0×10^{-3}
6	3×10^{-3}	6×10^{-3}	4.5×10^{-3}

Similarly, when [NO] is trebled (experiment 6 compared to experiment 4), the reaction rate is *not* trebled, *but* increases nine-fold (i.e. 3^2). These results show that

$$\text{Reaction rate} \propto [NO]^2$$

and

$$\text{Reaction rate} \propto (H_2]$$

∴ The order of reaction with respect to NO = 2 and the order of reaction with respect to H_2 = 1

We can combine these results in a single rate equation as

$$\text{Reaction rate} = k[NO]^2[H_2]$$

k, the rate constant, is constant for a given reaction at a particular temperature. It is important to realise that the rate equation can only be obtained by experiment; it cannot be deduced either theoretically or from the stoichiometric equation.

■ Using the concentrations of NO and H_2 and the initial rate of reaction in experiment 1, determine the value of k from the rate equation.

$$\text{Rate} = k[NO]^2[H_2]$$

■ What are the units of k in this case?

In experiment 2

$$[NO] = 6 \times 10^{-3}\,\text{mole dm}^{-3},$$

$$[H_2] = 2 \times 10^{-3}\,\text{mole dm}^{-3}$$

and

$$\text{rate} = 6 \times 10^{-3}\,\text{mole dm}^{-3}\,\text{s}^{-1}$$

Substituting these values in $\text{Rate} = k[NO]^2[H_2]$, we get

$$6 \times 10^{-3} = k(6 \times 10^{-3})^2 \times 2 \times 10^{-3}$$

$$\Rightarrow k = \frac{6 \times 10^{-3}}{(6 \times 10^{-3})^2} \times \frac{1}{2 \times 10^{-3}} = \frac{10^6}{12} = 8.33 \times 10^4$$

By substituting the units for reaction rate and concentrations in the rate equation, we can determine the units of k. For example, in the NO/H_2 reaction

$$k = \left(\frac{\text{rate}}{[NO]^2[H_2]} \right) \frac{\text{mole dm}^{-3}\,\text{s}^{-1}}{(\text{mole dm}^{-3})^2 . (\text{mole dm}^{-3})}$$

$$= \left(\frac{\text{rate}}{[NO]^2[H_2]} \right) \text{mole}^{-2}\,\text{dm}^6\,\text{s}^{-1}$$

As the rate constant is constant at a fixed temperature, our results show that $k = 8.33 \times 10^4\,\text{mole}^{-2}\,\text{dm}^6\,\text{s}^{-1}$ at 800°C.

Suppose we take the general case of a reaction,

$$x\text{A} + y\text{B} \rightarrow \text{products}$$

with a rate equation which can be expressed as

$$\text{Rate} = k[\text{A}]^m[\text{B}]^n$$

The indices m and n are known as the **orders of the reaction with respect to A and B**, respectively. The **overall order** of the reaction is described as $(m + n)$.

It is therefore important to be clear whether we are discussing the overall order of the reaction or the order with respect to an individual reactant.

The order of a reaction with respect to a given reactant is the power of that reactant's concentration in the experimentally determined rate equation.

The overall order of the reaction is the sum of the powers of the concentration terms in the rate equation.

The reaction between propanone and iodine in aqueous solution,

$$CH_3COCH_3(aq) + I_2(aq) \rightarrow CH_2ICOCH_3(aq) + H^+(aq) + I^-(aq)$$

is catalysed by H^+ ions. Experiments show that the rate law can be expressed as:

$$\text{Rate} = k[CH_3COCH_3](H^+]$$

■ What is the order of the reaction with respect to
   (a) propanone,
   (b) iodine,
   (c) H^+ ions?
■ What is the overall order of the reaction?

This particular reaction shows very clearly that the stoichiometric equation tells us nothing about the rate equation. I_2 appears in the stoichiometric equation, but it does not feature in the rate equation. Furthermore, H^+ ions which feature in the rate equation do not appear in the stoichiometric equation.

25.7 The half-life for a first-order reaction and for radioactive decay

Look closely at figure 25.7 once more.

■ How many seconds does it take for the concentration of bromine to fall from 0.010 M to 0.005 M?
■ How many seconds does it take for the concentration of bromine to fall from 0.005 M to 0.0025 M?
■ How many seconds does it take for the concentration of bromine to fall from 0.0025 M to 0.00125 M?

At any time during the reaction it takes 200 seconds for the concentration of bromine to halve (i.e. to fall from a concentration of say xM to $\frac{x}{2}$M).

The time taken for the concentration to fall to half its original value is called the half-life of the reactions.

All first-order reactions have constant half-lives.

Thus, the kinetics of a first-order reaction with its constant half-life are similar to the decay of a radioactive isotope which also has a constant half-life. (See figure 25.10 and section 7.4.)

Indeed, the constant half-lives of radioactive isotopes form the basis of archaeological dating. Archaeological dating can be carried out with organic remains containing radioactive $^{14}_{6}C$ or with mineral remains containing isotopes like radioactive $^{238}_{92}U$.

Half-lives range from minute fractions of a second (such as 10^{-21} seconds for $^{5}_{2}He$) to thousands of millions of years (such as 4.5×10^9 years for $^{238}_{92}U$). The half-life of a radioactive isotope is a fundamental property unaffected by chemical and physical changes.

Dating geological remains

From the half-life of a radioactive isotope, we can deduce the time it will take for a certain proportion of the isotope to decay. Thus, the decay of $^{238}_{92}U$ to $^{206}_{82}Pb$ provides a method of dating rocks in the earth's crust. We begin by assuming that the uranium-bearing rocks in the earth's crust originally contained $^{238}_{92}U$ but no $^{206}_{82}Pb$. The present ratio of $^{238}_{92}U : ^{236}_{82}Pb$ in the rocks can then be used to calculate the time which has elapsed since the rocks formed. Using this technique, the ages of different rocks vary from forty million to four thousand million years. Geologists often take the larger of these two values to be the age of the earth.

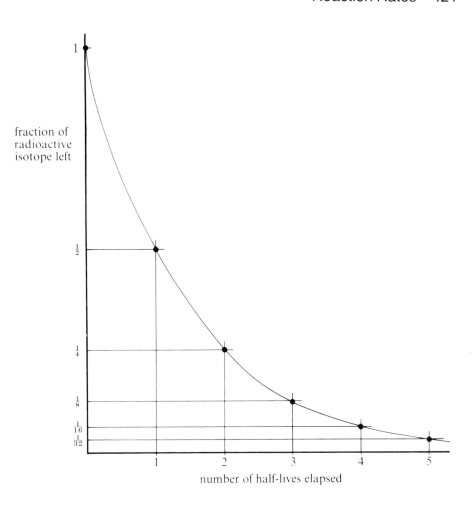

Figure 25.10
The fraction of a radioactive isotope remaining after 1, 2, 3, 4 and 5 half-lives.

Dating the remains of living things

The remains of living things are often dated from the amount of carbon-14 in them. The basis of carbon dating is the simultaneous production and disintegration of radioactive $^{14}_{6}\text{C}$. As high-energy cosmic rays pass through the earth's atmosphere, they strike atoms. This splits the atoms into sub-atomic fragments including neutrons. These neutrons (which are produced mainly in the upper atmosphere) react with nitrogen atoms producing $^{14}_{6}\text{C}$:

$$^{1}_{0}n + {}^{14}_{7}\text{N} \rightarrow {}^{14}_{6}\text{C} + {}^{1}_{1}\text{H}$$

At the same time as $^{14}_{6}\text{C}$ is being formed, the radioactive isotope is also decaying to $^{14}_{7}\text{N}$ by β-decay (section 7.3):

$$^{14}_{6}\text{C} \rightarrow {}^{14}_{7}\text{N} + {}^{0}_{-1}e$$

As a result of this simultaneous formation and decay of $^{14}_{6}\text{C}$, the atmosphere contains a constant concentration of $^{14}_{6}\text{C}$. The $^{14}_{6}\text{C}$ is present as $^{14}_{6}\text{CO}_2$. This $^{14}_{6}\text{CO}_2$ enters plants via photosynthesis. Eventually vegetable foods containing $^{14}_{6}\text{C}$ are eaten and metabolised by animals. Thus, all living things have a constant proportion of their carbon in the form of $^{14}_{6}\text{C}$. However, when the animal or plant dies, replacement of $^{14}_{6}\text{C}$ ceases while decay of $^{14}_{6}\text{C}$ continues.

Suppose that carbon-14 makes up $x\%$ of the carbon in living things and that the half-life of $^{14}_{6}\text{C}$ is 5800 years.

- What percentage of $^{14}_{6}\text{C}$ will the remains of a plant contain 5800 years after it has died?
- How long will it take for the percentage of $^{14}_{6}\text{C}$ in a dead plant to fall from $\frac{x}{2}\%$ to $\frac{x}{4}\%$?
- Approximately how old is an object which has $\frac{x}{8}\%$ of $^{14}_{6}\text{C}$?

Table 25.4

Investigating the effect of temperature on the reaction between HCl(aq) and $Na_2S_2O_3$(aq)

Temperature/K	Time for ink cross to disappear/s
296	135
298	119
302	90
307	62
311	45
317	37
320	34
326	24
332	20

By comparing the content of $^{14}_6C$ in archaeological specimens with that in similar materials living at the present time, it is possible to estimate the age of the specimens. Radiocarbon dating has been widely used to establish Egyptian chronology and to check the authenticity of ancient remains such as the Dead Sea scrolls.

25.8 Investigating the effect of temperature on the rate of a reaction

When dilute hydrochloric acid is added to sodium thiosulphate solution ($Na_2S_2O_3$(aq)), the solution becomes cloudy. The cloudiness is caused by a precipitate of sulphur. As the precipitate gradually thickens, its yellow colour becomes apparent.

$$S_2O_3^{2-}(aq) + 2H^+(aq) \rightarrow S(s) + SO_2(g) + H_2O(l)$$

How can we measure the reaction rate in this case?

Add $10cm^3$ of 1.0 M HCl to $50cm^3$ of 0.05 M $Na_2S_2O_3$(aq) and mix the contents thoroughly. Place the flask above an ink cross on white paper. Measure the interval between the addition of HCl and the obscuring of the ink cross as the precipitate thickens.

The results in table 25.4 were obtained when the experiment was carried out at different temperatures.

Notice that a rise in temperature of about 10 K causes the reaction rate to double. For example, the cross disappears in 119 seconds at 298 K, but in about half the time (62 s) at 307 K. The cross disappears in 90 seconds at 302 K and in half the time (45 s) at 311 K.

The question we must now ask is 'Why does a small rise in temperature cause such a large percentage increase in the reaction rate?'

25.9 Explaining the increase in reaction rate with temperature—the collision theory

As temperature rises, the average speeds of reacting particles increase. At high temperatures, there are more collisions per second and this results in an increase in the rate of reaction.

The increase in collision frequency explains the increase in reaction rate as temperature rises. But, does it also explain how *rapidly* the rate increases—the rates of many reactions double for a temperature rise of only 10 K?

From the kinetic theory, we can predict the relative increase in number of collisions when the temperature rises by 10 K.

The kinetic energy of a particle is proportional to its absolute temperature:

$$\tfrac{1}{2}mV^2 \propto T$$

but the mass of a given particle remains constant

$$\Rightarrow V^2 \propto T$$

$$\therefore \frac{V_1^2}{V_2^2} = \frac{T_1}{T_2} \qquad \text{(equation 1)}$$

where V_1 is the velocity at temperature T_1, and V_2 is the velocity at temperature T_2.

Now, suppose that the average speed of a particle is V at 300 K. What will its average speed be at 310 K?

Substituting in equation 1,

$$\frac{V_1^2}{V^2} = \frac{310}{300}$$

$$\therefore V_1 = \sqrt{\frac{310}{300}} \, V = \sqrt{1.033} \, V = 1.016V$$

$\therefore$ The average speed at 310 K is only 1.016 times greater than that at 300 K, i.e. it has increased by only 1.6%.

Assuming that the frequency of collisions depends on the average speed of the particles, we might expect the rate of collisions and hence the reaction rate to be 1.6% greater at 310 K than at 300 K. In practice, the reaction rate roughly doubles between 300 and 310 K, i.e. it increases by approximately 100%.

Clearly, the *simple* collision theory cannot account fully for the increase in reaction rate as temperature rises. How then do we explain the relatively large increase in reaction rate with temperature?

During a chemical reaction, bonds are first broken and then others are formed. Consequently, energy is required to break bonds and start this process, whether the overall reaction is exothermic or endothermic. Therefore, it is reasonable to assume that *particles do not always react when they collide*. They may not have sufficient energy for the necessary bonds to be broken.

A reaction will only occur if the colliding particles possess more than a certain minimum amount of energy. This minimum energy for a reaction to occur is known as the **activation energy**, E_A. The activation energy enables chemical bonds to stretch and break and rearrangements of atoms, ions and electrons to occur as the reaction proceeds.

The reaction can be imagined to proceed as shown in figure 25.11.

exothermic reaction

endothermic reaction

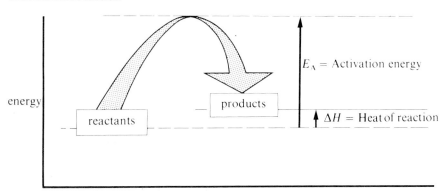

Figure 25.11
The progress of an exothermic and an endothermic reaction according to the collision theory.

The diagrams show the relationship between the activation energy, E_A, and the heat of reaction, ΔH, for an exothermic and an endothermic reaction. E_A is related to the rate constant of a reaction (see later in this section). It gives an indication of *how fast* the reaction will occur. If the activation energy is very large, only a small proportion of molecules have enough energy to react, so the reaction proceeds very slowly. If, however, the activation energy is very small, most of the molecules have sufficient energy to react and the reaction proceeds very fast.

In comparison to E_A, ΔH is related to the equilibrium constant of a reaction (section 21.7). ΔH gives information concerning *how far* a reaction goes towards completion.

The fact that a certain minimum energy is needed to initiate most reactions is well illustrated by fuels and explosives. These usually require a small input of energy to start their exceedingly exothermic reactions.

The idea of activation energies leads to the next important question.

What fraction of the particles have more than the activation energy – the minimum energy required for reaction? In the case of a gas, the energy of particles is largely kinetic energy ($\frac{1}{2}mv^2$). For, particles of a given mass this is determined by their speed. As we discovered in section 11.8, the distribution of speeds (and hence energies) can be determined by an apparatus similar to that used by Zartmann (figure 11.3).

The graph in figure 25.12 shows how the energies of particles are distributed. This graph is essentially a histogram. It shows the number of particles in each small range of kinetic energy. So, the area beneath the curve is proportional to the total number of particles involved. Furthermore, the number of particles with energy greater than E_A is proportional to the cross-hatched area beneath the curve at energies above E_A. Hence the *fraction* of particles with energy $> E_A$ is given by the ratio

$$\left(\frac{\text{cross-hatched area beneath curve}}{\text{total area beneath curve}} \right)$$

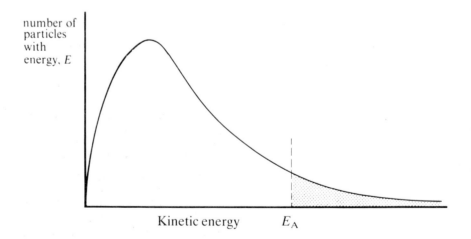

Figure 25.12
Distribution of the kinetic energies of particles in a gas.

Using probability theory and the kinetic theory of gases, Maxwell and Boltzmann derived equations for the distribution of kinetic energies amongst the molecules of a gas. From their equations, they calculated that the fraction of molecules with an energy greater than E_A J mole^{-1} is given by $e^{-E_A/RT}$. R is the gas constant (8.3 J K^{-1} mole^{-1}) and T is the absolute temperature.

This suggests that at a given temperature, T, the reaction rate $\propto e^{-E_A/RT}$.

Now, since the rate constant, k, is a measure of the reaction rate, we can write

$$k \propto e^{-E_A/RT}$$

$$\Rightarrow k = Ae^{-E_A/RT}$$

This last expression is sometimes called the **Arrhenius equation** because it was first predicted by the Swedish chemist, Svante Arrhenius in 1889.

In the Arrhenius equation, A (the Arrhenius constant), can be regarded as a **collision frequency** and **orientation factor** in the reaction rate, whilst $e^{-E_A/RT}$ represents an **activation state factor**. Thus, A is determined by the total number of collisions per unit time and the orientation of molecules when they collide (collision geometry). In contrast $e^{-E_A/RT}$ is determined by the fraction of molecules with sufficient energy to react.

If we take logarithms to base e in $k = Ae^{-E_A/RT}$

$$\ln k = \ln A + \ln e^{-E_A/RT}$$

$$\Rightarrow \ln k = \ln A - \frac{E_A}{RT}$$

Comparing the last equation with

$$y = c + mx$$

a graph of $\ln k$ against $1/T$ should be a straight line with gradient $-E_A/R$ (figure 25.13). E_A is the activation energy in J mole^{-1} and R is the gas constant (8.3 J K^{-1} mole^{-1}). Thus, by plotting a graph of the logarithm to base e of velocity constants against the reciprocal of the absolute temperature, we can measure the gradient $(-E_A/R)$. From this gradient, we can calculate the activation energy (E_A).

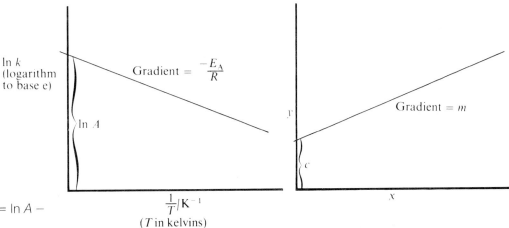

Figure 25.13
Comparing the equation $\ln k = \ln A - E_A/RT$ with $y = c + mx$.

Using the concept of activation energy, we can now return to our original question concerning the relatively large increase in reaction rate with temperature.

Look closely at figure 25.14. This shows how the kinetic energies of the molecules in a gas might be distributed at T K and $(T + 10)$ K. Let us suppose that colliding molecules must have a kinetic energy of E_A before a reaction takes place. Notice in figure 25.14 that only a small fraction of molecules (indicated by the dotted area) have sufficient energy to react at T K. However, when the temperature rises by 10 K, the fraction of molecules with sufficient energy to react (indicated by the vertically lined area) roughly doubles. This causes the reaction rate to double as well.

25.10 Catalysis

In a chemical reaction, existing bonds must break and new bonds must form as reactants are converted to products. In order to do this, energy is needed, usually in the form of heat. The 'energy barrier' (as we might call the activation energy) for the reaction between N_2 and H_2 to produce NH_3 is shown in figure

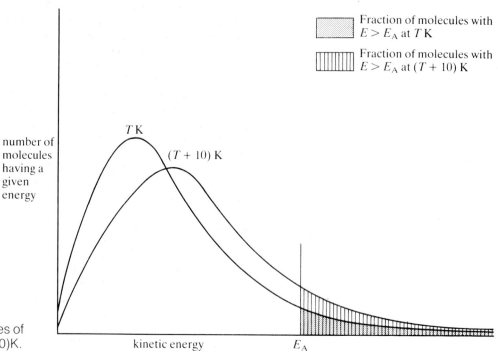

Figure 25.14
Distribution of the kinetic energies of
gas molecules at *T* K and (*T* + 10)K.

25.15. The horizontal axis of the diagram shows the progress of the reaction
from initial reactants to final products. The curve showing the energy of the
materials throughout the reaction is usually referred to as an **energy profile.**

Let us follow the progress of the reaction from left to right along the energy
profile in figure 25.15. As the molecules approach, there is little change in their
total energy until they get close to each other. Hence the flat part on the left of
the energy profile. When the molecules are within a few nanometres of each

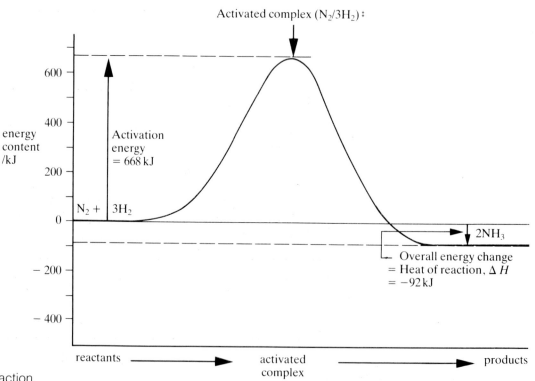

Figure 25.15
An energy profile for the reaction,
$N_2 + 3H_2 \rightarrow 2NH_3$.

other, repulsions between their nuclei and between their negative electron clouds begin to operate. Thus, the molecules must have sufficient speed and kinetic energy (668 kJ per mole of N_2) if they are to overcome these repulsive forces as the molecules get closer.

The peak of the energy profile represents an 'energy barrier'. This must be surmounted before bonds are stretched and sufficiently broken for products to form. The height of the energy barrier above the original height of the reactant molecules is the activation energy for the reaction. Only those molecules with enough energy to surmount the energy barrier will be able to form products. At the summit of the energy profile, the reactant molecules have a high energy content. They are described as the **activated complex** (designated by the superscript ‡). This activated complex can either break up and form the product molecules down the right-hand side of the energy profile or separate into the original molecules.

When nitrogen is mixed with hydrogen, no detectable reaction occurs even at high temperatures and high pressures. As the molecules approach each other, they have insufficient kinetic energy to overcome their mutual repulsions. So they never reach the activated state. They rise part of the way up the left-hand side of the energy profile, repel one another and separate again.

We can, however, speed up the reaction by using a **catalyst** (sections 19.6 and 22.3). *A catalyst can be defined as a substance which alters the rate of a reaction without itself undergoing any permanent chemical change.* Thus, a small amount of catalyst is capable of catalysing an infinite amount of reaction. Usually, the catalysed reaction is much faster than the uncatalysed reaction.

When the reaction

$$N_2(g) + 3H_2(g) \rightarrow 2NH_3(g)$$

is catalysed by tungsten, the activation energy is much lower than in the uncatalysed reaction (figure 25.16).

In the presence of tungsten, a greater proportion of molecules have sufficient energy to overcome the lower activation energy barrier. Therefore the reaction goes much faster (figure 25.17). Notice in figure 25.16 that the energy levels of reactants and products are the same in the catalysed reaction as in the uncatalysed reaction. The catalyst has not supplied any extra energy for the reactants, yet the reaction has been speeded up.

The catalyst has, in fact, provided a new reaction path for the breaking and rearrangement of bonds. The new reaction path has a lower activation energy,

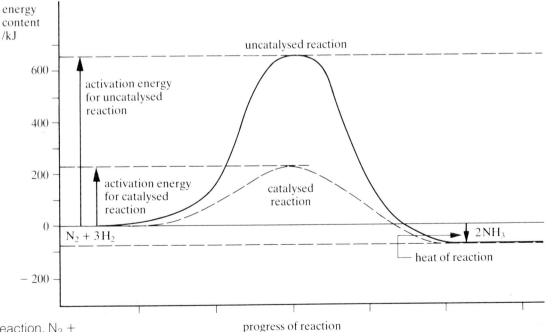

Figure 25.16
Energy profiles for the reaction, $N_2 + 3H_2 \rightarrow 2NH_3$, (a) uncatalysed and (b) catalysed by tungsten.

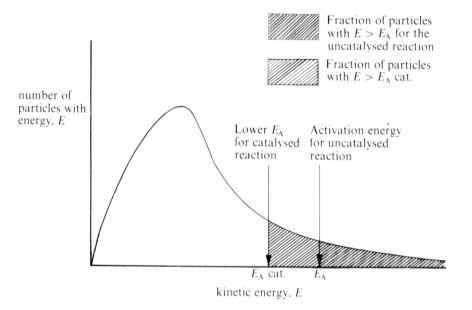

Figure 25.17
Distribution of the kinetic energies of reacting particles and the activation energies for catalysed and uncatalysed reactions. Notice the greater proportion of molecules which have an energy > the activation energy for the catalysed reaction.

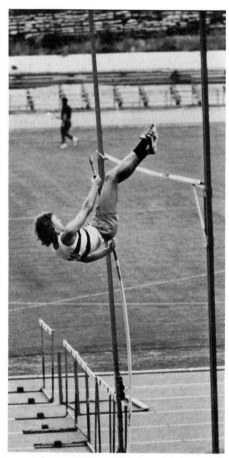

A strong pole-vaulter has sufficient height (potential energy) to clear the cross bar (activation energy barrier).

so that many more molecules can pass over the energy barrier. The situation in a catalysed reaction can be compared to a pole-vaulting event in which the bar has been lowered so that many more athletes can get over.

The reaction between nitrogen and hydrogen involves gases passing over a solid catalyst. Reactions of this kind in which the reactants are in a different physical state to the catalyst are said to involve **heterogeneous catalysis.**

In contrast to this, catalysed reactions in which reactants and catalyst are mixed together in the same state are said to involve **homogenous catalysis**.

Many important industrial processes involve heterogeneous catalysis. In contrast, the action of enzymes in biological systems usually involves homogenous catalysis.

■ Name the catalyst used in each of the following industrial processs;
   (a) the Haber Process,
   (b) the Contact Process,
   (c) the oxidation of ammonia to nitrogen oxide in the manufacture of nitric acid.
■ Do these processes involve heterogeneous or homogeneous catalysis?

Heterogeneous catalysis can reduce the atmospheric pollution from car exhaust fumes. The toxic chemicals in exhaust fumes include oxides of nitrogen, carbon monoxide and unburnt petrol. These are normally blown away and do not cause any problems. Under certain conditions, however, and particularly in traffic-dense city streets, these fumes cause a great deal of atmospheric pollution. Consequently, industrial chemists have developed a solid catalyst which can be used in the exhaust systems of all vehicles. This catalyses the conversion of these polluting chemicals to fully oxidised, harmless exhaust products. This is discussed further in section 27.6.

Another interesting development in the field of catalysis has been the increasing use of enzymes in industrial processes. These include the use of papain (a protein-hydrolysing enzyme) to remove the haze in beer and to tenderise beef. The manufacture of fruit juices, beer, vitamins and pharmaceutical products also uses enzymes extracted from animal tissues, plants, yeasts and fungi. The so-called 'biological' washing powders contain enzymes which attack animal and plant tissues. These are especially useful for removing the stains caused by biological materials such as food and blood.

Spear thistle (*Cirsium arvense*) before and after treatment with 2,4-D. 2,4-D and similar substances, act as plant hormones. They control the rate of metabolic processes and hence the rate of growth in plants. They do this by controlling the synthesis and activity of enzymes. 2,4-D is such a powerful growth promoter in broad-leaved plants that it can make them grow so rapidly that they exhaust their food supply and die. This has happened to the thistle in the lower photograph. Thin-leaved plants, such as grass surrounding the thistle, are unaffected. Hence 2,4-D can be used as a selective herbicide. Its full name is 2,4-dichlorophenoxyethanoic acid and its formula is

OCH_2COOH

Many catalysts are highly specific and the details of their operation are still not fully understood. The final choice of catalyst for many industrial processes may often be a combination of scientific deduction, trial and error or simply inspired guess-work. Indeed, the precise composition of an industrial catalyst is often a closely guarded secret.

25.11   The uses of reaction rate studies

Reaction rate studies give us useful information about the rates of chemical reactions. This is important for industrial processes in which time and the efficient use of resources are crucial. It is also useful for biological processes, archaeological dating and many other reactions, such as rusting and burning, which affect our everyday lives. Nevertheless, the most exhaustive studies of reaction rates have usually been done on industrial processes. In these cases chemists, engineers and economists endeavour to obtain maximum product from the minimum amount of raw material, using minimum fuel in the minimum possible time. When viewed in this light, it is not surprising that catalysts play such an important part in industrial processes.

Rate studies also allow us to interpret reactions on a molecular level. By considering the order of a reaction with respect to the different reactants, we can speculate about the sequence in which bonds break and atoms rearrange. From these ideas, it is possible to suggest a **reaction mechanism**.

As an example, let us consider the reaction between oxygen and hydrogen bromide at 700 K:

$$4\,HBr(g) + O_2(g) \rightarrow 2\,H_2O(g) + 2\,Br_2(g)$$

This equation indicates that four HBr molecules react with one O_2 molecule. If the reaction were to take place in a single step, these five molecules would need to collide with each other simultaneously. This is extremely improbable. Since the reaction occurs quite rapidly at 700 K, it is likely that it proceeds by a sequence of steps rather than by a single step involving the simultaneous collision of five molecules. In fact, *most chemical reactions which proceed at a measurable rate are believed to take place in a series of simple steps. This series of simple reactions is known as the reaction mechanism.*

Quantitative studies of the reaction between HBr and O_2 show that the reaction rate is proportional to both the concentration of HBr and the concentration of O_2. This means that the reaction is first order with respect to both HBr and O_2. We can summarise this information in the rate equation

$$\text{Rate} = k[\text{HBr}][\text{O}_2]$$

Notice that the stoichiometric equation involves HBr and O_2 in molar proportions of 4:1, yet the rate equation shows proportions of 1:1. How can we explain this?

The overall reaction must take place in a series of simple steps. These must satisfy both the rate equation and the stoichiometric equation.

The following mechanism has been proposed for the reaction:

	$HBr + O_2$	$\rightarrow$	~~HBrOO~~	Step 1	Slow
~~HBrOO~~ $+ HBr$		$\rightarrow$	~~2 HBrO~~	Step 2	Fast
~~HBrO~~	$+ HBr$	$\rightarrow$	$H_2O + Br_2$	Step 3	Fast
~~HBrO~~	$+ HBr$	$\rightarrow$	$H_2O + Br_2$	Step 4	Fast
Overall	$4\,HBr + O_2$	$\rightarrow$	$2\,H_2O + 2\,Br_2$		

Notice that each step in the reaction mechanism involves the collision of only two molecules. This is a much more likely event than the simultaneous collision of five molecules. Notice also the suggestion that the first step in the mechanism is slow whilst the others are fast. This explains why the reaction rate is proportional to both [HBr] and [O_2].

The first step producing HBrOO is very much a 'bottleneck' in the oxidation of hydrogen bromide. Thus, HBrOO forms slowly. Then, it is immediately consumed in the fast second step by reaction with HBr. Although the second, third, and fourth steps are very rapid, they produce water and Br_2 only as fast as the slowest stage in the sequence. Hence, those factors that determine the rate of formation of HBrOO determine the overall rate of reaction. The formation of HBrOO is the one step which dictates the rate because it is the slowest stage in the reaction mechamism. The slowest stage in a mechanism is called the **rate-determining step** or the **rate-limiting step**.

Iodine reacts with propanone in acid solution as follows:

$$I_2 + CH_3COCH_3 \rightarrow CH_2ICOCH_3 + H^+ + I^-$$

The reaction is first order with respect to CH_3COCH_3, zero order with respect to I_2 and first order with respect to H^+.

- Write a rate equation for the reaction.
- What is the overall order of reaction?
- Which substances are probably involved in the slow, rate-determining step of the reaction?
- Try to write a possible mechanism for the reaction.

The rate-determining step in the reaction involves propanone and H^+, but not iodine. Hence the suggested mechanism is:

The reaction rate is dictated by the first, slow stage which requires only the participation of CH_3COCH_3 and H^+. Once this first step is completed, the remaining steps take place rapidly. Consequently, the reaction rate is independent of the concentration of iodine and $[I_2]$ does not feature in the rate equation. If this mechanism is correct, then the reaction of Br_2 with propanone should take place at a similar rate to iodination. This is found to be so.

Both the reaction mechanisms that we have considered so far involve an initial slow step. This is not always the case. In order to illustrate this point, consider the reaction between bromide and bromate(v) ions in acid solution:

$$5\,Br^- + BrO_3^- + 6H^+ \rightarrow 3\,Br_2 + 3\,H_2O$$

Kinetic studies show that the reaction is fourth order overall; first order with respect to bromide, first order with respect to bromate(v) and second order with respect to H^+ ions.

$$Rate = k[Br^-][BrO_3^-][H^+]^2$$

The immediate deduction from this is that the rate-determining step involves one Br^-, one BrO_3^- and two H^+ ions. But, the simultaneous collision of four ions is most unlikely. A more likely explanation is that the slow rate-determining step is preceded by faster reactions. Hence, the suggested mechanism involves the initial formation of HBr and $HBrO_3$ in two fast reactions, followed by a reaction between these two substances in the slow rate-determining step.

$$H^+ + Br^- \rightarrow HBr \qquad \text{Step 1 Fast}$$

$$H^+ + BrO_3^- \rightarrow HBrO_3 \qquad \text{Step 2 Fast}$$

$$HBr + HBrO_3 \xrightarrow{\text{rate-determining step}} HBrO + HBrO_2 \quad \text{Step 3 Slow}$$

The HBrO and $HBrO_2$, produced in the slow step, then react rapidly with more HBr, forming bromine and water:

$$HBrO_2 + HBr \rightarrow 2\,HBrO \qquad \text{Step 4 Fast}$$

$$HBrO + HBr \rightarrow H_2O + Br_2 \qquad \text{Step 5 Fast}$$

Summary

1 Reaction rate $= \dfrac{\text{change in amount (or concentration) of a substance}}{\text{time taken}}$

2 The rate of a reaction can be affected by:
   (a) the availability of reactants and their surface area,
   (b) concentration (partial pressure for gases),
   (c) temperature,
   (d) catalysts,
   (e) light.

3 For the hypothetical reaction

$$a\text{A} + b\text{B} + c\text{C} \rightarrow \text{products}$$

it is found experimentally that the reaction rate can be expressed as

$$\text{Reaction rate} = k[\text{A}]^\alpha[\text{B}]^\beta[\text{C}]^\gamma$$

This is known as the rate equation for the reaction. k is known as the rate constant and its value is constant for a given reaction at a particular temperature. When all reactant concentrations are 1.0 M, k, the rate constant, is numerically equal to the reaction rate.

4 The rate equation can only be obtained experimentally. It cannot be deduced from the stoichiometric equation.

5 The order of a reaction with respect to a given reactant is the power of that reactant's concentration in the rate equation. For the hypothetical reaction just mentioned,

order of reaction with respect to A $= \alpha$,
order of reaction with respect to B $= \beta$,
order of reaction with respect to C $= \gamma$.

The overall order of a reaction is the sum of the powers of the concentration terms in the rate equation. Hence, the overall order of the hypothetical reaction $= \alpha + \beta + \gamma$.

6 The half-life of a first-order reaction is constant. The kinetics of a first-order reaction with its constant half-life are similar to the decay of a radioactive isotope. This also has a constant half-life.

7 A reaction will only occur if the colliding particles possess more than a certain minimum amount of energy known as the activation energy.

8 A catalyst is a substance which alters the rate of a reaction without itself undergoing any permanent chemical change. Catalysts usually speed up a reaction. They do this by introducing an entirely different reaction mechanism with a much lower activation energy than the uncatalysed reaction.

9 Most chemical reactions proceed by a sequence of simple steps, each of which involves only one or two particles. This series of simple steps is known as the reaction mechanism.

The slowest step in the reaction mechanism dictates the overall reaction rate. This slowest step is usually known as the rate-determining step.

10 The mechanism for a reaction is related to the rate equation. It cannot be deduced from the stoichiometric equation.

Study questions

1 (a) Give one example in each case of a reaction which takes place:
   (i) instantaneously,
   (ii) at a moderate rate,
   (iii) rapidly at a high temperature, but not at all room temperature.
   (b) Explain why the reactions you chose in (a) (i) and (a) (iii) behave as they do.

2 (a) Why is it difficult to hard boil an egg on the top of Mount Everest?
   (b) Is it difficult to fry an egg there? Explain.

3 For the gaseous reaction,

$A(g) + B(g) \rightarrow C(g) + D(g)$ it is found that,

Reaction rate $= k[A]^2[B]$

How many times does the rate increase or decrease if:
   (a) the partial pressures of both A and B are doubled,
   (b) the partial pressure of A doubles, but that of B remains constant,
   (c) the volume of the reacting vessel is doubled,
   (d) an inert gas is added, which doubles the overall pressure whilst the partial pressures of A and B remain constant.
   (e) the temperature rises by 30°C?

4 (a) Draw a sketch graph of the percentage reactant remaining against time, for a zero-order and a first-order reaction. (Assume, in each case, that it takes 10 minutes for the amount of reactant to fall from 100% to 50%.)
   (b) The half-life of radioactive $^{238}_{92}U$ is 4.5×10^9 years. It takes about 4.5×10^9 years for half of a given amount of uranium to disintegrate by radioactive emission and turn into lead. When uranium and lead are found together in rocks, the age of the rock may be deduced. In a particular sample, uranium and lead are found in molar proportions of 1:3.
   Estimate the age of the rock, stating any assumptions which you make.

5 The isotope $^{24}_{11}Na$ decays by β-particle emission to a stable nuclide. The half-life of $^{24}_{11}Na$ is 15 hours.
   (a) Write a nuclear equation for the decay process.
   (b) The rate of emission of β-particles from a freshly prepared sample of $^{24}_{11}Na$ is measured for 60 hours. Draw a sketch graph showing the rate of emission of β-particles (vertically) against time, during the 60 hours.
   (c) 2.0 g of $^{24}_{11}Na$ is allowed to decay. What mass of $^{24}_{11}Na$ will be left after 60 hours?

6 (a) Explain the following terms:
   (i) order of reaction,
   (ii) rate constant,

Table 25.5
The rate constants for the decomposition of hydrogen iodide at different temperatures

Rate constant, k /mole^{-1} dm^3 s^{-1}	Temperature/K
3.75×10^{-9}	500
6.65×10^{-6}	600
1.15×10^{-3}	700
7.75×10^{-2}	800

    (iii) half-life,
    (iv) activation energy,
    (v)   activated state.
(b) Rate constants (k) for the decomposition of hydrogen iodide at different temperatures are given in table 25.5.
    (i) Plot a graph of ln $k(\log_e k)$ against $1/T$.
    (ii) Use this graph to obtain a value for the activation energy for the decomposition of hydrogen iodide.
        (The gas constant, $R = 8.3$ J K^{-1} mole^{-1}.)

7 Hydrogen peroxide reacts with iodide ions in acid solution according to the following equation:

$$H_2O_2(aq) + 2I^-(aq) + 2H^+(aq) \rightarrow I_2(aq) + 2H_2O)l)$$

The rate of the reaction can be calculated by measuring the time for the first appearance of I_2 in the solution. When iodine first appears the concentration of I_2 is 10^{-5} mole dm^{-3}.

(a) For a particular experiment, the initial concentrations are
    $[H_2O_2] = 0.010$ M, $[I^-] = 0.010$ M and $[H^+] = 0.10$ M.
    Calculate the reaction rate if I_2 first appears after 6 seconds.
(b) In a second experiment, the initial concentrations are
    $[H_2O_2] = 0.005$ M, $[I^-] = 0.010$ M and $[H^+] = 0.10$ M.
    Calculate the reaction rate if I_2 first appears after 12 seconds.
(c) From these calculations show that the reaction is first order with respect to H_2O_2.
(d) Given the further information that the rate law is,
    Reaction rate $= k[H_2O_2][H^+][I^-]$, calculate the rate constant, k.
(e) What are the units of k?
(f) Predict the rate of reaction when
    $[H_2O_2] = 0.05$ M, $[H^+] = 0.10$ M and $[I^-] = 0.02$ M.

8 In an experiment to study the acid-catalysed reaction of propanone (CH_3COCH_3) with iodine, 50 cm^3 of 0.02 M I_2 were mixed with 50 cm^3 of acidified 0.25 M propanone solution. 10 cm^3 portions of the reaction mixture were removed at 5 minute intervals and added rapidly to excess $NaHCO_3(aq)$. The remaining iodine was then titrated against $Na_2S_2O_3(aq)$. The graph in figure 25.18 shows the volume of $Na_2S_2O_3(aq)$ required to react with the remaining iodine at different times from the start of the reaction.

(a) Why are the 10 cm^3 portions of the reaction mixture added rapidly to excess $NaHCO_3(aq)$ before titration with $Na_2S_2O_3(aq)$?
(b) What is the rate of reaction in terms of cm^3 of $Na_2S_2O_3(aq)$ min^{-1}?
(c) How does the *rate* of change of iodine concentration vary during the experiment?
(d) Is the reaction rate dependent on the concentration of iodine?
(e) What is the order of reaction with respect to iodine?
(f) Write an equation for the reaction between $S_2O_3^{2-}$ and I_2.
(g) What is the molarity of I_2 in the 100 cm^3 of reaction mixture at time $= 0$ min?
(h) Use the graph to predict the volume of $Na_2S_2O_3(aq)$ which reacts with 10 cm^3 of the reaction mixture at time $= 0$ min.
(i) What is the molarity of the $Na_2S_2O_3(aq)$ used in the titrations?
(j) Suppose the reaction is first order with respect to propanone. What would be the rate of reaction (in cm^3 $Na_2S_2O_3(aq)$ min^{-1}) if 0.50 M propanone were used in place of 0.25 M?

Figure 25.18
Volume of $Na_2S_2O_3$(aq) required to react with the remaining iodine at different times during the reaction of iodine with propanone.

9   Two gases, X and Y, react according to the equation

$$X(g) + 2Y(g) \rightarrow XY_2(g)$$

Experiments were performed at 400 K in order to determine the order of this reaction and the following results were obtained.

Experiment number	Initial concentration of X/ mole dm^{-3}	Initial concentration of Y/ mole dm^{-3}	Initial rate of formation of XY_2/ mole dm^{-3} s^{-1}
1	0.10	0.10	0.0001
2	0.10	0.20	0.0004
3	0.10	0.30	0.0009
4	0.20	0.10	0.0001
5	0.30	0.10	0.0001

(a) What is the order of this reaction with respect to (i) X, (ii) Y?
(b) Write a rate equation for the reaction of X with Y.
(c) Using the rate equation, predict a possible mechanism for this reaction.
(d) Using the results from experiment 1, calculate the numerical value of the rate constant, k.
(e) What are the units of k?
(f) What further experiments would you carry out to find the activation energy of the reaction between X and Y?
(g) Why are chemists interested in obtaining orders of reaction and rate equations?

10   Suggest experimental means by which the rates of the following reactions could be followed.

(a) $CaCO_3(s) \rightarrow CaO(s) + CO_2(g)$
(b) $2NO(g) + 2H_2(g) \rightarrow N_2(g) + 2H_2O(g)$
(c) $Cl_2(aq) + 2Br^-(aq) \rightarrow Br_2(aq) + 2Cl^-(aq)$

11 (a) Draw an energy profile curve for the reaction
$$H_2(g) + I_2(g) \rightarrow 2\,HI(g) \qquad \Delta H^{\ominus} = -10\,kJ$$
Put labelled arrows on your diagram to indicate the activation energy of the reaction and the enthalpy change of the reaction.

(b) The rate of this reaction is given by
$$Rate = k[H_2][I_2]$$
   (i)   What is the order of the reaction with respect to iodine?
   (ii) What is the overall order of the reaction?
   (iii) Name the constant, k.

(c) When 0.1 moles of hydrogen and 0.2 moles of iodine were mixed at 400°C in a $1\,dm^3$ vessel, the initial rate of formation of hydrogen iodide was 2.3×10^{-5} mole $dm^{-3}\,s^{-1}$.
   (i)   What is the value of k at 400°C?
   (ii) What are the units of k?

12 The following results were obtained from a study of the isomerisation of cyclopropane to propene in the gas phase at 433°C.

Time/hours	0	2	5	10	20	30
% of cyclopropane remaining	100	91	79	63	40	25

(a) Write an equation for the reaction involved.
(b) Show that the reaction is first order with respect to cyclopropane.
(c) Calculate the velocity constant of the reaction at 433°C.

26 Introduction to Carbon Chemistry

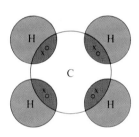

Figure 26.1
Electron sharing in methane.

Table 26.1
Some average bond energies

Bond	Bond energy /kJ mole^{-1}
C—C	346
C=C	610
C≡C	835
Si—Si	226
Si=Si	318 (estimated)
S—S	272
C—O	360
Si—O	464
C—H	413
Si—H	318

26.1 Carbon—a unique element

The number of compounds containing carbon and hydrogen whose formulae are known to chemists is about seven million. This is far more than the number of compounds of all the other elements put together. Why does carbon have this unique ability to form an enormous number of compounds? And how can we hope to begin to study more than a tiny fraction of them? There are three important properties of carbon that enable it to form so many stable compounds.

(a) Carbon has a fully shared octet of electrons in its compounds

For example, in methane the outer-shell electrons are shared as shown in figure 26.1. This means that the carbon atoms have no lone pairs or empty orbitals in their outer shells, so they are unable to form dative bonds. The inability of carbon to bond in this way once it has an octet of electrons is responsible for the kinetic stability of its compounds (see below).

(b) Carbon can form strong single, double and triple bonds to itself

The stability of the single C—C bond can be seen by comparing the bond energies in table 26.1.

It is worth comparing carbon and silicon here since silicon might be expected to show similarities to carbon, being the next member of group IV. Note the strength of the C—C bond compared with that of the Si—Si bond. Note too the high strength of the C—H bond: all but a handful of the vast range of carbon compounds also contain hydrogen. Of course, it is not enough to consider simply the strength of bonds between carbon atoms and hydrogen atoms. If carbon compounds are to be stable they must be stable under normal conditions, and that means in the presence of air. In fact, compounds containing carbon and hydrogen are *not* stable relative to their oxidation products, carbon dioxide and water. We would therefore expect them to react with oxygen exothermically. This, of course, they do. For example, methane:

$$CH_4(g) + 2O_2(g) \rightarrow CO_2(g) + 2H_2O(g); \quad \Delta H^{\ominus} = -890\,kJ$$

Most people are familiar with this reaction: it occurs in most gas-fired appliances in Britain, since North Sea gas is largely methane. It is a familiar fact that although methane is energetically unstable relative to its combustion products, it does not react with air until heated to quite high temperatures. In other words, it needs lighting before it burns. This is because the reaction between methane and oxygen has a high activation energy which must be supplied before the reaction

437

will proceed. Thus methane, like most compounds containing carbon and hydrogen, is energetically unstable in the presence of air, but kinetically stable (see section 12.7).

Compounds containing silicon and hydrogen are also energetically unstable relative to their combustion products:

$$SiH_4(g) + 2O_2(g) \rightarrow SiO_2(s) + 2H_2O(g); \quad \Delta H^{\ominus} = -1428\,kJ$$
silane

This reaction is much more exothermic than the corresponding one for methane, largely due to the very high energy of the Si—O bond. Unlike methane, silane is not kinetically stable in the presence of oxygen. The activation energy of the above reaction is quite low, and silane bursts into flame spontaneously in air. Other compounds containing silicon and hydrogen behave similarly, so there is no huge range of silicon compounds to compare with that of carbon. However, the high strength of the Si—O bond relative to the Si—Si and Si—H bonds means that silicon exists naturally as highly stable silicon(IV)oxide (sand) and silicates.

The ability of carbon to form strong bonds to itself means that it can form chains and rings of varying size. This is called **catenation**. These chains and rings are the basis of carbon's many stable compounds.

The kinetic stability of hydrocarbons in air is very important to society. It means they can be stored and, barring accidents, the energy of their oxidation can be released when it is required. This makes hydrocarbons, of which oil is our major source, the most important modern fuels.

In the presence of air, hydrocarbons are energetically unstable, but kinetically stable. They can, therefore, be safely stored for long periods at room temperature . . .

. . . but at high temperatures they undergo rapid combustion.

(c) Carbon can form four covalent bonds

The bond energies given in table 26.1 suggest that, like carbon, sulphur should be able to form reasonably stable bonds to itself. However, sulphur forms only two bonds, so a chain of sulphur atoms cannot have side-groups attached to it. By contrast, carbon can form *four* bonds. This means a chain of carbon atoms can have many different groups attached, leading to a wide diversity of compounds.

26.2 Organic Chemistry

The diversity of carbon chemistry is responsible for the diversity of life itself. The ability to form a virtually unlimited range of compounds has led to an almost unlimited range of living organisms constructed out of molecules containing carbon. You yourself are a unique individual because you contain unique proteins. Only carbon could form a range of compounds diverse enough to provide a different one for every individual.

The major source of compounds containing carbon and hydrogen is living or once-living material: animals, plants, coal, oil and gas. For this reason, it was originally thought that only living organisms could produce these compounds. This has since been shown to be untrue, but the name **organic** is still applied to that branch of chemistry concerned with the study of compounds containing C—H bonds. This includes the vast majority of carbon compounds, but traditionally compounds such as CO, CO_2 and carbonates have been considered to belong to the field of inorganic chemistry.

The position of carbon as the basis of the molecules of life means that the study of organic chemistry is of central importance in understanding the chemistry, and therefore the biology, of living systems. (The chemical study of living systems is called **biochemistry**.) A knowledge of organic chemistry enables chemists to develop and manufacture drugs, agricultural chemicals, anaesthetics and other chemicals whose effects on life processes are important to humans.

Many other organic chemicals are of prime importance to modern society. These include, for example, the many polymers (polythene, nylon) whose properties of flexibility and elasticity are a direct result of carbon's unique ability to form chains.

There can be little doubt of the importance of an understanding of organic chemistry, but with so many compounds to consider, we need some means of simplifying and systematising our study.

26.3 Functional groups

The ability of carbon to form strong bonds to itself and to hydrogen leads to the formation of stable compounds. Hydrocarbons contain *only* hydrogen and carbon. The simplest hydrocarbons, containing only single bonds, are the **alkanes**. Butane is an example:

$$\begin{array}{ccccccc} & H & H & H & H & \\ & | & | & | & | & \\ H- & C- & C- & C- & C & -H \\ & | & | & | & | & \\ & H & H & H & H & \end{array}$$

But consider another compound, butan-1-ol:

$$\begin{array}{ccccccc} & H & H & H & H & \\ & | & | & | & | & \\ H- & C- & C- & C- & C & -OH \\ & | & | & | & | & \\ & H & H & H & H & \end{array}$$

Butane and butan-1-ol, despite their structural similarities, have very different properties. Butane is a gas, butan-1-ol is a liquid. Butane has no effect on sodium, but butan-1-ol reacts with the evolution of hydrogen. Clearly, the OH group in butan-1-ol has a big effect on the properties of the unreactive butane skeleton to which it is attached.

The OH group in butan-1-ol is an example of a **functional group**. A given functional group, such as OH, has much the same effect whatever the size and shape of the hydrocarbon skeleton to which it is attached. This greatly simplifies the study of organic compounds because all molecules containing the same functional group can be considered as members of a family, with similar properties. As the hydrocarbon chain gets bigger it increasingly dominates the properties of the compound. Because of this, members of the family show a steady gradation of physical and chemical properties as the size of the hydrocarbon portion increases.

A family of compounds containing the same functional group is called a **homologous series**. Butan-1-ol is a member of the homologous series of **alcohols**, all of which contain the OH group. The first two alcohols are methanol and ethanol:

For any homologous series you can write a general formula in terms of the number of carbon atoms present. For example, the general formula of the alcohols is $C_nH_{2n+1}OH$.

The main functional groups and homologous series considered in this book are shown in table 26.2. The methods of naming the different compounds will be explained as we go along.

The idea of functional groups can also be applied to compounds containing more than one group. Thus, the properties of the molecule as a whole can be predicted by considering the modifying effect of each functional group.

Look at table 26.2 and answer these questions:

■ To what homologous series does propanal, CH_3CH_2CHO, belong?
■ To what homologous series does ethoxyethane, $CH_3CH_2OCH_2CH_3$, belong?

Table 26.2
Common functional groups

Functional group	Name of homologous series	Example
—OH	alcohols	CH_3OH methanol
—NH$_2$	amines	CH_3NH_2 methylamine
	carboxylic acids	CH_3COOH ethanoic acid
	alkenes	$H_2C{=}CH_2$ ethene
—C≡C—	alkynes	HC≡CH ethyne
—Halogen	halogeno compounds	CH_3Cl chloromethane
	aldehydes	CH_3CHO ethanal
	ketones	CH_3COCH_3 propanone
—O—	ethers	CH_3OCH_3 methoxymethane

26.4 Finding the formulae of organic compounds

To predict the properties of a compound, we need to know its structural formula, showing the position and nature of its functional groups.

A modern combustion analysis machine.

Finding empirical formulae

Chapter 1 explains how the empirical formula of a compound may be found from its percentage composition. The composition of organic compounds can be found by **combustion analysis**. A known mass of a compound is burned and the carbon dioxide and water formed are collected and measured. Other elements that may be present, such as nitrogen and halogens, can also be estimated. From the masses of the combustion products the empirical formula can be calculated. In modern laboratories this composition analysis is performed automatically by machines.

Example

A compound **X** containing only carbon, hydrogen and oxygen was subjected to combustion analysis. $0.1\,g$ of the compound on complete combustion gave $0.228\,g$ of carbon dioxide and $0.0931\,g$ of water. Calculate the empirical formula of the compound.

First, calculate the mass of C and H in $0.1\,g$ of the compound.

$$44\,g\ CO_2\ \text{contains}\ 12\,g\ C \Rightarrow \text{mass of C in}\ 0.1\,g\ \mathbf{X} = \frac{12}{44} \times 0.228 = 0.0621\,g$$

$$18\,g\ H_2O\ \text{contains}\ 2\,g\ H \Rightarrow \text{mass of H in}\ 0.1\,g\ \mathbf{X} = \frac{2}{18} \times 0.0931 = 0.0103\,g$$

Mass of C + H in $0.1\,g$ **X** $= 0.0621 + 0.0103 = 0.0724\,g$

The remainder of the $0.1\,g$ **X** must have been oxygen.

$\Rightarrow$ Mass of O in $0.1\,g = 0.1 - 0.0724 = 0.0276\,g$

$\therefore$ Ratio by mass C : H : O is $\quad 0.0621 \quad : \quad 0.0103 \quad : \quad 0.0276$

$\Rightarrow$ Ratio by moles C : H : O is $\dfrac{0.0621}{12} \quad : \quad \dfrac{0.0103}{1} \quad : \quad \dfrac{0.0276}{16}$

$$= \quad 0.00518 \ : \ 0.0103 \ : \ 0.00173$$

$$= \quad 3 \quad : \quad 6 \quad : \quad 1$$

$\therefore$ Empirical formula of **X** is C_3H_6O.

Finding molecular formulae

Once the empirical formula has been found, you can find the molecular formula of the compound if you know its relative molecular mass. Several methods are available for doing this. Traditional methods include measurement of gas density (chapter 10) and measurement of osmotic pressure. A quicker and more accurate modern method for finding relative molecular masses is mass spectrometry, described in chapter 1.

When the molecules of an organic compound pass through a mass spectrometer, they get broken up into fragments. This means that the mass spectrum of the compound contains several peaks, corresponding to the different fragments. But a few molecules pass through intact. These intact molecules give a peak showing the relative molecular mass of the compound.

The mass spectrum obtained for the compound **X** in the example above is shown in figure 26.2 on the next page.

The relative mass of the heaviest particle recorded in the spectrum is 58. We can assume that this corresponds to the intact molecule with a single positive charge, i.e. $\mathbf{X}^+$. This means that the relative molecular mass of **X** must be 58.

With an empirical formula of C_3H_6O, **X** could have molecular formula C_3H_6O, $C_6H_{12}O_2$, $C_9H_{18}O_3$ and so on. But since its relative molecular mass is known to be 58, the only possible molecular formula is C_3H_6O.

Finding structural fomulae

The molecular formula of a compound gives the number of atoms of the different elements in one molecule of the compound, but it gives no information

relative abundance

relative mass

Figure 26.2
Mass spectrum for compound X, empirical formula C_3H_6O.

about the way the atoms are arranged. A compound with molecular formula C_2H_6O, for example, could have one of two structural formulae:

$$CH_3OCH_3 \qquad CH_3CH_2OH$$
methoxymethane   ethanol

The structural formula of ethanoic acid is given in table 26.2.

■ What is its empirical formula?
■ What is its molecular formula?
■ Write one other structure with the same molecular formula.

To find the exact structural formula of a compound, more information is needed in order to decide which functional groups it contains. This information can be found by various methods. Some of them are described below.

(a) The mass spectrum of the compound

The fragments produced by an organic molecule in a mass spectrometer provide useful information. Each fragment gives a line in the mass spectrum. From the position of the line the relative mass of the fragment can be found and from this the nature of the fragment can be deduced. It is then possible, by piecing together the fragments, to deduce the structure of the parent molecule.

For example, if we look again at figure 26.2, we see that the mass spectrum of the compound **X** (molecular formula C_3H_6O) has prominent peaks at 15 and 43. These probably relate to CH_3^+ and CH_3CO^+, respectively. This suggests the structural formula CH_3COCH_3 (propanone). The other possible structure for **X**, CH_3CH_2CHO, propanal, is ruled out because this would give a very strong peak at mass 29, corresponding to the fragments $CH_3CH_2^+$ and CHO^+. Try to identify the other peaks in the mass spectrum of **X**.

(b) Other spectroscopic methods

The two modern techniques of nuclear magnetic resonance and infrared spectroscopy need sophisticated and expensive automated machinery not normally possessed by schools. The details of these methods are beyond the scope of this book and the briefest outline only is given below.

(i) Nuclear magnetic resonance (n.m.r.)

When atoms are placed in a strong magnetic field, their nuclei behave like tiny bar magnets and align themselves with the field. Electrons behave like this too, and for this reason both electrons and nuclei are said to possess 'spin', since any spinning electric charge has an associated magnetic field.

Once a nuclear 'magnet' has become aligned with a magnetic field, it can be made to change the orientation of its own magnetic field. Its field is then opposed to the external field (figure 26.3). To make this change occur, energy

Much of the work of investigating the structure of organic compounds is performed in modern laboratories by complex machines. This is a nuclear magnetic resonance (n.m.r.) instrument.

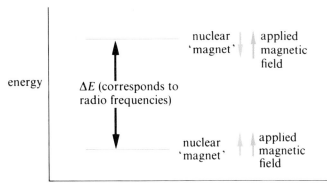

Figure 26.3
Two orientations of the nuclear 'magnet' in an external magnetic field. The energy difference between the two orientations is the basis of the technique of n.m.r.

Table 26.3
Characteristic absorption wavelengths in the infrared region for vibration of some bonds commonly encountered in organic compounds

Bond	Approximate maximum absorption wavelength/ 10^{-6} m
C—H	3.4
C=C	6.1
C≡C	4.5
C—Cl	14
C—Br	16
C—I	20
C=O	5.8

must be supplied to the nucleus. It happens that the energy absorbed corresponds to radio frequencies. The precise frequency of energy depends on the environment of the nucleus, that is, on the other nuclei and electrons in its neighbourhood. So by placing the sample being examined in a strong magnetic field and measuring the frequencies of radiation it absorbs, information can be obtained about the environments of nuclei in the molecule.

The technique of n.m.r. is used mainly to identify the positions of hydrogen atoms, whose nuclei lend themselves particularly well to the technique. For example, in methanol the three H atoms in the CH_3 group will give a different absorption frequency to the H atom in the OH group.

(ii) Infrared spectroscopy

Infrared radiation has a longer wavelength than visible light, occupying the wavelength range from about 2500 nm to about 25 000 nm. In infrared spectroscopy, infrared radiation containing a broad spectrum of wavelengths is shone through a sample of a compound. Radiation of a specific frequency is absorbed, causing the molecule to rotate and making bonds in the molecule vibrate.

The frequency, and therefore the wavelength, of radiation absorbed by the bonds in a particular functional group is characteristic of that group. Therefore, measurement of the wavelengths at which a compound absorbs infrared radiation provides information about its structural formula. Table 26.3 gives the characteristic absorption wavelengths of some bonds commonly encountered in organic chemistry.

Figure 26.4
Nuclear magnetic resonance (n.m.r) spectrum of ethylbenzene

Each major peak corresponds to a hydrogen atom in a particular environment (e.g. attached to the benzene ring). The area under each peak is proportional to the number of atoms of that type in the molecule.

Try to identify each peak and account for the relative sizes of the areas under the peaks.

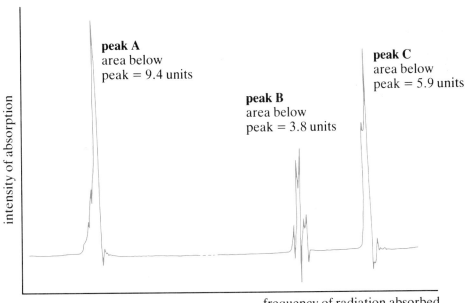

peak A
area below
peak = 9.4 units

peak B
area below
peak = 3.8 units

peak C
area below
peak = 5.9 units

intensity of absorption

frequency of radiation absorbed

(a)

(b)

(c)

(d)

(e)

(f)

(g)

(h) $CH_3CH_2CH_2CH_2CH_3$

Figure 26.5
Representations of the structural formula of pentane.

(c) Physical properties of the compound, such as boiling point

Physical properties are dependent on structure and provide information that can help us work out the structural formula. For example methoxymethane is a gas at room temperature, its boiling point being 248 K. In contrast, ethanol, with the same molecular formula, is a liquid with boiling point 341 K. All members of the homologous series of alcohols tend to have high boiling points relative to other compounds of comparable relative molecular mass.

(d) Chemical properties of the compound

Each functional group has certain chemical characteristics. For example, alcohols such as ethanol react with sodium to liberate hydrogen. Ethers such as methoxymethane do not. Thus a knowledge of the chemical properties of a compound can provide clues about the functional groups it contains.

26.5 Writing structural formulae

The structural formula of a compound shows how the atoms are joined together. This information is crucial in deciding the properties of the compound. The actual shape of the molecule also affects its properties. It is helpful if the structural formula can give some indication of the shape or **stereochemistry** of the molecule.

The methane molecule, as was shown in section 8.7, is tetrahedral in shape. This tetrahedral arrangement of bonds is common to all saturated carbon atoms (that is, carbon atoms bonded to four other groups). Unfortunately the tetrahedral shape creates a problem in representing structural formulae, because it is difficult to show a three-dimensional shape on two-dimensional paper. Figure 26.5 shows eight ways of representing the structural formula of pentane.

Possibly the most accurate representation is the **space-filling model** (figure 26.5(a)) and it would be useful if you could try building this model yourself. Space-filling models show the extent of the electron cloud of each atom accurately.

Figure 26.5(b) attempts to represent the tetrahedral carbon atom by showing the bonds in the plane of the paper as single lines and those projecting in front of or behind the paper as thicker, tapering lines. This is called a **sterochemical** formula. The idea is to give perspective by making the bond appear wider at the end nearer the observer. This system is of limited use, and the tapering lines are often omitted as in figure 26.5(c), which is called a **displayed** formula.

Occasionally a **skeletal structure** (figure 26.5(d)) is used, in which only carbon–carbon bonds and functional groups are shown.

A common and easy way of writing formulae is shown in figure 26.5(e), but this has drawbacks because it represents the bond angles at each carbon atom as 90° instead of 109°. This type of diagram shows its limitations when an attempt is made to represent rotation about a single bond: compare figure 26.5(g) with figure 26.5(f). It is easy to see that (f) can be obtained from (c) by rotating about the bond between the second and third carbon atoms. On the other hand, (g) appears to be a different compound to (e), though both in fact represent pentane.

Finally, figure 26.5(h) shows a shortened way of writing a structural formula.

These different ways of representing structural formulae will be used as appropriate in different parts of this book.

■ Write a full structural formula for $CH_3CH_2CHOHCH_2CH_3$.
■ Write a skeletal formula for

26.6 Isomerism

We saw in section 26.4 that a compound of molecular formula C_2H_6O could have two possible structural formulae, CH_3OCH_3 (methoxymethane) or CH_3CH_2OH (ethanol). Compounds such as these, possessing the same molecular formula but with their atoms arranged in different ways, are called **isomers**. Isomerism is very common in organic chemistry. All carbon compounds with four or more carbon atoms, and many with less, show isomerism. There are 4.11×10^9 isomers of molecular formula $C_{30}H_{62}$. Isomerism also occurs, though less commonly, in inorganic chemistry. In this book we are concerned with three main types of isomerism.

Structural isomerism

Structural isomers differ in which atom is attached to which. Methoxymethane and ethanol are structural isomers. In methoxymethane the oxygen atom is attached to two carbon atoms while in ethanol it is attached to a carbon and a hydrogen. Structural isomers can be members of different homologous series, like the pair just mentioned. Or they may be members of the same series, like the following two structural isomers of molecular formula C_4H_{10}, which are both alkanes:

butane   and   methylpropane

Note that

is *not* a third isomer of C_4H_{10}, because it can be formed from butane simply by rotating about a single bond. It is possible to rotate a structure freely about any single carbon–carbon bond. Structures that can be interconverted by rotating about a bond in this way are not isomeric. The structure above appears at first sight to be a separate isomer, but this is due to the short-comings of representing the carbon bond angles as 90° instead of 109°. A little experimenting with a molecular model will make this point clear.

- How many isomers are there of molecular formula C_5H_{12}?
- Write the structural formulae of all the alcohols (i.e. the compounds with the OH functional group) of molecular formula $C_4H_{10}O$.

Structural isomers usually show considerable differences in physical and chemical properties, even if they are members of the same homologous series. Thus, the boiling point of butane is 273 K while that of methylpropane is 261 K. It is not really surprising that structural isomers differ. Any functional group will be influenced by its environment and will therefore have its properties modified by the atoms to which it is attached.

Optical isomerism

Light is a form of electromagnetic radiation and can be regarded as consisting of waves. A ray of normal light consists of waves which vibrate in many directions at right angles to the direction of travel of the ray (figure 26.6(a)).

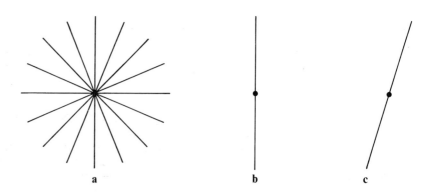

Figure 26.6
a Normal light ray travelling towards the observer: each line represents a wave seen 'end-on'. **b** Plane-polarised light: only waves in a single plane are present. **c** Plane-polarised light from **b** after passing through an optically active solution.

Certain materials have the ability to remove from normal light all waves except those vibrating in a single plane (see figure 26.6(*b*)). The light is then said to be **plane polarised**. It is rather like the light being combed as it passes through the polariser. A well known polariser is polaroid, which is used in the lenses of some sunglasses.

Certain chemical substances have the ability to rotate the plane of polarised light. That is, if polarised light is shone through a solution of the substance in a suitable solvent, the plane in which the polarised light vibrates will be rotated to the right or left (figure 26.6(*c*)). These substances are said to be **optically active**. An example is 2-hydroxypropanoic acid (lactic acid, $CH_3CHOHCOOH$), the substance responsible for the sour taste in sour milk.

Chemists have isolated two forms of lactic acid: one of these rotates polarised light to the right, the other rotates it to the left. For a given concentration of solution the two forms rotate light to exactly the same extent, but in opposite directions. In addition, crystals of the two forms are found to be mirror images of one another. Apart from these differences, the forms are physically and chemically identical. Since the two forms of lactic acid are chemically identical, they must contain exactly the same groups attached to each other in the same way. Their difference can only lie in the way the groups are arranged relative to each other in space. Furthermore they must be exact opposites in this respect since their effect on polarised light is equal and opposite.

A consideration of the molecular structure of different optically active substances shows they have one thing in common. They all have asymmetric molecules, that is to say, their molecules have no centre, axis or plane of symmetry. As a result they are different from their own mirror images. Optical activity is shown by all substances with asymmetric molecules. Such substances have two isomeric forms which are mirror images of one another and which rotate polarised light in opposite directions.

Objects that are asymmetric are said to be **chiral**. The simplest type of chiral molecule is one in which four different groups are attached to the same carbon atom—lactic acid is such a molecule. The asymmetric carbon atom is called a **chiral centre**. Figure 26.7 shows the two forms, called D-lactic acid and L-lactic acid, but you should try making models of each of the forms to satisfy yourself that they are indeed different molecules and that they are mirror images. You can show they are different by trying to superimpose the two molecules on one another: it is impossible to arrange them so that all the groups correspond in position. Note, though, that in both D- and L-lactic acid all the groups are

Figure 26.7
The enantiomeric forms of lactic acid. The two isomers can be seen to be images of one another reflected in the imaginary mirror.

OH

C

HOOC   CH₃ H

(a) D-lactic acid.

imaginary
mirror

OH

C

H   CH₃ COOH

(b) L-lactic acid

Much of the early work on optical isomerism was done by the great French scientist Louis Pasteur. These models, constructed for Pasteur, represent crystals of the two forms of ammonium 2-hydroxybutanedioate (ammonium malate). Like the molecules themselves, the two types of crystal are mirror images of one another. The model on the left is of one of the optically active forms of potassium 2,3-dihydroxybutanedioate (potassium tartrate).

attached together in the same way, and that the spacings of the various groups are the same in each isomer. This is why the two forms of lactic acid have identical chemical properties.

The relationship between optical isomers such as D- and L-lactic acid is like the relation between your right and your left hand. The lengths of the different fingers and the distances between them are the same for both hands, but they are mirror images of one another and cannot be superimposed on each other—try putting your right glove on the left hand. Structures that are mirror images of one another are called **enantiomers**.

Many organic compounds and some inorganic ones have chiral molecules and therefore show optical isomerism. Optically active compounds are very common in nature. Almost all amino acids are optically active, as are all the sugars. The interesting thing is that most naturally occurring optically active compounds occur as one isomer only. Thus all naturally occurring glucose is D-glucose, and all naturally occurring optically active amino acids are L-isomers. The enzymes

Early models constructed to illustrate optical isomerism. The two central models are of (left) L-lactic acid and (right) D-lactic acid.

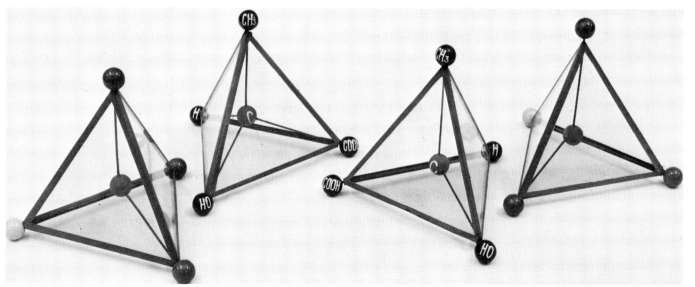

that produce and break down these substances are so specific that they fit only one type of molecular configuration. Perhaps if we went through the looking-glass like Alice all the glucose in our bodies would become L-glucose.

Many optically active compounds have complex structures and it is difficult to tell quickly whether or not their molecules are chiral. For simpler molecules it is safe to say that if the molecule contains a carbon atom to which four different groups are attached, it will show optical isomerism. Such a carbon atom is called a **chiral centre.**

- Would you expect (a) CH_2ClF (b) $CHClBrF$ to show optical isomerism?
- Which of the isomeric alcohols whose structures you wrote earlier in this section would show optical isomerism? Write the structures of the enantiomers.
- Write the structure of the first alkane to show optical isomerism.

Lactic acid can be extracted from natural sources such as sour milk or muscle tissue and the acid from these sources always shows optical activity. It is also possible to prepare lactic acid in the laboratory from fairly simple starting materials, but the acid prepared in this way shows no optical activity at all. This is because most laboratory methods for preparing lactic acid produce the D- and L- isomers in equal amounts. The two isomers of course cancel each other out in their effect on polarised light. This kind of mixture of optical isomers containing equal molar quantities of D- and L-isomers is called a **racemic mixture**. Because of the identical properties of the isomers it is very difficult to separate them from a racemic mixture, though methods do exist for doing so. The process of separation is called **resolution**.

Geometric isomerism

This type of isomerism is described in section 28.4.

Summary

1 The ability of carbon atoms to form strong bonds to four other atoms, including other carbon atoms, results in a huge range of organic compounds.

2 Organic compounds, though kinetically stable in air, are energetically unstable relative to their combustion products.

3 Organic compounds can be regarded as having a basic hydrocarbon skeleton with functional groups attached. Compounds containing the same functional group form a family with similar properties called a homologous series.

4 The empirical formula of an organic compound can be found by analysis of its combustion products.

5 The molecular formula can be found from the empirical formula once the relative molecular mass is known.

6 The structural formula shows the precise arrangement of atoms and can be found from a knowledge of some of the properties of the compound.

7 Compounds with the same molecular formula but with their atoms arranged in different ways are called isomers.

8 Structural isomers differ in the way the atoms are attached to one another.

9 Compounds with chiral (asymmetric) molecules exist in two forms whose molecules are mirror images or enantiomers. The two forms rotate polarised light in opposite directions and are called optical isomers.

Study questions

1 An organic compound was subjected to combustion analysis. 1.0 g of the compound formed 1.37 g carbon dioxide, 1.12 g water and no other products.
   (a) Calculate the percentage by mass of carbon and of hydrogen in the compound.
   (b) What other element must be present?
   (c) Calculate the empirical formula of the compound.
   (d) The mass spectrum of the compound is shown below.
      (i)   Use this to find the relative molecular mass of the compound, and thus its molecular formula.
      (ii) Using the fragments shown in the mass spectrum, deduce the structural formula of the compound.
      (iii) Give the formula of the fragment to which each peak can be attributed.

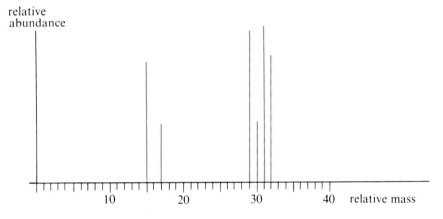

2 **A**, **B** and **C** are isomeric compounds of molecular formula C_3H_8O. Two of the compounds are members of the same homologous series. The table gives some data about **A**, **B** and **C**.

	A	**B**	**C**
Boiling point/K	370	284	356
Density/g cm^{-3}	0.80	0.72	0.79

   (a) Which two compounds are members of the same homologous series?
   (b) Write the structural formulae of all three possible isomers of C_3H_8O.
   (c) Use table 26.2 to decide to which homologous series each of the isomers you have drawn belongs.
   (d) Is it possible from the data given to say which of the isomers you have drawn is **A**, which is **B** and which is **C**? Explain your answer.

3 Use table 26.2 to decide to which homologous series each of the following compounds belongs.
   (a) $CH_3CH_2CH_2OH$
   (b) $CH_3CH_2COCH_3$
   (c) CH_3CH_2Cl
   (d) CH_3CH_2COOH
   (e) $CH_3CH{=}CHCH_3$
   (f) H_2NCH_2COOH
   (g) $CH_3CH_2OCH_2CH_2CH_3$

4 Write the full structural formulae of all the isomers of the following, stating which type of isomerism is involved:
   (a) C_3H_7Cl
   (b) C_6H_{14}
   (c) $C_2H_3Cl_2Br$

5 Silicones are unreactive polymers with the structure

$$
\begin{array}{ccccc}
CH_3 & CH_3 & CH_3 & CH_3 & CH_3 \\
| & | & | & | & | \\
-Si-O-Si-O-Si-O-Si-O-Si- \\
| & | & | & | & | \\
CH_3 & CH_3 & CH_3 & CH_3 & CH_3
\end{array}
$$

Look at the bond energies given in table 26.1. Suggest a reason why silicones are inert and unreactive while silanes (compounds analogous to alkanes but containing silicon atoms instead of carbon atoms) are unstable and ignite spontaneously in air.

6 Fluorocarbons are compounds analogous to alkanes but containing fluorine instead of hydrogen, e.g. CF_4, CF_3CF_3, etc. They are extremely unreactive, being quite stable in air even at high temperatures. Long-chain fluorocarbons such as 'Teflon' ($-CF_2-CF_2-CF_2-CF_2-CF_2-$) are used as unreactive corrosion-resistant materials for gaskets and protective coatings.
   Use the bond energies given below to explain why fluorocarbons are stable in air while hydrocarbons are energetically unstable.

Bond	Bond energy/kJ mole^{-1}
C—H	413
C—F	485
H—O	463
F—O	234

7 A compound containing carbon, hydrogen and nitrogen only was analysed. 0.1 g of the compound on combustion gave 0.228 g of carbon dioxide and 0.124 g of water. On reduction of the compound, all the nitrogen in it was converted to ammonia and it was found that 0.1 g of the compound gave ammonia equivalent to $17.2 \, cm^3$ of 0.1 M hydrochloric acid when titrated.
   (a) How many moles of carbon, hydrogen and nitrogen are there in 0.1 g of the compound?
   (b) What is the empirical formula of the compound?
   (c) The relative molecular mass of the compound was found to be 116. What is its molecular formula?
   (d) The infrared spectrum of the compound indicated the presence of the NH_2 group. Suggest one possible structure for the compound.

8 The following structural formulae represent only three different substances. Some of the formulae are equivalent to others. Which formulae are equivalent to which?

A
$$
\begin{array}{cccc}
H & H & H & H \\
| & | & | & | \\
H-C-C-C-C-H \\
| & | & | \\
H & H & H \\
& & | \\
& & H-C-H \\
& & | \\
& & H-C-H \\
& & | \\
& & H
\end{array}
$$

B

C $CH_3CH_2CH_2CH_2CH_2CH_3$

D
$$
\begin{array}{cccccc}
& & & H & & \\
& & & | & & \\
& & & H-C-H & & \\
H & H & H & | & H & H \\
| & | & | & | & | & | \\
H-C-C-C-C-C-C-H \\
| & | & | & | & | & | \\
H & H & H & H & H & H
\end{array}
$$

E

F $CH_3(CH_2)_5CH_3$

G $CH_3CH_2CH_2CHCH_2CH_3$
$\qquad\qquad\quad |$
$\qquad\qquad\quad CH_3$

H
$$
\begin{array}{ccc}
H & H & H \\
| & | & | \\
H-C----C----C-H \\
| & | & | \\
H-C-H \quad H \quad H-C-H \\
| & & | \\
H-C-H & & H-C-H \\
| & & | \\
H & & H
\end{array}
$$

9 The earth's crust contains only 0.036% by mass of carbon, compared with 49% oxygen and 26% silicon. Despite its relatively low abundance on earth, carbon is the element that forms the basis of all living things. What are the properties of carbon that make it so suitable for this rôle?

27 Petroleum and Alkanes

27.1 Crude oil

In 1973 the oil producers of the Middle East, who supply a large proportion of the crude oil used in the Western world, decided to cut back production of crude oil and at the same time greatly increased its price. Overnight, a cheap and abundant raw material became expensive and insecure.

The effects of this move were felt in every area of life in Britain, so dependent is it on crude oil. Motor fuels, lubricating oils and heating oils naturally increased in price. So did electricity, much of which was at that time generated in oil-fired power stations. But many other commodities from plastics and detergents to paints and antifreeze quickly became far more expensive than they had been in the days when oil was cheap. For as well as supplying a large part of our energy needs, crude oil is the source of 70% of Britain's organic chemicals. It is the most important of all modern raw materials.

Crude oil was formed from the remains of small marine animals and plants that were buried in the beds of the seas millions of years ago. The decay of these remains under the layers of overlaying rock formed the liquid known as crude oil or petroleum (from Greek words meaning 'rock oil'). Similar conditions led to the formation of natural gas. Natural gas is often found associated with crude oil as well as in deposits on its own, such as those in the North Sea.

Crude oil is a complex mixture of hydrocarbons. It has no uses in its raw form: to provide useful products its components must be partly separated and if necessary modified. Once crude oil has been located and extracted, it must therefore be transported to a refinery where it is processed. The fundamental process is primary distillation, the details of which will be familiar to most readers and are simply summarised here in diagrammatic form in figure 27.1.

Figure 27.1
Primary distillation of crude oil.

Figure 27.2
Variation of composition of crude oil from different sources.

Different crudes contain varying proportions of the different fractions. The output of the refinery can be arranged to suit market demands by blending different crudes and by converting heavy fractions into lighter ones by cracking.

The different fractions are used as follows.

Refinery gas (1–2% of crude oil) is similar in composition to natural gas, containing those hydrocarbons that are gases at normal temperatures. This includes the alkanes with one to four carbon atoms in their molecules, with methane as the major component. The main use of refinery gas is as a gaseous fuel. But like natural gas, it can also be used as a starting point for making petrochemicals, since most organic chemicals are built up from small molecules containing one, two or three carbon atoms.

Gasoline (15–30%) is a complex liquid mixture of hydrocarbons containing mainly C_5–C_{10} compounds whose boiling points range from 40°C to 180°C. The major use of gasoline is of course as a fuel in internal combustion engines (see section 27.6). In the USA, where the motor car is the most popular method of transport, fuel taxes are low and cars have traditionally been built to consume large quantities of fuel. Gasoline is therefore in great demand and American refineries use most of the gasoline fraction for blending in motor fuels. In

Western Europe and most of the rest of the world, the demand for gasoline is less. A considerable proportion of this fraction is used in the manufacture of chemicals, after the cracking processes described in section 27.6. The part of the gasoline fraction used to produce chemicals is called **naphtha**.

Kerosine (10–15%) consists mainly of C_{11} and C_{12} hydrocarbons, with boiling points from 160°C–250°C. It is used as a fuel in jet engines and for domestic heating purposes. It can also be cracked to produce extra gasoline (see section 27.6).

The major use of kerosine is as jet fuel.

Diesel oil or **gas oil** (15–20%) containing C_{13}–C_{25} compounds, boils in the range 220°–350°C. It is used in diesel engines, where the fuel is ignited by compression instead of by a spark and also in furnaces for industrial heating purposes. Like kerosine it can be cracked to produce extra gasoline.

Residue (40–50%) The residual oil from the primary distillation boils above 350°C and is a highly complex mixture of involatile hydrocarbons. Most of it is used as fuel oil in large furnaces such as those in power stations or big ships. A proportion of it, however, is used to make **lubricating oils** and **waxes**. Both these materials contain C_{26}–C_{28} hydrocarbons. When pure these hydrocarbons are solid, but lubricating oil is a complex mixture, each member of which depresses the melting point of the others so that the mixture is a liquid. In paraffin wax the components are similar enough in structure to form a solid.

To obtain lubricating oil and paraffin wax from the residue, the appropriate hydrocarbons must be distilled off. The distillation is done in a vacuum to avoid the high temperatures that would be needed for distillation at atmospheric pressure, since such temperatures would tend to crack the hydrocarbons. Paraffin wax is separated from lubricating oil by solvent extraction. The solid left after vacuum distillation is an involatile tarry material called **bitumen** or **asphalt**. It is used to surface roads and to waterproof materials.

Bitumen mixed with stone chippings is used to surface roads. It is applied hot, when it can be easily spread, but at normal temperatures its viscosity is so high that it hardly flows at all.

27.2 The composition of crude oil

The distillation of crude oil can be carried out on a small scale in the laboratory: you will probably be familiar with the experiment represented in figure 27.3 on the next page. Table 27.1 on the next page gives the properties of the fractions collected over different temperature ranges.

Notice that all these properties show a steady gradation: there are no sudden changes in the properties of the different fractions. This suggests that crude oil is a mixture of many components and that the components have similar properties, many of them perhaps belonging to the same homologous series.

mineral wool
soaked in
crude oil

distillate

heat

Figure 27.3
The laboratory distillation of crude oil.
Fractions are collected over several
different boiling ranges.

Table 27.1
Properties of fractions obtained in laboratory distillation of crude oil

Property	Room temp. to 70°C	70°C–120°C	120°C–170°C	170°C–220°C
Colour	pale yellow	yellow	dark yellow	brown
Viscosity	runny	fairly runny	fairly viscous	viscous
Behaviour when ignited	burns readily: clean yellow flame	quite easily ignited: yellow flame, some smoke	harder to ignite: quite smoky flame	hard to ignite: smoky flame

Hydrocarbons are the main and the most important components of crude oil. Three different homologous series of hydrocarbons are present: **alkanes**, which will be examined in more detail later in this chapter, **cycloalkanes**, considered in section 27.7, and **aromatics** (compounds containing benzene-type rings), considered in chapter 29. The proportions of the different classes present depend on the source of the oil. Alkanes and cycloalkanes form the large majority, with aromatics making only about 10% of the hydrocarbon total.

All the more volatile fractions of crude oil are made up of hydrocarbons. The increasing boiling ranges of the fractions correspond to the increasing size of the alkane molecules they contain. The involatile tarry residue left after vacuum distillation consists of compounds with large molecules called resins and asphalts. These contain mainly carbon and hydrogen, with some oxygen and sulphur.

Look again at table 27.1.

■ Why are the first fractions easiest to ignite?
■ Why do the higher fractions burn with the smokiest flames?
■ Suggest a reason why the fractions become increasingly viscous.

We will now consider in detail the largest group of compounds present in crude oil, the alkanes.

The pouring characteristics of different petroleum fractions. From left to right: gasoline, diesel oil, fuel oil.

Table 27.2
Straight-chain alkanes

Formula	Name
CH_4	methane
CH_3CH_3	ethane
$CH_3CH_2CH_3$	propane
$CH_3CH_2CH_2CH_3$	butane
$CH_3(CH_2)_3CH_3$	pentane
$CH_3(CH_2)_4CH_3$	hexane
$CH_3(CH_2)_8CH_3$	decane
$CH_3(CH_2)_{18}CH_3$	eicosane

27.3 Naming alkanes

The alkanes are **saturated hydrocarbons**. 'Saturated' means that they contain the maximum amount of hydrogen possible, with no double or triple bonds between carbon atoms. The general formula of the alkanes is C_nH_{2n+2}. As we have seen in chapter 26, it is possible to have alkanes with straight or with branched chains, for example

$$CH_3—CH_2—CH_2—CH_2—CH_3 \quad \text{straight-chain}$$

$$CH_3—CH_2—CH—CH_3 \quad \text{branched-chain}$$
$$|$$
$$CH_3$$

Table 27.2 gives the names of some simple straight-chain alkanes.

You can see that after the first four alkanes the name is formed by adding the suffix **-ane** to the Greek root (e.g. *pent-* five, *hex-* six) indicating the number of carbon atoms in the molecule. The first part of the name indicates the number of carbon atoms, the ending -ane indicates that it is an alkane. All compounds based on alkane skeletons are named by this method, with a stem to indicate the number of carbon atoms and a suffix to indicate the functional group.

Branched-chain alkanes are named by considering them as straight-chain alkanes with side groups attached. For example:

$$CH_3—CH_2—CH—CH_3$$
$$|$$
$$CH_3$$

is regarded as butane with CH_3 attached to the *second* atom. It is therefore called 2-methylbutane. The name **methyl** indicates the CH_3 group, which is just methane with a hydrogen atom removed so it can be attached to another atom. Similarly CH_3CH_2 is called **ethyl**, $CH_3CH_2CH_2$ **propyl**, and so on. Side groups of this kind are called **alkyl groups**; their general formula is C_nH_{2n+1}. The symbol **R** is often used to represent a general alkyl group.

Another example of a branched-chain alkane is:

$$CH_3—CH_2—CH—CH—CH_3$$

This molecule can be regarded as a five-carbon chain with a methyl and an ethyl group attached to the second and third carbon atoms, respectively. It is therefore called 3-ethyl-2-methylpentane, the side groups being arranged in alphabetical order. Note that if the carbon atoms were numbered from the left, the name would be 3-ethyl-4-methylpentane. This name is not used because the convention is to use the name that includes the lowest numbers. Note also that for each side group the name includes a number showing the carbon atom to which the group is attached. Thus

is called 2,2-dimethylbutane, and

is called 2,2-dimethyl-3-ethylheptane.

So, the procedure for naming an alkane from its structural formula is:

1 Look for the longest unbranched chain in the molecule.

2 Look for the side groups attached to the main chain and the numbers of the carbon atoms to which they are attached.

3 The name then consists of the name of the longest unbranched chain, prefixed by the names of the side groups and the numbers of the carbon atoms to which they are attached.

■ Name the following alkanes:

$$CH_3-\underset{\underset{CH_3}{|}}{CH}-\underset{\underset{CH_3}{|}}{CH}-\underset{\underset{CH_3}{|}}{\overset{\overset{CH_3}{|}}{C}}-CH_3 \qquad CH_3-\underset{\underset{CH_3}{|}}{\overset{\overset{CH_3}{|}}{C}}-CH_3 \qquad CH_3(CH_2)_6CH_3$$

■ Write the formulae of the following alkanes:
2,2,4-trimethylhexane,   methylpropane,   5-ethyldecane.

The system of nomenclature described above is called the IUPAC system (IUPAC stands for International Union of Pure and Applied Chemistry). The system can be extended to apply to all organic compounds. Using this system, it is possible to write the structural formula of any compound from its name.

The IUPAC system is gradually replacing the older, less systematic way of naming organic compounds. IUPAC names will be used wherever possible in this book. The only disadvantage of the system is that for complicated molecules the name becomes very cumbersome. In such cases systematic nomenclature is often abandoned in favour of a less informative but more easily spoken name is used. Thus the name 2,3,4,5,6-pentahydroxyhexanal is dropped in favour of 'glucose'.

Table 27.3
Physical properties of straight-chain alkanes

Number of carbon atoms	Formula	Name	State (at 298 K)	Boiling point/K	Melting point/K	Density /g cm^{-3}
1	CH_4	methane	g	112	90	0.424
2	C_2H_6	ethane	g	184	101	0.546
3	C_3H_8	propane	g	231	85	0.501
4	C_4H_{10}	butane	g	273	138	0.579
5	C_5H_{12}	pentane	l	309	143	0.626
6	C_6H_{14}	hexane	l	342	178	0.657
7	C_7H_{16}	heptane	l	371	182	0.684
8	C_8H_{18}	octane	l	399	216	0.703
9	C_9H_{20}	nonane	l	424	219	0.718
10	$C_{10}H_{22}$	decane	l	447	243	0.730
11	$C_{11}H_{24}$	undecane	l	469	247	0.740
12	$C_{12}H_{26}$	dodecane	l	489	263	0.749
15	$C_{15}H_{32}$	pentadecane	l	544	283	0.769
20	$C_{20}H_{42}$	eicosane	s	617	310	0.785

27.4   Physical properties of alkanes

The alkanes form a homologous series. As with all homologous series the members show a gradual change in physical properties as the number of carbon atoms in their molecules increases. This can be seen in the distillation of crude oil, which is after all largely composed of alkanes. Successive fractions from crude oil have successively higher boiling ranges and steadily increasing viscosity. The fractions range from gas (refinery gas), through liquids (gasoline, kerosine) to solids (paraffin wax, bitumen).

Some properties of individual straight-chain alkanes are shown in table 27.3. Notice that these properties show a gradual steady change as the number of carbon atoms in the molecules increases. Figure 27.4 shows graphically the smooth increase of boiling point with increasing number of carbon atoms.

Such steady variation in physical properties is characteristic of all homologous series. This makes it possible to predict the properties of a compound from the properties of other members of its homologous series.

Examining sea birds which became coated with oil following the accident to the tanker *Amoco Cadiz*.

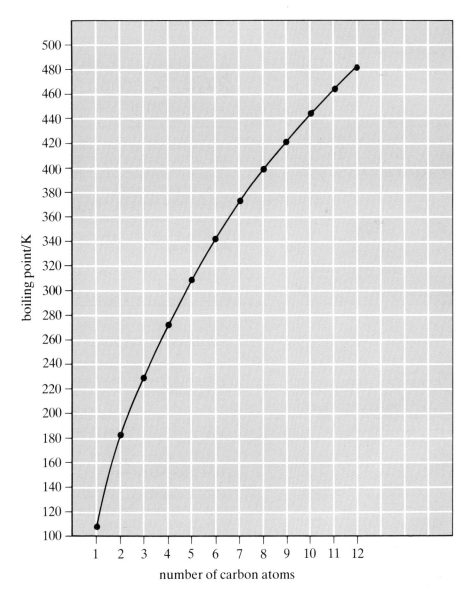

Figure 27.4
Variation of boiling point of straight-chain alkanes with number of carbon atoms.

Crude oil leaks from the hull of a damaged tanker. Hydrocarbons are less dense than water, so they float on the surface.

Use the data in table 27.3 to predict:

■ The density of tridecane ($C_{13}H_{28}$),
■ The boiling point of tetradecane ($C_{14}H_{30}$).
■ The formula of the first alkane to be solid at 20°C.

As table 27.3 shows, the first four straight-chain alkanes (C_1–C_4) are gases. The next twelve (C_5–C_{16}) are liquids, and the remainder are solids. Solid alkanes of high molecular mass are similar in nature to polythene, which is effectively composed of long-chain alkane molecules. Alkanes are all less dense than water and therefore float on it. Alkanes are colourless when pure. The viscosity of liquid alkanes increases with increasing molecular mass.

The steady change in physical properties of alkanes is caused by the steadily increasing molecular size. The increase in boiling point is due to the increasing forces of attraction between molecules of increasing size (see section 9.4). The higher viscosity of the higher alkanes is due to the tendency of the long molecules to become 'tangled up' with one another.

The trends in physical properties considered so far have been those among the straight-chain alkanes. Branched-chain alkanes do not show the same steady gradation of properties as straight-chain alkanes. The variation in molecular structures of branched alkanes is too great for trends to be clear. On the whole, the effect of branching is to increase volatility and reduce density. Thus the boiling point of pentane is 309 K while that of its highly branched isomer 2,2-dimethylpropane is 283 K.

27.5 Reaction mechanisms in organic chemistry

One of the things that makes organic chemistry a rewarding subject is the fact that although there are many, many different organic compounds, the reactions of these compounds are quite simply classified.

In the next section we will look at the particular reactions of alkanes. But in this section we will look at some of the general principles behind organic reactions. Some organic reactions go in a single step. But most reactions involve a series of steps. The sequence of steps in a reaction is called the **reaction mechanism**.

Types of organic reactions

In this book we will be concerned with three important types of organic reaction.

Substitution reactions

In this type of reaction one atom or group of atoms is substituted by another. For example, when bromoethane reacts with ammonia, bromine atoms are replaced by the NH_2 group:

$$CH_3CH_2Br + NH_3 \rightarrow CH_3CH_2NH_2 + HBr$$
bromoethane   ethylamine

Addition reactions

In these reactions, two substances react together to form a single substance. A double or triple bond is usually involved. For example, when bromine reacts with ethene, the bromine adds across the double bond to form a single product, 1,2-dibromoethane:

$$\begin{array}{ccc} \underset{H}{\overset{H}{\diagdown}}C=C\underset{H}{\overset{H}{\diagup}} + Br_2 & \longrightarrow & \begin{array}{c} H\ H \\ |\ \ | \\ H-C-C-H \\ |\ \ | \\ Br\ Br \end{array} \end{array}$$

Elimination reactions

In an elimination reaction, a small molecule is removed, or eliminated, from a larger one. This usually results in the formation of a double or triple bond. For example, when bromoethane is heated with a solution of hydroxide ions in ethanol, HBr is eliminated, forming ethene:

$$
\begin{array}{c}
\text{H}\ \ \text{H} \\
|\ \ \ | \\
\text{H—C—C—H} \\
|\ \ \ | \\
\text{H}\ \ \text{Br}
\end{array}
\xrightarrow{\text{OH}^-\ \text{(ethanol)}}
\begin{array}{c}
\text{H} \\
\diagdown \\
\text{C=C} \\
\diagup \\
\text{H}
\end{array}
\begin{array}{c}
\text{H} \\
\diagup \\
 \\
\diagdown \\
\text{H}
\end{array}
+\ \text{HBr}
$$

■ Look at the reactions, represented by the equations below. Classify each reaction as

A Substitution
B Addition
C Elimination

1 $CH_2 = CH_2 + H_2 \rightarrow CH_3CH_3$

2 $CH_3CH_2Br + OH^- \rightarrow CH_3CH_2OH + Br^-$

3 $CH_3CH_2OH \rightarrow CH_2 = CH_2 + H_2O$

4 $CH_4 + Cl_2 \rightarrow CH_3Cl + HCl$

Ways of breaking bonds

All reactions involve the breaking, and remaking, of bonds. Breaking bonds is sometimes called **bond fission**. The way bonds break has an important influence on reactions.

In a covalent bond, a pair of electrons is shared between two atoms. For example, in the HCl molecule:

$$\text{H}\dot{\vdots}\text{Cl}$$

When the bond breaks, these electrons get redistributed between the two atoms. There are two ways this redistribution can happen.

Homolytic fission

In this type of bond fission, one of the two shared electrons goes to each atom. In the case of HCl:

$$\text{H}\dot{\vdots}\text{Cl} \rightarrow \text{H}\cdot + \text{Cl}\cdot$$

The dot · beside each atom indicates the unpaired electron that the atom has gained from the shared pair in the bond. The atoms have no overall electric charge, because each has equal numbers of protons and electrons. But the atoms are highly reactive, because the unpaired electron has a strong tendency to pair up with another electron from another substance (figure 27.5). These highly reactive atoms or groups of atoms with unpaired electrons are called **free radicals**.

$$\text{H}^\bullet \quad \Rightarrow \quad \text{H}^\bullet \quad \text{Cl}\overset{\bullet}{\underset{\bullet}{\text{·}}}\text{Cl} \quad \Rightarrow \quad \text{H}\overset{\bullet}{\underset{\bullet}{\text{·}}}\text{Cl} \quad \text{Cl}^\bullet$$

A hydrogen radical. It has an unpaired electron which would be much more stable if paired with another electron.

The hydrogen radical meets a chlorine molecule.

The unpaired electron on the hydrogen pairs up with an electron on one of the chlorines. This forms a bond. It leaves the other chlorine as a reactive free radical.

Figure 27.5
Why free radicals are reactive.

Another example of free radical formation is when a C—H bond in methane is broken:

methane   methyl radical

Free radicals are most commonly formed when the bond being broken has electrons that are fairly equally shared. When the electrons are unequally shared, the bond is polar and heterolytic fission is more likely.

Heterolytic fission

When the bond breaks, both of the shared electrons go to just one of the atoms. This atom becomes negatively charged, because it has one more electron than protons. The other atom becomes positively charged. In the case of HCl:

$$H \div Cl \rightarrow H^+ + :Cl^-$$

Heterolytic fission is more common where a bond is already polar. For example, bromomethane contains a polar C—Br bond, and under certain conditions this can break heterolytically:

Notice that the CH_3^+ ion contains a positively charged carbon atom. It is an example of a **carbocation**. An ion containing a negatively charged carbon atom is called a **carbanion**. Both these types of ions tend to be unstable and highly reactive, so they only exist as short-lived reaction intermediates.

■ Look at the groups listed below. Classify each as

    A carbocation
    B carbanion
    C free radical
    D none of these

1 $CH_3CH_2\cdot$ 2 $CH_3CH_2CH_2^+$ 3 CH_3^+

4 $CH_3CH_2^-$ 5 OH^- 6 $OH\cdot$

■ Which type of bond fission, homolytic or heterolytic, would you expect for (a) a bond between identical atoms, (b) a bond between atoms whose electronegativities differ widely, (c) a bond between atoms whose electronegativities are similar?

Electrophiles and nucleophiles

Like all ions, carbocations and carbanions are attracted to groups which carry an opposite electric charge. Carbocations are **electrophiles** ('electron-lovers'): they are short of electrons, so they are attracted to groups which can donate electron pairs. Carbanions are **nucleophiles** ('nucleus-lovers'): they are electron-rich, so they are attracted to groups which can accept electron pairs. Figure 27.6 illustrates this idea.

Carbocations are electrophiles. They are attracted to groups which can donate electron pairs.

Carbanions are nucleophiles. They are attracted to groups which can accept electron pairs.

Figure 27.6

The terms electrophile and nucleophile do not apply only to ions. Many organic compounds are polar: they carry partial charges, though not full positive or negative charges. These partial charges can also make a group electrophilic or nucleophilic, as you can see in figure 27.7.

Figure 27.7
A polar molecule can have both an electrophilic and a nucleophilic part.

■ Look at the ions and groups below. Classify each as

    A electrophile
    B nucleophile
    C could act as electrophile or nucleophile

1 Cl^- **2** NO_2^+ **3** CH_3^+

4 HI   **5** CN^- **6** H^+

7 CH_3NH_2 **8** CH_3OH

Describing reaction mechanisms

We can use the ideas and terms outlined above to describe the mechanisms of the important classes of organic reactions that will be met in this book. You will find them in the following sections:

    Nucleophilic substitution: section 30.4

    Electrophilic addition: section 28.5

    Aromatic electrophilic substitution: section 29.6

    Nucleophilic addition: section 32.1

    Free radical substitution: section 27.6.

Curly arrows

When reaction mechanisms are being described, 'curly arrows' are sometimes used to show the **movement of pairs of electrons**. The beginning of the arrow shows where the electron pair starts from, and the arrow-head shows where the pair ends up. Figure 27.8 shows an example.

(a)

The arrow shows a pair of electrons moving from the Br^- ion to the region between the bromine and the carbon, where it forms a covalent bond between the two atoms.

(b)

The same diagram, with all the bonding electrons shown.

Figure 27.8
Using curly arrows.

27.6 Reactivity of alkanes

The effects of various reagents on hexane, a typical alkane, are shown in table 27.4.

Clearly hexane is unreactive. It is unaffected by acids, alkalis, dehydrating agents and aqueous oxidising agents. A close look at the substances in the table with which hexane *does* react, namely bromine and oxygen, shows two things. First, neither of these reagents has any centre of electrical charge in their molecule: they are non-polar. This is in contrast with the other substances in the table, which are all polar reagents. Second, before reaction can occur, energy must be supplied: heat in the case of oxygen, light in the case of bromine.

Table 27.4
The effect of common reagents on hexane

Reagent	Effect
air	no effect cold. Burns when heated
sodium hydroxide solution	no effect hot or cold
concentrated sulphuric acid	no effect hot or cold
potassium manganate(VII) (potassium permanganate) solution	no effect hot or cold
bromine	no effect in dark. Bromine slowly decolorised in sunlight

The failure of ions or polar molecules to have any effect on alkane molecules is due to the non-polarity of the C—H bond. Carbon and hydrogen are very close in electronegativity, so the electron pair in the covalent bond between carbon and hydrogen is fairly evenly shared: consequently there is little polarity in the C—H bond. The electron pair in the C—C bond is of course also evenly shared, so this bond is non-polar too. Thus there are no polar bonds in alkane molecules, and so no centres of electrical charge to act as electrophiles or nucleophiles and attract normally reactive species like H^+, OH^- and MnO_4^-.

Alkanes, then, are unreactive towards polar or ionic reagents, but will react with non-polar substances like oxygen or bromine. The mechanism of these reactions involves free radicals.

Free-radical reactions of alkanes

We have seen that hexane and bromine react in sunlight. A similar reaction occurs between methane and chlorine and we will use this as a simple example to look at in more detail.

Methane and chlorine do not react at all in the dark, but in sunlight an explosive reaction occurs, forming chloromethane and hydrogen chloride:

$$CH_4(g) + Cl_2(g) \xrightarrow{\text{light}} CH_3Cl(g) + HCl(g) \qquad \Delta H = -98\,kJ$$

This is a substitution reaction. Like all reactions between covalent molecules, it involves breaking some bonds, for which energy must be supplied, and making new bonds, when energy is released.

The reaction between methane and chlorine does not proceed in the dark at room temperature because there is not enough energy available to start the reaction by breaking the necessary bonds.

$$Cl_2 \rightarrow Cl\cdot + Cl\cdot \qquad \Delta H = +242\,kJ\,mole^{-1}$$

$$CH_4 \rightarrow CH_3\cdot + H\cdot \qquad \Delta H = +435\,kJ\,mole^{-1}$$

These figures show that the Cl—Cl bond is easier to break than the C—H bond. When light is shone on the reaction mixture, chlorine molecules are supplied with the energy necessary to split them into atoms. This stage is called **initiation**.

The chlorine atoms, being free radicals, are highly reactive. When they collide with a methane molecule they combine with one of its hydrogen atoms, forming a new free radical:

$$CH_4 + Cl\cdot \rightarrow CH_3\cdot + HCl \qquad \Delta H = +4\,kJ$$

The $CH_3\cdot$ free radical then reacts with another chlorine molecule:

$$CH_3\cdot + Cl_2 \rightarrow CH_3Cl + Cl\cdot \qquad \Delta H = -97\,kJ$$

and so the process continues. These two reactions enable a chain reaction to occur: they are **propagation** steps. Note that each propagation step involves both the breakage and the formation of a bond: the net energy change is therefore relatively small.

The reaction chain ends when two free radicals collide and combine; this is called **termination** and is highly exothermic:

$$Cl\cdot + Cl\cdot \rightarrow Cl_2 \qquad \Delta H = -242\,kJ$$

$$CH_3\cdot + Cl\cdot \rightarrow CH_3Cl \qquad \Delta H = -339\,kJ$$

$$CH_3\cdot + CH_3\cdot \rightarrow C_2H_6 \qquad \Delta H = -346\,kJ$$

Each chain may go through 100 to 10000 cycles before termination occurs. The processes are extremely rapid, hence the explosive nature of the reaction.

The net result of such reactions is the formation of large amounts of CH_3Cl and HCl and small amounts of C_2H_6. (In the presence of excess chlorine, further substitution may occur, forming CH_2Cl_2, $CHCl_3$ and CCl_4, as explained in the next section.) The overall energy change of $93\,kJ\,mole^{-1}$ evolved represents the difference between the energy released in forming C—Cl and H—Cl bonds and the energy absorbed in breaking C—H and Cl—Cl bonds. The energy supplied by the light which initiates the reaction is the activation energy.

Practically all the reactions of alkanes proceed by free-radical mechanisms, characterised by high activation energies and a tendency to proceed rapidly in the gas phase. Some of these reactions will now be considered.

The combustion of alkanes is a gas-phase reaction. Before it can burn, candle wax must vaporise. The wick provides a surface from which the molten wax vaporises.

27.7 Important reactions of alkanes

Alkanes show little reactivity towards all the common polar and ionic reagents. Indeed the alkanes were once known as the **paraffins**, from the Latin words *parum* (little) and *affinitas* (affinity). Hence there are only a few reactions of alkanes, but these are of great importance.

Burning

Although alkanes are kinetically stable in the presence of oxygen, they are energetically unstable with respect to their oxidation products. Therefore when the necessary activation energy is supplied, as when they are ignited, combustion occurs. The reaction has a free-radical mechanism, and it occurs rapidly in the gas phase. Because it is a gas-phase reaction, liquid and solid alkanes must first be vaporised. This is why less volatile alkanes burn less readily. If the oxygen supply is plentiful, the combustion products are carbon dioxide and water. For example:

$$C_7H_{16}(g) + 11\,O_2(g) \longrightarrow 7\,CO_2(g) + 8\,H_2O(l); \quad \Delta H = -485\,kJ$$
heptane

Car exhaust is a major source of pollution in modern cities.

Table 27.5
Gases in petrol engine exhaust

Gas	Percentage by volume
carbon dioxide	9
oxygen	4
hydrogen	2
carbon monoxide	4–9
hydrocarbons	up to 0.2
aldehydes	0.004
nitrogen oxides	0.05–0.4
sulphur dioxide	0.006
ammonia	0.0006
(lead (as solids)	$4\,\text{mg m}^{-3}$)

If the oxygen supply is limited, the products may include carbon monoxide and carbon.

- Write an equation for the complete combustion of octane, C_8H_{18}.
- Estimate the heat of combustion of octane given that the heats of combustion of hexane and heptane are $-4195\,\text{kJ mole}^{-1}$ and $-4854\,\text{kJ mole}^{-1}$, respectively.
- Write an equation for the combustion of octane to form carbon monoxide and water.

The combustion of hydrocarbons in general and alkanes in particular is of immense importance. It occurs in power stations, furnaces, domestic heaters, candles, gas heaters, internal combustion engines and many other devices essential to a technological society. An understanding of the nature of the combustion process is vital to the design of such devices.

An important example is the **petrol engine**. In the cylinder of a motor-car engine a mixture of gasoline vapour (which contains mostly C_5 to C_{10} alkanes) and air is ignited by an electric spark. This produces an explosive reaction which drives the piston down. The rate at which this reaction occurs and the ease with which it is initiated are very important for the efficiency of the engine. If the explosion occurs too rapidly, heat will be dissipated instead of being converted to useful kinetic energy. If the explosion starts too early the pistons are subjected to harmful jarring. 'Knocking' or 'pinking' of the engine then occurs. This is more likely to occur in the engines of high-performance cars, where high compression ratios tend to cause premature ignition.

Gasoline mixtures that are rich in straight-chain alkanes such as heptane ignite very readily and explode rapidly, causing 'knocking' and inefficient combustion. The combustion of branched-chain alkanes such as 2,2-4-trimethylpentane (iso-octane) is much smoother and more controlled. So gasoline mixtures rich in branched-chain alkanes are more efficient fuels and less likely to cause knocking. The **octane rating** of 2,2-4-trimethylpentane is set at 100, and that of heptane at 0. Thus gasoline mixtures rich in branched-chain alkanes burn smoothly and efficiently in high-performance engines and have high octane numbers. Octane numbers are usually represented by stars on petrol pumps. 'Four star' petrol has a higher octane number than 'two star'.

There are two ways of meeting the demand of modern high-performance engines for fuels with high octane numbers. One is to produce artificial gasoline mixtures that are rich in branched-chain alkanes (see below). The other is to add an anti-knock compound to gasoline. The anti-knock compound normally used is tetraethyllead(IV), $Pb(CH_3CH_2)_4$. When burned, this compound produces small particles of lead oxide which tend to combine with free radicals produced in the chain reaction of combustion. This slows the reaction down and makes it smoother.

To prevent lead accumulating in the engine, 1,2-dibromoethane (CH_2BrCH_2Br) is also added to the gasoline. This results in the formation of lead bromide, which is volatile and is swept away in the car exhaust. The unfortunate effect of this is to add lead to the atmosphere, whence it can be inhaled. Lead is a neurotoxin, or nerve poison. It can cause damage to the brain and nervous system, particularly in young children. There is considerable concern that leaded petrol may be increasing the concentration of lead in the blood of city-dwellers to near the point where symptoms of toxicity may occur. Several countries are now phasing out the use of lead in gasoline.

Lead is not the only pollutant produced by petrol engines. Table 27.5 gives the composition of typical petrol engine exhaust.

- What single gas omitted from table 27.5 makes up the remainder of car exhaust?
- Why does exhaust contain carbon monoxide?
- Why does the exhaust contain oxides of nitrogen?
- Why does exhaust contain hydrocarbons?

The most dangerous pollutants in car exhaust, apart from lead, are probably carbon monoxide, hydrocarbons and oxides of nitrogen. Carbon monoxide is very toxic because it forms a stable compound with haemoglobin in the blood, making the haemoglobin unable to transport oxygen. In a confined space the

Right. Acid fog in the Black Forest in Germany—one of several forms of acid deposition that is thought to be responsible for the death of trees.

A catalytic converter for fitting to car exhaust systems.

Tokyo police carry out spot checks on car exhausts in an attempt to control pollution. Owners of cars whose exhausts contain more than 5.5% carbon monoxide are ordered to tune their engines.

carbon monoxide in exhaust gases can be fatal. In the open the danger is less, though the carbon monoxide concentration in a busy street is undesirably high. It is worth noting, though, that the carbon monoxide level in a railway compartment full of smokers is twice as high as that in a busy London street.

It is thought that oxides of nitrogen and unburnt hydrocarbons may play an important role in the formation of **acid rain**. The biggest single cause of acid rain is probably sulphur dioxide, but these other pollutants may take part in the complicated series of reactions that turn air pollutants to **acid deposition.**

Acid deposition includes all the different ways that acid materials get deposited on the earth, including rain, snow and solid particles. Acid deposits mainly consist of sulphuric acid, nitric acid, sulphates and nitrates, and automobile exhausts are thought to play an important part in their formation. For example, oxides of nitrogen may help to catalyse the conversion of sulphur dioxide in the air to sulphuric acid. In the city of Los Angeles, California, oxides of nitrogen play an important part in the complex sequence of chemical reactions that leads to the formation of the notorious photochemical smog. The seven million inhabitants of Los Angeles own four million cars which between them consume thirty million dm^3 of petrol per day. The exhaust gases produced in this automobile-dominated city combine with particular geographical conditions to produce a unique smog of especially unpleasant character.

The Californian authorities have introduced legislation designed to reduce the level of pollutants in exhaust gases. One way of doing this is to install a catalyst unit in the exhaust system which converts the carbon monoxide, nitrogen oxides and hydrocarbons to harmless products. A problem associated with the use of such catalytic converters is that the catalysts are poisoned by lead.

Apart from pollution caused by car exhausts, oil is responsible for other forms of pollution, such as the smoke from diesel engines and spillages from oil tankers. As far as air pollution is concerned, though, in Britain coal is as much of a nuisance as oil. And as far as danger to health is concerned, neither of these sources can compare with the hazards caused by tobacco smoke.

Reaction with halogens

The reaction of methane with chlorine has already been considered in some detail. This reaction produces several products:

$$CH_4 + Cl_2 \rightarrow CH_3Cl + HCl$$
$$\text{chloromethane}$$

$$CH_3Cl + Cl_2 \rightarrow CH_2Cl_2 + HCl$$
$$\text{dichloromethane}$$

$$CH_2Cl_2 + Cl_2 \rightarrow CHCl_3 + HCl$$
$$\text{trichloromethane (chloroform)}$$

$$CHCl_3 + Cl_2 \rightarrow CCl_4 + HCl$$
$$\text{tetrachloromethane (carbon tetrachloride)}$$

The proportions of different products formed depend on the proportions of chlorine and methane used. All four products are useful in industry, but they have to be separated from one another by distillation. Chlorination of alkanes is therefore of limited use for preparing chloroalkanes, particularly in the case of the higher alkanes where the number of possible products is very large.

Similar reactions occur between other alkanes and fluorine, chlorine and bromine. The reaction with fluorine occurs in the absence of sunlight and is very vigorous.

Cracking

When alkanes are heated to high temperatures their molecules vibrate strongly enough to break and form smaller molecules. One of these molecules is usually an alkene. Such reactions are known as **cracking**. For example:

$$C_{11}H_{24} \longrightarrow C_9H_{20} + H_2C = CH_2$$
undecane $\qquad$ nonane $\qquad$ ethene

- In which petroleum fraction would $C_{11}H_{24}$ be found?
- In which fraction would C_9H_{20} be found?
- Why must one of the products of cracking always be unsaturated?

By using a catalyst, cracking can be made to occur at fairly low temperatures. This is known as **catalytic cracking**.

Cracking is very important in the petroleum industry. It is used for the following.

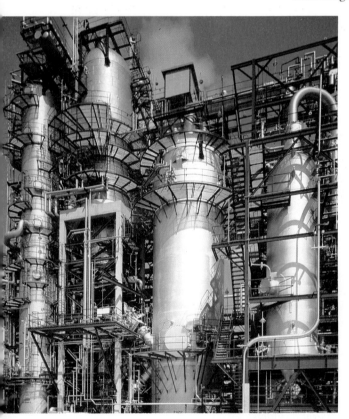

A catalytic cracking unit. The diagram on the right explains the working of the unit. The reaction actually occurs in the central unit: the catalyst is regenerated in the left-hand unit and the products are separated in the distillation tower on the right.

(a) To provide extra gasoline

The example above shows how undecane, a member of the kerosine fraction, can be cracked to produce nonane, a component of gasoline. Thus heavier fractions can be cracked to produce extra gasoline. Furthermore, cracking tends to produce branched-chain rather than straight-chain alkanes. So the gasoline produced this way has a high octane rating. A process similar to cracking is used to convert low-grade gasoline to high-grade fuel. This process is called **reforming**. It involves breaking up straight-chain molecules, then reassembling them as branched-chain molecules and aromatic rings.

(b) As a source of alkenes

Because they are so unreactive, alkanes are not a good starting point from which to make the many organic chemicals derived from crude oil. Alkenes, with their reactive double bonds, are a more suitable starting point. The petrochemical industry uses vast quantities of ethene and propene as units for building larger organic molecules.

We have seen that cracking reactions produce alkenes, and under the right conditions large yields of ethene and propene can be obtained. Alkenes are thus produced as by-products from cracking heavy fractions (such as $C_{11}H_{24}$) and by cracking some of the gasoline fraction (which is called naphtha when used as a source of alkenes). In countries such as the USA where gasoline is in high demand, alkenes are made by cracking gaseous alkanes (ethane, propane and butane).

The reactions of alkenes and the products obtained from them are discussed in the next chapter.

Which pump dispenses a higher proportion of straight-chain alkanes?

27.8 Cycloalkanes

The alkanes considered so far have all had open-chain molecules, that is molecules which come to an end at some point. It is also possible for alkane molecules to form rings: such compounds are named by using the prefix **cyclo-**. Some cycloalkanes are shown in figure 27.9.

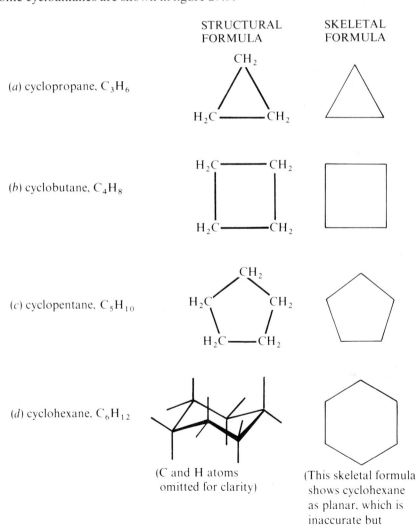

STRUCTURAL FORMULA   SKELETAL FORMULA

(a) cyclopropane, C_3H_6

(b) cyclobutane, C_4H_8

(c) cyclopentane, C_5H_{10}

(d) cyclohexane, C_6H_{12}

(C and H atoms omitted for clarity)

(This skeletal formula shows cyclohexane as planar, which is inaccurate but easier to draw)

Figure 27.9
Some cycloalkanes.

Cyclopropane, cyclobutane and cyclopentane all have planar molecules. The natural bond angles around a saturated carbon atom are 109° (the angle at the centre of a tetrahedron). In cyclopropane the bond angle is only 60°, since the carbon atoms within the molecule form an equilaterial triangle. Hence there is considerable bond strain in this molecule, so cyclopropane is unstable and reactive, tending to break open its ring. The bond angle in cyclobutane is 90°, so the bond strain is less, but nevertheless considerable, so cyclobutane is more reactive than open-chain alkanes. Cyclopentane, with a bond angle of 108°, has little bond strain and is therefore stable. Cyclohexane and the higher cycloalkanes can relieve bond strain by puckering, so that their rings are no longer planar. These cycloalkanes are therefore stable and very similar in properties to open-chain alkanes. Figure 27.9(*d*) shows the puckered structure of cyclohexane, in which all the carbon atoms have an unstrained bond angle of 109°. You should try building this molecule with models to see exactly how bond strain is eliminated.

The general formula of cycloalkanes is C_nH_{2n}, compared with C_nH_{2n+2} for open-chain alkanes. Although cyclopentane, cyclohexane and higher cycloalkanes behave similarly to open-chain alkanes their melting and boiling points tend to be somewhat higher. Cycloalkanes are present in considerable quantities in crude oil. Industrially the most important is cyclohexane, used in the manufacture of nylon.

Summary

1 Crude oil is a complex mixture of hydrocarbons consisting of alkanes, cycloalkanes and aromatic compounds. It is separated into useful fractions by fractional distillation.

2 Alkanes are saturated hydrocarbons with general formula C_nH_{2n+2}.

3 Straight-chain alkanes are named by adding the suffix -*ane* to a prefix indicating the number of carbon atoms in the molecule.

4 Branched-chain alkanes are named by considering their molecules as straight-chain alkanes with side groups attached.

5 Alkanes show a steady gradation in physical properties with increasing molecular size.

6 When a covalent bond breaks, it may form either uncharged free radicals (homolytic fission) or positive and negative ions (heterolytic fission).

7 Nucleophiles are electron-deficient and are attracted to negatively charged groups. Electrophiles are electron-rich and are attracted to positively charged groups.

8 Alkanes are chemically unreactive towards polar or ionic reagents, but they react with reagents such as oxygen or halogens by free-radical mechanisms.

9 Alkanes burn in a plentiful supply of oxygen to form carbon dioxide and water. They undergo substitution reactions with halogens in sunlight.

10 Alkane molecules break down to smaller molecules at high temperatures: this process is called cracking and is important in the petroleum industry.

11 Cycloalkanes are saturated hydrocarbon rings with general formula C_nH_{2n}.

Study questions

1   Name the following alkanes

(a) $CH_3-CH_2-CH_2-CH_2-CH_2-CH_3$

(b) $CH_3-CH-CH_2-CH-CH_2-CH_3$
        | |
        CH_3 CH_2CH_3

(c)

(d) $CH_3-CH_2-CH_2-CH_2$
                          |
                        CH_2
                          |
                        CH_3

(e) CH_2
        | $>CH-CH_3$
        CH_2

(f)

2   Consider the following alkanes

A $CH_3(CH_2)_{18}CH_3$ B   C $CH_3(CH_2)_6CH_3$

D $CH_3CH(CH_2)_4CH_3$ E $CH_3CH_2CH_3$
        |
        CH_3

(a) In which crude oil fraction would each be found?

(b) Which two alkanes are isomers?

(c) Which alkane might have been formed by cracking another alkane shown above?
     (Identify *both* alkanes.)

(d) Arrange the alkanes in order of increasing boiling point.

(e) Which alkane is a solid at 298 K?

(f) Which alkane's formula does *not* fit the general formula C_nH_{2n+2}?

3   Write (a) full structural formulae; (b) molecular formulae for the following alkanes:

(i)   ethylcyclohexane,

(ii)   1,2-dimethylcyclopentane,

(iii) 2,2,3-trimethylbutane,

(iv)   2,2-dimethyl-3,4-diethylheptane.

4   Study the table, which gives the heats of combustion and the relative molecular masses of some alkanes.

Alkane	No. of carbon atoms, n	Relative molecular mass	Heat of combustion $\Delta H_c/kJ\ mole^{-1}$
methane	1	16	−890
ethane	2	30	−1560
propane	3	44	−2220
butane	4	58	−2877
pentane	5	72	−3509
methylbutane	5	72	−3503
2,2-dimethylpropane	5	72	−3517
hexane	6	86	−4195

(a) Plot a graph of ΔH_c against n for the first six straight-chain alkanes.

(b) Does ΔH_c increase by the same amount for each extra carbon atom in an alkane chain?

(c) If so, what is the average increase in ΔH_c per carbon atom?

(d) Your answer to (c) represents the heat of combustion of which structural group?

(e) Where does your graph intercept the ΔH_c axis? What is the physical significance of this intercept?

(f) Compare the ΔH_c values for the three isomeric alkanes with five carbon atoms. Why are they so similar?

(g) Work out the heat of combustion per gram of butane, pentane and hexane, respectively. Comment on your result.

5   For the purposes of this question, assume that petrol has a density of $0.75\,\mathrm{g\,cm}^{-3}$ and is a mixture of hydrocarbons all of formula C_8H_{18}. Assume $0.5\,\mathrm{cm}^3$ of tetraethyllead(IV) are added to every dm^3 (litre) of petrol. Tetraethyllead(IV), $Pb(C_2H_5)_4$, is a liquid of density $1.6\,\mathrm{g\,cm}^{-3}$.

(a) Why is tetraethyllead(IV) added to petrol?

(b) Write an equation for the complete combustion of petrol in excess air.

(c) How many moles of C_8H_{18} are there in $1\,\mathrm{dm}^3$ of petrol?

(d) Calculate the volume of air needed to burn completely $1\,\mathrm{dm}^3$ of petrol (assume air is 20% oxygen by volume and that 1 mole of a gas occupies $24\,\mathrm{dm}^3$ under ordinary conditions).

(e) Calculate the volume of nitrogen emitted as exhaust when $1\,\mathrm{dm}^3$ of petroi burns. (Assume all the nitrogen in the air intake passes into the exhaust.)

(f) Calculate the mass of nitrogen emitted as exhaust when $1\,\mathrm{dm}^3$ of petrol burns.

(g) Calculate the mass of carbon dioxide produced when $1\,\mathrm{dm}^3$ of petrol burns.

(h) Assuming that all the lead in the tetraethyllead(IV) passes out in the exhaust, calculate the mass of lead passed out in the exhaust when $1\,\mathrm{dm}^3$ of petrol burns.

6   This question is about octane (C_8H_{18}) and cyclooctane (C_8H_{16}).

(a) Draw the structural formula of octane.

(b) Draw the structural formula of cyclooctane, showing the exact shape you would expect for the molecule.

(c) Suppose octane and cyclooctane were each reacted with chlorine so that only one hydrogen atom in each molecule was substituted by chlorine. Draw structural formulae for all the products you would expect in each case.

7   The table shows the heats of combustion of some cycloalkanes.

Name	Formula	Heat of combustion, $\Delta H_c/\mathrm{kJ\,mole}^{-1}$
cyclopropane	$(CH_2)_3$	-2090
cyclobutane	$(CH_2)_4$	-2740
cyclopentane	$(CH_2)_5$	-3320
cyclohexane	$(CH_2)_6$	-3948
cycloheptane	$(CH_2)_7$	-4635
cyclooctane	$(CH_2)_8$	-5310
cyclononane	$(CH_2)_9$	-5980
cyclodecane	$(CH_2)_{10}$	-6630

(a) For each compound, calculate the heat of combustion per CH_2 group.

(b) Would you expect the heat of combustion per CH_2 group to be constant among: (i) open-chain alkanes; (ii) unstrained cycloalkanes; (iii) cyclo alkanes with ring strain?

(c) Explain your answers and comment on the figures you obtained in (a).

8   If a few drops of bromine are added to hexane a deep red solution is formed. No reaction occurs if the mixture is kept in the dark, but in sunlight the red colour slowly disappears and a misty gas is given off.

(a) Identify the misty gas formed in this reaction.

(b) Write an overall equation for the reaction of one mole of bromine with one mole of hexane under these conditions.

(c) Why does the reaction only occur in sunlight?

(d) This is a free-radical chain reaction. Write equations for
   (i)   the initiation stage,
   (ii) a propagation stage,
   (iii) a termination stage.

9   Write the structural formulae of all the products you would expect to be formed when ethane reacts with excess chlorine in sunlight.

10 For each of the following, state whether or not you would expect reaction to occur to any significant extent. Write equations for any reactions which you think will occur.
   (a) Chlorine is bubbled through hexane in sunlight.
   (b) Sodium metal is added to warm hexane.
   (c) Hexane is boiled with acidified potassium dichromate(VI) solution.
   (d) Chlorine and hexane vapour are heated in the dark.
   (e) Hexane and hydrogen are heated in sunlight.
   (f) Hexane vapour is heated strongly on its own, in the absence of air.

11 Methane has been suggested and used as an alternative to gasoline as a fuel for motor cars. What modifications would be necessary to an ordinary car to enable it to run on methane? What would be the advantages of using methane instead of gasoline?

12 Natural gas from the Texas Panhandle has the percentage composition by volume shown in the table.

Component	Percentage by volume
Methane	80.9
Ethane	6.8
Propane	2.7
Butane and higher alkanes	1.6
Nitrogen	7.9
Carbon dioxide	0.1

The natural gas is processed in the following way:
Carbon dioxide is first removed, then ethane, propane, butane and higher alkanes are separated off and mainly used to make alkenes, especially ethene.
   (a) How might the carbon dioxide be removed from the gas?
   (b) How might ethane and the higher alkanes be separated off?
   (c) What would the remaining natural gas be used for?
   (d) Why is it not necessary to remove the nitrogen from the gas?
   (e) Why is ethene an important industrial chemical?
   (f) Ethene is produced from ethane by a cracking reaction that eliminates hydrogen. Write an equation for this reaction. What conditions would be needed to make the reaction occur?
   (g) In Europe ethene is manufactured from naphtha (part of the gasoline fraction from petroleum distillation) rather than ethane. Suggest a reason for this difference.

28 Unsaturated Hydrocarbons

Ethene is used to speed up the ripening of fruit such as bananas. This 'Ethylene Gas Generator' produces ethene by the dehydration of ethanol over a heated catalyst. What might the catalyst be?

28.1 Importance of alkenes

What do pvc raincoats and antifreeze have in common? Or polythene bottles and adhesives? Like many other things in everyday use, they are made from ethene, the simplest alkene and the most versatile organic compound in use today. In the UK, more than a million tonnes of ethene are produced and consumed every year. Ethene, $CH_2{=}CH_2$, with its reactive double bond, can be used as a building block to prepare complex organic molecules. Propene, $CH_3CH{=}CH_2$, is used in a similar way, though on a smaller scale. The necessary large quantities of these alkenes are manufactured by cracking processes described in the last chapter.

28.2 Nomenclature of alkenes

Ethene and propene are the first two members of the homologous series of **alkenes**. All members of this series contain a double carbon–carbon bond. They therefore have two atoms of hydrogen less than the corresponding alkane and their general formula is C_nH_{2n}. Because they contain less than the maximum amount of hydrogen they are said to be **unsaturated**.

Alkenes are named using the same general rules described for alkanes in the last chapter. The suffix **-ene** is used instead of -ane, together with a number indicating the position of the double bond in the chain. Thus the molecule $CH_3CH{=}CHCH_3$ is named but-2-ene. Note that although the double bond joins carbon atoms 2 and 3, the number 2 is used because this is the lower. Table 28.1 gives the formulae and names of some more alkenes.

Table 28.1
Nomenclature of alkenes

Formula	Name
$CH_3CH_2CH_2CH{=}CH_2$	pent-1-ene
$CH_3CH_2CH{=}CHCH_3$	pent-2-ene
$CH_3C{=}CHCH_3$ $\mid$ CH_3	2-methylbut-2-ene
$CH_2{=}CHCH_2CH{=}CH_2$	penta-1, 4-diene
⬡	cyclohexene

472

■ Name these alkenes:
(a) $CH_2\!\!=\!\!CHCH_2CH_3$
(b) $CH_2\!\!=\!\!C\!\!-\!\!CH_3$
$\qquad\;\;|$
$\qquad\;CH_3$

(c)

■ Write structural formulae for these alkenes:
(a) hex-2-ene
(b) buta-1, 3-diene
(c) 2,3-dimethylbut-2-ene.

28.3 The nature of the double bond

When ethene is bubbled through bromine, the bromine is decolorised. No sunlight is needed. A colourless liquid, immiscible with water, is formed, but no hydrogen bromide is produced. This reaction is clearly different in nature to the reaction of bromine with alkanes such as ethane, in which sunlight is needed to make the reaction occur and hydrogen bromide is produced. The reaction between ethene and bromine is an addition reaction. One of the two carbon–carbon bonds breaks, enabling bonds to be formed to bromine:

1,2-dibromoethane
(a colourless liquid)

Addition reactions are characteristic of all alkenes, which are much more reactive than alkanes. This reactivity is perhaps a little surprising at first: we might expect a double bond to be stronger than a single one, and therefore more stable.

Look at the bond energies in table 28.2. You can see that the bond energy of a double bond is greater than that of a single bond, though not twice as great. This suggests that the two bonds in C=C may not be identical. In fact, two kinds of covalent bond are thought to be involved.

Table 28.2
Bond energies and bond lengths of C—C, C=C, and C≡C

	C—C (in ethane)	C=C (in ethene)	C≡C (in ethyne)
bond energy/kJ mole^{-1}	346	598	813
bond length/nm	0.154	0.134	0.121

(a) A bond situated symmetrically between the two carbon atoms, formed by the overlap of two orbitals. This is called a **sigma-(σ) bond** (see figure 28.1(a) on the next page).

(b) A bond formed by the 'sideways' overlap of two $2p$ orbital. Because each p orbital has two lobes, this bond has two regions, one above and one below the plane of the molecule. It is called a **pi (π) bond** (see figure 28.1(b)).

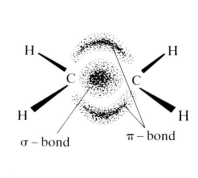

Figure 28.1
σ- and π-bonds in ethene. Note that the plane of the molecule is perpendicular to the page.

p orbitals, each containing one electron
(*a*) σ – bond only

(*b*) σ – bond and π – bond

The two electrons of the π-bond are not situated axially between the carbon atoms. This means they are not 'on average' as close to the nuclei of the atoms as the electron pair in the σ-bond. Therefore they do not attract the nuclei so strongly, so the π-bond is not as strong as the σ-bond. Nevertheless, the π- and σ-bonds together are stronger than the single σ-bond that links the carbon atoms in ethane. Consequently the C=C bond in ethene is stronger and shorter than the C—C bond in ethane.

If double bonds are stronger than single bonds, why are alkenes more reactive than alkanes? When bromine adds across the double bond in ethene, the π-bond is broken: this requires energy. The energy used in breaking the one π-bond, however, is more than repaid by the energy released when two new bonds are made to bromine atoms. Hence alkenes are *energetically* unstable relative to their products in an addition reaction. They are also *kinetically* unstable, because the high electron density in the double bond tends to attract electron-deficient groups (electrophiles), thus initiating addition reactions. Before considering the mechanism of these reactions in detail, there is another important consequence of the double bond to consider.

28.4 *Cis-trans* isomerism

X-ray diffraction evidence shows that the ethene molecule is planar. This fits in with our ideas on shapes of molecules (section 8.7) for ethene has no lone pairs. The three groups round each carbon atom are arranged trigonally, at approximately 120° to one another, so the ethene molecule is drawn thus:

$$\begin{array}{ccc} H & & H \\ \diagdown & & \diagup \\ & C=C & \\ \diagup & & \diagdown \\ H & & H \end{array}$$

Unlike ethane the ethene molecule cannot be rotated about the bond between the carbon atoms. It is possible to rotate about a σ-bond, because this does not affect the orbital overlap. With a double bond, though, rotation would involve breaking the π-bond and this requires more energy than is available at normal temperatures.

Now consider the compound 1,2-dibromoethene, BrCH=CHBr. This can be made by addition of bromine to ethyne (HC≡CH). The product is a colourless non-polar liquid, boiling point 381 K, melting point 266 K and immiscible with water.

However, another compound is known with the same formula BrCH=CHBr. This second substance is also a colourless liquid, immiscible with water, but its boiling point is 383 K, its melting point is 220 K and its molecule is appreciably polar. These two compounds exhibit a kind of isomerism called **cis-trans** or **geometric** isomerism. It arises from the lack of free rotation about a double bond. The first compound (T_b = 381 K, T_m = 266 K, non-polar) has the structure

$$\underset{Br}{\overset{H}{\diagdown}} C = C \underset{H}{\overset{Br}{\diagup}}$$

It is called *trans*-1,2-dibromoethene, because the bromine atoms are on opposite sides of the double bond. The second compound (T_b = 383 K, T_m = 220 K, polar) has the structure

$$\underset{Br}{\overset{H}{\diagdown}} C = C \underset{Br}{\overset{H}{\diagup}}$$

It is called *cis*-1,2-dibromoethene (*cis*, Latin: on the same side), both bromine atoms being on the same side of the double bond. Since free rotation about the double bond is prevented these isomers cannot readily be interconverted, unless the temperature is high enough to make available the energy needed to break the π-bond.

Cis-trans isomerism is common in compounds containing double bonds. The isomers normally have similar chemical properties but often their physical properties are markedly different. *Cis-trans* isomerism is not limited to compounds with C=C double bonds. It can arise wherever rotation about a bond is restricted, for example in ring compounds.

- Why is *cis*-1,2-dibromoethene polar while its *trans* isomer is not?
- Suggest a reason for the large difference in melting point between the two isomers.
- Would you expect but-2-ene, $CH_3CH=CHCH_3$, to have *cis* and *trans* isomers? If so, write their structures.
- Would you expect 1,1-dibromoethene, $Br_2C=CH_2$, to have *cis* and *trans* isomers? If so, write their structures.

- Why are $H-\underset{\underset{Br}{|}}{\overset{\overset{H}{|}}{C}}-\underset{\underset{H}{|}}{\overset{\overset{Br}{|}}{C}}-H$ and $H-\underset{\underset{Br}{|}}{\overset{\overset{H}{|}}{C}}-\underset{\underset{Br}{|}}{\overset{\overset{H}{|}}{C}}-H$ not considered to be isomers?

28.5 Mechanism of addition to a double bond

Ethene and bromine undergo an addition reaction to form 1,2-dibromoethane. The reaction occurs in the dark at room temperature, which suggests that a free-radical mechanism like that described in section 27.5 is not involved. The mechanism is believed to involve heterolytic rather than homolytic fission.

However, the ethene molecule has a region of high electron density caused by the π-electrons in its double bond. As the bromine molecule approaches the ethene, it becomes polarised by this negative charge:

A loose association is formed between the ethene and bromine molecules. Negative charge moves from the double bond towards the positively charged bromine atom. At the same time, electrons in the Br—Br bond are repelled towards the negatively charged Br atom. The net result is the formation of a C—Br bond and the production of two ions, a carbocation and a Br^- ion:

Notice the curly arrows in this equation. They are used to show the movement of electron pairs (section 27.5).

The carbocation is very unstable and quickly combines with the Br^- to form 1,2-dibromoethane:

This mechanism is accepted by most chemists, though the exact structure of the intermediate carbocation is not altogether clear. The mechanism is described as **electrophilic addition** because the Br_2 molecule acts as an electrophile when it is attracted to the electron-rich double bond.

Evidence for this mechanism comes from observation of the reaction in the presence of Cl^- ions. When ethene reacts with bromine in the presence of Cl^- ions, it is found that 1-chloro-2-bromoethane, $ClCH_2$—CH_2Br, is formed as well as 1,2-dibromoethane. This suggests that the intermediate carbocation has indeed been formed and has reacted with Cl^- as well as Br^- ions.

■ What product would you expect from an addition reaction between ethene and hydrogen chloride molecules? Write its formula.
■ In the reaction between ethene and HCl, which end of the HCl molecule would you expect to attack the double bond initially?
■ What would be the structure of the intermediate carbocation in this reaction?

28.6 Important reactions of alkenes

Nearly all the important reactions of alkenes are addition reactions. Many have mechanisms similar to that described for the reaction with bromine. The product of an addition reaction can often be predicted. You need to look carefully at the molecule which is being added to ethene. Try to decide how it will split into a positive and negative part. These two parts will then join on either side of the double bond.

Predict the structure of the products of addition reactions between ethene and:

■ H_2O,　　　　　　　■ HOCl　　　　　　　■ H_2.
■ H_2SO_4

Hint: think of H_2SO_4 as

You can check your answers below.

The reactions given below all involve ethene, which is industrially the most important alkene. The reactions of ethene are in any case typical of alkenes in general, for the alkenes are a well-graded homologous series.

Physically, alkenes are similar to alkanes, with boiling points generally a little lower (e.g. ethane 185 K, ethene 169 K; propane 231 K, propene 225 K).

Many of these reactions can be carried out in the laboratory, though some require special conditions normally only available on an industrial scale. Industrially, of course, ethene is produced by cracking light hydrocarbon fractions. In the laboratory it can be conveniently produced by the dehydration of ethanol (see section 31.7). More details of the preparation of alkenes are given in Practical 31 of the *Laboratory Manual and Study Guide*.

Reaction with halogens

We have already seen that bromine reacts with ethene under normal conditions to form 1,2-dibromoethane. This compound is added to gasoline as a scavenger for lead (see section 27.6). As expected, chlorine also reacts with ethene:

$$CH_2=CH_2 + Cl_2 \rightarrow CH_2Cl—CH_2Cl$$
$$\text{1,2-dichloroethane}$$

The reaction occurs under ordinary conditions. The product, 1,2-dichloroethane, is used to manufacture chloroethene ('vinyl chloride'), from which pvc is made. Fluorine reacts explosively with ethene, but the reaction with iodine is rather slow.

Decolorisation of bromine water is a useful test-tube reaction to detect a double bond.

Reaction with hydrogen

Hydrogen and ethene do not react under normal conditions, but in the presence of a finely divided metal catalyst, usually nickel at about 140°C, ethane is produced:

$$H_2C=CH_2 + H_2 \xrightarrow{\text{Ni}} CH_3CH_3$$
$$\text{ethane}$$

This reaction is useful in analysing organic compounds. By measuring the number of moles of hydrogen absorbed by one mole of a hydrocarbon, the number of double (or triple) bonds in its molecule can be established.

Hydrogenation of double bonds is used to convert edible oils into margarine. Vegetable oils such as palm oil consist of esters of long-chain carboxylic acids (see section 33.6) which contain double bonds. Treatment of these oils with hydrogen in the presence of a nickel catalyst saturates the carbon chains. This raises the melting point of the oil so that it becomes solid, i.e. a fat, at room temperature. By controlling the amount of hydrogenation, the margarine can be made as soft or hard as required. There is now considerable medical evidence that saturated fats, i.e. those containing few double bonds, are dangerous to health. In particular, they are thought to contribute to heart and circulatory disease. Animal fats such as cream, butter and pork fat tend to have a high proportion of saturated fats as well as containing cholesterol, which is also thought to be dangerous in large amounts. Vegetable fats, including soft margarine made from vegetable oil, contain more unsaturated fats and less cholesterol. They are therefore thought to be safer.

Above. Which contains more unsaturated molecules?

Right. Equipment used in the hydrogenation of oils in margarine manufacture.

- Predict the formula of the product of the reaction between propene and chlorine.
- Predict the formula of the product of the reaction between but-2-ene and hydrogen.

Reaction with hydrogen halides

Ethene reacts with concentrated aqueous solutions of hydrogen halides in the cold:

$$CH_2\!\!=\!\!CH_2 + HX \rightarrow CH_3CH_2X$$

Hydrogen ions act as electrophiles, attacking the double bond and forming an intermediate carbocation:

This ion then reacts with halide ions to form the product:

The reaction of hydrogen chloride with ethene produces chloroethane. When reacted with a sodium–lead alloy chloroethane produces tetraethyllead(IV), the petrol anti-knock additive.

When propene reacts with a hydrogen halide such as HCl, there are two possible products:

When this reaction is actually carried out, it is found that much more **B** is formed than **A**. The reason for this can be found by looking at the mechanism of the reaction.

When hydrogen ions attack the double bond in propene, two carbocations, **C** and **D**, can be formed:

D is the more stable of these two ions. This is because of a very important property of an alkyl group:

an alkyl group tends to donate electrons slightly to any carbon atom to which it is attached.

In ion **D** there are two methyl groups donating electrons to the positively charged carbon atom; in **C** there is only one ethyl group doing so. As a result the positive charge is stabilised slightly more in **D** than in **C**, because the donated electrons tend to cancel out the charge. **D** is therefore the more stable of the two possible intermediate carbocations, though still very unstable. It therefore tends to persist longer, making it more likely to combine with Cl⁻ to form the product

B. This means that **B** is the major product, though a certain amount of **A** is also formed.

- When 2-methylpropene (CH$_3$—C=CH$_2$) reacts with HCl, what are the

  $\qquad\qquad\qquad\qquad\qquad\qquad$ |

  $\qquad\qquad\qquad\qquad\qquad\qquad$ CH$_3$

  structures of the two possible intermediate carbocations?
- Which of these two ions is the more stable?
- What will be the major product of the reaction between 2-methylpropene and HCl?

The major products of asymmetric addition reactions can be predicted by a useful general rule, known as **Markovnikoff's Rule**. It says that:

when a molecule HA adds to an asymmetric alkene, the major product is the one in which the hydrogen atom attaches itself to the carbon atom already carrying the larger number of hydrogen atoms.

Thus in the example above,

$$CH_3-C{=}CH_2 + HCl \longrightarrow CH_3-\underset{\underset{CH_3}{|}}{\overset{\overset{Cl}{|}}{C}}-CH_3 + CH_3-\underset{\underset{CH_3}{|}}{CH}-CH_2Cl$$

this carbon atom is the one carrying the greater number of hydrogen atoms $\qquad$ major product $\qquad$ minor product

Predict the major products of reactions between

- CH$_3$CH$_2$CH=CH$_2$ and HBr
- CH$_3$C=CHCH$_3$ and HI

  $\quad$ |

  $\quad$ CH$_3$

Reaction with sulphuric acid: hydration

Sulphuric acid is another strong acid with which ethene undergoes addition. Concentrated sulphuric acid reacts with ethene in the cold:

sulphuric acid $\qquad\qquad\qquad$ ethyl hydrogensulphate

When added to water and warmed, ethyl hydrogensulphate is readily hydrolysed to ethanol:

ethyl hydrogensulphate $\qquad\qquad$ ethanol $\qquad\qquad$ sulphuric acid

The overall effect of these two reactions is to combine ethene with water to form ethanol. The sulphuric acid is regenerated. At one time this was the most important method of manufacturing ethanol from ethene. Nowadays most ethanol is manufactured by the direct catalytic hydration of ethene in the vapour phase:

$$CH_2{=}CH_2(g) + H_2O(g) \xrightarrow[330°C, 60\ atm]{H_3PO_4} CH_3CH_2OH(g)$$
$$\qquad\qquad\qquad\qquad\qquad\qquad\qquad\qquad ethanol$$

The reaction with sulphuric acid is still used industrially to produce propan-2-ol from propene.

Reaction with potassium manganate(VII)

Ethene will decolorise acidified potassium manganate(VII) (potassium permanganate). This reaction is a useful test for a double bond.

The reaction with manganate(VII) is complicated and involves both addition and oxidation. The product is ethane-1,2-diol, $HOCH_2CH_2OH$, which is itself further oxidised if excess manganate(VII) is present.

The equation for this reaction, like all redox reactions in organic chemistry, can be written using the half-equation method described in section 3.4:

$$CH_2{=}CH_2 + 2H_2O \rightarrow HOCH_2CH_2OH + 2H^+ + 2e^- \quad (i)$$
$$\text{ethane-1,2-diol}$$

$$MnO_4^- + 8H^+ + 5e^- \rightarrow Mn^{2+} + 4H_2O \quad (ii)$$

Multiplying (i) by 5 and (ii) by 2 and adding, we get

$$5CH_2{=}CH_2 + 2H_2O + 2MnO_4^- + 6H^+ \rightarrow 5HOCH_2CH_2OH + 2Mn^{2+}$$

Potassium manganate(VII) is much too expensive for this process to be useful for making ethane-1,2-diol industrially. In industry this important compound is made from epoxyethane (see below).

Reaction with oxygen

Like all hydrocarbons, ethene burns in air. This is an unimportant reaction, a waste of such a useful compound.

In the presence of a finely divided silver catalyst at 180°C, ethene can be made to combine with oxygen much more fruitfully, forming epoxyethane:

$$2CH_2{=}CH_2 + O_2 \xrightarrow[180°C]{Ag} 2\ CH_2{-}CH_2$$

epoxyethane

Epoxyethane, with its strained three-membered ring, is unstable and can be readily converted to a number of useful products. In particular it reacts with water to form ethane-1,2-diol (ethylene glycol):

$$CH_2{-}CH_2 + H_2O \rightarrow HOCH_2CH_2OH$$
$$\text{ethane-1,2-diol}$$

This is an important organic chemical, used in antifreeze and also for manufacturing polyester fibres.

Epoxyethane can also be made to join to itself, forming a polymer of structure $-OCH_2CH_2OCH_2CH_2O-$. This polymer, when added to the water used in fire hoses in the proportion 30 parts per million, reduces the friction between the water and the hose walls so successfully that the range of the hose is doubled.

Reaction with itself

Perhaps the most important industrial reaction of ethene is with itself, to form poly(ethene). This is described in section 28.8.

All the reactions above have referred to ethene, but they are applicable to alkenes in general.

Predict the outcome of the following reactions:

- $CH_3CH{=}CH_2 + H_2$ over Ni catalyst
- $CH_3CH{=}CHCH_3 + H_2SO_4$ (concentrated)
- excess $CH_3CH{=}CH_2 + KMnO_4$ in acid
- $CH_3CH{=}CHCH_3 + Br_2$

A polymer made from epoxyethane can be used to reduce the friction between water and the walls of fire hoses.

28.7 Polymers

About 80% of the world's output of organic chemicals is used to make polymers. The most important synthetic polymers are addition polymers, described in the next section. In this section we will look at the important general properties of polymers.

The use of polymers by humans is not new: wood, cotton, wool and rubber are all naturally occurring polymeric materials. However, the manufacture and use of *synthetic* polymers has only really got under way since the second world war. Since then their use has increased very rapidly and they are steadily replacing traditional natural materials. Synthetic polymers are usually cheaper than natural materials. They are often better suited to their particular function since chemists are now able to produce a polymer to suit most specifications.

One drawback of the increased use of synthetic polymers is the problem of their disposal. Synthetic polymer molecules, not being of natural origin, cannot be broken down by the enzymes in bacteria. They are said to be non-biodegradable. What is more, polymers are difficult to recycle economically. For these reasons their disposal presents a considerable problem: plastic containers are among the worst forms of modern litter.

Polymers are long-chain molecules made by joining together many small molecules. The small molecules from which the polymer is built are called **monomers**. Polymers may be natural or synthetic. Table 28.3 gives some examples of both types. Natural polymers include many important materials in common use. Many important biochemicals are polymeric, including the proteins in your hair, the DNA in your genes and the starch in your food. The structures of natural polymers are considered in other sections of this book (proteins, sections 9.8 and 34.8; nucleic acids, section 9.8; carbohydrates, sections 9.8 and 32.7).

Polymers vary widely in physical properties such as strength, flexibility and softening temperature. Some, such as polyesters, are very strong and not readily stretched. They are therefore suitable for use as **fibres**. Others, such as polythene, are more easily deformed and are classed as **plastics**.

The properties of a polymer are, of course, dependent on the properties of the molecular chains it contains. Four factors are particularly important.

Table 28.3
Examples of natural and synthetic polymers

	Polymer	Monomer	Where you find it
Natural	protein	amino acids	wool, silk, muscle, etc.
	starch	glucose	potato, wheat, etc.
	cellulose	glucose	paper, wood, dietary fibre
	DNA	nucleotides	chromosomes, genes
Synthetic	poly(ethene)	ethene	bags, washing-up bowls, etc.
	poly(chloroethene) (pvc)	chloroethene	fabric coatings, electrical insulation, etc.
	poly(phenylethene) (polystyrene)	phenylethene	toys, expanded polystyrene
	polyester	ethane-1, 2-diol and benzene-1, 2-dicarboxylic acid	skirts, shirts, trousers

(a) Chain length

The characteristic polymer properties of flexibility and tensile strength arise from the chain structure of polymer molecules. Strength and melting point increase with chain length. Typical polymer properties become apparent for average chain lengths of 50 monomer units and upwards. Mechanical strength increases up to about 500 units, after which it changes only slightly. It is necessary to talk in terms of *average* chain lengths where polymers are concerned, since individual chains vary in length even in a pure sample of polymer.

(b) Intermolecular forces

If the intermolecular forces between chains are high, the polymer will tend to be strong and difficult to melt. The size of the intermolecular forces depends on the nature of the side groups on the chain.

(c) Branching

Highly branched polymer chains cannot pack together so regularly as straight chains: highly branched polymers therefore tend to have lower tensile strength and to melt more easily (see figure 28.2). In most polymers there are regions in which the chains are regularly packed. These are called crystalline regions. The regions in which packing is irregular are called non-crystalline regions.

(d) Cross-linking

Some polymers have extensive cross-linking between chains. This forms a rigid network, making a hard material (figure 28.2). The more cross-linking, the more

(a) Polymer with few branched chains, e.g. high-density polythene.
Chains regularly packed with extensive crystalline regions. High tensile strength, high melting point.

(b) Polymer with many branched chains, e.g. low-density polythene.
Chains irregularly packed: largely amorphous with few crystalline regions. Lower tensile strength, lower melting point.

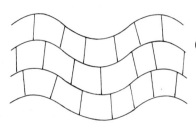

(c) Polymer with much cross-linking, e.g. bakelite. Rigid and hard, fairly brittle.

Figure 28.2
Physical properties of polymers.

rigid and brittle the polymer. Such polymers are called *thermosetting* (see section 31.7).

Forces between polymer molecules are considered further in section 9.4.

Addition and condensation polymerisation

When a polymer forms, the monomer chains join together. This joining together can happen in two ways: by addition and by condensation.

(a) *Addition polymerisation* involves addition reactions between monomer molecules. Usually only a single type of monomer molecule is involved. Most addition polymers are synthetic: for example, ethene forms an important synthetic polymer, poly(ethene). This type of polymerisation is considered further in the next section.

(b) *Condensation polymerisation* involves a **condensation** reaction between monomers. In this type of reaction, a small molecule, usually a water molecule, is eliminated from between two larger molecules. In the process a bond is formed between the two larger molecules:

$$\boxed{A} - \boxed{OH + H} - \boxed{B} \longrightarrow \boxed{A} - \boxed{B} + H_2O$$

Condensation polymerisation usually involves two different monomers joined in a chain by condensation:

$$H - \boxed{X} - [H + HO] - \bigcirc{Y} - [OH + H] - \boxed{X} - [H + HO] - \bigcirc{Y} - OH$$

$$\Downarrow$$

$$H - \boxed{X} - \bigcirc{Y} - \boxed{X} - \bigcirc{Y} - OH + 3H_2O$$

Most natural polymers are condensation polymers, and so are some synthetics. Table 28.4 gives examples of both types. There is more about condensation polymerisation in section 33.5.

Table 28.4

Addition and condensation polymers

Addition	Condensation
polythene	polyester
polystyrene	nylon
pvc	proteins
poly(propene)	starch
ptfe	cellulose
acrylic	nucleic acids

28.8 Addition polymerisation

Molecules containing double bonds are particularly useful monomers as they can usually be made to undergo addition reactions among themselves. The double bond in each molecule can be thought of as breaking open, enabling the free bonds to link with one another forming a chain.

Thus with ethene:

An insulated fish box made from high-density polythene.

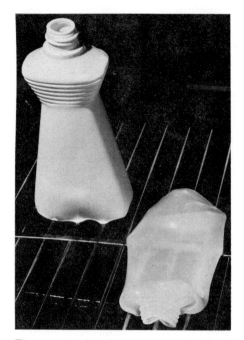

These two bottles, one made from high-density polythene and the other from low-density polythene, were heated in an oven. Which is which?

The product, **poly(ethene)** or **polythene**, is the most commonly used synthetic polymer. Note that the empirical formula of the polymer is the same as that of the monomer, CH_2. This is always the case with addition polymerisation, but is not so with condensation polymerisation. Details of some important addition polymers are now given.

(a) Polythene

Polythene was discovered by ICI in 1933. The method they used to make it involved heating ethene at about 200°C and a pressure of 1200 atm in the presence of traces of oxygen. The polythene produced in this way has branched chains which cannot pack closely together. It is therefore fairly readily melted and easily deformed. The polymer melts at about 105°C and softens in boiling water. The majority of polythene manufactured today is of this form: it is called **low-density polythene**. It is used for making film and sheeting for bags and wrappers and for making moulded articles such as washing-up bowls and 'squeezy' bottles.

Another method of making polythene was developed by Ziegler in the 1950s. This process uses catalysts at low temperatures and pressure (about 60°C and 1 atm). The molecules produced have little branching and can pack closely together. This form of polythene, called **high-density polythene**, is therefore more rigid and melts at a higher temperature (about 135°C) than the low-density form. High-density polythene is used for moulding rigid articles such as bleach bottles and milk bottle crates.

(b) Poly(propene) (polypropylene)

Propene, which like ethene is readily available from petroleum, can be polymerised by the Ziegler process:

Below. Drink crates made from poly-(propene).

Below right. Illuminated signs made from coloured Perspex.

Below far right. Non-stick frying pan and electric iron coated in ptfe

Poly(propene) produced in this way has the CH_3 side groups arranged in a highly regular fashion. Its chains pack together closely, producing a material similar to high-density polythene. It is used in mouldings and film and can be made into a fibre. It is increasingly used to make ropes, which is an example of a synthetic polymer replacing a natural one. Polypropene has the advantage that being non-biodegradable it does not rot.

(c) Poly(chloroethene) (pvc)

Chloroethene (vinyl chloride) polymerises to form poly(chloroethene) or polyvinyl chloride (pvc):

pvc-insulated cables.

The polar C—Cl bond results in considerable intermolecular attraction between the polymer chains, making pvc a fairly strong material. Its best feature is its versatility. On its own, pvc is a fairly rigid plastic, but additives called plasticisers can be used to make it highly flexible. It is used among other things for coating fabrics, for covering wires and cables and for making gramophone records.

(d) Other addition polymers

Table 28.5 shows some other addition polymers, their monomers and their uses.

Table 28.5
Addition polymers

Polymer: systematic name	poly(phenylethene)	poly(tetrafluoroethene)	poly(methyl 2-methyl-propenoate)	poly(propeno-nitrile)
Polymer: common name	polystyrene	ptfe, 'Teflon', etc.	'Perspex', acrylic	'Acrilan', etc.
Monomer: systematic name	phenylethene	tetrafluoroethene	methyl 2-methyl-propenoate	propenonitrile
Monomer: formula	$C_6H_5CH=CH_2$	$CF_2=CF_2$	$\begin{array}{c}CH_3\\[-4pt]\diagdown\\CH_2{=}\!\!\diagup C\!-\!COOCH_3\end{array}$	$CH_2=CHCN$
Monomer: common name	styrene	tetrafluoroethylene	methyl methacrylate	acrylonitrile
Properties	brittle (but cheap)	very stable. Low friction, anti-stick properties	transparent	strong, fibre properties
Uses	expanded polystyrene for insulation. Plastic toys, etc.	non-stick coatings on pans. Insulators	as a substitute for glass	making textiles (wool substitute)

Use table 28.5 to answer these questions.

■ Draw a section of the chain structures of poly(phenylethene) and poly-(propenonitrile).
■ Suggest a reason why poly(phenylethene) is brittle.
■ Suggest a reason why poly(propenonitrile) is a strong material, suitable for making fibres.

28.9 Rubber—a natural addition polymer

Rubbers are **elastomers**—polymers with elastic properties (figure 28.3). Many forms of synthetic rubber are now available, many of them cheaper and more useful than natural rubber, but the natural polymer is still much-used. Most naturally occurring polymers are condensation polymers. Natural rubber, however, can be regarded as an addition polymer. The monomer is 2-methylbuta-1, 3-diene (isoprene), whose formula is shown in the margin.

When dienes of this sort undergo addition polymerisation, one of the double bonds is retained in the resulting polymer. The structure of poly(2-methylbuta-1,3-diene) is shown in the margin.

$$CH_2\!=\!C\!-\!CH\!=\!CH_2$$
$$|$$
$$CH_3$$

2-methylbuta-1,3-diene

$$-\!\left(\!CH_2\!-\!\underset{\underset{CH_3}{|}}{C}\!=\!CH\!-\!CH_2\!\right)_{\!n}\!-$$

poly(2-methylbuta-1,3-diene)

Unstretched
Polymer chains tangled

PULL

Stretched
Chains less tangled. Crosslinks make sure the rubber goes back to its original shape when unstretched

Figure 28.3
Why rubber is elastic.

Because of the double bond, *cis-trans* isomerism is possible. Two isomeric polymers are known—natural rubber (*cis* form) and **gutta-percha** (*trans* form).

Natural rubber—*cis*

Gutta percha—*trans*

The structural difference between rubber and gutta-percha gives them different properties. Gutta-percha is much less elastic than rubber. The outer cover of golf balls is often made of gutta-percha.

CH₃
|
—CH—C=CH—CH₂—
|
S
|
S   CH₃
| |
—CH—C=CH—CH₂—

Cross-link between rubber chains.

Natural rubber comes from the Para rubber tree. When the outer bark is cut, a milky liquid called **latex** oozes out. This contains small globules of rubber suspended in water. Some adhesives (e.g. Copydex) are made from latex with ammonia added as a preservative. Raw rubber is readily precipitated from latex, but is sticky and soft. In 1832 Charles Mackintosh used raw rubber to waterproof cotton coats, but these became sticky in hot weather. It was not until Charles Goodyear discovered vulcanisation in 1839 that rubber became really useful for footwear, tyres, hoses, foams, etc.

When raw rubber is heated with sulphur it becomes harder, tougher and less temperature-sensitive. It is said to be **vulcanised**. Sulphur atoms form cross-links between rubber chains, as shown in the margin.

In ordinary soft rubber about 4% sulphur is used. This gives cross-links between 5–10% of the polymer units. By using more sulphur, harder rubbers can be made.

28.10 Alkynes

Alkynes are hydrocarbons containing triple bonds. The simplest is **ethyne**, commonly called acetylene, H—C≡C—H. Alkynes are named in a similar manner to alkenes, using the suffix **-yne** instead of -ene.

■ Write the structural formula of but-1-yne.
■ Name the compound with formula $CH_3CH_2C{\equiv}CCH_3$.
■ What is the general formula of the alkynes?

The triple bond in alkynes can be thought of as consisting of a σ-bond and two π-bonds (see section 28.3). As in ethene, the σ-bond is formed by the axial overlap of two orbitals; the π-bonds are each formed by the overlap of two $2p$ orbitals. Because the $2p$ orbitals are at right angles to one another and have two lobes each, the two π-bonds give rise to four regions of negative charge around the C—C bond. These regions tend to join up and form a cylinder of charge (see figure 28.4). Since ethyne has no lone pairs, its molecule is linear.

With a high density of electrons in the triple bond, ethyne may be expected to attract electrophiles and undergo addition reactions like those of ethene. This is indeed the case: ethyne reacts with most of the substances that undergo addition with ethene. With the two π-bonds, ethyne might be expected to be more reactive towards electrophiles than ethene. In fact, although ethyne is *energetically* less stable than ethene with respect to its addition products, it reacts rather more slowly—that is, it is *kinetically* slightly more stable. Thus while ethene decolorises bromine water almost instantaneously, with ethyne the colour takes a few minutes to disappear.

Figure 28.4
Bonding in ethyne.

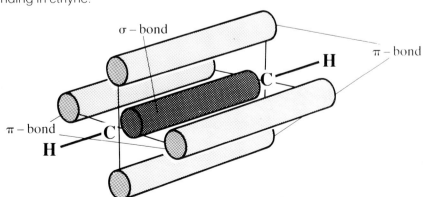

(*a*) Three-dimensional representation.
Limited regions of bonding orbitals shown.

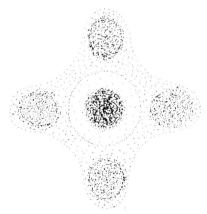

(*b*) End-on view. Depth of stippling represents electron density.

Because it has two π-bonds to add across, ethyne can give two different addition products, depending on the number of moles reacting. Thus, with hydrogen in the presence of a nickel catalyst:

$$H-C\equiv C-H + H_2 \xrightarrow[150°]{Ni} \begin{array}{c} H \\ \diagdown \\ C=C \\ \diagup \qquad \diagdown \\ H \quad \text{ethene} \quad H \end{array} \quad \text{(one π-bond broken)}$$

or

$$H-C\equiv C-H + 2H_2 \xrightarrow[150°]{Ni} H-\overset{\displaystyle H}{\underset{\displaystyle H}{C}}-\overset{\displaystyle H}{\underset{\displaystyle H}{C}}-H \quad \text{(both π-bonds broken)}$$

ethane

In practice, when ethyne reacts with hydrogen in the presence of a nickel catalyst, a mixture of ethane and ethene is formed. By carefully adjusting the quantities and conditions, one of these can be obtained in preference to the other. For example, using a palladium/barium sulphate catalyst in place of nickel gives almost entirely ethene.

Other addition reactions of ethyne

(a) With hydrogen halides

For example:

chloroethene   1,1-dichloroethane

The second stage of this addition obeys Markovnikoff's Rule (see section 28.6). The second molecule of HCl adds onto chloroethene in such a way that the hydrogen atom bonds to the carbon atom already carrying the larger number of hydrogen atoms.

(b) With halogens

Ethyne and chlorine explode spontaneously when mixed:

$$H-C\equiv C-H + Cl_2 \rightarrow 2C + 2HCl$$

This is a free-radical reaction. If, however, the reactants are mixed in the presence of an inert material such as kieselguhr (almost pure silicon(IV) oxide) the free radicals are absorbed and the expected addition reactions can take place:

1,2-dichloroethene   1,1,2,2-tetrachloroethane

Acetylene (ethyne) lamps like this one were once common on bicycles. The upper compartment contains water which drips onto calcium dicarbide contained in the lower compartment. The ethyne produced burns at the jet.

Predict the formulae of both possible products of each of the following reactions:

■ $CH_3C\equiv CH + H_2$ over nickel catalyst
■ $CH_3C\equiv CCH_3 + Br_2$
■ $CH_3CH_2C\equiv CH + HI$

Formation of salts from ethyne

Ethyne has one characteristic that is quite distinct from ethene. The hydrogen atoms in ethyne are slightly acidic and can be replaced by metal ions to form salts called **dicarbides**. Ethyne is an extremely weak acid: with a K_a of about 10^{-20} mole dm^{-3} it is much weaker than water, but about 10^{18} times stronger than ethene.

For example, if ethyne is bubbled through ammoniacal silver nitrate, a yellow precipitate of silver dicarbide is produced:

$$HC\equiv CH(g) + 2Ag^+(aq) \rightarrow \underset{\text{silver dicarbide}}{Ag-C\equiv C-Ag(s)} + 2H^+(aq)$$

This reaction is shown by all alk-1-ynes (alkynes with the triple bond at the end of the chain), and it provides a useful test for such compounds.

An important dicarbide is calcium dicarbide, CaC_2, often simply called calcium carbide. It is made by heating calcium oxide and coke at $2000°C$ in an electric furnace. Calcium dicarbide is ionic, containing the $(C\equiv C)^{2-}$ ion. It reacts readily with water to form ethyne:

$$\underset{\text{dicarbide ion}}{(C\equiv C)^{2-}} + 2H_2O \rightarrow \underset{\text{ethyne}}{HC\equiv CH} + 2OH^-$$

At one time this was a highly important reaction for manufacturing ethyne. With the decline of the coal industry and the high price of electricity, it has become too expensive and nowadays most ethyne is made from petroleum by cracking reactions. Ethyne used to be a very important material for the manufacture of organic chemicals, but it has lost this position to ethene, which is now a great deal cheaper to produce. If oil becomes very expensive it is possible that the carbide route to ethyne may again become economic and ethyne could regain its importance.

Ethyne is still used to manufacture some chloroethene (for making pvc) and 1,1,2-tetrachloroethane (a solvent). It is important as a fuel in 'oxy-acetylene' torches for cutting and welding metal. A flame temperature of about $2800°C$ can be obtained by the combustion of ethyne in pure oxygen.

Storage of ethyne under pressure presents a problem. The heat of formation of ethyne is $+277\,kJ\,mole^{-1}$ which means that it is energetically unstable with respect to carbon and hydrogen. It is therefore liable to decompose explosively when compressed on its own, so it must be stored dissolved in propanone under pressure.

Ethyne burning in pure oxygen gives a flame temperature high enough to melt most metals.

Summary

1 Alkenes are unsaturated hydrocarbons containing a carbon–carbon double bond in their molecules. Their general formula is C_nH_{2n} and they are named using the suffix -ene together with a number to indicate the position of the double bond.

2 Double bonds consist of a σ-bond situated axially between the carbon atoms and a weaker non-axial π-bond.

3 Cis and trans isomers differ in the way their substituents are arranged about a double bond.

4 The characteristic reaction of alkenes is electrophilic addition. In general:

$$\underset{}{\diagup}C=C\underset{}{\diagdown} + X-Y \longrightarrow -\overset{X}{\underset{|}{C}}-\overset{Y}{\underset{|}{C}}-$$

5 The mechanism of electrophilic addition involves initial attack on the π-electrons, often leading to the formation of an intermediate carbonium ion.

6   Some important reactions of a double bond are shown in figure 28.5.

Figure 28.5
Reactions of a double bond.

The importance of alkenes in the synthesis of organic compounds is discussed in chapter 35.

7   There are two types of polymerisation: addition polymerisation and condensation polymerisation.

8   The properties of polymers are determined by the properties of the chains they contain, in particular chain length, branching and cross-linking.

9   Many compounds containing double bonds readily undergo addition polymerisation. The most important case is ethene, which polymerises to polythene.

10 Alkynes are hydrocarbons containing a triple bond: they are named using the suffix -*yne*.

11 Alkynes undergo two-stage addition:

12 The terminal H atom in the alk-1-ynes is acidic and can be replaced by metal ions to form salts.

Study questions

1   (a) Name the following compounds:

(i) $CH_3CH_2CH_2CH_2CH=CH_2$

(ii)

(iii)

(iv) $CH_3C{\equiv}CCH_3$

(v) $CH_3C{\equiv}CCHCH_3$
            $|$
           CH_3

(b) Write structural formulae for these compounds:
   (i) propyne
   (ii) *cis*-pent-2-ene
   (iii) cycloocta-1,3,5,7-tetraene
   (iv) 2-methylpent-2-ene.

2 1,1-Dichloroethene, $Cl_2C{=}CH_2$, readily undergoes addition polymerisation.
(a) What is the systematic name of the polymer?
(b) Write the structure of a section of the polymer chain.
(c) Would you expect the polymer to have a higher or a lower melting point than poly(chloroethene) (pvc)? Explain your answer.

3 (a) For each of the following compounds, say whether you would expect them to show geometric (*cis-trans*) isomerism. For those compounds that show geometric isomerism, draw the structures of all the possible isomers.

(i)

(iii)

(ii)

(iv)

(b) 1,2-Dichlorocyclopropane,

, has *cis* and *trans* isomers.

Explain why this is so, and draw the structure of the two isomers.

4 A hydrocarbon **A** contains 87.8% carbon and 12.2% hydrogen by mass. Its relative molecular mass is 82. **A** decolorises bromine water and in the presence of a nickel catalyst it reacts with hydrogen to form **B**. 0.1 g of **A** was found to absorb 27.3 cm³ of hydrogen (measured at s.t.p.). **B** does not decolorise bromine water.
(a) What is the empirical formula of **A**?
(b) What is the molecular formula of **A**?
(c) How many moles of hydrogen react with 1 mole of **A**?
(d) How many double bonds does **A** have in its molecule?
(e) What is the molecular formula of **B**?
(f) Suggest structural formulae for **A** and **B**.

5 Propan-2-ol is an important alcohol. It is used as a solvent, as a de-icing fluid and for manufacturing propanone (acetone). Propan-2-ol is manufactured as follows:
Impure propene, containing traces of ethene, is passed up an absorption tower down which 85% sulphuric acid trickles. The propene reacts to form 2-propyl hydrogensulphate:

$$CH_3CH{=}CH_2 + H_2SO_4 \rightarrow CH_3\underset{\underset{HSO_4}{|}}{C}HCH_3$$

Virtually no 1-propyl hydrogensulphate is produced, and hardly any of the ethene present as impurity reacts. The 2-propyl hydrogensulphate produced in this way is added to water, when a highly exothermic reaction occurs. 2-Propyl hydrogensulphate is hydrolysed to propan-2-ol:

$$CH_3{-}\underset{\underset{HSO_4}{|}}{C}H{-}CH_3 + H_2O \rightarrow CH_3{-}\underset{\underset{OH}{|}}{C}H{-}CH_3 + H_2SO_4$$

(a) How is propene obtained industrially? Why is it contaminated with ethene?

(b) When propene reacts with sulphuric acid, initial attack is by H^+ on the π-electrons of the double bond. Write the structures of the two possible carbocations formed as a result of this attack.

(c) Which of the two carbocations formed in (b) is the more stable?

(d) Use your answer to (c) to explain why virtually no 1-propyl hydrogensulphate is formed in the reaction of propene with sulphuric acid.

(e) Write the structure of the carbocation formed when ethene is attacked by H^+.

(f) Is the carbocation in (e) more or less stable than the more stable carbocation formed by propene (refer back to your answer to (c))?

(g) Use your answer to (f) to explain why hardly any of the ethene contaminating the propene reacts with sulphuric acid, whereas the propene reacts readily.

(h) Why is it important that 2-propyl hydrogensulphate is virtually the only product of the reaction of the gases with sulphuric acid?

6 How would you distinguish between the compounds in each of the following pairs, using simple test-tube reactions?
(a) $CH_3CH_2C{\equiv}CH$ and $CH_2{=}CHCH{=}CH_2$
(b) hex-1-ene and cyclohexane
(c) $CH_3CH_2C{\equiv}CH$ and $CH_3C{\equiv}CCH_3$.

7 Some bond energies are given below.
C=C (in ethene) $598\,kJ\,mole^{-1}$
C—C (in ethane) $346\,kJ\,mole^{-1}$
C—H (general) $413\,kJ\,mole^{-1}$
H—H $436\,kJ\,mole^{-1}$
Consider the reaction of ethene with hydrogen:

(a) What conditions are needed for this reaction to occur?

(b) How much energy must be supplied to break the π-bonds in a mole of ethene?

(c) How much energy must be supplied to split a mole of hydrogen molecules into hydrogen atoms?

(d) How much energy is released when two new C—H bonds are formed by a mole of ethene?

(e) Calculate the energy change for the reaction of a mole of ethene with a mole of hydrogen.

8 Alkenes such as ethene and propene have been described as the building blocks of the organic chemical industry. Discuss this statement, giving examples. What particular features of the chemistry of alkenes make them suitable for this rôle and why are alkanes less suitable?

9 Predict the structures of the products, if any, of the following reactions.

(a) + $KMnO_4$ (aq)   (b) + HI

(c) $CH_3C{\equiv}CCH_3$ + ammonical silver nitrate
(d) $CH_3CH_2C{\equiv}CH$ + excess HBr
(e) $CH_3CH{=}CHCH_3$ reacts with itself, i.e. polymerises.

10 Consider the following compounds.

   A hex-2-ene $CH_3CH_2CH_2CH=CHCH_3$
   B hex-1-yne $CH_3CH_2CH_2CH_2C\equiv CH$
   C hex-2-yne $CH_3CH_2CH_2C\equiv CCH_3$
   D hexane $CH_3CH_2CH_2CH_2CH_2CH_3$

   E cyclohexane

The answers to the following questions may be one or more than one of the above compounds.

(a) Which would decolorise bromine in the absence of sunlight?
(b) Which would give a yellow precipitate with an ammoniacal solution of silver nitrate?
(c) Which would react with chlorine, but only when heated or exposed to light?
(d) Which would absorb 1 mole of hydrogen per mole of the compound in the presence of a nickel catalyst?
(e) Which has *cis* and *trans* isomers?
(f) Which are unsaturated?
(g) In which are there three or more carbon atoms arranged in a straight line?

29 Aromatic Hydrocarbons

A. Kekulé, 1829–1896.

Part of a German paper written by Kekulé describing his proposed structure for benzene.

29.1 Aromatic hydrocarbons

In the last chapter we considered unsaturated compounds with double and triple bonds. There is another important class of unsaturated compounds that needs to be considered separately because their properties are so different from the alkenes and alkynes.

This class of compounds is called the **aromatic hydrocarbons** and its simplest and most important member is **benzene, C_6H_6**. The name 'aromatic' was originally used because some derivatives of these hydrocarbons have pleasant smells. It is now known that just as many of them smell unpleasant, and in any case many of the aromatic vapours are toxic, so it is unwise to smell them. Benzene itself is particularly toxic. Long-term exposure to benzene may cause certain kinds of cancer. The name aromatic has been retained to indicate certain chemical characteristics rather than odorous properties.

29.2 The structure of benzene

Since 1834 the molecular formula of benzene has been known to be C_6H_6. The exact structural formula, however, posed a problem for many years. With such a high C:H ratio benzene must clearly be highly unsaturated. A possible structure might be hexatetraene, e.g. $CH_2{=}C{=}CH{-}CH{=}C{=}CH_2$. Such a structure would be expected to have two isomeric monosubstituted chlorocompounds, C_6H_5Cl:

$$ClCH{=}C{=}CH{-}CH{=}C{=}CH_2 \text{ and } CH_2{=}C{=}CCl{-}CH{=}C{=}CH_2.$$

Only one form of chlorobenzene has ever been isolated. The structure of benzene must therefore be one in which all six hydrogen atoms occupy equivalent positions.

The problem was solved by **Kekulé** in 1865. He proposed a ring structure in which alternate carbon atoms were joined by double bonds:

Here is Kekulé's description of how this structure occurred to him.

494

An article lampooning Kekulé's structure by proposing a ring made up of six monkeys. The monkeys join hands to form single bonds and link tails to make double bonds.

'I turned my chair to the fire and dozed. Again the atoms were gambolling before my eyes. This time the smaller groups kept modestly in the background. My mental eye, rendered more acute by repeated visions of this kind, could now distinguish larger structures, of manifold conformation; long rows, sometimes more closely fitted together; all twining and twisting in snakelike motion. But look. What was that? One of the snakes had seized hold of its own tail, and the form whirled mockingly before my eyes. As if by a flash of lightning I awoke.'

This structure, called the Kekulé or cyclohexatriene structure, explains many of the properties of benzene and was accepted for a long time. But even this structure leaves some problems, particularly concerning bond length and thermochemistry.'

Bond lengths

X-ray diffraction studies show that benzene is planar: this is to be expected from the Kekulé structure. X-ray diffraction also shows that all the C—C bonds in benzene are the same length:

carbon–carbon bond length in all bonds in benzene, 0.139 nm
carbon–carbon single bond length in cyclohexane, 0.154 nm
carbon–carbon double bond length in cyclohexene, 0.133 nm

The Kekulé model would suggest unequal carbon–carbon lengths, alternating between double and single bond values. In fact we find a constant bond length, somewhere between the value for a single and a double bond.

Thermochemistry of benzene

It is revealing to work out a theoretical value for the heat of formation of benzene on the basis of the Kekulé model, and to compare it with the experimental value obtained from the heat of combustion.

The heat of formation of gaseous benzene is the enthalpy change when a mole of gaseous benzene is formed from its elements:

$$6C(s) + 3H_2(g) \rightarrow C_6H_6(g)$$

Relevant data are:

heat of atomisation of C(s)   715 kJ (mole of C atoms)$^{-1}$
heat of atomisation of H_2(g)   218 kJ (mole of H atoms)$^{-1}$
bond energy of C=C (average)   610 kJ mole^{-1}
bond energy of C—C (average)   346 kJ mole^{-1}
bond energy of C—H (average)   413 kJ mole^{-1}

■ Work out the heat of formation of benzene by the following stages.

1 Calculate the energy needed to produce
 (a) six moles of gaseous carbon atoms from C(s)
 (b) six moles of gaseous hydrogen atoms from H_2(g)

2 Calculate the energy released when
 (a) three moles of C—C bonds are formed from gaseous atoms
 (b) three moles of C=C bonds are formed from gaseous atoms
 (c) six moles of C—H bonds are formed from gaseous atoms.

3 Use your answers to 1 and 2 to calculate the total energy change when a mole of gaseous benzene is formed from its elements.

4 Compare your answer with the experimental value of +82 kJ mole^{-1}.

5 Do your results suggest that 'real' benzene is more or less stable than the Kekulé structure?

The theoretical heat of formation of gaseous benzene based on the Kekulé structure is +252 kJ mole^{-1}, about 170 kJ mole^{-1} greater (i.e. more endothermic) than the experimental value of +82 kJ mole^{-1}. This implies that the actual structure of benzene is considerably more stable than the Kekulé structure. The result agrees reasonably closely with the stablisation energy of benzene obtained using heats of hydrogenation in section 12.11.

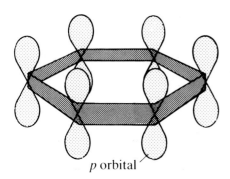

(a) σ–bonded skeleton
(C and H atoms
omitted for clarity)

(b) σ–skeleton with π–bonds

Figure 29.1
Bonding in benzene. Note that the plane of the molecule is perpendicular to the paper.

Electron delocalisation in benzene

The extra stability of benzene and the fact that its C—C bonds are all of equal length can be explained using the following model, which is currently accepted by most chemists.

The carbon atoms in the ring are bonded to one another and to their hydrogen atoms by σ-bonds. This leaves one unused p orbital on each carbon, each containing a single electron. These p orbitals are perpendicular to the plane of the ring, with one lobe above and one below this plane (figure 29.1(a)). Each p orbital overlaps sideways with the two neighbouring orbitals to form a single π-bond that extends as a ring of charge above and below the plane of the molecule (figure 29.1(b)).

The electrons in the π-bond cannot be said to 'belong to' any particular carbon atom. Each electron is free to move throughout the entire π system, so the electrons are said to be **delocalised**. It is this delocalisation that gives benzene its extra stability: any system in which electron delocalisation can occur is stabilised. The reason for this is not hard to see. Electrons tend to repel one another, so a system in which they are as far apart from one another as possible will involve minimum repulsion and will therefore be stabilised.

To conform with this model the structural formula of benzene is nowadays often written as ⬡(○) rather than ⬡ . As we shall see in section 29.5, the delocalisation of π-electrons has a profound effect on the chemical properties of benzene.

The word **aromatic** is used to describe any system that is stabilised by a ring of delocalised π-electrons. Non-aromatic compounds (such as alkanes and alkenes) are called **aliphatic**.

The structure described above means that the benzene molecule is planar, symmetrical and non-polar. Its lack of polarity results in benzene being a liquid at room temperature and makes it immiscible with water. Its boiling point is 80°C and its melting point 6°C. The surprisingly high melting point is due to the ease with which highly symmetrical benzene rings can pack into a crystal lattice. Compare it with that of the structurally similar but less symmetrical compound methylbenzene, $C_6H_5CH_3$, which melts at −95°C.

29.3 Nomenclature of aromatic compounds

Benzene is not the only aromatic hydrocarbon. Many others exist, either substituted forms of benzene or compounds containing different ring systems. Some other aromatic hydrocarbons are discussed in section 29.9. Aromatic hydrocarbons are sometimes known as **arenes**.

The name 'benzene' comes from *gum benzoin*, a natural product containing benzene derivatives. The name *phene*, derived from the Greek *pheno* 'I bear light', was at one time suggested as an alternative to 'benzene' (benzene was originally isolated from illuminating gas by Michael Faraday). It was not adopted, but it survives in the word **phenyl** used for the C_6H_5 group. The phenyl group is an example of an **aryl** group.

The hydrogen atoms in a benzene ring can be substituted by other atoms and groups, as we shall see in section 29.5. Some examples are given in table 29.1.

Where more than one hydrogen atom is substituted, numbers are used to indicate which of the six possible ring positions are concerned. Thus:

methyl-2-chlorobenzene

methyl-3-chlorobenzene

methyl-4-chlorobenzene

Table 29.1
Some derivatives of benzene

Substituent group	Systematic name	Other name
methyl, —CH_3	methylbenzene	toluene
chloro, —Cl	chlorobenzene	—
nitro, —NO_2	nitrobenzene	—
hydroxy, —OH	phenol	—
amino, —NH_2	phenylamine	aniline
carboxylic acid, —COOH	benzenecarboxylic acid	benzoic acid

The methyl group is regarded as occupying the 1 position and the ring is numbered clockwise as shown in the first formula. Two more examples are

3-nitrophenylamine   1,2-dimethylbenzene

Note that other numbers could be used (e.g. 3,4-dimethylbenzene instead of 1,2-dimethylbenzene), but the numbers actually employed are the lowest ones possible.

Use these rules and table 29.1 to write the structural formulae of the following compounds:

■ ethylbenzene
■ 2-methylphenol
■ 1,3-dinitrobenzene.

Name the following:

You may come across the terms *ortho, meta* and *para* in the names of disubstituted benzene derivatives, particularly in older books. Under this system, 1,2 derivatives are given the prefix *ortho-*, 1,3 derivatives are *meta-* and 1,4 derivatives *para-*. Thus the three methylchlorobenzenes shown on the previous page become *ortho*-methylchlorobenzene, *meta*-methylchlorobenzene and *para*-methylchlorobenzene, respectively.

MANSFIELD'S BENZENE STILL 1848

Early equipment for distilling benzene from coal tar.

Figure 29.2
The destructive distillation of coal.

29.4 The importance of benzene

Detergents, polystyrene, nylon and insecticides can all be made from benzene, which is industrially the most important arene. Most benzene is normally manufactured from oil by **catalytic reforming**. In the presence of a catalyst, C_6–C_8 hydrocarbons from the gasoline fraction rearrange their molecules, producing a variety of aromatic hydrocarbons including benzene. In the past, benzene was produced as a by-product of the destructive distillation of coal. Figure 29.2 shows a small-scale version of this process which can be carried out in the laboratory.

Coal is heated in the absence of air and the gases produced are bubbled through water. Coal tar condenses and floats on the water, leaving a gaseous fuel called coal gas. The solid residue is coke. On an industrial scale this process was once the basis of the gas industry in Britain, and also produced coke for boilers and for steel-making. The tar obtained contains a wide range of aromatic compounds which can be separated by distillation. At one time all aromatic hydrocarbons were produced from coal tar, but with the development of catalytic reforming and the decline of coal, most aromatic hydrocarbons now come from oil.

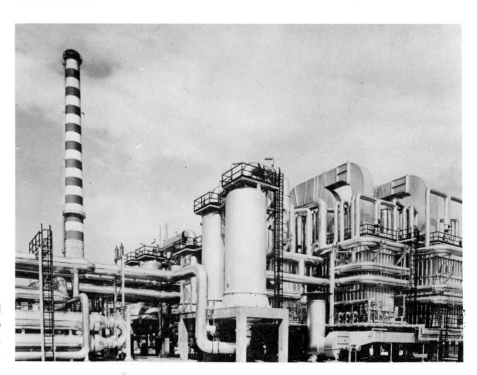

A catalytic reforming plant in an oil refinery, used to convert gasoline fractions to a mixture of aromatic hydrocarbons.

29.5 Chemical characteristics of benzene

Table 29.2 compares some reactions of cyclohexane, cyclohexene and benzene.

Table 29.2
Some reactions of cyclohexane, cyclohexene, and benzene

Reagent	cyclohexane	cyclohexene	benzene
bromine (in dark)	no reaction	bromine decolorised, no HBr evolved	no reaction with bromine alone. In presence of iron filings, bromine decolorised and HBr fumes evolved
acidified potassium manganate(VII) (potassium permanganate)	no reaction	manganate(VII) decolorised	no reaction
hydrogen over very finely divided nickel catalyst	no reaction	one mole absorbs one mole of hydrogen at room temperature	one mole absorbs three moles of hydrogen at 150°C
mixture of concentrated nitric and concentrated sulphuric acids	no reaction	oxidised	substitution reaction: yellow oil formed

A coal tar distillery, used to produce aromatic hydrocarbons. This rather messy method of production has been largely replaced by catalytic re-forming of oil fractions.

Look at the table and answer these questions.

■ Does the evidence suggest that (a) cyclohexene, (b) benzene undergo addition with bromine in the dark?
■ Does benzene undergo (i) addition, (ii) substitution with bromine in the presence of iron filings?
■ Does benzene undergo catalytic hydrogenation as readily as cyclohexene?

As table 29.2 shows, benzene undergoes addition reactions far less readily than we might expect for so unsaturated a compound. In fact, substitution reactions are more characteristic of benzene than addition reactions. You can see this by looking at the reaction of benzene with bromine in the presence of iron filings. HBr is evolved, which implies a substitution reaction.

If we remember the delocalisation of π-electrons in benzene, it is quite easy to see why addition reactions are difficult. For example, if a molecule of benzene underwent addition with a molecule of bromine, the ring of delocalised π-electrons would be broken:

This would require considerably more energy than is needed to break the one double bond in cyclohexene. In its reactions, therefore, benzene has a tendency to maintain its π-electron ring intact and to undergo substitution rather than addition (see study question 10 at the end of this chapter).

29.6 Mechanism of substitution reactions of benzene

The substitution reactions of benzene involve **electrophilic substitution**. The high density of negative charge in the delocalised electron system of the benzene ring tends to attract electrophiles. Consider as an example the nitration of benzene. Benzene reacts with a mixture of concentrated nitric acid and concentrated sulphuric acid (called a **nitrating mixture**) at 50°C. The product is nitrobenzene:

nitrobenzene
(a yellow oil)

This is a substitution reaction. Hydrogen has been substituted by a nitro group, NO_2.

The reaction of benzene with concentrated nitric acid alone is slow, whilst pure sulphuric acid at 50°C has practically no effect on benzene. This suggests that the sulphuric acid must somehow react with the nitric acid, producing a species that then reacts with benzene. There is good evidence that this species is NO_2^+, the **nitryl cation** formed by the removal of OH^- from HNO_3 by sulphuric acid:

$$HNO_3 + 2H_2SO_4 \rightarrow NO_2^+ + 2HSO_4^- + H_3O^+$$
nitryl cation

Here nitric acid is acting as a base in the presence of the stronger sulphuric acid. The NO_2^+ ion is a strong electrophile and tends to attack the negative π-electron system in benzene. First a loose association is formed:

The NO_2^+ then attacks one of the carbon atoms of the ring, forming a bond to it and disrupting the delocalised π system:

The formation of this intermediate requires the input of considerable energy to break the delocalised π system, so the reaction has a fairly high activation energy.

The intermediate cation then breaks down, either reforming benzene or producing nitrobenzene. The delocalised π system re-forms, releasing energy:

Notice that the first two stages of this mechanism, up to the formation of the intermediate cation, are similar to the stages in addition to an alkene double bond. The difference is that the aromatic cation loses its charge by loss of H^+, thus regaining aromatic character. With alkenes there is no delocalisation energy to be lost or gained, so the intermediate cation combines with an anion to form an addition product.

Polyester (section 33.5)

'Direct Red 39' – a red dye (Section 34.7)

$CH_3CH_2CH_2CH_2CH_2CH_2CH_2CH_2CH_2CH_2CH_2$—⬡—$SO_3^-Na^+$

A synthetic detergent (section 33.3)

Figure 29.3
Important aromatic compounds:
polyester to make clothes, dyes to
colour them, detergents to wash
them.

29.7 Important electrophilic substitution reactions of benzene

Nitration

The detailed mechanism of this reaction has already been considered in the last section. It is used industrially to manufacture nitrobenzene, from which phenylamine (aniline), $C_6H_5NH_2$, is produced by reduction. Phenylamine is used to manufacture dyes (see chapter 34).

Sulphonation

If benzene and concentrated sulphuric acid are refluxed together for several hours, benzenesulphonic acid is formed:

benzenesulphonic acid

This reaction is known as **sulphonation**. The electrophile that initially attacks the benzene ring is thought to be SO_3, which carries a large partial positive charge on the sulphur atom.

This theory is borne out by the fact that benzene is sulphonated in the cold by 'fuming sulphuric acid', which is a solution of SO_3 in concentrated sulphuric acid, but not by concentrated sulphuric acid alone unless heated.

Benzenesulphonic acid is important industrially as an intermediate in the manufacture of phenol (section 31.8). Phenol, C_6H_5OH, is a simple derivative of benzene, but it cannot be made by the direct reaction of OH^- with benzene. This is because the high electron density in the π system makes the ring unreactive towards negatively charged groups. When the sulphonate group is attached to a benzene ring, however, the electron-withdrawing effect of the group gives rise to a positive charge on the carbon atom to which it is attached.

This makes the carbon atom susceptible to attack by negatively charged groups such as OH^-, which can displace the SO_3^{2-} group. Consequently when benzene-sulphonic acid is heated with molten sodium hydroxide, phenol is formed:

benzenesulphonate ion   phenol

This is one of several methods used today to manufacture phenol. What other group might have a similar effect to the sulphonate group in making the carbon atom it is attached to susceptible to attack by negatively charged groups?

Polystyrene (section 28.8)
Made from phenylethene (styrene)(section 29.7)

Bakelite (section 31.7)
Made from phenol and methanal

Figure 29.4
Plastics manufactured from aromatic compounds.

Aspirin (section 33.6)

A sulphonamide (section 34.6)
Sulphonamides were among the earliest antibacterial drugs.

The insecticide DDT (section 32.6)

Figure 29.5
Some aromatic compounds of biological importance.

Halogenation

Benzene does not react with chlorine, bromine or iodine on their own in the dark. This is because the non-polar halogen molecule has no centre of positive charge to initiate electrophilic attack. However, in the presence of a catalyst such as iron filings, iron(III) bromide or aluminium chloride, benzene is substituted by chlorine or bromine:

chlorobenzene

The catalyst (called a **halogen carrier**) is thought to work by accepting a lone pair from one of the halogen atoms. This induces polarisation in the halogen molecule:

$$Cl\!-\!Cl\!: \rightarrow AlCl_3$$
$$\delta+ \quad \delta-$$

The positively charged end of the halogen molecule is now electrophilic and attacks the benzene ring.

- What would you expect to be the product of a substitution reaction between benzene and bromine(I) chloride, BrCl? (*Hint:* think about the way this molecule is polarised.)
- Iodine is too unreactive to substitute benzene even in the presence of a halogen carrier. Quite good yields of iodobenzene can however be obtained by reacting benzene with iodine(I) chloride, ICl. Explain why.

The electron-withdrawing tendency of the Cl group on chlorobenzene has a similar effect to that of the SO_3H group in benzenesulphonic acid. It makes the carbon atom to which it is attached susceptible to nucleophilic attack. For example, when boiled with hot, aqueous sodium hydroxide under pressure, it undergoes nucleophilic substitution to form phenol:

chlorobenzene   phenol

This reaction provides another method for the manufacture of phenol from benzene. Other reactions of chlorobenzene are described in chapter 30.

Interhalogen compounds such as BrCl will substitute benzene without a halogen carrier, since they are already polarised. BrCl reacts with benzene to form bromobenzene.

In the presence of sunlight, chlorine and bromine undergo *addition* reactions with benzene (see section 29.8).

Alkylation: Friedel–Crafts reaction

Aluminium chloride can be used as a catalyst to polarise halogen molecules and cause them to substitute a benzene ring. The same catalyst can be used to bring about the substitution of a benzene ring by halogenalkanes. For example, if benzene is warmed with chloromethane and aluminium chloride under anhydrous conditions, a substitution reaction occurs and methylbenzene is formed:

methylbenzene

As in the reaction with halogens, the aluminium chloride accepts an electron pair from the chlorine atom, polarising the chloromethane molecule:

$$\overset{\delta+}{CH_3} - \overset{\delta-}{Cl} : \rightarrow AlCl_3$$

Polarisation may reach the extreme where a carbocation is actually formed. The positively charged methyl group attacks the benzene ring and electrophilic substition occurs. This is an example of a **Friedel–Crafts** reaction. Such reactions occur between aromatic hydrocarbons and any combination of reagents that can give rise to a positively charged carbon atom. These include alkenes and alcohols as well as halogenoalkanes. For example, ethene and benzene undergo a Friedel–Crafts reaction in the presence of hydrogen chloride and aluminium chloride, to form ethylbenzene:

ethylbenzene

This reaction is used industrially to manufacture ethylbenzene, from which phenylethene (styrene) is made by catalytic dehydrogenation:

ethylbenzene   phenylethene

■ What important plastic is manufactured from phenylethene?
■ Ethylbenzene could be manufactured by a Friedel–Crafts reaction between chloroethane and benzene. Why is ethene used in preference to chloro-ethane?

Dodecylbenzene, important in the manufacture of detergents (see section 33.3) is made by a Friedel–Crafts reaction between benzene and dodecene:

dodecene   dodecylbenzene

A similar reaction between benzene and propene is used to make (1-methylethyl)benzene, which is used in the manufacture of phenol (see study question 7). The importance of electrophilic substition reactions in the synthesis of simple aromatic compounds is shown in chapter 35.

29.8 Other important reactions of benzene

The characteristic reactions of benzene and other aromatic hydrocarbons involve electrophilic substitution, because this type of reaction retains the delocalised π-electron system. Reactions involving the disruption of this system do occur, however.

Addition reactions

1 With hydrogen

Alkenes undergo addition with hydrogen in the presence of a nickel catalyst. Benzene also gives this reaction, but considerably higher temperatures are required than for aliphatic compounds. The higher temperature is needed because extra energy must be supplied to break up the delocalised π-electron system.

$$3H_2 + \text{[benzene ring]} \xrightarrow[150°C]{\text{Raney nickel}} \text{[cyclohexane ring]}$$

cyclohexane

The catalyst, Raney nickel, is a form of nickel with an extremely high surface area, and is very active,

If one mole of benzene and one mole of hydrogen are reacted in this way, one third of the benzene is converted into cyclohexane and the remainder is left unreacted. We might expect some cyclohexadiene, [structure], and cyclohexcne, [structure], to be formed, but this does not in fact happen. Can you suggest a reason why?

The catalytic hydrogenation of benzene is important industrially in the manufacture of cyclohexane, from which nylon is made (section 34.9).

2 With chlorine

Benzene can undergo addition as well as substitution reactions with chlorine. In ultraviolet light, chlorine adds to benzene to form 1,2,3,4,5,6-hexachlorocyclohexane. The need for light suggests a free-radical mechanism (section 27.5).

$$\text{[benzene ring]} + 3Cl_2 \xrightarrow{\text{light}} \text{[hexachlorocyclohexane structure]}$$

There are eight geometric isomers for 1,2,3,4,5,6-hexachlorocyclohexane due to the restricted rotation about the C—C bonds. Try drawing them or making models of them. One of these eight forms is a very effective insecticide, known commercially as Gammexane or BHC.

Burning

Benzene burns in air with a sooty, smoky flame. This sort of flame is characteristic of all hydrocarbons containing a high percentage of carbon.

29.9 Other arenes

Some examples of arenes other than benzene are shown in table 29.3. Naphthalene and anthracene are examples of **fused ring** arenes, in which two or more rings are joined.

Methylbenzene

The reforming reactions used to manufacture benzene from petroleum (see section 29.4) also produce considerable quantities of methylbenzene (commonly called toluene). This substance is used in the manufacture of plastics and explosives (see below), but much more is produced than can be used in this way. The majority of the methylbenzene produced is added to motor fuel to increase its octane rating: it does not have the pollution disadvantage of tetraethyl-lead(IV). Some methylbenzene is converted to benzene, which has more uses.

The remains of the Nypro plant at Flixborough after the 1974 explosion. As the result of a leak, cyclohexane, made from benzene and used in the manufacture of nylon, mixed with air and exploded.

Table 29.3
Some arenes

Systematic name	Other name	Molecular formula	Structural formula
methylbenzene	toluene	C_7H_8	
1,3-dimethylbenzene	*para*-xylene	C_8H_{10}	
naphthalene	—	$C_{10}H_8$	
anthracene	—	$C_{14}H_{10}$	

Reactions of methylbenzene

The properties of aromatic compounds are very different from those of aliphatic ones. Methylbenzene has an aromatic portion (the benzene ring) and an aliphatic portion (the CH_3 group). These two portions make different contributions to the properties of methylbenzene and have a modifying effect on one another.

1 The CH₃ group

The CH_3 group shows some reactions we would expect of an alkyl group: for example, it can be substituted by chlorine. This reaction occurs when chlorine is bubbled into boiling methylbenzene in sunlight:

methylbenzene   (chloromethyl)benzene

■ Write the structure of two other compounds that might form when the CH_3 group is substituted by chlorine.
■ Do the reaction conditions suggest that homolytic fission or heterolytic fission of the chlorine molecule is involved?
■ Under certain conditions chlorine will substitute the *ring* in methylbenzene instead of the CH_3 group. What will these conditions be?

The CH_3 group in methylbenzene does not behave as a typical alkyl group in all its reactions. The benzene ring with its regions of high electron density has a modifying effect on any group that is attached to it. We shall see later that a given functional group behaves differently depending on whether it is attached to an aliphatic or to an aromatic molecule.

One example of the way the ring modifies the properties of the CH_3 group in methylbenzene is the reaction with potassium manganate(VII). Alkanes are inert to oxidation, but the alkyl group in methylbenzene can be oxidised by alkaline manganate(VII) to give benzoic acid. The manganate(VII) is reduced to green manganate(VI).

manganate(VII)   benzoic acid   manganate(VI)

Note that the ring is not affected. This is another indication of its stability.

2 The aromatic ring

If chlorine is bubbled through methylbenzene in the absence of sunlight and in the presence of a halogen carrier such as $AlCl_3$, the ring is substituted instead of the CH_3 side-chain. This reaction proceeds by the electrophilic substitution mechanism described in section 29.6, rather than by the free-radical mechanism involved in side-chain chlorination. A mixture of two isomers is obtained:

$$\text{CH}_3 \quad + \text{Cl}_2 \xrightarrow{\text{AlCl}_3}$$

2-chloromethylbenzene
(58%)

4-chloromethylbenzene
(42%)

+ HCl

Virtually none of the other possible isomer, 3-chloromethylbenzene, is produced. Furthermore, the reaction proceeds at a considerably higher rate than the corresponding reaction of chlorine with benzene. Clearly the CH_3 group influences the aromatic ring, making it more susceptible to electrophilic substitution and dictating the positions in which it is substituted. This effect is discussed further in the next section.

An important substitution reaction of methylbenzene is nitration. When methylbenzene is heated with a mixture of concentrated nitric acid and concentrated sulphuric acid, it is substituted by one, two or three NO_2 groups, depending on the conditions. The main products are

2-nitromethylbenzene

4-nitromethylbenzene

2,4-dinitromethylbenzene

2,4,6-trinitromethylbenzene

Manufacturing high explosive shells in 1940. TNT was an important explosive in both world wars.

Blowing up a large building. High explosives such as TNT owe their explosive force to the production of large volumes of gas at high temperature.

Note the effect of the CH_3 group in determining that the main positions substituted are 2,4 and 6 rather than 3 or 5. 2,4,6-Trinitromethylbenzene (trinitrotoluene or TNT) is an important high explosive. It is fairly resistant to shock and so can be used in shells without risk of explosion under the shock of firing from a gun. When detonated it decomposes forming large volumes of CO, H_2O and N_2 at high temperature. The sudden expansion of these gases is responsible for the explosive force of TNT. TNT has the advantage that it is a solid which melts below 100°C, so it can be melted with steam and poured into its container.

29.10 Position of substitution in benzene derivatives

We have already seen that the methyl group in methylbenzene activates the ring towards electrophilic substitution. It also favours substitution in positions 2,4 and 6 rather than 3 or 5. In fact, *any substituent group* attached to a benzene ring affects the rate and the position at which further substitution occurs. Table 29.4 on the next page shows the main products of mononitration (i.e. substitution by one nitro group) of different benzene derivatives, and whether they are nitrated faster or slower than benzene.

Look at the table and answer these questions.

■ Which groups tend to direct substitution to the 2 or 4 position?
■ Which groups tend to direct substitution to the 3 position?
■ Is there any correlation between the position to which a group directs substitution and the rate at which it causes the ring to substitute?

Table 29.4
Mononitration products of benzene derivatives

Compound	Main products of monatrition		Rate of nitration relative to benzene
methylbenzene			Faster
phenol			Faster
nitrobenzene			Slower
phenylamine			Faster
benzoic acid			Slower

It can be seen from table 29.4 that benzene derivatives fall into two classes as far as further substitution is concerned:

A Those which substitute faster than benzene and in which the new substituent is directed to the 2 or 4 position, a mixture of the two isomers being obtained. Functional groups causing this behaviour include: CH_3 and all alkyl groups, OH, NH_2, OCH_3.

B Those which substitute slower than benzene and in which the new substituent is directed to the 3 position. Functional groups causing this behaviour include COOH, SO_3H, NO_2.

In practice, a mixture of all possible isomers is obtained, but these rules give the *main* products. Note that the rules apply whatever the nature of the new substituent, not just to nitration. Note too, that as far as monosubstitution is concerned, the 6 position is equivalent to the 2 position and the 5 to the 3.

■ Write structural formulae for the main products you would expect from the following substitution reactions. Assume monosubstitution occurs in each case.

(a) CH_3

$\xrightarrow[\text{AlCl}_3]{\text{CH}_3\text{Cl}}$

(b) NO_2

$\xrightarrow[\text{AlCl}_3]{\text{Cl}_2}$

(c) $CH_2CH_2CH_3$

$\xrightarrow[\text{c.H}_2\text{SO}_4]{\text{c.HNO}_3}$

(d)   COOH

$\xrightarrow{\text{c.H}_2\text{SO}_4}$

Summary

1 Aromatic hydrocarbons (arenes) are ring compounds stabilised by electron delocalisation.

2 Benzene, C_6H_6, has a symmetrical planar ring of six carbon atoms with a ring of delocalised π-electrons above and below the plane of the molecule.

3 Benzene is manufactured by catalytic reforming of petroleum fractions.

4 Disubstituted benzene derivatives are named by referring to numbered positions on the ring.

5 The C_6H_5 group is called phenyl.

6 Benzene undergoes addition considerably less readily than alkenes.

7 The characteristic reaction type of benzene is electrophilic substitution. Some important substitution reactions are shown on the following diagram:

8 A benzene ring considerably influences the properties of functional groups attached to it.

9 A functional group influences the position and rate of substitution of the benzene ring to which it is attached.

10 Methylbenzene shows some properties of an alkane and some of an arene, but the properties of the CH_3 group are substantially modified by the ring.

Study questions

1 Name the following benzene derivatives:

2   Write structural formulae for the following benzene derivatives:
   (a) 2,4,6-trinitrophenol
   (b) 1,4-dichlorobenzene
   (c) 4-nitrophenylamine
   (d) 2-methylbenzenesulphonic acid
   (e) 2-hydroxybenzoic acid
   (f) 2-chlorophenylamine.

3   Consider the catalytic hydrogenation of cyclohexene:

   (a) Suggest a suitable catalyst for this reaction.
   (b) Assuming benzene has the cyclohexatriene (Kekulé) structure, calculate the expected value of its heat of hydrogenation to cyclohexane, using the data above.
   (c) Compare your answer in part (b) with the experimental value:

   (d) Suggest why the two values differ.

4   Deuterium is an isotope of hydrogen containing a neutron in its nucleus in addition to the single proton present in normal hydrogen. The symbol for deuterium is ^2_1H or D, and its chemical properties are almost identical to those of normal hydrogen, ^1_1H.   If deuterium chloride, DCl, is dissolved in methylbenzene, no reaction occurs. If, however, anhydrous aluminium chloride is added to the solution, hydrogen atoms on the aromatic ring are rapidly substituted by deuterium atoms. If excess DCl is present, all five ring hydrogens are substituted, forming:

   (a) Write the structure of the complex formed between $AlCl_3$ and DCl.
   (b) What is the effect of the $AlCl_3$ on the polarisation of the DCl molecule?
   (c) Why is $AlCl_3$ effective in causing DCl to substitute the aromatic ring?
   (d) Why are none of the hydrogen atoms on the CH_3 side chain substituted, even in the presence of excess DCl?

5   How would you distinguish between the members of the following pairs of compounds, using simple chemical tests?

6 Predict the major products of the following reactions.

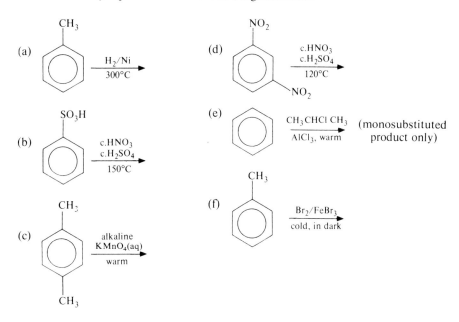

7 (1-Methylethyl)benzene, commonly called cumene, has the structure.

$$CH_3—CH—CH_3$$

It is an important intermediate in a process used to manufacture phenol and propanone (acetone).

(1-Methylethyl)benzene is manufactured by a Friedel–Crafts-type reaction between propene and benzene in the presence of an acid catalyst:

The reaction is believed to proceed via a carbocation intermediate.

(a) Write the structures of the two possible carbocations formed by the attack of H^+ on the double bond of propene.

(b) Which is the more stable of these two carbocations?

(c) Show how this carbocation can attack and substitute the benzene ring. Explain why (1-methylethyl)benzene is virtually the only product of this reaction, hardly any propylbenzene being produced.

Treatment of the product with air followed by dilute acid gives propanone and phenol:

(d) Why is this the most economic of the several methods available for manufacturing phenol?

8 (a) Write structural formulae for all the compounds of molecular formula C_8H_{10} containing one benzene ring.
   (b) For each of these compounds, write the formulae of all the possible mononitration products (not just the ones you would expect for the major product).
   (c) For one of the compounds, state which of the mononitration products you would expect to be produced in the majority.

9 Consider the following compounds,

(v) $CH_3CH_2CH{=}CHCH_2CH_3$

   (a) Which are aromatic hydrocarbons?
   (b) Which are cyclic compounds?
   (c) Which are unsaturated hydrocarbons?
   (d) Which have a planar ring in their molecule?
   (e) Which would decolorise bromine water in the dark?
   (f) Which would evolve fumes of HBr when treated with bromine and iron filings?
   (g) Which would react with alkaline potassium manganate(VII) solution?

10 Consider two possible reactions of chlorine with benzene.

A   [benzene]$_{(g)}$ + $2Cl_2(g)$ $\longrightarrow$ [dichlorobenzene with Cl and $Cl(g)$] + $2HCl(g)$

B   [benzene]$_{(g)}$ + $Cl_2(g)$ $\longrightarrow$ [cyclohexadiene with Cl and $Cl(g)$]

Some relevant bond energies (in kJ mole^{-1}) are: Cl—Cl 242; C—Cl (general value) 339; C—H (in benzene) 430; H—Cl 431.
   (a) Which of the two reactions is addition and which is substitution?
   (b) Calculate the energy change in reaction A by the following stages.
      (i) How much energy is needed to split two moles of chlorine molecules into atoms?
      (ii) How much energy is needed to break two moles of C—H bonds in benzene?
      (iii) How much energy is released when two moles of new C—Cl bonds form?
      (iv) How much energy is released when two moles of HCl molecules are formed?
      (v) What is the total energy change for reaction A?
   (c) Calculate the energy change in reaction B by the following stages, assuming that the energy needed to break one C—C bond in benzene, i.e. convert

      to   is 434 kJ mole^{-1}.

      (i) How much energy is needed to split one mole of chlorine molecules into atoms?
      (ii) How much energy is released when two moles of C—Cl bonds form?
      (iii) What is the total energy change in reaction B?
   (d) Which reaction is more likely to occur between chlorine and benzene, A or B? Give reasons for your answer.

30 Organic Halogen Compounds

Early surgery was primitive and performed without anaesthetics.

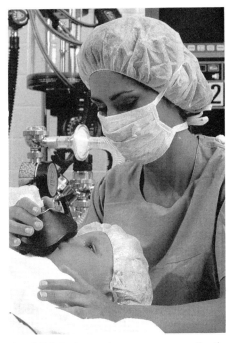

A child being given an anaesthetic prior to surgery.

30.1 Anaesthetics

Before the advent of anaesthetics, surgery was a savage and primitive affair. It was agony for the patient, and surgeons were therefore only prepared to operate if it was absolutely essential, for example the amputation of a damaged limb that would otherwise become gangrenous. Anaesthetics enabled surgery to develop from crude carpentry to its present precise and sophisticated form.

Three of the most important early anaesthetics were nitrous oxide (dinitrogen oxide, N_2O), ether (ethoxyethane, $CH_3CH_2OCH_2CH_3$) and chloroform (trichloromethane, $CHCl_3$). Nitrous oxide is non-toxic and non-flammable, but it only produces light anaesthesia. It is still used today in minor surgery such as tooth extraction. Ether is an effective anaesthetic but it is highly flammable and therefore dangerous. Chloroform produces deep anaesthesia and is non-flammable, but it is toxic and carries the risk of liver damage.

The ideal anaesthetic must be a gas or volatile liquid, so that it can be inhaled and absorbed via the lungs. It must be non-flammable. It must produce deep anaesthesia but must be non-toxic. In 1951, ICI began the search for the ideal anaesthetic. They decided to look at **halogenoalkanes**.

It was known that the substitution of chlorine atoms into an alkane chain gave it anaesthetic properties, but also made it toxic. Thus dichloromethane, CH_2Cl_2, is a fairly weak anaesthetic with little toxicity, trichloromethane (chloroform), $CHCl_3$, is stronger and more toxic, and tetrachloromethane is a very strong anaesthetic and also very toxic. The introduction of halogen atoms into an alkane skeleton also tends to make it non-flammable. Fluorine is useful in this respect as the C—F bond is very stable. Fluoroalkanes are inert, non-flammable and non-toxic. ICI therefore looked for a short-chain halogenalkane containing fluorine for inertness and chlorine for anaesthetic properties, with a suitable boiling point in the range 40°C to 60°C. They concentrated on two-carbon molecules, and after many trials produced the compound.

$$\begin{array}{ccc} F & & H \\ F-C & -C & -Cl \\ F & & Br \end{array}$$

The bromine atom was introduced to produce a substance with a sufficiently high boiling point to be conveniently stored. This substance, 2-bromo-2-chloro-1,1,1-trifluoroethane, was given the name Halothane. Since 1956 it has been in widespread use in hospitals.

Halothane illustrates several of the important properties of halogenoalkanes. It illustrates the increasing reactivity of the C—Hal bond as we go from F to I; the decreasing volatility of R—Hal as we move in the same direction; and the effect of halogen atoms in reducing the flammability of a hydrocarbon chain.

Although organic halogen compounds are uncommon in nature a study of their properties is important to chemists as they have many uses in industry and in the laboratory.

30.2 Nomenclature

Organic halogen compounds are named using the prefixes **fluoro-, chloro-, bromo-** and **iodo-**. Numbers are used if necessary to indicate the position of the halogen atom in the molecule. Thus,

CH_3CH_2Cl chloroethane,
$CH_3CHBrCH_3$ 2-bromopropane.

If the molecule contains more than one halogen atom of the same kind, the prefixes **di-, tri-**, etc., are used. Thus,

CH_2ClCH_2Cl 1,2-dichloroethane,
$CHCl_2CHClCH_3$ 1,1,2-trichloropropane.

Name these compounds.

- $CH_3CH_2CHICH_3$
- CH_3CHCl_2

Write the formulae for these compounds.

- 1,3,5-tribromobenzene
- 1,2-dibromo-3-chloropropane

30.3 The nature of the carbon–halogen bond

Unreactive halogenoalkanes: the C—F bond

The C—F bond is very strong: compare its bond energy of $485\,kJ\,mole^{-1}$ with 435 for C—H and 327 for C—Cl. The C—F bond is thus very unreactive. Consequently, fluorocarbons, compounds containing fluorine and carbon only, are extremely inert (see study question 6, chapter 26). The C—Cl bond is more reactive than C—F but nevertheless highly chlorinated compounds such as CCl_4 and $CHCl_3$ are fairly inert. In particular they are non-flammable.

An important group of compounds is the **chlorofluorocarbons**, often known by the trade name 'Freons'. Some examples are:

CCl_3F trichlorofluoromethane,
CCl_2F_2 dichlorodifluoromethane,
CCl_2F—$CClF_2$ 1,1,2-trichloro-1,2,2-trifluoroethane.

These compounds have some useful properties:

(a) They are very stable, and in particular are non-flammable.
(b) They have low toxicity.
(c) They are odourless.
(d) They are volatile, with a wide range of boiling points.

Some examples of their uses are:

(i) *Refrigerants.* Dichlorodifluoromethane, CCl_2F_2, has a boiling point of $-30°C$, which makes it a useful refrigerant fluid. In the refrigerator the fluid is liquefied by compression then vaporised by sudden expansion, which gives a cooling effect. CCl_2F_2 is particularly useful for this purpose because it is unreactive so it does not corrode the machinery, and it is non-toxic so there is no danger from possible leaks.

(ii) *Aerosol propellants.* Dichlorodifluoromethane is also suitable for use as an aerosol propellant, either alone or mixed with other chlorofluorocarbons. Under pressure in the aerosol can it is liquid, but when the valve is opened some of it vaporises, carrying with it the active components: insecticide, paint, hair lacquer or whatever.

The stability of CCl_2F_2 means that it does not interact with the active component, and its lack of toxicity is of course vital. Each year, hundreds of

Chlorofluorocarbons make excellent aerosol propellants, but they cause damage to the ozone layer.

millions of aerosol packs are sold in the United Kingdom. Between them these pump large quantities of chlorofluorocarbons into the atmosphere.

Once in the atmosphere, the inertness of these compounds is a disadvantage, for they tend to persist in the atmosphere for long periods. There is evidence that they interfere with the formation of the ozone layer that is vital in screening the Earth from excess ultraviolet radiation. Recently it has been discovered that there are large 'holes' in the ozone layer above the poles of the Earth, and scientists believe chlorofluorocarbons are to blame. More research needs to be done before the case against chlorofluorocarbons is clear, but the US govenment has banned the use of chlorofluorocarbons in 'non-essential' aerosols, and other countries are likely to follow suit.

(iii) *Solvents.* Chlorinated hydrocarbons tend to be very good solvents for non-polar materials. For example, dichloromethane will partially dissolve old paint and is therefore used as a paint-stripper. Fluorocarbons tend to be less effective than chlorocarbons as solvents. Chlorofluorocarbons have intermediate solvent properties and have many applications where a moderate solvent is needed, for example in cleaning and degreasing machinery and electronic circuits. Here a powerful solvent might damage plastic components in the machinery or in the circuit. Chlorofluorocarbons will dissolve grease but are not powerful enough solvents to affect components. Particularly useful for these purposes is 1,1,2-trichloro-1, 2,2-trifluoroethane, a liquid that boils at 48°C. It is also used for dry-cleaning delicate materials such as suede or fur.

Degreasing machinery using a halogenoalkane solvent.

Reactive halogenoalkanes: the C—Br bond

The C—F bond is not typical of the halogenoalkanes as a whole, which tend to be much more reactive than fluoroalkanes. To get some idea of the nature of the C—Hal bond, we will consider some bromine-containing compounds.

Table 30.1 compares the effect of adding silver nitrate solution to sodium bromide, 1-bromobutane and bromobenzene. In the case of the two organic liquids, vigorous shaking is necessary since they do not mix appreciably with the aqueous phase.

Table 30.1
Effect of aqueous silver nitrate on bromine-containing compounds at room temperature

Sodium bromide	1-Bromobutane	Bromobenzene
pale yellow precipitate appears immediately	no reaction at first: faint precipitate appears after several minutes	no reaction even after several hours

Look at table 30.1 and answer these questions.

■ What is the pale yellow precipitate produced in the reaction between silver nitrate and sodium bromide?
■ Write an ionic equation for this reaction.
■ Why does silver nitrate produce no immediate precipitate with 1-bromobutane, even though it contains bromine?
■ Suggest a reason why a precipitate appears after several minutes.
■ Compare the reactivity of bromobenzene with that of 1-bromobutane.

The C—Br bond in 1-bromobutane is covalent. 1-Bromobutane therefore contains no Br^- ions, so it does not produce a precipitate of silver bromide with silver nitrate. The slow appearance of a precipitate of silver bromide suggests that Br^- ions are slowly being produced. Why?

Bromine is more electronegative than carbon and the C—Br bond is thus polar:

The partial positive charge on the carbon atom tends to attract nucleophiles, such as NH_3, OH^- and H_2O, with their lone pairs of electrons. Water molecules from the aqueous silver nitrate can act as nucleophiles since their oxygen atoms carry two lone pairs and a partial negative charge:

The water attacks the partially positive carbon atom in 1-bromobutane and a substitution reaction takes place, releasing bromide ions:

This is an example of a **nucleophilic substitution** reaction. Such reactions are typical of halogenoalkanes and their mechanism is discussed further in the next section. The failure of bromobenzene to react under these conditions suggests that the C—Br bond is stronger in aromatic than in aliphatic compounds.

Physical properties of halogenoalkanes

The C—Hal bond is polar, but not polar enough to have an appreciable effect on the physical properties of organic halogen compounds. They are all immiscible with water. As table 30.2 shows, their volatility is determined more by the size and number of halogen atoms they contain than by the polarity of the bonds. Thus iodomethane is a liquid at room temperature while chloromethane is a gas, even though chloromethane has a more polar molecule.

Table 30.2
Boiling points of some halogen compounds

Compound	State at 298 K	Boiling point/K
CH_3F	g	195
CH_3Cl	g	249
CH_3Br	g	277
CH_3I	l	316
CH_2Cl_2	l	313
$CHCl_3$	l	335
CCl_4	l	350
C_6H_5Cl	l	405

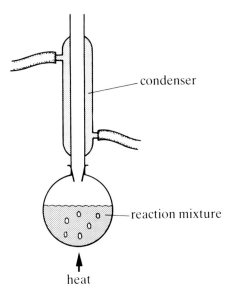

Figure 30.1
Reflux apparatus used for boiling an organic reaction mixture. The condenser prevents escape of volatile reagents.

30.4 Nucleophilic substitution

We have seen that water can act as a nucleophile towards halogenoalkanes. In general

$$RHal + H_2O \rightarrow ROH + HHal$$

This reaction is slow, as the slow appearance of the AgBr precipitate in the last section showed. If OH^- is used as a nucleophile instead of H_2O the reaction is quicker. Thus bromoethane forms ethanol quite rapidly when heated under reflux with aqueous sodium hydroxide (figure 30.1)):

$$CH_3CH_2Br + OH^- \rightarrow CH_3CH_2OH + Br^-.$$
bromoethane   ethanol

The mechanism of reactions between halogenoalkanes and hydroxide ions has been extensively studied by chemists. As suggested in the last section, it involves nucleophilic attack by the OH^- ion, e.g.

The actual mechanism is more complex than the simplified version shown here. In some cases, the halide ion may leave before the OH^- actually attacks, so that a carbocation (section 27.5) is formed as an intermediate. In other cases, the C—Hal bond may break at the same time as the C—OH bond forms. Further details of the mechanism of this substitution reaction are given in Practical 35 of the *Laboratory Manual and Study Guide.*.

The substitution reaction outlined above between CH_3CH_2Br and OH^- is general for all halogenoalkanes. It can be used as a method for preparing alcohols:

$$RHal + OH^- \rightarrow ROH + Hal^-$$

For a given alkyl group R, the iodo compound reacts most readily, the bromo compound less so and the chloro compound reacts least readily. This is because the C—Hal bond becomes progressively stronger passing from I to Cl.

As we saw in the last section, if the halogen is attached to a benzene ring, substitution is more difficult. It is thought that lone pairs of electrons on the halogen atom interact with the delocalised π-electron system of the benzene ring, strengthening the C—Hal bond. In addition, the high electron density on the aromatic ring tends to repel the approaching negatively charged OH^- ion. Consequently, chlorobenzene reacts with aqueous sodium hydroxide at a reasonable rate only at 300°C under a pressure of 200 atmospheres. (This reaction provides one way of manufacturing phenol—see section 29.7.) Compare this with the behaviour of chlorobutane, which is readily converted to butanol by refluxing with aqueous sodium hydroxide under ordinary conditons.

■ Predict the organic product of the reactions when aqueous sodium hydroxide is boiled with
  (a) 2-chloropropane   (c)
  (b) 1,2-dibromoethane

30.5 Important substitution reactions of halogenoalkanes

Organic halogen compounds have important uses. Several of these have already been mentioned, and their importance in insecticides is considered in section 17.12. Other uses are discussed in chapter 27.

Organic halogen compounds occur hardly at all in nature so they must be synthesised. They are usually manufactured from alkenes or alkanes. Chloroethane, for example, used in the manufacture of tetraethyllead(IV) (section 27.6), is made by the addition reaction between ethene and hydrogen chloride. Tetrachloromethane is made by the substitution reaction between chlorine and methane. In the laboratory, halogenoalkanes are usually prepared from alcohols (see section 31.6 and Practical 34 in the *Laboratory Manual and Study Guide*).

As well as being useful in their own right, halogenoalkanes are important intermediates in synthesis. They can be used to introduce a reactive site into a hydrocarbon molecule. The reactive halogen can then be substituted by another group which could not be introduced directly. Examples of groups that can be introduced in this way are discussed later in this section.

This kind of synthesis is important in small-scale preparations such as those carried out in the laboratory or in the manufacture of pharmaceuticals. Many drugs have complicated organic molecules and their synthesis involves building up a complex molecule from a simple starting compound. Such synthesis often involves many steps. For example, the synthesis of phenobarbitone (a sleep-inducing drug) from methylbenzene involves eight steps (see figure 30.2). Note in figure 30.2 how a chlorine atom is used to provide an active site in the relatively unreactive methylbenzene molecule so that other groups can then be introduced.

Right. Small-scale synthesis of a pharmaceutical. Laboratory methods are used and the reagents are often expensive.

Above. Large-scale synthesis of ethene. On the large scale the emphasis is on expensive plant and cheap reagents.

The synthetic methods described in this chapter are used for small-scale operations, but they are relatively expensive and therefore not used for large-scale, high-tonnage industrial operations. On this scale high-pressure, high-temperature catalytic processes, very different from those used in the laboratory, tend to be used. These processes involve expensive plant and equipment, but cheap reagents. In the long run they are cheaper than small-scale laboratory processes. The subject of organic synthesis is considered in more detail in chapter 35.

Halogenoalkanes undergo substitution reactions with a wide range of nucleophiles. Species that can act as nucleophiles include not only those carrying a full negative charge (OH^-, CN^-, CH_3COO^-, $CH_3CH_2O^-$) but also neutral molecules carrying an unshared pair of electrons (H_2O, NH_3).

■ Predict the structure of the compound formed by the reaction of bromoethane with each of the nucleophiles mentioned above. The answers are given opposite, together with the conditions needed for the reaction. Do not look at these until you have predicted the structures yourself.

Figure 30.2
Synthesis of phenobarbitone from methylbenzene. Phenobarbitone is a hypnotic, used in sleeping pills.

(a) *Reaction with hydroxide ion.* This has already been dealt with (p. 517).

(b) *Reaction with cyanide ion.* When bromoethane is heated under reflux with a solution of potassium cyanide in ethanol, propanenitrile is formed:

$$CH_3CH_2Br + CN^- \rightarrow CH_3CH_2CN + Br^-$$
$$\text{propanenitrile}$$

This reaction is useful in synthesis as a means of increasing the length of a carbon chain (section 35.3).

(c) *Reaction with ethanoate ion.* If bromoethane is warmed with dry silver ethanoate, ethyl ethanoate, an ester, is formed:

$$CH_3COO^-Ag^+ + CH_3CH_2Br \rightarrow CH_3COOCH_2CH_3 + AgBr$$
$$\text{silver ethanoate} \qquad\qquad \text{ethyl ethanoate}$$

The reaction is encouraged by using the silver salt which removes bromide ion as insoluble silver bromide.

(d) *Reaction with ethoxide ion.* Bromoethane reacts with an ethanolic solution of sodium ethoxide, $Na^+{}^-OCH_2CH_3$ (see section 31.5) to form an ether, ethoxyethane:

$$CH_3CH_2Br + CH_3CH_2O^- \rightarrow CH_3CH_2OCH_2CH_3 + Br^-$$
$$\text{ethoxide ion} \qquad\qquad \text{ethoxyethane}$$

The ethoxide ion can also act as a base rather than as a nucleophile under these conditions, in which case a different product, ethene, is formed (see next section).

Just as both water and hydroxide ion can act as nucleophiles, so both ethanol and ethoxide ion give nucleophile substitution reactions with halogenoalkanes. Ethanol reacts considerably slower than ethoxide ion as we would expect, but the product is the same:

$$CH_3CH_2Br + CH_3CH_2OH \rightarrow CH_3CH_2OCH_2CH_3 + HBr$$

(e) *Reaction with water*. This has already been dealt with (p. 516). In the case of bromoethane the product is ethanol, though the reaction is slow even when heated:

$$CH_3CH_2Br + H_2O \rightarrow CH_3CH_2OH + HBr$$
$$\text{ethanol}$$

(f) *Reaction with ammonia*. Ammonia is a nucleophile because of the lone pair of electrons on its nitrogen atom. When bromoethane is heated with a concentrated aqueous solution of ammonia in a sealed tube, ethylamine is formed:

$$CH_3CH_2Br + NH_3 \rightarrow CH_3CH_2NH_2 + HBr$$
$$\text{ethylamine}$$

The nitrogen atom in ethylamine still possesses a lone pair, so it can still act as a nucleophile. With excess bromoethane, further substitution therefore occurs:

$$CH_3CH_2NH_2 + CH_3CH_2\ Br \rightarrow (CH_3CH_2)_2NH + HBr$$
$$\text{ethylamine} \qquad\qquad\qquad \text{diethylamine}$$

and so on. In practice a mixture of products is obtained, so this reaction is of little use as a preparative method.

The reactions mentioned above have all been illustrated by reference to bromoethane, but they are applicable to halogenoalkanes generally. Many different compounds can be synthesised from halogenoalkanes and this is considered further in chapter 35.

Describe the reagents and conditions you would use to prepare:

- propane-1,2-diol from 1,2-dibromopropane,
- 2-methylpropanenitrile from 2-iodopropane.
- methoxyethane from chloroethane.

30.6 Elimination reactions

In the reactions we have considered so far, OH^- has acted as a nucleophile in its reactions with halogenoalkanes. For example, when 2-bromopropane is refluxed with aqueous sodium hydroxide, the hydroxide ion acts as a nucleophile and substitution occurs:

$$CH_3CHBrCH_3 + OH^- \rightarrow CH_3CHOHCH_3 + Br^-$$
$$\text{2-bromopropane} \qquad\qquad \text{propan-2-ol}$$

Under certain conditions, OH^- can act as a *base* instead of a nucleophile. When OH^- behaves in this way, it removes H^+ from a halogenoalkane. The C—Br bond breaks at the same time as OH^- removes H^+ from the neighbouring carbon atom, and an alkene is formed. For example, when 2-bromopropane is refluxed with a solution of sodium hydroxide in ethanol instead of in water, propene is formed:

i.e. $CH_3CH_2\ Br + OH^- \longrightarrow CH_2{=}CH_2 + H_2O + Br^-$

The overall effect of this reaction is to eliminate HBr from the molecule of 2-bromopropane, so this is an elimination reaction. Notice that the reagent, sodium hydroxide, is the same as for the substitition reaction outlined at the beginning of this section. In the substitution reaction, OH^- acted as a nucleophile; in the elimination reaction it acts as a base. By altering the conditions, we can alter the manner in which it acts: in aqueous solution it behaves as a nucleophile, in ethanolic solution it behaves as a base. Under each set of conditions, both substitution and elimination will occur, but by controlling the conditions we can ensure that one particular reaction occurs to a greater extent than the other.

Try to predict the effect of an elimination reaction in a halogenoalkane that contains *two* halogen atoms. For example, what would be the product if 1,2-dibromoethane were heated under reflux with a solution of sodium hydroxide in ethanol?

30.7 Acyl halides

So far we have considered molecules in which a halogen atom is attached directly to an alkyl or aryl group. An important group of halogenated compounds is the **acyl halides**, in which a halogen atom is attached to a $\diagdown C=O$ group.

An example is ethanoyl chloride, $CH_3—C—Cl$, usually written CH_3COCl.
$$\underset{O}{\overset{\|}{}}$$

Acyl halides have the general formula **RCOHal**, where R is an alkyl or aryl group. They are named using the suffix **-oyl halide** after a stem indicating the number of carbon atoms in the molecule. Thus CH_3CH_2COBr is called propanoyl bromide and $CH_3CH_2CH_2COCl$ is called butanoyl chloride.

Like halogenoalkanes, acyl chlorides undergo nucleophilic substitution reactions in which the halogen is substituted by a nucleophile. Acyl halides undergo nucleophilic substitution much more readily than the halogenoalkanes. For example, when water is added to ethanoyl chloride (which is a liquid), a violent reaction occurs. The liquid boils and clouds of hydrogen chloride fumes are evolved:

$$CH_3COCl + H_2O \rightarrow CH_3COOH + HCl$$
$$\text{ethanoic acid}$$

The reaction of acyl halides with nucleophiles involves a different mechanism to that of halogenoalkanes. The first stage of the reaction involves *addition* of the nucleophile across the $C=O$ double bond. (Nucleophilic addition is described further in section 32.3.) The double bond is reformed at a later stage of the reaction. This different reaction mechanism enables acyl halides to react with nucleophiles much more rapidly than halogenoalkanes do. Acyl chlorides react with all the nucleophiles that will substitute halogenoalkanes.

■ Predict the structure of the product of the reaction between ethanoyl chloride, CH_3COCl, and each of the nucleophiles NH_3 and CH_3CH_2OH. The answers appear below.

Like halogenoalkanes, acyl halides are important in synthesis. Just as RHal is used to attach an alkyl group R to a nucleophile, RCOHal is used to attach an acyl group RCO. Because of their greater reactivity, acyl halides will react rapidly with those nucleophiles, such as water and alcohols, that react only slowly with halogenoalkanes.

Some reactions of ethanoyl chloride are as follows:

(a) *With ammonia.* Ethanoyl chloride reacts violently with an aqueous solution of ammonia at room temperature. The product is ethanamide, an amide. Amides have the general formula $RCONH_2$ and should not be confused with amines, RNH_2.

$$CH_3COCl + NH_3 \rightarrow CH_3CONH_2 + HCl$$
$$\text{ethanamide}$$

The $C=O$ group in ethanamide has a strong electron-withdrawing effect. It tends to withdraw electrons from the NH_2 group, making the nitrogen atom much less nucleophilic than in NH_3 or in amines such as $CH_3CH_2NH_2$.

Ethanamide is therefore not a strong nucleophile, despite carrying a lone pair on its nitrogen atom. As a result, further substitution does not occur and ethanamide is the only product. Compare this with the reaction of chloroethane with ammonia, which gives several products.

(b) *With alcohols*. Ethanoyl chloride reacts vigorously with alcohols at room temperature to form esters. For example, with ethanol:

$$CH_3CH_2OH + CH_3COCl \rightarrow CH_3COOCH_2CH_3 + Hcl$$
ethyl ethanoate

As expected, the reaction occurs much more readily than the corresponding reaction of chloroethane with ethanol.

The reactions of ethanoyl chloride described above are of course quite general and apply to all acyl chlorides.

Summary

1 Halogenoalkanes are named using a prefix (*chloro-*, etc.) to indicate the nature of the halogen atom and a number to indicate its position on the alkane chain.

2 The C—F bond is very strong. Fluorocarbons are very stable.

3 The C—H bond is polar. The positively charged carbon atom tends to be attacked by nucleophiles. Nucleophilic substitution is the characteristic reaction of halogenoalkanes. Some examples are shown in this diagram.

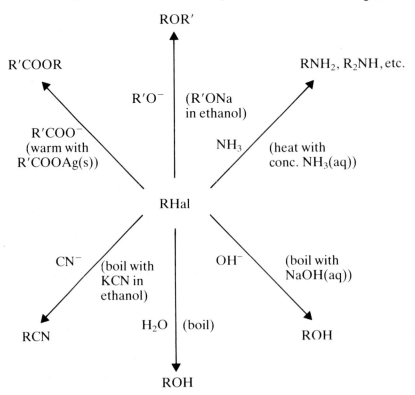

4 The Ar—Hal (Ar = aryl group) bond is considerably less reactive than the R—Hal bond.

5 In the presence of a strong base, halogenoalkanes can undergo elimination reactions, forming alkenes.

6 Acyl halides contain the COCl group. They undergo nucelophilic substitution considerably more readily than halogenoalkanes.

Study questions

1 The table below gives formulae of some halogenoalkanes and their boiling points.

Formula	Boiling point/:C
CCl_3F	24
$\begin{matrix} F_2C-CF_2 \\ \mid \qquad \mid \\ F_2C-CF_2 \end{matrix}$	-6
$CCl_2F-CClF_2$	48
$CBrF_3$	149

(a) Name each compound.
(b) Which would be suitable for use as an aerosol propellant?
(c) Which would be suitable as a degreasing solvent?
(d) Which would be a suitable fire extinguisher?

2 Consider the following compounds.

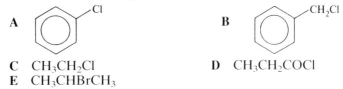

A (benzene ring with Cl)   **B** (benzene ring with CH_2Cl)

C CH_3CH_2Cl **D** CH_3CH_2COCl
E $CH_3CHBrCH_3$

(a) Which are halogenoalkanes?
(b) Which is an acyl halide?
(c) Which would react most readily with cold water?
(d) Which would react least readily with aqueous sodium hydroxide?
(e) Which would form an amide when reacted with aqueous ammonia?
(f) Which would have the lowest boiling point?

3 This question concerns the hydrolysis of three different halogenoalkanes, 1-chlorobutane, 1-bromobutane and 1-iodobutane.
Four drops of each halogenoalkane are added separately to three separate tubes standing in a water-bath at 60°C. Each tube contains $1\,cm^3$ of $0.1\,M$ silver nitrate solution. The results are as follows:
1-Chlorobutane—slight cloudiness after three minutes. Still only slightly cloudy after 15 minutes.
1-Bromobutane—slightly cloudy after one minute, opaque after three minutes, coagulation and precipitation after six minutes.
1-Iodobutane—immediately opaque, yellow precipitate within first minute.
(a) What is the precipitate formed in each case?
(b) Why does the precipitate form?
(c) In each case a substitution reaction is occurring between the halogenoalkane and a nucleophile. What is the nucleophile involved?
(d) Which halogenoalkane undergoes substitution most readily and which least readily?
(e) Explain your answer to (d), given the following bond energies in kJ mole^{-1}: C—Cl (in chloroethane) 339; C—Br (in bromoethane) 284; C—I (in iodoethane) 218.

4 Predict the products of reactions between the following pairs of substances.

(a) $CH_3CH_2CH_2Br$ and NH_3

(b) $CH_2ICH_2CH_2CH_2I$ and $CH_3CH_2CH_2COO^-Ag^+$

(c) (benzene ring with CH_2Br) and KCN (ethanolic)

(d)

and NaOH(aq)

(e)

and NaOH(ethanolic)

(f)

$$CH_3-\underset{\underset{Br}{|}}{\overset{\overset{Br}{|}}{C}}-CH_3 \quad \text{and} \quad CH_3O^-Na^+$$

(g)

and CH_3CH_2OH

5 Use your knowledge of nucleophilic substitution reactions to predict the structural formulae of the products of reactions between bromoethane and the following.
  (a) sodium hydrogensulphide (Na^+SH^-)
  (b) sodium ethynide ($HN{\equiv}C^-Na^+$)
  (c) potassium nitrite ($K^+NO_2^-$)
  (d) potassium chloride
  (e) lithium tetrahydridoaluminate, $LiAlH_4$ (nucleophile = H^-)
  (f) sodamide ($Na^+NH_2^-$).

6 An organic liquid **A** is immiscible with water. When 1.00 g of **A** was refluxed for several hours with excess aqueous sodium hydroxide it slowly dissolved in the aqueous layer. The resultant solution was neutralised with nitric acid and excess silver nitrate solution was added, when a precipitate of silver bromide was formed. The precipitate was filtered off and dried and its mass was found to be 2.00 g. The filtrate was distilled and a high-boiling liquid **B** was obtained. **B** contained 39% C, 9.5% H and 51.5% O by mass. Suggest structural formulae for **A** and **B** and explain the above reactions.

$$(C = 12, H = 1, O = 16, Br = 80, Ag = 108)$$

7 Two compounds **X** and **Y** have the same molecular formula, C_3H_5OCl. **X** reacts vigorously with cold water to give a strongly acidic solution, but **Y** has no visible reaction with cold water. Suggest structural formulae for **X** and **Y**.

8 The reaction of 3-chloro-3-ethylpentane with aqueous sodium carbonate at room temperature yields two products, one an alcohol and one an alkene.
  (a) Write the structural formula of 3-chloro-3-ethylpentane.
  (b) Consider the reaction that produces the alcohol.
      (i)   What type of reaction is this?
      (ii) Write the structural formula of the alcohol.
      (iii) Write an equation for the reaction.
  (c) Consider the reaction that produces the alkene.
      (i)   What type of reaction is this?
      (ii) Suggest a possible structural formula for the alkene.
      (iii) Write an equation for the reaction.
  (d) What changes in reaction conditions might favour the formation of the alkene?

9 Give reagents, reaction conditions and equations to show how you would convert
  (a) $CH_2{=}CH_2$ to CH_3CH_2CN (2 steps)
  (b) $CH_2{=}CH_2$ to $CH_3OCH_2CH_2OCH_3$ (2 steps)

(c)

(1 step)

COCl   to   CONH$_2$

(d)

(2 steps)

CH$_3$ to   CH$_2$OH

10 Identify the compounds **A** to **D** and the reagents **P** and **Q** in the reaction scheme below.

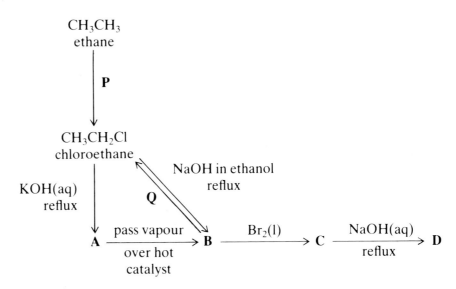

CH$_3$CH$_3$
ethane

P

CH$_3$CH$_2$Cl
chloroethane

KOH(aq)
reflux

Q   NaOH in ethanol
reflux

A   pass vapour
over hot
catalyst   **B**   Br$_2$(l)   **C**   NaOH(aq)
reflux   **D**

11 Suggest explanations for each of the following observations.
   (a) Iodobenzene is more reactive than chlorobenzene towards nucleophiles, but much less reactive than iodoethane.
   (b) Tetrachloromethane, CCl$_4$, is much less reactive towards nucleophiles than chloromethane, CH$_3$Cl.
   (c) Iodo- and bromo- compounds are more useful as intermediates in synthesis than chloro- compounds.
   (d) When chloroethanoyl chloride, ClCH$_2$COCl, is added to water, one chlorine atom is quickly substituted to give ClCH$_2$COOH, but the other atom is only substituted slowly even on boiling.

Alcohols, Phenols and Ethers

31.1 Fermentation

When Noah left the Ark he promptly planted a vineyard and was soon drinking its produce. People have been fermenting grape juice for at least ten thousand years, and probably fermenting honey for even longer. The reason for doing this, of course, is that fermentation of sugar by yeast produces ethanol, the well-known intoxicant. Ethanol is a member of the homologous series of alcohols and is commonly just called 'alcohol'.

$$C_6H_{12}O_6 \rightarrow 2CH_3CH_2OH + 2CO_2$$
$$\text{a sugar} \qquad \text{ethanol}$$

This is an exothermic reaction that provides the yeast with energy.

Alcoholic fermentation occurs naturally wherever sugar-containing materials such as fruit are allowed to decay. In autumn it is quite common to see drunken (and therefore dangerous) wasps which have been feeding off fermented fruit.

In the manufacture of alcoholic drinks, fermentation is carried out under more controlled conditions. The source of sugar varies: it may be grapes (for wine), honey (for mead), malted grain (for beer), apples (for cider) or indeed any sugar-containing fruit or plant. From the point of view of yeast, the ethanol produced in fermentation is a toxic waste product, which kills the yeast at concentrations greater than about 15% by volume. It is therefore impossible to produce alcoholic drinks containing more than 15% alcohol by fermentation alone. There are reasons for producing beverages with greater alcoholic content than this. For one thing they can be stored longer, because alcohol in high concentration is toxic to bacteria. Another reason is simply that some people like drinks containing alcohol in high concentrations.

To satisfy these requirements, fermented liquids are distilled to increase their alcoholic concentration and produce *spirits*. For example, distillation of wine produces brandy. A typical spirit contains about 40% ethanol. In Britain alcoholic strength is measured in degrees proof. 'Proof spirit' is an ancient standard, defined as the weakest mixture of pure alcohol and water which when poured over gunpowder and ignited will allow the powder to burn. 100% proof spirit corresponds to about 57% ethanol, so typical spirits are about 70% proof.

All the members of the homologous series of alcohols are toxic to a greater or lesser extent. The first member of the series, methanol, is much more toxic than ethanol. It is added to industrial alcohol (on which no excise duty is charged) to make it undrinkable. It is then called methylated spirits. Unfortunately a few people drink it nevertheless, the eventual result being blindness or death.

The higher alcohols are moderately toxic, unpleasant-tasting compounds. A mixture of these compounds is produced in small amounts during fermentation: the mixture is called *fusel oil* (*fusel* is German for 'bad liquor').

Yeast on a vat of fermenting beer. The yeast has multiplied ten-fold and the best is retained for the next brewing. The remainder is sold to the manufacturers of yeast extract.

Liber de arte Distil
landi de Compositis.
Das büch der waren kunst zu distillieren die
Composita vñ simplicia/vnd dz Büch thesaurus pauperis/Ein schatz d arme ge-
na_ Dicariū/die bedsamlin gefallen vō dē büchern d Artzny/vnd durch Experimēt
vō mir Jheronimo brüschwick vff geclubt vñ geoffenbart zū trost dené die es begeré.

An early method for the distillation of brandy.

Ethanol itself is intoxicating in small amounts but toxic in large amounts. In small amounts it has the effect of making people relaxed and for this reason it undoubtedly has its social uses. In large amounts it has a serious effect on mental and physical performance and its use can be very dangerous, particularly to drivers of motor vehicles. It is addictive if regularly taken in large quantities. Ethanol is firmly established as an accepted social drug with many advantages, but there is little doubt that if it were newly introduced today and its properties were known, it would be banned as a dangerous drug.

Fermentation is used throughout the world to produce alcoholic drinks. At one time it was also a major source of organic chemicals. Ethanol, produced by the fermentation of cheap carbohydrate such as grain or molasses, was once a major feedstock in the production of aliphatic compounds. Its place as a feedstock has now been taken by ethene, which is much cheaper. Most industrial ethanol is now produced from ethene. Nevertheless, ethanol still has important uses as a solvent, and other alcohols are of industrial importance too.

31.2 Nomenclature of alcohols

In this chapter we shall be considering organic **hydroxy compounds**, containing the OH group. A study of the properties of the OH group is important to the chemist because of the industrial importance of compounds containing this functional group and because of its wide occurrence in biological molecules.

Aliphatic and aromatic hydroxy compounds differ considerably in their properties. They are, therefore, regarded as two distinct groups of compounds: aliphatic hydroxy compounds are called **alcohols** and aromatic ones are called **phenols.**

Alcohols are named using the suffix **-ol**, preceded if necessary by a number to indicate its position in the carbon skeleton. Thus, CH_3OH is methanol, $CH_3CH_2CH_2OH$ is propan-1-ol, $CH_3CHOHCH_3$ is propan-2-ol

and $CH_3\!-\!\overset{\displaystyle CH_3}{\underset{\displaystyle OH}{\overset{|}{\underset{|}{C}}}}\!-\!CH_3$ is 3-methylpropan-2-ol.

Alcohols with structures of the form RCH_2OH, where R is an alkyl or aryl group or hydrogen, are called **primary** alcohols. Methanol and propan-1-ol are

primary alcohols. Those with the structure $\overset{\displaystyle R_1}{\underset{\displaystyle R_2}{>}}CHOH$ are called **secondary**

alcohols, for example propan-2-ol. Alcohols with the structure $\overset{\displaystyle R_1}{\underset{\displaystyle R_3}{R_2\!-\!}}COH$ are

called **tertiary** alcohols, for example 2-methylpropan-2-ol. Primary, secondary and tertiary alcohols have some important differences in chemical reactivity, as we shall see.

Some alcohols, particularly biologically occurring ones, contain more than one OH group in their molecule. They are known as **polyhydric** alcohols. They are named using the suffixes **-diol, -triol**, etc., depending on how many OH groups they contain. Thus,

CH_2OHCH_2OH ethane-1,2-diol,
$CH_2OHCHOHCH_2OH$ propane-1,2,3-triol.

Compounds in which the OH group is attached to an aromatic ring are called phenols. The simplest and most important is phenol itself:

Figure 31.1
Which of these two compounds would be classed as a phenol?

Phenols have the OH group attached *directly* to the ring. One of the two compounds shown in figure 31.1 is a phenol. Which one?

- Name the following:
  (a) $CH_3CHOHCH_2CH_3$

  (b)
$$H\!-\!\underset{\underset{OH}{|}}{\overset{\overset{H}{|}}{C}}\!-\!\underset{\underset{OH}{|}}{\overset{\overset{CH_3}{|}}{C}}\!-\!CH_3$$

- Write formulae for the following:
  (a) butane-1,2,4-triol
  (b) 2-methylpentan-2-ol
- From the previous questions, classify those alcohols containing a single OH group as primary, secondary or tertiary.

31.3 Alcohols as a homologous series

Reference has already been made to the industrial and biological importance of compounds containing the OH group. Industrially, alcohols are manufactured by the hydration of alkenes (see section 28.6). In the laboratory they can be made by nucleophilic substitution reactions between halogenoalkanes and hydroxide ions as described in section 30.4.

The alcohols illustrate very well the steady change in physical properties that occurs when a homologous series is ascended. The OH group has a big effect on the physical properties of any molecule of which it is a part. Alcohols show marked physical differences from the organic compounds we have met so far.

Consider three compounds of similar relative molecular mass: propane ($M_r = 44$), chloromethane ($M_r = 50.5$) and ethanol ($M_r = 46$). Ethanol is far less volatile than the other two: it is a liquid at room temperature while the others are gases. (Boiling points 231 K, 249 K and 351 K, respectively.) Furthermore, ethanol is soluble in water in all proportions, while the other two are practically insoluble.

The cause of ethanol's anomalous properties is hydrogen bonding (see section 9.5). Hydrogen bonds form between OH groups of adjacent ethanol molecules. This gives ethanol relatively high intermolecular forces and therefore relatively low volatility (see figure 31.2). Hydrogen bonding between ethanol molecules and water molecules explains why these two liquids are miscible in all proportions.

Pure ethanol has a strong affinity for water and tends to absorb it from the atmosphere. Furthermore, it is impossible to separate pure water from an ethanol–water mixture by distillation alone. The two liquids form a constant-boiling mixture containing 95.6% ethanol which distils over unchanged.

To produce 100% ethanol (called **absolute alcohol**) the remaining 4.4% water must be removed by a chemical drying agent such as calcium oxide.

Figure 31.3 shows the boiling points of the first seven straight-chain primary alcohols and the first seven straight-chain alkanes. Look at the graph and answer the following questions.

- Why are the boiling points of alcohols higher than those of the corresponding alkanes?
- Why do the differences in boiling points between corresponding alcohols and alkanes get less as the number of carbon atoms increases?
- Where would the two graphs intersect, and what is the physical significance of the point of intersection?

As the homologous series of alcohols is ascended, the influence of the OH group becomes less and less important compared to that of the increasingly large hydrocarbon portion. The properties of the higher alcohols therefore tend more and more towards those of the corresponding alkane. This trend can be seen in solubility as well as volatility, as table 31.1 shows.

Figure 31.2
Hydrogen bonding between ethanol molecules. Hydrogen bonds are represented by dotted lines.

Table 31.1
Solubility of alcohols in water

Name	Formula	Solubility/g per 100g of water
methanol	CH_3OH	infinite (miscible in all proportions)
ethanol	CH_3CH_2OH	infinite
propan-1-ol	$CH_3CH_2CH_2OH$	infinite
butan-1-ol	$CH_3CH_2CH_2CH_2OH$	8.0
pentan-1-ol	$CH_3CH_2CH_2CH_2CH_2OH$	2.7
hexan-1-ol	$CH_3CH_2CH_2CH_2CH_2CH_2OH$	0.6

The lower alcohols such as ethanol tend to be good solvents for polar as well as non-polar solutes. This is because they contain both a highly polar OH group and a non-polar hydrocarbon portion. For example, both sodium hydroxide (ionic) and hexane (molecular) dissolve well in ethanol. This property makes ethanol, and to a lesser extent methanol and propanol, valuable solvents in the laboratory and in industry. One example is the use of ethanol as a base for perfumes and after-shave lotions: the ethanol dissolves both the water-insoluble oils which provide the aroma and the water which makes up the bulk of the preparation. In Germany, propan-2-ol is often used for this purpose instead of ethanol.

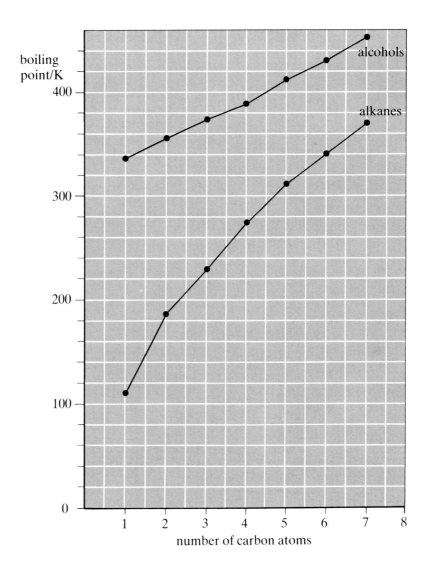

Figure 31.3
Boiling points of the first seven straight-chain alkanes and straight-chain primary alcohols.

(a)

(b)

(c)

The effect of hydrogen bonding on viscosity. (a) Identical ball bearings are held magnetically above (from left to right) ethanol, ethane-1,2-diol, propane-1,2,3-triol. (b) and (c) the ball bearings are released simultaneously and fall through the liquids.

Ethanol   **Ethane-1,2-diol**   **Propane-1,2,3-triol**

Hydrogen bonding between OH groups also has an effect on the viscosity of alcohols, particularly those with more than one OH group in their molecule. Thus ethanol has a viscosity of $1.06 \times 10^{-3}\,N\,s\,m^{-2}$ at 298 K, about the same as water. But propane-1,2,3-triol (commonly called glycerine) is very thick and sticky, with a viscosity of $942 \times 10^{-3}\,N\,s\,m^{-2}$ at the same temperature. This is because of extensive interaction between its molecules which carry three OH groups each.

In phenol, the large, non-polar benzene ring to some extent dominates the OH group. Thus phenol is only partially soluble in water (9.3 g in 100 g water at 20°C). It is a crystalline solid, melting point 43°C.

31.4 The amphoteric nature of hydroxy compounds

Alcohols can be regarded as being derived from water, by replacing one hydrogen atom by an alkyl group. If we replace both hydrogens by alkyl groups, we get an **ether** (see section 31.9).

$$H—O—H \qquad R—O—H \qquad R—O—R$$
$$\text{water} \qquad\quad \text{an alcohol} \qquad \text{an ether}$$

Alcohols might therefore be expected to show some similarity to water, and we have already seen this is true of their physical properties.

Water is an amphoteric compound. It can act as an acid, donating a proton, or as a base, accepting a proton:

as an acid: $H_2O \rightarrow OH^- + H^+$

as a base: $H_2O + H^+ \rightarrow H_3O^+$

overall: $H_2O + H_2O \rightarrow OH^- + H_3O^+$
$\qquad\qquad\quad$ acid $\quad$ base

Alcohols also show amphoteric behaviour:

as an acid: $ROH \rightarrow RO^- + H^+$

as a base: $ROH + H^+ \rightarrow ROH_2^+$

overall: $ROH + ROH \rightarrow RO^- + ROH_2^+$
$\qquad\qquad\quad$ acid $\quad$ base

When it acts as an acid, the alcohol cleaves at the O—H bond. When it acts as a base, it can, as we shall see, subsequently cleave at the R—O bond. Both these forms of bond cleavage are characteristic of hydroxy compounds, and we will consider them separately. In doing so we will take ethanol as a typical aliphatic hydroxy compound and phenol as a typical aromatic one.

31.5 Reactions involving cleavage of the O—H bond

Acidity

Table 31.2 shows how sodium reacts with water, with ethanol and with a solution of phenol in ethanol.

Table 31.2
Reaction of a small piece of sodium with different hydroxy compounds

Water	Ethanol	Solution of phenol in ethanol
sodium floats, melts, rushes about on surface, rapid evolution of hydrogen	sodium sinks, steady evolution of hydrogen, does not melt	sodium sinks, rapid evolution of hydrogen, does not melt

Look at the table and answer these questions.

- Why does sodium float in water but sink in ethanol?
- The general reaction in each case is the reduction of H^+ ions by sodium. Write a general ionic equation.
- Judging from these reactions, which contains a higher concentration of H^+ ions, water or ethanol?
- Which contains a higher concentration of H^+ ions, ethanol or a solution of phenol in ethanol?
- Which is the stronger acid, water or ethanol?
- Which is the stronger acid, phenol or ethanol?
- Write a full equation for the reaction of ethanol with sodium.

The fact that ethanol reacts with sodium liberating hydrogen suggests that the alcohol contains some hydrogen ions and that these are being reduced by sodium:

$$2Na + 2H^+ \rightarrow 2Na^+ + H_2$$

The fact that the reaction is slower with ethanol than with water suggests that ethanol contains a lower concentration of H^+ than water, i.e. it is a weaker acid. In fact, the K_a of ethanol is 10^{-18} whereas the value for water is 10^{-16}. Remember, the higher the K_a, the stronger the acid (see section 23.9). We can see why ethanol is a weaker acid if we compare the equilibria:

$$CH_3CH_2OH + H_2O \rightleftharpoons \underset{\text{ethoxide ion}}{CH_3CH_2O^-} + H_3O^+$$

$$H_2O + H_2O \rightleftharpoons \underset{\text{hydroxide ion}}{HO^-} + H_3O^+$$

The ethoxide ion is more basic than the hydroxide ion, because of the tendency of alkyl groups to donate electrons. The CH_3CH_2 group donates electrons to the O, increasing the negative charge density on that atom and thus making it more ready to accept protons. Thus the equilibrium above is further to the left in the case of ethanol than with water. Therefore ethanol is the weaker acid. Both water and ethanol are of course extremely weak acids compared with the substances we normally call acids. Correspondingly, the hydroxide and ethoxide ions are strong bases.

In its reaction with sodium, ethanol is converted to ethoxide ion, so the two products are hydrogen and sodium ethoxide:

$$2Na + 2CH_3CH_2OH \rightarrow \underset{\text{sodium ethoxide}}{2CH_3CH_2O^-Na^+} + H_2$$

The reactions with sodium indicate that phenol is a stronger acid than ethanol. In fact it is also a stronger acid than water (K_a for phenol $= 10^{-10}$). Nevertheless, sodium reacts more vigorously with water than with an ethanolic solution of phenol, because the latter only contains a fairly low concentration of phenol.

If we consider the equilibrium:

we can see why phenol is a stronger acid than aliphatic alcohols. In the phenoxide ion, the negative charge on the O can to some extent be delocalised round the ring. This reduces the tendency of the phenoxide ion to attract protons, in other words it reduces its strength as a base. Consequently, phenol is a stronger acid than aliphatic alcohols, though it is still weak.

The antiseptic action of 'Coal Tar Soap' depended on the phenol and related compounds it contained.

2,4,6-trichlorophenol. 23 times as effective a germicide as phenol in a given concentration.

4-chloro-3,5-dimethyl-phenol. 280 times as effective as phenol.

Figure 31.4
Derivatives of phenol used as antiseptics.

Because of its acidic nature, phenol is much more soluble in sodium hydroxide solution than in water. Hydroxide ions from sodium hydroxide remove hydrogen ions, displacing the above equilibrium to the right so that the phenol dissolves as sodium phenoxide.

Phenol used to be known as 'carbolic acid' and was used as one of the earliest antiseptics. The Edinburgh doctor Joseph Lister first used it in the 1860s to prevent wounds going septic after surgery. Unwittingly, it had been used as an antiseptic even before this, because the old-fashioned way of treating an amputation wound was to cover it with coal-tar, which contains phenol. Phenol is effective in killing bacteria, but it is also very corrosive to the skin. It has been largely replaced by other antiseptics. Many of these are derivatives of phenol and are better germicides. Two examples are shown in figure 31.4.

An early example of the use of anti-septics. In this operation a spray of phenol is being directed onto the wound.

Esterification

In the presence of an acid catalyst, alcohols react with carboxylic acids to form esters. Water is eliminated. The alcohol cleaves at the O—H bond:

$$CH_3-\underset{\underset{O}{\|}}{C}-OH + CH_3CH_2-O-H \xrightarrow{H^+} CH_3-\underset{\underset{O}{\|}}{C}-O-CH_2CH_3 + H_2O$$

ethanoic acid   ethanol   ethyl ethanoate (an ester)

This reaction is considered more fully in chapter 33.

Phenol is again rather different from aliphatic alcohols: it does not react with acids directly. Phenyl esters can however be made by reacting phenol with an acyl halide (see section 30.7). For example:

$$CH_3{-}\underset{\underset{O}{\|}}{C}{-}Cl + \underset{phenol}{\text{(benzene ring with OH)}} \longrightarrow CH_3{-}\underset{\underset{O}{\|}}{C}{-}O{-}\text{(benzene ring)} + HCl$$

ethanoyl chloride   phenol   phenyl ethanoate

Predict the formulae of the products of reactions between

- propan-2-ol and sodium,
- propan-1-ol and propanoic acid, in the presence of an acid catalyst.

31.6 Reactions involving cleavage of the C—O bond

Reaction with halide ions

In the presence of a strong acid, the OH group in an alcohol bonds to H^+ ions, and becomes protonated.

$$\underset{}{\overset{}{\gg}}C{-}\overset{H}{\underset{H^+}{O}} \longrightarrow \gg C{-}\overset{H}{\underset{H}{O^+}}$$

The oxygen atom thus carries a positive charge, and tends to attract electrons very strongly from the carbon atom next to it (figure 31.5). This causes a large positive charge on the carbon, making it attractive to nucleophiles. Thus, *in the presence of concentrated sulphuric acid*, ethanol reacts with Br^- to form bromoethane:

$$CH_3CH_2{-}OH + H^+ \longrightarrow CH_3CH_2\overset{\delta+}{\underset{\underset{Br^-}{H}}{{-}OH}} \longrightarrow CH_3CH_2 \underset{\underset{Br}{|}}{} + \underset{\underset{H}{}}{OH}$$

Overall, this amounts to the reaction of the alcohol with HBr:

$$CH_3CH_2OH + HBr \rightarrow CH_3CH_2Br + H_2O$$
bromoethane

In some cases an intermediate carbocation may actually be formed before the Br^- attacks. Cleavage of the C—O bond is greatly aided by protonation because it is much easier for the molecule to lose a neutral H_2O molecule than a charged OH^- ion. This is another example of a nucleophilic substitution reaction. Compare it with the substitution reactions of halogenoalkanes (section 30.4). The reaction is normally carried out by heating ethanol under reflux with potassium bromide and concentrated sulphuric acid. Similar reactions occur between alcohols and Cl^-, Br^- and I^- in acid conditions. Cl^- is the least reactive and I^- the most.

Other halogenation reactions

There are several other reagents that are useful for replacing an OH group with a halogen atom. Some of the more important are:

    phosphorus pentachloride: $ROH + PCl_5 \rightarrow RCl + HCl + POCl_3$
    phosphorus tribromide or triiodide (in practice a mixture of red phosphorus and bromine or iodine is used): $3ROH + PBr_3 \rightarrow 3RBr + H_3PO_3$

Phenol is again different in that it cannot be halogenated by normal reagents.

Figure 31.5
Effect of protonation on polarisation of the C—O bond in alcohols. The arrow indicates the direction of displacement of electrons in the bond.

31.7 Reactions involving the carbon skeleton

So far we have considered reactions of alcohols in which part or all of the OH group is replaced. Alcohols also undergo several reactions which involve *both* the carbon skeleton and the OH group.

Dehydration

(i) Alkene formation

Consider again the protonated form of ethanol that we met in the last section. This ion can readily lose water, forming a carbocation:

The square brackets indicate that the carbocation is an unstable intermediate. In the reaction with HBr, this ion rapidly reacts with Br⁻ to form bromoethane. In the absence of any nucleophile like Br⁻, however, a different reaction may occur. The ion may lose H⁺ and form ethene:

Thus in the presence of strong acids, ethanol forms ethane in what amounts to an overall elimination reaction:

Since water is eliminated, this can also be thought of as a dehydration reaction. In practice the intermediate carbocation may never form: the H^+ and the H_2O may leave simultaneously.

The reaction is normally carried out by heating ethanol at 170°C with excess concentrated sulphuric acid. Ethene is evolved and can be collected over water. The concentrated sulphuric acid can be thought of as a dehydrating agent, removing water from ethanol, though the actual mechanism as we have seen is more complex than this. Another way of preparing ethene from ethanol is by catalytic dehydration of ethanol vapour. A heated catalyst of aluminium oxide or pumice stone is used, in the apparatus shown in figure 31.6.

Figure 31.6
Catalytic dehydration of ethanol.

(ii) Ether formation

The carbocation formed by ethanol in the presence of an acid is open to attack by nucleophiles. Ethanol, with lone pairs on its oxygen atom, is itself a nucleophile, so we might expect it to attack the carbocation:

$$\left[CH_3CH_2^+ \right] \quad \ddot{O}-CH_2CH_3 \longrightarrow \left[CH_3CH_2-\overset{+}{O}-CH_2CH_3 \right]$$
$$\qquad\qquad\qquad\quad | \qquad\qquad\qquad\qquad\qquad\qquad |$$
$$\qquad\qquad\qquad\quad H \qquad\qquad\qquad\qquad\qquad\qquad H$$

ethyl carbocation ethanol

This new ion can then lose an H^+ ion:

$$\left[CH_3CH_2-\overset{+}{\underset{H}{O}}-CH_2CH_3 \right] \longrightarrow CH_3CH_2-O-CH_2CH_3 + H^+$$

ethoxyethane

The product is ethoxyethane, an ether. Overall, this too is effectively a dehydration reaction:

$$\begin{matrix} CH_3CH_2OH \\ + \\ CH_3CH_2OH \end{matrix} \longrightarrow \begin{matrix} CH_3CH_2 \\ \diagdown \\ CH_3CH_2 \end{matrix} O + H_2O$$

ethanol ethoxyethane

The dehydration of ethanol, then, can give two different products. Both this reaction and the one in which ethene is formed occur when ethanol is heated with concentrated sulphuric acid. By adjusting the reaction conditions we can largely determine which product is formed. The formation of a molecule of the ether involves two molecules of ethanol, but the formation of a molecule of the alkene involves only one. Ether formation is therefore favoured by having an excess of ethanol present. Compare this with the excess of sulphuric acid used in the preparation of ethene.

Ethoxyethane is prepared by heating concentrated sulphuric acid with excess ethanol at 140°C. The ethoxyethane distils off. The properties of ethers are considered further in section 31.9.

The dehydration reactions here are applicable to alcohols in general, but not to phenols.

■ Predict the formulae of the main products of the following reactions.
   (a) Butan-2-ol is heated with excess concentrated sulphuric acid.
   (b) Excess propan-1-ol is heated with concentrated sulphuric acid.
   (c) Propan-2-ol vapour is passed over heated aluminium oxide.

Oxidation

Table 31.3 shows the effect of warming three different alcohols with acidified potassium dichromate(VI).

Look at the table and answer these questions.

■ Classify the three alcohols as primary, secondary and tertiary.
■ When potassium dichromate(VI) is reduced in acid solution, green Cr^{3+} ions are formed. Which of the three alcohols is oxidised by acidified potassium dichromate(VI)?

Primary and secondary alcohols are readily oxidised by a variety of oxidants. Acidified dichromate(VI), acidic or alkaline manganate(VII) or air in the presence of a catalyst are all suitable oxidants. The initial product is a **carbonyl** compound (chapter 32) which in the case of a primary alcohol is an **aldehyde**:

propan-1-ol → propanal, an aldehyde + $2H^+ + 2e^-$ electrons accepted by oxidiser

Table 31.3
Effect of warming alcohols with acidified potassium dichromate(VI)

Name	Formula	Observation
propan-1-ol	$CH_3CH_2CH_2OH$	orange dichromate (VI) slowly turns green
propan-2-ol	$CH_3CHOHCH_3$	orange dichromate(VI) slowly turns green
2-methyl-propan-2-ol	$CH_3 - \overset{\overset{\displaystyle CH_3}{\mid}}{\underset{\underset{\displaystyle OH}{\mid}}{C}} - CH_3$	no change

Aldehydes are themselves readily oxidised to acids, thus:

The product of oxidising a primary alcohol is therefore usually an acid, unless the aldehyde is distilled from the reaction mixture as it forms.

In the case of a secondary alcohol the product of oxidation is a **ketone**:

Ketones are not readily oxidised, so the reaction stops at this point.

Tertiary alcohols cannot be readily oxidised since they have no hydrogen atom to be removed from the carbon atom carrying the OH group. Strong oxidisers cause their molecules to be broken up, giving a mixture of oxidation products.

■ Predict the products, if any, of oxidising the following alcohols with acidified dichromate(VI):
  (a) ethanol (product distilled off immediately),
  (b) ethanol (reagents heated together under reflux for some time),
  (c) 2-methylbutan-2-ol,
  (d) butan-2-ol.

The oxidation of alcohols is important in industry and in living organisms. Methanal (commonly called formaldehyde and used in the production of plastics, see the next section) is manufactured by passing a mixture of methanol vapour and air over a silver catalyst at 500°C:

Home-made beer fermenting in a glass vessel. The curved tube at the top is a trap to keep out bacteria. It allows carbon dioxide to escape from the fermenting beer, but it does not allow entry to bacteria which might turn the beer to vinegar.

Propanone (acetone) is manufactured from propan-2-ol by a similar process, using air and a copper catalyst at 500°C. In these industrial processes, high-temperature catalytic reactions are used in which air is the oxidiser, rather than aqueous oxidisers like dichromate(VI). This is for economic reasons. Air is much cheaper than dichromate(VI) and although the catalyst is expensive it has a long life before it needs renewing.

The bacterial oxidation of ethanol to ethanoic acid (acetic acid) has long been a problem for wine producers. The bacterium, *Acetobacter*, uses air to oxidise ethanol in wine, producing a weak solution of ethanoic acid called vinegar. The bacterium uses this oxidative process as a source of energy. Once a bottle of wine has been opened it will turn to vinegar fairly quickly because of the considerable number of these bacteria in the air. One way of preventing this is to add extra alcohol to the wine so that the concentration of ethanol is too high for the bacteria to tolerate. Wine treated in this way is said to be *fortified*: sherry and port are examples.

Ethanol is disposed of by the body by oxidation: it is oxidised in the liver, first to ethanal and eventually to CO_2 and water. The liver can oxidise about 8g of ethanol an hour, but very high concentrations are too much for it to cope with. Excessive amounts of ethanol can damage the liver and cause a condition called cirrhosis.

Ethanol, like all alcohols, undergoes complete oxidation to CO_2 and H_2O when heated in the presence of air. In other words, it burns. Ethanol has a clean, smokeless flame and is occasionally used as a fuel (as in methylated spirit stoves). Its use as a fuel is limited by its cost in the United Kingdom. But in some tropical countries where sugar is easily grown, ethanol is cheaply produced by fermentation. In Brazil, for example, ethanol is an important motor fuel because it can be produced at a cost comparable to that of gasoline.

Phenol, like tertiary alcohols, cannot be oxidised to a carbonyl compound. However, the aromatic ring with its high electron density is susceptible to attack by oxidisers.

A spirit burner. Ethanol burns with a very clean flame.

31.8 Substitution reactions of the aromatic ring in phenol

If bromine water is added to a solution of phenol the bromine is immediately decolorised and a white precipitate is formed. This is a substition reaction, the white precipitate being 2,4,6-tribromophenol:

2,4,6-tribromophenol

Benzene does not react with bromine except in the presence of a halogen-carrier catalyst (section 29.7). Phenol is more susceptible than benzene to attack by electrophiles. This is because the lone pairs of electrons on the oxygen of the OH group become partially delocalised round the ring: this increases the electron density there and makes it more susceptible to attack by electrophiles. The 2,4 and 6 positions are preferentially substituted (see section 29.10).

Phenol is nitrated far more readily than benzene. Dilute nitric acid alone is sufficient, the products being 2- and 4-nitrophenol. A mixture of concentrated nitric acid and concentrated sulphuric acid readily converts it to 2,4,6-tri-nitrophenol, commonly known as picric acid:

2,4,6-trinitrophenol

This substance is used as a high explosive. It decomposes spontaneously and exothermically, producing large volumes of CO, steam and N_2 which expand and produce an explosive shock.

Plastics from phenol

When phenol is heated with methanal in the presence of an acid or alkali catalyst, a hard, brittle plastic is formed. The initial stage of this reaction is substitution of the phenol in the 2 or 4 position:

This product then undergoes a condensation reaction in which a molecule of water is eliminated between it and another molecule of phenol.

The product may then be substituted by another methanal molecule:

Condensation then occurs with a further molecule of phenol and the process continues until all the aromatic rings are substituted in the 2,4 and 6 positions. A molecular network like that shown in figure 31.7, on the next page, builds up.

This polymer is very hard and rather brittle, because of its extensively cross-linked three-dimensional network. It is a dark brown material, known as 'Bakelite'. It has the disadvantage that it sets hard on heating and cannot be remelted, which makes it difficult to mould. Its hardness and cheapness are however a considerable advantage and it was once widely used for such things as electric plugs and motor-car distributor caps. Plastics of this kind, which set hard and cannot be remelted, are called **thermosetting**. Plastics like polythene which can be moulded by melting are called **thermosoftening**.

TCP antiseptic contains phenol derivatives. What do you think the initials stand for?

A radio with 'Bakelite' case, manufactured just after the War. 'Bakelite' was the earliest synthetic plastic.

Figure 31.7
Condensation polymer of phenol and methanal.

etc.

31.9 Ethers

Ethers are compounds with the general structure ROR′, where R and R′ are alkyl or aryl groups. They are named by regarding them as alkanes substituted by alkoxy groups. Thus,

CH_3OCH_3 methoxymethane,
$CH_3CH_2OCH_2CH_3$ ethoxyethane,
$CH_3OCH_2CH_3$ methoxyethane.

Ethoxyethane is an example of a **symmetrical** ether; methoxyethane is an **unsymmetrical** ether.

Since they have no capacity for hydrogen bonding, ethers are far more volatile than alcohols. Their volatility corresponds closely to alkanes of comparable relative molecular mass. Thus the most commonly encountered ether, ethoxyethane (commonly called simply 'ether'), is a very volatile liquid of boiling point 35°C. This volatility makes ethoxyethane a dangerously flammable substance. Since its relative molecular mass is 74, ethoxyethane vapour is much denser than air. It tends to diffuse slowly and stay at floor or bench level in a laboratory, where it is likely to be ignited by naked flames. Serious explosions and fires have been caused in this way.

The molecule of ethoxyethane is analogous in shape to that of water:

$$CH_3CH_2-\ddot{O}:$$
$$\diagdown CH_2CH_3$$

It is therefore slightly polar, but not polar enough to have much effect on the properties of ethoxyethane. Like all ethers, it is only slightly soluble in water and is a good solvent for non-polar substances. It is frequently used as a solvent in practical work because it is volatile and easily distilled off, but its use must be carefully controlled because of the fire risk. The small polarity in the molecule is not enough to make it significantly reactive towards nucleophiles or electrophiles. Ethers are not much more reactive than alkanes.

Ethers do, however, show some reactivity in acid solution. The lone pairs on the oxygen atom enable ethers to act as bases. For example, ethoxyethane is protonated by concentrated sulphuric acid:

$$CH_3CH_2\ddot{O}CH_2CH_3 + H^+ \rightarrow CH_3CH_2\overset{+}{\ddot{O}}CH_2CH_3$$
$$H$$

Ethoxyethane will therefore dissolve in concentrated sulphuric acid though it is almost insoluble in water.

WONDERFUL EFFECTS OF ETHER IN A CASE OF SCOLDING WIFE.

Patient.—" THIS IS REALLY QUITE DELIGHTFUL—A MOST BEAUTIFUL DREAM."

Ethoxyethane (ether) was at one time used as an anaesthetic.

The monument erected at Boston Common, USA, by a grateful patient, Thomas Lee.

Protonation of the ether makes it more liable to cleavage. The carbon atoms now carry a considerable positive charge and are open to attack by nucleophiles. Thus ethers are split by heating with concentrated hydriodic acid, HI:

$$CH_3CH_2-\overset{..}{O}-CH_2CH_3 \rightarrow CH_3CH_2-\overset{+}{O}-CH_2CH_3 \rightarrow CH_3CH_2 + O-CH_2CH_3$$

$$\overset{|}{H^+} \qquad\qquad \overset{|}{H} \qquad\qquad \overset{|}{I}$$

$$I^- \qquad\qquad\qquad\qquad \text{iodoethane ethanol}$$

Overall: $CH_3CH_2OCH_2CH_3 + HI \rightarrow CH_3CH_2I + CH_3CH_2OH$

An interesting example of a **cyclic ether** is epoxyethane, CH_2-CH_2. This

$$\diagdown \diagup$$
$$O$$

compound has considerable ring strain (section 27.8) so it is a good deal more reactive than straight-chain ethers. Its chemical reactivity makes it an important intermediate in industrial synthesis (see section 28.6).

Summary

1 Alcohols are aliphatic compounds containing the OH group. They are named using the suffix *-ol*. Phenols are aromatic compounds with the OH group attached directly to the ring.

2 Primary alcohols are of the form RCH_2OH; secondary alcohols are of the form $\overset{R_1}{\underset{R_2}{\diagdown}}CHOH$; tertiary alcohols are of the form $\overset{R_1}{\underset{R_3}{R_2-C-OH}}$.

3 The physical properties of alcohols, especially the early members of the series, are strongly influenced by hydrogen bonding.

4 Alcohols can act as proton donors but are weaker acids than water. Phenols are considerably stronger acids.

5 Alcohols can act as bases, becoming protonated on the oxygen atom. This can lead to the formation of a carbocation and subsequent dehydration or substitution reactions.

6 Primary alcohols can be oxidised to aldehydes, then to acids. Secondary alcohols can be oxidised to ketones. Tertiary alcohols cannot be oxidised without fragmenting the molecule.

7 Some reactions of ethanol are shown in the diagram below.

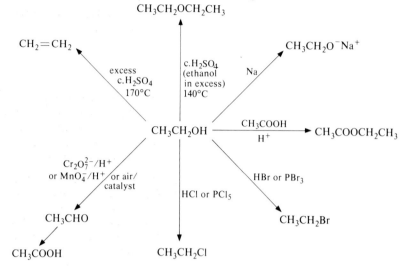

The importance of alcohols in the synthesis of other organic compounds is considered in chapter 35.

8 Some reactions of phenol are shown below.

9 Ethers are compounds of the general form ROR', where R and R' are alkyl or aryl groups. They are rather unreactive.

Study questions

1 Consider the following compounds.

              CH₃
               |

A CH₃CH₂—C—CH₃

               |
              OH

B CH₃CH₂CHOHCH₃

C $CH_3OCH_2CH_2CH_3$

D $CH_3CH_2CH_2CH_2OH$
  OH

E

(a) Name each compound.
(b) Which is a primary alcohol?
(c) Which is a tertiary alcohol?
(d) Which is a phenol?
(e) Which is an ether?

2 Refer again to the compounds in question 1.
   (a) Which react with sodium metal?
   (b) Which could be oxidised to an aldehyde?
   (c) Which could be oxidised to a ketone?
   (d) Which would be the strongest acid?
   (e) Which would form an alkene when heated with excess concentrated sulphuric acid?
   (f) Which has the lowest boiling point?

3 Table 31.4 gives some physical properties of water, ethanol and ethoxyethane.

Table 31.4
Some physical properties of water, ethanol, and ethoxyethane

Name	Formula	M_r	Boiling point/°C	Density at 273 K/ g cm^{-3}	Surface tension at 293 K/N m^{-1}
water	H_2O	18	100	1.00	7.28
ethanol	CH_3CH_2OH	46	78	0.79	2.23
ethoxyethane	$CH_3CH_2OCH_2CH_3$	74	35	0.71	1.69

Suggest explanations in terms of intermolecular forces, why
(a) the boiling point of ethanol is greater than that of ethoxyethane,
(b) the density of water is greater than that of ethanol,
(c) the surface tension of water is greater than that of ethoxyethane.

4 An organic liquid **A** contains carbon, hydrogen and oxygen only. On combustion 0.463 g of **A** gave 1.1 g of carbon dioxide and 0.563 g of water. When vaporised, 0.1 g of **A** occupy 54.5 cm^3 at 208°C and 98.3 kPa (740 mm Hg).
   Standard pressure = 101 kPa (760 mm Hg); 1 mole of a gas occupies 22.4 dm^3 at s.t.p.
   (a) What is the percentage composition of **A**?
   (b) Find the empirical formula of **A**.
   (c) Calculate the relative molecular mass of **A**.
   (d) Give the structures of possible non-cyclic isomers of **A**.
   (e) Which isomers will react with sodium to give hydrogen?
   (f) Which isomers will reduce acidified dichromate(VI) ion to green Cr^{3+}?

5 Suggest explanations for the following observations.
   (a) Butan-1-ol is much more soluble in 5 M hydrochloric acid than in water.
   (b) 2,4,6-Trinitrophenol (picric acid) is a much stronger acid than phenol ($K_a = 10^{-1}$ and 10^{-10}, respectively).
   (c) Phenol can be nitrated by dilute nitric acid to give a mixture of 2-nitrophenol and 4-nitrophenol, whereas benzene is only nitrated by a mixture of concentrated nitric and concentrated sulphuric acids.
   (d) Ethanol is more acidic than 2-methylpropan-2-ol.
   (e) Heating butan-2-ol with excess concentrated sulphuric acid produces a mixture of three isomeric alkenes.

6   Give the reagents and conditions you would use to carry out the following conversions. One or more steps may be involved in the conversions.
   (a) Ethanol to ethyl ethanoate (using ethanol as the only organic starting material).
   (b) Ethanol to 1,2-dibromoethane.
   (c) Ethene to ethanoic acid.
   (d) Ethanol to propanenitrile.
   (e) Bromoethane to ethoxyethane.

7   Suggest structures for the following compounds **X**, **Y** and **Z**. Explain the reactions involved in each case.
   (a) **X**, C_7H_8O, burns with a smoky flame. **X** is only slightly soluble in water, but very soluble in sodium hydroxide solution.
   (b) **Y**, $C_6H_{14}O$, is insoluble in water but soluble in concentrated sulphuric acid. **Y** has no reaction with sodium.
   (c) **Z**, $C_5H_{12}O$, gives hydrogen when reacted with sodium metal but has no effect on acidified potassium dichromate(VI).

8   Predict the structure of the organic products of the following reactions.
   (a) Butan-2-ol is warmed with acidified potassium manganate(VII).
   (b) An aqueous solution of chlorine is added to phenol.
   (c) Propan-2-ol is warmed with excess concentrated sulphuric acid.
   (d) 2-Methylphenol is treated with propanoyl chloride.
   (e) Methanol is treated with phosphorus pentachloride.

9   Primary alcohols can be thought of as compounds derived by replacing one hydrogen atom in a water molecule by an alkyl group. Ethers can be thought of as resulting from the replacement of both hydrogen atoms in water by alkyl groups. To what extent do the physical and chemical properties of water, alcohols and ethers fit in with this idea? Illustrate your answer by reference to ethanol and ethoxyethane.

10   For each of the common uses of alcohols given below, explain which property or properties of the alcohols involved make them suitable for that use.
   (a) Ethanol is used to clean surfaces before applying adhesive.
   (b) Ethane-1,2-diol is used as an antifreeze.
   (c) Ethanol is used as the base for many perfumes.
   (d) Vinegar is manufactured from a weak solution of ethanol.
   (e) Fruits, such as peaches, are sometimes preserved in brandy.
   (f) 'Breathalysers' containing potassium dichromate(VI) can be used to test for ethanol.

11   Five different bottles contain the following liquids

   A propan-1-ol $CH_3CH_2CH_2OH$

   B butoxybutane $CH_3CH_2CH_2CH_2-O-CH_2CH_2CH_2CH_3$

   C cyclopentanol   ⬠—OH

   D 2-methylbutan-2-ol $CH_3CH_2\overset{\overset{\displaystyle OH}{|}}{C}CH_3$
                                           $\underset{|}{}$
                                           CH_3

   E methanol CH_3OH

   If the bottles were unlabelled, what *chemical* tests would you use to find out which liquid was which?

32 Carbonyl Compounds

32.1 The carbonyl group

In chapter 28 we looked at compounds containing C=C double bonds and saw that their typical reactions involved electrophilic addition. In chapter 31 we looked at compounds containing the C—O single bond, which is polar and tends to bring about substitution reactions. We will now turn our attention to the **carbonyl** group, C=O, which is the functional group in aldehydes and ketones. We might expect the reactions of this group to show similarity to the reactions of both

$$ \diagdown C=C \diagup \text{and} -\overset{\diagup}{\underset{\diagdown}{C}}-O-.$$

The double bond between C and O in the carbonyl group, like the double bond in alkenes, can be considered to consist of a σ-bond and a π-bond. Unlike the C=C group, however, the carbonyl group does not have an even electron distribution between the two atoms. There is a greater electron density over the more electronegative oxygen atom, as shown in figure 32.1.

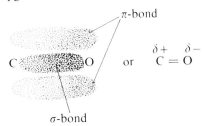

Figure 32.1
Electron distribution in the C=O bond.

This electron distribution makes the carbon atom attractive towards nucleophiles. Nucleophiles tend to attack and bond to this carbon, breaking the π-bond and resulting eventually in addition. With a general nucleophile $\ddot{X}$—Y:

The exact mechanism depends on the nature of X and Y. In some cases, Y may bond to the O atom before X attacks the carbon. In most cases, Y is hydrogen. This sort of reaction, which is typical of the carbonyl group, is called **nucleophilic addition**: compare it with the *electrophilic* addition which is typical of alkenes. There is more about nucleophilic addition in section 30.3.

The reactions of the carbonyl group are important because the group is common in biological molecules, particularly carbohydrates (section 32.7). Carbonyl compounds also have considerable industrial significance, for example as solvents and in the manufacture of thermosetting plastics (section 31.8).

32.2 Aldehydes and ketones—nature and nomenclature

Aldehydes and ketones both contain the carbonyl group, but they differ in its position in the hydrocarbon skeleton. **Aldehydes** have the carbonyl group at the end of a chain. Their general formula is therefore R—C—H, usually written $\overset{\|}{O}$

RCHO, where R is an alkyl or aryl group or hydrogen. **Ketones** have the carbonyl group in a non-terminal position in the chain. The general formula is R—C—R′, usually written RCOR′, where R and R′ are alkyl or aryl groups. $\overset{\|}{O}$

Thus aldehydes and ketones are structurally quite similar, but their properties differ considerably and they are considered as different homologous series.

Aldehydes are named using the suffix **-al** after a stem indicating the number of carbon atoms (including the one in the carbonyl group). Thus CH_3CHO is called ethanal. Ketones are named using the suffix **-one** after a stem indicating the number of carbon atoms, together with a number, if necessary, to indicate the position of the carbonyl group in the chain. Thus $CH_3COCH_2CH_2CH_3$ is called pentan-2-one. Table 32.1 gives the names and formulae of some important aldehydes and ketones.

- Name the compounds:
  (a) $CH_3CH_2CH_2CHO$,
  (b) $CH_3CH_2COCH_2CH_2CH_3$.
- Give the formula of hexan-2-one.
- Why is there no such compound as ethanone?

Table 32.1
Some important aldehydes and ketones

Formula	Systematic name	Other name	State at room temp.	Boiling point/K	Solubility in water
HCHO	methanal	formaldehyde	g	254	soluble
CH_3CHO	ethanal	acetaldehyde	l	294	infinite
CH_3CH_2CHO	propanal	propionaldehyde	l	321	soluble
CH_3COCH_3	propanone	acetone	l	329	infinite
$CH_3COCH_2CH_3$	butanone	methylethyl ketone	l	353	very soluble
$CH_3CH_2COCH_2CH_3$	pentan-3-one		l	375	very soluble
⬡—CHO	benzaldehyde		l	451	slightly soluble
⬡—COCH₃	phenylethanone	acetophenone	l	475	insoluble

Physical properties

The polarity of the C=O group has a big influence on the physical properties of aldehydes and ketones. The earlier members of both series are considerably less volatile than alkanes of corresponding relative molecular mass. Thus ethanal

(CH$_3$CHO), with a boiling point of 20°C, is a liquid (though a very volatile one) at room temperature. Propane, CH$_3$CH$_2$CH$_3$, with the same relative molecular mass is a gas at room temperature (boiling point, −42°C).

The polar C═O group has less effect on intermolecular forces than the OH group, since the latter is able to participate in hydrogen bonding. Compare the boiling points of ethanal and ethanol, which are 20°C and 78°C, respectively.

The early members of the aldehydes and ketones are soluble in water, and will themselves dissolve both polar and non-polar solutes. Propanone (acetone), for example is a widely used industrial solvent. As expected, the polar C═O group has less and less influence on the physical properties of carbonyl compounds as the homologous series are ascended. Table 32.1 gives some physical properties of carbonyl compounds.

Making aldehydes and ketones

In the laboratory these compounds can be made by the oxidation of alcohols, as described in section 31.7 and in Practical 37 in the *Laboratory Manual and Study Guide*. Aldehydes are made by oxidising primary alcohols, ketones by oxidising secondary alcohols. This method is also used industrially for making methanal (from methanol) and propanone (from propan-2-ol). Most propanone, however, is now produced as a by-product from the manufacture of phenol (see study question 7, chapter 29). Some ethanal is still manufactured by the oxidation of ethanol, but most is made by direct oxidation of ethene (see study question 3 at the end of this chapter).

A 40% solution of methanal in water, called formalin, is used for preserving biological specimens.

32.3 Addition reactions of carbonyl compounds

The characteristic reaction of compounds containing the carbonyl group is nucleophilic addition:

$$\begin{array}{c} R_1 \\ \diagdown \overset{\delta+}{} \quad \overset{\delta-}{} \\ C{=}O + \ddot{X}{-}Y \longrightarrow \\ \diagup \\ R_2 \end{array} \qquad \begin{array}{c} R_1 \quad OY \\ \diagdown \diagup \\ C \\ \diagup \diagdown \\ R_2 \quad X \end{array}$$

The readiness with which such an addition reaction occurs is partly determined by the size of the partial positive charge on the carbon atom of the carbonyl group.

■ Consider the cases when R$_1$ and R$_2$ are (a) both H, (b) CH$_3$ and H, (c) both CH$_3$. Place (a), (b) and (c) in order according to the size of positive charge you would expect on the carbonyl carbon in each of the three compounds. (Remember that alkyl groups like CH$_3$ tend to donate electrons to groups to which they are attached.)
■ Place (a), (b) and (c) in order of readiness to undergo nucleophilic addition.
■ Name compounds (a), (b) and (c).
■ How would you expect benzaldehyde (C$_6$H$_5$—C—H) to compare with
$$\quad\quad\quad\quad\quad\quad \overset{\displaystyle \|}{O}$$
ethanal in its readiness to undergo nucleophilic addition?

Aldehydes tend to be more reactive than ketones in nucleophilic addition reactions. Methanal is the most reactive aldehyde. Benzaldehyde is less reactive than aliphatic aldehydes because of the tendency of the benzene ring to delocalise the positive charge on the carbonyl carbon. On the whole, though, the reactions of aromatic aldehydes and ketones (i.e. those with the carbonyl group attached directly to the benzene ring) are quite similar to those of aliphatic ones.

The important stage of addition to C═O is attack by a nucleophile. Compare this with C═C, where the important stage of addition is attack by an electrophile. Carbonyl compounds do react with some of the compounds that undergo addition reactions with alkenes, but many of their reactions are different. Some examples are now given.

Reaction with HCN

In most addition reactions of carbonyl compounds, the molecule adding across the C=O double bond is of the form HX. A good example is HCN.

ethanal   2-hydroxypropanenitrile

The reaction occurs with ketones as well as aldehydes:

propanone   2-hydroxy-2-methylpropanenitrile

2-Hydroxy-2-methylpropanenitrile is an intermediate in the manufacture of methyl 2-methylpropenoate, the monomer for Perspex.

Reaction with sodium hydrogensulphite

Carbonyl compounds undergo addition reactions when shaken with saturated aqueous sodium hydrogensulphite ($NaHSO_3$). As expected, ethanal reacts more readily than propanone.

ethanal   sodium hydrogen-   Reaction 88% complete
               sulphite   after 30 min at 25°C

propanone   Reaction 47% complete
                             after 30 min at 25°C

The products of these reactions are crystalline ionic compounds, soluble in water.

Reduction

Like the C=C bond, the C=O bond undergoes addition with hydrogen in the presence of a metal catalyst such as platinum or nickel. Aldehydes give primary alcohols, ketones give secondary alcohols.

ethanal   ethanol

propanone   propan-2-ol

- Write the formulae of the products you would expect when propanal reacts with (a) H_2 in the presence of a nickel catalyst, (b) HCN.
- Write the formula of the product you would expect when (a) butanone, (b) benzaldehyde reacts with $NaHSO_3$.

'Self-addition': polymerisation

Just as alkenes can be made to undergo addition polymerisation, carbonyl compounds too can form polymers, by addition across the C=O double bond. Ketones are not reactive enough to polymerise easily, but aldehydes can readily be converted to a variety of addition polymers. As expected, methanal polymerises most readily. Some of the polymers formed by methanal and ethanal are shown in table 32.2.

A plastic made from poly(methanal) is used to replace metal in machine parts such as gear wheels and clips.

32.4 Condensation reactions of carbonyl compounds

In some cases the addition reactions outlined above are followed by elimination of a molecule of water. Many of these elimination reactions involve derivatives of ammonia of the general form $X—NH_2$:

Table 32.2
Polymers of methanal and ethanal

Method of preparation	Name	Structure	Uses
1 From methanal evaporate aqueous solution of methanal	poly (methanal)	$—O—CH_2—O—CH_2—O—CH_2—O—$	high-strength plastic (making gear wheels, etc.), marketed as Delrin and Kematal
distil methanal from acidic solution	methanal trimer		
2 From ethanal add dilute acid to ethanal	ethanal trimer		hypnotic (sleep-inducing) drug
add acid to ethanal below 0°C	ethanal tetramer		slug poison, solid fuel in portable stoves (commonly called metaldehyde)

This is a **condensation** or **addition–elimination** reaction (section 28.7). Important examples are shown in table 32.3

The products of condensation reactions between 2,4-dinitrophenylhydrazine and carbonyl compounds are all orange crystalline solids with well-defined melting points. They are useful for identifying individual carbonyl compounds. The condensation product is prepared, its melting point is measured accurately and the compound is identified from tabulated known melting points.

Table 32.3
Condensation products of carbonyl compounds

Reactant	Product
NH_2OH hydroxylamine	(from propanone)
NH_2NH_2 hydrazine	(from benzaldehyde)
(2,4-dinitrophenylhydrazine)	(from ethanal)

32.5 Oxidation of carbonyl compounds

Table 32.4 shows the effect of some oxidising agents on different carbonyl compounds.

Look at table 32.4 and answer the questions at the top of the next page.

Table 32.4
Effect of warming different carbonyl compounds with oxidising agents

	Oxidising agent	
Carbonyl compound	Complexed Ag^+ in alkaline solution (Tollen's reagent)	Complexed Cu^{2+} in alkaline solution (Fehling's solution)
methanal	silver mirror formed on walls of tube	copper metal formed together with copper(I) oxide
ethanal	silver mirror formed on walls of tube	red precipitate of copper(I) oxide formed
propanone	no reaction	no reaction

■ Write a half-equation for the reduction of Ag^+ to Ag.
■ Which of the carbonyl compounds in the table are oxidised by Ag^+?
■ Suggest a compound to which ethanal might be oxidised.
■ To what oxidation state is Cu^{2+} reduced by (a) methanal (b) ethanal?
■ Which is the more powerful reducer, methanal or ethanal?

Aldehydes carry a hydrogen atom next to their carbonyl group. This hydrogen is activated by the carbonyl group and is readily oxidised to OH. Aldehydes are therefore readily oxidised to carboxylic acids. For example, the oxidation of ethanal by Ag^+:

$$CH_3-\overset{\overset{\textstyle ||}{\textstyle O}}{C}-H + H_2O \rightarrow CH_3\overset{\overset{\textstyle ||}{\textstyle O}}{C}-OH + 2H^+ + 2e^-$$

ethanal ethanoic acid

$$Ag^+ + e^- \rightarrow Ag$$

Overall:

$$CH_3\overset{\overset{\textstyle ||}{\textstyle O}}{C}-H + H_2O + 2Ag^+ \rightarrow CH_3\overset{\overset{\textstyle ||}{\textstyle O}}{C}-OH + 2H^+ + 2Ag$$

Methanal has two hydrogen atoms attached to the carbonyl group, both of which are oxidisable. Methanal is very readily oxidised and is consequently a more powerful reducer than other aldehydes:

$$H-\overset{\overset{\textstyle ||}{\textstyle O}}{C}-H + H_2O \longrightarrow H-\overset{\overset{\textstyle ||}{\textstyle O}}{C}-OH + 2H^+ + 2e^-$$

methanal methanoic acid

accepted by oxidiser

Then:

$$H-\overset{\overset{\textstyle ||}{\textstyle O}}{C}-OH + H_2O \longrightarrow \left[HO-\overset{\overset{\textstyle ||}{\textstyle O}}{C}-OH\right] + 2H^+ + 2e^-$$

methanoic acid carbonic acid

accepted by oxidiser

$$\downarrow$$
$$CO_2 + H_2O$$

Thus methanal can reduce $Cu(II)$ to $Cu(0)$, while other aldehydes reduce it only to $Cu(I)$.

Methanal is also different from most other aldehydes in being able to **disproportionate** (see section 17.10). In the presence of aqueous sodium hydroxide, molecules of methanal simultaneously oxidise and reduce one another. The products are methanol (from reduction of methanal) and methanoic acid (from oxidation of methanal).

$$HCHO + HCHO \longrightarrow CH_3OH + HCOOH$$

2 molecules of methanal methanol methanoic acid

This is known as **Cannizaro's reaction**. Most other aldehydes do not give the reaction because in the presence of sodium hydroxide they polymerise instead (section 32.3). The only aldehydes giving Cannizaro's reaction are those with no H atom on the carbon atom next to the carbonyl group. Benzaldehyde comes into this category, so it gives Cannizaro's reaction. When heated with sodium hydroxide solution, benzaldehyde gives phenylmethanol and benzoic acid.

Ketones are not readily oxidised at all. They have no effect on mild oxidising agents like those in table 32.4. This is because they have no oxidisable hydrogen atom joined to the carbonyl group. Strong oxidisers like hot concentrated nitric acid can oxidise ketones, but the effect is to break up the molecule, forming at least two molecules of carboxylic acid.

The oxidisers in table 32.4 are commonly used to test for the aldehyde group.

Fehling's solution is made by mixing copper(II) sulphate solution with an alkaline solution containing 2,3-dihydroxybutanedioate ions (tartrate ions).

These complex the Cu^{2+} ions and prevent precipitation of copper(II) hydroxide from the alkaline solution.

Tollen's reagent is made by adding excess ammonia solution to a solution of silver(I) ions. It contains the $[Ag(NH_3)_2]^+$ ion in alkaline solution. Complexing of Ag^+ by NH_3 prevents precipitation of silver hydroxide.

Many other oxidisers can be used to convert aldehydes to carboxylic acids. In the laboratory, acidified dichromate(VI) is commonly used to prepare acids from aldehydes. Cataytic air oxidation of aldehydes was at one time used industrially to manufacture carboxylic acids, but nowadays they are mainly produced directly from alkanes (see section 33.1).

32.6 Effect of the carbonyl group on neighbouring atoms

Figure 32.2
Polarising effect of the carbonyl group in ethanal.

The carbonyl group is polar with a considerable partial positive charge on the carbon atom. This charge has the effect of withdrawing electrons from neighbouring carbon atoms, as shown in figure 32.2.

The result is to make the C—H bond of the neighbouring carbon more polar than normal. As a result, the hydrogen atoms are more readily replaced than those in alkanes. For example, ethanal reacts readily with chlorine even in the dark, forming trichloroethanal:

$$CH_3CHO + 3Cl_2 \longrightarrow CCl_3CHO + 3HCl$$
$$\text{ethanal} \qquad\qquad \text{trichloroethanal}$$

Trichloroethanal is used in the manufacture of the insecticide DDT (see figure 32.3). Section 18.12 gives more information about DDT and particularly the dangers involved in its use.

Figure 32.3
Right. Manufacture of DDT.

The iodoform reaction

If chlorine and ethanal are warmed together in the presence of a base, trichloroethanal is initially formed. This then reacts with the base to form trichloromethane, commonly known as chloroform:

$$CCl_3CHO + OH^- \longrightarrow CHCl_3 + HCOO^-$$
$$\text{trichloroethanal} \qquad \text{trichloromethane} \quad \text{methanoate ion}$$

A similar reaction occurs when ethanal is warmed with iodine and a base. The product in this case is triiodomethane (iodoform), CHI_3. This kind of reaction is not limited to ethanal. Any compound of the general formula CH_3COR, where R is an alkyl group or hydrogen, will form CHI_3 when warmed with iodine and a base. Triiodomethane is a yellow solid with a characteristic smell. Being insoluble in water, it appears as a yellow crystalline precipitate. This reaction, known as the **iodoform reaction**, is a useful test for compounds of the form CH_3COR. The reaction is also given by compounds of the form CH_3CHOHR, since these are themselves oxidised by the iodine to CH_3COR. For example, ethanol, CH_3CH_2OH, gives the iodoform reaction.

(i)

(ii)

Figure 32.4
Structure of glucose: (i) full structural formula, showing the puckered shape of the ring; (ii) skeletal formula. The hexagonal ring should be thought of as *perpendicular* to the paper, with the OH groups projecting above or below it. Hydrogen atoms attached to ring carbon atoms are not shown.

Which of the following compounds would give a yellow precipitate of triiodomethane (iodoform) when heated with a solution of iodine in aqueous sodium carbonate?

■ CH_3CH_2CHO ■ $CH_3CHOHCH_3$ ■ $CH_3CH_2CH_2OH$
■ $CH_3CH_2COCH_3$ ■ HCHO

32.7 Sugars—naturally occurring carbonyl compounds

Sugars are sweet-tasting soluble carbohydrates. Carbohydrates derive their name from the fact that they are composed of carbon, hydrogen and oxygen with H and O in the ratio of 2:1, as in water. **Monosaccharides** like glucose (figure 32.4) are usually **pentoses** or **hexoses**, i.e. they contain 5 or 6 carbon atoms in their molecules. **Disaccharides** like sucrose consist of two monosaccharide molecules joined by the elimination of a molecule of water. **Polysaccharides** like starch are made up of many monosaccharide units joined together. The structure of polysaccharides is considered further in section 9.8. Table 32.5 gives some examples of common carbohydrates. Notice that the monosaccharides all have asymmetric molecules: they therefore exhibit optical isomerism.

The most obvious feature of the structures of the monosaccharides and disaccharides shown in table 32.5 is the presence of large numbers of OH groups. These give them a large capacity for hydrogen bonding, so they are involatile solids, soluble in water. The presence of OH groups on several adjacent carbon atoms in the molecule is thought to be responsible for the sweet taste of sugars.

Table 32.5
Some common carbohydrates

Name	Type	Structure	Occurrence
glucose	monosaccharide, aldose, hexose		occurs abundantly in plants and animals
fructose	monosaccharide, ketose, hexose		in fruit and honey
ribose	monosaccharide, aldose, pentose		component of the molecules of ribonucleic acid (RNA) and vitamin B12
sucrose	disaccharide	glucose – fructose	sugar cane, sugar beet (commonly simply called 'sugar')
maltose	disaccharide	glucose – glucose	malt
lactose	disaccharide	glucose – galactose	milk
starch	polysaccharide	chains of glucose units	plant storage organs, e.g. potato, wheat grain
cellulose	polysaccharide	chains of glucose units (linked differently to those in starch)	structural material of plants

As well as showing the properties of polyhydroxy compounds, sugars show many properties in solution that are typical of carbonyl compounds. For example, glucose gives a crystalline condensation compound with 2,4-dinitrophenylhydrazine (section 32.4). This is surprising since the structure of glucose shown in figure 32.4 contains no carbonyl group.

The carbonyl propeties possessed by glucose arise from the fact that in addition to its normal ring form it can exist as an 'open-chain' form.

ring form   open-chain form

The two forms are readily interconverted and in aqueous solution about 1% of glucose molecules exist in the open-chain form. This form can be clearly seen to carry an aldehyde group. Because of this group, glucose has several properties typical of an aldehyde and is sometimes called an **aldose**. Thus, in addition to the condensation reaction already mentioned, glucose shows the reducing properties typical of an aldehyde. The reduction of Fehling's solution (or Benedict's solution, which is very similar) is a standard test for glucose and other reducing sugars.

The open-chain form of fructose is

Fructose is therefore a **ketose**.

The tendency of the open-chain form of glucose and other sugars to revert to the ring form results from the tendency of the carbonyl group to undergo nucleophilic addition. The nucleophile involved is the oxygen atom of one of the OH groups of the molecule. An internal nucleophilic addition reaction occurs, forming a ring.

glucose: open-chain form   glucose: ring form

Worker bees filling a comb with honey. Honey is a concentrated solution of sugars including glucose and fructose.

Right. One lump or two? Sucrose dissolves because its OH groups form hydrogen bonds.

Far right. Carbohydrates—a mono-saccharide, a disaccharide and a polysaccharide.

This reaction occurs spontaneously. Under normal conditions in aqueous solution the two forms exist in equilibrium, with the ring form predominating.

An understanding of the chemistry of carbohydrates is vital to an understanding of biology. These molecules occur in all living organisms, as structural materials (e.g. cellulose), energy storage compounds (e.g. starch) and primary energy sources (e.g. glucose). The biological importance of carbohydrates is considered further in section 9.8.

Consider the following sugars. In each case the normal ring form is shown, with the open-chain form below it.

A   B   C

- Which is a heptose?
- Which would reduce Tollen's reagent to silver?
- Which is a ketose?
- Which would undergo a condensation reaction with 2,4-dinitrophenyl-hydrazine?

Summary

1 The carbonyl group, C=O, is present in aldehydes and ketones. In aldehydes it is in a terminal position in the carbon chain. In ketones it is in a non-terminal position.

2 Aldehydes and ketones are named using the suffixes *-al* and *-one*, respectively.

3 Aldehydes are prepared by oxidising primary alcohols, ketones by oxidising secondary alcohols.

4 The carbonyl group readily undergoes nucleophilic addition. This is sometimes followed by the elimination of a molecule of water, resulting in a condensation reaction.

5 Aldehydes are generally more reactive than ketones.

6 The tendency of aldehydes to undergo nucleophilic addition makes them polymerise readily.

7 Aldehydes can be oxidised to carboxylic acids by a variety of reagents. Ketones are not readily oxidised.

Table 32.6
Important reactions of ethanal and propanone

Reagent	Reaction type	Ethanal	Propanone
HCN	addition	$\longrightarrow CH_3CH\begin{smallmatrix}OH\\CN\end{smallmatrix}$	$\longrightarrow CH_3CCH_3$ with OH and CN
NaHSO$_3$	addition	$\longrightarrow CH_3CH$ with OH and $SO_3^-Na^+$	$\longrightarrow CH_3CCH_3$ with OH and $SO_3^-Na^+$
H$_2$/Ni	addition/reduction	$\longrightarrow CH_3CH_2OH$	$\longrightarrow CH_3CHOHCH_3$
dilute acid	addition/polymerisation	$\longrightarrow (CH_3CHO)_3$ above 0°C $\longrightarrow (CH_3CHO)_4$ below 0°C	does not polymerise
NH$_2$OH	condensation	$\longrightarrow CH_3CH$ ‖ NOH	$\longrightarrow CH_3CCH_3$ ‖ NOH
NH$_2$NH$_2$	condensation	$\longrightarrow CH_3CH$ ‖ NNH$_2$	$\longrightarrow CH_3CCH_3$ ‖ NNH$_2$
Ag$^+$ (complexed) (Tollen's reagent)	oxidation	$\longrightarrow CH_3COOH$ Ag$^+$ reduced to Ag	no reaction
Cu^{2+} (complexed) (Fehling's solution)	oxidation	$\longrightarrow CH_3COOH$ Cu^{2+} reduced to Cu$_2$O	no reaction
I$_2$/base	iodoform reaction	yellow ppte of CHI$_3$	yellow ppte of CHI$_3$

8 The carbonyl group activates the hydrogen atoms on neighbouring carbon atoms, making them more readily substituted than those in alkanes.

9 Monosaccharides like glucose or fructose are polyhydroxy compounds usually containing five or six carbon atoms. They show many properties typical of carbonyl compounds.

10 Table 32.6 gives some of the important reactions of ethanal and propanone.

Study questions

1 Consider the following compounds:

A $CH_3COCHCH_2CH_3$
  |
  CH_3

B $CH_3CH_2CH_2CH_2CHO$

C

D

E (cyclohexanone structure)

(a) Which are aldehydes?
(b) Which are ketones?
(c) Which is a hexose?
(d) Name compounds **A, B, D** and **E**.
(e) Which would produce a red precipitate of copper(I) oxide when boiled with Fehling's solution?
(f) Which would give a yellow crystalline precipitate when reacted with iodine in basic solution?
(g) Which would be reduced to a secondary alcohol by hydrogen in the presence of a nickel catalyst?

2 'The C=C and C=O groups might be expected to show chemical similarity, but in fact show very little.' Is this true? Illustrate and discuss the statement, referring to a wide range of examples and to the underlying mechanistic principles.

3 This question is about the manufacture of ethanal. Read the passage below and then answer the questions on it.

Ethanal used to be made from ethyne, CH≡CH, which is readily manufactured from coke. This process is now being replaced by a method based on ethene. The overall process amounts to oxidation. Ethene and oxygen are bubbled together through an aqueous solution containing $CuCl_2$ and $PdCl_2$ catalysts:

$$CH_2{=}CH_2 + \tfrac{1}{2}O_2 \xrightarrow[\text{CuCl}_2]{\text{PdCl}_2} CH_3CHO$$

The mechanism of the reaction involves initial attack on the ethene molecule by water. Since H_2O is a nucleophile, it is necessary to render the ethene, which is normally subject only to electrophilic attack, attractive to nucleophiles. This is done by the Pd^{2+} catalyst, which forms a complex with the ethene, decreasing the electron density in the double bond and making the molecule prone to nucleophilic attack.

$$H_2\ddot{O}: \quad \begin{matrix} CH_2 \\ \| \\ CH_2 \end{matrix} \text{------}Pd^{2+}$$

The complex thus formed breaks down by a number of stages to form ethanal and palladium metal. The palladium metal is oxidised back to Pd^{2+} by oxygen. The $CuCl_2$ is a catalyst for this oxidation.

(a) Why is ethanal an important industrial chemical?

(b) Why do you think ethene is replacing ethyne as the starting material for ethanal manufacture?

(c) The suggested mechanism involves the formation of a complex between ethene and Pd^{2+}. How do you think ethene is bonded to Pd^{2+} in this complex?

(d) Suggest a reason why the formation of this complex renders ethene attractive to nucleophiles rather than electrophiles.

(e) Could this method be used to manufacture *propanone*? If so, what starting material would be used instead of ethene?

4   Predict the formulae of the products of the following reactions.

(a) $CH_3COCH_2CH_3 + H_2 \xrightarrow{Ni}$

(b) $C_6H_5COCH_3 + NH_2OH \longrightarrow$

(c) $CH_3CH_2COCH_2CH_3 + HCN \longrightarrow$

(d) $C_6H_5CHO + KMnO_4 \longrightarrow$

(e) $CH_3CH_2CHO + Cl_2 \longrightarrow$

5   Write structural formulae for all compounds of molecular formula C_4H_8O containing a carbonyl group. How would you distinguish between the different compounds, using simple chemical tests?

6   (a) A compound **A** has molecular formula $C_4H_6O_2$. **A** reacts with HCN to form compound **B**, $C_6H_8O_2N_2$. **A** is readily oxidised by acidified potassium dichromate(VI) to an acidic compound **C**, $C_4H_6O_4$. When 1.0 g of **C** is dissolved in water and titrated with 1.0 M sodium hydroxide, 16.9 cm^3 of sodium hydroxide are required for neutralisation.
Suggest structural formulae for **A**, **B** and **C** and explain the above reactions.

(b) A compound **X** contains 64.3% C, 7.1% H, and 28.6% O by mass. Its relative molecular mass is 56. **X** reduces Fehling's solution to copper(I) oxide. **X** reacts with hydrogen in the presence of a nickel catalyst: 0.1 g of **X** was found to react with 80 cm^3 of hydrogen (measured at s.t.p.).
Suggest a structural formula for **X** and explain the above reactions.

7   Suggest explanations for the following.

(a) Five different oxidation products of ethane-1,2-diol are known (excluding carbon dioxide and water).

(b) When ethanal is added to heavy water (D_2O) containing a small amount of base, trideuteroethanal (CD_3CHO) is formed.

(c) Trichloroethanal undergoes addition reactions far more readily than ethanal.

8   This question is about **Grignard reagents**, compounds that are of great use in organic synthesis for forming carbon–carbon bonds.
Grignard reagents are compounds of general formula RMgX where X = Br or I. They are very reactive, giving rise to the highly nucleophilic ion R^-. When a solution of a Grignard reagent in dry ether is added to a carbonyl compound, the R^- attacks the carbonyl group:

If water is now added to this product, an alcohol is formed:

Overall: $\overset{\diagdown}{\diagup}C{=}O + RMgBr + H_2O \longrightarrow \underset{\diagup}{\overset{\diagdown}{}}C\overset{\diagup OH}{\underset{\diagdown R}{}} + MgBrOH$

Predict the formulae of the compounds formed when the following are treated with the Grignard reagent methyl magnesium bromide, followed by water.

(a) Methanal
(b) Ethanal
(c) Propanone
(d) Carbon dioxide.

33 Carboxylic Acids and their Derivatives

33.1 Carboxylic acids

Why is an ant like a pickled onion?

They both owe their powerful effect to carboxylic acids: in the case of the ant to methanoic acid injected into the skin when the ant bites; in the case of the pickled onion to the ethanoic acid present in vinegar. Carboxylic acids and their derivatives occur widely in nature. They are also present in many manufactured products, such as soaps and polyesters.

Carboxylic acids contain the **carboxyl** group, $-\overset{\displaystyle \|}{\underset{\displaystyle O}{C}}-OH$. Their general formula is usually written RCOOH, where R is an alkyl or aryl group or hydrogen. They are named using the suffix **-oic acid** after a stem indicating the number of carbon atoms (including the one in the carboxyl group). The first two members of the series, systematically named methanoic acid and ethanoic acid, are often called by their older names, formic acid and acetic acid.

Where two carboxyl groups are present, the suffix **-dioic acid** is used. Table 33.1 further illustrates the naming of carboxylic acids.

Table 33.1
Some important carboxylic acids

Formula	Systematic name	Other name	Occurrence and uses
HCOOH	methanoic acid	formic acid (from Latin *formica*, an ant)	used in textile processing and as a grain preservative; ants use it as a poison, injecting it when they bite their victims
CH_3COOH	ethanoic acid	acetic acid (from Latin *acetum*, vinegar)	in vinegar; used in making artificial textiles
CH_3CH_2COOH	propanoic acid	propionic acid	calcium propanoate is used as an additive in bread manufacture
COOH (on benzene ring)	benzoic acid		food preservative
COOH | COOH	ethanedioic acid	oxalic acid	in rhubarb leaves

The C=O and O—H of the carboxyl group are so close that they modify one another's properties a great deal. Carboxylic acids therefore have many reactions that are different from those of both alcohols and carbonyl compounds.

The properties of the COOH group are modified only slightly when it is attached to a benzene ring. Aromatic carboxylic acids as a result have many properties in common with aliphatic ones.

Making carboxylic acids

In the laboratory, carboxylic acids are normally prepared by oxidising primary alcohols (section 31.7) or aldehydes (section 32.5). Details of the preparation of ethanoic acid from ethanol are given in practical 37, *Laboratory Manual and Study Guide*. Carboxylic acids can also be prepared by hydrolysis of nitriles (section 34.2). The hydrolysis is carried out by heating with strong acid or base, e.g.

$$RCN + 2H_2O + HCl \rightarrow RCOOH + NH_4Cl$$

Nitriles are themselves prepared from halogenoalkanes (section 30.5).

The oxidation of alcohols and aldehydes was at one time the major industrial method of preparing carboxylic acids. Nowadays the most important aliphatic acids are produced by direct oxidation of alkanes. A volatile petroleum fraction is oxidised in the liquid phase by passing air through it under pressure in the presence of a catalyst. Ethanoic acid is the major and most important product. By-products include methanoic and propanoic acids.

Benzoic acid is made industrially and in the laboratory by the oxidation of methylbenzene (chapter 29).

Consider the following acids:

A

$$CH_2 < ^{COOH} \atop CH_2 < _{COOH}$$

B

$$CH_3CH_2CH_2COOH$$

C

$$CH_3 - \bigcirc - COOH$$

D

$$CH_3(CH_2)_{16} COOH$$

- Which is octadecanoic acid?
- Which is a dibasic acid?
- Which is butanoic acid?
- Which could be prepared by oxidation of butane-1,4-diol?
- Which is an aromatic acid?

Physical properties

Ethanoic acid melts at 17°C and boils at 118°C. This means that although it is normally a liquid in British laboratories, it freezes in cold weather. In fact it is a fairly good rule that when the ethanoic acid freezes it is time to switch the heating on. Because of the readiness with which it freezes, and the similarity of solid ethanoic acid to ice, pure ethanoic acid is often described as 'glacial'. Dilute solutions of the acid, of course, freeze at about the same temperature as water.

The boiling point of ethanoic acid is higher than that of either ethanol (78°C), which has the same number of carbon atoms, or propan-1-ol (97°C), which has the same relative molecular mass. The relatively high boiling point of ethanoic acid and other carboxylic acids reflects a considerable degree of hydrogen bonding. Carboxylic acids form stronger hydrogen bonds than alcohols. This is because their OH group is more polarised due to the presence of the electron-withdrawing C=O group:

compare

$$R - C < ^{O} \atop O - H \atop {\delta-} \quad {\delta+}$$

with

$$H - C(R)(H) - O - H \atop {\delta-} \quad {\delta+}$$

Both these bottles contain glacial ethanoic acid. Pure ethanoic acid freezes at 17°C.

Table 33.2
Some properties of carboxylic acids

Acid	Formula	State at 25°C	Boiling point/°C	K_a at 25°C/mole dm^{-3}	Solubility in water
methanoic	HCOOH	l	101	1.6×10^{-4}	infinite
ethanoic	CH_3COOH	l	118	1.7×10^{-5}	infinite
propanoic	CH_3CH_2COOH	l	141	1.3×10^{-5}	infinite
butanoic	$CH_3CH_2CH_2COOH$	l	164	1.5×10^{-5}	infinite
octanoic	$CH_3(CH_2)_6COOH$	l	237	1.4×10^{-5}	slightly soluble
chloroethanoic	$ClCH_2COOH$	s	189	1.3×10^{-3}	very soluble
dichloroethanoic	$Cl_2CHCOOH$	l	194	5.0×10^{-2}	infinite
trichloroethanoic	Cl_3CCOOH	s	196	2.3×10^{-1}	very soluble
benzoic	C_6H_5COOH	s	249	6.4×10^{-5}	slightly soluble
ethanedioic	$(COOH)_2$	s		3.5×10^{-2} (first dissociation) 4.0×10^{-5} (second dissociation)	soluble

Furthermore, carboxylic acids have the possibly of forming doubly hydrogen-bonded dimers:

$$R—C \begin{matrix} O ---H—O \\ \\ O—H---O \end{matrix} C—R$$

Carboxylic acids in the liquid and solid states exist mostly in this form (see section 10.8).

Because of their capacity for hydrogen bonding, the early members of the series are miscible with water in all proportions. As with all homologous series solubility decreases with increasing molecular size. Table 33.2 gives physical properties of some carboxylic acids.

Methanoic, ethanoic and propanoic acids have strong, sharp vinegary odours. The C_4 to C_8 acids (butanoic to octanoic) have very strong, unpleasant odours. Butanoic acid is responsible for the smell of rancid butter and is present in human sweat. Its smell can be detected at concentrations of 10^{-11} mole dm^{-3} by humans and at concentrations of 10^{-17} mole dm^{-3} by dogs. The ability of a dog to track a person is due to its ability to detect carboxylic acids in the sweat from the person's feet. Each person's sweat glands produce a characteristic blend of carboxylic acids which can be detected and recognised by the sensitive nose of the dog.

A dog can be trained to track people by recognising the characteristic blend of carboxylic acids in the sweat from a particular person's feet.

33.2 The carboxyl group

In chapter 31 we discussed the acidity of the OH group in alcohols, and found that alcohols were even less acidic than water. Carboxylic acids also contain the OH group, but they are much stronger acids than alcohols and water. Most of the acids we taste in our food, such as ethanoic acid in vinegar and citric acid in lemons, are carboxylic acids. They appear to have a very strong acid taste;

nevertheless, they are only present in low concentrations (vinegar is about 7% ethanoic acid) and they are weak acids compared to inorganic ones like hydrochloric acid.

When a carboxylic acid dissociates in water, it forms a carboxylate anion $RCOO^-$:

$$H_2O + R-C\overset{\displaystyle O}{\underset{\displaystyle OH}{<}} \longrightarrow H_3O^+ + R-C\overset{\displaystyle O}{\underset{\displaystyle O^-}{<}}$$

carboxylic acid   carboxylate anion

X-ray diffraction studies of the anion show the two carbon–oxygen bonds to be of equal length. This implies that the two bonds are identical. It suggests that the negative charge and double bond character are distributed evenly over the whole carboxylate group, delocalising the negative charge:

$$R-C\overset{\displaystyle O}{\underset{\displaystyle O}{<}}^-$$

Thus each oxygen effectively carries half a negative charge. This delocalisation of charge on the anion makes it less likely to become protonated again. As a result, the equilibrium

$$RCOOH + H_2O \rightleftharpoons RCOO^- + H_3O^+$$

is much further to the right than in the case of alcohols, so carboxylic acids are stronger acids. Nevertheless, the majority of the acid is in the un-ionised form and 1.0 M ethanoic acid is only 0.3% ionised.

The nature of the group to which COOH is attached can have a considerable effect on the strength of a carboxylic acid. In general, an electron-withdrawing group reduces the density of negative charge on the anion, and so increases the strength of the acid. An electron-donating group does the reverse.

Consider the following acids:

$$CH_3COOH \quad CH_2ClCOOH \quad CCl_3COOH \quad HCOOH$$

Which would be the stronger acid:

- CH_3COOH or $CH_2ClCOOH$?
- CCl_3COOH or $CH_2ClCOOH$?
- CH_3COOH or $HCOOH$?

methanoate ion

ethanoate ion

chloroethanoate ion

Figure 33.1
Anions formed from methanoic, ethanoic, and chloroethanoic acids.

Table 33.2 gives the K_a values for different carboxylic acids (remember, the higher the K_a, the stronger the acid). The electron-donating CH_3 group tends to increase the negative charge density on the carboxylate group in the ethanoate ion. Ethanoic acid is therefore weaker than methanoic, which carries no electron-donating alkyl group. On the other hand, the electron-withdrawing chlorine atom reduces the charge density on the chloroethanoate ion so chloroethanoic acid is stronger than ethanoic (figure 33.1).

The electron-donating effect of all alkyl groups is roughly equal, so unsubstituted aliphatic carboxylic acids with more than one carbon atom (ethanoic, propanoic, etc.) are roughly equal in strength.

33.3 Salts of carboxylic acids

Like all acids, carboxylic acids react with bases to form salts. Table 33.3 gives some experimental data about benzoic acid.

Look at the table, then answer the questions which follow it.

Table 33.3
Behaviour of benzoic acid in basic and acidic solution

Experiment		Result
1	solid benzoic acid is shaken with water	most of the benzoic acid is undissolved
2	sodium hydroxide is added to the aqueous suspension from experiment (1)	benzoic acid dissolves
3	hydrochloric acid is added to the solution from experiment (2)	precipitate of benzoic acid appears

■ Why is benzoic acid not very soluble in water?
■ Why does it dissolve in sodium hydroxide solution?
■ Why does hydrochloric acid precipitate benzoic acid from its solution in sodium hydroxide?

Carboxylic acids are weak acids and so only slightly dissociated in aqueous solution:

$$H_2O + RCOOH \rightleftharpoons RCOO^- + H_3O^+$$
$$\text{carboxylic acid} \qquad \text{carboxylate}$$
$$\text{ion}$$

Addition of a base removes H_3O^+, shifting this equilibrium to the right. This converts the acid to the carboxylate ion and forms the salt. Benzoic acid is much more soluble in basic solution than in water because in the presence of a base it is converted almost entirely to soluble benzoate ion. The addition of a strong acid to a solution of the carboxylate salt reverses the equilibrium, reforming free acid. The anions of carboxylic acids are themselves basic: a molar solution of sodium ethanoate has a pH of about 9.5.

An interesting application of the formation of salts by carboxylic acids occurs with ethanedioic acid (oxalic acid). This acid is poisonous, because of the toxic nature of its anion, the ethanedioate ion. The normal way of treating people who have swallowed an acid is to give them an alkali like sodium hydrogencarbonate to neutralise it. In the case of ethanedioic acid, though, this only makes the poisoning worse. This is because the addition of an alkali encourages the formation of the poisonous anion by removing H^+. The correct way to treat this kind of poisoning is to give an alkali which can form an insoluble salt with ethanedioic acid. For example, magnesium hydroxide can be used, because magnesium ethanedioate is insoluble.

Salts of carboxylic acids show marked differences in properties from the free acids, because they are ionic rather than molecular substances. Thus all carboxylate salts are solids, and most are soluble in water. Not surprisingly, as the hydrocarbon chain increases in length, the properties of these salts tend more and more to those of molecular hydrocarbons rather than ionic salts.

Salts of carboxylic acids with medium-length chains have the characteristics of both hydrocarbons and salts. This makes them valuable detergents.

Detergents

Detergents are substances that improve the cleaning properties of water. The most common 'dirt' is that on human clothes and bodies . Typically this consists of a mixture of natural fats, long-chain carboxylic acids, skin debris, silicon(IV) oxide and carbon. The main problem is caused by greasy fats and carboxylic acids, which tend to bind the other materials to the skin or fabric surface. This prevents them being washed away by water. A detergent works by enabling water to mix with and remove these greasy materials.

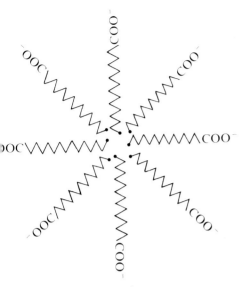

Figure 33.2
Arrangement of soap chains in a micelle, shown in cross-section. The charged ends are on the outside, since they are attracted by water. (Each micelle contains far more chains than shown here.)

Soap was the first detergent, dating from Roman times, and it is still one of the commonest. Soap is a mixture of the salts of medium- and long-chain carboxylic acids. Its manufacture is described in the next section. A typical salt present in soap is sodium octadecanoate (sodium stearate):

$$CH_3(CH_2)_{16}COO^- Na^+$$

or

$COO^- Na^+$

- Which end of this molecule would you expect to mix well with water?
- Which end would you expect to mix well with grease?
- Can you see how soap helps water to remove grease from a surface?
- Why is hot water better than cold for washing in?

The charged COO^- group at the end of a soap chain enables it to dissolve in water. When soap dissolves, it does so in the form of spherical clusters called **micelles** (figure 33.2). These micelles are large enough to cause the solution to scatter light: this is why soapy water is cloudy. The non-polar hydrocarbon chain mixes well with greasy substances. When soapy water comes into contact with grease, the non-polar end of the molecule mixes with the non-polar grease (figure 33.3). This leaves the grease surrounded by an outer sheath of polar COO^- groups which are attracted to water. The forces between water and grease are thus much increased, so that the grease is lifted off the surface in the form of small globules. These can then be rinsed away. The sheath of negatively charged groups causes the globules to repel one another, so they do not coagulate and redeposit on the surface.

(a)

Figure 33.3
Effect of detergent on grease.

(c)

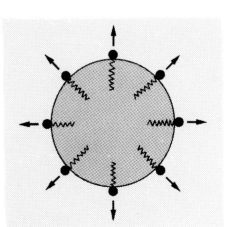

Grease micelle washed away by water

(b)

Soapless detergents

Soap is not the only detergent. Any molecule with a fairly large non-polar 'tail' and a charged or polar 'head' will have detergent properties. The head may carry a negative charge (**anionic** detergents are the most common), a positive charge (**cationic**), or may consist of a polar, uncharged group (**non-ionic**).

The effect of soap on the wetting ability of water. The droplet on the right is pure water; on the left is a soap and water solution.

Soaps have the disadvantage that they form a 'scum' with hard water. Hard water contains calcium ions. These ions react with the soap to form the insoluble calcium salt of the carboxylic acid. This is the scum. For example:

$$2\,CH_3(CH_2)_{16}COO^-(aq) + Ca^{2+}(aq) \rightarrow (CH_3(CH_2)_{16}COO)_2Ca(s)$$

octadecanoate ions   scum
from soap

The scum is itself undesirable and its formation leads to the use of excessive quantities of soap.

Soapless detergents have the advantage that they form no scum with hard water. In addition, they are made from by-products of oil refining. Soap, on the other hand, is made from vegetable or animal fats which are expensive and could be used for human consumption.

A typical and very common soapless detergent has the structure:

$$CH_3(CH_2)_{11}\text{—}\bigcirc\text{—}SO_3^- Na^+ \quad \text{i.e.} \quad \diagdown\!\diagup\!\diagdown\!\diagup\!\diagdown\!\diagup\!\diagdown\!\diagup\!\diagdown\!\diagup\text{—}\bigcirc\text{—}SO_3^- Na^+$$

Which are the water-attracting and water-repelling parts of this compound?

This detergent is manufactured by the sequence shown in figure 33.4. Look back to section 29.7 to remind yourself of the reactions involved. This kind of soapless detergent is present in well-known and well-advertised washing powders. Earlier synthetic detergents included branched-chain compounds with the structure:

These are cheap to make. Unfortunately because of their side chains they cannot be degraded by bacteria, unlike the synthetic detergent shown earlier. They therefore tended to persist in water after sewage treatment and caused foaming in rivers and streams. They have now been replaced by detergents with unbranched side chains.

Synthetic detergents are gradually replacing soap. Many different compounds with detergent properties are now known and manufactured. They are used not only for cleaning purposes but also as wetting agents, foam stablisers and emulsifying agents. All these uses depend on their ability to mix both with water and with water-repelling materials. Some more examples are given in table 33.4.

$$CH_3(CH_2)_9CH = CH_2 + \bigcirc$$
dodecene

AlCl$_3$, HCl
(Friedel-Craft's reaction)

$$CH_3(CH_2)_{11}\text{—}\bigcirc$$
dodecylbenzene

c.H$_2$SO$_4$

$$CH_3(CH_2)_{11}\text{—}\bigcirc\text{—}SO_3H$$

NaOH

$$CH_3(CH_2)_{11}\text{—}\bigcirc\text{—}SO_3^- Na^+$$

Figure 33.4
Manufacture of a soapless detergent.

Table 33.4
Some synthetic detergents

Type	Non-polar portion	Polar portion	Uses
Non-ionic	C_9H_{19}——⬡——	$O(CH_2CH_2O)_x CH_2CH_2O$ (x = 5 to 10)	liquid detergents emulsifying agents wetting agents
Cationic	$C_{15}H_{31}$———————	$CH_2\overset{+}{N}(CH_3)_3Br^-$	hair conditioner
Anionic	$C_{12}H_{23}$———————	$COOCH_2CHOHCH_2OSO_3^-Na^+$	toothpastes shampoos

Non-biodegradable detergents are responsible for the persistent foam to be seen in some rivers.

33.4 Some important reactions of carboxylic acids

All these products contain detergents of one sort or another.

Reaction with bases

This has already been considered (section 33.3).

Esterification

This important reaction is considered in the next section.

Reaction with phosphorus pentachloride

You may recall from section 31.6 that phosphorus pentachloride reacts with the OH group in alcohols. What is the product of such a reaction? Given that carboxylic acids contain the OH group, what product would you expect from the reaction of, say, ethanoic acid with phosphorus pentachloride?

Carboxylic acids react with phosphorus pentachloride to form acyl chlorides (section 30.7). For example

$$CH_3\underset{\underset{O}{\|}}{C}OH + PCl_5 \longrightarrow CH_3\underset{\underset{O}{\|}}{C}Cl + HCl + POCl_3$$

ethanoic acid   ethanoyl
                       chloride

Other halogenating agents such as $SOCl_2$ and PBr_3 react similarly.

Oxidation

As a rule, carboxylic acids are not readily oxidised. This is because they are themselves the end products of the oxidation sequence

primary alcohol → aldehyde → carboxylic acid.

There are however two exceptions to this general rule.

(i) Methanoic acid and methanoates are readily oxidised, for example by potassium manganate(VII), to carbon dioxide:

$$HCOOH \longrightarrow CO_2 + 2H^+ + \quad 2e^-$$

methanoic acid   accepted
                                        by oxidiser

(ii) Ethanedioic acid and its salts are oxidised to carbon dioxide by warm potassium manganate(VII):

$$(COOH)_2 \longrightarrow 2CO_2 + 2H^+ + \quad 2e^-$$

ethanedioic acid   accepted
                                          by oxidiser

Reduction

Carboxylic acids are rather hard to reduce. Powerful reducing agents like lithium tetrahydridoaluminate, $LiAlH_4$, will convert them to the corresponding primary alcohol:

$$R\underset{\underset{O}{\|}}{C}OH + 4H^+ + \quad 4e^- \longrightarrow RCH_2OH + H_2O$$

carboxylic acid   from   alcohol
                       reducer

Esters (section 33.5) are somewhat easier to reduce than acids. They can be converted to alcohols by the use of high-pressure hydrogen and a catalyst.

Dehydration

Methanoic and ethanedioic acids are exceptional as the only carboxylic acids that can be readily dehydrated.

■ What products would you expect to be formed when a molecule of water is eliminated from (a) a molecule of methanoic acid, HCOOH, (b) a molecule of ethanedioic acid, $(COOH)_2$?

■ What reagent could you use as a dehydrating agent for these two compounds?

Methanoic acid is dehydrated to carbon monoxide when warmed with concentrated sulphuric acid:

$$HCOOH \xrightarrow{\text{conc. } H_2SO_4} CO + H_2O$$

Ethanedioic acid is dehydrated by the same method to carbon monoxide and carbon dioxide:

$$(COOH)_2 \xrightarrow{\text{conc. } H_2SO_4} CO + CO_2 + H_2O$$

Use your knowledge of the reactions of carboxylic acids to write the structural formulae of the products formed when

- benzoic acid is treated with phosphorus pentachloride,
- butanoic acid is treated with lithium tetrahydridoaluminate,
- ethanedioic acid is treated with excess sodium hydroxide.

33.5 Esters

Figure 33.5 illustrates an experiment in which ethanol and ethanoic acid are heated in the presence of concentrated sulphuric acid. Study it, then answer the following questions.

Figure 33.5
Reaction between ethanol and ethanoic acid in the presence of concentrated sulphuric acid.

- Does (a) ethanol, (b) ethanoic acid, (c) sulphuric acid, mix with water?
- Do any of the substances above have a sweet, 'gluey' smell?
- What can you say about the product of this reaction between ethanol and ethanoic acid?
- Why do you think the reaction mixture is poured into cold water before smelling?

Ethanol and ethanoic acid react together to form an ester, ethyl ethanoate:

$$CH_3C\!\!-\!\!OH \; + \; CH_3CH_2OH \longrightarrow CH_3C\!\!-\!\!OHC_2CH_3 + H_2O$$
$$\qquad \overset{||}{O} \qquad\qquad\qquad\qquad\qquad \overset{||}{O}$$

ethanoic acid   ethanol   ethyl ethanoate

This reaction proceeds extremely slowly under ordinary conditions. However, it goes at an appreciable rate in the presence of a strong acid catalyst such as sulphuric or hydrochloric acid. This is a general method of preparing esters, and can be applied to any combination of acid and aliphatic alcohol:

$$R\!\!-\!\!C\!\!-\!\!OH + R'OH \rightarrow R\!\!-\!\!C\!\!-\!\!OR' + H_2O$$
$$\quad \overset{||}{O} \qquad\qquad\qquad \overset{||}{O}$$

Notice that the bridging oxygen atom between the RCO and R′ groups in the ester could have come either from the alcohol or from the acid. As the equation stands, there is no way of telling from which molecule it originated. To put the problem another way, does the oxygen atom in the alcohol molecule end up in the ester or in the water?

The answer to this question was found by two American chemists, Roberts and Urey, using a technique known as **isotopic labelling**. The method is outlined in figure 33.6.

This experiment shows that the bridging O in fact comes from the alcohol. This information enabled chemists to propose a mechanism for the esterification reaction. Isotopic labelling is often used to investigate the mechanism of a chemical reaction and has advanced our knowledge considerably, especially in the field of biochemistry.

---

1 Prepare methanol 'labelled' with the oxygen isotope ^{18}O.

$$CH_3 \, ^{18}OH$$

2 React this with benzoic acid in the presence of an acid catalyst.

— COOH + $CH_3 \, ^{18}OH$

3 Two sets of products are possible according to which bond breaks in $CH_3 \, ^{18}OH$.

— $CO \, ^{18}OCH_3 + H_2O$ ($M_r = 138$)   or   — $COOCH_3 + H_2 \, ^{18}O$ ($M_r = 136$)

4 Separate the ester and measure its relative molecular mass in a mass spectrometer.

5 M_r of the ester is found to be 138 corresponding to — $CO \, ^{18}OCH_3$. Hence the bridging O must come from the alcohol.

Figure 33.6
Investigation of the fate of oxygen atoms in an esterification reaction.

The nature of esters

Esters are the products of condensation reaction between alcohols and carboxylic acids. They can also be prepared by reacting alcohols with acyl chlorides (section 30.7) or acid anhydrides (section 33.7).

Esters are named by regarding them as alkyl (or aryl) derivatives of carboxylic acids. Thus the name is obtained from a stem indicating the alcohol from which the ester is derived, with the suffix **-yl**, followed by a stem indicating the acid, with the suffix **-oate**. So the ester derived from methanol and propanoic acid is called methyl propanoate. The name of an ester is most easily worked out from its formula by these stages:

(i) Divide the formula into two portions by mentally drawing a line after the bridging O of the COO group, e.g.

$$CH_3CH_2C\!-\!O\!-\!\!\!+\!\!\!CH_2CH_2CH_2CH_3$$
$$\|$$
$$O$$

(ii) Name the portion that does *not* carry the COO group. This portion is an alkyl group—*butyl* in this example.
(iii) Name the portion that carries the COO group. This portion is a carboxylate radical—*propanoate* in this example.
(iv) Combine (ii) and (iii) to give the name of the ester—*butyl propanoate*.

Try these questions to test your understanding of how esters are named and prepared.

■ Give the name of
  (a) $C_6H_5COOCH_3$,
  (b) $HCOOCH_2CH_3$.
■ Write the formula of
  (a) heptyl decanoate,
  (b) phenyl benzoate.
■ How would you prepare propyl propanoate?
■ Write the formula of the product formed when ethanoic acid and butan-2-ol are warmed together in the presence of an acid catalyst.

Unlike the acids and alcohols from which they are derived, esters have no free OH groups so they cannot form hydrogen bonds. They are therefore volatile compared with acids and alcohols of similar molecular mass and they are not very soluble in water. Ethyl ethanoate, for example, is a liquid, boiling point 77°C. At 25°C its solubility is 8.5 g per 100 g of water.

Volatile esters have characteristically pleasant, fruity smells. The flavour and fragrance of many fruits and flowers are due to mixtures of compounds, many of them esters. Artificial fruit flavourings are made by mixing synthetic esters to give the approximate flavour (raspberry, pear, cherry, etc.) required. Organic acids are usually added to give the sharp taste characteristic of fruit. Artificial fruit flavours can only approximate to the real thing, because it would be too costly to include all the components of the complex mixture of compounds present in the real fruit.

The smell of some adhesives is very like that of ethyl ethanoate. Polystyrene cement, for example, consists of polystyrene dissolved in ethyl ethanoate. When the cement is applied the ethyl ethanoate evaporates, leaving behind the solid plastic which binds together the surfaces being joined.

Polyesters

Polyesters are synthetic fibres used as substitutes for cotton and wool. They are marketed under such trade names as 'Terylene' and 'Dacron'. As the name 'polyester' suggests, they are polymers joined by an ester linkage. By far the commonest polyester is 'Terylene', poly(ethane-1,2-diyl benzene-1,4-dicarboxylate). This is made by esterifying ethane-1,2-diol with benzene-1,4-dicarboxylic acid (terephthalic acid) (figure 33.7 on the next page).

Having two OH groups on the alcohol and two COOH groups on the acid makes possible the formation of the polymer. This is an example of condensation polymerisation.

Polyesters have great tensile strength. They are used as the bonding resin in glass fibre plastics as well as being widely used to make fabrics.

Esterification as an equilibrium reaction

The reaction of a carboxylic acid with an alcohol to form an ester is fairly slow, even in the presence of an acid catalyst. As well as being kinetically quite slow, esterification is also an equilibrium reaction that does not normally reach completion.

If known quantities of ethanol, ethanoic acid and hydrochloric acid catalyst are sealed together and left for two or three weeks, an equilibrium is reached.

Most yachts today have sails made of polyester. It is stronger than the natural fibres formerly used, and it does not rot.

Polyester can withstand higher temperatures than most plastics, and this makes it suitable for the manufacture of food roasting bags.

Monomers
HOCH$_2$CH$_2$OH
ethane -1,2-diol

HOOC —⬡— COOH
benzene – 1,4 – dicarboxylic acid

Polymerisation

....+ HOCH$_2$CH$_2$OH + HOOC—⬡—COOH + HOCH$_2$CH$_2$OH + HOOC—⬡—COOH +—....

↓

....—OCH$_2$CH$_2$OOC—⬡—COOCH$_2$CH$_2$OOC—⬡—CO—....

+   +   +   +
H$_2$O   H$_2$O   H$_2$O   H$_2$O

Figure 33.7
The formation of a polyester.

Ethanol, ethanoic acid, ethyl ethanoate and water are all present, as well as unchanged acid catalyst.

$$CH_3CH_2OH + CH_3COOH \underset{}{\overset{H^+}{\rightleftharpoons}} CH_3COOCH_2CH_3 + H_2O$$
ethanol   ethanoic acid   ethyl ethanoate

If the reaction mixture is now titrated with standard sodium hydroxide, the amount of ethanoic acid present at equilibrium can be found.

From this the amounts of all the other components of the equilibrium mixture can be worked out. From these results the equilibrium constant K_c for the esterification reaction can easily be found: its value is about 4 at 25°C.

■ Write an expression for the equilibrium constant for the esterification of ethanol and ethanoic acid as in the last equation.
■ If 1 mole each of ethanoic acid and ethanol are allowed to reach equilibrium at 25°C, what will be the number of moles of ethyl ethanoate present at equilibrium? (Assume $K_c = 4$ at 25°C.)
■ What would you expect to happen if ethyl ethanoate were refluxed with an excess of water for a long time?

A mixture of 1 mole of ethanol and 1 mole of ethanoic acid brought to equilibrium at 25°C contains 2/3 mole each of ethyl ethanoate and water and 1/3 mole each of ethanol and ethanoic acid. The esterification reaction is thus far from complete, and it is readily reversed.

The reverse of esterification is **ester hydrolysis**, in which an ester reacts with water to form an alcohol and a carboxylic acid. Like esterification, ester hydrolysis is a slow reaction, but it is speeded up by an acid catalyst. It is also catalysed by alkali, but in this case the carboxylic acid formed by hydrolysis reacts with excess alkali to form the carboxylate salt. This removes the carboxylic acid from the equilibrium mixture as it is formed, which means that ester hydrolysis can proceed to completion in the presence of alkali. This cannot happen when an acid catalyst is used.

For example:

$$CH_3COOCH_2CH_3 + OH^- \longrightarrow CH_3COO^- + CH_3CH_2OH$$
ethyl ethanoate   ethanoate ion   ethanol

or

$$CH_3COOCH_2CH_3 + NaOH \longrightarrow CH_3COO^-Na^+ + CH_3CH_2OH$$

Esters are therefore hydrolysed more effectively in alkaline than in acidic solution. But in both cases the reaction is still quite slow and the mixture must be boiled under reflux. Strictly speaking the alkali is not acting catalytically in this

Fats are energy storage compounds

reaction because, as the equation shows, the OH⁻ is used up. The hydrolysis of an ester by alkali is sometimes called **saponification**, a word derived from Latin words meaning 'soap-making'. The reason for the use of this word will become clear shortly.

33.6 Fats

Fats and oils are naturally occurring esters which are used as energy storage compounds by plants and animals. They are members of a larger class of naturally occurring compounds called **lipids**, which are insoluble in water but soluble in organic solvents.

Fats are derived from propane-1,2,3-triol, $CH_2OH—CHOH—CH_2OH$ (commonly known as glycerol or glycerine). This molecule has the capacity to combine with one, two or three molecules of carboxylic acid. In practice, most fats are *triesters* derived from propane-1,2,3-triol and a variety of long-chain carboxylic acids, sometimes called **fatty acids**. For example, a simple fat molecule is derived from a molecule of propane-1,2,3-triol and three molecules of octadecanoic acid:

$$CH_2OH \qquad HOOC(CH_2)_{16}CH_3 \qquad CH_2OOC(CH_2)_{16}CH_3$$
$$| \qquad\qquad\qquad\qquad\qquad\qquad\qquad |$$
$$CHOH \quad + \quad HOOC(CH_2)_{16}CH_3 \longrightarrow CHOOC(CH_2)_{16}CH_3 + 3H_2O$$
$$| \qquad\qquad\qquad\qquad\qquad\qquad\qquad |$$
$$CH_2OH \qquad HOOC(CH_2)_{16}CH_3 \qquad CH_2OOC(CH_2)_{16}CH_3$$

1 mole of   3 moles of   1 mole of propane-
propane-1,2,3-triol   octadecanoic acid   1,2,3-triyl trioctadecanoate

Table 33.5 gives some common fatty acids. Fifty or so are found in nature, the vast majority of them having an even number of carbon atoms in their molecule.

Fats containing a large proportion of unsaturated acids tend to have low melting points. Many are liquid at room temperature and these are called **oils**. They can be converted to solid fats by hydrogenation (section 26.6).

Like all esters, fats can be hydrolysed. The products are propane-1,2,3-triol and a mixture of carboxylic acids. If the hydrolysis is carried out by boiling the fat with an alkali such as sodium hydroxide, the products are propane-1,2,3-triol and a mixture of the sodium salts of fatty acids. For example:

What oils do you think this soap was made from?

$$CH_2OOC(CH_2)_{16}CH_3 \qquad\qquad\qquad\qquad\qquad CH_2OH$$
$$| \qquad\qquad\qquad\qquad\qquad\qquad\qquad\qquad\qquad\qquad |$$
$$CHOOC(CH_2)_{16}CH_3 + 3NaOH \longrightarrow 3CH_3(CH_2)_{16}COO^-Na^+ + CHOH$$
$$+$$
$$\qquad\qquad\qquad\qquad\qquad\qquad\qquad\qquad sodium\ octadecanoate \qquad CH_2OH$$
$$| \qquad\qquad\qquad\qquad\qquad\qquad\qquad\qquad\qquad\qquad\qquad\qquad propane-$$
$$CH_2OOC(CH_2)_{16}CH_3 \qquad\qquad\qquad\qquad\qquad\qquad\qquad 1,2,3-triol$$
propan-1,2,3-triyl
trioctadecanoate

Sodium octadecanoate is a soap.

This reaction is the basis of **soap-making**, one of the oldest chemical manufacturing processes. Soap is manufactured by boiling fat with sodium or

Table 33.5
Some common fatty acids

Structure	Systematic name	Common name	Occurrence
$CH_3(CH_2)_{16}COOH$	octadecanoic acid	stearic acid	mainly in animal fats
$CH_3(CH_2)_{10}COOH$	dodecanoic acid	lauric acid	coconut oil, palm-kernel oil
$CH_3(CH_2)_{14}COOH$	hexadecanoic acid	palmitic acid	most fats, especially palm oil
$CH_3(CH_2)_7CH{=}CH(CH_2)_7COOH$	octadec-9-enoic acid	oleic acid	most fats, especially olive oil

potassium hydroxide. The fat is usually tallow (storage fat of cattle and sheep) or vegetable oil such as coconut oil. Because fats contain a mixture of fatty acids, soaps are in practice a mixture of the salts of fatty acids. Potassium soaps are milder than sodium ones, so potassium hydroxide is the alkali used in the manufacture of toilet soaps. Propane-1,2,3-triol (glycerol) is a useful by-product of soap manufacture. It is used to make paints and the explosive 'nitroglycerine'.

Making soap, *c.*1870. Hot liquid soap is being poured into a mould.

33.7 Acid anhydrides

Acid anhydrides can be regarded as the products of a dehydration reaction between two molecules of carboxylic acid. For example, ethanoic anhydride:

ethanoic acid   ethanoic anhydride

Despite their name, acid anhydrides are not readily formed by direct dehydration of carboxylic acids. Ethanoic anhydride is normally made in the laboratory by the reaction of ethanoyl chloride with sodium ethanoate. As shown in section 30.7, ethanoyl chloride reacts readily with nucleophiles. The ethanoate ion is a nucleophile, so:

ethanoyl chloride   ethanoate ion   ethanoic anhydride

(The actual mechanism of this reaction involves more stages than shown in this outline.)

Industrially, ethanoic anhydride is made from ethanal or from ethanoic acid (in the latter case, indirectly, not by direct dehydration).

Methanoic anhydride does not exist, since methanoic acid dehydrates to give carbon monoxide.

The oxygen atoms in ethanoic anhydride withdraw electrons from the carbons to which they are attached, giving them a large partial positive charge:

Ethanoic anhydride, like ethanoyl chloride, is readily attacked by nucleophiles such as H_2O, NH_3, ROH, RNH_2, etc. Representing a general nucleophile as $\ddot{X}$–Y, we can show the reaction as

In most cases, Y is hydrogen.

In general, acid anhydrides, $(RCO)_2O$, act as **acylating agents**, joining the acyl group RCO onto a nucleophile. In this respect they behave like acyl chlorides which are also acylating agents, though acyl chlorides react more vigorously than acid anhydrides. Compare the general acylating reactions of acid anhydrides and acyl chlorides:

Acid anhydrides react with all those nucleophiles which react with acyl chlorides.

In the following reactions, ethanoic anhydride is used as an example, though the reactions apply quite generally to all acid anhydrides.

With **water** to give the carboxylic acid:
e.g. $(CH_3CO)_2O + H_2O \rightarrow 2\,CH_3COOH$

With **alcohols** to give esters:
e.g. $(CH_3CO)_2O + CH_3CH_2OH \rightarrow CH_3COOCH_2CH_3 + CH_3COOH$
                              ethyl ethanoate

With **ammonia** to give amides:
e.g. $(CH_3CO)_2O + NH_3 \rightarrow CH_3CONH_2 + CH_3COOH$
                        ethanamide

With **amines** to give substituted amides:
e.g. $(CH_3CO)_2O + CH_3CH_2NH_2 \rightarrow 2\ CH_3CONHCH_2CH_3 + CH_3COOH$

■ Predict the formulae of the products of reactions between the following pairs of substances:
(a) ethanoic anhydride and phenol,
(b) propanoic anhydride and methylamine,
(c) butanoic anhydride and ammonia.

In industry, ethanoic anhydride is used in preference to ethanoyl chloride where an **ethanoylating agent** is required. This is because it is cheaper to make and less vigorous in its reactions. Aspirin, for example, is manufactured by ethanoylating 2-hydroxybenzoic acid:

2-hydroxybenzoic acid + (CH$_3$CO)$_2$O → aspirin + CH$_3$COOH

The artificial fibre cellulose triethanoate (sold under the trade name 'Tricel') is made by ethanoylating cellulose with ethanoic anhydride. Three of the OH groups on each glucose molecule of the cellulose chain are ethanoylated. This gives a fibre with good 'wash and wear' properties.

Certain dicarboxylic acids can form **internal anhydrides** by eliminating a molecule of water from one rather than two molecules of the acid. For example, *cis*-butenedioic acid (maleic acid) readily eliminates water when heated, forming its anhydride:

cis-butenedioic acid (maleic acid) → anhydride + H$_2$O

However, the *trans* isomer of this compound, commonly called fumaric acid,

does not readily dehydrate on heating. This is because with the *trans* isomer it is impossible to form a cyclic internal anhydride. This is an unusual example of a pair of *cis* and *trans* isomers having a major difference in chemical properties.

Summary

1   Carboxylic acids possess the carboxyl group, COOH. They are named using the suffix *-oic acid*.

2   Carboxylic acids are prepared by oxidising primary alcohols or aldehydes or by the hydrolysis of nitriles.

3 The COOH group is strongly acidic compared to the OH group, but nevertheless carboxylic acids are weak compared with mineral aicds such as hydrochloric acid. The strength of a carboxylic acid is increased by the presence of electron-withdrawing substituents.

4 Some important reactions of ethanoic acid are shown in the following diagram:

The importance of carboxylic acids in the synthesis of organic chemicals is developed further in chapter 35.

5 Esters have the general formula RCOOR′. They are formed by a condensation reaction between an alcohol and a carboxylic acid, usually in the presence of an acid catalyst.

6 Esterification is a reversible reaction. Esters can be hydrolysed by boiling with dilute acid or alkali.

7 Acid anhydrides have the general formula $(RCO)_2O$. They are acylating agents, with the ability to join the RCO group to nucleophiles.

8 Detergents are substances that improve the cleaning of water. They have both a polar and a non-polar portion in their molecules. Soap is a detergent; soaps are the sodium or potassium salts of long-chain carboxylic acids.

9 Fats are esters derived from propane-1,2,3-triol and long-chain carboxylic acids. Alkaline hydrolysis of fats produces soaps.

10 Methanoic acid and ethanedioic acid differ from other carboxylic acids in being readily oxidised and dehydrated.

Study questions

1 Consider the following compounds:

    A $CH_3CHCOOH$
             |
             CH_3

    B [benzene ring]—COOH
              —CH_3

    C $CH_3CH_2COOCH_3$
    D $HOOCCH_2CH_2COOH$
    E CH_3CH_2CO
                     O
      CH_3CH_2CO

(a) Which is an ester?
(b) Which is a dibasic acid?
(c) Which is an acid anhydride?
(d) Name each compound.
(e) Which would be almost insoluble in water, but would slowly dissolve when boiled with sodium hydroxide solution?
(f) Which would be almost insoluble in water, but soluble in cold sodium hydroxide solution?
(g) Which would form a pleasant-smelling liquid when warmed with ethanol and concentrated sulphuric acid?
(h) Which would react with ammonia to form a mixture of propanamide and propanoic acid?

2   Predict the formulae of the products of the following reactions.
(a) $CH_3CH_2COOH + PBr_3 \longrightarrow$
(b) $C_6H_5COOH + LiAlH_4 \longrightarrow$
(c) $CH_3COO(CH_2)_4CH_3 + NaOH(aq) \xrightarrow{boil}$
(d) $CH_3COOH + Ca(OH)_2(aq) \longrightarrow$
(e)

3   (a) When a carboxylate salt (e.g. $RCOO^-Na^+$) is heated with 'soda lime' (a mixture of sodium and calcium hydroxides) a *decarboxylation* reaction occurs. In this reaction, CO_2 is effectively eliminated from the molecule, forming an alkane:

$$RCOO^-Na^+ + NaOH \rightarrow RH + Na_2CO_3$$
$$\text{an alkane}$$

What do you think would be formed when *soap* is heated with soda lime?
(b) Suggest explanations for the following observations.
   (i) Butane, propan-1-ol, propanal and ethanoic acid all have approximately the same relative molecular mass, but their boiling points are 273 K, 370 K and 391 K, respectively.
   (ii) When dilute hydrochloric acid is added to an aqueous solution of soap, a white insoluble substance is formed.

4   Three isomeric acids **A**, **B** and **C** have molecular formula $C_8H_6O_4$ and all contain a benzene ring. In each case, one mole of the acid will react with 2 moles of sodium hydroxide. Suggest structures for the acids.
When the three acids are separately heated, **A** and **B** melt without decomposing, but **C** loses a molecule of water at about 250°C to form **D**, $C_8H_4O_3$. Suggest structures for **C** and **D**.

5   How would you carry out the following conversions in the laboratory? One or more steps may be involved in each case.
(a) CH_3CHO to $CH_3COOCH_2CH_3$
(b) $CH_2{=}CH_2$ to CH_3COOH
(c) CH_3CH_2COOH to $(CH_3CH_2CO)_2O$
(d)

(e) CH_3COOH to CH_3CONH_2

6   What simple chemical tests would you use to distinguish one compound from the other in the following pairs:
(a) CH_3CH_2CHO and CH_3CH_2COOH,
(b) $HCOOH$ and CH_3COOH,
(c) $CH_3COCH_2CH_3$ and $CH_3COOCH_2CH_3$,
(d) CH_3COOCH_3 and $HCOOCH_2CH_3$,
(e) $CH_2FCOO^-Na^+$ and $CH_3COO^-Na^+$.

7   To what extent does the carboxyl group, —C—OH, show properties typical

                                                      $\|$

                                                      O

    of (a) the —C— group, (b) the OH group? To what extent are its properties

                        $\|$

                        O

    different from those of either of these groups?

8   Write structural formulae for all acids and esters of molecular formula $C_4H_8O_2$. What simple chemical tests would you use to distinguish between the different esters?

9   Consider the following compounds:

    A $CH_2OOC(CH_2)_{14}CH_3$

            $|$

          $CHOOC(CH_2)_{16}CH_3$

            $|$

          $CH_2OOC(CH_2)_{14}CH_3$

    B $HOCH_2CH_2OH$

    C $CH_3(CH_2)_{15}COOH$

    D $CH_3(CH_2)_{15}SO_3{}^-Na^+$

    E $HOOC(CH_2)_4COOH$

    (a) Which is a fatty acid?

    (b) Which is an anionic detergent?

    (c) Which would form soap on boiling with sodium hydroxide solution?

    (d) Which pair of compounds could together be used to make a polyester?

    (e) Which is a fat?

    (f) Which would be oxidised to carbon dioxide and water on boiling with acidified potassium manganate(VII)?

    (g) Which might form a cyclic internal anhydride?

10 This question is about an investigation into the mechanism of hydrolysis of ethyl ethanoate:

                                   HCl

$$CH_3COOCH_2CH_3 + H_2O \xrightarrow{\text{catalyst}} CH_3COOH + CH_3CH_2OH$$

       ethyl ethanoate   ethanoic   ethanol

                                                   acid

    Two experiments, **A** and **B**, were carried out. Read the accounts of the experiments, then answer the questions following.

    Experiment A Ethyl ethanoate was refluxed with deuterated water, D_2O, containing deuterium chloride, DCl ($D = {}_1^2H$). The two hydrolysis products were separated and purified and their relative molecular masses were measured using a mass spectrometer.

    The alcohol formed had $M_r = 47$, and the acid had $M_r = 61$.

    (a) Compare M_r for the alcohol formed in this experiment with the value expected from the equation above. Account for any difference.

    (b) Compare M_r for the acid formed in this experiment with the value expected from the equation above. Account for any difference.

    (c) What information if any, does experiment **A** give about the mechanism of the hydrolysis reaction?

    Experiment B Ethyl ethanoate was refluxed with $H_2{}^{18}O$ containing HCl. The two hydrolysis products were separated and purified and their relative molecular masses were measured using a mass spectrometer. The alcohol formed had $M_r = 46$, and the acid had $M_r = 62$.

    (d) and (e) Repeat steps (a) and (b) above, this time using the results from experiment **B**.

    (f) What information, if any, does experiment **B** give about the mechanism of the hydrolysis reaction?

    (g) Compare your answer with the information given on pp. 569–70.

34 Organic Nitrogen Compounds

Sir William Perkin, discoverer of the dye mauve.

34.1 Dyestuffs and the development of the organic chemical industry

---

The structure of the organic chemical industry is very different from its inorganic counterpart. The inorganic chemical industry is rather static. It tends to stick for long periods to established processes producing a small number of essential chemicals in very large quantities—for example, sulphuric acid, sodium hydroxide and steel. In organic chemistry, on the other hand, the number of possible compounds and the number of routes by which they can be made is far greater. As a result, the industry is more dynamic, constantly discovering new compounds to do a particular job, and new ways of making them. Pharmaceuticals and polymers are just two areas of the organic chemical industry that show this rapid evolution. With new compounds and new processes to be investigated, the research chemist is more closely involved.

How it all began

The evolution of the organic chemicals industry started with the discovery of synthetic dyestuffs. Until the 1850s dyes were all produced from living materials, usually plants. Consequently, they were expensive, limited in colour range and tended to fade.

In 1856 in London, W.H. Perkin, an 18-year-old student of the German chemist Hofmann, was attempting to prepare the drug quinine from aniline. Although he failed, he managed, quite by accident, to produce a brilliant purple dye that was named *mauve*. This was the first synthetic dyestuff. It was the forerunner of many other dyes produced from aniline, or phenylamine as it is now called. The presence of the NH_2 group attached to a benzene ring makes phenylamine especially suited to the preparation of coloured materials, particularly **azo dyes**.

Germany pulls ahead

The synthetic dyestuffs industry was dominated by Britain, its country of origin, until the 1870s. However, many German chemists had gained experience of synthetic dyestuffs in Britain, and Germany slowly began to draw ahead and to develop an extraordinarily dynamic dyestuffs industry. This was partly due to the German education system, which placed great importance on the training of scientists in universities. Germany thus had the large reserve of trained scientists needed to carry out the research necessary to develop the new industry. Then, as now, the key to the development of science and technology was education.

By the First World War Britain was importing most of its dyes from Germany, and producing only 20% of its own needs. Indeed, even after the outbreak of war, dyes had to be imported secretly from Germany to dye British soldiers' uniforms.

The Antwerp plant of the giant German chemical company BASF. The initials stand for Badische Anilin und Soda Fabrik, indicating the origin of the company in the dye industry.

The experience gained by the German chemical industry in the development of dyestuffs helped it diversify into pharmaceuticals (aspirin was first used in Germany in 1898), synthetic polymers, photographic chemicals and many other fields. By the eve of the First World War the German chemical industry was superior to that of any other country.

The organic chemical industry has evolved a long way since its start with dyestuffs. These materials now account for only a very small proportion of the total tonnage of organic chemicals produced. Dyestuffs are manufactured on a fairly small scale by relatively lengthy and complex processes. Their production tends to resemble scaled-up laboratory methods rather than the giant processes used in the manufacture of petrochemicals. Nevertheless, dyestuffs are essential materials and their high price makes them commercially important. Section 34.7 gives further details of the production of azo dyes.

34.2 Important organic nitrogen compounds

In this chapter we shall be mainly concerned with compounds containing the NH_2 group, but we will briefly meet other nitrogen compounds. The range of nitrogen compounds covered and the methods used to name them are dealt with in this section.

Amines

Amines can be considered as compounds formed from ammonia by substituting hydrogen atoms with alkyl or aryl groups. If one of the hydrogen atoms in NH_3 is substituted, we get a compound of the form RNH_2, called a **primary amine**. Such compounds are named using the suffix **-ylamine** after a stem indicating the number of carbon atoms in the molecule. Thus CH_3NH_2 is methylamine. Substitution of two of the hydrogens of NH_3 gives compounds of the form R_1R_2NH, called **secondary amines**, named in a similar way to primary amines. Thus $(CH_3)_2NH$ is dimethylamine. You should be able to work out the general structure and nomenclature of **tertiary amines**. Table 34.1 gives some examples of amines and other nitrogen compounds together with their systematic and other names.

The prefix **amino-** can be used to indicate the presence of an NH_2 group in molecules containing more than one functional group, e.g. aminoethanoic acid.

Table 34.1
Some important organic nitrogen compounds

Formula	Type of compound	Name	Other name
$CH_3CH_2NH_2$	primary amine	ethylamine	aminoethane
⬡—NH_2	primary amine	phenylamine	aniline
$(CH_3)_3N$	tertiary amine	trimethylamine	—
CH_3CONH_2	amide	ethanamide	acetamide
⬡—NO_2	nitro-compound	nitrobenzene	—
CH_3CN	nitrile	ethanenitrile	acetonitrile

$CH_3CHCOOH$
   |
  NH_2
	amino acid	2-aminopropanoic acid	alanine

Amides

Not to be confused with amines, **amides** have the general structure $RCNH_2$. The presence of the $C{=}O$ group 'next door' has a big effect on the properties of the NH_2 group, so amides behave very differently from amines. Amides are named using the suffix **-amide** after a stem indicating the number of carbon atoms in the molecule, including that in the $C{=}O$ group. Thus CH_3CONH_2 is ethanamide.

Nitro-compounds

Compounds containing the NO_2 group are called **nitro-compounds**. Only aromatic ones need concern us here. They are named using the prefix **nitro-**. Thus $C_6H_5NO_2$ is called nitrobenzene.

Nitriles (cyano-compounds)

Compounds containing the cyano group $C{\equiv}N$ are called **nitriles** or **cyano-compounds**. They are named using the suffix **-nitrile** after the stem indicating the number of carbon atoms, including that in the CN group. Thus CH_3CN is ethanenitrile.

Amino acids

As their name implies, **amino-acids** contain both the NH_2 and COOH groups. By far the most important are those in which the NH_2 and the COOH are both attached to the same carbon, thus:

$$R{-}CH{-}COOH$$
$$|$$
$$NH_2$$

Most of the amino acids found in nature are of this form. Amino acids are named as amino derivatives of carboxylic acids. Thus

$$CH_3CHCOOH$$
$$|$$
$$NH_2$$

is called 2-aminopropanoic acid. Since many natural amino acids have quite complex structures, they are often referred to by trivial names.

- Name the following compounds:
  (a) $CH_3CH_2CH_2CONH_2$
  (b) CH_3CH_2CN
  (c) $(CH_3CH_2)_2NH$
- Give the formulae of the following compounds:
  (a) 2-nitrophenylamine
  (b) aminoethanoic acid
  (c) propylamine
  (d) propanamide

The peculiar smell of fish is mainly due to amines.

34.3 The nature and occurrence of amines

The NH_2 group is very widespread in biological molecules, especially proteins. Normally the group is associated with other functional groups. Free amines are relatively rare in nature, though they do occur in decomposing protein material. Amines are formed by the action of bacteria on amino acids. For example, $NH_2(CH_2)_4NH_2$ and $NH_2(CH_2)_5NH_2$ are found in decaying animal flesh, as their common names, putrescine and cadaverine, evocatively suggest. Dimethylamine and trimethylamines are found in rotting fish and are partly responsible for its peculiar smell.

The physical and chemical properties of the early members of the amine series resemble those of ammonia. Their smell is very similar to that of ammonia, though with a slightly fishy character. Like ammonia, the early amines are gaseous and very soluble in water. Their high solubility is due to hydrogen bonding between the NH_2 group and water molecules.

Table 34.2
Physical properties of ammonia and some amines

Name	Formula	State at 25°C	Boiling point/°C
ammonia	NH_3	g	-33
methylamine	CH_3NH_2	g	-6
ethylamine	$CH_3CH_2NH_2$	g	17
propylamine	$CH_3CH_2CH_2NH_2$	l	49
butylamine	$CH_3CH_2CH_2CH_2NH_2$	l	78
phenylamine	$C_6H_5NH_2$	l	184
dimethylamine	$(CH_3)_2NH$	g	7
trimethylamine	$(CH_3)_3N$	g	3
triethylamine	$(CH_3CH_2)_3N$	l	90

Table 34.2 gives some physical properties of amines. Can you suggest a reason why trimethylamine has a lower boiling point than dimethylamine even though its relative molecular mass is higher?

Phenylamine, with its large hydrocarbon portion, is only sparingly soluble in water. Phenylamine dissolves well in organic solvents. Its ability to dissolve in fats means that it is readily absorbed through the skin. Since it is toxic, great care must be taken when using phenylamine.

34.4 Making amines

There are many ways of making amines. Four of the most important methods are listed here.

From ammonia and halogenoalkanes

Halogenoalkanes undergo substitution reactions with ammonia to form a mixture of primary, secondary and tertiary amines (section 30.5):

$$RCl + NH_3 \longrightarrow RNH_2 + HCl$$

$$RCl + RNH_2 \longrightarrow R_2NH + HCl$$

$$RCl + R_2NH \longrightarrow R_3N + HCl$$

In practice the product is the salt of the amine, formed by combination of the HCl with the free amine (section 34.5). The problem with this method is that it produces a mixture of products which must be separated by distillation. Nevertheless, the method is widely used both in the laboratory and industrially, because of the ready availability of the starting materials.

By reduction of nitro-compounds

This method is normally used only for aromatic amines.

Aromatic nitro-compounds are readily prepared by the nitration of aromatic hydrocarbons (see section 29.6). Reduction of nitro-compounds in acid solution produces amines:

In the laboratory the reducing agent used is normally tin in concentrated hydrochloric acid:

$$Sn \longrightarrow Sn^{4+} + 4e^-$$

This method is also used industrially to manufacture phenylamine, the reducing agent in this case being iron in concentrated hydrochloric acid.

By reduction of nitriles

Primary amines can be produced by the reduction of nitriles using powerful reducing agents such as lithium tetrahydridoaluminate ($LiAlH_4$):

$$R\!-\!C\!\equiv\!N + 4H^+ + 4e^- \longrightarrow RCH_2NH_2$$
$$\text{from}$$
$$\text{reducer}$$

This method is useful in the laboratory but because of the relatively high cost of the reagents it is not used industrially.

By reduction of amides

Amides, like nitriles, are reduced by $LiAlH_4$ to primary amines:

$$RCONH_2 + 4H^+ + 4e^- \rightarrow RCH_2NH_2 + H_2O$$

■ How would you prepare:
   (a) propylamine from chloroethane (2 stages),
   (b) ethylamine from chloroethane,
   (c) 2-methylphenylamine from methylbenzene (2 stages)?

34.5 Amines as bases

Figure 34.1 illustrates an experiment in which acid is added to a solution of ethylamine.

Answer these questions.

■ What evidence is there for a chemical reaction between ethylamine and hydrochloric acid?
■ Why does the smell of ethylamine disappear when hydrochloric acid is added?
■ Why does the smell reappear when sodium hydroxide is added?

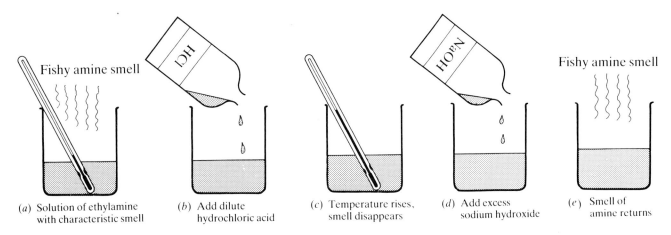

(a) Solution of ethylamine with characteristic smell

(b) Add dilute hydrochloric acid

(c) Temperature rises, smell disappears

(d) Add excess sodium hydroxide

(e) Smell of amine returns

Figure 34.1
Effect of acid on aqueous ethylamine.

Like ammonia and all primary amines, ethylamine carries a lone pair of electrons on its nitrogen atom. This enables it to form a dative bond to a hydrogen ion:

Ethylamine is therefore a base like ammonia.

When an acid is added to a solution of ethylamine, a salt is formed—in this case **ethylammonium chloride**. $CH_3CH_2NH_3^+Cl^-$. Like all salts it is involatile and therefore has no smell. When a strong base such as sodium hydroxide is added to this salt, protons are removed from it. This reforms the free amine:

$$CH_3CH_2NH_3^+ + OH^- \rightarrow CH_3CH_2NH_2 + H_2O$$
ethylammonium ion   ethylamine

Compare this series of reactions with the familiar corresponding reactions of ammonia:

$$NH_3 + H^+ \longrightarrow NH_3^+$$
$$NH_4^+ + OH^- \longrightarrow NH_3 + H_2O$$

Amine salts are soluble white crystalline solids, similar to ammonium compounds.

Figure 34.2 shows another experiment that can be carried out with ethylamine and HCl. What would you expect to happen? Why?

glass tube

Figure 34.2
An experiment with ethylamine and HCl. The apparatus is set up as shown. What happens?

cotton wool soaked in concentrated ethylamine solution

cotton wool soaked in concentrated hydrochloric acid

Is ethylamine a stronger or weaker base than ammonia?

To answer this question, look at the following equilibria:

$$CH_3CH_2NH_2 + H^+ \rightleftharpoons CH_3CH_2NH_3^+$$
$$NH_3 + H^+ \rightleftharpoons NH_4^+$$

The stronger the base, the further the equilibrium is to the right. The position of equilibrium depends on the stability of the cation on the right-hand side. The more stable the cation, the further the equilibrium will be to the right. Recalling the electron-donating effect of alkyl groups, which will be the more stable cation, NH_4^+ or $CH_3CH_2NH_3^+$?

Primary amines are stronger bases than ammonia. By considering the electron-donating effect of alkyl groups and the electron-withdrawing effect of the C=O group, try to arrange the following in order of basic strength:

(a) NH_3 (b) CH_3CNH_2 (c) $(CH_3CH_2)_2NH$ (d) $CH_3CH_2NH_2$
      ‖
      O

Table 34.3 gives K_b values for different compounds containing the NH_2 group. Remember, the higher the K_b value, the stronger the base (section 23.9). Note that phenylamine is a much weaker base than aliphatic amines. This is due to the partial delocalisation round the benzene ring of the lone pair of electrons from the nitrogen atom.

Table 34.3
K_b values of ammonia, amines and ethanamide

Formula	Name	K_b at 25°C/mole dm^{-3}
NH_3	ammonia	1.8×10^{-5}
$CH_3CH_2NH_2$	ethylamine	5.4×10^{-4}
$(CH_3CH_2)_2NH$	diethylamine	1.3×10^{-3}
CH_3CONH_2	ethanamide	10^{-15}
$C_6H_6NH_2$	phenylamine	5×10^{-10}

34.6 Other reactions of amines

With acyl chlorides

The lone pair of electrons on the nitrogen atom gives amines nucleophilic character. Thus they tend to react readily with electrophiles like ethanoyl chloride. For example:

phenylamine   ethanoyl chloride   N-phenylethanamide (m.p. 114°C)

The products of this type of reaction are called acyl derivatives. Unlike the amines from which they are made, they are crystalline solids with sharp melting points. They are therefore useful for identifying (*characterising*) unknown amines. The acyl derivative is prepared and its melting point is taken. This is then checked against the accepted melting points of known amides in published tables.

With nitrous acid

Nitrous acid (HNO_2) is rather unstable, so when it is used as a chemical reagent it is usually generated on the spot from sodium nitrite and hydrochloric acid.

When ethylamine is reacted with nitrous acid, nitrogen is rapidly evolved and a mixture of ethanol, ethene and other products is formed. However, when phenylamine is reacted with nitrous acid at low temperatures (below 10°C), a clear solution is formed, but no nitrogen appears. If, however, this solution is warmed, nitrogen is evolved and phenol is formed.

These observations are explained as follows. When amines react with nitrous acid, **diazonium compounds** are formed:

$$RNH_2 + HNO_2 \longrightarrow R-\overset{+}{N}\equiv N + OH^- + H_2O$$
$$\text{diazonium}$$
$$\text{ion}$$

The $-\overset{+}{N}\equiv N$ group is rather unstable. In the case of ethyldiazonium ion, it decomposes at once:

$$CH_3CH_2-\overset{+}{N}\equiv N \longrightarrow N_2 + \left[CH_3\overset{+}{C}H_2 \right]$$

$$\xrightarrow{H_2O} CH_3CH_2OH + H^+$$

$$\longrightarrow CH_2{=}CH_2 + H^+$$

ethyldiazonium ion

When the $-\overset{+}{N}\equiv N$ group is attached to a benzene ring, though, the ion is stabilised to some extent by the delocalised electrons of the ring. Benzenediazonium ion is therefore much more stable than its aliphatic counterparts. Nevertheless it decomposes readily above 10°C. Aromatic diazonium compounds are considered further in the next section.

Reactions of the aromatic ring in phenylamine

The lone pair of electrons on the nitrogen atom in phenylamine tends to get partly delocalised round the ring. We have already seen how this reduces the basic strength of phenylamine relative to aliphatic amines. Another result is that the electron density round the ring in phenylamine is considerably increased. This makes it undergo electrophilic substitution much more readily than benzene. For example, phenylamine reacts with bromine water even in the absence of a halogen carrier catalyst:

$$+ 3Br_2 \longrightarrow \text{2,4,6-tribromophenylamine} + 3HBr$$

The high electron density in the phenylamine molecule makes it very easy to oxidise. Pure phenylamine is colourless, but it quickly darkens due to atmospheric oxidation. Chemical oxidisers can convert it to a host of different products, including the important pigment Aniline Black (figure 34.3).

An important ring-substituted derivative of phenylamine is 4-aminobenzenesulphonamide:

Figure 34.3
The pigment Aniline Black is a mixture of several substances. The structure of one of them is believed to be that shown here (the + and − charges are effectively delocalised over the molecule, which is electrically neutral overall).

Figure 34.4
Maternal mortality death rate from childbed fever. Deaths per 100 000 total births, 1850–1964, in England and Wales (logarithmic scale). Sulphonamides were introduced in 1935.

This compound and other compounds derived from it are the active agents in the **sulphonamides**, one of the earliest groups of anti-bacterial drugs. The sulphonamide drugs played an important part in the decline of puerperal or child-bed fever, once a major cause of death in childbirth. They were first used in 1935 (see figure 34.4). They also helped to achieve the vast reduction in deaths from pneumonia and tuberculosis over the last 40 years. They are still being used today though they have been largely superseded by antibiotics such as penicillin.

34.7 Diazonium salts

The only stable diazonium salts are aromatic ones (section 34.6), and even these are not particularly stable. In aqueous solution benzenediazonium chloride decomposes above about 5–10°C, and the compound is explosive when solid. Its reactive nature makes it useful in synthesis, however.

Benzenediazonium chloride is prepared by adding a cold solution of sodium nitrite to a solution of phenylamine in concentrated hydrochloric acid below 5°C:

$$C_6H_5NH_2 + HNO_2 + HCl \longrightarrow C_6H_5N_2^+Cl^- + 2H_2O$$

phenylamine   benzene diazonium chloride

Owing to the explosive nature of the solid, the compound is always used in solution.

Consider the structure of the benzenediazonium ion, $C_6H_5{-}\overset{+}{N}{\equiv}N$.

- What stable molecule can be eliminated from the ion?
- What would be the product left when this molecule is eliminated?
- What kind of reaction would this product tend to undergo?

The benzenediazonium ion reacts with nucleophiles:

The introduction of the diazonium group is therefore a way of making the aromatic ring susceptible to *nucleophilic* substitution. (Remember the characteristic mode of reaction of an aromatic ring is normally *electrophilic* substitution.) This makes diazonium compounds useful in synthesis. For example:

$$C_6H_5N_2^+ + I^- \longrightarrow C_6H_5I + N_2 \quad \text{(warm benzenediazonium}$$
iodobenzene   chloride with KI solution)

$$C_6H_5N_2^+ + H_2O \longrightarrow C_6H_5OH + N_2 + H^+ \quad \text{(warm the aqueous}$$
phenol   solution)

$$C_6H_5N_2^+ + Cl^- \longrightarrow C_6H_5Cl + N_2 \quad \text{(warm benzenediazonium}$$
chlorobenzene   chloride with CuCl catalyst)

Coupling reactions

The positive charge on the N_2^+ group of benzenediazonium ion means that this group is itself a strong electrophile. Thus we might expect it to attack another benzene ring, particularly one that has an electron-donating group such as OH attached to it:

(4-hydroxyphenyl)azobenzene

This is an example of a **coupling reaction**. If a cold solution of benzenediazonium chloride is added to a cold solution of phenol in sodium hydroxide, a bright orange precipitate is immediately formed. This is (4-hydroxyphenyl)-azobenzene. It is an example of an **azo compound**. Many different azo compounds can be formed by coupling reactions between diazonium compounds and activated aromatic rings. They are all brightly coloured. Their colour arises from the extensive delocalised electron systems they possess, delocalisation extending from one ring through the N=N group to the next ring.

Coupling reactions are important in the dyestuffs industry for making azo dyes. Many different colours can be obtained by adjusting the structure of the azo compound. These can be quite complex as figure 34.5 shows.

Unlike diazonium compounds, azo compounds are quite stable. Azo dyes do not fade or lose their colour.

■ Write the structures of the products you would expect to be formed at each stage when:
(a) phenylamine is dissolved in excess concentrated hydrochloric acid,
(b) sodium nitrite solution is added to the cooled solution from the first stage,
(c) the product from the second stage is added to a fresh solution of phenylamine.

'Acid Orange 7' (bright reddish-orange)

'Direct Red 39' (bluish-red)

Figure 34.5
Structures of some azo dyes.

'Direct Brown 57' (reddish-brown)

34.8 Amino acids and proteins

All proteins are chemically similar, but they perform many different functions in living things. They act as structural materials (e.g. skin and fingernails). They make up the muscle fibres that enable animals to move. They are the basic material of the enzymes that catalyse the many chemical reactions that are the driving force of all organisms. Protein molecules are chemically very similar, yet their diverse functions mean there must be many different ways of putting such molecules together.

Manufacture of a dye, c. 1925.

Proteins are long-chain molecules made by linking together relatively small molecules called amino acids. There are 22 different amino acids found widely in nature and many proteins contain several thousand amino acid units. For a protein containing 5000 such units the number of possible arrangements using 22 different amino acids is 22^{5000}, i.e. about 10^{6700}. So it is not surprising that proteins come in many different forms. Indeed, the surprising thing is that a given organism can control precisely the form of protein it produces.

Amino acids were introduced in section 34.2. Table 34.4 gives some examples of naturally occurring amino acids. All amino acids except aminoethanoic acid have four different groups attached to the central carbon atom (see figure 34.6). Because of this, they exhibit optical isomerism (section 26.6). All naturally occurring amino acids exist as the L isomer.

Amino acids carry at least two functional groups, NH_2 and COOH. They therefore show the properties of both amines and carboxylic acids. Consider the simplest amino acid, aminoethanoic acid (glycine), H_2NCH_2COOH.

■ Write the structure of the ion formed by aminoethanoic acid when it reacts with a base.
■ Write the structure of the ion formed by aminoethanoic acid when it reacts with an acid.
■ Could aminoethanoic acid form an ion under neutral conditions?

In neutral solution and in the solid state, aminoethanoic acid exists as a dipolar ion:

$$H_3\overset{+}{N}—CH_2—COO^-$$

The ion is formed as a result of an *internal* acid–base reaction. The COOH group donates a proton to the NH_2 group. This kind of ion is called a **zwitterion**, from the German word 'zwei' meaning two. The ionic character of aminoethanoic acid accounts for its high solubility and high melting point (234°C).

Aminoethanoic acid can thus exist in three forms, depending on the pH:

$$H_3\overset{+}{N}—CH_2—COOH \qquad H_3\overset{+}{N}—CH_2—COO^- \qquad H_2N—CH_2—COO^-$$

in acid conditions / in neutral conditions / in basic conditions

Table 34.4

Some naturally occurring amino acids. In this table X stands for the

$$—CH\overset{\displaystyle NH_2}{\underset{\displaystyle COOH}{}}$$

group common to all these amino acids

Formula	Common name
X—H	glycine
X—CH₃	alanine
X—CH₂COOH	aspartic acid
X—CH₂⬡	phenylalanine
X—CH₂OH	serine
X—(CH₂)₄NH₂	lysine
X—CH₂SH	cysteine

Figure 34.6
The asymmetry of amino acid molecules.

Buttons made from casein, a protein obtained from milk.

Other amino acids behave similarly, though the pH ranges in which the three forms exist differ for different amino acids. Some amino acids have extra NH_2 and COOH groups elsewhere in their molecule, and this affects their behaviour towards acids and bases.

The peptide link

The amino acids in a protein chain are linked by the elimination of a molecule of water between the NH_2 of one amino acid and the COOH of the next. Such a link is called a **peptide link**:

peptide link

$+ H_2O$

This kind of linkage is not easily formed under laboratory conditions, but it forms readily in living systems under the influence of enzymes. Long chains of amino acids, called **polypeptides**, can be formed in this way. Once formed, these polypeptide chains can take up very precise three-dimensional arrangements, leading to the formation of proteins. The structure of proteins is considered further in section 9.8.

The formation of a peptide link between amino acids is reversible, so proteins can be hydrolysed to reform amino acids. Hydrolysis can be achieved by boiling with aqueous acid, but it is carried out more effectively by digestive enzymes like pepsin and trypsin.

Hydrolysis is used to identify the amino acids present in a protein. The protein is hydrolysed, then subjected to paper chromatography. The individual amino acids separate on the chromatogram and can be identified.

34.9 Nylon

Naturally occurring fibres like wool, cotton and silk are gradually being replaced by synthetic fibres for many uses. Synthetic fibres are cheaper and often have better wearing and washing characteristics. However, consumers tend to prefer synthetic fibres that feel and look like natural ones. It is not therefore surprising that several synthetic fibres are structurally similar to natural ones.

Wool and silk are protein fibres. Cotton is cellulose. Many of the artificial fibres used in Britain are in fact only semi-synthetic, being based on natural cellulose. These are the fibres known as rayon (viscose) and cellulose triacetate ('Tricel'). They are made by using chemical reactions to convert cellulose (wood pulp, straw, etc.) into a form suitable for use as a textile. (See section 9.8. for the structure of cellulose.)

Nylon is one of the major wholly synthetic fibres. Nylon has some structural similarity to protein fibres like wool and silk. Like proteins, nylon is a condensation polymer. The monomer units are joined by a link similar to the peptide link. There are several forms of nylon. One of the commonest is nylon 66, which is made by a condensation reaction between 1,6-diaminohexane and hexanedioic acid. The product is named nylon 66 because both monomers contain 6 carbon atoms.

Extruding nylon through spinnarets.

Notice the structural similarity of this polymer to a polypeptide (page 591). The CONH grouping is also present in amides (see section 34.2), so nylons are sometimes given the general name **polyamides**. The 1,6-diaminohexane and hexanedioic acid used in this process are obtained from phenol or cyclohexane by the route outlined in figure 34.7.

The polymer produced in this way is melted and squeezed (extruded) through small holes called spinnarets. As the jets emerge they are cooled and they solidify. The fibres produced in this way contain molecular chains that are tangled and disarrayed. The fibres are now stretched or **drawn**, with the result that the molecules become more aligned and the fibre is strengthened (see figure 34.8).

Figure 34.7
Outline of the manufacture of 1,6-diaminohexane, hexanedioic acid, and hence nylon 66.

Nylon 66 can be more readily produced on a laboratory scale from hexanedioyl chloride instead of hexanedioic acid. You will recall from chapter 30 that acyl chlorides are more reactive than the carboxylic acids from which they are produced. Hexanedioyl chloride reacts readily with 1,6-diaminohexane at room temperature to form nylon 66, with the elimination of HCl.

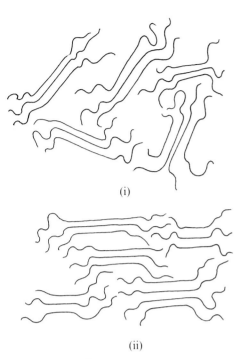

(i)

(ii)

Figure 34.8
Arrangement of polymer chains, (i) before, and (ii) after drawing.

1,6-diaminohexane   hexanedioyl chloride   1,6-diaminohexane

$$\longrightarrow HN-(CH_2)_6-NH-\underset{O}{\overset{}{C}}-(CH_2)_4-\underset{O}{\overset{}{C}}-NH-(CH_2)_6-NH-$$

+ HCl   + HCl

This reaction forms the basis of the well-known 'nylon rope trick' (see figure 34.9). The reaction is not used industrially because of the high cost of hexanedioyl chloride.

Despite its structural similarity to wool, nylon lacks the softness and moisture-absorbing properties of the natural fibre. It is, however, harder wearing and has good 'wash and wear' characteristics. One of the earliest uses of nylon was as a substitute for silk in the manufacture of stockings. The elasticity and strength of nylon make it ideal for making stockings and tights. About 75% of UK nylon consumption goes into clothing. Other uses include tufted carpets, tyre cords and machine gear wheels and bearings.

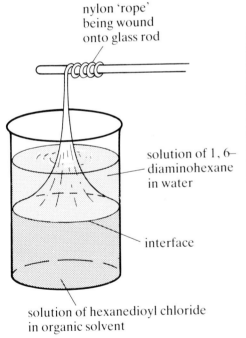

nylon 'rope' being wound onto glass rod

solution of 1,6-diaminohexane in water

interface

solution of hexanedioyl chloride in organic solvent

Figure 34.9
Above. The nylon rope trick. The hexanedioyl chloride and 1,6-diaminohexane react at the interface between the aqueous and organic layers, and a nylon 'rope' can be continuously pulled out.
Right. In the 1940s nylon was newly discovered and nylon stockings were a luxurious novelty.

Summary

1 Primary amines have the structure RNH_2. They are named using the suffix *-ylamine*.

2 Four ways of making amines are summarised below.

    (a) $RCl \xrightarrow{NH_3} RNH_2$

    (b) $RNO_2 \xrightarrow{reduce} RNH_2$ (R = aryl)

    (c) $RCN \xrightarrow{reduce} RCH_2NH_2$

    (d) $RCONH_2 \xrightarrow{reduce} RCH_2NH_2$

3 Amines, particularly the early members of the series, show considerable physical and chemical similarity to ammonia.

4 Amines are basic. Alkyl amines are stronger bases than ammonia, aryl amines are weaker than ammonia.

5 This diagram shows some important reactions of amines.

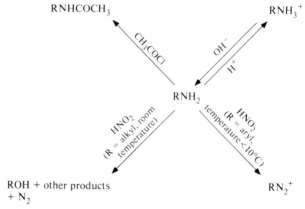

6 This diagram shows some important reactions of benzenediazonium chloride.

7 Amino acids contain both NH_2 and $COOH$ groups. They can behave as bases and as acids.

8 Proteins contain chains of amino acids joined by a condensation linkage between the $COOH$ of one amino acid and the NH_2 of the next. This is called a peptide link.

9 Nylons are condensation polymers containing the $CONH$ group. Nylon 66 is made by a condensation reaction between $H_2N(CH_2)_6NH_2$ and $HOOC(CH_2)_4COOH$.

Study questions

1 Consider the following compounds:

A CH₃CHCH₃
         |
         CN

B CH₃CH₂CH₂NH₂

C O₂N   NO₂

D CH₃
         \
           NH
         /
   CH₃CH₂

E CH₃CH₂CONH₂
F (CH₃CH₂CH₂)₃N

(a) Which is a primary amine?
(b) Which is a nitrile?
(c) Which is an amide?
(d) Which is a tertiary amine?
(e) Name each compound.

2 (a) Place the following in expected order of basic strength. Give reasons for your answer.

(i)   NH₂   (ii)   CH₂NH₂   (iii)   NHCH₃

     CH₃

(b) Place the following in expected order of boiling point. Give reasons for your answer.
(i) CH₃CH₂CH₂NH₂   (ii) CH₃CH₂CH₂CH₃   (iii) CH₃CH₂CH₂OH

3 Draw up a table comparing the principal physical and chemical properties of methylamine and ammonia. Do the two compounds in general behave similarly?

4 Consider the following compounds:

A

CH₃

NH₂

B CH₃CH₂NH₂
C (CH₃CH₂)₂NH
D H₂N(CH₂)₅NH₂

(a) Which would be among the products of the reaction of chloroethane with ammonia?
(b) Which would be converted to a diazonium compound by the action of nitrous acid below 10°C?
(c) Which could be made by reduction of ethanenitrile?
(d) Which would react with hydrochloric acid in the ratio one mole of the compound to two moles of hydrochloric acid?
(e) Which is the weakest base?
(f) Which could be one of the reagents in the manufacture of a nylon (polyamide)?

5 Suggest explanations for the following.
(a) Ethylamine can be readily made from chloroethane and ammonia, but it is difficult to make phenylamine from chlorobenzene and ammonia.

(b) If electrodes are placed in a solution of aminoethanoic acid at pH 0 the aminoethanoic acid migrates to the cathode. If the same experiment is repeated at pH 14 the aminoethanoic acid migrates to the anode.

(c) Phenylamine is much more soluble in dilute hydrochloric acid than in water.

6   In the following, explain how you would convert the first compound to the second. Each conversion may involve one or more steps.

(a) $CH_2\!=\!CH_2$ to $H_2NCH_2CH_2NH_2$

(b)

to

(c) $CH_2\!=\!CH_2$ to $CH_3CH_2CH_2NH_2$

(d) $CH_3CH_2NH_2$ to $CH_3CH_2NH\!-\!C(=\!O)\!-\!C_6H_5$

(e)

to

7   A compound **X** containing carbon, hydrogen and nitrogen only has relative molecular mass 88. When reacted with nitrous acid, 0.1 g of **X** released $50.9\,cm^3$ of nitrogen gas, measured at s.t.p.

(a) How many moles of nitrogen gas are produced by 1 mole of **X** when it reacts with nitrous acid?

(b) How many NH_2 groups are there in one molecule of **X**?

(c) Write three possible structural formulae for **X**.

(d) What volume of 0.1 M hydrochloric acid would be needed to neutralise $50\,cm^3$ of a 0.2 M solution of **X**?

8   Copy out the following reaction sequences, inserting the formulae of the products formed in the blank spaces.

(a) $CH_3COCl \xrightarrow{NH_3} \quad \xrightarrow{LiAlH_4} \quad \xrightarrow{HCl}$

(b) $CH_3CH_2CH_2CH_2Br \xrightarrow{NH_3}$

(c)

8   Copy out the following reaction sequences, inserting the formulae of the products formed in the blank spaces.

9   Suppose a molecule of each of the amino acids serine, lysine and alanine are joined together by peptide links to form a tripeptide.

(a) How many different tripeptides are possible?

(b) Write the structural formula of each different tripeptide. (Refer to table 34.4 for the structures of these amino acids.)

10 A compound **A** of molecular formula C_7H_8 was treated with a mixture of concentrated nitric acid and concentrated sulphuric acid to form two isomeric compounds. One of these isomers was **B**, $C_7H_7O_2N$. **B** could be converted to **C**, C_7H_9N by reduction. **C** reacted with dilute hydrochloric acid to form **D**, $C_7H_{10}NCl$. When cold sodium nitrite solution was added to a cold solution of **D** in hydrochloric acid, a solution of **E**, $C_7H_7N_2Cl$, was produced. When a cold solution of **E** was added to a cold solution of phenol, a brightly coloured substance **F** was produced. When the solution **E** was warmed alone an unreactive gas was evolved and **G**, C_7H_8O, was formed. Give possible structural formulae for **A** to **G**.

Synthetic Routes for Organic Chemicals

STEVENSON COLLEGE EDINBURGH LIBRARY

35.1 Introduction

One of the most important applications of organic chemistry is the synthesis of new compounds. These new compounds have known structures and properties that can be accurately predicted. This approach is particularly important in the synthesis of drugs, dyes, anaesthetics and pesticides.

Once an active compound has been discovered, a wide range of similar compounds can be synthesised. These are then tested to find the most effective compound with the required properties.

For example, one of the earliest pain-killers was 2-hydroxybenzoic acid (salicylic acid) (figure 35.1). Unfortunately this acid irritated the patient's mouth and stomach. In severe cases, it caused internal bleeding. Between 1850 and 1900, various derivatives of salicylic acid were prepared in order to find an equally therapeutic but less toxic form of the drug. In 1899, the German chemist Hofmann synthesised the ethanoyl (acetyl) derivative of salicylic acid. This compound was found to have the beneficial properties of the original drug without causing the same irritation to the mouth and stomach. It is still used today as **aspirin** (figure 35.1).

Synthetic highly selective pesticide being sprayed onto crops from an aircraft.

salicylic acid

acetylsalicylic acid (aspirin)

Figure 35.1
The structures of 2-hydroxybenzoic acid (salicyclic acid) and 2-ethanoyloxybenzoic acid (acetylsalicylic acid, aspirin).

35.2 Starting materials for synthesis

Many of the organic chemicals synthesised in industry rely on crude oil, vegetable oils or fats for their starting materials. The cracking of alkanes in crude oil produces large quantities of short-chain alkenes. These alkenes have anything from two to eight carbon atoms. They include ethene ($CH_2{=}CH_2$) and propene ($CH_3CH{=}CH_2$) and can be converted into alcohols, ketones, acids and other derivatives by appropriate reactions.

The hydrolysis of naturally occurring fats and oils provides an important source of carboxylic acids. These carboxylic acids contain between eight and eighteen carbon atoms but they are restricted to compounds with an even number of carbon atoms.

Figure 35.2 shows some important synthetic routes based on the availability of alkenes from petrochemicals and the availability of carboxylic acids from naturally occurring fats and oils. This overall scheme contains many of the reactions included in the summaries at the end of the organic chemistry chapters.

Look closely at figure 35.2. Notice that alkenes, alcohols and carboxylic acids are the starting points for several reactions. These compounds are often the starting point of many organic syntheses. The cracking of crude oil fractions provides a ready supply of alkenes for industry while fats and oils can be used to

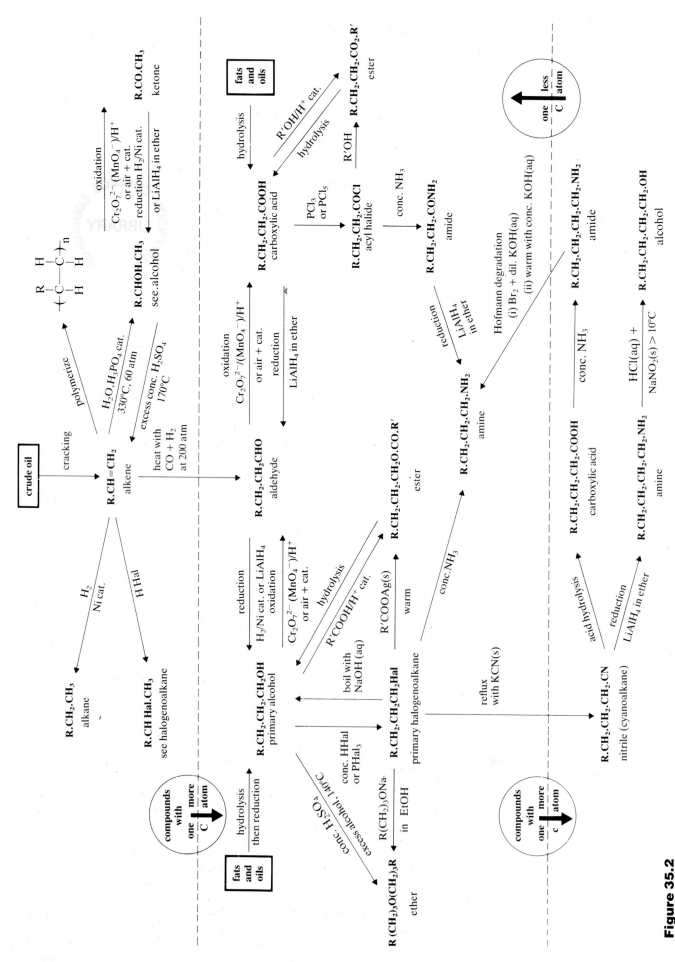

Figure 35.2
Important synthetic routes in organic chemistry.

C$_2$H$_5$OH
ethanol

↓ conc. HBr

C$_2$H$_5$Br
bromoethane

↓ reflux with KCN in alcohol

C$_2$H$_5$CN
propanenitrile

↓ LiAlH$_4$ in ether

CH$_3$CH$_2$CH$_2$NH$_2$
propylamine

↓ HCl(aq) + NaNO$_2$(s) >10 °C

CH$_3$CH$_2$CH$_2$OH
propan-1-ol

Figure 35.3
Ascending a homologous series.

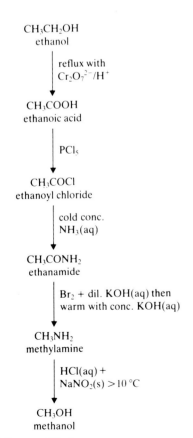

CH$_3$CH$_2$OH
ethanol

↓ reflux with Cr$_2$O$_7^{2-}$/H$^+$

CH$_3$COOH
ethanoic acid

↓ PCl$_5$

CH$_3$COCl
ethanoyl chloride

↓ cold conc. NH$_3$(aq)

CH$_3$CONH$_2$
ethanamide

↓ Br$_2$ + dil. KOH(aq) then warm with conc. KOH(aq)

CH$_3$NH$_2$
methylamine

↓ HCl(aq) + NaNO$_2$(s) >10 °C

CH$_3$OH
methanol

Figure 35.4
Descending a homologous series.

manufacture alcohols and carboxylic acids. Primary alcohols and carboxylic acids can, of course, be interconverted via aldehydes using suitable oxidants and reductants.

35.3 Ascending and descending a homologous series

The other key reagents in preparing organic aliphatic chemicals are halogenoalkanes and nitriles. Indeed, the conversion of halogenoalkane to nitrile provides an important means of increasing the number of carbon atoms in a molecule. So, nitriles allow us to ascend a homologous series. For example, ethanol can be converted to propan-1-ol via propanenitrile along the synthetic route shown in figure 35.3.

Another important process for ascending a homologous series is the 'oxo' reaction. Alkenes, on heating under pressure with carbon monoxide and hydrogen, give aldehydes with one more carbon atom.

$$RCH{=}CH_2 + CO + H_2 \rightarrow RCH_2CH_2CHO$$

The aldehyde can then be readily reduced to a primary alcohol. Consequently, this route provides an important industrial process for the manufacture of primary alcohols.

The last few paragraphs have focused our attention on means of ascending a homologous series. How can we reverse this process and *descend* a homologous series? The crucial reaction in this case is called Hofmann's Degradation. In this reaction an amide is first treated with bromine and dilute potassium hydroxide and then warmed with concentrated alkali. The amide is converted to an amine with one less carbon atom.

$$RCONH_2 + Br_2 + 4KOH \rightarrow RNH_2 + 2KBr + K_2CO_3 + 2H_2O$$

Overall, the process involves removing the C=O group from the amide to give the amine

$$RCONH_2 \rightarrow RNH_2$$

Figure 35.4 shows a sequence of reactions which could be used to descend a homologous series from ethanol to methanol.

35.4 Synthetic routes for aromatic compounds

The reactions shown in figure 35.2 relate principally to aliphatic substances in which R represents either H or an alkyl group such as CH$_3$. In many cases, R could also represent an aromatic group such as phenyl (C$_6$H$_5$). However, there are some reactions of aromatic chemicals (arenes) without parallels in aliphatic chemistry. In particular, these include the various substitutions of an aromatic ring. Thus, there is a separate range of synthetic routes for arenes. Some of the conversions linking the simpler aromatic chemicals such as benzene, methylbenzene (toluene) and phenol are shown in figure 35.5.

Drugs, detergents and dyes can all be made from benzene, which is unquestionably the most important arene. At one time, benzene, toluene and phenol were produced from coal tar. With the development of catalytic reforming, almost all benzene and toluene now come from oil.

There is one last point. In the industrial manufacture of chemicals, cost is all-important. Hence, cheap reagents such as water, hydrogen, sodium hydroxide and sulphuric acid are used in the production of 'heavy' organic chemicals whenever possible. Expensive reagents such as potassium dichromate(VI), lithium tetrahydridoaluminate(III) (LiAlH$_4$) and silver compounds are only used in small-scale laboratory preparations when the cost is relatively unimportant.

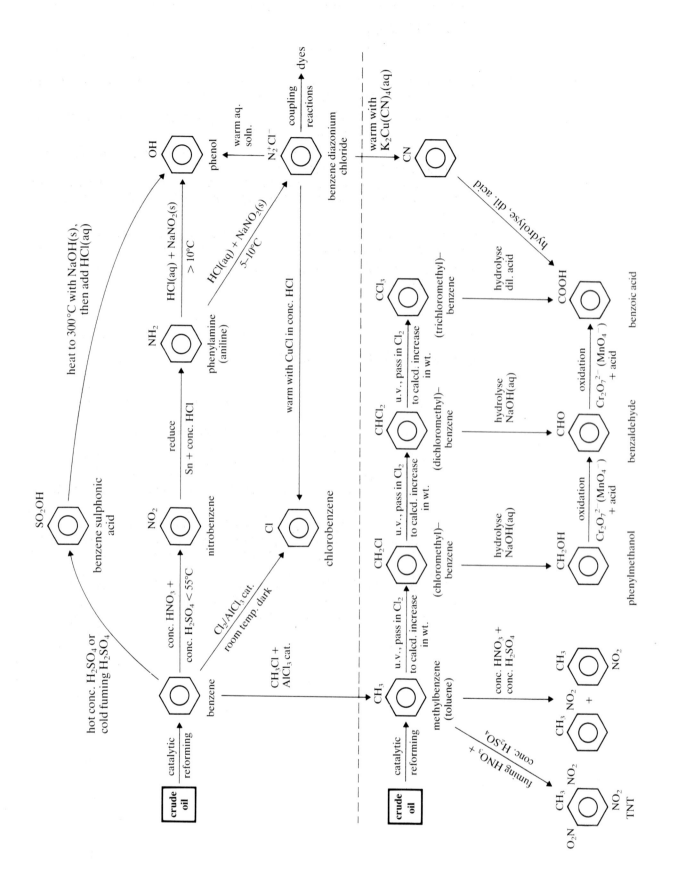

Figure 35.5
Important synthetic routes for arenes.

Summary

1 Chemists have found ways of synthesising new compounds with known structures and with properties that can be predicted.

2 The natural resources for many of the organic chemicals synthesised in industry are crude oil, fats and vegetable oils.

3 Alkenes, alcohols and carboxylic acids are the starting materials for the synthesis of many organic compounds.
   The cracking of crude oil fractions provides a ready supply of alkenes. Fats and vegetable oils can be hydrolysed to produce alcohols and carboxylic acids.

4 There are two important reactions which can be used to ascend a homologous series: the conversion of halogenoalkane to nitrile, or the 'oxo' reaction.

5 The crucial reaction in descending a homologous series is the Hoffman Degradation.

Study questions

1   Use the information in figure 35.2 to construct synthetic routes for the manufacture of
   (a) butanoic acid ($CH_3(CH_2)_2COOH$) from crude oil,
   (b) hexylamine ($CH_3(CH_2)_5NH_2$) from crude oil,
   (c) propanone (CH_3COCH_3) from crude oil,
   (d) decanal ($CH_3(CH_2)_8CHO$) from naturally occurring fats,
   (e) 1-bromoduodecane ($CH_3(CH_2)_{10}CH_2Br$) from naturally occurring fats.

2   Write synthetic routes for the laboratory preparation of
   (a) propylamine ($CH_3(CH_2)_2NH_2$) from butylamine ($CH_3(CH_2)_3NH_2$),
   (b) 1-bromopentane ($CH_3(CH_2)_4Br$) from 1-bromobutane ($CH_3(CH_2)_3Br$),
   (c) butylbutanoate   ($CH_3(CH_2)_2COO(CH_2)_3CH_3$)   from   butanal ($CH_3(CH_2)_2 CHO$),
   (d) propan-2-ol ($CH_3CHOHCH_3$) from propan-1-ol ($CH_3CH_2CH_2OH$),
   (e) benzoic acid (C_6H_5COOH) from benzene (C_6H_6),
   (f) phenol (C_6H_5OH) from benzene (C_6H_6).

Table of relative atomic masses

The table gives the relative atomic masses of elements correct to one decimal place.

Element	Symbol	A_r	Element	Symbol	A_r
Aluminium	Al	27.0	Molybdenum	Mo	95.9
Antimony	Sb	121.8	Neodymium	Nd	144.2
Argon	Ar	39.9	Neon	Ne	20.2
Arsenic	As	74.9	Nickel	Ni	58.7
Barium	Ba	137.3	Niobium	Nb	92.9
Beryllium	Be	9.0	Nitrogen	N	14.0
Bismuth	Bi	209.0	Osmium	Os	190.2
Boron	B	10.8	Oxygen	O	16.0
Bromine	Br	79.9	Palladium	Pd	106.4
Cadmium	Cd	112.4	Phosphorus	P	31.0
Caesium	Cs	132.9	Platinum	Pt	195.1
Calcium	Ca	40.1	Potassium	K	39.1
Carbon	C	12.0	Praseodymium	Pr	140.9
Cerium	Ce	140.1	Rhenium	Re	186.2
Chlorine	Cl	35.5	Rhodium	Rh	102.9
Chromium	Cr	52.0	Rubidium	Rb	85.5
Cobalt	Co	58.9	Ruthenium	Ru	101.1
Copper	Cu	63.5	Samarium	Sm	150.4
Dysprosium	Dy	162.5	Scandium	Sc	45.0
Erbium	Er	167.3	Selenium	Se	79.0
Europium	Eu	152.0	Silicon	Si	28.1
Fluorine	F	19.0	Silver	Ag	107.9
Gadolinium	Gd	157.3	Sodium	Na	23.0
Gallium	Ga	69.7	Strontium	Sr	87.6
Germanium	Ge	72.6	Sulphur	S	32.1
Gold	Au	197.0	Tantalum	Ta	180.9
Hafnium	Hf	178.5	Tellurium	Te	127.6
Helium	He	4.0	Terbium	Tb	158.9
Holmium	Ho	164.9	Thallium	Tl	204.4
Hydrogen	H	1.0	Thorium	Th	232.0
Indium	In	114.8	Thulium	Tm	169.9
Iodine	I	126.9	Tin	Sn	118.7
Iridium	Ir	192.2	Titanium	Ti	47.9
Iron	Fe	55.8	Tungsten	W	183.9
Krypton	Kr	83.8	Uranium	U	238.0
Lanthanum	La	138.9	Vanadium	V	50.9
Lead	Pb	207.2	Xenon	Xe	131.3
Lithium	Li	6.9	Ytterbium	Yb	173.0
Lutetium	Lu	175.0	Yttrium	Y	88.91
Magnesium	Mg	24.3	Zinc	Zn	65.4
Manganese	Mn	54.9	Zirconium	Zr	91.2
Mercury	Hg	200.6			

The modern periodic table (wide form)

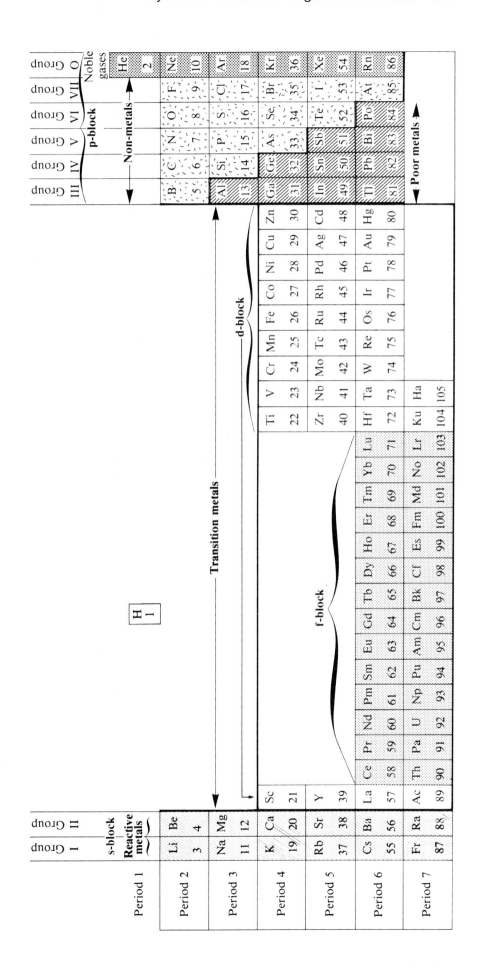

Answers to study questions

Chapter 1 Atoms, Atomic Masses and Moles

1 (a) 16g (b) 31g (c) 32g (d) 124g
(e) 44g (f) 216g (g) 12.8g (h) 120g
2 (a) 0.1 (b) 0.1 (c) 0.1 (d) 0.2
(e) 1.0 (f) 0.125 (g) 3.0 (h) 0.002
3 (a) 1.2×10^{24} (b) 6.0×10^{22}
(c) 1.8×10^{24} (d) 1.8×10^{22}
5 (a) 0.567 moles X + 0.85 moles O
(b) 1.5 (c) X_2O_3
6 (a) 0.02 (b) 0.02 (c) 267
(d) 85.5 (e) Rb
7 (a) 0.96g C, 0.16g H, 1.28g O
(b) e.f. CH_2O, m.f. $C_2H_4O_2$

Chapter 2 Reacting Quantities and Equations

1 (a) 0.01 (b) 0.03 (c) $AlBr_3$
(e) Al_2Br_6
2 C and D
3 (b) 16g (c) 16g
4 (e) 0.85g
7 A, B and D are true. C and E are false.
8 (b) 1:1
9 (a) 7.93g (b) 2.07g (c) 0.057 moles
(d) 0.115 moles (e) 2 moles.

Chapter 3 Redox

1 (a) $-1, +1, +5, -1, -1, -1$
(b) $+1, +2, +4, +5, -2, -3$
3 (b) and (d) are redox reactions
6 (a) A +6, B +6, C +3, D+6, E+6, F +3
7 (a) A +5, B +4, C +4, D +3, E +3,
F +2, G +5, H +5, I +4
9 $3.6cm^3$
10 (b) 5 (c) 0.005 (d) 2.8g (e) 28%

Chapter 4 Patterns and Periodicity

1 (a) (i) Li, Be (ii) B, N
(iii) LiF, BeO, BN, CO, NO
(iv) O, F, N
2 (a) (i) and (ii) C, Si
(b) (i) and (ii) He, Ne, Ar
(c) (i) molecular lattices
(ii) monatomic gases
7 (a) 238 (b) U
(c) Y = UH_3, Z = UCl_4

Chapter 5 Atomic Structure

3 (a) (i) 9 (ii) 19
(b) 3.17×10^{-23}g
(c) 6.0×10^{16}kg m^{-3} (6.0×10^{13}g cm^{-3})
(e) 10^{12}
4 (a) 14 (b) 28, 29, 30 (c) 28.11
5 (b) 2,3,4 (c) 12% $^2_1$H, 88% $^1_1$H
6 (a) (i) 28.0062 (ii) 28.0172
(iii) 27.9949 (iv) 28.0312
(v) 28.0282
(b) DCN
7 (b) $CH_3CH_2CH_2OH$, $CH_2(OH)CHO$

Chapter 6 Electronic Structure

2 (a) A (b) B and D are in group 2
(c) group 3 (d) C
3 (b) 16 (c) sulphur

Chapter 7 Nuclear Structure and Radioactivity

2 (b) 25.7°C above surroundings
3 (b) $^{208}_{82}$Pb (d) 200
4 (a) 8.8%

Chapter 9 Intermolecular Forces

1 (a) ICl, CH_2Cl_2
(c) $C_2H_5NH_2$, CH_3OH, H_2SO_3
(d) CO_2, BF_3
4 (b) hydrogen bonding in NH_3
(c) 13.5kJ mole^{-1} (d) 9.9kJ mole^{-1}
5 (b) 22kJ (c) 27%
6 (a) B (b) 64, B 64, C 128
7 (a) $RCH(NH_2)CO_2H$ (c) 100
(c) 1000

Chapter 10 Structure, Bonding and Properties—the Solid State

4 (a) A (b) D (c) C (d) B
(e) E
5 (a) D (b) C (c) B (d) E
(e) D (f) E
10 (a) 3.177×10^{-10}m (b) 28.9°
(c) (i) 3.207×10^{-29}m^3 (ii) 0.5
(iii) 6.414×10^{-29}m^3
(d) 3.73×10^{-5}m^3 (e) 5.8×10^{23}
13 (b) $(\sqrt{2})-1$ (c) LiCl
14 (b) 4 (c) 2.73g cm^{-3}

11 The Gaseous State

2 The volume indoors would be 1045cm^3—the balloon bursts.
4 (a) 200mm Hg (b) isotopes
5 (b) 59.7
6 4cm^3 ethyne, 6cm^3 ethene
7 25% CO_2
8 (c) PH_3
9 (c) 10cm^3 (d) 10cm^3
(e) 29cm^3
(f) (i) 5cm^3 (ii) 24cm^3
(g) (i) 16cm^3 (ii) 14cm^3
10 C_2H_2
11 C_4H_8

Chapter 12 Energy Changes and Bonding

1 (a) 16.8kJ (b) 840kJ
2 (d) -276kJ mole^{-1}
3 (a) 335cm^3
5 (b) -84kJ mole^{-1}
7 (a) -621kJ mole^{-1}
8 (c) -672kJ mole^{-1}
9 (a) -3401kJ mole^{-1}

Chapter 13 Patterns Across the Periodic Table

4 (a) N, F (b) B (c) B, N
(d) Ne (e) Be (f) Li, Be, N
8 (a) $1s^2 2s^2 2p^6 3s^2 3p^6 3d^{10} 4s^2 4p^1$
(b) p-block (c) +3 (d) Y_2O_3
(e) amphoteric
9 P is Na or K, Q is Cr

Chapter 14 Competition Processes

2 (a) 0.5 (b) 1.75×10^{-3}
(c) 8.75×10^{-4} (d) 8.75×10^{-3}
(e) 8.74×10^{-2} (f) 10
5 (c) 1.10V
8 (d) 1.21V
9 (a) 16.5mg per 100cm^3

Chapter 15 Groups I and II—the Alkali Metals and the Alkaline-Earth Metals

5 (b) (ii) $^{42}_{20}$Ca (iii) 0.015g
8 (a) 3×10^{-4} (b) 3.78×10^{-2}g
(c) 378 p.p.m. (e) 0.070g

Chapter 16 Aluminium

3 (a) r.m.m. = 268 at 200°C, 193 at 600°C, 132 800°C
(b) Al_2Cl_6 at 200°C, $AlCl_3$ at 800°C
5 (a) A is C_3H_9Al, B is CH_4, C is Al_2O_3
(c) $Al(CH_3)_3$

Chapter 17 Group VII—the Halogens

5 (b) 1.25×10^{-3} (c) 2.5×10^{-2}
(d) 80% Cu

Chapter 18 Group IV—Carbon to Lead, Non-Metal to Metal

4 (c) 15.68g
5 (c) $CO_2(g) + C(s) \rightarrow 2CO(g)$ $\Delta H = +172$kJ
$CO(g) + \frac{1}{2}O_2(g) \rightarrow CO_2(g)$ $\Delta H = -283$kJ

Chapter 20 Metals and the Activity Series

4 (b) 5 (c) 3.75×10^{-3} (d) 2.1g
(e) 21%

Chapter 21 Equilibria

1 (a) 10.0% (b) 3.31% (c) 1.56%
2 (c) (i) 4.55g (ii) 4.86g
3 (c) 3 moles of ethanol, 8 moles of ethyl ethanoate and 6 moles of water
(d) 4 (e) (ii) 1.33 (ii) 1.33
(iii) 2.67 (iv) 2.67
4 (a) 2.67×10^4Pa (0.267 atm) (b) $5.34 \times$ 10^{5}Pa (5.34 atm)
5 (a) (i) $K_c = K_p.RT$
(ii) $K_c = K_p.(RT)^{\frac{1}{2}}$

6 (v) 4 (unitless)
(c) partial pressures are 3.75×10^3 Pa (0.0375 atm) for H_2, 3.375×10^4 Pa (0.3375 atm) for I_2, 2.25×10^4 Pa (0.225 atm) for HI
(d) partial pressures are 1.5×10^4 Pa (0.15 atm) for H_2, 1.5×10^4 Pa (0.15 atm) for I_2, 3×10^4 Pa (0.30 atm) for HI.

Chapter 22   Factors Affecting Equilibria

3 (c) (i) 10^{-10} (ii) 10^{20}
4 (c) $125\,dm^3\,mole^{-1}$
(f) 0.02 mole dm^{-3} for both
5 (a)
$$K_p = \frac{\alpha^2 P}{1 - \alpha^2}$$
(b) (i) 1.41×10^{-3} (ii) 1.0×10^{-3}

Chapter 23   Ionic Equilibria in Aqueous Solution

1 4.9×10^{-13}
2 (a) 1.08×10^{-4} mole dm^{-3}
3 (a) (i) 1.26×10^{-4} mole dm^{-3}
(ii) 1.6×10^{-7} mole dm^{-3}
(iii) 1.6×10^{-6} mole dm^{-3}
4 (b) (i) $1.0M$ HA (ii) $1 \times 10^{-3}\,M$
5 (a) $[H^+] = 2 \times 10^{-5}\,M$, $[OH^-] = 5 \times 10^{-10}\,M$
(b) 4.7
7 (b) $1.1 \times 10^{-3}\,M$ (c) 3 (d) 4.6
(e) New pH is 4.62, a change of 0.03
8 1.61
9 (a) 4 (b) 10.3
(c) pH = 0 if $H_2X \rightleftharpoons 2H^+ + X^{2-}$ but pH = 0.3 if $H_2X \rightleftharpoons H^+ + HX^-$
(d) 3.42 (e) 11.62

Chapter 24   Entropy and Free Energy

1 (a) To the right; iodine does dissolve
(b) 10
(c) 1
(d) Among the CCl_4 molecules
(e) Yes
(f) 6×10^{23}
(g) 1. It is not likely to
2 (a) $270.3\,JK^{-1}\,mole^{-1}$
(b) $53.6\,JK^{-1}\,mole^{-1}$
(c) $-216.7\,JK^{-1}\,mole^{-1}$
(d) No
(e) $+4040\,JK^{-1}\,mole^{-1}$
(f) +3824
(g) Yes
3 (a) A (b) C (c) A (d) C
(e) A (f) B (g) B (h) A
5 (a) For $CaCO_3$:
1. $+128\,830\,J\,mole^{-1}$ 2. No
3. (a) Yes (b) No 4. 1079K
5. 1079K
(b) For $MgCO_3$:
1. $+48\,209.6\,Jmole^-$ 2. No
3. (a) Yes (b) No 4. 574K
5. 574K
(c) $MgCO_3$ less stable than $CaCO_3$

Chapter 25   Reaction Rates

3 (a) increase $\times 8$ (b) increase $\times 4$
(c) decrease $\div 8$ (d) no change
(e) increase $\times 8$
4 (b) 9×10^9 years, assuming all U decays to Pb
5 (c) 0.125 g
6 (ii) $186\,kJ\,mole^{-1}$
7 (a) $1.67 \times 10^{-6}\,mole^{-1}$
(b) $8.33 \times 10^{-7}\,mole^{-1}$
(d) 1.67×10^{-1}

(e) $mole^{-2}\,dm^6\,s^{-1}$
(f) 1.67×10^{-5} mole s^{-1}
8 (b) $-0.1\,cm^3\,min^{-1}$ (g) 0.01 M
(h) $20\,cm^3$ (i) 0.01 M
(j) $-0.2\,cm^3\,min^{-1}$
9 (a) (i) 0 (ii) 2 (b) rate = $k[Y]^2$
(c) $Y + Y \rightarrow Y_2$ (slow), $Y_2 + X \rightarrow XY_2$ (fast)
(d) 0.01 (e) $mole^{-1}\,dm^3\,s^{-1}$
11 (b) (i) 1st order (ii) 2nd order
(c) (i) $1.15 \times 10^{-3}\,mole^{-1}\,dm^3\,s^{-1}$
(ii) $mole^{-1}\,dm^3\,s^{-1}$
12 (a)
$$\underset{H_2C-CH_2}{\overset{CH_2}{\diagdown\diagup}} \rightarrow CH_3CH{=}CH_2$$
(c) $1.28 \times 10^{-5}\,s^{-1}$

Chapter 26   Introduction to Carbon Chemistry

1 (a) C 37.4%, H 12.4%
(b) oxygen (c) CH_4O
(d) (i) 32 (ii) CH_3OH
2 (a) A and C
(c) A and C are alcohols, B is an ether
(d) A is propan-1-ol (higher boiling point), B is methoxyethane, C is propan-2-ol
3 (a) alcohol (b) ketone
(c) chloroalkane (d) carboxylic acid
(e) alkene (f) amino acid
(g) ether
7 (a) 5.18×10^{-3} moles C, 1.38×10^{-2} moles H 1.72×10^{-3} moles N
(b) C_3H_8N (c) $C_6H_{16}N_2$
(d) $NH_2(CH_2)_6NH_2$
8 A = C, B = F = H, D = E = G

Chapter 27   Petroleum and Alkanes

2 (a) A in diesel oil; B in gasoline; C in gasoline; D in gasoline; E in refinery gases
(b) D and C
(c) several possibilities, e.g. C from A
(d) E, B, D, C, A, (e) A (f) B
4 (c) $360\,kJ\,mole^{-1}$ (d) CH_2
(g) butane $-49.6\,kJ\,mole^{-1}$, pentane $-48.7\,kJ\,mole^{-1}$, hexane $-48.8\,kJ\,mole^{-1}$
5 (c) 6.58 (d) $9870\,dm^3$
(e) $7896\,dm^3$ (f) 9212 g
(g) 2316 g (h) 0.51 g
7 Values of ΔH_c/kJ mole^{-1} per CH_2 group
cyclopropane −697
cyclobutane −685
cyclopentane −664
cyclohexane −658
cycloheptane −662
cyclooctane −664
cyclononane −664
cyclodecane −663

Chapter 28   Unsaturated Hydrocarbons

4 (a) C_3H_5 (b) C_6H_{10} (c) one
(d) one (e) C_6H_{12}
(f) A could be cyclohexene, B cyclohexane
7 (a) heat with a Ni catalyst
(b) $252\,kJ\,mole^{-1}$
(c) $436\,kJ\,mole^{-1}$
(d) $826\,kJ\,mole^{-1}$
(e) $-138\,kJ\,mole^{-1}$
10 (a) A, B, C (b) B (c) D, E
(d) A (e) A (f) A, B, C (g) B, C

Chapter 29   Aromatic Hydrocarbons

3 (b) $-360\,kJ\,mole^{-1}$
9 (a) (i), (ii) (b) (i), (ii), (iii), (iv)
(c) (i), (ii), (iv) (d) (i), (ii)
(e) (iv), (v) (f) (i), (ii)
(g) (i), (iv), (v)
10 (a) A is substitution, B is addition
(b) (i) $484\,kJ\,mole^{-1}$
(ii) $860\,kJ\,mole^{-1}$
(iii) $-678\,kJ\,mole^{-1}$
(iv) $-862\,kJ\,mole^{-1}$
(v) $-196\,kJ\,mole^{-1}$
(c) (i) $242\,kJ\,mole^{-1}$
(ii) $-678\,kJ\,mole^{-1}$
(iii) $-2\,kJ\,mole^{-1}$
(d) A (more exothermic process)

Chapter 30   Organic Halogen Compounds

2 (a) C, E (b) D (c) D (d) A
(e) D (f) C
6 A is CH_2BrCH_2Br, B is $CH_2(OH)CH_2OH$
7 X is CH_3CH_2COCl, Y is CH_2ClCH_2CHO
10 A ethanol B ethene C 1,2-dibromoethane D ethane-1,2-diol P chlorine in UV light Q hydrogen chloride

Chapter 31   Alcohols, Phenols and Ethers

1 (b) D (c) A (d) E (e) C
2 (a) A, B, D, E (b) D (c) B
(d) E (e) A, B, C (f) C
4 (a) C 64.8%, H 13.6%, O 21.6%
(b) $C_4H_{10}O$ (c) 74.4

Chapter 32   Carbonyl Compounds

1 (a) B, D (b) A, E (c) C
(e) B, C, D (f) A (g) A, E
6 (a) A is $(CH_3CHO)_2$, B is $(CH_3CH(OH)CN)_2$, C is $(CH_2CO_2H)_2$
(b) X is $CH_2 = CHCHO$
8 (a) CH_3CH_2OH
(b) $CH_3CH(OH)CH_3$
(c) $(CH_3)_3COH$ (d) CH_3CO_2H

Chapter 33   Carboxylic Acids and their Derivatives

1 (a) C (b) D (c) E (e) C
(f) B (g) A, B, D, E (h) E
9 (a) C (b) D (c) A, C
(d) B and E (e) A (f) B (g) E

Chapter 34   Organic Nitrogen Compounds

1 (a) B (b) A (c) E (d) F
2 (a) increasing base strength (i), (iii), (ii)
(b) increasing boiling point (ii), (i), (iii)
4 (a) B, C (b) A (c) B (d) D
(e) A (f) D
7 (a) 2 (b) 2 (d) $200\,cm^3$
10 B must be 2-nitromethylbenzene

Index